# NATURE OF PROMINENCES AND THEIR ROLE IN SPACE WEATHER

IAU SYMPOSIUM No. 300

*COVER ILLUSTRATION:*

Solar image: SDO/NASA satellite in HeII 30.4 nm
Composit image with Eiffel Tower by S. Cnudde

# IAU SYMPOSIUM PROCEEDINGS SERIES

INTERNATIONAL ASTRONOMICAL UNION

UNION ASTRONOMIQUE INTERNATIONALE

International Astronomical Union

# NATURE OF PROMINENCES AND THEIR ROLE IN SPACE WEATHER

## PROCEEDINGS OF THE 300th SYMPOSIUM OF THE INTERNATIONAL ASTRONOMICAL UNION HELD IN PARIS, FRANCE JUNE 10–16, 2013

Edited by

**BRIGITTE SCHMIEDER**
*Observatoire de Paris, MEUDON*

**JEAN-MARIE MALHERBE**
*Observatoire de Paris, MEUDON*

and

**S.T.WU**
*University of Alabama, HUNTSVILLE*

CAMBRIDGE
UNIVERSITY PRESS

CAMBRIDGE UNIVERSITY PRESS
The Edinburgh Building, Cambridge CB2 2RU, United Kingdom
32 Avenue of the Americas, New York, NY 10013–2473, USA
10 Stamford Road, Oakleigh, Melbourne 3166, Australia



First published 2013

Printed in the United Kingdom by CPI Group (UK) Ltd, Croydon, CR0 4YY

Typeset in System LaTeX $2_\varepsilon$

*A catalogue record for this book is available from the British Library*

*Library of Congress Cataloguing in Publication data*

This journal issue has been printed on FSC-certified paper and cover board. FSC is an independent, non-governmental, not-for-profit organization established to promote the responsible management of the worlds forests. Please see www.fsc.org for information.

ISBN 9781107045194 hardback
ISSN 1743-9213

# Table of Contents

## SESSION II -2.1 Prominence destabilization, CMEs, 3D reconstructions



## SESSION II -2.2 CMEs and Magnetic clouds in the Heliosphere and their impacts on Earth's Environment

## SESSION III Stellar Ejecta and Impact on Exoplanets



## SESSION IV Instrumentation, Missions and Techniques



## CONCLUSION

## POSTERS SESSION I

## POSTERS SESSION II

## POSTERS SESSION III



## POSTERS SESSION IV

# Preface

IAU Symposium 300, "Nature of solar prominences and their Role in Space Weather" was coordinated through Division II, with the strong support of Division IV and several commissions. It was held in Paris, France, from 10 to 14 June 2013.

This symposium was dedicated to Einar Tandberg Hanssen. We started the symposium with memories about his career by inviting his two daughters to Paris. Jean Claude Pecker (from the Académie des Sciences), S.T. Wu, R. Moore from Hunstville and B. Schmieder from LESIA (Observatoire de Paris) presented his work from all along his long career as a specialist in solar prominences and Principal Investigator of the UV instrument (UVSP) on-board the Solar Maximum Mission satellite (SMM). 175 scientists from 30 countries attended the meeting at the Ecole Nationale Supérieure de Chimie Paris Tech (Paris Sciences et Lettres Research University). We had 36 participants from France, 25 from the US, 17 from Spain, 15 from the UK, and 14 from China. More than 6 scientists came from each of the following countries : Russia, South Korea, Germany, Belgium and India. Four came from Japan and the Czech Republic. Between 1 and 3 participants came from : Iran, Poland, Argentina, Norway, Brazil, Costa Rica, Tajikistan, Slovenia, Austria, Sweden, Romania, Slovakia, Mexico, Portugal, Serbia, Italy and Canada. Many young researchers attended the meeting and three of them received a prize for the best poster selected by the SOC members. The conference dinner aboard a cruise ship on the Seine during the sunset left all the participants with good memories.

The meeting was divided in 4 Sessions : Prominences (I), Coronal Mass Ejections and Space Weather (II), Ejections from Stars (III) and Instrumentation (IV).

There were 28 invited reviews, 48 contributions and 98 posters. The topics were very interesting. The aim of this IAU Symposium was to present a review of the state-of-the-art theoretical and numerical modeling, and space-borne (Hinode, STEREO and SDO) and ground-based observational studies of prominences and their role in the dynamics of Sun-Earth relations. It also opened new perspectives for people, and especially young ones, working in the field. Prominences have an active role in Space Weather. Magnetic clouds and Interplanetary Coronal Mass Ejections (ICME) associated with erupting prominences can produce severe perturbations in the Earth's environment. Moreover, huge prominences and CMEs have been detected in solar-type stars (and others) and exoplanets. It was interesting to put the properties of solar prominences in a broader perspective, on one hand, and to present the status of the sophisticated solar analysis to the stellar community on the other hand. Eric Priest made a very lively summary detailing all the sessions and the keynote talks at the end of the meeting (see this issue).

Two american scientists were not allowed to come because of the NASA restrictions and five participants could not come because of visa problems. They were replaced on time and finally all the talks were given on the right schedule.

We would like to thank the meeting sponsors (IAU, KLSA/CAS from China, SCOSTEP, ESA) and from France (SF2A, CNES, Observatoire de Paris, LESIA, IAS, PNST). They allowed us to support financially more than 50 participants.

The editors are also indebted to all the LOC members and particularly E.Pariat and E. Buchlin. We are very grateful to the following reviewers who helped a lot for improving the papers: Chae J., Dasso S., Heinzel P., Jardine M., MacKay D., Schmieder B, Srivastava N., van Driel L., Webb D. assisted by Aulanier G., Démoulin P., Bommier V., Koutchmy S., Ballester J.L., Gopalswamy N, Gilbert H., Gunar S., and Vial J.C. Please notice that many of the papers contain color figures, which are printed in black and white but which can be viewed online in color.

The editors of IAUS300: B. Schmieder, J.-M. Malherbe and S.T. Wu 25 September 2013

**IAUS300, Nature of Prominences and their role in Space Weather**

## SCIENTIFIC ORGANIZING COMMITTEE (SOC)

Brigitte Schmieder, LESIA, Observatoire de Paris, France, Chairman
Shi T. Wu, University of Alabama, Hunstville, US, Co-Chairman
and
Jongchul Chae, Seoul National University, Korea
Sergio Dasso, IAFE, Argentina
Lidia van Driel-Gesztelyi, LESIA, Observatoire de Paris, France-UK-Hungary
Holy Gilbert, GSFC/NASA, USA
Nat Gopalswamy, NASA Goddard Space Flight Center, USA
Petr Heinzel, Astronomical Institute, Czech Republic
Moira Jardine, St. Andrews University, UK
Duncan MacKay, St. Andrews University, UK
Valentin Martínez Pillet, Instituto de Astrofsica de Canarias, Spain
Nandita Srivastava, Udaipur Solar Observatory, India
David F. Webb, Boston College, USA
Yihua Yan, NAO, Beijing, China

## LOCAL ORGANIZING COMMITTEE (LOC)

Brigitte Schmieder (LESIA, Observatoire de Paris, France), Co-chairman
Jean-Marie Malherbe (LESIA, Observatoire de Paris, France), Co-chairman
and
Eric Buchlin (IAS, Orsay, France)
Etienne Pariat (LESIA, Observatoire de Paris, France)
Marie Pierre Issartel (LESIA, Observatoire de Paris, France)
Sylvain Cnudde (LESIA, Observatoire de Paris, France)
Isabelle Bualé (LESIA, Observatoire de Paris, France)
Frédéric Dauny (LESIA, Observatoire de Paris, France)
Guillaume Aulanier (LESIA, Observatoire de Paris, France)
Miho Janvier (LESIA, Observatoire de Paris, France)
Kevin Dalmasse (LESIA, Observatoire de Paris, France)

# CONFERENCE PHOTOGRAPH

DE CHIMIE
Chimie ParisTech
PSL

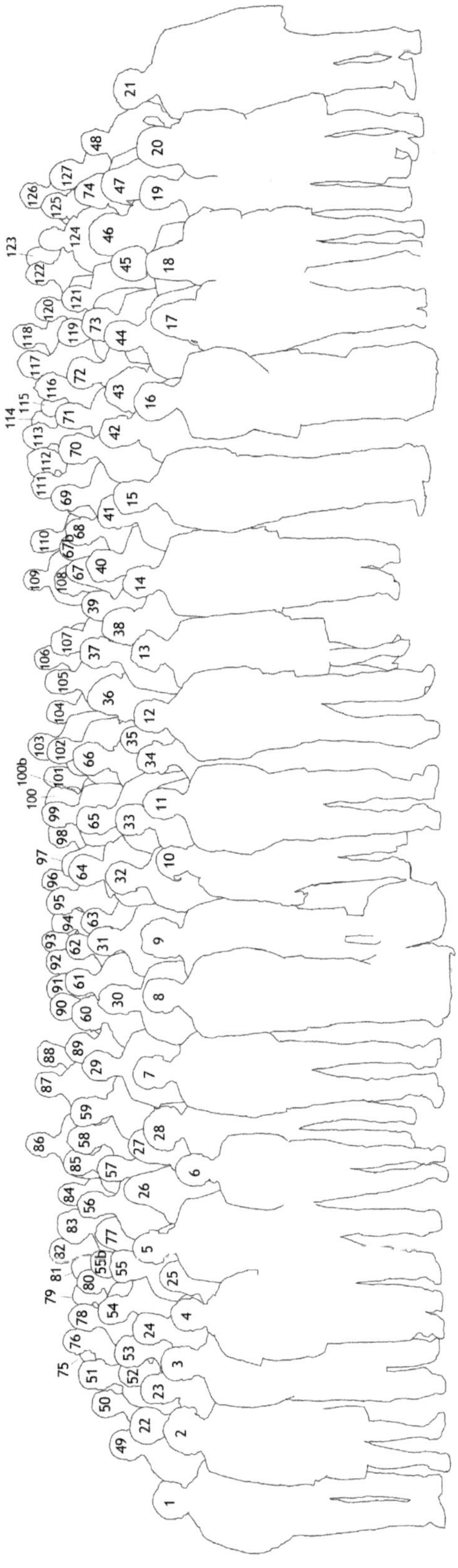
1
2
3
4
5
6
7
8
9
10
11
12
13
14
15
16
17
18
19
20
21
22
23
24
25
26
27
28
29
30
31
32
33
34
35
36
37
38
39
40
41
42
43
44
45
46
47
48
49
50
51
52
53
54
55
55b
56
57
58
59
60
61
62
63
64
65
66
67
67b
68
69
70
71
72
73
74
75
76
77
78
79
80
81
82
83
84
85
86
87
88
89
90
91
92
93
94
95
96
97
98
99
100
100b
101
102
103
104
105
106
107
108
109
110
111
112
113
114
115
116
117
118
119
120
121
122
123
124
125
126
127

## List of Participants

The number refers to the photography

Aarnio Alicia 67bis
Agueda Neus 43
Ajabshirizadeh Ali 121
Alexeeva Irina
Amory-Mazaudier Christine
Arregui Inigo 99
Asensio Ramos Andres 60
Auchère Frédéric
Aulanier Guillaume 88
Bak-Steslicka Urszula 36
Baker Deborah 71
Ballester José 76
Balmaceda Laura 27
Bazin Cyrille
Berger Thomas 15
Bilenko Irina 74
Bocchialini Karine 91
Bogachev Sergey 72
Bommier Véronique 63
Bonnin Xavier
Brown Gerrard 114
Buchlin Eric 118
Carbonell Marc
Carlsson Mats 93
Carlyle Jack 113
Chae Jongchul
Chen Hua-dong 49
Cheng Xin 70
Chian Abraham
Chifu Iulia
Cho Kyungsuk 59
Cid Consuelo 83
Dal Lago Alisson 24
Dalmasse Kévin 56
Damé Luc
Dasso Sergio 84
Démoulin Pascal 65
Diaz Antonio
Ding Adalbert 95
Dudik Jaroslav 82
Dwivedi Bhola
Dzifcakova Elena 89
Ehrenreich David
Fang Xia 64

Farnik Frantisek 87
Filippov Boris 44
Fontaine Dominique 52
Forland Blake
Gary Gilmer 37
Gibson Sarah 38
Golub Leon 119
Green Lucie
Gunár Stanislav 42
Gutiérrez Heidy 62
Habbal Shadia 32
Haerendel Gerhard 55
Hanaoka Yoichiro 79
Hasan Sirajul
Heinzel Petr
Hillier Andrew 61
Hu Qiang 33
Hussain Gaitee 46
Ibadov Subhon 34
Janvier Miho 18
Jardine Moira 97
Jejcic Sonja
Joshi Bhuwan 1
Joshi Anand 80
Khodachenko Maxim
Khomenko Elena
Kim Iraida 19
Knizhnik Kalman 54
Kontar Eduard 111
Koutchmy Serge 116
Kuckein Christoph 86
Kumar Pankaj 7
Labrosse Nicolas 115
Lavraud Benoit 98
Leibacher John
Lepri Susan 67
Li Leping 3
Li Ting 6
Lima Joao José 122
Lites Bruce 96
Liu Zhong 53
Lopez Ariste Arturo 29
Lugaz Noe 101bis
Luna Manuel
Ma Suli 20
Mackay Duncan 94
Malherbe Jean-Marie 12
Martinez Gonzalez Maria Jesus 28
Mashnich Galina
McCauley Patrick

Mein Nicole
Mein Pierre 104
Mierla Marilena 108
Milic Ivan 103
Miteva Rositsa
Moore Ron 107
Mouradian Zadig 21
Muglach Karin 26
Nishizuka Naoto 66
Oliver Ramon
Olmedo Oscar 57
Ontiveros Veronica
Orozco Suárez David 110
Pagano Paolo 58
Palacios Judith 48
Panesar Navdeep 10
Pariat Etienne
Park Sung-Hong
Park Hyungmin 90
Pecker Jean-Claude
Pick Monique 2
Pinter Teodor
Pinto Rui 73
Poletto Giannina
Pötzi Werner 109
Priest Eric 14
Rachmeler Laurel 35
Rimmele Thomas 101
Rodríguez-Gasén Rosa 17
Rozelot 120
Ruffenach Alexis
Salas-Matamoros Carolina 127
Schmieder Brigitte 13
Schmit Donald 55 bis
Schwartz Pavol 78
See Victor
Sharma Rahul 123
Shen Yuandeng 22
Shibata Kazunari 16
Shimojo Masumi 40
Soler Roberto
Srivastava Nandita 55 ter
Stere Oana
Strugarek Antoine 112
Suiunova Elvira 75
Taliashvili Lela 81
Tavabi Ehsan 117
Terradas Jaume 102
Title Alan
Trichtchenko Larisa

Turc Lucile 25
Valori Gherardo 126
van Ballegooijen Adriaan 30
van Driel-Gesztelyi Lidia 39
Vial Jean-Claude 92
Vidotto Aline 47
Vilmer Nicole 9
Warnecke Joern
Williams David 69
Wimmer-Schweingruber Robert 5
Wu Shican 11
Xia Chun 31
Xu Zhi 125
Yan Yihua 8
Yan Xiaoli 23
Yang Shuhong 51
ZapiÓr Maciej 105
Zerbo Jean Louis 124
Zhang Qingmin 4
Zhang Jun
Zhao Jie 85
Zuccarello Francesco 68

# Opening session

## INTRODUCTION OF THE SYMPOSIUM

dedicated to the memory of Einar Tandberg Hanssen

*Nature of Prominences and their role in Space Weather*
*Proceedings IAU Symposium No. 300, 2013*
*B.Schmieder J.-M. Malherbe & S. T. Wu, eds.*

doi:10.1017/S1743921313011861

# Einar Tandberg-Hanssen

**Figure 1.** Einar Tandberg-Hanssen portrait (@Julia Gary).

*Nature of Prominences and their role in Space Weather*
*Proceedings IAU Symposium No. 300, 2013*
*B. Schmieder J.-M. Malherbe & S. T. Wu, eds.*

doi:10.1017/S1743921313010685

# Einar Tandberg-Hanssen

**Brigitte Schmieder[1], Jean-Claude Pecker[2], Allen Gary[3], S.T. Wu[3], Ronald Moore[4] and Else Biesmann[5]**

[1]Observatoire de Paris, LESIA, Meudon, France, [2]Académie Française, Paris, [3]University of Alabama, Hunstville, AL, US, [4]NASA/Marshall Space Flight Center, Hunstville, AL, US, [5]10160 Canyon Place, Rapid City, SD 57702, US.
email: brigitte.schmieder@obspm.fr

## 1. Einar Tandberg-Hanssen and Solar Physics (Brigitte Schmieder's speech)

Dear Colleagues,

I would like to report first on the scientific career of Einar Tandberg-Hanssen: how he became a Solar Physicist particularly interested in prominences. In the second part of my talk I will show what he brought to the French community from the science perspective.

Einar Tandberg Hanssen was born on the 6 August 1921 in Bergen in Norway.

After WWII he studied Astrophysics, Stellar, and Solar Physics. He worked with S. Rossland and E. Jensen in Oslo. He built the telescope of the Observatory in Blindern (University of Oslo). In 1951 he went to the Astrophysics Institute (IAP) in Paris and started his collaboration with the French scientists: Chalonge, Dollfus, Martres, Michard, Pecker, Schatzman, Steinberg ... From 1952-1959, with Zirin and Pecker in California, he worked on the abundance of Beryllium in stellar atmospheres and on the classification of prominences according to their spectra.

For a short time during this period he visited England and worked on radio astronomy. Finally he came back to Oslo as a professor and gave lectures on Solar Physics for two years between 1959-1961. In 1960 he presented his thesis on "An Investigation of the Temperature Conditions in Prominence".

He was interested in the role of the magnetism in the exchange between cool prominences and the hot corona. He followed the polarization measurements made by Ratier at Pic du Midi in 1975 and the theory of the Hanle effect (Sahal, Bommier, Leroy 1977), which allows the derivation of the angle between the vector magnetic field in prominences and the local vertical on the solar surface. The conclusion of the latter study was that in prominences the magnetic field lines are horizontal and not, as they looked, vertical. MHD models appeared to take account of this discovery (the models of Kuperus and Raadu, 1974 and Kieppenhan-Schluter, 1957)

I personally started to know Einar just after my thesis on the propagation of waves in the solar atmosphere, when the Solar Maximum Mission (SMM) was launched in 1980. Einar was the Principal Investigator of the spectrometer on board SMM called UVSP. He travelled between Hunstville (MSFC) and Washington (GSFC) and organized several workshops (Figure 1 top left panel). The French scientists: J.-C. Pecker, G. Simon, B. Schmieder, P. and N. Mein, proposed a program on the "Detection of the Moreton waves in UV (CIV line)" to be run on this instrument. The program was accepted by the UVSP team. This lead to the theses of J.-M. Malherbe, P. Démoulin on the "Dynamics of filaments" and 14 refereed publications of Tandberg-Hanssen co-signed with the Meudon group. He had many other collaborators in different institutes in France: S. Koutchmy, J.-C. Vial, Z. Mouradian, J-C. Hénoux, P. Lantos, and J.L. Leroy.

**Figure 1.** Einar Tandberg-Hanssen (*top left* ) in Airlie (US) during a SMM workshop in 1989 (Art Poland, Einar, Brigitte Schmieder, Guy Simon, Jean Marie Malherbe); (*bottom left*) Einar Tandberg-Hanssen in Oslo with Pierre Foukal, David Rust, Brigitte Schmieder, Oddbjorn Engvold and his spouse, Einar: (*right panels*) Einar in front of his cabin in Sandefjord in 1996. (*Figure adapted from Tandberg-Hanssen 2011*)

**Figure 2.** The daughters and grand-daughters of Einar in Paris on June 10, 2013 (from left to right: S.T.Wu, Jacquelyn, Karin, and Jennifer Brock, Brigitte Schmieder, Ronald Moore, Else Biesmann, Jean Claude Pecker).

Einar brought many things to our group in Meudon. We learned how to work in a team on a project and not to work alone. He was very convivial and also very well organized. He transmitted to us this capability. He opened our department of the Observatory of Paris to Space Research and showed how it is important to observe in different wavelengths. With him we acquired a great number of collaborations in the United States and in Norway. One Norwegian PhD student (Jun Elin Wiik) worked on her PhD in our laboratory and get a French thesis in 1994. I got a part-time fellowship of Professor in Oslo between 1996 and 2006, working with Oddbjorn Engvold and his students on fine structures in prominences (Figure 1 bottom left panel). For us, and particularly for me, Einar was a very good friend. He invited me to Sandefjord to visit his "Cabin" along the fjord (Figure 1 right panels).

I knew his wife Erna very well and remember her last sentences at the IAU General Assembly in La Hague in 1994, announcing to me her future death "I have achieved what I wanted to do in my life, it is fine for me to leave this world". I was also very surprised to receive an email transmitted by J.L. Leroy from Einar six months before his death. "I have been diagnosed with ALS, also called Lou Gehrig's disease, so we will not meet again. I would appreciate if you would let the Schmieders know this and all my old friends at Meudon. I am doing all right, but weak". He died on the 22 July 2011 in Huntsville.

We lost a great scientist and it is a great honor for me to be allowed to organize this IAU Symposium on prominences in memory of him. He loved Paris. He was in Paris every year "aux Ides" of March for the IAU FAG meetings. He would have like to attend this meeting. He attended the last one in Aussois in 1997 (IAU S167) . His daughters are here and for us it is a part of him.

References :
Tandberg-Hanssen 2011, Solar Physics, 269, 237

## 2. Einar Tandberg-Hanssen in the 60's (Jean-Claude Pecker speech)

Dear Colleagues;

I am amongst you the one who knew Einar and Erna Tandberg-Hanssen for the longest time, more than sixty years! Allow me to say a few words about the many occasions during which we met and became friends.

They were Norwegians. Actually, in the post-war years, remember, between-countries trips were quite difficult, even within Europe. But, due to the diligent action of Colonel F.J.M. Stratton, General Secretary of the IAU, and eager to organize the "exchange" of astronomers, some international exchanges were organized. And this created indeed excellent relations, in particular, between the French astronomers, the young ones especially, and their colleagues in Scandinavia, particularly in Norway. Following Vova Kourganoff, I spent some time at Oslo, in order to be acquainted with Svein Rosseland, then a key figure of the theory of stellar-solar atmospheres. During the same period, Einar and Erna spent two years in Paris, principally at the Institut d'Astrophysique, where the attractive figure of Daniel Chalonge acted as a magnet to young people.

Very soon, we felt a great sympathy for each other. The quiet and smiling behaviour of Einar, his competence, and seriousness, impressed me, as did the bright smile of the charming Erna, also a very competent scientist. I could not separate one from the other in my tribute to this talented and happy couple.

Later on, I had the pleasure to be at the HAO, in Boulder, at the time when Walter Roberts was its powerful and cheerful Director, and at the same time as the Tandberg-Hanssens. He was by then becoming "the" expert in the physics of solar prominences. That field of research was the symbol of his life-track, observing and understanding better and better the physics of matter within the magnetic fields of the solar corona. I remember a long discussion about spiraling motions in prominences, about which I had published a paper with Don Billings, a good friend of both Einar and myself.

After that time, we met of course many times, at IAU meetings, or elsewhere. And we kept a very solid and quiet friendship. I always admired, along the years, Einar's serenity, his deep concern for scientific rigour, but also his very subtle sense of humour. We miss him. We miss them.

I am glad to have been able to say these few words of respect and friendship in front of their children. His daughter will now speak about her parents.

Thank you for your attention. Jean Claude Pecker (Figure 2).

## 3. Einar Tandberg-Hanssen in Hunstville (Allen Gary, S.T. Wu and Ronald Moore)

In 1974, Dr. Einar Tandberg-Hanssen left the High Altitude Observatory at the National Center for Atmospheric Research in Boulder, Colorado to join the Space Sciences Laboratory at NASAs Marshall Space Flight Center (MSFC) in Huntsville, Alabama. There, he was a Senior NASA Research Scientist and later served as Deputy Director. He served as Director from 1987 until his retirement from NASA in 1993. He promptly took a part-time post at the Center for Space Plasma and Aeronomic Research at The University of Alabama in Huntsville where he worked until his death.

During his tenure at NASA, Einar along with Dr. Mona Hagyard and Dr. S. T. Wu, built up a substantial, internationally-based group of solar physicists at MSFC and UAH. He was lead investigator on two instruments aboard NASA spacecraft: the S-056 X-ray Event Analyzer on the "Skylab" Apollo Telescope Mount (which provided pioneering, high-time-cadence temperature and density information on solar X-ray-emitting regions) and the Ultraviolet Spectrometer and Polarimeter on the Solar Maximum Mission (which carried out sweeping new studies of EUV emission from solar active regions and flares). He received the NASA Exceptional Service Medal.

Dr. Tandberg-Hanssen's books about various aspects of solar activity, viz. Solar Activity (Blaisdell, 1967), Solar Prominences (Reidel, 1974), The Physics of Solar Flares (with A. G. Emslie) Cambridge, 1988, and The Nature of Solar Prominences (Reidel, 1995) have become international standard works within the discipline of solar physics.

Dr. Tandberg-Hanssen's Solar Physics Memoir paper entitled "Solar Prominences An Intriguing Phenomenon" was recently published before his death and can be downloaded via http:www.springerlink.com/content/1166j74k577kv332/. The article starts with an autobiographical account, where the author relates how his several study trips abroad gradually led him to the study of solar physics in general, and prominences in particular. Einar's residence as a research fellow at the Instut d'Astrophysique in Paris in the 1950s laid the foundation for a lifelong interest in France and French culture. His great interest and knowledge of French medieval churches, as well as the Norwegian state churches is reflected in two books "Letters to My Daughters" (Ivy House Pub. Group, 2004), and "The Joy of Travel: More Letters to My Daughters" (Pentland Press, 2007), which serve as a review, tourist guide and history book shaped in the form of letters home to his two daughters from his many travels in Norway and France.

This image (Figure 3) of Einar with a unicycle was captured at the end of a photographic session with Julia Gary. Einar had requested a portrait for his book, "The Joy of Travel: More Letters to My Daughters". After the formal session, Allen Gary brought out his unicycle, placed it next to Einar, whose quizzical look quickly changed into a whimsical grin as he posed in his usual aristocratic manner. This photographic event in 2006 was the culmination of a dinner which occurred in the early 1980s and was hosted by Einar and Erna in Huntsville, Alabama for Brigitte Schmieder, one of the editors of this symposium. At that time Einar was engaged as the principal investigator for NASA Solar Maximum Missions Ultraviolet Spectrometer/Polarimeter experiment working with Bruce Woodgate, Grant Athay, Jacques Beckers, and John Brandt, Dick Shine, Art Poland, and groups at Marshall Space Flight Center, Goddard Space Flight Center, University of Colorado, The University of Alabama in Huntsville, and the Observatoire de Paris. At this dinner, Julia Gary mentioned that she had just come from a local French group. The Tandberg-Hanssens were excited to learn about the Huntsville group and decided to join. The French group frequently met at the Gary's home and after Ernas death in 1994, the tradition of inviting Einar to remain for dinner was begun. The

**Figure 3.** Einar Tandberg-Hanssen with a unicycle In Hunstville in 1997 (@Julia Gary)

dinner discussions included solar science, photography, the art and history of cathedrals, and the culture and literature of France, as always, accompanied by Einars insightful, aristocratic Norwegian intellect and twinkle, and, of course, *une bouteille de vin rouge français.* The enthusiastic discussions of Einar's letters to his daughters in the Huntsville French group and in the dinner conversations led to the publication of his two books and the photographic portrait session.

Einar was a true scholar and true gentleman. As evidence by his papers, his books and his dealings with others, he was always seeking not only to expand his own knowledge and understanding, but also to find new ways of communicating his remarkable insight to others. He is survived by his daughters, Else and Karin and grandchildren.

## 4. Einar Tandberg-Hanssen ( Else Biesmann's Speech)

Bonjour, God Dag, Hello,

It is our privilege to be invited to join you here in one of our fathers favorite cities and to share in this event with you. We are honored and pleased to be included.

Our father, as most of you may already know, was a very special gentleman. He taught us to enjoy nature and God's wondrous surroundings as well as the incredible creations of mere man. He seemed to know everything about everything but never came across as a know-it-all. His gentle and wise demeanor seemed to give everyone who met him the feeling that you were the most important person in the world to him at that moment.

He was a Francophile to the highest degree and would have chosen France as his home if circumstances would have permitted. Indeed, he spent the last few years of his life making sure that Karin and I had an opportunity to experience as much of France as possible in the way he had learned to love it - to see her many wonders and beautiful sights from north to south and east to west. Our annual expeditions taught us what a diverse and wonderful country France truly is. One of his favorite areas was Provence where we would end each of our French Adventures. Most likely his love for Provence had to do with his early years spent there with our mother when he was earning his Doctorate degree. He made life-long friends wherever he went, even with the hosts of several of the places to which we returned, including one of his favorites in Saintes Maries de la Mer to which he returned year after year. We always explored new areas but he made sure that we returned to two or three of the same places as anchor spots - the places that he liked the best. It seemed that everywhere we went we would find people that knew him and would welcome him with warm smiles and open arms. This welcome would of course extend to us as well as soon as he introduced us, which gave us a warm and happy feeling.

I hope I will not be considered blasphemous when I say that being in my father's presence was always like sitting at the feet of God and soaking in his radiance. I always held my father is such high regard. He was always on a pedestal for me and could do no wrong. Perhaps it is similar for most little girls, but I never outgrew this feeling for my father. He was always so wise and loving and forgiving because I could not possibly live up to what I thought were his standards. As it turned out, he did have high expectations of us but he accepted us as we were. But my impression of what he expected sure gave me a good goal to work for, although I am still fairly certain I have not yet attained it. I remember a time as a little girl, perhaps eight or nine years old, when I ran away from home after misbehaving and arguing with our mother and not being willing to admit my guilt. I ran and walked about a block from our home and found a big rock in an empty field where I decided I would stay. I remember sitting on my rock and pouting and wondering if anyone would care enough to find me. After a while, which seemed like a very long time, my father came walking by as I was sitting there on my rock, my new home. He found me sitting there and came and asked if he could sit on my rock. And then he sat with me in silence for a while. Eventually he began to talk to me about my new surroundings and where I had chosen to run away to - he loved rocks and knew exactly what kind of rocks and minerals we were sitting on and, although I don't remember his exact words, as he was explaining the feldspar and mica and granite in the rock, I knew he had forgiven me and it wasn't long before he persuaded me to come home with him. He assured me that we could come back and look at my rock again another time.

Our father often spoke of the times when we were growing up in Boulder, Colorado, as some of his best years. He was a devout, disciplined and caring father and spent a great deal of time making sure we were learning and gaining the best experiences. We were both given instruments to play and lessons in music and dancing and we were each required to practice each instrument for half an hour every morning before school. He would reward us by reading to us in the evenings after our family suppers. Karin and I would snuggle up on either side of him on the sofa so we could see any pictures in the books and eventually as we grew older we could follow along with the words. But our father did not read traditional childrens books to us - he would read things like the Time-Life series "The American Revolution", so our evening readings were not only a cozy family time, they were also educational, although sometimes gory and a little scary for little girls. As I looked back on this later, I am certain he was trying to teach us some history of the country we had chosen as our new home.

Our father was always very active in the church and one of my fondest and most secure memories of growing up was sitting next to my father in the church pew leaning on his shoulder with my hand slipped up into his crossed arms and holding his hand. Karin always sat on his other side between him and my mother - perhaps Karin was holding his other hand? Those times were the most serene and perfect days a child could possibly have to look back on. But there was the one morning driving home after church when to our surprise and alarm our father was pulled over to the side of the road by a policeman - our perfect father had been driving faster than the speed limit and the policeman wanted to give him a ticket. I believe this is the only time I remember seeing my father feel shame or guilt, yet even this situation he handled in a calm and dignified, yet humbled, manner. He has always set a good example for us.

The importance of planning was something our father tried over and over to teach us. And it seems that planning trips was one of our father's favorite things to do. When Karin and I were just becoming teenagers we were fortunate to be included in another trip to explore Europe on our way to our beloved Norway because Pappa had an IAU conference to attend. We were indeed fortunate - every three years while we were growing up Pappa would bring us somewhere where the IAU meetings were being held and share the world's wonders with us. These trips were always planned to include visits to castles and the magnificent cathedrals created centuries before in the area where each conference was held as well as cathedrals all along our trip from the conference location to Norway. On this particular trip, when we were about 12 and 14 years old, we were taking several detours on our journey to see what we had begun to consider a few too many cathedrals. Being teenagers we had perhaps not yet come to appreciate the true magnificence of each these individual structures - each of them had unique features that he wanted to us to see and appreciate. The flying buttresses, a particularly fine stained glass window, an unusual baptismal font. The special features list was seemingly endless and in his view he would have loved to see every last one of them. We had not yet learned to understand the splendor of each of these exceptional works of art and finally we rebelled and told him "Not another cathedral!" To his great surprise we had seen enough cathedrals. In his wisdom he did not let us see his full disappointment, but we did note his surprise at our intolerance. We spent many years afterwards joking about our rebellion and Pappa tried to pick only the finest cathedrals for our future trips together. It became a family tradition to joke about our father's love of cathedrals and fortunately he was able to find others who shared his great love of cathedrals. When he traveled without us he would take in as many of them as he could. Our father loved to tell the story of one of his solo trips when he met a gentleman surveying a cathedral who walked with a limp. They struck up a conversation about the particular features of the cathedral they were in and soon my father learned that this gentleman, Mr. Dodge from Michigan, had received his limp by walking through a cathedral while looking up and admiring the soaring ceilings and since he was not looking where he was going he tripped and hurt his leg. The two of them exchanged information and were soon good friends and would get together periodically over the decades to discuss their latest travels and which cathedrals they should recommend to each other. Our father had found another kindred spirit. But our father was not only determined to teach us about ancient cultures and the finest manmade things in the world. He also wanted us to enjoy what God has given us and teach us to care for nature and those around us. From the time we were born we were taken for hikes and skiing trips. To begin with we were in backpacks on his back and then we were fitted with hiking boots or skis depending on the season and were given the love of the great outdoors. I was put in my first ski race at the ripe old age of two when we were still living in Norway. And both Karin and I were taken to a local downhill ski area when we were five and seven years old respectively after our move to Colorado. On our hikes Pappa taught us about the way fallen trees would normally twist their trunks to the right when they died but once in a while we would find a fallen tree that twisted to the left it was a marvel that became something we would learn to hunt for. He taught us about the mosses growing in the moist, shady areas along the normally dry, Colorado mountain trails, and about the lichens that grew in various colors on the rocks in the sun. There were the sounds of the creatures in the forest, around the mountain lakes and in the high rocky passes that we soon learned to identify. We would bring little books on our hikes teaching us about the beautiful birds and flowers all along our paths throughout the beautiful and rugged Colorado Rocky Mountains. On our American vacations we explored the great southwestern part of the US touring many American Indian ruins before it became fashionable to do so. Our parents had found a great appreciation for the handicrafts of the southwestern Indians. They had found another fascinating ancient culture that they wished to share with us and it was located in another beautiful, unusual landscape that our father found intriguing. It was another mixture of manmade and God-made beauty that he wished to share with us. During our vacations to our beloved Norway we were taught about the ancient Viking culture while exploring the woods, gently rolling mountains and farmlands of our native land. It is in our native land where we find the most peaceful paradise our father could share with us and preserve for us: our family cabin which is perched along a gently sloping fjord in the southern part of Norway where a few of his colleagues have also had an opportunity to visit. Our father gave us this treasure to love and cherish so we may maintain our family ties and continue our family gatherings. For our father's 80th birthday, our entire family gathered at the cabin. Once again our father did some excellent planning as there were so many of us arriving from America that we had to borrow rooms in the neighboring cabins, which our neighbors were happy to help Pappa with. The planning for the event was meticulous and everything went smoothly. A few years earlier we had all gathered for our mother's burial and Pappa had also planned things meticulously. Mamma and Pappa's friends and our Norwegian relatives joined us for a big luncheon following the interment in an elegant, old restaurant in town near the cabin. The day afterwards all the Americans went to Oslo for some sightseeing. Pappa had spent time in Oslo a few weeks beforehand preparing where we were to stay and how our large contingent of all ages would travel from place to sight-seeing place and enjoy the sights of Oslo. His planning included actually tracing the steps we would take and which trams we could all take from the hotel including the tram time-tables. He had always been an excellent tour guide and we were all looking forward to our sight-seeing adventure. After arriving in Oslo and checking into our hotel, we all gathered in the hotel lobby and proceeded to follow him like a long line of ducklings to the nearest tram stop. As we approached the spot where the tram was to stop to take us to town, we noticed our father's face acquire a priceless look - a combination of shock and dismay - as the tracks had been dug up and the train would no longer be coming to pick us up! But, as always, our ever resourceful father came up with Plan B and, after a little hiccup in his plans, we proceeded to tour Oslo with the best, most knowledgeable tour guide imaginable.

Thanks to our fathers enthusiasm for exploring the world, which was truly contagious, we were both smitten with a love of nature and bitten very early in life by a severe case of the travel bug. I now live in the woods watching the wildlife from our home and my jobs have been helping me to explore the world. Karin has lived for many years by the ocean and she creates beautiful things from shells and things that are found by the sea. We are both truly pleased to be included in this symposium honoring our father and feel very honored to be able to share a little bit about him from his daughters perspectives. Thank you again for inviting us.

Else Biesmann and Karin Brock the daughters of Einar, Jacquelyn and Jennifer the grand-daughters of Einar in Paris (Figure 2).

# Session I - 1.1

## PROMINENCE FINE STRUCTURE, DYNAMICS AND SEISMOLOGY

*Nature of prominences and their role in space weather*
*Proceedings IAU Symposium No. 300, 2013*
*B. Schmieder, J.-M. Malherbe & S. T. Wu, eds.*

doi:10.1017/S1743921313010697

# Solar Prominence Fine Structure and Dynamics

**Thomas Berger**

National Solar Observatory,
P.O. Box 62, Sunspot, NM, 88349 USA
email: tberger@nso.edu

**Abstract.** We review recent observational and theoretical results on the fine structure and dynamics of solar prominences, beginning with an overview of prominence classifications, the proposal of possible new "funnel prominence" classification, and a discussion of the recent "solar tornado" findings. We then focus on quiescent prominences to review formation, down-flow dynamics, and the "prominence bubble" phenomena. We show new observations of the prominence bubble Rayleigh-Taylor instability triggered by a Kelvin-Helmholtz shear flow instability occurring along the bubble boundary. Finally we review recent studies on plasma composition of bubbles, emphasizing that differential emission measure (DEM) analysis offers a more quantitative analysis than photometric comparisons. In conclusion, we discuss the relation of prominences to coronal magnetic flux ropes, proposing that prominences can be understood as partially ionized condensations of plasma forming the return flow of a general magneto-thermal convection in the corona.

**Keywords.** Sun: prominences, Sun: filaments, Sun: magnetic fields, Sun: atmospheric motions, Sun: chromosphere, instabilities, Sun: oscillations, Sun: UV radiation

---

## 1. Introduction

It is an honor to be asked to review our state of knowledge of prominence fine structure and dynamics at a meeting dedicated to Professor Einar Tandberg-Hanssen. The number of fundamental results in this field that were directly made or influenced by Prof. Tandberg-Hanssen is truly impressive. Since the last edition of his book (Tandberg-Hanssen 1995), and particularly since the recent launches of the *Hinode* Solar Optical Telescope (SOT; Tsuneta *et al.* 2008) and SDO/AIA instruments (Lemen *et al.* 2011), there has been an explosion of results too numerous to cover in this brief review. Thus I will focus on key results from the last several years and in particular on recent findings in quiescent prominences, my own field of expertise. For broader reviews see Labrosse *et al.* (2010) and Mackay *et al.* (2010).

This article, being based on the first talk of the conference, begins by introducing prominences and filaments†. Section 2 briefly reviews current definitions of prominences and the distinctive structures and dynamics in various types. We also discuss whether a new type of coronal condensation, thus far called "funnel prominences", constitutes a real prominence or is perhaps better understood as a unique condensation event more akin to coronal rain. We conclude the discussion with a review of the so-called "tornado" phenomenon and its relation to other prominence structures. Section 3 then focuses on quiescent prominence dynamics, in particular prominence formation, downflow dynamics, and the enigmatic prominence bubble instability. We conclude by proposing the unifying

† As usual we use the term "filaments" to refer synonymously to prominences seen on the disk.

hypothesis that prominences can be understood as largely neutral plasma condensations within magnetic flux ropes in the solar corona, with the observed differences in structure and dynamics caused by variations in the magnetic field strength, plasma ionization degree, and height of the flux rope in the atmosphere.

It will be noticed that the theoretical and measured magnetic structure of prominences is not covered in detail in this review. References to "structure" in this review concentrate primarily on plasma morphology and not on magnetic field topology, although the two are of course linked to varying degrees in different prominence types. Bommier *et al.* (1994) summarizes the pioneering efforts to measure quiescent prominence magnetic fields using the Hanle effect at the Pic du Midi observatory through the 1990s, while Casini *et al.* (2003), López Ariste and Casini (2003), Merenda *et al.* (2006), and Schmieder *et al.* (2013) present more recent measurements. Also notably lacking will be details of radiative transfer models; see Gunár (2013), Berlicki *et al.* (2011), Heinzel *et al.* (2008) for recent progress in this sub-field of prominence research. As with any brief review, omission of key papers in the field is inevitable, due only to my own limitations in following the literature and the required brevity of the review, and not to any intentional neglect.

## 2. What are prominences and filaments?

Since Prof. Tandberg-Hanssen's book, there has been a narrowing of the definition of "prominence" to include only those structures composed of relatively low-temperature, $\sim 10^4$ K, plasma in the solar corona that are associated with large-scale, $O(10\text{–}100)$ Mm, magnetic polarity inversion lines (PILs) in the photosphere (Engvold 1998). Equivalently, prominences all lie above chromospheric "filament channels" that form in association with PILs in the photosphere (Lites *et al.* 2010; Gaizauskas *et al.* 1997). Thus so-called "surge prominences" are now classified as active region jet phenomena. Similarly "loop prominences" are now recognized as post-flare loop arcades, quite distinct from any prominence or filament structures on the Sun.

In the following sections we employ the common "active region", "intermediate", and "quiescent" prominence classifications to delineate basic variations in prominence characteristics. More sophisticated schemes based on PIL topologies can be found, e.g., in Mackay *et al.* (2008). It should also be noted that there are common structural characteristics in all prominence types, e.g., the "barb" structures that are seen as extensions of filaments reaching downwards (Martin 1998), the "spine" regions which comprise the major axis of the structure (particularly evident in long filaments seen on the disk), and the associated characteristic of "chirality" that is observationally related to the direction of barb formation relative to the PIL (Martin *et al.* 2008).

### 2.1. *Active region prominences*

Active region (AR) prominences occur adjacent to sunspots, typically over PILs in associated plage regions. The key structures in AR prominences are long thin threads that occur in groups of relatively horizontal "bundles". The bundles are generally not straight and can exhibit upward or downward curvature. Thread lengths are typically 5–30 Mm with typical thicknesses of 350–650 km in *Hinode*/SOT images (Okamoto *et al.* 2007). Figure 1 shows a typical AR prominence. Note that the prominence is only 2–3 times higher than the chromospheric spicules, i.e. about 15 Mm in maximum height.

AR prominences are highly dynamic with flows along the thread structures on the order of $10\,\mathrm{km\,s^{-1}}$ (Okamoto *et al.* 2007) and perpendicular (possibly erupting) motions up to $70\,\mathrm{km\,s^{-1}}$ (Kucera *et al.* 2003). The thread groups themselves sometimes rise or appear to twist in impulsive events, with plasma continuing to flow along the threads during these

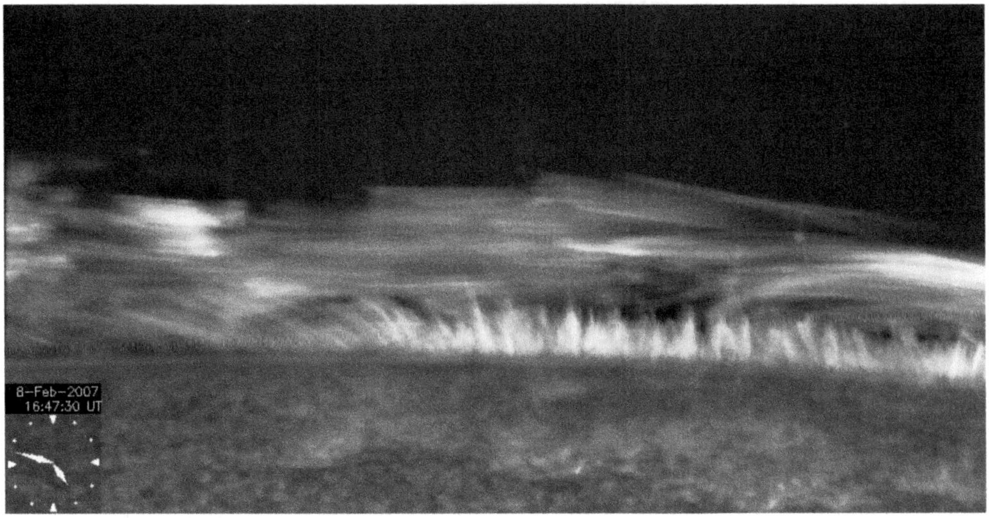


**Figure 1.** A typical active region prominence seen at the limb in the *Hinode*/SOT Ca II 396.8 nm bandpass on 08-Feb-2007. The image has been rotated to place the solar limb horizontal. Courtesy of J. Okamoto, ISAS/JAXA.

transients. Doppler velocities in AR filaments show only small mostly upward velocities of $\sim 1\,\mathrm{km\,s^{-1}}$ (Kuckein 2012), perhaps due to localized heating events (Filippov and Koutchmy 2002). AR prominences are the most eruptive of prominences with typical time between eruptions or major "activation" events measured in hours. In comparison, intermediate and quiescent prominence erupt on time scales of days or weeks.

Evidence of Alfvén waves in AR prominences is found with periods of 120–250 s and wavelengths of 250 Mm (Okamoto *et al.* 2007). Prominence oscillation events have been related to pre-eruption dynamics as well (Bocchialini *et al.* 2011). A recent review of prominence "seismology" is given by Ballester (2006).

### 2.2. *Intermediate prominences*

Intermediate prominences form outside of active regions, typically in the mid-latitude regions between remnant plage regions that have been sheared by differential rotation to form elongated PILs, sometimes extending for 500 Mm or more. Intermediate prominences do not usually occupy the entire PIL, occurring in shorter segments of order 100 Mm in the extended filament channel.

Like AR prominences, intermediates are composed of a multitude of thin, ~300–500 km, threads. However intermediate prominence threads are shorter in length and occur in upward-arcing "dips", at least when seen at the limb along some sight lines (prominence appearance is highly dependent on the angle of the line-of-sight to the local prominence axis at the limb). Figure 2 shows a typical intermediate prominence at the limb as well as the associated mid-latitude filament. Magnetostatic models of prominences based on extrapolations of measured photospheric magnetic fields appear most similar to intermediate prominences due to their assumption that prominence plasma collects only in concave horizontal dips of the magnetic field (Dudík *et al.* 2008; Aulanier and Demoulin 1998; Aulanier *et al.* 1998; Demoulin *et al.* 1987).

Ahn *et al.* (2010) finds large scale flows with velocities of $\sim 10\,\mathrm{km\,s^{-1}}$, with evidence of counterstreaming similar to that found in the observations of Zirker *et al.* (1998), in a large intermediate prominence, but more often one observes only localized flows or large scale "rippling" oscillations in these structures. For example, oscillations have been

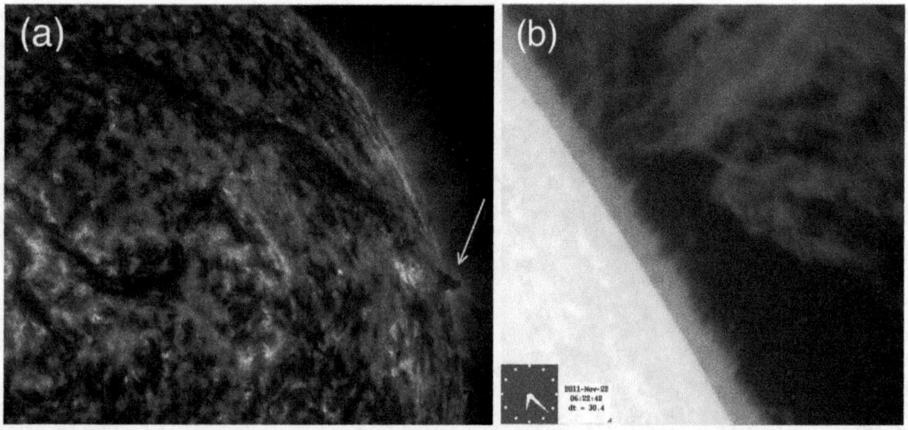


**Figure 2.** (a) A typical long intermediate filament complex observed in the SDO/AIA He II 304 Å bandpass on 20-Nov-2011. The arrow points to the location of the *Hinode*/SOT observation in Panel b on the following day. (b) *Hinode*/SOT H$\alpha$ line-center 656.3 nm prominence image on 21-Nov-2011. Note the short horizontal dip segments comprising the majority of the structure.

studied in intermediate prominence sheets, showing evidence of transverse magnetosonic wave propagation with power spectrum peaks at 200–300 s (Schmieder *et al.* 2013).

Intermediate filaments exhibit the same activation events as AR structures but with lower frequency. Intermediate filament eruptions typically involve only individual segments with neighboring segments often remaining unaffected. Su and Van Ballegooijen (2012) model the magnetic field configuration of a large erupting intermediate filament to show that the pre-eruption structure is consistent with that of a magnetic flux rope anchored at one end in an active region that injects destabilizing plasma into the prominence. Recent SDO observations (Schrijver and Title 2011) establish that large intermediate filament eruptions, perhaps themselves triggered by waves generated in AR flares, can trigger nearby filaments to erupt. Török *et al.* (2011) have successfully modeled this cascade of eruptions as an interacting flux rope/arcade field system.

### 2.3. *Quiescent Prominences*

Quiescent prominences are typically found in high latitude regions (>50°) far from active regions, are generally shorter in latitudinal extent, and associated with weaker photospheric fields. Quiescent prominences are the tallest prominences, sometimes extending 50 Mm or more above the limb. All quiescent prominences have overlying coronal cavities, however the cavity may not be visible in typical EUV filtergrams for some sight angles. Figure 3 shows a typical quiescent prominence extending to about 35 Mm above the photospheric limb.

Typical quiescent prominence structures in visible light passbands are long, predominately quasi-vertical threads. These threads are thicker and more complex than active or intermediate prominence threads and do not appear to me to be structured on horizontally dipped magnetic field lines. Typical thread thickness is 500–700 km in *Hinode*/SOT observations with lengths on the order of 10–20 Mm (Berger *et al.* 2008). However when seen on the disk, quiescent filament threads appear much thinner and more horizontal than limb prominence threads (see e.g., Lin *et al.* 2005). The difference between the on-disk and off-limb appearance of quiescent prominences remains one of the key mysteries in prominence studies. Oscillations in quiescent filament threads have been analyzed by

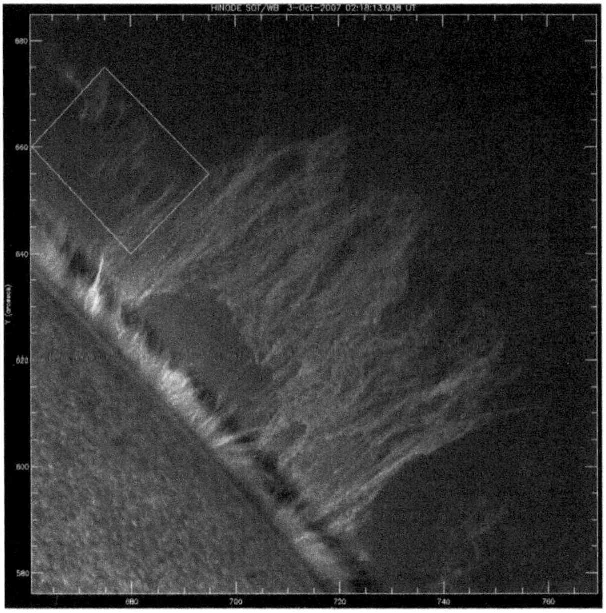

**Figure 3.** *Hinode*/SOT Ca II 396.8 nm image of a quiescent prominence on the NE limb on 03-October-2007. The box on the left outlines a region of prominence formation with droplet downflows (Haerendel and Berger 2011). The large gap in the lower region of the prominence demarcates the location of an earlier prominence bubble passage. A radial gradient filter is applied to the image causing the dark band above the spicules. Major tickmarks are 20 arcseconds apart.

Ning *et al.* (2009) and Lin *et al.* (2007), showing periods of 200–400 s with a dominant period of 300 s and typical amplitudes of ∼1000 km.

In EUV passbands, quiescent prominences can exhibit "horn" structures emanating from the top of the prominence (Berger *et al.* 2012; Berger 2012; Vourlidas *et al.* 2012; Plunkett *et al.* 2000). These upwardly curved extensions are not seen in visible light images of quiescent prominences and may represent channels of hot plasma comprising the "prominence corona transition region" (PCTR, Parenti and Vial 2007) connecting the choromospheric prominence to the hotter (Reeves *et al.* 2012) overlying coronal cavity flux rope. In eruptions of quiescent prominences these horns appear to conform exactly to the bottom contour of the rising cavity portion of the flux rope (Régnier *et al.* 2011).

Quiescent prominences are dominated by plasma downflows in the quasi-vertical threads with speeds on the order of 10 km s$^{-1}$ (Schmieder *et al.* 2010; Berger *et al.* 2008). In some cases, isolated "knots" or droplets of plasma are observed to be ejected from quiescent prominences and to fall at near free-fall speeds of ∼100 km s$^{-1}$ (Haerendel and Berger 2011; Hillier *et al.* 2012b). Recent observations have also revealed the "prominence bubble" phenomenon to be common in quiescent prominences (Dudík *et al.* 2012; Hillier *et al.* 2011; Schmieder *et al.* 2010; Berger *et al.* 2010; Ryutova *et al.* 2010; de Toma *et al.* 2008; Berger *et al.* 2008). Prominence bubbles have not yet been identified in intermediate or AR prominences. We defer further discussion of these dynamics to Sec. 3.

### 2.4. *Funnel Prominences*

Continuous, full-Sun, multi-spectral, data from SDO/AIA have made it clear that condensation events resulting in "chromospheric" plasma in the corona are common. Prominences and coronal rain are well-known examples, but another type of event has recently been termed "funnel prominences" (Liu 2013). These events initiate at shallow dips in long coronal loop systems and are usually first detected in the He II 304 Å bandpass as they form into large drainage flows within hours of first nucleation. Superficially, funnel flows resemble quiescent prominences but they can appear at lower latitudes and have not yet been identified with PIL/filament channel systems. Figure 1 of Liu (2013) shows an SDO/AIA composite image that distinguishes between established prominence types and funnel events.

It has been suggested that funnel prominences are equivalent to "cloud prominences" described in earlier ground-based observations (Allen *et al.* 1998), but the latter appear closer in character to "coronal rain" events with high-altitude condensations subsequently flowing at high speeds along coronal loop field lines (Antolin and Verwichte 2011; Schrijver 2001). In contrast, funnel prominence downflows are constrained to narrow conical regions (hence the name) with the plasma apparently flowing *across* coronal magnetic loops. Funnel flow speeds are thus slower than coronal rain speeds, typically only about $30\,\mathrm{km\,s^{-1}}$. However this is still 2–3 times the typical flow speed in quiescent prominence downflows.

### 2.5. *Prominence pillars and solar "tornados"*

Recent SDO/AIA observations in the EUV show apparent rotational motion in so-called prominence "pillars". Pillars appear as narrow absorption features in the EUV, extending more or less vertically at semi-regular intervals along a filament. They are visible in emission in visible lines as well, but are usually wider than the associated EUV structures. Earlier observations (Pettit 1932) suggest vertical axis rotational motions in prominences, but the continuous full-Sun SDO observations establish these pillars as ubiquitous features of intermediate and quiescent prominences on the Sun†. Figure 4 shows the typical appearance of these structures in a variety of spectral bandpasses.

The apparent rotation of a predominately vertical structure has resulted in these structures being termed "solar tornados" (Su *et al.* 2012). However measurement of continual rotation in these structures is not yet definitive. Several analyses rely solely on AIA movies showing sinusoidal patterns in "time slices" through the structures and such patterns could conceivably be caused by rapid transverse oscillations or counterstreaming flows rather than rotation. However ground-based observations of pillars utilizing doppler velocity measurements show red and blue shifts of $\sim 6\,\mathrm{km\,s^{-1}}$ (Orozco Suárez *et al.* 2012) to $\sim 20\,\mathrm{km\,s^{-1}}$ (Wedemeyer-Böhm *et al.* 2013; Schmieder *et al.* 1991) on alternate sides of the pillar, consistent with a rotational motion of the structure. Wedemeyer-Böhm *et al.* (2013) discuss the possible link between this apparent rotation and vortical motions in the photosphere and lower chromosphere (Wedemeyer-Böhm *et al.* 2012; Zhang and Liu 2011; Brandt *et al.* 1988).

The relation of these possibly rotating prominence pillars to filament barbs is unclear (see however recent observations of Li and Zhang (2013)). High resolution images of filament barbs (e.g., Fig. 1 in Lin *et al.* 2005) do not appear anything like the pillars seen in EUV images, with no evidence of rotational motion in the accompanying movies. The detailed relation of solar "tornados" to other prominence structures such as barbs and spines remains to be determined. Also, whether there is a consistent plasma flow *along* the pillars, either upwards or downwards, is unknown. Finally, the term "tornado" has recently also been used to describe large scale rotational flow in a coronal cavity

† To my knowledge, pillars and the tornado phenomenon have not yet been identified in AR prominences.

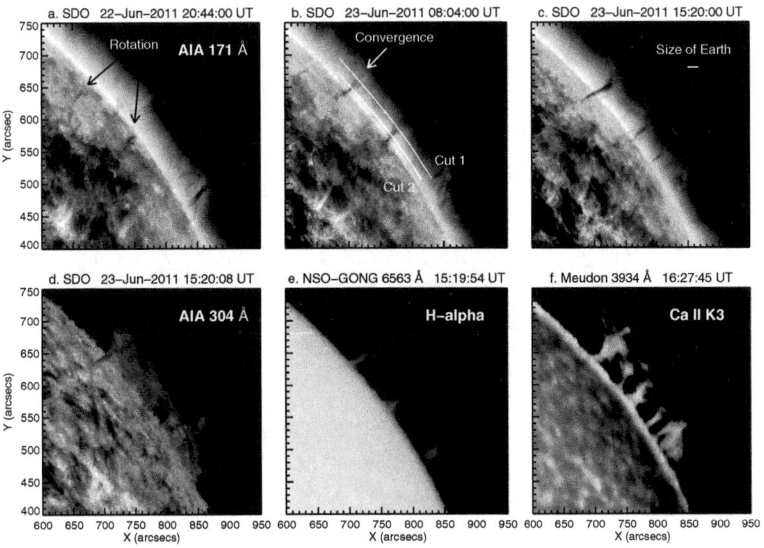


**Figure 4.** Multispectral view of solar tornados from Su *et al.* (2012). The prominence is a large intermediate complex that extends around the limb. Note that the dark apparently rotating pillars in the SDO/AIA EUV bandpasses are significantly narrower than the structures seen in the visible H$\alpha$ and Ca II K-line images. The pillars fan out into the spine region of the prominence at high altitudes to give the appearance of plasma "trees" in the corona.

(Panesar *et al.* 2013), but this is likely distinct from the pillar tornados, perhaps related to spinning motions reported by Wang and Stenborg (2010).

## 3. Quiescent Prominence Fine Structure and Dynamics

We concentrate now on recent findings on quiescent prominences, with an obvious bias towards my own research and interpretations in this area. Quiescent prominences have the advantage over intermediate and AR types of relative freedom from interference by surrounding active region structures, although they have weaker surface field strengths that can make connections to the lower atmosphere less clear. For example, the three-part structure of CMEs consisting of the prominence, coronal cavity, and overlying streamer fields is best seen in polar crown CMEs (Low and Hundhausen 1995; Low 2001). Engvold (1998) and Zirker (1989) give earlier reviews of quiescent prominences. Priest (1989) presents a summary of knowledge to the late-1980s on these objects.

### 3.1. *Formation and downflow dynamics*

Mechanisms of prominence formation typically discussed are (1) direct injection of chromospheric plasma into the corona via siphon flows (Pikel'Ner 1971; Engvold and Jensen 1977), magnetic reconnection jets (Litvinenko and Martin 1999; Wang 1999), or flux emergence (Hu and Liu 2000; Okamoto *et al.* 2009); (2) condensation of coronal plasma injected at the footpoints of prominence magnetic field lines (so-called "thermal non-equilibrium" formation) (Xia *et al.* 2012; Luna *et al.* 2012; Karpen and Antiochos 2008; Antiochos and Klimchuk 1991); and (3) direct, or *in situ*, condensation of coronal plasma.

Saito and Tandberg-Hanssen (1973) argue against *in situ* condensation on the grounds that typical prominence densities imply the entire corona contains insufficient mass for the number of observed prominences. However Schmieder *et al.* (1984), following on the suggestions of Malherbe *et al.* (1983), suggest that if the process is dynamic, i.e., if the corona is continually resupplied with plasma, then *in situ* condensation would be possible.

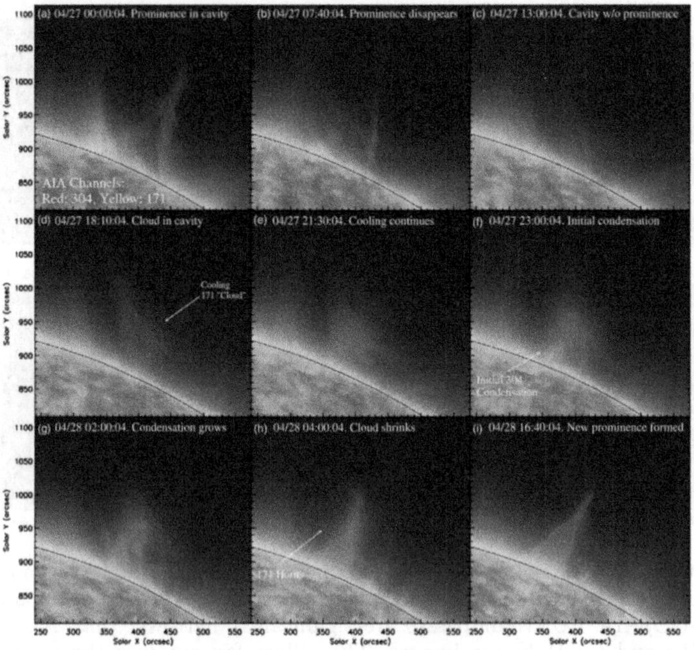


**Figure 5.** *In situ* condensation of a polar crown quiescent prominence in a coronal cavity from Berger *et al.* (2012). Emission in the $\sim 10^6$ K bandpass dominated by Fe IX 171 Å; chromospheric He II 304 Å emission is shown in orange. Time advances from left to right, top to bottom. The upper right frame shows the empty coronal cavity following the drainage disappearance of the prominence; the lower right shows the prominence fully reformed approximately 27 hours later. The time from first appearance of the hot cloud in the cavity to first evidence of prominence condensation is approximately 5 hours.

Liu *et al.* (2012) observe an intermediate prominence formation via condensation but it is unclear whether it is an *in situ* or a transport condensation event. However Berger *et al.* (2012) analyze multi-wavelength SDO/AIA observations of a coronal cavity and find clear evidence of quiescent prominence formation via *in situ* condensation with no evidence of flows of hot plasma as would be implied by thermal non-equilibrium models. Figure 5 shows a series of SDO/AIA images of the event including the initial disappearance of the prominence due to drainage from the coronal cavity.

Quiescent prominence formation via *in situ* condensation results in vertical drainage of plasma across the apparently horizontal magnetic field lines of the system. For fully ionized plasma such cross-field transport is prevented by Lorentz forces. Low *et al.* (2012a,b) address the physics of cross-field transport of partially ionized prominence plasma. Using analytic radiative cooling functions, they show that a thermal instability can set in that condenses plasma to much lower ionization states and significantly higher densities than typically assumed values (Hirayama 1985). This mechanism is an evolution of the "dragging reconnection" downflows discussed in Low and Petrie (2005) and cited by Chae *et al.* (2008) in his observational study of *Hinode*/SOT prominence downflows. Gilbert *et al.* (2007) also present evidence of cross-field diffusion of neutral He in intermediate prominences, following on initial work by Mercier and Heyvaerts (1977).

Haerendel and Berger (2011) empirically model "magnetic droplet" formation that also allows cross-field transport of condensed prominence plasma. Hillier *et al.* (2012b) analyze *Hinode*/SOT observations and a numerical prominence model to show that reconnection can produce such detached "droplets", falling at nearly free-fall velocities.

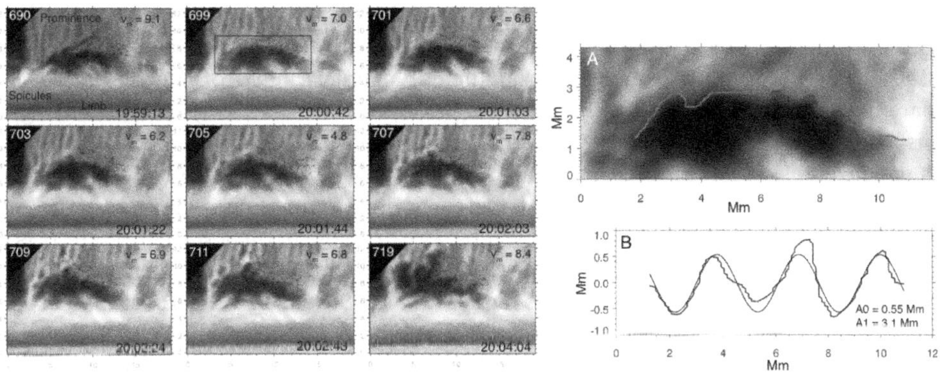

**Figure 6.** Left: sequence of images in the *Hinode*/SOT Hα bandpass showing the development of the RT instability in a prominence bubble. Arrows show the plasma flow as derived by the NAVE code (courtesy of J. Chae). The box in Frame 699 shows the region enlarged on the right. Right: enlargment of the prominence bubble just prior to RT instability onset. The red curve traces the boundary and the blue segment is fit below with a sine function to show perturbations of approximately 3 Mm in wavelength.

Finally Leonardis *et al.* (2012) analyze *Hinode*/SOT time series observations of a quiescent prominence to show that the flow character is consistent with a turbulent cascade, in support of mechanisms that imply a tangled/chaotic magnetic field in quiescent prominence downflows (van Ballegooijen and Cranmer 2010).

### 3.2. *Quiescent prominence bubbles*

Prominence bubbles are large ~10 Mm, dark (in visible light images) cavities rising into quiescent prominences from below. Bubbles sometimes stagnate at heights of 5–10 Mm before going unstable to form either a single large plume or a series of smaller ones. The plumes are apparently turbulent flows in the bright prominence body that rise with typical speeds of 10–30 km $s^{-1}$ (Berger *et al.* 2010, 2008), with a horizontal component to the velocity as well (Berger *et al.* 2011; Schmieder *et al.* 2010). Smaller plumes may rise only 10 Mm or so to "fade" into the prominence body while larger plumes may pass entirely through the prominence.

Quiescent prominence bubbles were first observed by Stellmacher and Wiehr (1973). In hindsight, many quiescent prominence images show evidence of bubbles, but single images cannot reveal the dynamics. Following the launch of the *Hinode*/SOT instrument, the phenomenon was rediscovered. Simultaneously, de Toma *et al.* (2008) observed the phenomenon in Mauna Loa Solar Observatory (MLSO) Hα patrol images. Ryutova *et al.* (2010) first suggested that the flows are consistent with a Rayleigh-Taylor (RT) buoyancy instability, implying that the bubbles are underdense relative to the prominence.

The RT instability occurs at the interface of fluids of disparate densities in an accelerated frame, typically a static gravitational one. Analysis of prominence bubbles shows that plasma flows *along* the boundary of the bubble can develop Kelvin-Helmholtz (KH) instabilities to form waves that act as the initial perturbations for the RT instability onset (Berger *et al.*, in preparation). Figure 6 shows a local correlation tracking analysis of flows along the boundary of a small quiescent prominence bubble. Note that there is a significant component of the flow upwards and to the left, along the bubble boundary, in the plane of the sky. This lateral flow generates waves on the "upwind" side of the boundary that subsequently grow into the RT instability plumes. This is perhaps the first observation of triggering of the RT instability via KH waves in an astrophysical setting.

Three-dimensional MHD simulations based on the KS prominence model (Kippenhahn and Schlüter 1957) verify that the observed flows are consistent with a magnetic RT "interchange" instability triggered by the presence of a hot, under-dense, region introduced within the cooler prominence plasma (Hillier *et al.* 2011). This simulation was criticized on the grounds that the perpendicular magnetic field of the KS model is in apparent contradiction to measurements (e.g. Bommier *et al.* 1994) of so-called "guide fields" along the prominence axis. Guide fields supposedly suppress the RT instability due to magnetic tension forces resisting small "undular" mode perturbations. In 2-D analytic models, magnetic tension suppresses all perturbation components parallel to the magnetic field below a critical wavelength given by $\lambda_c = B^2 \cos^2\theta / g\mu_0(\rho_2 - \rho_1)$ where $B$ is the prominence magnetic flux density, $\theta$ is the angle between the magnetic field and the perturbation wave vector, $\rho_2$ and $\rho_1$ are the higher and lower density values of the superposed fluids, respectively, and $g$ is the acceleration of gravity (Chandrasekhar 1981). Assuming $B = 10^{-3}$ Tesla and $\theta \sim 30\,\mathrm{deg}$ as implied by measurements, $\rho_2 = 10^{-9}\,\mathrm{kg\,m^{-3}} >> \rho_1$ and $g_\odot = 274\,\mathrm{m\,s^{-2}}$, the critical wavelength is ~5 Mm, significantly larger than the perturbations shown in Fig. 6 that develop into plumes.

This discrepancy shows that the simple 2-D magnetic tension analysis does not apply to real prominences which are complex, 3-D, partially ionized systems. Hillier *et al.* (2012a) modifies the KS model to include a guide field and shows that the instability is indeed not suppressed, but that the resulting plumes are larger and less turbulent. Khomenko (2013) studies the effects of partial ionization on the magnetized RT instability to show that the critical wavelength approaches zero as ionization decreases. While issues of line-tying boundary conditions may effect the model outputs, the general conclusion I draw is that in a real 3-D system the magnetic field can only effect certain perturbation modes for any given geometry and there will always arise other perturbations (e.g. those perpendicular to the local field direction) that cannot be suppressed (see, e.g., Stone and Gardiner 2007, who show 3-D MHD simulations of multi-mode RT instabilities.).

The buoyancy of prominence bubbles begs the question of their internal composition. Since they are always dark in H$\alpha$ and Ca II images, the bubbles must either contain plasma heated above the ionization temperatures of these lines (~138,000 K for Ca II) or they must be empty voids that allow the background corona to "shine through". In the latter case, the buoyancy is strictly magnetic; in the former, it could be both magnetic and thermal. Dudík *et al.* (2012) analyze a single bubble and compare SDO/AIA Fe XII 193 Å interior intensity to the surrounding corona. By defining the reference coronal intensity outside of the prominence cavity complex, they show that the bubble interior intensity is less than this background, implying that the bubble is either empty or contains traces of absorbing cooler plasma. Low-resolution spectroscopic studies of prominences apparently support this finding, but these studies have not demonstrated clear discrimination of bubble events in their data (Berlicki *et al.* 2011; Labrosse *et al.* 2011). However, Berger *et al.* (2011) analyze SDO/AIA data to find relative emission (thus implying a hotter-than-background plasma) in two separate events viewed in the 171 Å passband, shown by Parenti *et al.* (2012) to be due to temperatures of at least $\log T = 5.6$ K. In the second event analyzed by Berger *et al.* (2011), the bubble is very large and there is no foreground confusion, nor is there any "prominence" plasma that could lead to PCTR emission in the bubble. Figure 7 confirms that background reference choice is critical to determining the outcome of photometric comparisons: by choosing a nearby background rather than the distant one in Dudík *et al.* (2012), the comparison shows no relative difference in emission, thus implying only that the bubble interior plasma is not *hotter* than the background plasma emission in the Fe XII 193 Å bandpass.

The lower panels of Fig. 7 show a preliminary differential emission measure (DEM) analysis of one of the prominence bubbles and the nearby background corona from Berger *et al.* (2011). Using all six SDO/AIA EUV bandpasses, the DEM analysis implies that

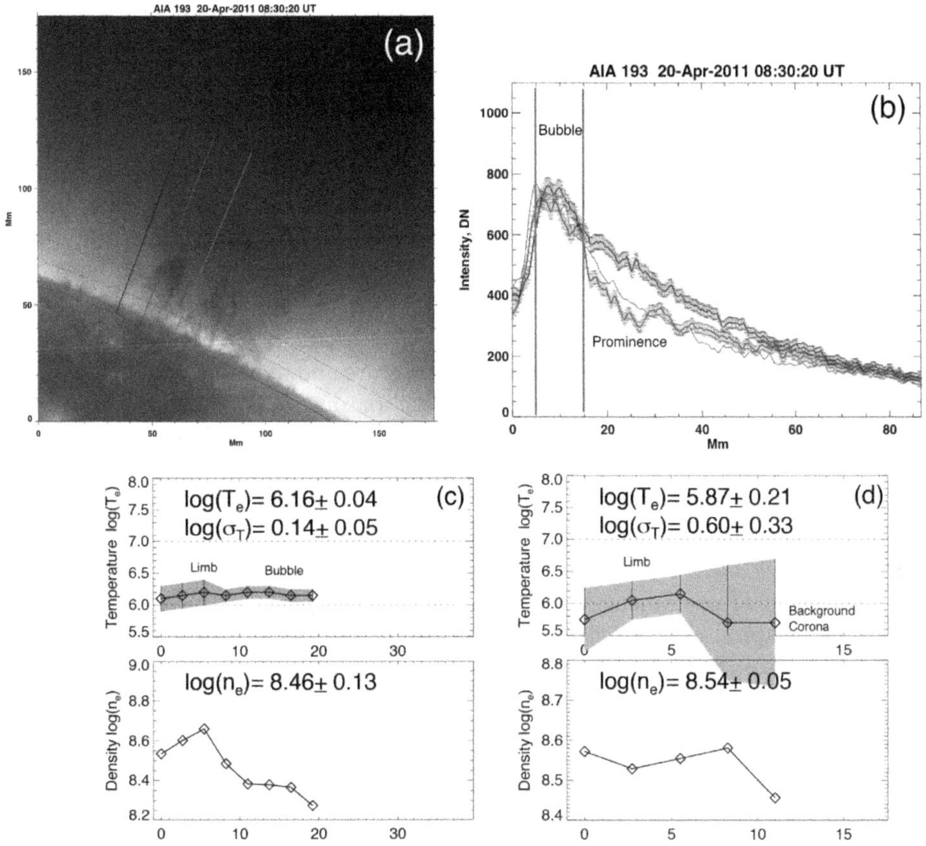


**Figure 7.** (a) Prominence bubble event on 20-Apr-2011 analyzed in Dudík *et al.* (2012). Colored vertical lines show photometric measurements of the background corona (blue), the bubble (red), and the coronal cavity central region (green). (b) Plots of SDO/AIA intensity in the 193 Å bandpass along the cuts shown in (a). Error bars are $\pm 1\sigma$ deviations due to Poisson statistics. The bubble is indistinguishable from the background in this EUV bandpass. (c) DEM analysis of a prominence bubble from 22-Jun-2010. The top frame show temperature as a function of altitude in Mm from the solar limb; the bottom panel shows density. Limb brightening emission accounts for the first 5–7 Mm of the measurement. (d) Same DEM analysis as in (c) applied to a background coronal region near the prominence. The DEM results show that the bubble is both hotter and lower in density than the background corona. Courtesy M. Aschwanden, LMSAL.

the prominence bubble is both hotter and lower in density than the nearby background plasma, even though the intensity in the bubble in both the Fe XII 193 Å and Fe XIV 211 Å bandpasses is not above background. This analysis supports the conclusion of Berger *et al.* (2011) that prominence bubbles could contain plasma with temperatures up to $10^6$ K. While line-of-sight integration is still an issue with DEM analyses of prominences at the limb, it avoids the selectivity biases of photometric comparisons.

## 4. Conclusion

Although the variation in prominence appearance and dynamics is large, we propose that all prominences can be understood as the partially ionized plasma signatures of coherent magnetic flux ropes in the solar corona. The distinct morphologies of AR, intermediate, and quiescent prominences may thus be due to the relative magnetic field strengths in the associated flux ropes. AR flux ropes are apparently compact 5–10 Mm structures with field strengths in the kilogauss range, while intermediate flux ropes extend to heights of 20–30 Mm with field strengths of 10–100 gauss, and quiescent polar crown flux ropes extend to heights of 50 Mm or more and have field strengths of 5–10 gauss. It follows that AR prominences are compact horizontal structures with highly constrained flows on long narrow threads forming directly in the core of the flux rope, while intermediate prominences are larger structures with plasma concentrated on shorter horizontal dips in the field (indicating formation in the lower region of the flux rope), and quiescent prominences are extended structures with mostly vertical flows suggesting formation below the flux rope in a current sheet region. Quiescent prominence "horns" also imply that these prominences form below the concave lower portion of the flux rope that manifests in the EUV as a coronal cavity (Gibson 2013; Gibson *et al.* 2004).

This suggests that differences in prominence structure and dynamics can be understood as plasma-$\beta$ variations in the associated magnetic flux ropes, with quiescent prominences exhibiting possible high-$\beta$ flow characteristics. Although the suggestion of relatively high-$\beta$ plasma in any coronal structure is controversial, we point out that *Hinode*/SOT movies of quiescent prominences consistently show flows that appear more "hydrodynamic" in character, as supported by analysis of quiescent prominence plume bow-waves implying plasma-$\beta$ values of $\sim$1 or larger (Hillier 2013; Hillier *et al.* 2012c).

The association of prominences with magnetic flux ropes explains observed characteristics such as spiral and doppler flows in coronal cavities (Panesar *et al.* 2013; Bak Steślicka *et al.* 2013; Schmit *et al.* 2009), capture and subsequent condensation of coronal plasma in complex field topologies, and energy storage and eruption of the entire flux rope/prominence system due to accumulation of magnetic flux and helicity (Roussev *et al.* 2012; Fan and Gibson 2007; Zhang *et al.* 2006; Gibson and Low 1998). The finding that prominence bubbles may be caused by magnetic flux emergence across the PIL, as modeled by Dudík *et al.* (2012), and that they may contain hot plasma, implies a mechanism for transport of plasma, magnetic flux, and helicity via the Rayleigh-Taylor instability into the overlying flux rope. One can then envision a form of convection in the outer solar atmosphere with hot plasma transported upward and prominences representing the condensation return flow of the system. Similar cyclic/convective processes have been proposed for the quiet Sun (McIntosh *et al.* 2012; Antolin and Rouppe van der Voort 2012; Marsch *et al.* 2008) and coronal rain (Landi *et al.* 2009; Schrijver 2001).

Quoted in the frontispiece of Prof. Tandberg-Hanssen's book, Secchi (1877) astutely states that "Protruberances [prominences] present all the bizarre and capricious aspects that are absolutely impossible to describe with any exactitude". We agree and hope that this brief review and personal viewpoint has managed to convey some of the capricious aspects of prominences to yet another generation of solar researchers who will go on to make even more bizarre discoveries about these fascinating objects.

## Acknowledgements

We thank the referees for helpful comments and suggestions to improve this review. Dr. Berger was supported by the National Solar Observatory which is operated by the Association of Universities for Research in Astronomy, Inc. (AURA) for the National Science Foundation. *Hinode* is a Japanese mission developed and launched by ISAS/JAXA, collaborating with NAOJ as a domestic partner, NASA (US) and STFC (UK) as international partners. Support for the post-launch operation is provided by JAXA and NAOJ (Japan), STFC (UK), NASA, ESA, and NSC (Norway). SDO/AIA data are available to the public without restriction.

## References


Ahn, K., Chae, J., Cao, W., & Goode, P. R.: 2010, *ApJ* **721**, 74

Allen, U. A., Bagenal, F., & Hundhausen, A. J.: 1998, in New Perspectives on Solar Prominences, IAU Colloquiium 167, D. F. Webb, D. M. Rust, and B. Schmieder (eds.), *Astron. Soc. of the Pacific Conf. Series* **150**, p. 290

Antiochos, S. K. & Klimchuk, J. A.: 1991, *ApJ* **378**, 372

Antolin, P. & Rouppe van der Voort, L.: 2012, *ApJ* **745**, 152

Antolin, P. & Verwichte, E.: 2011, *ApJ* **736**, 121

Aulanier, G. & Demoulin, P.: 1998, *A & A* **329**, 1125

Aulanier, G., Demoulin, P., van Driel-Gesztelyi, L., *et al.*: 1998, *A & A* **335**, 309

Bak Steślicka, U., Gibson, S. E., Fan, Y., *et al.*: 2013, *ApJ* **770**, L28

Ballester, J. L.: 2006, *Space Sci Rev* **122**, 129

Berger, T.: 2012, in The Second ATST-EAST Meeting: Magnetic Fields from the Photosphere to the Corona, T. Rimmele, A. Tritschler, F. Woeger, *et al.* (eds.), *Astron. Soc. of the Pacific Conf. Series* **463**, p. 147

Berger, T. E., Liu, W., & Low, B. C.: 2012, *ApJ* **758**, L37

Berger, T., Testa, P., Hillier, *et al.*: 2011, *Nature* **472**, 197

Berger, T. E., Slater, G., Hurlburt, *et al.*: 2010, *ApJ* **716**, 1288

Berger, T. E., Shine, R. A., Slater, G. L., *et al.*: 2008, *ApJ* **676**, L89

Berlicki, A., Gunár, S., Heinzel, P., Schmieder, B., & Schwartz, P.: 2011, *A & A* **530**, A143

Bocchialini, K., Baudin, F., Koutchmy, S., Pouget, G., & Solomon, J.: 2011, *A & A* **533**, 96

Bommier, V., Landi Degl'Innocenti, E. *et al.*: 1994, *Sol Phys* **154**, 231

Brandt, P. N., Scharmer, G. B., Ferguson, S. *et al.*: 1988, *Nature* **335**, 238

Casini, R., López Ariste, A., Tomczyk, S., & Lites, B. W.: 2003, *ApJ* **598**, L67

Chae, J., Ahn, K., Lim, E.-K., Choe, G. S., & Sakurai, T.: 2008, *ApJ* **689**, L73

Chandrasekhar, S.: 1981, *Hydrodynamic and hydromagnetic stability*, Dover, 3rd edition

de Toma, G., Casini, R., Burkepile, J. T., & Low, B. C.: 2008, *ApJ* **687**, L123

Demoulin, P., Malherbe, J. M., Schmieder, B., & Raadu, M. A.: 1987, *A & A* **183**, 142

Dudík, J., Aulanier, G., Schmieder, B., & Bommier, V.; Roudier, T.: 2008, *Sol Phys* **248**, 29

Dudík, J., Aulanier, G., Schmieder, B., Zapiór, M., & Heinzel, P.: 2012, *ApJ* **761**, 9

Engvold, O.: 1998, in New Perspectives on Solar Prominences, IAU Coll. 167, D. F. Webb, D. M. Rust, and B. Schmieder (eds.), *Astron. Soc. of the Pacific Conf. Series* **150**

Engvold, O. & Jensen, E.: 1977, *Sol. Phys.* **52**, 37

Fan, Y. & Gibson, S. E.: 2007, *ApJ* **668**, 1232

Filippov, B. & Koutchmy, S.: 2002, *Sol. Phys.* **208**, 283

Gaizauskas, V., Zirker, J. B., Sweetland, C., & Kovacs, A.: 1997, *ApJ* **479**, 448

Gibson, S.: 2013, in these proceedings

Gibson, S. E., Fan, Y., Mandrini, C., Fisher, G., & Demoulin, P.: 2004, *ApJ* **617**, 600

Gibson, S. & Low, B.: 1998, *ApJ* **493**, 460

Gilbert, H., Kilper, G., & Alexander, D.: 2007, *ApJ* **671**, 978

Gunár, S.: 2013, in these proceedings

Haerendel, G. & Berger, T.: 2011, *ApJ* **731**, 82

Heinzel, P., Schmieder, B., Fárník, F., Schwartz, P., *et al.*.: 2008, *ApJ* **686**, 1383

Hillier, A.: 2013, in these proceedings

Hillier, A., Hillier, R., & Tripathy, D.: 2012c, *ApJ* **761**, 106

Hillier, A., Isobe, H., Shibata, K., & Berger, T.: 2012b, *ApJ* **756**, 110

Hillier, A., Berger, T., Isobe, H., & Shibata, K.: 2012a, *ApJ* **746**, 120

Hillier, A., Isobe, H., Shibata, K., & Berger, T.: 2011, *ApJ* **736**, L1

Hirayama, T.: 1985, *Sol. Phys.* **100**, 415

Hu, Y. Q. & Liu, W.: 2000, *ApJ* **540**, 1119

Karpen, J. T. & Antiochos, S. K.: 2008, *ApJ* **676**, 658

Kippenhahn, R. & Schlüter, A.: 1957, *Zeitschrift für Astrophysik* **43**, 36

Khomenko, E.: 2013, in these proceedings

Kucera, T., Tovar, M., & de Pontieu, B.: 2003, *Sol. Phys.* **212**, 81

Kuckein, C., Martinez Pillet, V., & Centeno, R.: 2012, *A & A* **542**, A112
Labrosse, N., Schmieder, B., Heinzel, P., & Watanabe, T.: 2011, *A&A* **531**, 69
Labrosse, N., Heinzel, P., Vial, J. C., *et al.*: 2010, *Space Sci Rev* **151**, 243
Landi, E., Miralles, M. P., Curdt, W., & Hara, H.: 2009, *ApJ* **695**, 221
Lemen, J. R., Title, A. M., Akin, *et al.*: 2011, *Sol. Phys.* **275**, 17
Leonardis, E., Chapman, S. C., & Foullon, C.: 2012, *ApJ* **745**, 185
Li, T. & Zhang, J.: 2013, *ApJ* **770**, L25
Lin, Y., Engvold, O., Rouppe Van Der Voort, L. H. M., *et al.*: 2007, *Sol. Phys.* **246**, 65
Lin, Y., Engvold, O., Van Der Voort, L. H. M., *et al.*: 2005, *Sol. Phys.* **226**, 239
Lites, B. W., Kubo, M., Berger, T., *et al.*: 2010, *ApJ* **718**, 474
Litvinenko, Y. E. & Martin, S. F.: 1999, *Sol. Phys.* **190**, 45
Liu, W.: 2013, these proceedings
Liu, W., Berger, T. E., & Low, B. C.: 2012, *ApJ* **745**, L21
López Ariste, A. & Casini, R.: 2003, *ApJ* **582**, L51
Low, B.: 2001, *J. Geophys. Res.* **106**, 25
Low, B. C., Berger, T., & Casini, R.: 2012a, *ApJ*
Low, B. C. & Hundhausen, J. R.: 1995, *ApJ* **443**, 818
Low, B. C., Liu, W., Berger, T., & Casini, R.: 2012b, *ApJ* **757**, 21
Low, B. C. & Petrie, G. J. D.: 2005, *ApJ* **626**, 551
Luna, M., Karpen, J. T., & DeVore, C. R.: 2012, *ApJ* **746**, 30
Mackay, D. H., Gaizauskas, V., & Yeates, A. R.: 2008, *Sol. Phys.* **248**, 51
Mackay, D. H., Karpen, J. T., Ballester, J. L., *et al.*: 2010, *Space Sci Rev* **151**, 333
Malherbe, J. M., Priest, E. R., Forbes, T. G., & Heyvaerts, J.: 1983, *A & A* **127**, 153
Marsch, E., Tian, H., Sun, J., Curdt, W., & Wiegelmann, T.: 2008, *ApJ* **685**, 1262
Martin, S. F.: 1998, *Sol. Phys.* **182**, 107
Martin, S. F., Lin, Y., & Engvold, O.: 2008, *Sol. Phys.* **250**, 31
McIntosh, S. W., Tian, H., Sechler, M., & De Pontieu, B.: 2012, *ApJ* **749(1)**, 60
Mercier, C. & Heyvaerts, J.: 1977, *A&A* **61**, 685
Merenda, L., Bueno, J. T., Degl'Innocenti, E. L., & Collados, M.: 2006, *ApJ* **642**, 554
Ning, Z., Cao, W., Okamoto, T. J., Ichimoto, K., & Qu, Z. Q.: 2009, *A & A* **499**, 595
Okamoto, T. J., Tsuneta, S., Berger, T. E., *et al.*: 2007, *Science* **318**, 1577
Okamoto, T. J., Tsuneta, S., Lites, *et al.*: 2009, *ApJ* **697**, 913
Orozco Suárez, D., Asensio Ramos, A., & Trujillo Bueno, J.: 2012, *ApJ* **761**, L25
Panesar, N. K., Innes, D. E., Tiwari, S. K., & Low, B. C.: 2013, *A & A* **549**, A105
Parenti, S., Schmieder, B., Heinzel, P., & Golub, L.: 2012, *ApJ* **754(1)**, 66
Parenti, S. & Vial, J.-C.: 2007, *A & A* **469**, 1109
Pettit, E.: 1932, *ApJ* **76**, 9
Pikel'Ner, S. B.: 1971, *Sol. Phys.* **17**, 44
Plunkett, S. P., Vourlidas, A., Šimberová, S., *et al.*: 2000, *Sol. Phys.* **194**, 371
Priest, E. R.: 1989, *Dynamics and Structure of Quiescent Solar Prominences* Kluwer Academic Publishers: Dordrecht
Reeves, K. K., Gibson, S. E., Kucera, T. A., Hudson, H. S., & Kano, R.: 2012, *ApJ* **746**, 146
Régnier, S., Walsh, R. W., & Alexander, C. E.: 2011, *A & A* **533**, L1
Roussev, I. I., Galsgaard, K., Downs, C., *et al.*: 2012, *Nature Physics* **8**, 1
Ryutova, M., Berger, T., Frank, Z., Tarbell, T., & Title, A.: 2010, *Sol. Phys.* p. 170
Saito, K. & Tandberg-Hanssen, E.: 1973, *Sol. Phys.* **31**, 105
Schmieder, B., Kucera, T. A., Knizhnik, K., Luna, M., Lopez-Ariste, A., Toot, D. *ApJ*, **777**, 108
Schmieder, B., Chandra, R., Berlicki, A., & Mein, P.: 2010, *A & A* **514**, 68
Schmieder, B., Malherbe, J. M., Mein, P., & Tandberg-Hanssen, E.: 1984, *A & A* **136**, 81
Schmieder, B., Raadu, M. A., & Wiik, J. E.: 1991, *A & A* **252**, 353
Schmit, D. J., Gibson, S. E., Tomczyk, S., *et al.*: 2009, *ApJ* **700**, L96
Schrijver, C. J.: 2001, *Sol. Phys.* **198**, 325
Schrijver, C. J. & Title, A. M.: 2011, *J. Geophys. Res.* **116**, A04108

Secchi, A.: 1877 *Le Soleil* Gauthier-Villars: Paris, vols. 1 & 2
Stellmacher, G. & Wiehr, E.: 1973, *A & A* **24**, 321
Stone, J. M. & Gardiner, T.: 2007, *Phys. Fluids* **19**, 4104
Su, Y. & Van Ballegooijen, A.: 2012, *ApJ* **757**, 168
Su, Y., Wang, T., Veronig, A., Temmer, M., & Gan, W.: 2012, *ApJ* **756**, L41
Tandberg-Hanssen, E.: 1995, *The Nature of Solar Prominences*, Kluwer Academic: Dordrecht
Török, T., Panasenco, O., Titov, V. S., *et al.*: 2011, *ApJ* **739**, L63
Tsuneta, S., Ichimoto, K., Katsukawa, *et al.*: 2008, *Sol. Phys.* **249**, 167
van Ballegooijen, A. A. & Cranmer, S. R.: 2010, *ApJ* **711**, 164
Vourlidas, A., Lynch, B. J., Howard, R. A., & Li, Y.: 2012, *Sol. Phys.* **284**, 179
Wang, H., Chae, J., Gurman, J. B., & Kucera, T. A.: 1998, *Sol Phys* **183**, 91
Wang, Y.-M.: 1999, *ApJ* **520**, L71
Wang, Y.-M. & Stenborg, G.: 2010, *ApJ* **719**, L181
Wedemeyer-Böhm, S., Scullion, E., Rouppe van der Voort, L., *et al.*: 2013, *ApJ* **774**, 123
Wedemeyer-Böhm, S., Scullion, E., Steiner, O., *et al.*: 2012, *Nature* **486**, 505
Xia, C., Chen, P. F., & Keppens, R.: 2012, *ApJ* **748**, L26
Zhang, J. & Liu, Y.: 2011, *ApJ* **741**, L7
Zhang, M., Flyer, N., & Low, B. C.: 2006, *ApJ* **644**, 575
Zirker, J. B.: 1989, *Sol. Phys.* **119**, 341
Zirker, J. B., Engvold, O., & Martin, S. F.: 1998, *Nature* **396**, 440

*Nature of prominences and their role in Space Weather*
*Proceedings IAU Symposium No. 300, 2013*
*B. Schmieder, J.-M. Malherbe & S. T. Wu, eds.*

doi:10.1017/S1743921313010703

# Prominence Seismology

**J. L. Ballester**

Departament de Física, Universitat de les Illes Balears,
E - 07122, Palma de Mallorca, Spain
email: joseluis.ballester@uib.es

**Abstract.** Quiescent solar prominences are cool and dense plasma clouds located inside the hot and less dense solar corona. They are highly dynamic structures displaying flows, instabilities, oscillatory motions, etc. The oscillations have been mostly interpreted in terms of magnetohydrodynamic (MHD) waves, which has allowed to perform prominence seismology as a tool to determine prominence physical parameters difficult to measure. Here, several prominence seismology applications to large and small amplitude oscillations are reviewed.

**Keywords.** Prominence Oscillations, MHD Waves

## 1. Introduction

Seismology refers to the process of deriving the physical conditions of a medium by analysing properties of the oscillations or the waves travelling through the medium. Helioseismology was invented almost 50 years ago, and became an efficient tool to probe the internal structure of the Sun, while thanks to space-based high-resolution observations, which provided with strong evidence about the presence of oscillations and waves in coronal magnetic structures, Solar Atmospheric Seismology, proposed by Rosenberg (1970), Uchida (1970) and Roberts *et al.* (1984), became a reality about 15 years ago.

MHD waves and oscillations provide with an indirect path to determine physical parameters [coronal magnetic field, transport coefficients (viscosity, resistivity, thermal conductivity, etc.), heating function, filling factors, inhomogeneity scale] of the solar corona. Therefore, MHD seismology is a method of remote diagnostics of plasma structures combining observations of oscillatory motions with an interpretation in terms of MHD waves.

MHD seismology involves the solution of the forward and inverse problems. In the forward problem, a theoretical model is built and used to predict the oscillations of many different modes. If the predictions do not agree with observations, then, we modify the model somehow and start again the comparison. On the contrary, in the inverse problem, instead of computing frequencies from a theoretical model, we construct the model from the observed frequencies. MHD seismology is similar to helioseismology, but with a few differences:

(*a*) Only a local diagnostics of the oscillating structures and their nearest vicinity (local seismology) is obtained

(*b*) Three wave modes are available: fast, slow magnetoacoustic and Alfvén

Prominence seismology was initially suggested by Roberts & Joarder (1994):"*...the possibility of using data on the modes of oscillation of a prominence to derive seismic information about the structure of the prominence...*" and by Vial (1998):"*...we may well be en route towards prominence seismology*", and its main aim is to determine physical parameters difficult to measure by other methods. The observational tools available for prominence seismology are:

(*a*) Large amplitude oscillations, with velocity amplitudes greater than 20 km/s. [Tripathi *et al.* (2009)]

(*b*) Small amplitude oscillations, whose velocity amplitudes range from the noise level up to a few km/s. [Arregui *et al.* (2012)]

## 2. Prominence seismology using large amplitude oscillations

The first seismological study was made by Hyder (1966) using observational data about large amplitude oscillations of 11 filaments reported by Ramsey & Smith (1966). Assuming that the filaments are located in a depressed magnetic field, these observations were interpreted in terms of vertical oscillations damped by the viscosity of the surrounding coronal plasma, and estimates of the radial magnetic field in the range 2 – 30 G were obtained. Furthermore, the coronal coefficient of viscosity was also determined. Kleczek & Kuperus (1969) re-interpreted the above mentioned observations in terms of horizontal (transverse) oscillations of filaments. They assumed that a line-tied magnetic field was directed along the filament, that the restoring force was provided by magnetic tension and that the oscillations were damped by the emission of acoustic waves. Then, from the equation of motion of a damped harmonic oscillator, the period is given by

$$P = 4\pi L B^{-1}\sqrt{\pi\rho_p}$$

Knowing the period of oscillation ($P$), measuring the length of the filament ($L$) and assuming a typical density ($\rho_p$), the strength of the magnetic field ($B$) can be determined. Nowadays, this model is still being used to explain prominence oscillations and to perform prominence seismology.

Transverse oscillations have been also observed in a pre-erupting filament (Isobe & Tripathi (2006)). These oscillations were present along one foot of a polar crown filament, prior to the eruption of the full structure. The oscillating part of the filament moved like a rigid body with a line-of-sight velocity amplitude > 20 km/s, a displacement amplitude $\sim$ 20 - 30 arcsec, and a period $\sim$ 2 hr. Following the theoretical model by Kleczek & Kuperus (1969), the magnetic field strength (9.8 G) and the Alfvén speed (87 km/s) were determined.

Longitudinal oscillations consist of periodic motions along the axis of a filament. Jing *et al.* (2003) and Jing *et al.* (2006) analysed several events reporting periods of 80, 160, 150, 100 min.; damping times of 210, 600, 450 min., maximum velocity amplitudes of 80, 50, 30, 100 km/s and maximum displacement amplitudes of 140, 160, 100, 80 Mm. These oscillations seem to be triggered by subflares, flares, filament eruption and reconnection events which happen close to the filaments. Vršnak *et al.* (2007) reported H$\alpha$ observations of longitudinal oscillations in a filament with an initial displacement of 24 Mm, an initial velocity amplitude of 51 km/s, a period of 51 min. and a damping time of 115 min. Assuming that the filament was embedded in a flux rope, Vršnak *et al.* (2007) suggested that the oscillations were triggered by additional poloidal flux injected at one of its legs creating a magnetic pressure gradient along the filament, which is the restoring force. After linearising the equation of motion, the expression for the longitudinal displacement ($X$), in dimensionless form, is:

$$\ddot{X} = -\frac{2v_{A\varphi}^2}{L^2}X$$

which provides with an expression for the period, $P \approx 4.4L/v_{A\varphi}$, as a function of the poloidal Alfvén speed ($v_{A\varphi}$) and the length of the filament ($L$). Then, knowing the period and the length of the filament, the poloidal Alfvén speed can be determined ($v_{A\psi} - 100$ km/s). Furthermore, assuming a set of prominence densities, the poloidal magnetic field strengh is in the range 5-15 G, and measuring the pitch angle, the longitudinal magnetic field strength can be also obtained (10 - 30 G).

Recently, a theoretical model for large amplitude longitudinal oscillations in filaments has been proposed [Luna & Karpen (2012), Luna *et al.* (2012), Zhang *et al.* (2013)]. The scenario is the following: When an energetic event (a subflare, for instance) happens close to a filament, the injected energy evaporates plasma at the fluxtube footpoint closest to the energetic event. Then, the flow of hot plasma pushes the cold plasma condensations (threads) located at the dips of a sheared double arcade, and the longitudinal oscillations start. After some time, they lose coherence due to period differences. The restoring force is the projected solar gravity directed towards the bottom of the dip and since the magnetic tension in the dip must be larger than the weight of the threads, we have,

$$\frac{B^2}{R} - mng_0 \geqslant 0$$

where $R$ is the radius of curvature, $m$ the particle mass, $n$ the particle density, and $g_0$ the gravity. On the other hand, since the oscillation is gravity driven:

$$\omega = \sqrt{\frac{g_0}{R}}$$

and combining the above two expressions, we obtain,

$$B \geqslant \sqrt{\frac{g_0^2 mn}{4\pi^2}} P$$

Then, again, knowing the period (P) and assuming a typical density (n), the strength of the magnetic field (B) can be determined (31 - 75 G for the threads considered). In this case, the damping of the oscillations has been attributed to thermal and mass accretion effects.

On the other hand, this model points out that if longitudinal oscillations are gravity driven, the seismological determination of the magnetic field would not be influenced by the radius of curvature of the magnetic field dips.

## 3. Prominence seismology using small amplitude oscillations

### 3.1. *Seismology of prominence slabs*

Seismology of prominence slabs was performed in order to infer prominence physical properties and the methodology was based on the identification of observed oscillations with theoretical eigenmodes. Régnier *et al.* (2001) observed an active region filament with SUMER/SoHO detecting oscillations covering different ranges of periods: $< 5$ min.; $6 - 20$ min.; $> 40$ min. and considered the possible theoretical modes that could explain their observations. The slab model with a uniform and inclined magnetic field by Joarder & Roberts (1993) was chosen. The dispersion relations for Alfvén modes and magnetoacoustic modes were considered, providing the frequency of six fundamental modes: the symmetric Alfvén, slow and fast kink modes and the antisymmetric Alfvén, slow and fast sausage modes, as a function of the prominence parameters. Observations provided with estimates for the width (8000 km) and length (63,000 km) of the filament, and assumptions on other parameters, such as the temperature of the filament (8000 K) and of its environment ($10^6$ K), the density of the slab ($10^{12}$ cm$^{-3}$), the magnetic field strength (20 G) and for the angle between the magnetic field and the long axis of the slab (25°), were made. The dispersion relations were then solved by using these parameters and the corresponding periods were obtained and classified.

From the comparison between the observed and calculated frequencies, an identification method of the oscillation modes in the observed filament was presented. The method makes use of the fact that the frequency ratio of the fundamental even Alfvén mode to the fundamental odd Alfvén mode only depends on the ratio of the half-with of the slab to the half-length of the filament, which is a measurable quantity. The same applies to the frequency ratios involving the slow kink/sausage and fast kink/sausage modes. Parametric calculations for the frequencies as a function of the magnetic field strength and the inclination angle, while keeping the slab density constant, were next performed. A diagnostic of the observed filament is obtained by looking for the parameters values that enable the matching of theoretical and observed frequencies. By following this method, the angle between the magnetic field and the long axis of the slab is estimated to be $18^{\circ}$. Using this value, an algebraic relation for the magnetic field strength as a function of the slab density is derived.

A more involved and ambitious diagnostic, using the slab model, was performed by Pouget *et al.* (2006). The long duration and high temporal resolution observations with CDS/SoHO enabled these authors to detect and measure the entire range of periodicities theoretically expected in a filament. In particular both the short (less that 10 min) and the long ones (more than 40 min) are detected.

The detailed analysis of three filaments is presented. The seismic inversion technique closely follows that of Régnier *et al.* (2001), in the sense that the first step towards the diagnostic is the use of frequency ratios between fundamental even/odd (kink/sausage) modes. These ratios only depend on the ratio of the filament half-width to its half-length. Once this ratio is measured, with a given uncertainty, Pouget *et al.* (2006) assume that their 16 hours long observation had allowed them to observe the six modes of interest, since the slowest mode is expected at a period of 5 hours, for standard prominence parameters.

The inversion method first assigns a possible triplet of measured frequencies to the 3 odd fundamental frequencies (odd Alfvén, slow sausage, and fast sausage modes). The coherence of each choice is examined against two tests. The first requires to find three corresponding even frequencies, with the condition that the even/odd frequency ratios are consistent with the measured half-width to half-length ratio. The second involves the inferred values for the density, temperature, magnetic field inclination angle, and magnetic field strength to be consistent with typical values reported in the literature. For each test, if the test was negative, the full triplet was changed and the series started again. On the contrary, if the tests succeeded, they considered that the six fundamental modes were identified.

The three filament observations led to coherent diagnostics and a single possible set of frequencies was found for each observation. The importance of this study is its ability to simultaneously determine the values of the inclination angle, temperature, and Alfvén speed for the same prominence. The drawback is that the modeling does not permit to capture the highly inhomogeneous nature of prominences.

Gravity is not considered in the above theoretical model, however, its inclusion produces a negligible effect on the prominence slab oscillations and seismological determinations.

### 3.2. *Seismology of filament threads*

Although quiescent prominences observed at the limb seem to be made of vertical threads (See Gunar's review in this volume), when observed on the disk as filaments they seem to be made by a myriad of fine structures which seem to be field aligned, outlining magnetic flux tubes [Engvold, (1998), Lin (2004), Lin *et al.* (2005), Lin *et al.* (2007),

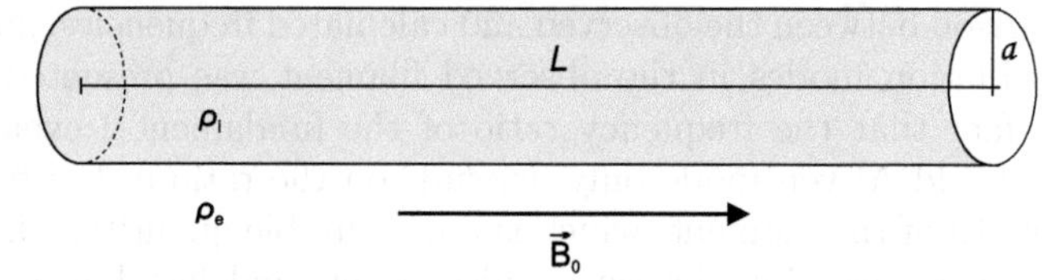


**Figure 1.** Sketch of an infinitely long thread immersed in the solar corona [Lin *et al.* (2009)].

Engvold (2008), Lin *et al.* (2008)]. These magnetic flux tubes are fully or partially filled with cold plasma condensations called threads [Lin (2004), Okamoto *et al.* (2007)], and the detected transverse thread oscillations, interpreted in terms of kink MHD waves, are used to perform thread seismology.

3.2.1. *Seismology using the period of filament thread oscillations*

Transverse thread oscillations observed by Lin *et al.* (2009) show evidence of waves propagating along individual threads. Ten of the swaying threads were chosen by Lin *et al.* (2009) for further investigation, and for each selected thread two or three perpendicular cuts were made in order to measure the properties of the propagating waves. Periods and amplitudes of the waves, as well as their phase velocity, were derived for each thread. Lin *et al.* (2009) interpreted the observed events as propagating MHD kink waves supported by the thread body. This mode is the only one producing a significant transverse displacement of the cylinder axis. In addition, it also produces short-period oscillations of the order of minutes, compatible with the observed periods. This interpretation also implies that the measured phase velocity is equal to the kink speed.

If an infinitely long, straight, cylindrical thread model, with the tube fully filled with cool and dense material, is assumed (Figure 1), a comparison between the observed wave properties and the theoretical prediction can be made. This enabled Lin *et al.* (2009) to obtain estimates for some physical parameters of interest, namely the Alfvén speed and the magnetic field strength in the studied threads. Assuming the thin tube approximation, length of the magnetic flux tube, $L$, much greater than the radius, $a$, and that inside and outside the magnetic flux tube the density is given by,

$$\rho_0(r) = \begin{cases} \rho_f, & r \leqslant a \\ \rho_c, & r > a \end{cases}$$

the kink speed, $c_k$, is,

$$c_k = \sqrt{\frac{2B_0^2}{\mu(\rho_f + \rho_c)}}$$

where $\mu$ is the magnetic permittivity, and the above expression can be written as,

$$c_k = v_{Af}\sqrt{\frac{2c}{c+1}}$$

with

$$\xi = \frac{\rho_f}{\rho_c}$$

and when this ratio becomes very large, then

$$c_k \sim \sqrt{2}v_{Af}$$

where $v_{Af}$ is the thread Alfvén speed. The results for the internal Alfvén speed show a strong dispersion, suggesting that the physical conditions in different threads were very different in spite of belonging to the same filament. This result clearly reflects the highly inhomogeneous nature of solar prominences. Once the Alfvén speed in each thread was determined, the magnetic field strength could be computed after a value for the thread density was assumed.

### 3.2.2. *Seismology using the period and damping time of filament thread oscillations*

A feature clearly observed by Lin *et al.* (2009) is that the amplitudes of the waves passing through two different cuts along a thread are notably different. Apparent changes can be due to damping of the waves in addition to noise in the data. The damping of prominence oscillations is a common feature in many observed events and damping time-scales provide with an additional source of information that can be used when performing parameter inference using seismology inversion techniques. Among the different damping mechanisms which can be considered, resonant absorption in the Alfvén continuum seems a very plausible one and has been used to perform prominence thread seismology, using the damping as an additional source of information.

The model considered here is a one-dimensional infinitely long thread of mean radius $a$ surrounded by a thin transition sheath of thickness $l$ in which a smooth transition from the thread to the coronal density takes place. The thread is located inside a flux tube, in static equilibrium, gravity is neglected, the magnetic field is uniform inside and outside and the low-$\beta$ approximation is considered. The thread plasma density is $\rho_f$, the coronal plasma density $\rho_c$, and the density contrast is $\zeta = \frac{\rho_f}{\rho_c}$ (Figure 2). Since $l/a \neq 0$ the kink MHD mode is resonantly coupled to Alfvén continuum modes and is damped in time.

For standing kink waves, and without using the thin tube and thin boundary approximation, the normal mode period and damping ratio are functions of the relevant equilibrium parameters,

$$P = P(k_z, \zeta, l/a, v_{Af}), \qquad \frac{P}{\tau_d} = \frac{P}{\tau_d}(k_z, \zeta, l/a),$$

with $v_{Af}$ the thread Alfvén speed. In the thin tube and thin boundary approximations (TTTB), the period does not depend on $l/a$ and the damping ratio is independent of the wavelength. For prominence threads, the wavelength of oscillations needs to be measured and the above relations indicate that, if no assumption is made on any of the physical parameters of interest, there are infinite different equilibrium models that can equally well explain the observations (namely the period and damping ratio). The parameter values that define these valid equilibrium models are displayed in Figure 3, where the analytical algebraic expressions in the TTTB approximations by Goossens *et al.* (2008) have been used to invert the problem. It can be appreciated that, even if an infinite number of solutions is obtained, they define a rather constrained range of values for

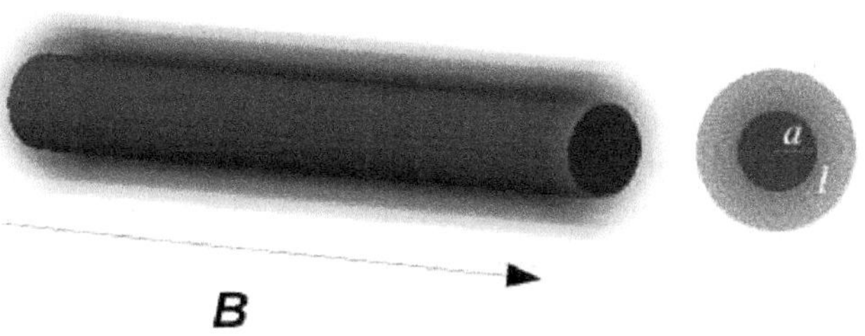


**Figure 2.** Radially non-uniform filament fine structure [Arregui *et al.* (2008)].

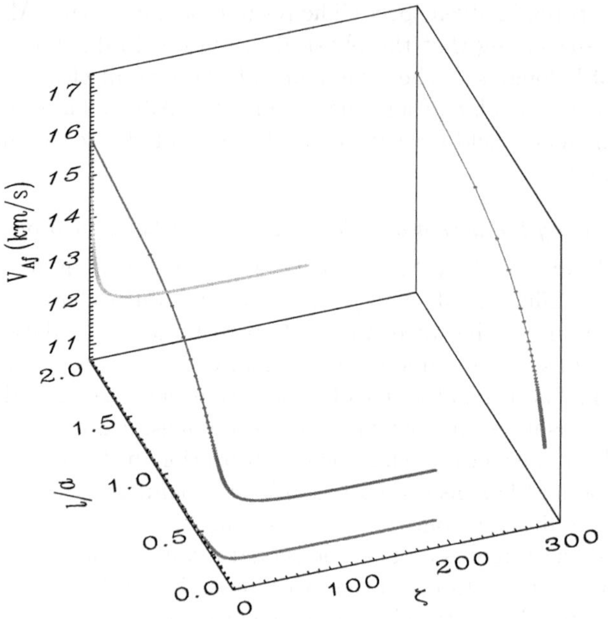


**Figure 3.** Analytic inversion of physical parameters in the $(\zeta, l/a, v_{Af})$ space for a filament thread with $P = 3$ min., $\tau_d = 9$ min. and a wavelength $\lambda = 3000$ km. [Arregui *et al.* (2012)]

the thread Alfvén speed. Because of the insensitiveness of the damping rate with the density contrast for the typically large values of this parameter in prominence plasmas, the obtained solution curve displays an asymptotic behaviour for large values of $\zeta$. This makes possible to obtain precise estimates for the thread Alfvén speed, $v_{Af} \simeq 12$ km/s and the transverse inhomogeneity length scale, $l/a \simeq 0.16$. The computation of the magnetic field strength from the obtained seismological curve requires the assumption of a particular value for either the filament or the coronal density. In the inversion curve displayed in Figure 3, a change in the period produces a vertical shift of the solution curve, hence the period influences the inferred values for the Alfvén speed.

The main shortcoming of this technique is the use of thread models in which the full magnetic tube is filled with cool and dense plasma. An example of the inversion of physical parameters for different values of the thread length $(L_p)$ was presented by Soler *et al.* (2010). When partially filled threads, i.e., with the dense part occupying a length shorter than the total length of the tube $L$, are considered, one curve is obtained for each value of the length of the thread. Even if each curve gives an infinite number of solutions, again, each of them defines a rather constrained range of values for the thread Alfvén speed, and the ratio $L_p/L$ is a fundamental parameter in order to perform an accurate seismology of prominence threads, since different curves produce different estimates for the prominence Alfvén speed. Because of the insensitiveness of the damping ratio with respect to the length of the thread, all solution curves for different lengths of the threads produce the same projection onto the $(\zeta,\, l/a)$-plane. Hence, the same precise estimates of the transverse inhomogeneity length scale obtained from infinitely long thread models are valid, irrespective of the length of the thread. The computation of the magnetic field strength from the obtained seismological curve requires the assumption of a particular value for either the filament or the coronal density.

Recently, and using Bayesian inference, a consistent solution for the inversion problem and the correct propagation of errors from observations to inferred parameters has been

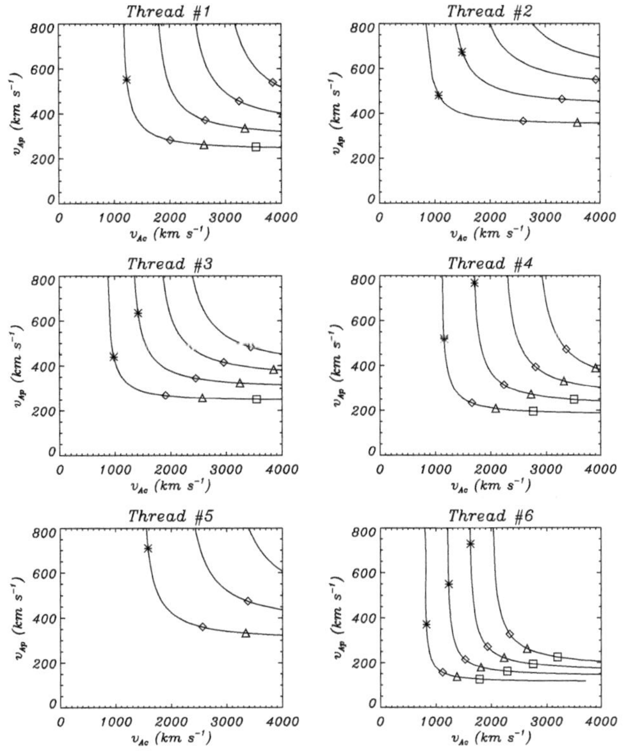


**Figure 4.** Dependence of the Alfvén velocity in the thread as a function of the coronal Alfvén velocity for the six threads observed by Okamoto *et al.* (2007). In each panel, from bottom to top, the curves correspond to a length of magnetic field lines of 100,000 km, 150,000 km, 200,000 km and 250,000 km, respectively. Asterisks, diamonds, triangles and squares correspond to ratio of the thread to the coronal one $\zeta \simeq 5, 50, 100, 200$. From Terradas *et al.* (2008).

proposed [Arregui *et al.* (2013), Arregui & Asensio Ramos, (2011)]. Using the same values as before and considering an uncertainty in time-scales of the order of 0.1 min. [Lin *et al.* (2009)], they obtained

$$v_{Ap} = 11.9 \pm 0.4\ km/s; \quad l/a = 0.21 \begin{cases} +0.001 \\ -0.002 \end{cases}$$

3.2.3. *Seismology of flowing and oscillating prominence threads*

The first application of prominence seismology using Hinode observations of flowing and transversely oscillating threads was presented by Terradas *et al.* (2008) using observations obtained in an active region filament by Okamoto *et al.* (2007).

The observations show a number of threads that flow following a path parallel to the photosphere while they oscillate in the vertical direction. The relevance of this particular event is that the coexistence of waves and flows can be firmly established, so that there is no ambiguity about the wave or flow character of a given dynamic feature: both seem to be present in this particular event.

In their seismological analysis of these oscillations Terradas *et al.* (2008) started by neglecting the mass flows. Then, they interpreted these events in terms of the standing kink mode of a finite-length thread in a magnetic flux tube. By using theoretical results by Díaz *et al.* (2002) and Dymova & Ruderman *et al.* (2005), Terradas *et al.* (2008) found that a one-to-one relation between the thread Alfvén speed and the coronal Alfvén speed could be established. This relation comes in the form of a number of curves relating the two Alfvén speeds for different values of the length of the magnetic flux tube and the density contrast between the filament and coronal plasma (Figure 4). An interesting property of the obtained solution curves is that they display an asymptotic behaviour for large values of the density contrast, which is typical of filament to coronal plasmas, and hence a lower limit for the thread Alfvén speed can be obtained. Take for instance thread 6. Considering a magnetic flux tube length of 100 Mm, a value of 120 km/s for the thread Alfvén speed is obtained.

Terradas *et al.* (2008) next incorporated mass flows into their analysis by considering the numerical solution of the non-linear, ideal, low-$\beta$ MHD equations. The numerical results indicate that the effect of the flow on the obtained periods is weak. In fact, using the flow velocities measured by Okamoto *et al.* (2007) slightly shorter kink mode periods than the ones derived in the absence of flow are obtained. Differences are small, however, and produce period shifts between 3 and 5% and, therefore, on the derived Alfvén speed values. We must note that in this case, and because of the small value of the flow speeds measured by Okamoto *et al.* (2007) in this particular event, there are no significant variations of the wave properties, and hence of the inferred Alfvén speeds.

More extensive information about seismological techniques can be found in Arregui *et al.* (2012)

## 4. Conclusions

Up to now, MHD seismology has not been used to interpret large amplitude oscillations. On the contrary, these oscillations have been explained in terms of oscillators whose restoring forces are magnetic tension, magnetic pressure gradient or projected gravity, and the performed seismology has been based in the analysis of these oscillations. On the other hand, following the analysis by Ruderman & Goossens (2013) about nonlinear kink oscillations in coronal magnetic loops, the ratio between the displacement and the radius (thickness) of a magnetic flux tube (slab) determines whether the transverse oscillations are linear or not. Taking this analysis into account, most of the large amplitude transverse oscillations observed in filaments could be nonlinear, therefore, this should be taken into account when setting up theoretical models to explain these oscillations.

In the case of small amplitude oscillations, prominence seismology is based on evidence of MHD waves and highly idealized theoretical models, in particular with respect to the magnetic configuration. Therefore, it is still a young science. However, and in spite of this, tentative values for unknown physical parameters (Alfvén speed, transverse inhomogeneity length scale) in prominence fine structures can be obtained.

An important step ahead for prominence seismology would be to couple radiative transfer with magnetohydrodynamic waves as a mean to establish a relationship between velocity, density, magnetic field and temperature perturbations, and the observed signatures of oscillations like spectral line shift, width and intensity (See Zapior *et al.* in this volume). Also, partial ionization is another topic of interest since, apart from influencing the behaviour of magnetohydrodynamic waves, it poses an important problem for prominence equilibrium models since cross-field diffusion of neutral atoms can give place to flows and drain prominence material. However, the study of the oscillatory behaviour of a fully three-dimensional and dynamical prominence model involving magnetic equilibrium, radiative transfer, etc., seems to be still a dream.

**Acknowledgements** JLB acknowledges the support from MINECO and FEDER Funds through grant AYA2011-22846 and from CAIB through the "grups competitius" scheme and FEDER Funds.

## References


Arregui, I., Terradas, J., Oliver, R., & Ballester, J. L. 2008, *ApJ*(Letters), **682**, L141

Arregui, I. & Asensio Ramos, A. 2011, *ApJ* (Letters), **740**, 44

Arregui, I., Oliver, R., & Ballester, J. L. 2012, *Living Reviews in Solar Physics*, **9**, 2

Arregui, I., Asensio Ramos, A., & Pascoe, D. J. 2013, *ApJ*, **769**, L34

Díaz, A. J., Oliver, R., & Ballester, J. L. 2002, *ApJ*, **580**, 550

Dymova, M. V. & Ruderman, M. S. 2005, *Solar Phys.*, **229**, 79

Engvold, O. 1998, in D. F. Webb, B. Schmieder, & D. M. Rust (eds.) *New Perspectives in Solar Prominences*, Proc. IAU Colloq. No. 167 (San Francisco, ASP), p. 23

Engvold, O. 2008, in R. Erdélyi, & C. A. Mendoza-Briceño (eds.) *Waves & Oscillations in the Solar Atmosphere*, Proc. IAU Symp. No. 247 (San Francisco, ASP), p. 152

Goossens, M., Arregui, I., Ballester, J. L., & Wang, T. J. 2008, *A & A*, **484**, 851

Hyder, C. L. 1966, *ZfA*, **63**, 78

Isobe, H. & Tripathi, D. 2006, *A & A*, **449**, L17

Jing, J., Lee, J., Spirock, T. J., Xu, Y., Wang H., & Choe, G. S. 2003, *ApJ* (Letters), **584**, L103

Jing, J., Lee, J., & Spirock, T. J., Wang H. 2006, *Solar Phys.*, **236**, 97

Joarder, P. & Roberts, B. 1993, *A & A*, **277**, 225

Kleczek, J. & Kuperus, M. 1969, *Solar Phys.*, **6**, 72

Lin, Y. 2004, *Ph.D. Thesis*, University of Oslo, Norway

Lin, Y., Engvold, O., Rouppe Van der Voort, L. H. M., Wiik, J. E., & Berger, T. E. 2005, *Solar Phys.*, **225**, 229

Lin, Y., Engvold, O., Rouppe Van der Voort, L. H. M., & Van Noort, M. J. 2007, *Solar Phys.*, **246**, 65

Lin, Y., Martin, S. F., Engvold, O., Rouppe Van der Voort, L. H. M., & Van Noort, M. J. 2003, *Adv. Sp. Res.*, **42**, 803

Lin, Y., Soler, R., Engvold, O., Ballester, J. L., Langagen, O., Oliver, R., & Rouppe Van der Voort, L. H. M. 2009, *ApJ*, **704**, 870

Luna, M. & Karpen, J. 2012, *ApJ* (Letters), **750**, L1

Luna, M., Díaz, A. J., & Karpen, J. 2012, *ApJ* (Letters), **757**, L98

Okamoto, T. J., Tsuneta, S., Berger, T. E., Ichimoto, K., Katsukawa, Y., Lites, B. W., Nagata, S., Shibata, K, Shimizu, T., Shine, R. A., Suematsu, Y., Tarbell, T. D., & Title, A. M. 2007, *Science*, **318**, 1577

Pintér, B., Jain, R., Tripathi, D., & Isobe, H. 2008, *ApJ* **680**, 1560

Pouget, G., Bocchialini, K., & Solomon, J. 2006, *A & A* **450**, 1189

Ramsey, H. E. & Smith, S. F. 1966, *AJ*, **71**, 197

Régnier, S.,Solomon, J., & Vial, J. C. 2001, *A & A*, **376**, 292

Roberts, B. & Joarder, P. 1994, in G. Belvedere, M. Rodon, & G. M. Simnett (eds.) *Advances in Solar Physics*, Proc. seventh European Meeting on Solar Physics (Lecture Notes in Physics 432, Springer), p. 173 *ARAA*, **41**, 241

Roberts, B., Edwin, P. M., & Benz, A. O. 1984, *ApJ*, **279**, 857

Rosenberg, H. 1970, *A & A*, **9**, 159

Ruderman, M. & Goossens, M. 2013, *Solar Phys.* (In press)

Soler, R., Arregui, I., Oliver, R., & Ballester, J. L. 2010, *ApJ*, **722**, 1778

Terradas, J., Arregui, I., Oliver, R., & Ballester, J. L. 2008, *ApJ*(Letters), **978**, L153

Tripathi, D., Isobe, H., & Jain, R. 2009, *Space Sci. Revs*, **149**, 283

Uchida, Y. 2004, *PASJ*, **22**, 341

Vial, J. C. 1998, in D. F. Webb, B. Schmieder, & D. M. Rust (eds.) *New Perspectives in Solar Prominences*, Proc. IAU Colloq. No. 167 (San Francisco, ASP), p. 175

Vršnak, B., Veronig, A. M., Thalman, J. K., & Žic, T. 2007, *ApJ*, **647**, 676

Zhang, Q. M., Chen, P. F., Xia, C., Keppens, R., & Ji, H. S. 2013, *A & A*, **554**, A124

*Nature of Prominences and their Role in Space Weather*
*Proceedings IAU Symposium No. 300, 2013*
*B. Schmieder, J.-M. Malherbe & S. T. Wu, eds.*

doi:10.1017/S1743921313010715

# Formation and evolution of an active region filament

**Christoph Kuckein[1,2], Rebeca Centeno[3] and Valentín Martínez Pillet[4]**

[1]Leibniz-Institut für Astrophysik Potsdam (AIP),
An der Sternwarte 16, 14482, Potsdam, Germany.
email: ckuckein@aip.de

[2]Instituto de Astrofísica de Canarias (IAC), Vía Láctea s/n, 38205, La Laguna, Tenerife, Spain

[3]High Altitude Observatory (NCAR), Boulder, CO 80301, USA

[4]National Solar Observatory (NSO), Sunspot, NM 88349, USA

**Abstract.** Several scenarios explaining how filaments are formed can be found in literature. In this paper, we analyzed the observations of an active region filament and critically evaluated the observed properties in the context of current filament formation models. This study is based on multi-height spectropolarimetric observations. The inferred vector magnetic field has been extrapolated starting either from the photosphere or from the chromosphere. The line-of-sight motions of the filament, which was located near disk center, have been analyzed inferring the Doppler velocities. We conclude that a part of the magnetic structure emerged from below the photosphere.

**Keywords.** Sun: filaments, prominences, Sun: photosphere, Sun: chromosphere, Sun: magnetic fields, techniques: polarimetric

## 1. Introduction

Owing to new observations and a continuous improvement of simulations, in the last years the solar community has published several works concerning the formation and evolution of filaments. Active region (AR) filament studies are still scarce in literature, and it is not clear how similar their formation process is to that of the quiescent (QS) ones. Generally speaking, there are two proposed scenarios that try to explain the formation of filaments: (1) the sheared arcade (SA) model and (2) the flux rope emergence (FRE) model. The difference between both models mainly exists in the atmospheric height of filament formation. On the one hand, in the SA model the filament is formed in the corona by shearing motions in addition to converging flows at the polarity inversion line (PIL) and reconnection processes. As a result, the magnetic structure can be a flux rope (e.g., van Ballegooijen & Martens 1989) or a dipped arcade (e.g., Antiochos *et al.* 1994). On the other hand, the FRE model assumes that a flux rope emerges from below the photosphere ascending into the corona (e.g., Okamoto *et al.* 2008). Kuckein *et al.* (2012a) provide an extensive discussion and references related to these two models. In this work, we present a multi-height study of the magnetic structure of an AR filament observed in July 2005 and investigate whether this filament fits into the SA or FRE model.

## 2. Observations

The present analysis is mainly based on spectropolarimetric observations acquired with the Tenerife Infrared Polarimeter attached to the Vacuum Tower Telescope (VTT) in Tenerife (TIP-II; Collados *et al.* 2007). The data sets of an AR filament belonging to NOAA 10781 were taken close to disk center ($\mu \sim 0.91$ and 0.95) on 2005 July 3 and

5. However, H$\alpha$ images from Big Bear Solar Observatory (BBSO; Denker *et al.* 1999) show that the filament was already present a few days before, on July 1. Line-of-sight (LOS) magnetograms from the Michelson Doppler Imager (MDI, Scherrer *et al.* 1995) show that below the filament an extensive facular region is seen with two polarities clearly separated by the PIL. Between July 1–7 the opposite polarities became more compact, almost touching each other at the PIL, and then broadened again. During this event, with characteristics of the "sliding door" effect (an effect firstly described by Okamoto *et al.* 2008), new pores and orphan penumbrae emerged at the PIL (see Kuckein *et al.* 2009, 2012a, for a detailed description and images). Okamoto *et al.* (2008) related this effect to the emergence of a flux rope below an AR filament.

The spectral region observed with TIP-II comprised the photospheric Si I 10827 Å line, the chromospheric He I 10830 Å triplet, and two telluric lines with a spectral sampling of $\sim$11.04 mÅ px$^{-1}$. Therefore, with this instrument it is possible to simultaneously analyze the vector magnetic field at two different heights in the solar atmosphere. In this work, we will concentrate on the second set of spectropolarimetric data taken on July 5 between 7:36 and 14:51 UT. The observing strategy was to scan the filament with the slit parallel to the PIL with a scanning step size of 0.3″. The pixel size along the slit was 0.17″.

The vector magnetic field was inferred by carrying out inversions of the four Stokes parameters with two different inversion codes: (1) for the Si I line we used the SIR code (Stokes Inversion based on Response functions; Ruiz Cobo & del Toro Iniesta 1992) and (2) for the He I triplet we used a Milne-Eddington-based inversion code (MELANIE; Socas-Navarro 2001). The 180°-ambiguity was solved using the AZAM code (Lites *et al.* 1995). The LOS velocities were determined from the inversions and converted to an absolute scale, i.e., corrections for orbital motions and the gravity shift were made (see Kuckein *et al.* 2012b, for a full description of the velocity calibration).

## 3. Results

The inferred magnetic field strength in this AR filament is 600–800 G (Kuckein *et al.* 2009, 2012a), which was the strongest value detected so far inside AR filaments. Recently, other authors have also reported similar strong magnetic fields inside filaments, (e.g., Guo *et al.* 2010; Xu *et al.* 2012). Therefore, it seems to be rather common to find these strong fields inside AR filaments.

The set of inferred chromospheric and photospheric vector magnetograms indicate that the magnetic structure which supports the filament is a flux rope. To substantiate this result, non-linear force-free (NLFF) field extrapolations, starting from the photosphere and chromosphere, were carried out. The extrapolations confirmed the presence of a flux rope structure, which lay surprisingly low in the atmosphere and had its axis located at $\sim$1.4 Mm above the solar surface (see Yelles Chaouche *et al.* 2012, for a complete description of the extrapolations and results).

Based on the vector magnetograms, we constructed the sketch presented in Fig. 1. The cartoon shows a gray-scale continuum image from the Dutch Open Telescope (DOT) on 2005 July 5. In the middle of the image, the aforementioned pores and orphan penumbrae that started to appear on July 4 are seen. The reconstruction of the filament is represented as a black structure which was extracted from the corresponding DOT H$\alpha$ image. The filament follows the PIL. The positive (negative) polarity is at the right (left) side of the PIL when viewing the figure from the lower right corner. There are three different magnetic field lines represented in the cartoon. (1) The dark red field lines are based on the photospheric vector magnetograms. These field lines are aligned along the PIL where pores and orphan penumbrae are seen. However, outside the orphan-penumbrae

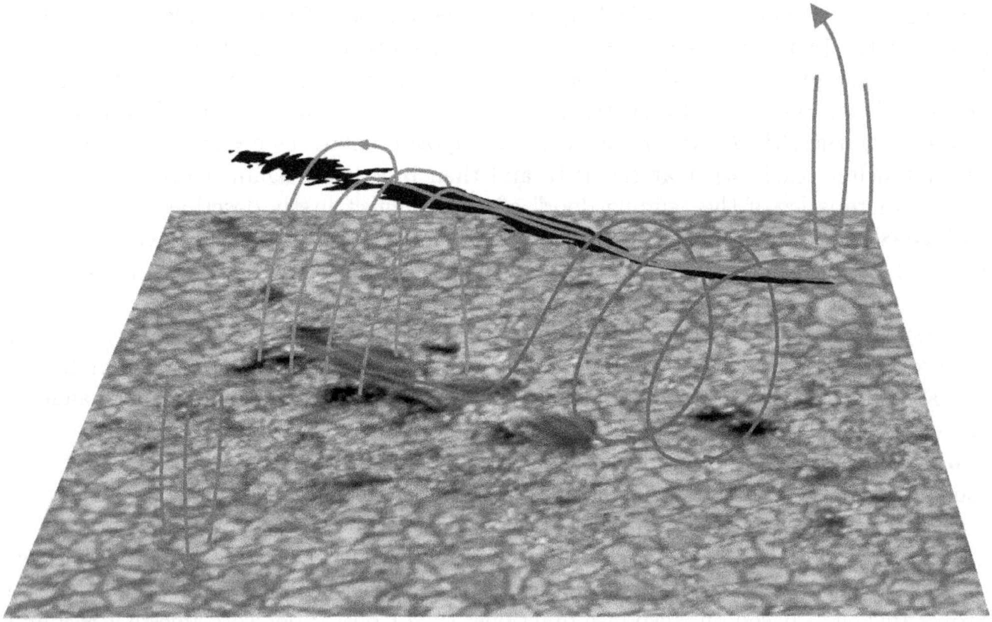

**Figure 1.** Sketch showing the inferred magnetic structure of the AR filament. The surface corresponds to a DOT continuum image. The dark red field lines are representative of what was inferred from the Si I vector magnetic field whereas the light blue lines represent the He I vector magnetic field. The filament is outlined by the black structure taken from the DOT H$\alpha$ image. The light green field lines on both sides of the filament represent the fields in the positive (upward arrow) and negative (downward arrow) faculae. This figure is from Kuckein (2012).

region, the field lines have an inverse polarity configuration (pointing from negative to positive polarity). (2) The light blue field lines were derived from the chromospheric vector magnetograms. In this case, the field lines are parallel to the filament axis outside the orphan-penumbrae region, whereas inside, the field lines show a normal configuration (the field lines point from positive to negative polarity). (3) The green field lines, close to the corners of the image, represent the positive (upward pointing arrow) and negative (downward pointing arrow) polarity of the faculae.

To shed light on the formation of the flux rope it was crucial to infer the LOS velocities. Calibrations using two telluric lines and corrections related to orbital motions and the gravity shift were carefully carried out (see Kuckein *et al.* 2012b).

The motions of the transverse magnetic fields in the photosphere, in the orphan-penumbrae area, show, on a seven-hour average, a slow upward trend. Above, in the chromosphere, the filament moves on average downward. Nevertheless, there are clearly localized upflow areas (see Fig. 3 in Kuckein *et al.* 2012b). This indicated, when looking at the orphan-penumbrae area in Fig. 1, that the axis of the flux rope is slowly rising. Half of the flux rope is below the surface, which seems to be responsible for the formation of the orphan penumbrae. The upper part of the flux rope reaches the chromosphere where groups of field lines produce the upflow patches detected in the chromosphere. The portion of the filament that does not have orphan penumbrae below behaves differently. Out of seven maps, the first four show on average upward motions of the transverse fields. The other three show on average velocities close to zero. The chromospheric counterpart shows downward motions.

## 4. Discussion

The initial formation phase of this AR filament cannot be described with the presented data sets because the filament was already present prior to our observations. However, we have shown that AR filaments can have extremely low-lying flux rope structures (even as low as the photosphere) that support the filament. This structure can eventually emerge from below the photosphere, as seen in our data sets, generating pores and orphan penumbrae along the PIL. Therefore, AR filaments can have a photospheric counterpart. In the chromosphere, the filament's plasma is being pushed upward by the magnetic field that expands from the emerging flux rope structure. At the same time, the emerging flux rope supports the filament material against gravity. Altogether, the present results favor a flux rope emergence scenario.

**Acknowledgements**

CK greatly acknowledges the travel support received from the IAU. The VTT is operated by the Kiepenheuer-Institute for Solar Physics in Freiburg, Germany, at the Spanish Observatorio del Teide, Tenerife, Canary Islands. The National Center for Atmospheric Research (NCAR) is sponsored by the National Science Foundation (NSF). The authors would like to thank C. Denker for carefully reading the manuscript.

## References

Antiochos, S. K., Dahlburg, R. B., & Klimchuk, J. A. 1994, *ApJ* 420, L41
Collados, M., Lagg, A., Díaz García, J. J., *et al.* 2007, *ASP-CS* 368, 611
Denker, C., Johannesson, A., Marquette, W., *et al.* 1999, *Solar Phys.* 184, 87
Guo, Y., Schmieder, B., Démoulin, P., *et al.* 2010, *ApJ* 714, 343
Kuckein, C. 2012, *PhD Thesis*, Universidad de La Laguna (Tenerife, Spain)
Kuckein, C., Martínez Pillet, V., & Centeno, R. 2012b, *A&A* 542, A112
Kuckein, C., Martínez Pillet, V., & Centeno, R. 2012a, *A&A* 539, A131
Kuckein, C., Centeno, R., Martínez Pillet, V., *et al.* 2009, *A&A* 501, 1113
Lites, B. W., Low, B. C., Martinez Pillet, V., *et al.* 1995, *ApJ* 446, 877
Okamoto, T. J., Tsuneta, S., Lites, B. W., *et al.* 2008, *ApJ* 673, L215
Ruiz Cobo, B. & del Toro Iniesta, J. C. 1992, *ApJ* 398, 375
Scherrer, P. H., Bogart, R. S., Bush, R. I., *et al.* 1995, *Solar Phys.* 162, 129
Socas-Navarro, H. 2001, *ASP-CS* 236, 487
van Ballegooijen, A. A. & Martens, P. C. H. 1989, *ApJ* 343, 971
Xu, Z., Lagg, A., Solanki, S., & Liu, Y. 2012, *ApJ* 749, 138
Yelles Chaouche, L., Kuckein, C., Martínez Pillet, V., & Moreno-Insertis, F. 2012, *ApJ* 748, 23

*Nature of prominences and their role in Space Weather*
*Proceedings IAU Symposium No. 300, 2013*
*B. Schmieder, J.-M. Malherbe & S. T. Wu, eds.*

doi:10.1017/S1743921313010727

# The formation and disappearance of filament barbs observed by SDO

**Leping Li and Jun Zhang**

Key Laboratory of Solar Activity, National Astronomical Observatories, Chinese Academy of Sciences, Beijing 100012, China
email: `lepingli@nao.cas.cn`

**Abstract.** Employing six-day (August 16-21, 2010) SDO/AIA observations, we systematically investigate the formation and disappearance of 58 barbs of a northern (∼N60) polar crown filament. Three different ways of barb formation are discovered, including (1) the convergence of surrounding moving materials (55.2%), (2) the flows of materials from the filament (37.9%), and (3) the material injections from neighboring brightening regions (6.9%). We also find three different types of barb disappearance, involving: (i) the bi-lateral movements (44.8%), and (ii) the outflowing (27.6%) of barb material resulting in the barb disappearance, as well as (iii) the barb disappearance associated with neighboring brightenings (27.6%). We propose that barbs exchange materials with the filament, surrounding atmosphere, and nearby brightening regions, causing the barb formation and disappearance.

**Keywords.** Sun: filaments, Sun: prominences, Sun: evolution, Sun: corona

---

## 1. Introduction

Solar filaments are characterized by relatively cool and dense plasma (Babcock & Babcock 1955, Schmieder *et al.* 2008). They typically have two linked substructures: the 'spine' and the 'barbs', between which the latter are made when some threads of the spine smoothly bend outwards (Lin 2008).

For the formation of the barbs, Wang (1999) presented that magnetic reconnection between the parasitic and the neighboring dominant polarities plays a key role. In contrast, Martens & Zwaan (2001) suggested that the barbs arise as a result of failed cancelation. Heinzel & Anzer (2001) suggested that barbs could be formed by the evaporation-condensation process if they are composed of vertically aligned dips. Aulanier *et al.* (1998) and Mackay & van Ballegooijen (2009) also proposed that the barbs are cool matter suspended in local dips.

Most of the studies about barbs are based on H$\alpha$ observations. After the launch of the Solar Dynamics Observatory (SDO), its high-quality observations provide us the opportunity to study the barbs in extreme ultraviolet (EUV) wavelengths. In this work, we present our systematic study of the formation and disappearance of barbs on a polar crown filament in six days. The observations are described in Section 2. We present the results and conclusions in Section 3, as well as the discussion, in Section 4.

## 2. Observations

The SDO/AIA is designed to acquire images of the solar atmosphere in a series of wavelength bands. In this study, we use the AIA 171 Å and 304 Å data to investigate the evolution of barbs. The time cadence and spatial resolution of the AIA images are 12 s and 1.2″, respectively. We also employ the HMI line of sight magnetograms, with spatial resolution of 1.0″ and time cadence of 45 s, to study the evolution of magnetic fields surrounding the barbs.

## 3. Results and Conclusions

From August 16 to 21, 2010, a northern (N50∼N70) polar crown filament was observed by SDO. Using six-day observations, we identified 69 barbs, and chose 58 of them, which formed away from the western solar limb (⩽W60), as our sample to study the barb formation and disappearance in detail.

### 3.1. *The formation of filament barbs*

Three different types of barb formation are detected (Li & Zhang 2013), involving: (1) the convergence of moving materials, (2) the flows of material from the filament main body, and (3) the material injections from neighboring brightening regions.

32 (55.2%) barbs display the first type of barb formation. An example on August 16, 2010, is shown in the upper panels of Fig. 1. A barb is presented at 12:44 UT in Fig. 1a3. We overlay the barb on Fig. 1a1 as red contours, and notice that there was no barb at 11:36 UT in and surrounding the contour area. However, several materials moved around. Sometimes, some of the materials, denoted by three red arrows in Fig. 1a2 with heads showing the moving directions, converged together. Finally, a new barb formed. In short, the ambient moving materials converge together forming a new barb.

22 (37.9%) barbs manifest the second type of barb formation. We display an example on August 16, 2010, in the middle panels of Fig. 1. Figure 1b3 illustrates a barb at 9:23 UT. We overly the barb as black contours on Fig. 1b1, and find that there was no barb existed at 9:07 UT. However, to the northwest, some small-scale materials flowing along the spine were detected. From 9:13 UT, the horizontal flows transformed into southward

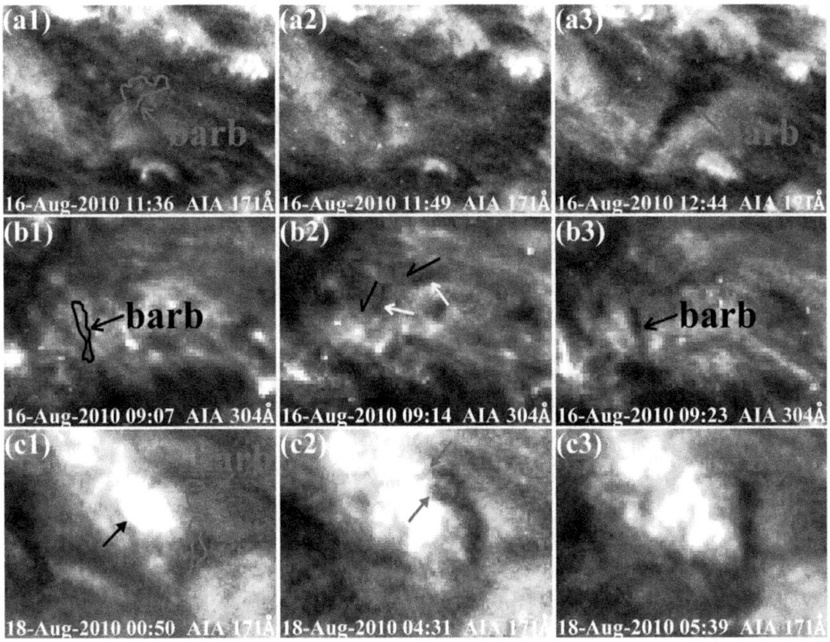


**Figure 1.** A series of SDO/AIA 171 Å (a1-a3), 304 Å (b1-b3), and 171 Å (c1-c3) images displaying the formation of a filament barb on August 16, 16, and 18, 2010, respectively. The contours in (a1), (b1), and (c1) separately enclose the barb in (a3), (b3), and (c3). Three red arrows in (a2) and two white arrows in (b2) denote the moving materials, and two black arrows in (b2), the moving directions. A black arrow in (c1) indicates a brightening region, and two red arrows in (c2), the connections between the brightening regions and the barb. The field of view (FOV) of (a1-a3), (b1-b3), and (c1-c3) is $96'' \times 72''$, $48'' \times 36''$, and $80'' \times 45''$, respectively.

flows (see the black arrows in Fig. 1b2). Finally, a new barb formed. In a word, the transition of the flows along the spine into downward flows forms a new barb.

In 4 (6.9%) cases of our sample, another type of barb formation is noted. An example on August 18, 2010, is displayed in the lower panels of Fig. 1. A barb is indicated in Fig. 1c3 at 5:39 UT. We overlay the barb as red contours on Fig. 1c1, and notice that there was no barb at 0:50 UT. Nevertheless, several neighboring brightening regions existed to the east of the contour regions. From 0:25 UT, the brightening regions began to inject small-scale materials into the contour area to form a new barb. Two connections between the brightening regions and the new formed barb appeared, labeled by two red arrows in Fig. 1c2. Finally, these connections disappeared, and the barb formed. In brief, the materials injected from the neighboring brightening regions form a new barb.

### 3.2. *The disappearance of filament barbs*

Barbs are seen to disappear in three different ways (Li & Zhang 2013), involving: (i) the barb moves bi-laterally and disappears into the surrounding atmosphere, and (ii) the outflowing of barb materials to the chromosphere results in the barb disappearance, as well as (iii) the barb disappearance is associated with neighboring brightenings.

In our sample, 26 (44.8%) barbs illustrate the first way to disappear. An example on August 18, 2010, is displayed in the upper panels of Fig. 2. Figure 2a1 shows a barb at 8:10 UT. Thereafter, the barb moved bi-laterally (see the green arrows in Fig. 2a2), and disappeared rapidly into the atmosphere. We overlay the barb as red contour on Fig. 2a3, and find that the barb completely disappeared at 12:39 UT. In brief, the bi-lateral movements of the barb lead to the barb disappearance.

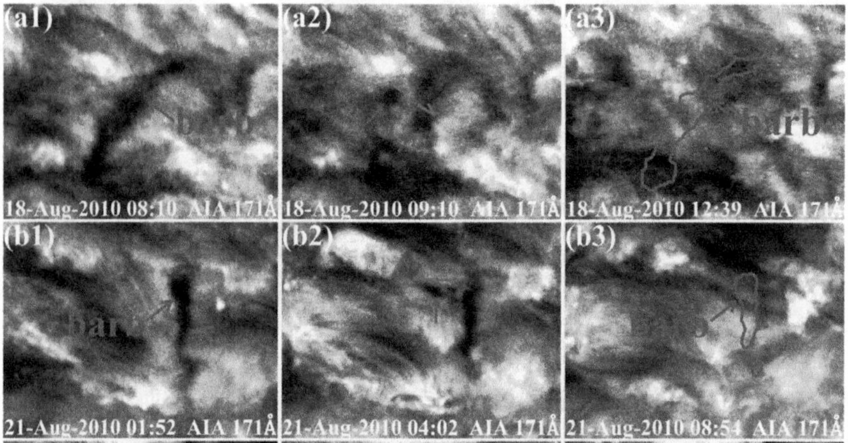

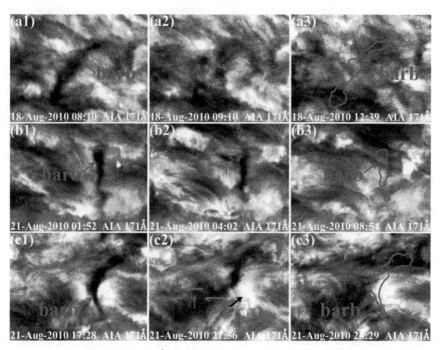


**Figure 2.** A series of SDO/AIA 171 Å images separately showing the disappearance of a filament barb on August 18 (a1-a3), 21 (b1-b3), and 21 (c1-c3), 2010. The red contours in (a3), (b3), and (c3) enclose the barb in (a1), (b1), and (c1), respectively. The red arrows in (a2), (b2), and (c2) separately mark the moving materials, and green arrows, the moving directions. A black arrow in (c2) denotes a brightening region. The FOV of (a1-a3), (b1-b3), and (c1-c3) is $96'' \times 72''$, $120'' \times 90''$, $120'' \times 90''$, respectively.

16 (27.6%) barbs display the second way to disappear. We demonstrate an example on August 21, 2010, in the middle panels of Fig. 2. A barb is presented at 1:52 UT in Fig. 2b1. From 2:43 UT, the barb materials began to outflow toward east, as denoted by a green arrow in Fig. 2b2. Consequently, the barb materials became less and less, and the barb disappeared. The red contours in Fig. 2b3 enclose the barb in Fig. 2b1. There was no barb existed at 8:54 UT in the contour area. In short, the outflowing of the barb material leads to the barb disappearance.

16 (27.6%) barbs show the third type of barb disappearance. An example on August 21, 2010, is demonstrated in the lower panels of Fig. 2. A barb is marked at 17:28 UT in Fig. 2c1. We overlay the barb as red contours on Fig. 2c3, and note that the barb completely disappeared at 23:29 UT. During the disappearance process, a brightening region, indicated by a black arrow in Fig. 2c2, was detected to the west of the barb. The barb materials (denoted by red arrows in Fig. 2c2) moved toward the east one by one (see the green arrows in Fig. 2c2). In a word, the barb disappearance is associated with the nearby small-scale solar brightening.

## 4. Discussion

About the barb formation, two aspects are considered: the formation of (1) the barb magnetic structures, supporting the material, and (2) the barb material. We investigate the evolution of magnetic fields, and find that magnetic emergences may form the barb magnetic structures. About the origination of the barb material, three different types of formation are found, and the third type seems to be consistent with the injection models (Wang 1999) of the filament material.

Similar to the formation, two aspects are also regarded during the barb disappearance, including: the disappearance of (1) the barb magnetic structures, and (2) the barb material. After investigation of magnetic field evolutions, we find that the magnetic field motions, emergences, cancelations, and disappearances may deform/destroy the barb magnetic structures.

In a word, the evolution of the photospheric magnetic fields may be the reason for the formation and disappearance of the barb magnetic structures. Moreover, the barbs exchange materials with the surrounding atmosphere, the filament spine, and the neighboring small-scale brightening regions, caused by the magnetic cancelations, can lead to the formation and disappearance of the barb material.

### Acknowledgements

The work is supported by the National Natural Science Foundations of China (G11003026, 11025315, 11303050, 11221063, 41074123, 11003024), the National Basic Research Program of China under grant G2011CB811403, and the CAS KJCX2-EW-T07.

### References

Aulanier, G., Démoulin, P., van Driel-Gesztelyi, L., *et al.* 1998, *Astron. Astrophys.*, 335, 309
Babcock, H. W. & Babcock, H. D. 1955, *Astrophys. J.*, 121, 349
Heinzel, P. & Anzer, U. 2001, *Astron. Astrophys.*, 375, 1090
Li, L. P. & Zhang, J. 2013, *Solar Phys.*, 287, 147
Lin, Y. 2011, *Space Sci. Rev.*, 158, 237
Mackay, D. & van Ballegooijen, A. 2009, *Solar Phys.*, 260, 321
Martens, P. C. & Zwaan, C. 2001, *Astrphys. J.*, 538, 872
Schmieder, B., Bommier, V., Kitai, T., *et al.* 2008, *Solar Phys.*, 247, 321
Wang, Y. M. 1999, *Astrophys. J. Lett.*, 520, L71

*Nature of Prominences and their role in Space Weather*
*Proceedings IAU Symposium No. 300, 2013*
*B. Schmieder, J.-M. Malherbe & S. T. Wu, eds.*

doi:10.1017/S1743921313010739

# The damping of transverse oscillations of prominence threads: a comparative study

**Roberto Soler, Ramon Oliver, and Jose Luis Ballester**
Departament de Física, Universitat de les Illes Balears, E-07122 Palma de Mallorca, Spain
email: roberto.soler@uib.es

**Abstract.** Transverse oscillations of thin threads in solar prominences are frequently reported in high-resolution observations. The typical periods of the oscillations are in the range of 3 to 20 min. A peculiar feature of the oscillations is that they are damped in time, with short damping times corresponding to few periods. Theoretically, the oscillations are interpreted as kink magnetohydrodynamic waves. However, the mechanism responsible for the damping is not well known. Here we perform a comparative study between different physical mechanisms that may damp kink waves in prominence threads. The considered processes are thermal conduction, cooling by radiation, resonant absorption, and ion-neutral collisions. We find that thermal conduction and radiative cooling are very inefficient for the damping of kink waves. The effect of ion-neutral collisions is minor for waves with periods usually observed. Resonant absorption is the only process that produces an efficient damping. The damping times theoretically predicted by resonant absorption are compatible with those reported in the observations.

**Keywords.** Sun: oscillations – Sun: corona – Sun: prominences – Sun: filaments – Waves

## 1. Introduction

Oscillations of small amplitude are frequently observed in solar prominences (see reviews by, e.g., Oliver & Ballester 2002, Engvold 2008, Arregui *et al.* 2012). High-resolution observations often show that individual prominence threads, i.e., the building blocks of prominences, oscillate transversely (e.g., Lin *et al.* 2005, 2007, 2009, Okamoto *et al.* 2007, Ning *et al.* 2009). Some features of transverse thread oscillations are that the periods are in the range of 3 to 20 min, that the velocity amplitudes are lower than 3 km s$^{-1}$, and that the oscillations are quickly damped. Typically, the oscillation amplitude decreases to the noise level in less than 10 periods (Ning *et al.* 2009).

Transverse thread oscillations are theoretically interpreted as kink magnetohydrodynamic (MHD) waves propagating along the magnetic flux tube that supports the thread (e.g., Díaz *et al.* 2002, Terradas *et al.* 2008, Lin *et al.* 2009, Soler *et al.* 2010). However, the mechanism responsible for the damping is not well known and several possible physical processes have been suggested (e.g., Ballai 2003, Arregui *et al.* 2008, Oliver 2009). Here, we compare the efficiency of various mechanisms that may damp kink MHD waves in prominence threads, namely radiative cooling, thermal conduction, ion-neutral collisions, and resonant absorption. This brief work is based on the results extensively discussed in Soler (2010).

## 2. Model and method

The prominence thread model is schematically represented in Figure 1. It is composed of a straight cylindrical magnetic flux tube of radius $a$, filled with prominence plasma with density $\rho_{\rm i}$ and temperature $T_{\rm i}$, and embedded in a coronal environment with density $\rho_{\rm e}$ and temperature $T_{\rm e}$. Between the prominence and coronal plasma there is a transverse

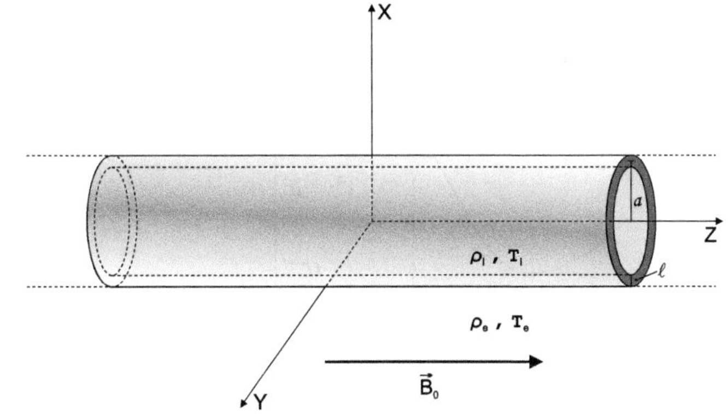


**Figure 1.** Schematic representation of the prominence thread model used in this work.

transitional layer of thickness $l$, where the physical conditions change continuously. The magnetic field strength, $B_0$, is constant. We use the following values for the various physical parameters: $\rho_{\rm i} = 5 \times 10^{-11}$ kg m$^{-3}$, $T_{\rm i} = 8000$ K, $\rho_{\rm e} = 2.5 \times 10^{-13}$ kg m$^{-3}$, $T_{\rm e} = 10^6$ K, $a = 100$ km, and $B_0 = 5$ G. These parameters verify the pressure balance condition at the boundary of the thread. The thickness of the transitional layer, $l$, is considered a free parameter. The prominence plasma is assumed to be partially ionized and the ionization degree is a free parameter as well.

We study linear MHD waves superimposed on the static equilibrium. The governing equations for nonadiabatic perturbations are

$$\frac{\partial \rho'}{\partial t} = -\mathbf{v} \cdot \nabla \rho - \rho \nabla \cdot \mathbf{v}, \tag{2.1}$$

$$\rho \frac{\partial \mathbf{v}}{\partial t} = -\nabla p' + \frac{1}{\mu} \left[ (\nabla \times \mathbf{B}') \times \mathbf{B}_0 \right], \tag{2.2}$$

$$\frac{\partial p'}{\partial t} = -\gamma p \nabla \cdot \mathbf{v} - (\gamma - 1) \left[ \rho \frac{\partial L}{\partial \rho} \rho' + \rho \frac{\partial L}{\partial T} T' - \kappa_{\parallel} \frac{\partial^2 T'}{\partial z^2} \right], \tag{2.3}$$

$$\frac{\partial \mathbf{B}'}{\partial t} = \nabla \times (\mathbf{v} \times \mathbf{B}_0) - \nabla \times (\eta \nabla \times \mathbf{B}') + \nabla \times \left\{ \frac{\eta_{\rm C} - \eta}{B_0^2} \left[ (\nabla \times \mathbf{B}') \times \mathbf{B}_0 \right] \times \mathbf{B}_0 \right\}, \tag{2.4}$$

$$\frac{p'}{p} = \frac{\rho'}{\rho} + \frac{T'}{T}, \tag{2.5}$$

where $\rho$, $p$, $T$, $\mathbf{B}_0$ are the equilibrium density, gas pressure, and temperature, $\rho'$, $p'$, $T'$, $\mathbf{B}'$ are their corresponding perturbations, $\mathbf{v}$ is the velocity perturbation, and $\mu$ is magnetic permittivity. In addition, $\partial L/\partial \rho$ and $\partial L/\partial T$ are the partial derivatives of the the heat-loss function $L$ with respect to density and pressure, $\eta$ and $\eta_{\rm C}$ are Ohm's and Cowling's resistivities, and $\kappa_{\parallel}$ is the parallel thermal conductivity. We refer the reader to Soler (2010) for the expressions of these quantities.

The perturbations are put proportional to $\exp(ik_z z + im\varphi - i\omega t)$, where $k_z$ and $m$ are the longitudinal and azimuthal wavenumbers ($m = 1$ for kink waves) and $\omega$ is the frequency. After fixing $k_z$ and $m$, Equations (2.1)–(2.5) form an eigenvalue problem where $\omega$ is the eigenvalue. The eigenvalue problem is solved numerically with the PDE2D code (Sewell 2005). Due to the presence of damping mechanism, $\omega$ is complex. The period, $P$, and exponential damping time, $\tau_{\rm D}$, are computed from the real and imaginary parts of

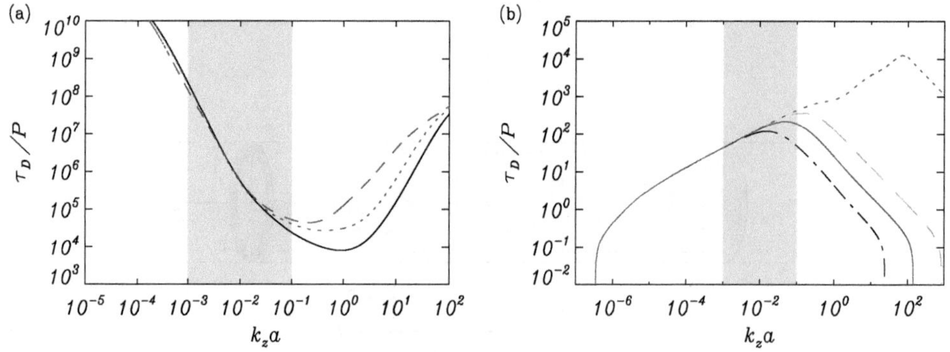


**Figure 2.** (a) Kink wave $\tau_D/P$ vs. $k_z a$ for damping due to nonadiabatic effects. The various lines are for different parameterizations of the heat-loss function (see details in Soler 2010). (b) Same as panel (a) but for damping due to Cowling's and Ohm's diffusion. The various lines are for different ionization degrees from almost fully ionized (upper line) to almost fully neutral (lower line) thread. In both panels, the shaded area is the realistic range $k_z a \in [10^{-3}, 10^{-1}]$. Results for $l/a = 0$.

$\omega$ as $P = 2\pi/\mathrm{Re}(\omega)$ and $\tau_D = 1/|\mathrm{Im}(\omega)|$. Values of $\tau_D/P \lesssim 10$ are needed for damping to be compatible with the observations.

## 3. Results

We compare the efficiency of the various damping mechanisms by computing the kink wave damping ratio, $\tau_D/P$, produced by each process as function of the dimensionless wavenumber, $k_z a$, whose realistic values are in the interval $k_z a \in [10^{-3}, 10^{-1}]$.

*Damping due to nonadiabatic effects.* We compute the damping ratio when thermal conduction and radiative cooling are considered (Fig. 2a). We obtain $\tau_D/P \gtrsim 10^4$, meaning that nonadiabatic effects are very inefficient damping mechanisms for kink waves and cannot explain the observed rapid attenuation.

*Damping due to ion-neutral collisions.* The effect of ion-neutral collisions is here included in the single-fluid approximation. Cowling's diffusion term in the induction equation (third term on the right-hand side of Equation (2.4)) contains the effect of collisions. For consistency, we also consider the effect of classic Ohm's diffusion. Figure 2b shows that $\tau_D/P \gtrsim 10^2$ in the realistic range of $k_z a$, which points out that damping due to Cowling's and Ohm's diffusion is inefficient. Although efficient damping due to Ohm's/Cowling's diffusion is obtained for $k_z a$ smaller/larger than realistic values, this result has no relevance for the observed waves.

*Damping due to resonant absorption.* Resonant damping takes place in the nonuniform transitional layer where the global kink wave couples to local Alfvén and slow waves. An approximate expression for the kink mode $\tau_D/P$ in the thin tube ($k_z a \ll 1$) and thin boundary ($l/a \ll 1$) approximations is (e.g., Ruderman & Roberts 2002, Goossens *et al.* 2002)

$$\frac{\tau_D}{P} \approx \frac{2}{\pi}\frac{a}{l}\frac{\rho_i + \rho_e}{\rho_i + \rho_e}\left[1 + (k_z a)^2 \left(\frac{v_s^2}{v_s^2 + v_A^2}\right)^2\right]^{-1}, \tag{3.1}$$

where $v_s$ and $v_A$ are the sound and Alfvén velocities. The first term within the square brackets of Equation (3.1) is due to the Alfvén resonance and the second term is due to the slow resonance. For typical parameters the slow resonance can be neglected. Using $l/a = 0.2$ and $k_z a = 10^{-2}$, Equation (3.1) gives $\tau_D/P \approx 3$, which suggests that

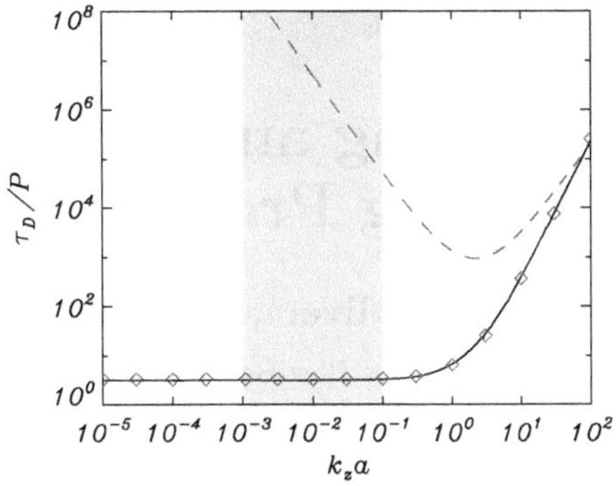


**Figure 3.** Same as Figure 2 but for damping due to resonant absorption. The solid line is the full result, the dashed line is the slow resonance damping, and symbols are the Alfvén resonance damping. We used $l/a = 0.2$.

resonant absorption is an efficient damping mechanism. This result is confirmed by the full numerical computations displayed in Figure 3.

We conclude that, among the considered damping mechanisms, resonant coupling to Alfvén waves is the only process that produces kink wave damping times compatible with those observed.

## Acknowledgements

We acknowledge support from MINECO and FEDER Funds through grant AYA2011-22846 and from CAIB through the 'grups competitius' program and FEDER Funds.

## References

Arregui, I., Terradas, J., Oliver, R., & Ballester, J. L. 2008, *ApJL*, 682, L141
Arregui, I., Oliver, R., & Ballester, J. L. 2012, *Living Rev. Sol. Phys.*, 9, 2
Ballai, I. 2003, *A&A*, 410, L17
Díaz, A. J., Oliver, R., & Ballester, J. L. 2002, *ApJ*, 580, 550
Engvold, O. 2008, *Waves & Oscillations in the Solar Atmosphere: Heating and Magneto-Seismology, Proceedings of the International Astronomical Union, IAU Symposium*, 247, 152
Goossens, M., Andries, J., & Aschwanden, M. J. 2002, *A&A*, 394, L39
Lin, Y., Engvold, O., Rouppe van der Voort, L., Wiik, J E., & Berger, T. E. 2005, *Sol. Phys.*, 226, 239
Lin, Y., Engvold, O., Rouppe van der Voort, L., & van Noort, M. 2007, *Sol. Phys.*, 246, 65
Lin, Y., Soler, R., Engvold, O., *et al.* 2009, *ApJ*, 704, 870
Ning, Z., Cao, W., Okamoto, T. J., Ichimoto, K., & Qu, Z. Q. 2009, *A&A*, 499, 595
Okamoto, T. J., Tsuneta, S., Berger, T. E., *et al.* 2007, *Science*, 318, 1577
Oliver, R. & Ballester, J. L. 2002, *Sol. Phys.*, 206, 45
Oliver, R. 2009, *SSR*, 149, 175
Ruderman, M. S. & Roberts, B. 2002, *ApJ*, 577, 475
Sewell, G. 2005, *The Numerical Solution of Ordinary and Partial Differential Equations, Pure and Applied Mathematics Series* (New York: Wiley)
Soler, R., Arregui, I., Oliver, R., & Ballester, J. L. 2010, *ApJ*, 722, 1778
Soler, R. 2010, *PhD thesis, Universitat de les Illes Balears*, available at `http://www.uib.es/depart/dfs/Solar/thesis_roberto_soler.pdf`
Terradas, J., Arregui, I., Oliver, R., & Ballester, J. L. 2008, *ApJL*, 678, L153

*Nature of Prominences and their role in Space Weather*
*Proceedings IAU Symposium No. 300, 2013*
*B. Schmieder, J.-M. Malherbe & S. T. Wu, eds.*

doi:10.1017/S1743921313010740

# Non-LTE Modeling and Observations of Oscillating Prominences

**M. Zapiór[1,2], P. Heinzel[2], R. Oliver[1], J. L. Ballester[1] and P. Kotrč[2]**

[1]Departament de Física, Universitat de les Illes Balears, E-07122 Palma de Mallorca, Spain
email: maciej.zapior@uib.es; ramon.oliver@uib.es; joseluis.ballester@uib.es

[2]Astronomical Institute, Academy of Sciences, 25165 Ondřejov, Czech Republic
email: pheinzel@asu.cas.cz; pkotrc@asu.cas.cz

**Abstract.** Prominence oscillations have been mostly detected using Doppler velocity, although there are also claimed detections by means of the periodic variations of half-width or line intensity. Our main aim here is to explore the relationship between spectral indicators such as Doppler shift, line intensity and line half-width and the linear perturbations excited in a simple prominence model.

**Keywords.** Sun: oscillations, Sun: filaments, Sun: prominences

---

## 1. Model and methods

For MHD simulations we used equilibrium model described by Oliver *et al.* (1993). It consists of a bounded, vertical, homogeneous slab, permeated by a transverse magnetic field, having prominence-like physical properties. We assumed linear perturbations. Then the dispersion relation for fast and slow modes has been derived. Perturbations of temperature, gas pressure and velocity along the line of sight (LOS) axis (see Figure 1) calculated by MHD simulations were used as input parameters for one-dimensional non-LTE radiative transfer code with 5 levels and continuum of hydrogen (see Heinzel 1995; Labrosse *et al.* 2010). Considered geometry was the same as for the MHD modeling. We assumed hydrogen-helium plasma with partially ionized hydrogen and neutral helium, microturbulent velocity equal to zero and slab thickness 12000 km. Incident radiation was carefully taken into account.

## 2. Results

As a result of calculations we obtained the full spectral profiles of the hydrogen H$\alpha$ and H$\beta$ lines in different phases of oscillations for four oscillatory modes (see Figures 2 and 3). For each step we calculated peak intensity of the spectral profile, full width at half maximum (FWHM) and the Doppler shift (see Figure 4). Detectable variations of the Doppler velocity with the peak-to-peak amplitude of the order 2 km s$^{-1}$ were found for the fundamental slow (fS) mode only. Other modes have variations below 0.1 km s$^{-1}$. Different behaviour of FWHM and maximum intensity variations (number of peaks and their relative maximum over the whole cycle) for fundamental slow (fS), first slow harmonic (1S) and fundamental fast (fF) modes brings the possibility to distinguish between modes from observational data. Asymmetry was detected for fS mode in the H$\alpha$ line profile only. Only 1F mode is practically non-detectable. All variations are summarized in Table 1.

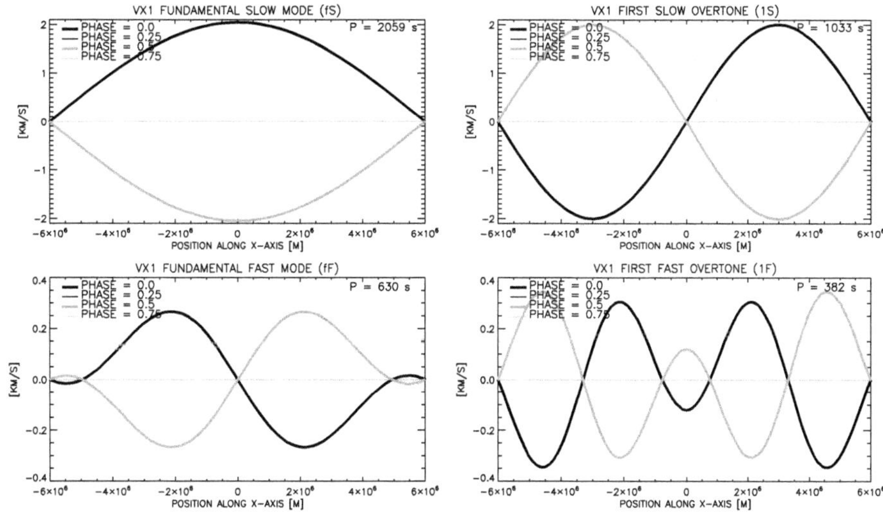


**Figure 1.** Velocity perturbations across the prominence slab for the fundamental slow mode (fS), the first slow overtone (1S), the fundamental fast mode (fF) and the first fast overtone (1F). The $x$-axes of plots are scaled in kilometers. Note the different velocity scale. Different lines represent different phases (i.e. time divided by period), labeled at the top left corner of each plot. For phase = 0.25 and phase = 0.75 the lines are merged. Oscillatory periods are shown at the top right corner of each plot.

| Line | Mode | $\Delta v_D$ | FWHM | $\max(I(\lambda))$ |
|---|---|---|---|---|
| H$\alpha$ | fS | 2.241 | 0.593 – 0.622 (4.8%) | 3.531 – 3.671 (3.9%) $\times 10^{-6}$ |
| | 1S | 0.026 | 0.600 – 0.625 (4.1%) | 3.521 – 3.651 (3.6%) $\times 10^{-6}$ |
| | fF | 0.004 | 0.581 – 0.609 (4.6%) | 3.491 – 3.590 (2.8%) $\times 10^{-6}$ |
| | 1F | 0.090 | 0.593 – 0.594 (0.3%) | 3.531 – 3.531 (0.0%) $\times 10^{-6}$ |
| H$\beta$ | fS | 2.175 | 0.298 – 0.303 (1.6%) | 5.760 – 6.551 (12.8%) $\times 10^{-7}$ |
| | 1S | 0.005 | 0.300 – 0.304 (1.2%) | 5.690 – 6.790 (17.6%) $\times 10^{-7}$ |
| | fF | 0.018 | 0.293 – 0.301 (2.6%) | 5.570 – 6.130 (9.6%) $\times 10^{-7}$ |
| | 1F | 0.098 | 0.296 – 0.297 (0.2%) | 5.780 – 5.820 (0.7%) $\times 10^{-7}$ |

**Table 1.** Variations of spectral indicators for H$\alpha$ and H$\beta$. Abbreviations of modes are the same as described in the main text. $\Delta v_D$ stands for peak-to-peak Doppler velocity amplitude. In brackets relative changes of each value with respect to the mean are presented.

## 3. Summary and future prospects

The first numerical simulations of the non-LTE radiative transfer in an oscillatory prominence slab were performed. To perform prominence seismology, analysis of the H$\alpha$ and H$\beta$ spectral line parameters could be a good tool to detect and identify oscillatory modes. Figure 5 presents results of analysis of exemplary observations of prominence oscillations performed with Ondřejov Multi-channel Spectrograph (see Kotrč, 2009). In the future we will investigate a grid of models with different physical conditions and different oscillatory parameters. Calculation of spectral profiles for different spectral lines may lead to investigation of velocity field in different optical depths and may be directly compared with observations. Ca II lines are detectable from ground, MgII h and k lines from space by the IRIS mission. Performing 2D radiative transfer simulations will bring more realistic results.

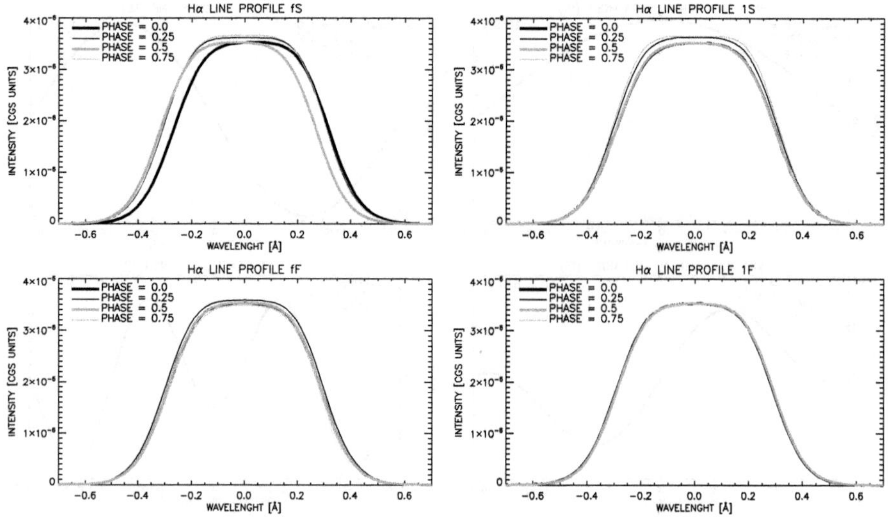


**Figure 2.** Time variation of the Hα spectral line profile for consecutive phases (labelled in the top left corner of each plot) and different modes. Specific line intensities are in cgs units erg $\sec^{-1}$ $\mathrm{cm}^{-2}$ $\mathrm{sr}^{-1}$ $\mathrm{Hz}^{-1}$.

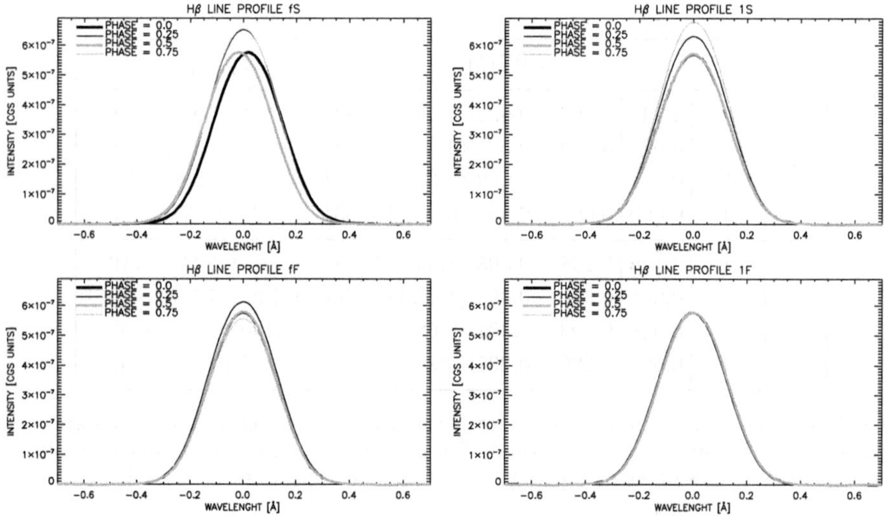


**Figure 3.** Same as Fig. 2, but for Hβ.

## Acknowledgements

JLB, RO and MZ acknowledge the financial support provided by MICINN and FEDER funds under grant AYA2011-22846. JLB and RO acknowledge the financial support from CAIB and Feder Funds under the ,,Grups Competitius scheme”. PH was supported by the grant P209/12/0906 of the GAČR.

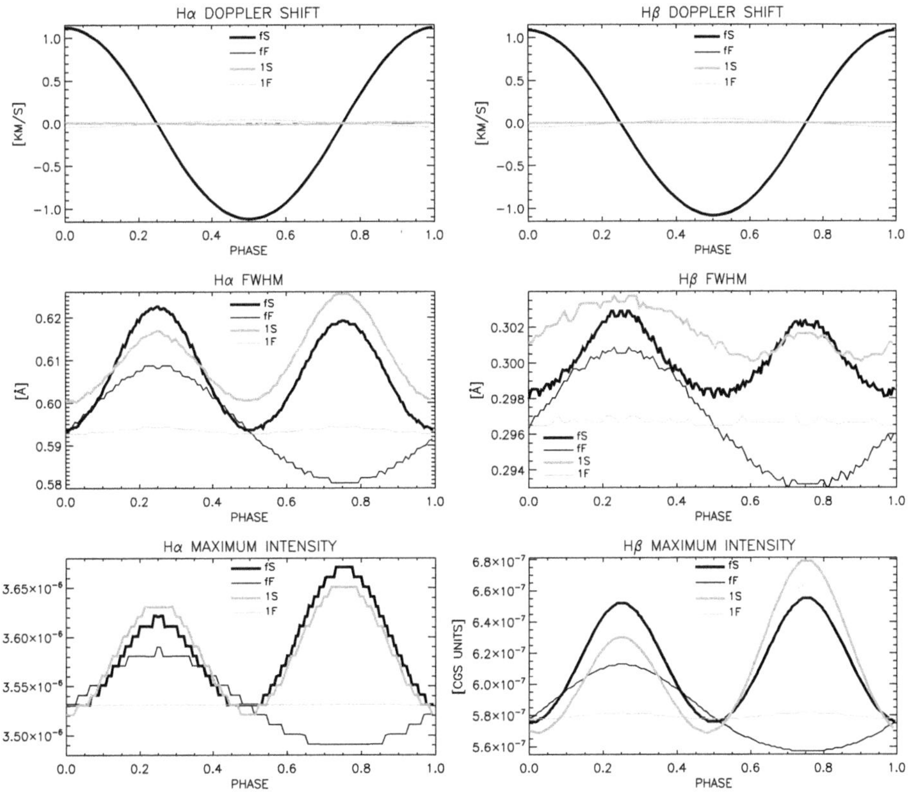


**Figure 4.** Time variation as a function of phase of the spectral line parameters for all modes. Different lines labelled in the plots correspond to different modes.

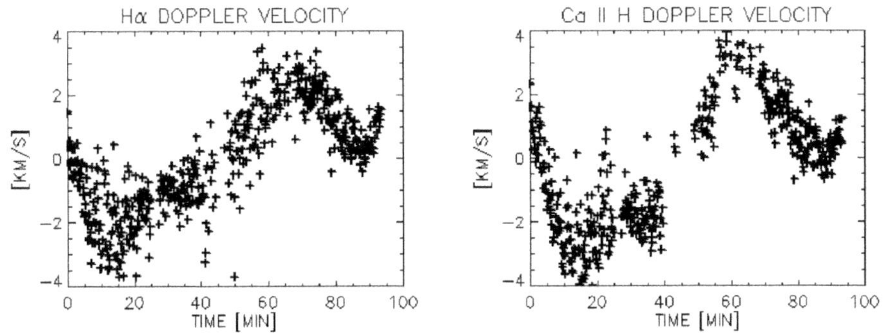


**Figure 5.** Variations of Doppler velocity of Hα and Ca II H spectral lines of the prominence of September 23, 2010 observed by Ondřejov Spectrograph.

## References


Heinzel, P. 1995, *A&A*, 299, 563

Kotrč, P. 2009, *Central European Astrophysical Bulletin*, 33, 327

Labrosse, N., Heinzel, P., Vial, J.-C., Kucera, T., Parenti, S., Gunár, S., & Schmieder, B. and Kilper, G. 1993, *Space Sci. Revs*, 151, 243

Oliver, R., Ballester, J. L., & Hood, A. W. and Priest, E. R. 1993, *ApJ*, 400, 369

# Session I - 1.2

## PROMINENCE PLASMA

*Nature of prominences and their role in Space Weather*
*Proceedings IAU Symposium No. 300, 2013*
*B. Schmieder, J.-M. Malherbe & S. T. Wu, eds.*

doi:10.1017/S1743921313010752

# Modelling of quiescent prominence fine structures

**S. Gunár**

Astronomical Institute of the Academy of Sciences of Czech Republic,
251 65 Ondřejov, Czech Republic
email: stanislav.gunar@asu.cas.cz

**Abstract.** We review here the current status and the latest results of the modelling of quiescent prominence fine structures. We begin with the simulations of the prominence magnetic field configurations, through an overview of the modelling of the fine structure formation and dynamics, and with the emphasis on the radiative transfer modelling of the realistic prominence fine structures. We also illuminate the future directions of the field that lie in the combining of the existing approaches into more complex multi-disciplinary models.

**Keywords.** Sun: prominences, Sun: filaments, magnetic fields, radiative transfer

## 1. Introduction

In this paper we review the state-of-the-art of the modelling of the prominence fine structures. We focus solely on the quiescent solar prominences and we consider only the simulations and models that include individual or multiple small-scale prominence structures.

Quiescent solar prominences are cool dense regions of plasma which lie in a near-equilibrium in a much hotter and rarer coronal environment. Their large-scale structure remains stable for days or weeks while their fine structures with dimensions as small as 100 km exhibit rather dynamical behaviour on time scales of a few minutes. The cool core of prominence fine structures have a temperature below 10,000 K, while the surrounding corona is over 1MK. The nature of the transition region between cool prominence plasma and hot corona (the prominence-corona transition region - PCTR) has a significant impact on the formation, energy equilibrium, and stability of prominences. Many of their important characteristics are given in the monograph by Tandberg-Hanssen (1995). The existence of prominences is mainly due to the coronal magnetic fields. Firstly, the magnetic fields define the framework configuration of prominences and support the dense prominence plasma against gravity. Secondly, they insulate the cool prominence fine structure material from the hot coronal plasma. Prominences, even the quiescent, may also become unstable, producing Coronal Mass Ejections - violent eruptions that may directly affect the Earth.

The exceptionally large amount of observations with ever-increasing spatial and temporal resolution, obtained by space-borne missions like SOHO, Stereo, Hinode, and SDO, together with a plethora of ground-based observatories facilitated a rapid advancement in our understanding of prominences in the last decade. Indeed, the synergy between the wealth of observations and the prominence modelling is still the main driving force of the prominence research. The contribution of these observations to the prominence research was reviewed by Patsourakos & Vial (2002) and Schmieder & Aulanier (2012). The fine structure of the solar prominences was comprehensively reviewed by Heinzel (2007) and the state-of-the-art of prominence physics was reviewed by Labrosse *et al.* (2010) and Mackay *et al.* (2010).

The modelling of the prominence fine structures relies on three main components. These are the simulations of the prominence magnetic field configurations, models of the prominence plasma and its dynamics, and the modelling of the radiative transfer in the prominence plasma. In the following sections we review the current status and the latest results of these approaches, beginning with the whole-prominence magnetic field simulations (Sect. 2). In Sections 3 and 4 we give a brief overview of the fine structure formation and oscillations, respectively. In Sect. 5 we describe the state-of-the-art of the radiative transfer modelling of prominence fine structures and in Sect. 6 we give our conclusions.

## 2. Whole-prominence magnetic field simulations

It is generally assumed that the majority of the dense cool plasma of the prominence fine structures lies in dips of the predominantly horizontal magnetic field (see e.g. observational findings of Schmieder *et al.* 2010 and López Ariste *et al.*2006). These dips can be either strongly affected or indeed be caused by the weight of the prominence fine-structure plasma or they can occur as the result of for example the force-free nature of the magnetic field configurations. For discussion of the force-free assumption validity in prominences see Anzer & Heinzel (2007).

In this section, we briefly review the whole-prominence magnetic field simulations under the assumption of the force-free field. This means that neither the whole magnetic field configuration nor the shape of the individual dipped magnetic field lines are affected by the prominence plasma weight (low plasma $\beta$ conditions). These simulations do not assume any representation of the prominence fine structure plasma (except for visualization purposes) and therefore also do not treat the radiative transfer. A comprehensive review of the prominence magnetic field simulations and the details of the techniques and methods used to construct them are presented by van Ballegooijen (2013) in this volume (see also the review by Mackay *et al.* 2010). In this review we place the emphasis on the connection between the prominence fine structures and recent prominence magnetic field simulations based on either the assumption of the sheared magnetic arcade or the magnetic flux rope forming the structure accommodating the magnetic dips. The **sheared arcade models** such as those of Antiochos, Dalhburg & Klimchuk (1994), DeVore & Antiochos (2000), and Aulanier, DeVore & Antiochos (2002) incorporate a magnetic arcade formed by shearing photospheric footpoint motions. More recently, Aulanier, DeVore & Antiochos (2006) simulated the interaction and merging of two sheared magnetic arcades. On the other hand the **flux rope models** contain a weakly twisted flux ropes overlying the polarity-inversion line and are based on the extrapolation of the photospheric magnetic flux distribution into the corona. Such models were constructed by Aulanier & Démoulin (1998) assuming the *linear force-free fields* and by Aulanier *et al.* (1998, 1999), Aulanier, Srivastava & Martin (2000), Aulanier & Démoulin (2003), and Dudík *et al.* (2008, 2012) assuming the *linear magneto-hydrostatic fields.* Another successful prominence magnetic field extrapolation models, assuming the *non-linear force-free fields,* were developed by van Ballegooijen (2004) and by Mackay & van Ballegooijen (2009) who considered the evolution of the structure of the magnetic dips caused by the advection of a single parasitic polarity bipole.

The common feature of all these magnetic field simulations is the existence of the dipped field lines lying above the polarity-inversion line that can accommodate the prominence fine structure plasma but are not caused or affected by its weight. The spatial variation of the simulated magnetic field along these dipped field lines is comparable with the largest geometrical dimensions of the observed prominence fine structures, especially with the length of the filament fibrils (several thousand km) that are probably aligned with the field lines. However, the scale on which the simulated magnetic field varies in the plane perpendicular to the field lines is much larger than the smallest observed dimensions of the prominence fine structures that can be as small as 100 km. In other words, the localized, small-scale 3D magnetic field elements, such as 3D magnetic dips, that could constrain the prominence fine structure plasma in the 3D space do not exist in these simulations. Rather, one can imagine the spatial magnetic field configuration produced by these simulations as a set of larger-scale nearly identical 2D valleys lying one above the other. On the other hand, the small-scale 3D magnetic field dips are produced by the tangled field models of van Ballegooijen & Cranmer (2010) that, however, assume a relatively simple cylindrical geometry and are not able to reproduce the larger-scale configurations of the prominence magnetic field.

Position of the magnetic dips within a 3D prominence magnetic field configuration produced by simulations can be visualized by commonly used techniques based on the drawing of portions of the dipped field lines. These are usually drawn to the geometrical extend that represents the dip height equivalent to one pressure scale-height (approximately 300 km if assuming an isothermal plasma with a temperature of 10,000 K). Such visualization shows that the force-free magnetic field simulations are highly successful in reproducing the appearance of the global features of prominences and filaments. For example, Dudík *et al.* (2008) performed a linear magneto-hydrostatic field extrapolation based on the observed photospheric flux distribution and compared the resulting magnetic field configuration with the corresponding H$\alpha$ observation of filament obtained by THEMIS on Oct 6 2004. The position of dips in the magnetic field configuration produced by simulations is in a remarkable agreement with the general structure of this filament (see Fig. 3, therein). Another striking example of the agreement between the magnetic field simulations (in this case the non-linear force-free fields) of the prominence observed by the SDO/AIA on Dec 6, 2010 is presented by Su & van Ballegooijen (2012) (see Fig. 13, therein). To show the position of the magnetic dips in these cases, precisely those dipped field lines are drawn that pass through the grid-points in the given simulation box. Such a technique is suitable for indication of position of dips in the particular simulation, but introduces several arbitrary effects into the visualization of the individual prominence fine structures. The apparent clusters of the drawn magnetic dips (see e.g. Fig. 13 in Su & van Ballegooijen 2012) are a product of the filling factor dependent on a number of grid-points in the simulation box and also on the boldness of the lines used for drawing of the dipped portions of the magnetic field lines. Moreover, the use of the non-transparent color bars as representations of the prominence fine structures is in fact equivalent to the observation in a spectral line with an infinitely large optical thickness, which is in contrast with the optical thickness of the H$\alpha$ line commonly used for prominence observations that is typically around unity in the whole prominence. This leads to the disregarding of any effects of integration of radiation along the line of sight that might play a significant role in the appearance of the observed prominence and filament fine structures. Thus, in the absence of a realistic visualization of the magnetic dips produced by current prominence magnetic field simulations these cannot be truly compared with the high-resolution observations of the prominence fine structures.

## 3. Fine-structure formation modelling

In this section we give a brief overview of the quiescent prominence formation modelling with focus only on the models that allow for the formation of the individual prominence fine structures. More thorough reviews of prominence formation modelling can be found in review by Mackay *et al.* (2010).

Presently the most developed and successful prominence fine structure formation models are the *thermal non-equilibrium models* of Karpen *et al.* (2006), see also Luna *et al.* (2012). The mechanism of lifting of the chromospheric plasma upwards along the field lines used in these **evaporation-condensation models** is based on the heating near the footpoints of the magnetic flux tubes extended into the corona that leads to an increase of the density in the flux tube. Subsequent cooling of the plasma located inside the flux tube in the corona leads to the condensation and thus to creation of the cool prominence structures. Luna *et al.* (2012) successfully demonstrated the ability of such models to predict observable signatures such as the thermal properties, speed, and mass of moving fine structures and showing that these models are consistent in many ways with the SDO/AIA observations. Thse models are based on the detailed 3D whole-prominence magnetic field structure provided by the double sheared arcade simulations of DeVore, Antiochos & Aulanier (2005). The asymmetrical heating is applied at the footpoints of the selected flux tubes leading to the chromospheric evaporation and thus increase of the plasma density. The thermal non-equilibrium processes govern the evolution of the prominence plasma modeled individually along each selected flux tube in the 1D geometry. They are shaped as elongated threads or compact blobs, depending on the geometry of the individual flux tubes and the location of a condensation along each flux tube. At the dipped flux tubes, inside the magnetic dip, plasma condensations tend to form rather massive cool elongated threads which remain in the dips for long time. On the other hand, condensations that form outside the dipped portions of the flux tubes tend to be compact and rapidly fall to the chromosphere. The dynamical behaviour of these two populations of the cool prominence fine structures is in a good agreement with the prominence observations. The drawback of these present thermal non-equilibrium prominence formation models is their inability to adequately simulate the plasma cooling below temperatures of 30,000 K, where the optically thick radiative transfer effects (radiative losses) start to play a significant role (see also the study of the prominence radiative equilibrium by Heinzel & Anzer 2012).

A novel idea of the plasma transport into the prominences is the **magneto-thermal convection** in which hot mass moves upwards (in essence perpendicularly to the field lines) in the form of fine structure plumes and returns to the photosphere in the form of cool fine structure plasma blobs. Such dynamical behaviour of the prominence fine structures often seen in the Hinode/SOT high-resolution observations (see Berger 2013 in this volume) seems to be caused by the magnetic Rayleigh-Taylor instability modeled by Hillier *et al.* (2012a,b). These authors used a 3D ideal MHD simulation of the isothermal prominence plasma with the pressure balance governed by the Kippenhahn-Schlüter-type equilibrium forming the magnetic dips. Fragmentation of the initially uniform plasma distribution creating numerous prominence fine structures and their dynamical motions are consequences of the Rayleigh-Taylor instability, caused by the introduction of a low-density region into the simulation. However, these simulations might be strongly affected by the stabilizing effect of the field line tying or by the radiative cooling/heating effects, that are not considered in the present simulations.

Low *et al.* (2012) recently suggested a mechanism that might lead to the formation of the vertically aligned sets of the prominence fine structures, due to a chain of successive break-downs of the frozen-in-magnetic-field conditions of the cold prominence plasma located in the dipped field lines. Such an approach relies on a radiative collapse of the 1D prominence plasma sheet into an infinitely narrow layer with near zero temperature due to the radiative losses. However, as was shown by several authors (e.g. Heinzel & Anzer

2012), at sufficiently low temperatures (around 4500 – 8000 K) net radiative losses can be significantly diminished, or even equal to zero when the prominence plasma reaches the radiative equilibrium (radiative losses are balanced by the radiative heating effect of the incident radiation). The radiative equilibrium temperatures are well above those needed for the breaking of the frozen-in-magnetic-field conditions, rendering this mechanism unrealistic in the quiescent prominence conditions.

An unconventional mechanism of support of the prominence fine structures against the gravity (but not their formation) in the predominantly vertical magnetic field (that might exist e.g. in the prominence feet) is the **plasma levitation** due to the weakly damped MHD waves modeled by Pécseli & Engvold (2000).

## 4. Fine-structure oscillations modelling

Recent models of the prominence oscillations successfully develop the theory of the prominence seismology utilizing the observations of the oscillations of the prominence fine structures for the diagnostics of their physical conditions. These models generally assume simplified geometry of the magnetic field and often 1D models of the prominence plasma including a simplified representation of the magnetic field but without any radiative transfer treatment. Modelling of the prominence oscillations is covered in depth in Ballester (2013) in this volume and was reviewed also by Mackay *et al.* (2010). Here we briefly mention only models explicitly dealing with the prominence fine structures.

Luna & Karpen (2012) (see also Luna 2013 in this volume) constructed a prominence fine structure model for investigation of the **large-amplitude longitudinal oscillations**. It is based on the prominence formation model described in the above section (Luna *et al.* 2012) and utilizes the geometrical shape of selected dipped flux tubes produced by the double sheared arcade magnetic field simulations (DeVore, Antiochos & Aulanier 2005). The 1D model is used to describe the prominence fine structure plasma located in the magnetic dips. These authors selected an ensemble of flux tubes representing the whole prominence volume. The prominence fine structures located in these flux tubes respond to an outside trigger (e.g. a nearby flare) by a collective oscillation along the flux tubes. After several periods the initially coherent oscillations of individual fine structures become increasingly out-of-phase, which is in agreement with observations. The damping of these longitudinal oscillations caused by the force of gravity is dependent on the curvature of the dipped magnetic field. This might allow us to indirectly investigate the internal structure of the prominence magnetic field.

Another type of prominence fine structure oscillations was studied by Arregui *et al.* (2008) who used a simplified model with a straight magnetic flux tube and a 1D cylindrical fine-structure plasma representation for the investigation of the **small-amplitude transverse oscillations** of filament fine structures (see also contribution by Soler 2013 in this volume). These authors identified the resonance absorption due to the inhomogeneities of the fine structure plasma as the most likely source of the damping of these transverse (perpendicular to the magnetic field) oscillations of the plasma structures embedded in the horizontal magnetic field. This might allow us to investigate the internal plasma properties of the prominence fine structures.

Although the results of the prominence fine structure oscillations modelling are promising, the full utilization of the prominence seismology potential requires an inclusion of the realistic magnetic field simulations and realistic models of the prominence fine structure plasma including the radiative transfer calculations. Only recently Heinzel *et al.* (2013) attempted for the first time to model the time-dependent synthetic spectra of the hydrogen Balmer lines arising from the oscillating prominence. However, these authors used the 1D whole prominence slab model and did not consider the fine structures.

## 5. Fine-structure radiative transfer modelling

The main topic of this review are models of the prominence fine structures employing the radiative transfer computations to obtain the synthetic spectra emerging from the plasma structure that realistically describes the physical conditions of prominences. This topic was recently reviewed in Labrosse *et al.* (2010), here we present the most recent developments and results.

The non-LTE (i.e. departures from local thermodynamic equilibrium) radiative transfer modelling of the prominence fine structures was considered already several decades ago when Morozhenko (1978) developed a multi-thread fine structure model composed of 1D plane-parallel isothermal and isobaric slabs and solved the radiative transfer for a two-level hydrogen atom to obtain the synthetic H$\alpha$ line spectra. Later, Fontenla & Rovira (1983) used the two-level hydrogen atom to obtain the Lyman-$\alpha$ synthetic spectra from individual 1D plane-parallel isobaric slabs with the temperature structure determined by the energy balance equation considering the conductive heat flux and radiative losses. This was further improved by Fontenla & Rovira (1985) who used the three-level plus continuum hydrogen atom to obtain the synthetic Lyman-$\alpha$, Lyman-$\beta$, and H$\alpha$ lines and the Lyman continuum. These authors assumed a similar energy balance equation. These early efforts were reviewed by Heinzel (1989) who also studied the mutual radiative interaction between individual 1D prominence fine structures. Later, Gouttebroze, Heinzel & Vial (1993) used a sophisticated 20 level plus continuum hydrogen atom and a large grid of 1D isothermal isobaric slab models. The partial frequency redistribution (Heinzel, Gouttebroze & Vial 1987) was considered for the Lyman-$\alpha$ and Lyman-$\beta$ lines. Fontenla *et al.* (1996) used the five-level plus continuum hydrogen atom and a collection of 1D isobaric plane-parallel models in the energy balance, including the effects of the ambipolar diffusion. Energy balance of the prominence fine structures was also studied by Anzer & Heinzel (1999) using a 1D slab model in magneto-hydrostatic equilibrium including the empirically prescribed temperature structure of the prominence-corona transition region (PCTR). Labrosse & Gouttebroze (2001) used the 1D isobaric isothermal slab models and a complex 33-level plus continuum helium atom to synthesize the He I and He II spectra. Later, Labrosse *et al.* (2002) used 1D slab models in the magneto-hydrostatic equilibrium (including the PCTR) to calculate the synthetic spectra of hydrogen, helium and calcium.

Although such 1D models still represent a useful and a computationally efficient approach for a number of situations, proper understanding of the prominence fine structures requires the use of a more general 2D or 3D models. The 2D vertically infinite models with the cylindrical cross-section were used by Gouttebroze (2006). The plasma parameters and the radiation field in these models varies with the cylinder radius and with the azimuth which allows for the inclusion of the anisotropic incident radiation. Gouttebroze (2007) used these models to study the temperature relaxation of the prominence fine structure plasma.

The first self-consistent 2D radiative-magneto-hydrostatic model of the prominence fine structures was developed by Heinzel & Anzer (2001). These authors generalized previously 1D prominence model of Anzer & Heinzel (1999) that is based on the work of Heasley & Mihalas (1976). The model of Heinzel & Anzer (2001) represents the quasi-vertical fine structure threads in a 2D vertically infinite geometry with a cross-section parallel to the solar surface. This means that all quantities vary in the $x$-$y$ plane parallel to the solar surface but are uniform along the vertical $z$-axis. The prominence plasma is suspended in the magnetic dips which are the product of the local 2D magnetohydrostatic (MHS) equilibrium of the Kippenhan-Schlüter type. These gravity-induced dips are caused by the weight of the prominence mass acting on the initially horizontal magnetic field lines, as opposed to the force-free magnetic dips occurring in the whole-prominence magnetic field simulations (see Sect. 2). Temperature structure is prescribed semi-empirically and accommodates two different forms of the PCTR. In the direction perpendicular to the magnetic field temperature steeply rises from the central cool part within a very narrow layer (typically few tens of km), while along the field lines the temperature gradient is shallow and the PCTR layer is much more extended. The gas pressure variation and the extension of the thread along the $x$-axis results from the MHS equilibrium. The multi-level non-LTE radiative transfer is solved in this 2D plasma structure assuming a 5-level plus continuum hydrogen atom and the partial frequency redistribution for the Lyman-$\alpha$ and Lyman-$\beta$ lines. The details of the method used for the radiative transfer computations are given in Heinzel & Anzer (2001), along with an example of the resulting synthetic spectra. These are consistent with the typically observed Lyman spectra obtained by the SOHO/SUMER and also with the H$\alpha$ line observations. An adaptive MHS grid was introduced into these 2D models by Heinzel & Anzer (2003). Further, Heinzel, Anzer & Gunár (2005) introduced the 12-level plus continuum hydrogen atom and analyzed a set of different models with focus on the dependence of the resulting synthetic Lyman spectra on the choice of the model input parameters and on the orientation of the magnetic field with respect to the line-of-sight. These authors also used the 2D contribution functions to indicate the place of formation of individual spectral lines inside the fine-structure threads. This work was complemented by Gunár, Heinzel & Anzer (2007), who studied the formation of the Lyman continuum. Later, Gunár *et al.* (2007) implemented a multi-thread model where a set of identical 2D threads (without mutual radiative interaction) is stochastically distributed with a given line-of-sight intersecting multiple threads. These authors then compared the resulting synthetic Lyman spectra with the observations. Such direct profile-to-profile analysis showed that the 2D multi-thread fine-structure models produce the synthetic spectra in a very good agreement with the observed spectra. Gunár *et al.* (2008) further improved the 2D multi-thread prominence fine structure models by introduction of randomly distributed line-of-sight velocities of individual threads. This allowed for investigation of the asymmetries of the hydrogen Lyman line profiles observed by the SOHO/SUMER (see also Vial, Ebadi & Ajabshirizadeh (2007)). The observed Lyman line profiles exhibit rather large asymmetries which, if attributed simply to a Doppler shift, would indicate velocities of the order of 100 km s$^{-1}$. However, the prevailing velocities in the quiescent prominences are only around 10 km s$^{-1}$ or below. In addition, the asymmetries of the neighbouring spectral lines, especially Lyman-$\alpha$ and Lyman-$\beta$ (observed nearly simultaneously) often show an opposite character at the same place in the prominence. The synthetic Lyman line profiles obtained by the 2D multi-thread model exhibit similar substantial asymmetries as the observed ones, even though the LOS velocities of individual threads are between $\pm$ 10 km s$^{-1}$. The synthetic Lyman line profiles also exhibit the same opposite character of asymmetries as the observed profiles. The ability of these 2D multi-thread prominence fine structure models to reproduce the observed spectra was further demonstrated by Gunár *et al.* (2010). These authors performed an extensive statistical comparison of a large data-set of observed Lyman lines obtained by the SOHO/SUMER and the synthetic Lyman spectra obtained by 2D modelling. The synthetic spectra resulting from a model with a realistic set of input parameters showed a very good agreement with the observations for most of the studied statistical parameters, including the asymmetries of the line profiles. This was further complemented by Gunár *et al.* (2012), who used the statistical comparison of the synthetic and observed H$\alpha$ line profiles (obtained by the Meudon/MSDP instrument) to analyze the prominence velocity fields. The aforementioned works use the 2D multi-thread models without the mutual radiative interaction between individual threads. An illustrative study of the effects of the mutual radiative interaction on the synthetic Lyman line profiles was done by Heinzel, Anzer & Gunár (2010) using a simple configuration of three identical fine structure threads.

To study the plasma structure produced by the 2D MHS equilibrium used in the 2D fine-structure thread models, Gunár, Heinzel & Anzer (2011) developed a method for obtaining the synthetic differential emission (DEM) measure curves from 2D multi-thread models. These were compared with the DEM curves derived from the SOHO/SUMER observations of a prominence by Gunár *et al.* (2011) and proved to be in a very good agreement with the observed prominence DEM curves within the temperature range covered by the model (up to 100,000 K).

## 6. Models combining whole-prominence magnetic field simulations and radiative transfer modelling

Present 3D whole-prominence magnetic field simulations (summarized in Sect. 2) provide realistic large-scale configurations of the prominence magnetic field containing regions of magnetic dips that correspond to the general structure of the observed prominences/filaments. However, details of the localized prominence fine structures cannot be fully determined from such simulations, as these do not assume any representation of the prominence fine structure plasma. Neither can the visualizations of these configurations (by drawing portions of the dipped field lines as color bars) be directly compared with the high-resolution observations such as those obtained by the Hinode/SOT. On the other hand, 2D radiative transfer models of individual prominence fine structures (see Sect. 5) are able to reproduce the observed prominence spectra with a high degree of accuracy, but their localized nature does not allow us to study a large-scale configurations of fine structures. To overcome these limitations we have to, in the future, combine models by integrating the 3D magnetic field simulations with a realistic description of the prominence fine structure plasma and self-consistent treatment of the radiation transfer.

The first step in this direction was taken by Gunár *et al.* (2013a). These authors use the whole-prominence magnetic field configurations produced by the 3D non-linear force-free simulations of Mackay & van Ballegooijen (2009) and extract the resulting force-free magnetic dips. These are then filled with the realistic 2D representation of the prominence fine-structure plasma by a newly developed technique. The 2D non-LTE radiative transfer is self-consistently solved in the resulting plasma structure using the method of Heinzel & Anzer (2001). The synthetic spectrum produced by these force-free dip models is in a qualitative agreement with a range of typical quiescent prominence observations. Moreover, their plasma structure is similar to that of the gravity-induced 2D models of Heinzel & Anzer (2001).

The availability of realistic models of individual prominence fine structures located in the 3D whole-prominence magnetic field configuration will allow us in the future to consistently visualize entire sets of prominence/filament fine structures with any given line of sight, either as a prominence in emission above the solar limb, or as a dark filament seen in absorption against the solar disk. This might help us to understand the hitherto puzzling relation between the quasi-vertical fine structures often seen in quiescent prominences observed on the solar limb and the horizontally aligned dark fibrils representing the fine structures observed in absorption against the solar disk. This topic was recently discussed by Gunár *et al.* (2013b).

## 7. Concluding remarks

Presently, the research on the quiescent solar prominences approaches a point, where most of the basic blocks of our understanding of these spectacular solar features are sufficiently well developed. In the near future the upcoming space-borne missions and ground-based telescopes will provide even wider range of prominence observations with very high spatial, temporal, spectral, and polarimetric resolution compared to what we have now. These include the IRIS, CLASP, Proba-3, Solar Orbiter, and Solar-C missions, and the large ground-based solar telescopes such as Gregor, ATST, EST, and even the radio-interferometer ALMA. This unprecedented amount of detailed observations of the prominence fine structures will bring further significant challenges for the prominence modelling. Such a situation, however, creates also new opportunities in enabling us to combine the existing approaches of prominence fine structure modelling into more complex models, such as those of Gunár *et al.* (2013a). The combined multi-disciplinary fine structure models will help us to accelerate our understanding of the quiescent prominences and solve some of the still open questions about their true nature.

## 8. Acknowledgements

S.G. acknowledges the support from grant P209/12/0906 of the Grant Agency of the Czech Republic. Work of S.G. was supported by the project RVO: 67985815. Participation of S.G. at the IAUS300 was supported by the ESA-PRODEX project 4000102852. This research has made use of NASA's Astrophysics Data System. The author thanks to U. Anzer, P. Heinzel, D. Mackay, B. Schmieder, and J. Štěpán for their valuable comments and suggestions.

### References

Antiochos, S. K., Dahlburg, R. B., & Klimchuk, J. A. 1994, *Astrophys. J.* 420, 41
Anzer, U. & Heinzel, P. 1999, *Astronomy & Astrophysics* 349, 974
Anzer, U. & Heinzel, P. 2007, *Astronomy & Astrophysics* 467, 1285
Arregui, I., Terradas, J., Oliver, R., & Ballester, J. L. 2008, *Astrophys. J.* 682, 141
Aulanier, G. & Démoulin, P. 1998, *Astronomy & Astrophysics* 329, 1125
Aulanier, G., Demoulin, P., van Driel-Gesztelyi, L., Mein, P., & Deforest, C. 1998, *Astronomy & Astrophysics* 335, 309
Aulanier, G., Demoulin, P., Mein, N., van Driel-Gesztelyi, L., Mein, P., & Schmieder, B. 1999, *Astronomy & Astrophysics* 342, 867
Aulanier, G., Srivastava, N., & Martin, S. F. 2000, *Astrophys. J.* 543, 447
Aulanier, G., DeVore, C. R., & Antiochos, S. K. 2002, *Astrophys. J.* 567, 97
Aulanier, G. & Démoulin, P. 2003, *Astronomy & Astrophysics* 402, 769
Aulanier, G., DeVore, C. R., & Antiochos, S. K. 2006, *Astrophys. J.* 646, 1349
Ballester, J. L. 2013, *The IAU Symposium 300 Proceedings* (this volume)
Berger, T. 2013, *The IAU Symposium 300 Proceedings* (this volume)
DeVore, C. R. & Antiochos, S. K. 2000, *Astrophys. J.* 539, 954
DeVore, C. R., Antiochos, S. K., & Aulanier, G. 2005, *Astrophys. J.* 629, 1122
Dudík, J., Aulanier, G., Schmieder, B., Bommier, V., & Roudier, T. 2008, *Solar Phys.* 248, 29
Dudík, J., Aulanier, G., Schmieder, B., Zapiór, M., & Heinzel, P. 2012, *Astrophys. J.* 761, 9
Fontenla, J. M. & Rovira, M. 1983, *Solar Phys.* 85, 141
Fontenla, J. M. & Rovira, M. 1985, *Solar Phys.* 96, 53
Fontenla, J. M., Rovira, M., Vial, J.-C., & Gouttebroze, P. 1996, *Astrophys. J.* 466, 496

Gouttebroze, P., Heinzel, P., & Vial, J. C. 1993, *Astronomy & Astrophysics Suppl.* 99, 513
Gouttebroze, P. 2006, *Astronomy & Astrophysics* 448, 367
Gouttebroze, P. 2007, *Astronomy & Astrophysics* 465, 1041
Gunár, S., Heinzel, P., & Anzer, U. 2007, *Astronomy & Astrophysics* 463, 737
Gunár, S., Heinzel, P., Schmieder, B., Schwartz, P., & Anzer, U. 2007, *Astronomy & Astrophysics* 472, 929
Gunár, S., Heinzel, P., Anzer, U., & Schmieder, B. 2008, *Astronomy & Astrophysics* 490, 307
Gunár, S., Schwartz, P., Schmieder, B., Heinzel, P., & Anzer, U. 2010, *Astronomy & Astrophysics* 514, A43
Gunár, S., Heinzel, P., & Anzer, U. 2011, *Astronomy & Astrophysics* 528, A47
Gunár, S., Parenti, S., Anzer, U., Heinzel, P., & Vial, J.-C. 2011, *Astronomy & Astrophysics* 535, A122
Gunár, S., Mein, P., Schmieder, B., Heinzel, P., & Mein, N. 2012, *Astronomy & Astrophysics* 543, 93
Gunár, S., Mackay, D. H., Anzer, U., & Heinzel, P. 2013a, *Astronomy & Astrophysics* 551, A3
Gunár, S., Heinzel, P., Anzer, U., & Mackay, D. H. 2013b, *J. Phys.: Conf. Ser.* 440, 012035
Heasley, J. N. & Mihalas, D. 1976, *Astrophys. J.* 205, 273
Heinzel, P. 1989, *Hvar Observatory Bulletin* 13, 317
Heinzel, P. 2007, in *The Physics of Chromospheric Plasmas, ASPC* 368, 27
Heinzel, P. & Anzer, U. 2001, *Astronomy & Astrophysics* 375, 1082
Heinzel, P. & Anzer, U. 2003, in *Stellar Atmosphere Modeling, ASPC* 288, 441
Heinzel, P. & Anzer, U. 2012, *Astronomy & Astrophysics* 539, 49
Heinzel, P., Gouttebroze, P., & Vial, J.-C. 1987, *Astronomy & Astrophysics* 183, 351
Heinzel, P., Anzer, U., & Gunár, S. 2005, *Astronomy & Astrophysics* 442, 331
Heinzel, P., Anzer, U., & Gunár, S. 2010, *Mem. S.A.It.* 81, 654
Heinzel, P., Zapior, M., Oliver, R., & Ballester, J. L. 2013, *Astronomy & Astrophysics*, submitted.
Hillier, A., Berger, T., Isobe, H., & Shibata, K 2012, *Astrophys. J.* 746, 120
Hillier, A., Isobe, H., Shibata, K., & Berger, T. 2012, *Astrophys. J.* 756, 110
Karpen, J. T., Antiochos, S. K., & Klimchuk, J. A. 2006, *Astrophys. J.* 637, 531
Labrosse, N. & Gouttebroze, P. 2001, *Astronomy & Astrophysics* 380, 323
Labrosse, N., Gouttebroze, P., Heinzel, P., & Vial, J.-C. 2002, in *Solar Variability: From Core to Outer Frontiers, ESA SP* 506, 451
Labrosse, N., Heinzel, P., Vial, J.-C., Kucera, T., Parenti, S., Gunár, S., Schmieder, B., & Kilper, G. 2010, *Space Sciences Reviews*, 151, 243
López Ariste, A., Aulanier, G., Schmieder, B., & Sainz Dalda, A. 2006, *Astronomy & Astrophysics* 456, 725
Low, B. C., Berger, T., Casini, R., & Liu, W. 2012, *Astrophys. J.* 755, 34
Luna, M. & Karpen, J. *Astrophys. J.* 750, 1
Luna, M., Karpen, J. T., & DeVore, C. R. 2012, *Astrophys. J.* 746, 30
Luna, M. 2013, *The IAU Symposium 300 Proceedings* (this volume)
Mackay, D. H., Karpen, J. T., Ballester, J. L., Schmieder, B., & Aulanier, G. 2010, *Space Sciences Reviews*, 151, 333
Mackay, D. H. & van Ballegooijen, A. A. 2009, *Solar Phys.* 260, 321
Morozhenko, N. N. 1978, *Solar Phys.* 58, 47
Patsourakos, S. & Vial, J.-C. 2002, *Solar Phys.* 208, 253
Pécseli, H. & Engvold, O. 2000, *Solar Phys.* 194, 73
Schmieder, B., Chandra, R., Berlicki, A., & Mein, P. 2010, *Astronomy & Astrophysics* 514, A68
Schmieder, B. & Aulanier, G. 2012, *EAS Publications Series* 55, 149
Soler, R. 2013, *The IAU Symposium 300 Proceedings* (this volume)
Tandberg-Hanssen, E. 1995, *The nature of solar prominences*, Kluwer, Dordrecht
van Ballegooijen, A. A. 2004, *Astrophys. J.* 612, 519
van Ballegooijen, A. A. & Cranmer, S. R. 2010, *Astrophys. J.* 711, 164
van Ballegooijen, A. A. 2013, *The IAU Symposium 300 Proceedings* (this volume)
Vial, J.-C., Ebadi, H., & Ajabshirizadeh, A. 2007, *Solar Phys.* 246, 327
Su, Y. & van Ballegooijen, A. 2012, *Astrophys. J.* 757, 168

*Nature of prominences and their role in Space Weather*
*Proceedings IAU Symposium No. 300, 2013*
*B. Schmieder, J.-M. Malherbe & S. T. Wu, eds.*

doi:10.1017/S1743921313010764

# On the nature of the prominence - corona transition region

**Susanna Parenti**[1] **and Jean-Claude Vial**[2]

[1]Royal Observatory of Belgium - STCE,
3 Av. Circulaire, Brussels, Belgium
email: s.parenti@oma.be

[2]Institut d'Astrophysique Spatiale, Université Paris Sud - CNRS,
Orsay Cedex, France
email: jean-claude.vial@ias.u-psud.fr

**Abstract.** Due to the complexity of their environment, prominences properties are still a matter of controversy. Prominences cool and dense plasma is suspended in the hot corona by a magnetic structure poorly known. Their thermal insulation from the corona results in a thin geometrical interface called prominence-corona-transition-region (PCTR). Here we will review the main properties of such a region as derived primarily from observations. We will introduce the thermal structure properties, describe the fine structure together with the Doppler-shift and width properties of lines of the emitting plasma. We will introduce the proposed interpretations of such observations and the limits of our knowledge imposed by the present instrumentation. We will conclude with a perspective for the future observations of the PCTR.

**Keywords.** prominences, UV-EUV, transition region, spectroscopy

---

## 1. Introduction

Solar prominences have the peculiar property of being made of cool plasma (mainly at chromospheric temperatures) embedded in a magnetic structure suspended in the corona. As in the case of the chromosphere, this implies the existence of an interface region with the corona, where the temperature of the plasma rises from about 7000 K to $1 \times 10^6$ K. Such a region is called prominence-corona transition region (PCTR).

Prominences are found above some of the photospheric neutral lines. At large scales, it is known that the magnetic field is at about 30 deg angle with the underlying neutral line, which itself is parallel to the main body of the prominence. This means that the overall lateral PCTR is at about 30 deg with the magnetic field.

Prominences are formed by a collection of thin magnetic flux tubes (about 200 km wide) filled by cool plasma. Looking at, for example, H$\alpha$ images (Figure 1), we see that a single flux tube is only partially filled by the prominence plasma. The rest may be filled by coronal material. This means that there exists also a PCTR inside the flux tube which has the temperature gradient almost aligned with the magnetic field. Considering that the efficiency of the thermal conduction depends on the angle between the magnetic field and the temperature gradient directions, it is clear that the large and small scale PCTRs may have different properties (see for example Heinzel and Anzer 2001). At present the only information we have on the field-aligned temperature gradient comes from modeling, as the observations are limited in spatial resolution. The PCTR is, in fact, observed by the EUV instruments which, until now, could not resolve single threads. With the launch of the new NASA/IRIS mission, such discrepancy is reduced and new results on this topic are expected.

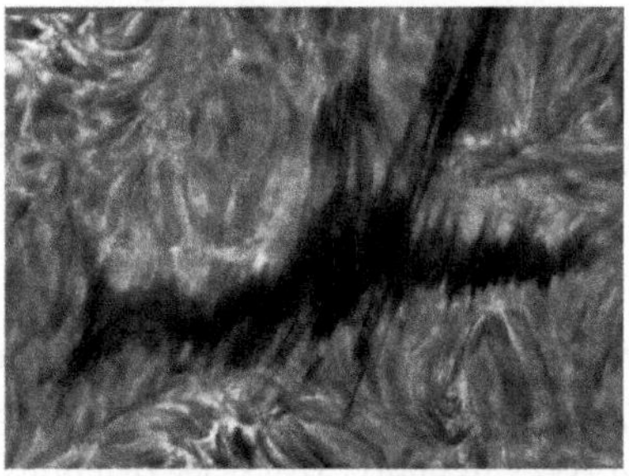

**Figure 1.** A segment of a filament observed in H$\alpha$ by the Swedish 1-m Solar Telescope (SST). The fine structure is only partially filled by cool plasma. Courtesy O. Engvold.

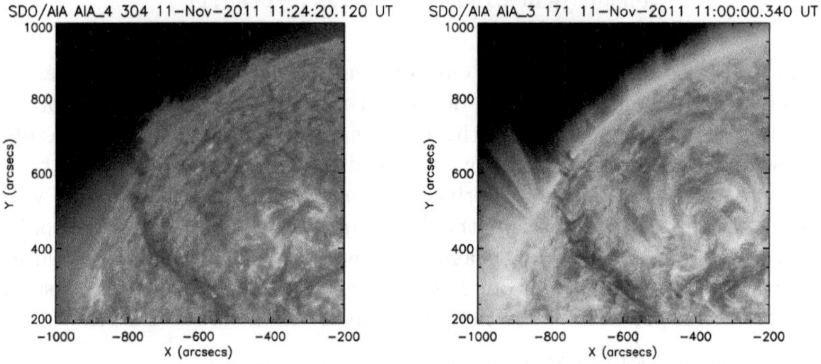


**Figure 2.** A prominence (left) and its transition region (right) observed by SDO/AIA in, respectively, the EUV He II 304 Å and Fe IX 171 Å wavebands.

The first property we discuss is the general appearance of the PCTR. Figure 2 shows the same prominence observed co-temporally in the wavebands He II 304 Å (sampling the low transition region, left plot) and Fe IX Å (sampling the high transition region-corona, right plot). Off-limb the two images are clearly different: in the He II image the prominence is brighter than the background, indicating that the PCTR is emitting between 2 and $8 \times 10^4$ K; in the Fe IX image the prominence is faintly visible in emission, suggesting that if present, the prominence emission at about 0.8 MK is quite low or absent, as the background corona emission is already low. In Sect. 2 we will further discuss the temperature structure of the PCTR and mention a few recent theoretical works on this topic. On the disk, the two images show a dark prominence. This aspect is mainly due, for the 304 Å channel, to He II self absorption, but also by the absorption of the spectral lines by the hydrogen and helium continua produced by the neutral H and He, and singly ionized He plasma of the prominence (see e.g. Chiuderi Drago *et al.* 2001, Chiuderi Drago and Landi 2002) and, at least for the 171 Å case, by the absence or weak emission of the PCTR with respect to the background quiet Sun. In Sect. 3 we will further discuss the PCTR properties as derived from on-disk filament observation. From these first properties of the PCTR emission we already understand that the mass of this region should be reduced with respect to that of the prominence core. However, the PCTR has its role to maintain the stability of the whole prominence. As we will see (Sect. 3), it is a dynamic region and its local and global emissions contribute to the radiative losses of the structure. We will conclude this paper with Sect. 4, with a few important recommendations for the future studies of the PCTR.

## 2. The thermal structure

The PCTR at the interface with the background corona is probably a thin layer with a steep gradient of temperature, as the thermal conduction perpendicular to the magnetic field is not efficient to thermalize the plasma. As mentioned, deriving the PCTR properties from the observations is important for understanding the physics of the whole prominence. As it has been shown by several works, the emerging emission of H (see e.g. Heinzel *et al.* 2001) and He (see e.g. Labrosse *et al.* 2002) from a prominence is affected by the presence or not of a transition region.

The prominence is an extremely complex environment which, in certain cases, imposes the use of some simplifications in the modeled physics. The close interaction between observation and modeling works is a key to improve this issue. Observers have the role of providing modelers with the best suitable information to properly constrain and simplify their work. We give some examples of such interaction in this section.

To infer the thermal structure of the PCTR we need to have observations in a wide temperature range, typically from about $10^4$ to $10^6$ K. The inversion of the data is done using the differential emission measure (DEM) technique applied to the optically thin UV-EUV emission of the PCTR. The DEM is a measure of the amount of plasma at each temperature ($DEM(T) = n_e^2 dh/dT$, where $T$ is the electron temperature, $n_e$ the electron density and $h$ the line of sight). Details of this technique are given in Labrosse *et al.* (2010).

Due to the properties of this technique, only off-limb prominence DEMs have been derived so far. For a better constrain of the inversion, a large number of spectral lines has to be used and only few results are present in the literature (e.g. Wiik, Dere, and Schmieder 1993, Cirigliano, Vial, and Rovira 2004, Parenti *et al.* 2010, Parenti and Vial 2007, Parenti, Vial, and Lemaire 2008, Gunár *et al.* 2011).

Figure 3 left shows one of the latest inferred prominences DEM (Parenti and Vial 2007). It was obtained inverting more than forty spectral lines intensity from elements in different stages of ionization measured by SOHO/SUMER on October 18 1999. On the same day a similar observation was obtained on the quiet Sun, with the purpose of comparing the DEM profiles of the two regions. The DEM of the quiet Sun is shown on Figure 3 right. As the DEM profile in temperature is shaped, at a given time, by the physical process acting inside a structure, this comparison is necessary to understand if and how much the PCTR behaves differently from the corona-chromosphere transition region (CCTR). As we see from the figure this is the case: the PCTR has smaller DEM values almost everywhere (beside the coronal part, which will be discussed later) and a different DEM gradient at similar temperatures (including a different value of temperature where the DEM minimum is located). The interpretation of these differences is still under study and some results are presented hereafter.

In addition to the comparison of the thermal structure of the PCTR to the CCTR, further investigations need to work out if and how the thermal structure (DEM) of different quiescent prominences differ from each other. This would allow to identify those properties typical to all prominences, which then can be used in constraining their modeling. From the few results published on this topic, the answer is not ready. This is because of the limitations, for example, in the data: the temperature coverage, the lack of corona background subtraction. Figure 3 left shows no data below $logT = 4.4$ (at this temperature the emission starts to become optically thick in most lines). Other data may have different limitations (see for example Gunár *et al.* 2011) which would produce a DEM having a different gradient at such temperature. At present we also often lack the suitable data needed for the estimation of the coronal contribution surrounding the prominence. This was the case for the structure of Parenti and Vial (2007): the peak of the DEM at 1MK in Figure 3 left is almost certainly the background corona, while the gradient of the high transition region part of the DEM is contaminated by the background corona. These limitations need to be overcome in order to advance in prominences investigations.

*Thermal structure and energy balance.* One of the motivations to reduce the uncertainties on the DEM constraints is its use to estimate the total radiative losses of the PCTR (e.g. Parenti and Vial 2007). The stable emission of a prominence in quiescent conditions suggests an almost continuous input of energy into the structure. At present there are several candidate processes for such sources (thermal conduction, waves, chromospheric irradiation, see Heinzel and Anzer 2012 for the cool core), but none of them seems to be sufficient to compensate the radiative losses from the prominence core to the low PCTR (up to about $10^5$ K, at higher temperature the thermal conduction is by now the best candidate for this compensation). Particularly critical is the energy budget in the core, where most of the emission is concentrated. To make the problem more complicated, the core emission should be treated using non-local thermodynamic equilibrium (NLTE) treatment of the radiative transfer. The estimate of the core's radiative losses is still not complete today, and the bridge with the radiative losses of the PCTR is missing. For this reason an extension of the DEM at lower temperature ($logT < 4.5$, where possible) and a better estimate of the core losses would allow to have the full temperature information on the energy balance.

A possible solution to this problem is the use of the H-Ly lines (e.g. Heinzel *et al.* 2001, Stellmacher and Wiehr 2008, Gunár *et al.* 2011, Vial *et al.* 2012, Schwartz *et al.* 2012) as they are formed at different heights from the core to the PCTR. Unfortunately most of the available data miss the profile of the H-Ly$\alpha$ line, which can put better constraints to the results. Including this line in the future observations is strongly recommended.

In addition, the newly launched NASA/IRIS spectrometer is observing the intense Mg II line, which until now was not included in the core losses budget known (however, Mg II modeling can be found, for example, in Vial 1982, Paletou, Vial, and Auer 1993). As far as radiative losses are concerned, according to Heinzel *et al.* 2013, the inclusion of Mg II losses decreases dramatically the radiative-equilibrium temperature at very low pressures. This situation is much different at high pressures (opacities) where the Mg II contribution is negligible, as it is the case in the CCTR.

We also point out that the high transition region and coronal part of the thermal emission of prominences is scarcely known due to the difficulty of isolating it from the

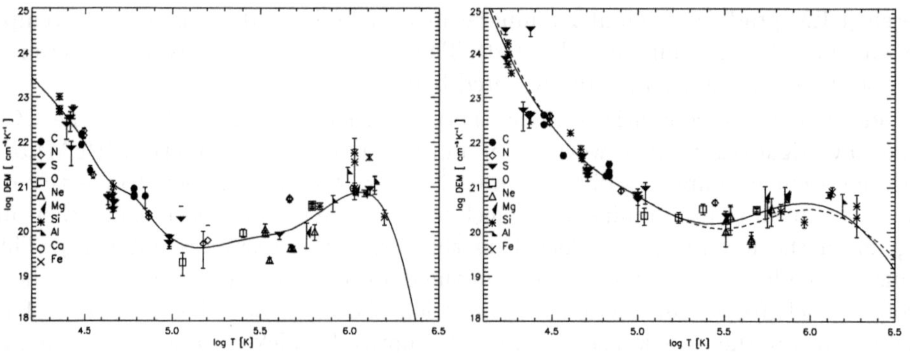


**Figure 3.** Logarithm of the DEM as function of the logarithm of temperature for the PCTR (left) and the quiet Sun (right, the double curve originate from different data). From Parenti and Vial (2007).

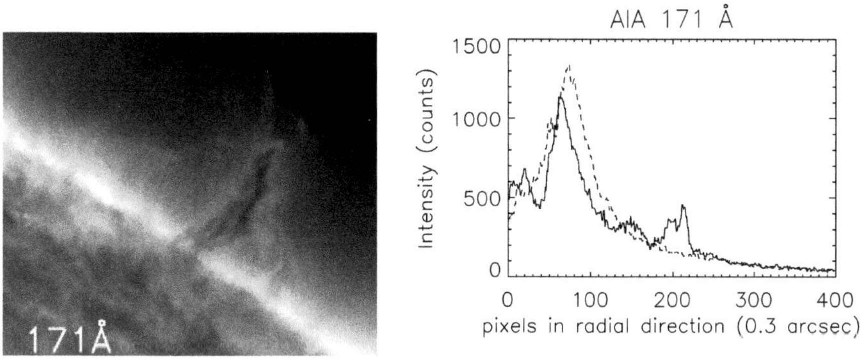


**Figure 4.** Left: prominence observed by SDO/AIA 171 Å on the 22 June 2010. A weak but clear emission is seen inside the structure surrounding the central part seen in absorption. Right: radial cut across the prominence (solid line) and off-limb corona at the side of the structure (dotted line). An emission above the background is clearly visible around pixel 200. From Parenti *et al.* (2012).

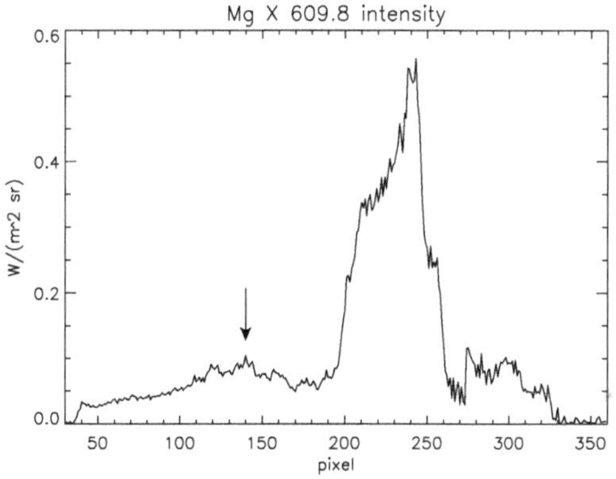


**Figure 5.** Mg X 609 Å line intensity across a prominence observed on November 11, 2011 with SOHO/SUMER. The limb is located at around pixel 240. The off-limb signal is affected by absorption by neutral hydrogen inside the denser part of the prominence (around pixels 150-200). A weak emission is seen between the pixels 120-150 as indicated by the vertical arrow. Parenti *et al.* 2013.

background/foreground off-limb emission. On this topic, the new images produced by the NASA/SDO high sensitivity AIA EUV instrument have revealed that the PCTR emits also at quite high temperature and, consequently it could be more massive than previously thought (Figure 4). Parenti *et al.* (2012) have investigated this aspect using a combination of AIA data analysis and simulation. They concluded that the emission seen in the AIA 171 Å channel is due to Fe IX, and not from cooler lines which also fall into the instrument waveband. They estimated a PCTR emission up to $4 \times 10^5$ K. Kucera *et al.* (2012) found signatures of emission in a prominence even at hotter temperatures (Fe X, Fe XI), suggesting that more work on this topic should be done.

These new results stimulate further studies which will better quantify the thermal distribution of the high temperature part of the PCTR. Preliminary results on spectroscopic data from SOHO/SUMER (Parenti *et al.*, 2013) of the off-limb prominence shown in Figure 2, seems to indicate a weak PCTR emission above the background level at temperature close to 1MK, as shown in Figure 5. Further investigations are under way.

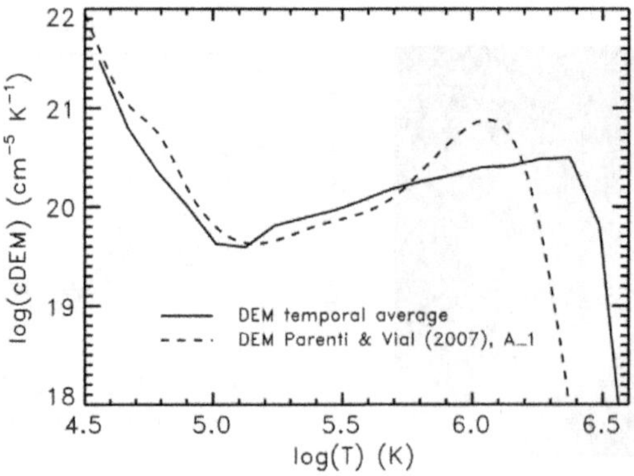


**Figure 6.** Solid line: Logarithm of the DEM as function of the logarithm of temperature for the PCTR, averaged over the time simulation and space. The dashed line is the DEM derived from observations by Parenti and Vial (2007). From Luna, Karpen, and DeVore (2012).

*Thermal structure and prominence formation.* The observed thermal structure of the PCTR has been used to study and constrain models of the formation of prominences. For instance, Luna, Karpen, and DeVore (2012) have recently simulated the formation of the fine scale of a prominence by the thermal non-equilibrium process inside a 3D shared arcade (see also DeVore, Antiochos, and Aulanier 2005). Their outputs included the DEM which was compared to that derived from observations by Parenti and Vial (2007). The result is well constrained in all the transition region, as it is shown in Figure 6. Some differences between the two DEMs are located at coronal temperature. But as mentioned earlier, this is probably because the Parenti and Vial (2007) DEM includes the background emission while, as pointed out by Luna, Karpen, and DeVore (2012), the high temperature part of the theoretical DEM they show was not completely representative of the whole coronal arcade they had in the model.

## 3. Fine structure and dynamics

*PCTR from on-disk filament observations.* In view of its (supposedly) extremely small thickness, the PCTR is difficult to detect tangentially when the prominence is seen at the limb (it would require a sub-arcsec resolution, still not achieved in the UV from space, except from VAULT, see below). This direct observation is equally difficult when the prominence is seen on the disk as a filament. But one can use the fact that one sees the fibril/thread structuring from above, which allows for precise cuts accross the aligned structures. For instance, the unique filament image obtained by VAULT (Vial *et al.* 2012, Figure 7) in the L$\alpha$ line (formed at the bottom of the PCTR) displays a fine fibril (or thread) structure at the level of the spatial resolution (0.3 arcsec). This does not mean that there is no fibril or thread with a smaller cross-section. But this also could indicate that we have a relatively large PCTR common for smaller numerous (cool) threads within the 0.3 arcsec span. Until the advent of the results from IRIS and its slit-jaw images (e.g. Si IV), there is no way to firmly conclude. With the VAULT L$\alpha$ data, Vial *et al.* (2012) found a typical thread cross-section of one arcsec with a minimum of 0.4 arcsec, slightly larger than the 0.3 arcsec cross-sections of the H$\alpha$ fibrils (Lin *et al.* 2005, Lin *et al.* 2008). For this limiting case, one could conclude that the individual thread PCTR is 0.05 arcsec (or about 40 km). But for the bulk of the observed threads, the large cross-section can also imply that we have a PCTR common to a set of threads.

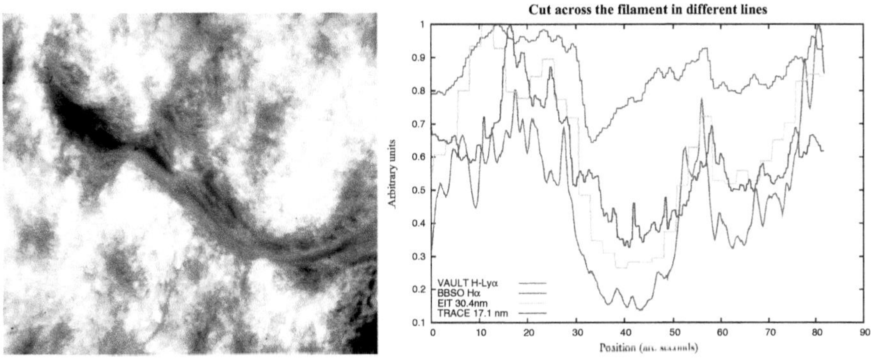


**Figure 7.** Left: a detail of the filament observed by VAULT. Right: A cut of the filament in different spectral lines: VAULT H-L$\alpha$ is the bottom curve, BBSO H$\alpha$ is the upper curve, EIT 304 Å and TRACE 171 Å are the middle curves. The arbitrary units are obtained by dividing the cut intensity by its maximum value. From Vial *et al.* (2012).

High spatial IRIS observations (launched in 2013) of prominences and filaments should bring more light on this issue.

A major feature of the VAULT observations (Vial *et al.* 2012) was the confirmation of filament extensions, as shown in Figure 7 (see e.g. Heinzel, Schmieder, and Tziotziou 2001). Such extensions are of the order of 10 arcsec (and less) and can be noticed in the UV and EUV absorbed lines. At this stage, it is tempting to consider them as the bottom of the global PCTR (up to $3 \times 10^4$ K). It should also be remembered that the VAULT observations concern an active region filament whose properties may differ from a quiescent one.

*Filling factor.* There are different ways for overcoming the difficulties of "direct" observations of the PCTR. One is to derive a (volume) filling factor through the measurement of the EM (and consequently the average of the square of the density) and the derivation of the local density from appropriate line ratios. Values of the order of 0.02 have been obtained by Mariska, Doschek, and Feldman (1979) and even lower ones by Cirigliano, Vial, and Rovira (2004) with the implication of a very high number of threads along the line-of-sight (up to 35). Similar results were obtained from the interpretation of the intensity, shift and width distributions : Wiik, Dere, and Schmieder (1993) also found 30 threads in the PCTR and later on (Wiik *et al.* 1999) 20 threads at the N V temperature. However, one should be aware of the many assumptions relevant to these different methods. EM and density derivations are often limited by photon noise and the intensity vs. velocity distributions can lead to different results.

*Modeling the small scale PCTR.* Beside the observational limitation of the PCTR, important progress has been made in the past years by the effort of combining observation and modeling works. For example, Gunár *et al.* (2011) have developed a 2D multi-thread model of prominences which includes small line-of-sight velocity and solves a multi-level NLTE problem for the hydrogen spectral lines. They tested different prominence conditions by comparing simulated and observed Lyman line profiles and DEMs. Using these two complementary quantities, it was possible to link the properties at the base of the transition region with those at higher temperatures. Their results showed good agreement between data and simulation, even though multiple solutions were still compatible with the data. An example of the DEMs comparison is shown in Figure 8. Even if this work gives a set of solutions, it goes in the right direction to better constrain prominence fine scale modeling with a PCTR (see also Gunár paper in this volume).

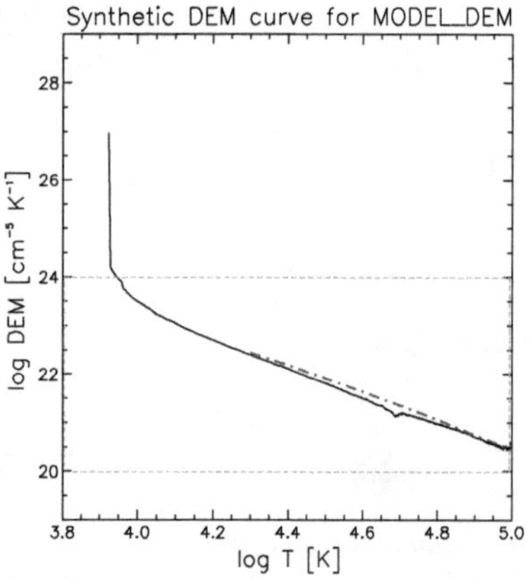


**Figure 8.** Synthetic DEM curve (solid black line) for one realization of the multi-thread prominence with 10 threads. The red dash-dotted line represents the DEM derived from the observations of the June 8, 2004 prominence. From Gunár *et al.* (2011)

Similarly to what is observed in the prominence core the PCTR is dynamic showing, for example, bulk motion and turbulence. We have to remember that we cannot reach the same spatial information as the dynamics observed for the core. This opens the questions if the two are related and if they are governed by the same mechanisms. This will have implications also in the question of the fine scale structure: is each thread surrounded by its own PCTR or a common PCTR exists for a bunch of threads (e.g. Pojoga 1994, Vial *et al.* 2012).

*Bulk motions*. The bulk motions of the PCTR have been studied both by following the paths of absorption features (due to H, He I and He II continua, e.g. Schmieder *et al.* 2008) and Doppler-shifts (e.g. Labrosse *et al.* 2010). The feature tracking technique found motions with similarities to the counter-streaming flow observed with the H$\alpha$ instruments. However, in general both techniques give variable results ($5 < v < 40$ km/s) which are affected by limitations of the data inversion methods and instruments. Again, we expect the new IRIS spectrometer to have better performances.

*Non-thermal velocities*. At the spatial resolution of unresolved threads, the spectral profiles of the optically thin UV-EUV lines have a width greater than their thermal Doppler width. This property is also present in the CCTR but, as the latest results show, their path in temperature are not exactly the same (and they can change from prominence to prominence, see e.g. Stellmacher, Wiehr, and Dammasch 2003). We show this in Figure 9 where are plotted the non-thermal velocities of the prominence on the left, and the quiet Sun on the right. This extra width of the spectral line could be interpreted as turbulence, waves or/and the superposition of line-of-sight flows of unresolved structures. We still have no answer for that, but the high spatial resolution of IRIS will certainly contribute to this topic. The different runs of the velocities with temperature shown in Figure 9, once properly interpreted, will let us understand better in what and why the CCTR and PCTR are different.

The small scale dynamics observed in the core have been recognized to have their role in the formation and stability of the prominence. What about the observed dynamics of the PCTR? One step forward could be done once we understand if there is a link between what are observed in the core and in the PCTR. Similarly to the case of the DEM, for the non-thermal velocities the complete path in temperature is missing. Only few km/s are derived in the coolest part of the structure, while about 10 km/s are derived at the base of the transition region. Recent results by Park *et al.* (2013) obtained Ca II and H$\alpha$

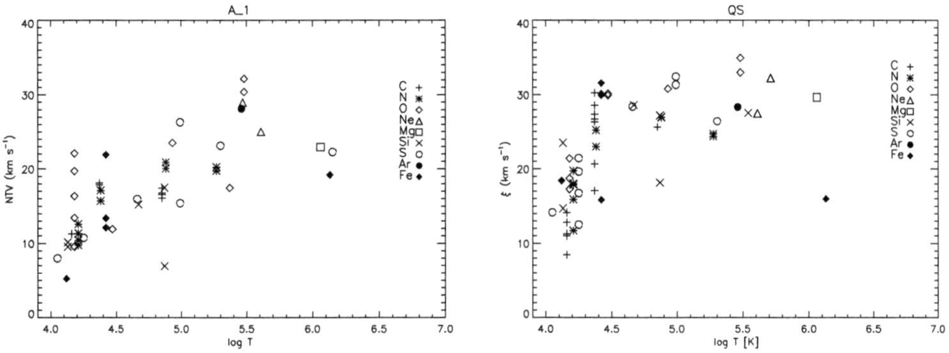


**Figure 9.** Non-thermal velocities as function of the logarithm of the temperature for the PCTR of the same prominence and quiet Sun of Figure 3. From Parenti and Vial (2007).

non-thermal velocities between 4 and 11 km/s, in agreement with the picture already proposed (for further recent results see also Ramelli *et al.* 2012) of non-thermal velocities increasing when moving outside the core.

## 4. Conclusion

In this paper we have briefly gone through some of open questions concerning the properties of the PCTR. We are aware that this layer is more complex than what we previously thought and to progress we need a common effort from the observation and modeling sides. The lack of high resolution observations contributes to limit our investigations concerning the thermal structure across and along the fine prominence threads, together with a continuous link with the thermal properties of the prominence core. This is also true for the dynamics studies, which use data affected by spatial and temporal integration. We are looking forward for the IRIS first results, but also for the future Solar Orbiter ESA mission, which will resolve about 100 km on the Sun and fraction of seconds in exposure time. To better understand which are the dominant physical processes governing the PCTR, to understand if they are common to the core, it would require to have co-temporal spectroscopic observations which cover a wide temperature range. This is still not completely achieved.

Even if we haven't gone in detail through the progress in the modeling of the PCTR, we have seen that the new multi-threads models appear to be promising. But more effort should be put by modelers, by extending the complexity of the physics they use, starting with including a PCTR reaching higher temperatures than the typical $10^5$ K. In addition, observations reveal that threads are not static and a variety of prominence internal motions exists. These are scarcely taken into account in modeling, beside the aspect of waves propagation, which is giving interesting results (see other works in this book).

Here we have not discussed the magnetic field properties in the PCTR. We could summarize saying that at present there are no observations able to detect such a weak field. However there are proposals for using, e.g. the Hanle effect in the H-L$\alpha$ line which, eventually, will sample the $2-3\times10^4$ K part of the PCTR. The access to such measurements will already be a step forward in the PCTR knowledge.

## References


Chiuderi Drago, F., Alissandrakis, C. E., Bastian, T., Bocchialini, K., & Harrison, R. A. 2001, *Solar Phys.* 199, 115.

Chiuderi Drago, F. & Landi, E. 2002, *Solar Phys.* 206, 315.

Cirigliano, D., Vial, J.-C., & Rovira, M. 2004, *Solar Phys.* 223, 95.

DeVore, C. R., Antiochos, S. K., & Aulanier, G. 2005, *ApJ* 629, 1122.

Gunár, S., Parenti, S., Anzer, U., Heinzel, P., & Vial, J.-C. 2011, *A&A* 535, A122.

Gunár, *et al.* 2013, this volume.

Heinzel, P., Schmieder, B., Vial, J.-C., & Kotrč, P. 2001, *A&A* 370, 281.

Heinzel, P., Schmieder, B., & Tziotziou, K. 2001, *ApJ* 561, L223.

Heinzel, P. & Anzer, U. 2001, *A&A* 375, 1082.

Heinzel, P., Schmieder, B., Fárník, F., Schwartz, P., Labrosse, N., Kotrč, P., Anzer, U., Molodij, G., Berlicki, A., DeLuca, E. E., Golub, L., Watanabe, T., & Berger, T. 2008, *ApJ* 686, 1383.

Heinzel, P. & Anzer, U. 2012, *A&A* 539, A49.

Kucera, T. A., Gibson, S. E., Schmit, D. J., Landi, E., & Tripathi, D. 2012, *ApJ* 757, 73.

Labrosse, N., Heinzel, P., Vial, J.-C., Kucera, T., Parenti, S., Gunár, S., Schmieder, B., & Kilper, G. 2010, *Space Sci. Revs* 151, 243.

Labrosse, N., Gouttebroze, P., Heinzel, P., & Vial, J.-C. 2002, *Solar Variability: From Core to Outer Frontiers*, 506, 451.

Lin, Y., Engvold, O., Rouppe van der Voort, L., Wiik, J. E., & Berger, T. E. 2005, *Solar Phys.* 226, 239.

Lin, Y., Martin, S. F., Engvold, O., Rouppe van der Voort, L. H. M., & van Noort, M. 2008, *Adv. Sp. Res.* 42, 803.

Luna, M., Karpen, J. T., & DeVore, C.R. 2012,*ApJ* 746, 30.

Mariska, J. T., Doschek, G. A., & Feldman, U. 1979, *ApJ* 232, 929.

Paletou, F., Vial, J.-C., & Auer, L. H. 1993, *A&A*, 274, 571.

Parenti, S., Vial, J.-C., & Lemaire, P. 2008, *Adv. Sp. Res.*, 41, 144.

Parenti, S. & Vial, J.-C. 2007, *A&A*, 469, 1109.

Parenti, S., Schmieder, B., Heinzel, P., & Golub, L. 2012, *ApJ*, 754, 66.

Parenti, S., Vial, J.-C., Schmieder, B., & Heinzel, P. 2013, in preparation.

Park, H., Chae, J., Song, D., Maurya, R. A., Yang, H., Park, Y.-D., Jang, B.-H., Nah, J., Cho, K.-S., Kim, Y.-H., Ahn, K., Cao, W., & Goode, P. R. 2013, *Solar Phys.* 72.

Pojoga, S. 1994, *IAU Colloq. 144: Solar Coronal Structures*, 357.

Ramelli, R., Stellmacher, G., Wiehr, E., & Bianda, M. 2012, *Solar Phys.* 281, 697.

Schmieder, B., Bommier, V., Kitai, R., Matsumoto, T., Ishii, T. T., Hagino, M., Li, H., & Golub, L. 2008, *Solar Phys.*, 247, 321.

Schwartz, P., Schmieder, B., Heinzel, P., & Kotrč, P. 2012, *Solar Phys.* 281, 707.

Stellmacher, G., Wiehr, E., & Dammasch, I. E. 2003, *Solar Phys.* 217, 133.

Stellmacher, G. & Wiehr, E. 2008, *A&A* 489, 773.

Vial, J. C. 1982, *ApJ* 254, 780.

Vial, J.-C., Olivier, K., Philippon, A. A., Vourlidas, A., & Yurchyshyn, V. 2012, *A&A* 541, A108.

Wiik, J. E., Dere, K., & Schmieder, B. 1993, *A&A* 273, 267.

Wiik, J. E., Dammasch, I. E., Schmieder, B., & Wilhelm, K. 1999, *Solar Phys.* 187, 405.

*Nature of Prominences and their role in Space Weather*
*Proceedings IAU Symposium No. 300, 2013*
*B. Schmieder, J.-M. Malherbe, & S. Wu, eds.*

doi:10.1017/S1743921313010776

# Plasma properties in eruptive prominences

**Nicolas Labrosse**

SUPA, School of Physics and Astronomy, University of Glasgow, Glasgow G12 8QQ, UK
email: Nicolas.Labrosse@glasgow.ac.uk

**Abstract.** Prominence eruptions are one of the most spectacular manifestations of our Sun's activity. Yet there is still some mystery surrounding their relevant physical conditions. What are their plasma parameters? How different are they from those of quiescent prominences? How do they relate to those within coronal mass ejections? We briefly review some recent results in non-LTE radiative transfer modelling which contribute to our knowledge of the plasma properties in eruptive prominences. We discuss in particular how these results, combined with observational data analysis, can help us in determining the plasma parameters in eruptive prominences.

**Keywords.** Sun: activity, Sun: prominences

## 1. Introduction

Recent efforts in non-LTE (i.e. out of local thermodynamic equilibrium) modelling have been focused towards understanding how the prominence plasma evolves during an eruption. This is an important question to answer as it can shed light on CME initiation mechanisms, and on the link between the chromosphere and the heliosphere.

In this paper we first describe the effect of plasma motions on the emergent radiation, and detail the main aspects of the non-LTE modelling we used. Section 3 presents observations in the He II 304 Å channel of SDO/AIA. Section 4 shows how the modelling results compare with these observations, and the ensuing diagnostic. Finally, we give conclusions in Section 5.

## 2. Effects of plasma motions on emitted radiation

Theoretical calculations have shown that when solar prominences move away from the surface of the Sun, their radiative output is affected via the Doppler dimming or brightening effects (Hyder & Lites 1970; Heinzel & Rompolt 1987; Gontikakis *et al.* 1997; Labrosse *et al.* 2007, 2008).

Doppler dimming (brightening) is the decrease (increase) in intensity of a line formed by resonant scattering of the incident radiation from the Sun due to the motion of the plasma. This effect is most commonly seen in strong resonance lines, and is widely used to diagnose the solar wind speed (Kohl & Withbroe 1982).

In order to study the effects of radial motions on the spectral lines emitted by moving or erupting prominences, we adopt the simple case of a 1D plane-parallel vertical slab standing above the solar surface (Fig. 1). The code used is the same as described by Labrosse *et al.* (2007, 2008). The following equations are solved: hydrostatic equilibrium, ionisation and statistical equilibrium, and radiative transfer for Hydrogen and Helium. The prominence atmosphere model is defined by the choice of the gas pressure, temperature, column mass along the line-of-sight, altitude above the limb, and the radial velocity. The latter is used to define velocity-dependent boundary conditions for the radiative transfer equation (Heinzel & Rompolt 1987). Pressure and temperature variations

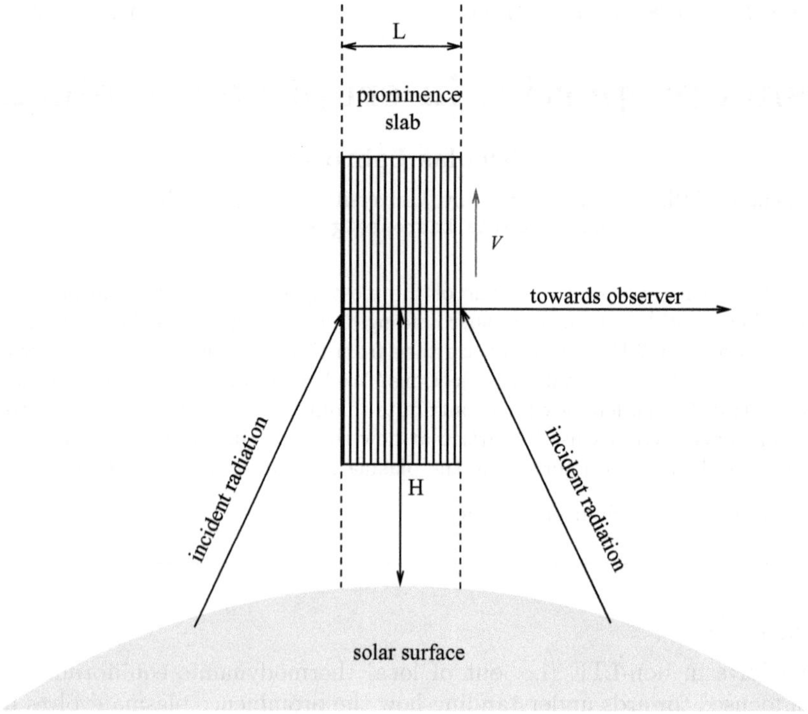


**Figure 1.** Sketch of the 1D plane-parallel prominence slab of geometrical width $L$, altitude $H$ above the solar surface, moving radially at a velocity $V$.

in the slab are described in Labrosse *et al.* (2008) and were taken from Anzer & Heinzel (1999).

Similar multi-level, multi-line non-LTE modelling set-ups have been used by various authors to study radial motions in moving or erupting prominences, and the results are summarised in Labrosse *et al.* (2010). In short, the characteristics of the Doppler dimming or brightening effects on lines emitted by prominence plasma moving in the radial direction depend on:

- the line formation mechanisms, namely the relative importance of the resonant scattering mechanism with respect to other mechanisms such as collisional excitation
- the shape of the incident radiation, i.e. whether the incident line is in emission or in absorption, and its strength
- any coupling with other energy states or with other lines arising from statistical equilibrium and radiation transfer (see e.g. the case of the H$\alpha$ line, well explained by Heinzel & Rompolt 1987; Gontikakis *et al.* 1997).

Emergent line profiles are affected by the Doppler dimming or brightening effects to various degrees as per the above criteria. In the case of a strongly sensitive line affected by partial redistribution in frequency, asymmetries will arise between the red and the blue wings of the profile (Gontikakis *et al.* 1997).

It is worth noting that the relative importance of the resonant scattering mechanism, with respect to other excitation mechanisms, will depend on the actual radial velocity and plasma parameters. For example, when a significant amount of hot material is present in the prominence-to-corona transition region (hereafter PCTR), collisional processes may become more important in the line formation mechanisms, and compete with the resonant scattering mechanism.

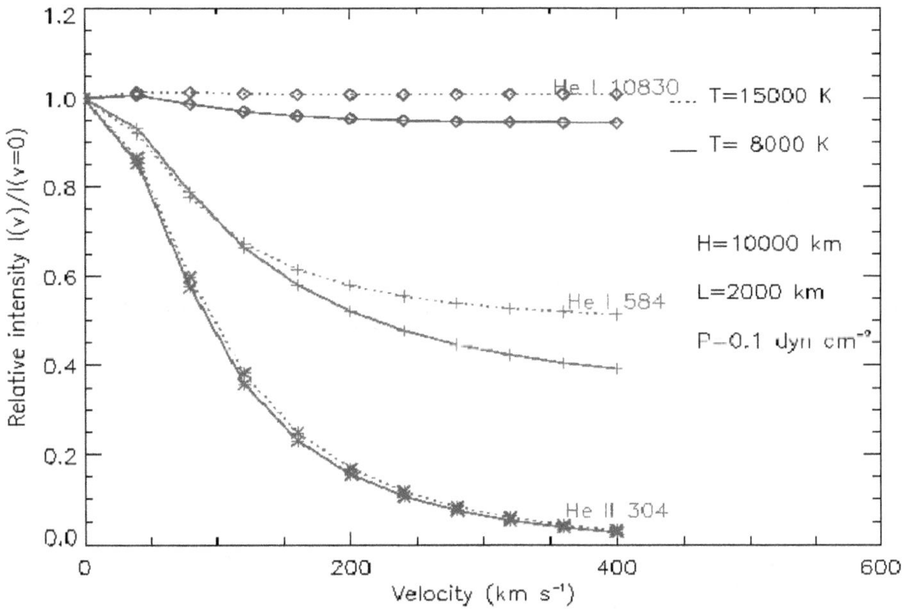


**Figure 2.** Relative intensity as a function of velocity at 8000 K (solid line) and 15 000 K (dotted line) for He I 584 (+) and 10 830 Å (◊) and He II 304 Å (*). Adapted from Labrosse *et al.* (2007).

The investigations performed by Labrosse *et al.* (2007, 2008) showed that the He II 304 Å line is very sensitive to the Doppler dimming effect. Under the prominence conditions considered in those studies, the main formation mechanism of that line is the scattering of the strong He II 304 Å emission line coming from the disk (Fig. 2). Hence it was shown that the Doppler dimming effect on the He II 304 line is stronger when most of the prominence plasma is cool, not too dense, or a large temperature gradient exists in the PCTR. When the physical conditions deviate from this situation, the relative contribution from the thermal processes becomes more and more important in the formation of the line – the relative contribution of the resonant scattering becomes smaller and smaller – and the sensitivity to the Doppler dimming effect is weaker.

It is important to stress that in all the previously referenced studies, the modelled prominence plasma parameters are kept constant while the radial velocity is changed in order to investigate the effects of radial motions on the radiation emitted by eruptive prominences.

## 3. Observations

Taking the theoretical results discussed in the previous section, the natural question to ask is then: Is Doppler dimming observed in eruptive prominences? A first step in answering this question was taken in Labrosse & McGlinchey (2012). In that paper, the authors asked whether observational signatures of the changes in the radiative output of eruptive prominences could be found in EUV observations of the He II 304 Å line. The authors studied the variation of intensity of arbitrarily selected features in the 304 channel from SDO/AIA observations of prominence eruptions as a function of velocity in the plane of the sky. These results were then compared with non-LTE radiative transfer calculations of the intensity of the He II 304 resonance line.

It was found that the observed intensity variations in various parts of the four eruptive prominences studied were not always consistent with the Doppler dimming effect on the

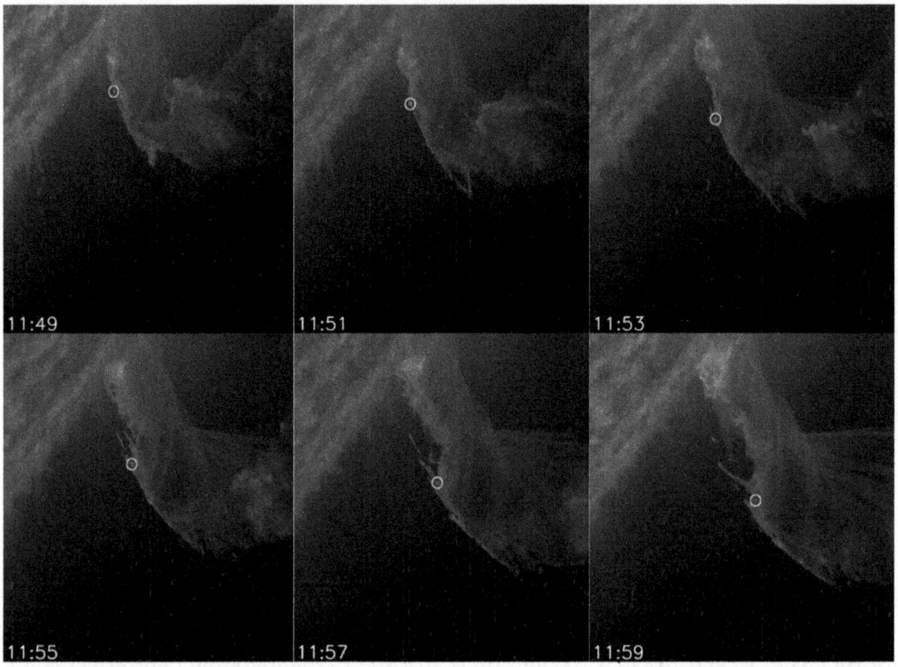


**Figure 3.** Evolution of the 2011-03-19 prominence eruption. The white circle marks the part of the prominence which was tracked and used to calculate the intensities as a function of altitude and velocity. The field of view is 390″×390″.

He II 304 Å line presented thus far. Namely, in some cases one observes an increase in intensity in the 304 channel with velocity, in contradiction to what is expected from the Doppler dimming effect alone. The use of new non-LTE models allowed the authors to explain the different behaviour of the intensity by changes in the plasma parameters inside the prominence, in particular the column mass of the plasma. The non-LTE models used in that study were more realistic than those used in previous calculations and discussed in the previous section. The main improvement is that they took into consideration the variation of the plasma parameters during the eruption (as opposed to keeping the plasma parameters constant, as in Fig. 2).

## 4. Prominence plasma diagnostic

We can now investigate whether these observations can be used to perform a diagnostic of the plasma of the eruptive prominence. We focus on one of the events presented in Labrosse & McGlinchey (2012) – the 19 March 2011 eruption observed by SDO/AIA (Fig. 3). We follow the same method as in Labrosse & McGlinchey (2012), although we have included some improvements in the comparison between observations and simulations.

### 4.1. *Normalisation of the model intensities*

The observed intensity data from AIA is normalised to the intensity that corresponds to the lowest observed radial velocity. In Labrosse & McGlinchey (2012), a model with zero radial velocity was designated as the reference model, and the modelled intensities were normalised to this reference model. This led to a radial velocity difference between the lowest observed and the reference model radial velocity. Additionally, the normalisation

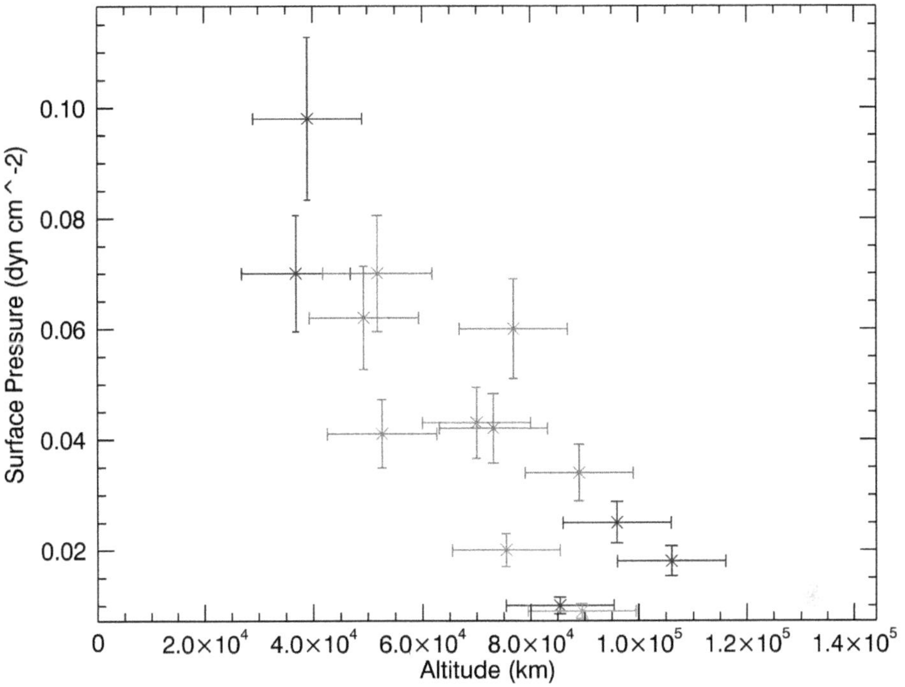


**Figure 4.** Surface pressure as a function of altitude for the selected models matching observations of the 19 March 2011 eruption.

did not account for the observed altitude, which led to differences in observed and reference model altitudes.

We improve upon this technique by creating a set of models with radial velocities fixed to the value of the lowest observed radial velocity. This results in multiple appropriate radial velocity models in which we can search for the model with the closest altitude to the observed altitude. This model is then designated as the reference model corresponding to the chosen observation data set.

This new normalisation procedure is far closer to the normalisation undergone by the observed intensities. This allows for accurate comparison between observation data and model data, an essential step in the prominence diagnostic process.

### 4.2. *Model search*

Each model is defined by the three observed parameters (intensities, radial velocities, and altitudes), along with other prominence plasma parameters. It is therefore possible to search the models for those which are probable matches for each observation point. We take each observation point and search for models which have altitude, intensity at 304 Å and radial velocity within the error boundaries of those observed parameters. Acceptable models are then combined across the lifetime of the prominence, and analysed.

### 4.3. *Results*

Figure 4 shows the results of this diagnostic method. It suggests that the surface pressure in the erupting prominence decreases as the eruption unfolds. This is consistent with the decrease in pressure with altitude in the corona.

## 5. Conclusions

We have shown here that we can obtain plasma parameters in eruptive prominences by comparing and minimising the differences between observations and non-LTE model results. As an example, we observe the decrease of the pressure at the surface of an eruptive prominence as its altitude in the corona increases, going from 0.1 dyn $cm^{-2}$ at the start of the observation to less than 0.02 dyn $cm^{-2}$ when the tracking has stopped.

This technique can be applied to any instrument that will monitor prominence eruptions in a resonantly scattered line such as Hydrogen and He II Lyman lines. In the future, the EUI instrument on Solar Orbiter will monitor the low atmosphere counterparts of large-scale solar eruptive events such as CMEs (see Auchere's paper in this issue). It will be ideally suited to study prominence eruptions. Using both H Ly-$\alpha$ and He II 304 lines will add constraints on the models. In addition, combining EUI observations with *in-situ* data out of the ecliptic plane will bring invaluable insights on the physical conditions within EPs, and a better understanding of the links between eruptive prominences and the dynamic heliosphere.

*Acknowledgement:* The author acknowledges the help of Scott Forrest in this work.

### References

Anzer, U. & Heinzel, P. 1999, A&A, 349, 974
Gontikakis, C., Vial, J.-C., & Gouttebroze, P. 1997, A&A, 325, 803
Heinzel, P. & Rompolt, B. 1987, Sol. Phys., 110, 171
Hyder, C. L. & Lites, B. W. 1970, Sol. Phys., 14, 147
Kohl, J. L. & Withbroe, G. L. 1982, ApJ, 256, 263
Labrosse, N., Gouttebroze, P., & Vial, J.-C. 2007, A&A, 463, 1171
Labrosse, N., Heinzel, P., Vial, J., *et al.* 2010, Space Science Reviews, 151, 243
Labrosse, N. & McGlinchey, K. 2012, A&A, 537, A100
Labrosse, N., Vial, J.-C., & Gouttebroze, P. 2008, Annales Geophysicae, 26, 2961

*Nature of prominences and their role in space weather*
*Proceedings IAU Symposium No. 300, 2013*
*B. Schmieder, J.-M. Malherbe & S. T. Wu, eds.*

doi:10.1017/S1743921313010788

# Determination of Temperature in Solar Prominences/Filaments Using FISS Observations

**Jongchul Chae, Hyungmin Park and Donguk Song**
Department of Physics and Astronomy, Seoul National University, Seoul 151-742, Korea
email: jcchae@snu.ac.kr

**Abstract.** Using the Fast Imaging Solar Spectrograph of the 1.6 meter New Solar Telescope at Big Bear, we simultaneously took the spectral profiles of the H$\alpha$ line and the Ca II line at 854.2 nm from prominences beyond the solar limb and filaments on the disk. The spectral data were fitted by the slab model of radiative transfer with constant source function, either with zero background intensity profile (in prominences) or with carefully constructed background intensity profile (in filaments). These observations with different perspectives and different analyses produced consistent results: temperature inside prominences/filaments ranges from 4000 to 20000 K with a mean of about 9500 K. We expect that this kind of observation and analysis with higher spatial resolution and higher temporal resolution will allow us to study in detail the thermal structure and evolution of plasma in prominences.

**Keywords.** line: profiles, plasmas, instrumentation: spectrographs, techniques: spectroscopic, Sun: prominences

## 1. Introduction

Determining the distribution and variation of temperature inside a prominence is fundamental to understanding its thermal structure and the relevant heating/cooling processes. One can infer temperature from the profiles of the H$\alpha$ line and the Ca II 854.2 nm line supposing these lines sense the same volume of a prominence. Since the mass of hydrogen atom much lighter than Ca II ion, the profile of the H$\alpha$ line is more subject to thermal broadening than the Ca II line, and it is possible to determine both temperature $T$ and non-thermal speed $\xi$ from the Doppler absorption widths $\Delta\lambda_D$ of the two lines. Here we present the results of the determination of temperature from the simultaneously taken H$\alpha$ line spectra and Ca II line spectra of limb prominences and disk filaments. The details of these works were described by Park *et al.* (2013) and Song *et al.* (2013).

## 2. Observation and Analysis

The observations were done with the Fast Imaging Solar Spectrograph (FISS) of the 1.6 meter New Solar Telescope at Big Bear Solar Observatory. The FISS is a dual-band Echelle spectrograph that can record the H$\alpha$ band and the Ca II 854.2 nm band simultaneously (Chae *et al.* 2013a). The spectral range of 0.97 nm is covered with the sampling of 0.019 nm/pixel for the H$\alpha$ band, and the one of 1.31 nm, covered with 0.025 nm/pixel for the Ca II band. The spectral resolving power is $1.4 \times 10^5$ and $1.3 \times 10^5$, respectively. The imaging is done with the slit scan. The quality of imaging mostly depends on the performance of the adaptive optics (AO) feeding the seeing-corrected light into the

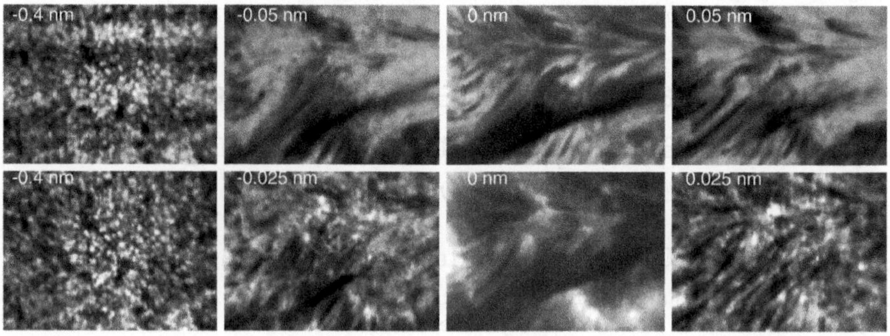


**Figure 1.** Raster images of a filament constructed at different wavelengths of H$\alpha$ line (upper row) and the Ca II 854.3 nm line (lower row) from FISS observations on 2012 July 11.

instrument. The imaging quality of the FISS observations has been improved with the advance of the AO project in Big Bear. Fig. 1 illustrates the imaging quality of FISS observations done in 2012. Our results presented here came from the earlier observations done in 2010 that have poorer imaging quality. The data processing and calibration are done in the ways described by Chae *et al.* (2013a, b).

The Doppler width $\Delta\lambda_D$ of line-emitting/absorbing plasma can be determined from the observed line profile $I_{\lambda,\mathrm{obs}}$ by fitting it with a simple slab model of radiative transfer with source function $S$ being assumed constant over wavelength and over position along the line of sight:

$$I_{\lambda,\mathrm{obs}} = S \cdot (1 - \exp(-\tau_\lambda)) + I_{\lambda,\mathrm{in}} \exp(-\tau_\lambda) \tag{2.1}$$

where $\tau_\lambda$ is the optical thickness profile

$$\tau_\lambda = \tau_0 \exp\left(-\left[\frac{\lambda - \lambda_0}{\Delta\lambda_\mathrm{D}}\right]^2\right) . \tag{2.2}$$

Note that the profile of light incident from the background $I_{\lambda,\mathrm{in}}$ is zero in prominence observations, but is not zero and has to be estimated in filament observations. The solution in this case of non-zero background intensity is reduced to the Becker's cloud model (Beckers 1964). The background intensity profile was determined by taking average of all the profiles taken from outside the filament that have the same intensities at wavelengths away from the cores as the observed profile of interest taken from the filament.

## 3. Results

Fig. 2 shows the example of a prominence observed through the H$\alpha$ line and the Ca II line. Even the prominence is optically thinner in the Ca II line than the H$\alpha$ line, but the two images constructed at the cores of the two lines look very similar to each other, which supports the assumption that the two lines are emitted by the same plasma of the prominence. We find that the fittings are fairly good as illustrated in Fig. 3. The determined Doppler widths of 0.034 nm (H$\alpha$) and 0.023 nm (Ca II) yield the estimates: $T = 10900$ K and $\xi = 7.8$ km s$^{-1}$. Fig. 3 also presents the scatter plot of Doppler widths

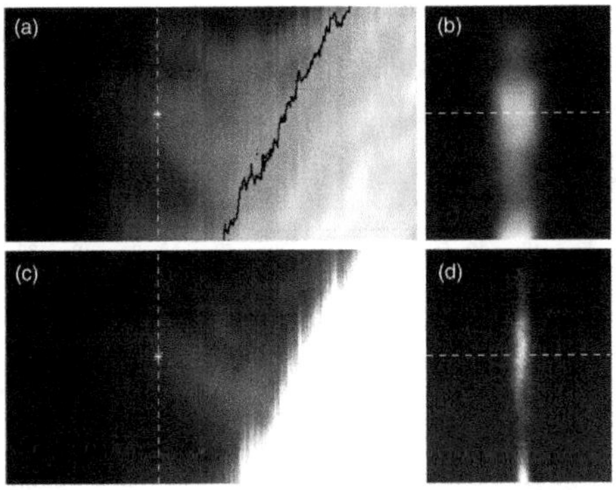


**Figure 2.** Left: raster images of the prominence observed on 30 June 2010 constructed at the center wavelengths of Hα line (upper) and the Ca II line (lower). Right: the spectrograms of the Hα line (upper) and the Ca II line (lower) at the slit position indicated by the dashed vertical lines.

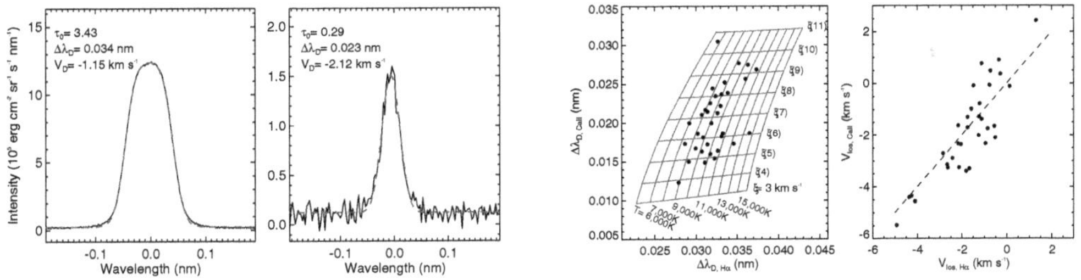


**Figure 3.** Left two: sesults of model fitting to the Hα line profile and the Ca II line profile. Right two: scatter plots of $\Delta\lambda_D$ (Hα) vs $\Delta\lambda_D$ (Ca II), and $V_{\rm los}$ (Hα) vs $V_{\rm los}$ (Ca II) for the prominence shown in Fig. 2. The overplotted iso-$T$ curves and the iso-$\xi$ curves allows an easy estimate of $T$ and $\xi$ from the plot.

with more number of data points. We find that $T$ here ranges from 7000 K to 15000 K with the mean of 10200K and the standard deviation of 1580 K. The right panel of Fig. 3 shows that the line-of-sight velocities determined from the two lines are close to each other, which is a support that the two lines sense the same volumes. We determined $T$ at many different parts of another prominence observed the same day, and obtained the mean of 9400 K and the standard deviation of 2200 K. The third prominence was observed on a different day, 15 August 2011. The mean temperature in this prominence was 8700 K and the standard deviation, 1100 K.

Fig. 4 shows the images of the filament observed on 29 July 2010. The filament is less prominent and optically thinner in the Ca II line than in the Hα line, like in the limb observations. Fig. 5 illustrates the goodness of the fittings of the contrast profiles $C_\lambda \equiv I_{\lambda,\rm obs}/I_{\rm in} - 1$ with the cloud model. The model fitting of the profiles marked by the diamond symbol, for example, produces $\Delta\lambda_D$ of 0.027 nm for the Hα line and 0.016 nm for the Ca II line, which can be explained by the parameter values: $T$ = 7500 K and $\xi$ = 5.3 km s$^{-1}$. In the same way, we obtain $T$ = 12600 K and $\xi$ = 4.4 km s$^{-1}$ for the data point marked by the circle symbol. We find that $T$ varies much with position inside the filament, with the mean of 9800 K and the standard deviation of 3100 K.

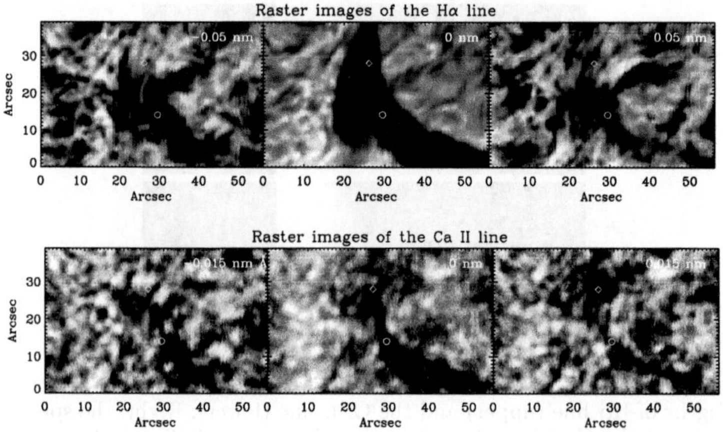


**Figure 4.** Raster images of the filament observed on 29 July 2010 constructed at different wavelengths of the Hα line (top) and the Ca II line (btottom).

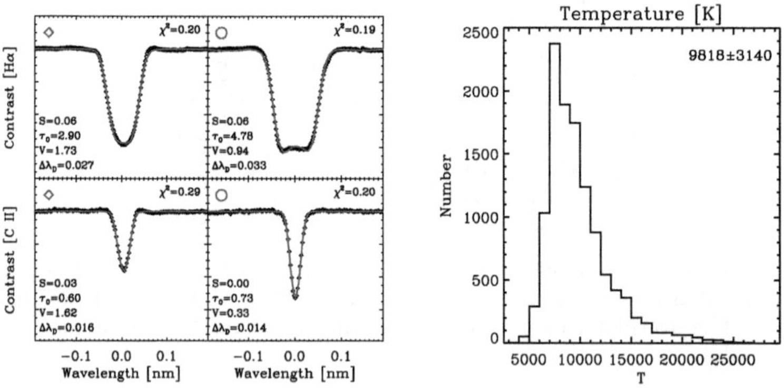


**Figure 5.** Left: examples of fitting of the spectra at two selected points. Right: The number distribution of temperature determined inside the filament.

## 4. Conclusion

Our limb observations of the prominences and disk observations of the filament produced consistent results on the values of temperature in prominences/filaments; the prominences/filaments have mean temperatures in the narrow range between 8700 to 10200 K while the variation of temperature inside each individual prominence/filament is large. The lowest temperature is around 4000 K, and the highest temperature is 20000 K or higher, being in agrement with previous studies (e.g. Hirayama 1985). This consistency between limb observations and disk observations is quite encouraging to us, since the analysis of disk observations requires the determination of background intensity, while limb observations do not. The consistency indicates that our analysis of disk observations is reasonable, so that our analysis can be extended to other disk observations. It is definitely worthwhile to investigate the spatial and temporal variation of $T$ inside a prominence/filament in detail.

This work was supported by the National Research Foundation of Korea (NRF - 2012R1A2A1A03670387).

## References

Beckers, J. M. 1964, A Study of the Fine Structures in the Solar Chromosphere, Ph.D. thesis, University of Utrecht

Chae, J., Park, H.-M., Ahn, K., *et al.* 2013a, *Solar Phys.*, in press

Chae, J., Park, H.-M., Ahn, K., *et al.* 2013b, *Solar Phys.*, in press

Hirayama, T. 1985, *Solar Phys.*, 100, 415

Park, H., Chae, J., Song, D., *et al.* 2013, *Solar Phys.*, in press

Song, D., Chae, J., Park, H., *et al.* 2013, *Solar Phys.*, submitted

*Nature of Prominences and their role in Space Weather*
*Proceedings IAU Symposium No. 300, 2013*
*B. Schmieder, J.-M. Malherbe & S. T. Wu, eds.*

doi:10.1017/S174392131301079X

# Rayleigh–Taylor instability in partially ionized prominence plasma

**E. Khomenko[1,2], A. Díaz[1,2], A. de Vicente[1,2], M. Collados[1,2] and M. Luna[1,2]**

[1]Instituto de Astrofísica de Canarias, 38205 La Laguna, Tenerife, Spain
email: khomenko@iac.es

[2]Departamento de Astrofísica, Universidad de La Laguna, 38205, La Laguna, Tenerife, Spain

**Abstract.** We study Rayleigh–Taylor instability (RTI) at the coronal–prominence boundary by means of 2.5D numerical simulations in a single-fluid MHD approach including a generalized Ohm's law. The initial configuration includes a homogeneous magnetic field forming an angle with the direction in which the plasma is perturbed. For each field inclination we compare two simulations, one for the pure MHD case, and one including the ambipolar diffusion in the Ohm's law, otherwise identical. We find that the configuration containing neutral atoms is always unstable. The growth rate of the small-scale modes in the non-linear regime is larger than in the purely MHD case.

**Keywords.** Instabilities; Numerical simulations; Chromosphere; Magnetic fields; Prominences.

---

Solar prominences are blocks of cool and dense chromospheric-type plasma remaining stable for days, or even weeks, in the solar corona. Despite the global stability, prominences are extremely dynamic at small scales, showing a variety of shapes, moving with vertical and horizontal threads. Berger *et al.* (2010) find large-scale 20–50 Mm arches, expanding from underlying corona into the prominences. At the top of these arches, at the border between the corona and the prominence, there are observed dark turbulent upflowing channels of 4-6 Mm maximum width with a profile typical for the Rayleigh–Taylor (RT) and Kelvin–Helmholtz (KH) instabilities (Berger *et al.* 2010; Ryutova *et al.* 2010). The upflows rise up to 15–50 Mm, with an average speed of 13–17 km s$^{-1}$, decreasing at the end. Lifetimes of the plumes are about 300–1000 sec. From the theoretical point of view, the existence of the instabilities at the interface between the prominence and the corona is easily explained since the two media have clearly different densities, temperatures and relative velocities. Recent numerical simulations of the RTI by Hillier *et al.* (2012), including a rising buoyant tube in a Kippenhahn–Schlüter prominence model show a good agreement with observations. Nevertheless, prominence plasma is expected to be only partially ionized. The presence of a large number of neutrals must affect the overall dynamics of the plasma. Linear theory of RTI and KHI in the partially ionized plasma has been recently developed by Soler *et al.* (2012); Díaz *et al.* (2013) showing that there is no critical wavelength as in the purely MHD case ($\lambda_c = B_0^2 \cos^2\theta/(\rho_2 - \rho_1)/g$; $B_0$ and $\theta$ are magnetic field strength and inclination, $\rho_2 - \rho_1$ is density contrast between the two media, $g$ is gravity), and perturbations in all the wavelength range are always unstable. The aim of our work is to model the dynamics of RTI in the partially ionized prominence plasma in the non-linear regime.

We simulate a small portion of the interface between prominence and corona, of the size of 1×1 Mm. The initial stratification of the pressure and density is in hydrostatic equilibrium for a given temperature. The equilibrium magnetic field, $\vec{B}_0$, is homogeneous and does not influence the force balance. The plasma is perturbed in the $XZ$ plane, the

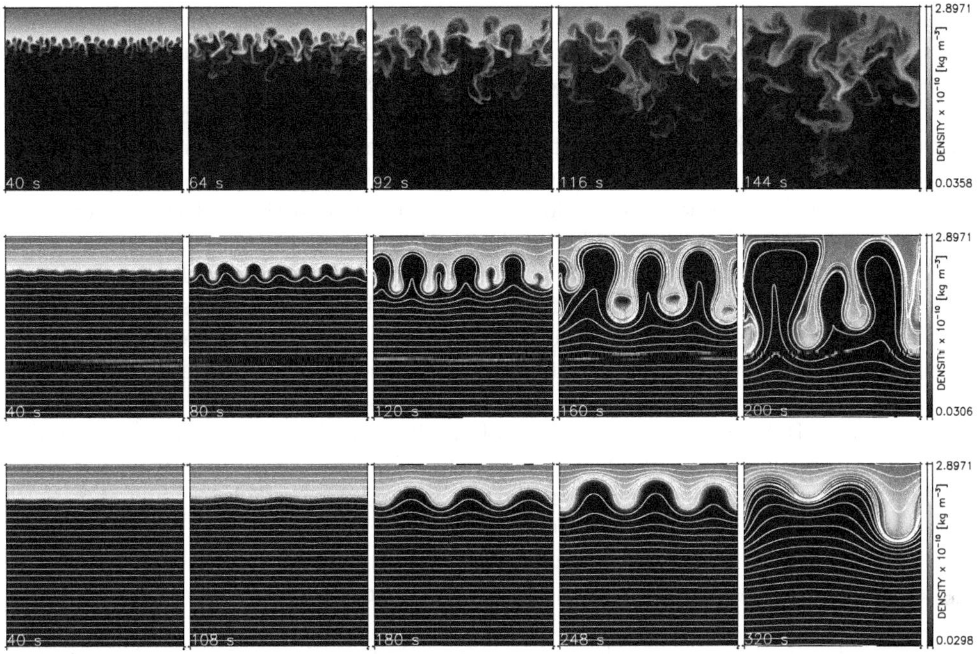


**Figure 1.** Top: Time evolution of density in the "ambipolar" simulation with $\theta = 90°$. Middle: same for the simulation with $\theta = 89°$. White lines are projections of magnetic field lines into $XZ$ perturbation plane. Bottom: same for the simulation with $\theta = 88°$.

magnetic field vector is initially in $XY$ plane, making an angle with $X$ axis. We consider 3 simulation runs with $\vec{B}_0$ at $\theta = 90°$ to $X$ axis (i.e. normal to the $XZ$ plane), $\theta = 89°$ and 88°. The following parameters of the equilibrium configuration were taken: $T_{\rm cor}$ = 400 kK, $T_{\rm prom}$ = 5 kK; $\rho_{\rm cor} = 3.7 \times 10^{-12}$ kg m$^{-3}$, $\rho_{\rm prom} = 2.9 \times 10^{-10}$ kg m$^{-3}$, and $B_0 = 10$ G. This provides prominence neutral fraction (the ratio of neutral density to the total density, $\rho_n/\rho$) of 0.9 and the ambipolar diffusion coefficient of $\eta_A = 2.3 \times 10^8$ m$^2$ s$^{-1}$. To initiate the instability we perturbed the position of the interface by a multi-mode perturbation containing 25 modes with wavelengths in the range $\lambda = 10 \div 250$ km. The resolution of the simulations was 1 km in each spatial direction. Such high resolution is expected to be necessary to study non-ideal plasma effects.

After subtracting the equilibrium conditions, we solve numerically the quasi-MHD equations of conservation of mass, momentum, internal energy, and the induction equation by means of our code MANCHA (Felipe *et al.* 2010; Khomenko & Collados 2012) with the inclusion of the physical ambipolar diffusion term in the equation of energy conservation and in the induction equation. We evolve the ionization fraction in time by Saha equation, as a first approximation. Our code uses hyperdiffusivity for stabilizing the numerical solution. To assure that the numerical diffusivity (whose action resembles the physical diffusivity) does not affect the results of the simulations, we kept the amplitude of artificial hyperdiffusive terms 2-3 orders of magnitude lower than the physical ambipolar diffusion so that the characteristic time scales of action of the former are orders of magnitude large. For each field inclination we perform two identical simulations, one for the purely MHD case and one with the ambipolar term switched on.

Figure 1 gives the time evolution of density perturbation in the "ambipolar" simulations for the three field inclinations. The case of the magnetic field normal to the plane of the instability ($\theta = 90°$) is analogous to a purely hydrodynamical case since there is not cut-off wavelength. Indeed, the simulation shows the development of very small scales. The comparison of the "MHD" and "ambipolar" simulations (not shown in the Figure) reveals different (but statistically equivalent, as is shown below) particular form of the turbulent flows, since the ambipolar diffusion, acting on small scales, forces their different evolution. The density variations have a pronounced asymmetry between the large-scale upward rising bubbles and small-scale downflowing fingers, caused by the mass conservation. Such asymmetric behavior can possibly explain the preferred detection of the larger-scale upward rising bubbles in observations (Berger *et al.* 2010).

By just rotating the field by 1° in $XY$ plane the scales developed in the simulation are significantly changed (middle panels of Fig. 1, $\theta = 89°$), small scales disappear and only few big drops are developed after about 200 sec of the simulation. The small scales can not develop the instability because of the cut-off induced by the magnetic field, $\lambda_c \approx 38$ km, for our equilibrium configuration and $\theta = 89°$. While not completely damped as in the purely MHD case, the growth rate of small scales is still very low compared to large scales, see the results by Díaz *et al.* (2013). One can appreciate from the figure that at $t = 40$ sec the dominant wavelength is around $\lambda \approx L/9$ km, while at $t = 120$ sec it becomes $\lambda \approx L/6$ and finally after $t = 200$ sec the dominant wavelength is equal to $\lambda \approx L/3$, being $L = 1$ Mm the size of the simulation box. During the evolution of the instability, the horizontal magnetic field component increases below and above the drops as the field gets compressed by the flows. The increase of the field produces a local increase of the cut-off wavelength, contributing additionally to the damping of small scales. Moreover, small scales tend to disappear with time due to non-linear interaction of harmonics with larger scales, as was already demonstrated by other numerical works on the RTI (Jun *et al.* 1995). Similar trend is observed in the simulation with $\theta = 88°$, in the bottom panels of Fig. 1. In this case, the critical wavelength is $\lambda_c \approx 155$ km, and, accordingly, the developed scales are even larger. Note that, at lower field inclinations, the evolution takes progressively more time, since the tension force by the magnetic field slows down the development of the instability, in agreement with the linear theory.

The velocity of flows lie in the range of 10–20 km s$^{-1}$, similar to observations. The downflows are observed in about 2/3 of all points, and upflows occupy the remaining 1/3 but the upflow velocities are, on average, larger. The comparison of the flows in the "ambipolar" and the "MHD" simulations in approximately linear stage of the RTI (first few tens of sec of the simulation), reveals slightly larger velocities in the "ambipolar" simulations, i.e. this case is slightly more unstable.

To analyze the stability of different scales in the non-linear regime of the RTI, at each time moment $t$ we calculated the power as a function of the horizontal wave number $k_x$ by fourier-transforming in space a portion of the snapshot of pressure variations around the discontinuity, and by averaging the power for the vertical $k_z$ wave number to decrease the noise. We then divided the "ambipolar" power map by the corresponding "MHD" map for each pair of the simulations, $\theta = 90°$, 89° and 88°. The result is given at the left panels of Fig. 2, while the right panel shows the time-averages of the relative power maps. In the case of the field directed normal to the perturbation plane ($\theta = 90°$), we do not observe any significant change in power between the "ambipolar" and "MHD" cases. The relative power fluctuates in time, but the average remains around one for all harmonics (upper left panel). This behavior is already anticipated from the linear analysis (Díaz *et al.* 2013). The relative power is significantly different to that in the cases with $\theta = 89°$ and 88°. Now there is a clear increase in the growth rate of the small-scale harmonics in the "ambipolar" simulations compared to the "MHD" ones. The growth rate of the large-scale harmonics is the same in both cases. The change in the behavior happens at about $\lambda \sim 30$ km for the $\theta = 89°$ case, and at $\lambda \sim 100$ km for the $\theta = 88°$ case, both

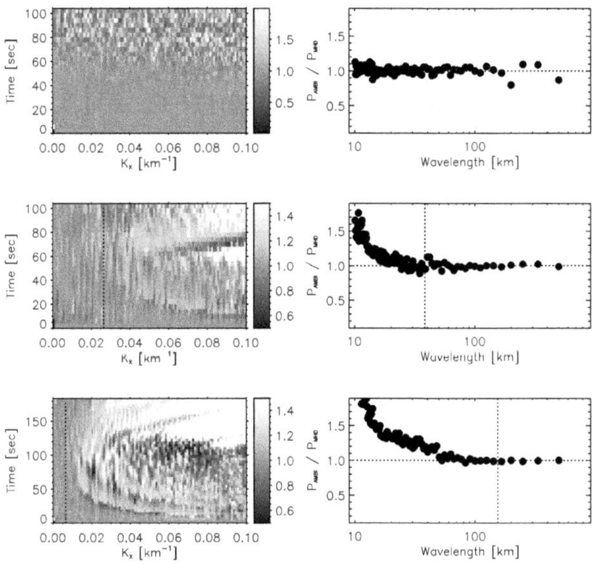


**Figure 2.** Fourier analysis of scales developed in the simulations. Left: relative power of the "ambipolar" vs "MHD" simulation as a function of horizontal wave number along the discontinuity, $k_x$, and time. Whiter colors mean that the "ambipolar" simulation has more power. Right: time average of the relative power from the panels on the left. Top: $\theta = 90°$. Middle: $\theta = 89°$. Bottom: $\theta = 88°$. In the last two cases, the vertical dotted lines at the right panels mark the cut-off wavelength $\lambda_c = 38$ and 155 km, correspondingly.

numbers being close to the corresponding cut-off wavelengths $\lambda_c$. This result confirms and extends the conclusion from the linear theory that all RTI modes become unstable when the presence of neutral atoms is accounted for in the analysis. We obtain up to 50% increase of the small-scale harmonics growth rate for the $\theta = 89°$ case and up to 90% increase for the $\theta = 88°$ case. Our simulations show that partial ionization of prominence plasma measurably influences its dynamics and must be taken into account in future models.

## Acknowledgements

This work contributes to the deliverables identified in FP7 European Research Council grant agreement 277829, "Magnetic connectivity through the Solar Partially Ionized Atmosphere", whose PI is E. Khomenko.

## References

Berger, T. E., Slater, G., Hurlburt, N., *et al.* 2010, *ApJ*, 716, 1288
Díaz, A. J., Khomenko, E., & Collados, M. 2013, *A&A*
Felipe, T., Khomenko, E., & Collados, M. 2010, *ApJ*, 719, 357
Hillier, A., Berger, T., Isobe, H., & Shibata, K. 2012, *ApJ*, 746, 120
Jun, B.-I., Norman, M. L., Stone, & J. M. 1995, *ApJ*, 453, 332
Khomenko, E. & Collados, M. 2012, *ApJ*, 747, 87
Ryutova, M., Berger, T., Frank, Z., Tarbell, T., & Title, A. 2010, *Solar Phys.*, 267, 75
Soler, R., Díaz, A. J., Ballester, J. L., & Goossens, M. 2012, *ApJ*, 749, 163

*Nature of prominences and their role in space weather*
*Proceedings IAU Symposium No. 300, 2013*
*B. Schmieder, J.-M. Malherbe & S. T. Wu, eds.*

doi:10.1017/S1743921313010806

# Determination of Prominence Plasma $\beta$ from the Dynamics of Rising Plumes

**Andrew Hillier[1], Richard Hillier[2] and Durgesh Tripathi[3]**

[1]Kwasan Observatory, 17-1 Kitakazan-ohmine-cho, Yamashina-ku, Kyoto, 607-8471, Japan
email: andrew@kwasan.kyoto-u.ac.jp

[2]Department of Aeronautics, Imperial College, London, United Kingdom

[3]Inter-University Centre for Astronomy and Astrophysics, Post Bag 4, Ganeshkhind, Pune 411 007, India

**Abstract.** Observations of quiescent prominences show rising plumes, dark in chromospheric lines, that propagate from large bubbles. In this paper we present a method that may be used to determine the plasma $\beta$ (ratio of gas pressure to magnetic pressure) from the rising plumes. Using the classic fluid dynamic solution for flow around a circular cylinder, the compression of the prominence material can be estimated. Application to a prominence gave an estimate of the plasma $\beta$ as $\beta = 0.47$–$1.13$ for a ratio of specific heats of $\gamma = 1.4$–$1.7$.

**Keywords.** Sun: prominences, magnetic fields, MHD, instabilities

## 1. Introduction

Understanding the strength of the magnetic field of prominences is of great importance for the study of prominence dynamics and space weather forecasting. Measurements of the magnetic field for quiet sun prominences and filaments gives a range of 3 to 30 G (e.g. Leroy 1989; Casini *et al.* 2009; Schmieder *et al.* 2013). However, the sample of prominences that have had their magnetic field strength measured is still very small, meaning that other methods are often employed to estimate the strength of a prominence's magnetic field. Commonly used methods include prominence seismology (e.g. Arregui *et al.* 2012; Ballester 2013) and extrapolation of the photospheric magnetic field (e.g. Aulanier & Démoulin 2003; van Ballegooijen 2013).

Since the launch of the Hinode satellite, there has been great focus on the dynamics of the rising plumes (see Figure 1) (Berger *et al.* 2008; de Toma *et al.* 2008; Berger *et al.* 2010) created by the magnetic Rayleigh-Taylor instability (Berger *et al.* 2010; Hillier *et al.* 2011, 2012a). As any instability, wave or flow that is observed in an Magnetohydrodynamic (MHD) system should contain information about the system itself, these rising plumes present a great opportunity to investigate the plasma parameters of quiescent prominences. In this paper, we present a method that uses observations by the Solar Optical Telescope (SOT; Tsuneta *et al.* 2008) on the Hinode satellite (Kosugi *et al.* 2007) of rising plumes to determine the plasma $\beta$ of a quiescent prominence. A full explanation of the methodology and results can be found in Hillier *et al.* (2012b)

## 2. Necessary Information to Determine Prominence Plasma $\beta$

The method to determine the plasma $\beta$ of a prominence, as presented in (Hillier *et al.* 2012b), is formulated using a set of observational information about prominences and their dynamics. Before presenting the methodology, first it is necessary to quickly review these observational characteristics of prominences, and in particular their plumes, that are important for this analysis. The points we need to remember are:

- The plumes rise at an approximately constant velocity (Berger *et al.* 2010)
- The plumes have an elliptical head (Berger *et al.* 2010)
- The intensity of the prominence material at the top of the plume is greater than the average prominence intensity (Hillier *et al.* 2012b)
- Flows of material can be observed along the head of the plume (Berger *et al.* 2011)
- The prominence magnetic field is mainly horizontal (e.g. Schmieder *et al.* 2013)

First we must relate the prominence intensity to the density of the prominence. The prominence intensity is dominated by scattering, giving the intensity as proportional to the column density. Hence the bright top of the plume would imply that there has been an increase in the density through compression of the prominence.

Next we can use the approximately constant velocity and elliptical plume head to perform two coordinate transforms that greatly simplify the situation under consideration. First we remove the plume stem, treating it as a rising ellipse in the prominence material. As this ellipse is moving at constant velocity through (what we assume to be) a constant medium, a shift in reference frame gives a stationary ellipse in a constant flow. Now we must consider the 3D nature of the geometry under investigation. The observed plumes are known to be caused by the magnetic Rayleigh-Taylor instability (Berger *et al.* 2010; Hillier *et al.* 2012a), which creates filamentary structures aligned with the magnetic field (Isobe *et al.* 2005, 2006b; Stone & Gardiner 2007; Hillier *et al.* 2012a) (for no shear in the magnetic field). Therefore, we can view the problem as being similar of an elliptical cylinder in a constant flow. Performing a conformal transform of the coordinate system to change to a circular cylinder creates a situation analogous to that of a constant flow around a circular cylinder.

For simplicity, we will now make the assumption that the plasma $\beta$ is approximately uniform. Based on this assumption, and modifying the derivation for a compressible flow around a circular cylinder (van Dyke 1975), we find:

$$\rho = \left(1 - \frac{\gamma - 1}{2} {M_*}^2 (v_r^2 + v_\theta^2 - 1)\right)^{1/(\gamma-1)} \tag{2.1}$$

$$v_r(r,\theta) = \left(1 - \frac{1}{r^2}\right)\cos\theta \tag{2.2}$$
$$- {M_*}^2\left[\left(-\frac{13}{12}\frac{1}{r^2} + \frac{3}{2}\frac{1}{r^4} - \frac{5}{12}\frac{1}{r^6}\right)\cos\theta + \left(\frac{1}{4}\frac{1}{r^4} - \frac{1}{4}\frac{1}{r^2}\right)\cos^3\theta\right] + O({M_*}^4)$$

$$v_\theta(r,\theta) = -\left(1 + \frac{1}{r^2}\right)\sin\theta \tag{2.3}$$
$$+ {M_*}^2\left[\left(\frac{13}{12}\frac{1}{r^2} - \frac{1}{2}\frac{1}{r^4} + \frac{1}{12}\frac{1}{r^6}\right)\sin\theta - \left(\frac{1}{12}\frac{1}{r^4} - \frac{1}{4}\frac{1}{r^2}\right)3\cos^2\theta\sin\theta\right] + O({M_*}^4)$$

$$M_* = \sqrt{\frac{\gamma\beta_\infty}{\gamma\beta_\infty + 2(\gamma - 1)}} M_\infty \tag{2.4}$$

Equation 2.1 can now be used to investigate the compression around a plume head.

## 3. Application to a Prominence Plume

In this section we apply the method presented above to determine the plasma $\beta$ of the plume displayed in the panel on the right of Figure 1. Due to the short timescales of

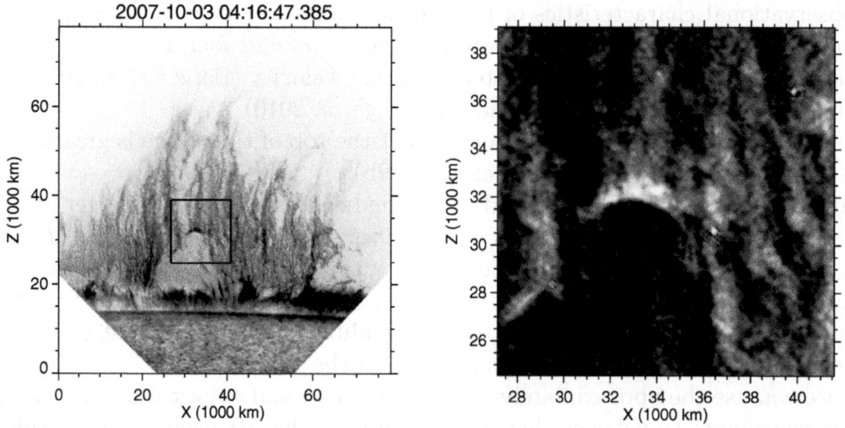


**Figure 1.** The left-hand panel shows a quiescent prominence observed on the 3-Oct-2007 at 04:16UT. The right-hand panel shows a zoom-in of the region in the box in the left-hand panel. The plume and the bright prominence plasma above it are clearly visible.

these compressions ($\sim 100\,\mathrm{s}$), we assume that the increased intensity comes through an increased column density resulting from compression.

Using the plume observations shown in Figure 1, it is possible to determine the Mach number of the rising plume and the plume dimensions. This prominence was observed on the NW solar limb on 2007-10-03 observed by the SOT with the Ca II H filter at a cadence of 30 s. The velocity of the plume was found to be approximately constant, with velocity and the velocity error calculated to be $v_{plume} = 12.3 \pm 0.6\,\mathrm{km\,s^{-1}}$. The sound speed of the prominence material is $C_s = [\gamma(R/\mu)T_{prom}]^{1/2} = 11\,\mathrm{km\,s^{-1}}$ for $\gamma = 5/3$, $\mu = 0.9$ and $T_{prom} = 8000\,\mathrm{K}$ giving a Mach number of the flow of $M_\infty = 1.12 \pm 0.05$. The dimensions of the elliptical plume head are 1.3 arcsec and 2.35 arcsec for the short and long axes, these are used for the conformal transform of the plume head.

Figure 2 shows the normalised intensity, which is used as a proxy for the column density. The fit is shown by the dotted line. The fitting error is determined as the standard deviation of the difference between the normalised intensity and the fit at each pixel and the intensity error is determined by the intensity fluctuations of the stray light in the corona. The error bars equivalent to $2\hat{\sigma}_{prom}$ and $2\hat{\sigma}_{fit}$ are shown in Figure 2.

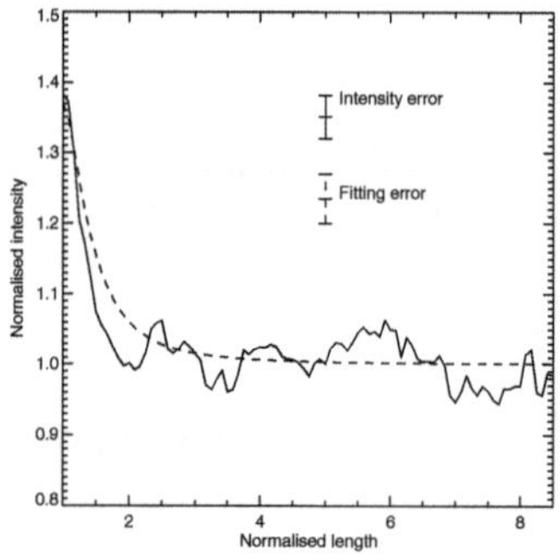


**Figure 2.** The graph shows a plot of intensity along the slit (solid line) and the fitting assuming the intensity proportional to density (dashed line) for $\gamma = 1.65$. Error bars equivalent to twice the standard deviation are presented for the observed intensity and the fit.

Using the value of $M_*$ from the fitting of the intensity and the definition of $M_*$ as given in Equation 2.4, it is possible to solve for the plasma $\beta$ of the prominence, where $\beta$ is given by:

$$\beta = \frac{2(\gamma - 1)}{\gamma} \frac{M_*^2}{M_\infty^2 - M_*^2} \tag{3.1}$$

Assuming the range $\gamma = 1.4 - 1.7$, we find a plasma $\beta$ of $\beta = 0.47 \pm 0.079$ to $1.13 \pm 0.080$.

## 4. Conclusion

In this paper, we presented a method to determine a prominence's plasma $\beta$. Applying Equation 2.1 to observations of a prominence we found that for the range $\gamma = 1.4$ - $1.7$ the observed prominence has a plasma $\beta$ of $\beta = 0.47 \pm 0.079$ to $1.13 \pm 0.080$. This method can be applied to any plume observation to provide an estimate of the plasma $\beta$ of the prominence in which the plumes are formed.

One interesting point to note is that we have equations that define the velocity around the plume head. Therefore, through analysis of the velocities at the observed plume head, the projection of the plume onto the plane-of-sky can be determined. Using this information, an estimation of the direction of the magnetic field in the prominence may be possible.

Hinode is a Japanese mission developed and launched by ISAS/JAXA, with NAOJ as domestic partner and NASA and STFC (UK) as international partners. It is operated by these agencies in co-operation with ESA and NSC (Norway). AH is supported by KAKENHI Grant-in-Aid for Young Scientists (B) 25800108.

### References

Arregui, I., Oliver, R., & Ballester, J. L. 2012, Living Reviews in Solar Physics, 9, 2
Aulanier, G. & Démoulin, P. 2003, *A & A*, 402, 769
Ballester, J. L. 2013, The IAU Symposium 300 Proceedings 2013, in press
Berger, T. E., *et al.* 2008, *ApJ* Letters, 676, L89
Berger, T. E., *et al.* 2010, *ApJ*, 716, 1288
Berger, T., *et al.* 2011, *Nature*, 472, 197
Casini, R., López Ariste, A., Paletou, F., & Léger, L. 2009, *ApJ*, 703, 114
Chae, J. 2010, *ApJ*, 714, 618
de Toma, G., Casini, R., Burkepile, J. T., & Low, B. C. 2008, ApJ Letters, 687, L123
Hillier, A., Isobe, H., Shibata, K., & Berger, T. 2011b, ApJ Letters, 736, L1
Hillier, A., Berger, T., Isobe, H., & Shibata, K. 2012a, ApJ, 746, 120
Hillier, A., Hillier, R., & Tripathi, D. 2012b, ApJ, 761, 106
Isobe, H., Miyagoshi, T., Shibata, K., & Yokoyama, T. 2005, Nature, 434, 478
Isobe, H., Miyagoshi, T., Shibata, K., & Yokoyama, T. 2006b, PASJ, 58, 423
Kosugi, T., *et al.* 2007, Sol. Phys., 243, 3
Leroy, J. L. 1989, Dynamics and Structure of Quiescent Solar Prominences, 150, 77
López Ariste, A. 2013, The IAU Symposium 300 Proceedings 2013, in press
Schmieder, B., Kucera, T. A., Knizhnik, K., *et al.* 2013, arXiv:1309.1568
Stone, J. M. & Gardiner, T. 2007, *ApJ*, 671, 1726
Tsuneta, S., *et al.* 2008, Sol. Phys., 249, 167
van Ballegooijen, A. 2013, The IAU Symposium 300 Proceedings 2013, in press
van Dyke, M. 1975, NASA STI/Recon Technical Report A, 75, 46926

# Session I - 1.3

## MAGNETIC FIELD: MEASUREMENTS AND MODELS

*Nature of prominences and their role in space weather*
*Proceedings IAU Symposium No. 300, 2013*
*B. Schmieder, J.-M. Malherbe & S. T. Wu, eds.*

doi:10.1017/S1743921313010818

# The inference of the magnetic field vector in prominences

**Bruce W. Lites**

High Altitude Observatory, National Center for Atmospheric Research,
P.O. Box 3000, Boulder, CO 80307-3000, USA
email: `lites@ucar.edu`

**Abstract.** Prominences owe their existence to the presence of magnetic fields in the solar corona. The magnetic field determines their geometry and is crucial to their stability, energetics, and dynamics. This review summarizes techniques for measurement of the magnetic field vector in prominences. New techniques for inversions of full Stokes spectro-polarimetry, incorporating both the Zeeman and Hanle mechanisms for generation and modification of polarization, are now at the forefront. Also reviewed are measurements of the magnetic fields in the photosphere below prominences, and how they may be used to infer the field geometry in and surrounding the prominence itself.

**Keywords.** magnetic fields, polarization, scattering, data analysis, polarimetric, Sun: prominences, Sun: filaments

## 1. Introduction

The major, but still incompletely understood aspects of the nature of prominences – their origin, topology, stability, and relationship to dynamic events in the corona – are intimately connected to the magnetic field. The ultimate observational objective would be to obtain precision measures of the magnetic field *vector* in the three-dimensional volume from the photosphere below the prominence, within the prominence itself, and to the larger coronal volume above and surrounding it. Were we able to attain this goal, our understanding of this phenomenon would be far more complete. Alas, this ideal observational picture of the field is beyond our current capability. However, within the last few years significant progress has been realized toward this ultimate goal – remote sensing of the magnetic field vector in and surrounding prominences – through new observations and development of advanced data analysis methods. Furthermore, the prospect of a dramatic advance of observational capability for precision polarimetry, using ground- and space-based facilities either under development or in the planning stages, promises to bring us much closer to this ultimate goal.

It is possible to infer some properties of the magnetic field vector using indirect methods, examples of which are intensity tracers for field alignment (see Lin *et al.* 2005), line profile diagnostics at various wavelengths including the ultraviolet (Schmieder *et al.* 2007), and dynamical response to external perturbations (i.e. the "winking filaments" of Hyder 1966). Indirect techniques provide important information regarding the magnetic field; information that is often needed to properly interpret direct methods (e.g. in resolution of the 180° azimuth ambiguity), but due to limitations of space and scope, this review concentrates on "direct" inference of the magnetic field vector, primarily through polarimetric means.

Despite the importance of the magnetic field, over the years there have been relatively few attempts to measure it in prominences, likely owing both to the difficulty of the measurements and the interpretation of the data. One source of the observational challenge is the paucity of suitable spectral diagnostics. There are only a handful of spectral features in the solar spectrum that present good polarimetric diagnostics of the magnetic field at (chromospheric) temperatures and densities typical of the conditions in prominences. Precision polarimetry is needed to make reliable field measurements, so one requirement of a spectral diagnostic is that there be significant optical depth in the body of the prominence (or filament if seen against the disk) in order to detect a sufficient photon flux. In practice this requirement has limited the diagnostics to spectra of the abundant elements H and He. The atomic processes governing the generation of polarized radiation in the spectra of H and He, especially at the lower chromospheric densities typical of prominences where scattering dominates the radiative transport, only recently have been fully understood. With the advent of new observational facilities and diagnostic tools grounded upon rather complete treatments of the atomic physics, a new era in measurement of the magnetic field in prominences is now upon us.

Section 2 of this review presents a brief summary of the history of inference of magnetic fields in prominences and filaments. Section 3 outlines results of measurement of the photospheric fields below prominences. Section 4 is a discussion of the possibilities for future measurement of prominence magnetic fields.

## 2. A Perspective on Measurement of Prominence Magnetic Fields

To date there have been few reviews of magnetic field measurements in prominences. The reader is referred to reviews by Paletou & Aulanier (2003), López Ariste & Aulanier (2007), Paletou (2008), and Mackay *et al.* (2010). Table 1 presents a selection of milestones for prominence field measurements to date. Early attempts at measurement of the prominence field used standard longitudinal magnetometry: considering the circular polarization signal only. These studies assumed that the polarization signal results from the Zeeman effect in H$\alpha$, H$\beta$, and/or He I $D_3$ (Zirin & Severny 1961, Rust 1967, Harvey 1969, Tandberg-Hannsen 1970). These observations did not resolve the Stokes $V$ line spectrally, so that the authors could not know that the profiles often had anomalous shapes (López Ariste *et al.* 2005); that is, rather than being antisymmetric about line center, the Stokes $V$ profiles can exhibit a symmetric shape arising a result of a scattering process. As pointed out much later by Brown, López Ariste, & Casini (2003), in spite of the interpretative mechanism being in error these early magnetographic inferences yielded plausible field strengths because both the Zeeman effect and scattering in the presence of an anisotropic radiation field induce approximately the same Stokes $V$ polarization levels.

### 2.1. *Early Application of the Hanle Effect to Prominence Measurements*

Hyder (1965) reported measurements of linear polarization in the H$\alpha$ line for 16 prominences. In that work he was able to measure the orientation of the linear polarization, but not its magnitude, and found that the orientation was not strictly tangential to the solar limb as would be expected from scattering in the non-magnetic case. He then interpreted the observed orientation of the linear polarization as arising from scattering in the presence of a magnetic field (the Hanle effect). In the Hanle effect, both the degree of polarization and and its orientation depend on the strength of the field. Without precision measurements of the degree of polarization, accompanied by a sophisticated analysis, no definitive conclusions could be drawn regarding the magnetic field strength and its orientation.

Much later, Leroy and co-workers (Leroy 1977; Leroy 1978; Bommier, Sahal-Brechot, & Leroy 1981; Leroy, Bommier, & Sahal-Brechot 1983, 1984; Bommier *et al.* 1994; Bommier

**Table 1.** Milestones in Measurement of Prominence Magnetic Fields.

| Reference | Method | Field Attributes |
|---|---|---|
| Zirin 1961 | Zeeman magnetometry Hα | [*few* ×100 G, < 2 G] for [active,quiet] |
| Hyder 1964, 1965 | scattering rotation of linear pol. (Hanle effect) Hα | *few*×10 G |
| Rust 1967 | Zeeman magnetometry Hα | 5-60 G |
| Harvey 1969 | Zeeman magnetometry Hα, He I $D_3$ | 0-15 G |
| Tandberg-Hanssen 1970 | Hα, He I $D_3$, others, Zeeman magnetometry | *few* G, aligned to prom. axis |
| Leroy 1977b, 1978, 1983<br>Leroy *et al.* 1984 | Hanle effect $I,Q,U$ He I $D_3$<br>Hanle magnetometry He I $D_3$(+Hα) | 0-10 G, horiz. field<br>small angle to prom. axis |
| House & Smart 1982<br>Landi degl'Innocenti 1982<br>Athay *et al.* 1983 | Hanle spectro polarimetry He I $D_3$<br>(allows meas. of all 3 field comp.)<br>inversion code for complete $I,Q,U,V$ | horiz. field<br>tendency for inverse config.? |
| Querfeld *et al.* 1985 | spectro-polarimetry He I $D_3$<br>Gaussian fits to profiles | Stokes $V$ is necessary |
| Bommier *et al.* 1994, 1998 | Hanle magnetometry He I $D_3$ (+Hα) | < 10 G, horizontal, inverse config. |
| H. Lin *et al.* 1998 | spectro-polarimetry He I 10830 Å<br>Stokes $U$ analysis only | filament on disk |
| López Ariste & Casini 2002, 2003, Casini *et al.* 2003<br>Brown *et al.* 2003<br>Schmieder *et al.* 2013 | spectro-polarimetry He I $D_3$<br>quantum interferences in incomplete<br>Paschen-Back regime, PCA inversion | spectrally resolved $I,Q,U,V$<br>$\|\mathbf{B}\| \sim$ 10-20 G<br>but some $\|\mathbf{B}\| >$ 50 G |
| Trujilo Bueno *et al.* 2002 | spectro-polarimetry He I 10830 Å<br>lower level atomic polarization | *a few* G, highly inclined |
| López Ariste *et al.* 2005 | $I,Q,U,V$ spectro-polarimetry Hα | scattering polariz.<br>Stokes $V$ non-Zeeman, electric fields? |
| Merenda *et al.* 2006 | spectro-polarimetry He I 10830 Å<br>complete treatment Hanle + Zeeman | vertical fields in polar<br>crown prominence |
| Kuckein *et al.* 2012 | spectro-polarimetry He I, Si I 1083nm<br>Zeeman effect only | active region filament<br>3-D structure → flux rope |
| Orozco Suárez *et al.* 2013 | spectro-polarimetry He I 10830 Å<br>complete Hanle-Zeeman analysis | $\|\mathbf{B}\|$ 5-30 G, incl. 65 − 75° |

& Leroy 1998) carried out analyses of quantitative measures of the linear polarization observed primarily in the He I $D_3$ line in quiescent prominences (Leroy 1977). Their analysis is based on computations of the modification of scattering polarization due to the Hanle effect (Sahal-Brechot, Bommier, & Leroy 1977, Bommier & Sahal-Brechot 1978, Bommier 1980). The observations did not resolve the spectral lines, so that the basic data consisted of two pieces of information: the degree of polarization and its angular orientation in the plane of the sky. Three independent measures are needed to fully specify the magnetic field vector, the ambiguity associated with the polarization orientation notwithstanding. Adopting the assumption that the prominence field is largely horizontal (Leroy 1978) allowed Sahal-Brechot, Bommier, & Leroy (1977), Leroy (1977) and Leroy (1978) to infer the field strength and the angle of the field in the horizontal plane. The analysis proceeds from a forward synthesis of the line polarization, resulting in a *Hanle effect polarization diagram* (for example, Fig. 5 of Sahal-Brechot, Bommier, & Leroy 1977) in which separate sets of contours of constant field strength and constant inclination to the line-of-sight are plotted against the angle of linear polarization in the plane of the sky (on the abscissa) and polarization degree (on the ordinate). Each of the observed data may then directly indicate the field strength and inclination.

Using simulations of lines, Bommier, Sahal-Brechot, & Leroy (1981) reported that the most effective method to augment the Hanle effect observations to allow a complete determination of the magnetic field vector was to obtain simultaneous measurements in two or more different lines having differing sensitivities to the Hanle effect. Following this reasoning, Leroy, Bommier, & Sahal-Brechot (1984) then interpreted measurements of many prominences, a significant fraction of which included simultaneous measurements in He I $D_3$ plus either H$\alpha$ or H$\beta$. Although the hydrogen line data were not included explicitly in their analysis, they independently confirmed that fields are close to horizontal in quiescent prominences. More detailed analysis of simultaneous measurements in He I $D_3$ and H$\beta$ (Bommier, Sahal-Brechot, & Leroy 1986), then in He I $D_3$ plus H$\alpha$ (Bommier *et al.* 1994), indicate that most quiescent prominences have fields that are nearly horizontal, and have "inverse configuration" with respect to the photospheric field below (that is, the sense of the field component perpendicular to the prominence axis is negative toward positive polarity, as opposed to the potential field case).

In the foregoing two-line analyses the authors confronted the complications of generation of scattering polarization in the hydrogen Balmer lines. In addition to treatment of the scattering process in the optically-thick non-LTE transfer of the H$\alpha$ line, it was necessary to treat the detailed quantum electrodynamics formulation of scattering polarization and collisional interaction via the density matrix formulation. The formulation of the problem and its forward solution for the prominence case are outlined in Landi Degl'Innocenti, Bommier, & Sahal-Brechot (1987). They present Hanle effect polarization diagrams for H$\alpha$. The unavoidable ambiguity of the magnetic field azimuth arising from any polarization diagnostic takes a different form when the line becomes optically thick. For the Zeeman effect and optically-thin Hanle effect diagnostics the ambiguity of the inferred field vector is symmetric with respect to the line-of-sight, but in the optically-thick scattering case the two-fold ambiguous field vectors no longer display that symmetry. It is noted that this broken symmetry provides an observational basis for resolution of this fundamental ambiguity (Bommier *et al.* 1994).

### 2.2. *The Modern Era of Magnetic Field Measurement in Prominences*

The pioneering studies summarized in Sect. 2.1 demonstrated the importance of the Hanle effect in measurement of prominence magnetic fields. Limitations of these early observations and their accompanying diagnostic techniques have been occulted in recent years as a result of the following advances.

*Spectral Resolution and Sampling:* House & Smartt (1982) reported the first systematic full-Stokes spectral profile measurements of He I $D_3$ multiplet in prominences. This spectral line contains several spectral components differing in sensitivity to the Hanle effect. When this added information is incorporated into the analysis it becomes possible to extract the full magnetic field vector. The earlier polarimetric measurements determined Stokes $Q/I$ and $U/I$ only, but as noted above and as pointed out by Landi Degl'Innocenti (1982), the Stokes $V$ profile also provides information regarding the strength and orientation of the magnetic field. For stronger fields, the Stokes $V$ profile will show the spectrally anti-symmetric signature arising from the Zeeman effect (splitting in wavelength of the Zeeman $M$-sublevels). The $M$-sublevels may also differ from their natural populations (atomic "orientation" and/or "alignment") due to radiative effects, particularly from level crossings arising in the incomplete Paschen-Back effect (see Fig. 4 of Sahal-Brechot, Bommier, & Leroy 1977) and another "more subtle effect" mentioned by Landi Degl'Innocenti (1982).

*Magnetic Field Measurements in the Infrared He* I *10830Å line:* Advancements in infrared detectors in recent times have permitted polarimetry of prominences in the He I infrared line at 10830Å. Like its counterpart He I $D_3$, the 10830Å line is in fact a multiplet whose spectral components also differ in sensitivity to the Hanle effect.

Several instruments permit polarimetric measurements in He I 10830 Å, most notably the Tenerife Infrared Polarimeter (TIP, Collados *et al.* 2007). TIP observations of prominence magnetic fields have been reported by Trujillo Bueno *et al.* (2002), Merenda *et al.* (2006), Orozco Suárez, Asensio Ramos, & Trujillo Bueno (2013).

*More Complete Quantum Mechanical Treatment of Polarization:* Subtle effects, among them atomic orientation, give rise to observable signatures in all the polarization profiles, thereby providing unique new information regarding the magnetic field that would be unavailable from measurements of Stokes $Q$ and $U$ alone. The full treatment of the incomplete Paschen-Back effect and its influence on the populations of the Zeeman $M$-subtlevels is a hallmark of many of the more recent studies (López Ariste & Casini 2002; López Ariste & Casini 2003; Casini *et al.* 2003; Merenda *et al.* 2006; Orozco Suárez, Asensio Ramos, & Trujillo Bueno 2013, Schmieder *et al.* 2013). Another effect is the possibility atomic polarization of the lower level of a transition influencing the observed polarization through selective absorption (Trujillo Bueno *et al.* 2002). The upper level of the weak blue component of He I 10830Å line has total angular momentum quantum number $J = 0$, thus this line is intrinsically unpolarizable, so scattering polarization in this line must arise from *lower level polarization.* This polarization is observed in absorption in filaments seen against the solar disk, so the imbalance of the lower-level $M$-sublevels leads to this selective absorption of one polarization direction. Note that the lower term of He I $D_3$ is the upper term of He I 10830 Å. Being optically thin, the He I $D_3$ line will not show any polarization arising from a selective absorption process, but an imbalance in the lower level $M$-sublevels may, through statistical equilibrium, influence the upper levels. Through fitting observed profiles, Trujillo Bueno *et al.* (2002) provide an example of prominence polarization measurements in 10830Å where the selective emission is influenced by lower level polarization. It is likely, then, that He I $D_3$ emission polarization will be influenced by polarization of its lower levels. It must be stressed that the complex physics involved in the scattering process leading to the Hanle effect needs to be treated in its full quantum mechanical generality, otherwise significant errors in the inference of magnetic fields will be encountered (Casini 2002).

*Innovative Inversion Procedures:* Quite a few assumptions needed to be invoked in order to produce Hanle effect diagrams for the optically-thick non-LTE, multi-dimensional H$\alpha$ line (Bommier *et al.* 1994). Furthermore, such diagrams fail to incorporate the polarization generated as a result of the Zeeman effect, even though it has been demonstrated that many prominences show Stokes $V$ profiles that have the Zeeman-like anti-symmetry (Trujillo Bueno *et al.* 2002). Recent years have seen the development of inversion codes that incorporate most of the complex quantum electrodynamics effects shown to be important to the scattering of polarized light in prominences. Among these is the HAZEL code (Asensio Ramos, Trujillo Bueno, & Landi Degl'Innocenti 2008) that embraces both the Hanle and Zeeman effects for various geometries of arbitrary optical depth. The latter is especially important for the formation of the red blend of the He I 10830 Å line. The HAZEL code uses the standard Levenberg-Marquardt least squares minimization augmented with an algorithm to invoke a search of the entire parameter space in order to select the global minimum. The main drawback of this standard inversion scheme is that the procedure must do at least one forward computation of the radiative transfer for each iteration of the least-squares procedure, so the computations become very time-consuming. Another approach is the application of pattern recognition techniques to identify the best fit to observations of Stokes profiles derived from a physical model. López Ariste, Casini, and co-workers have successfully applied principal components analysis (PCA, see for example Rees *et al.* 2000) to invert prominence observations (López

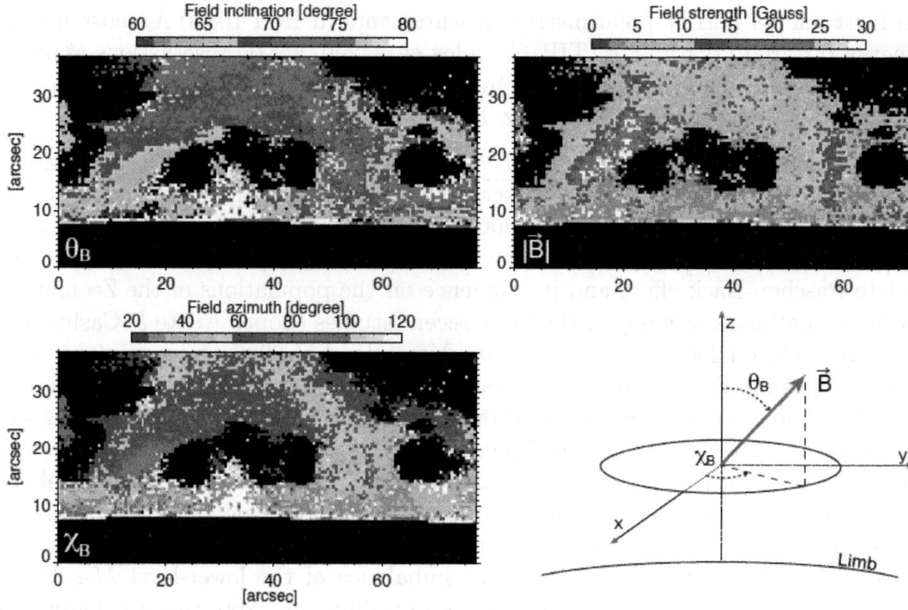


**Figure 1.** A two-dimensional map of the magnetic field vector within a prominence as inferred using the HAZEL procedure (see Sect. 2.2) is shown. This figure is adapted from Figure 3 of Orozco Suárez, Asensio Ramos, & Trujillo Bueno (2013). Shown as color images are the field strength and two angles defining the field orientation. Note that the field inclination to the local vertical $\theta_B$ is consistently large indicating fields close to horizontal. See onine version of this article for color display.

Ariste & Casini 2002; Casini *et al.* 2003; Brown, López Ariste, & Casini 2003, Casini *et al.* 2009; Casini *et al.* 2013, Schmieder *et al.* 2013). A considerable initial investment in computation is needed at the outset to develop a suitably chosen database of synthetic Stokes profiles spanning physically realistic ranges of parameters describing the physical model, but once this "training data set" is in hand, the method allows a very rapid search for the model best fitting any observed set of Stokes profiles. Furthermore, PCA inversions *always* select the global best fit.

*Measurements that Map the Solar Scene:* Information on the spatial variation of the vector magnetic field within a prominence is of paramount importance to understanding the prominence phenomenon. Unlike vector field maps of the photosphere, authors often reported inference of a few scattered points within a prominence, or even more commonly measurements at only one spatial location. In order to acquire a S/N adequate for Hanle effect analysis, it was necessary to carry out observations with long integration times, thereby preventing measurements at high spatial and/or temporal resolution. Also, not until the modern era were spectro-polarimeters capable of two-dimensional maps. Casini *et al.* (2003) were among the first to report such a map. Figure 1 shows results from a recent application of the HAZEL code to prominence polarimetry in He I 10830 Å. The authors of that paper show that the strength of the field and its orientation do not vary rapidly from point-to-point, but there are significant variations of larger scale across the prominence. They also demonstrate that, like the earlier observational studies using the Hanle effect, the prominence fields are nearly horizontal. Recently, a similar result has been reached by Schmieder *et al.* (2013) from maps of the magnetic field of a prominence observed in He I $D_3$. Their analysis also reveals that the prominence fields are nearly horizontal with orientation varying little from parallel to the plane-of-the-sky.

*Magnetic Field Measurements of Filaments Seen Against the Disk:* The 10830Å line has substantial optical depth in prominences, and for this reason it has been useful as a Hanle effect diagnostic of filaments seen against the solar disk. For quiescent filaments the Hanle effect provides an excellent diagnostic in He I 10830 Å because it affects the degree of line polarization for field strengths of a few Gauss. For stronger field strengths (in the saturation regime of the Hanle effect – between 10 - 100 Gauss for He I 10830 Å) the direction of linear polarization is an indicator of the orientation of the magnetic field in the plane of the sky, as is also the case for forbidden coronal emission lines. Of course, as the field strength increases, the Zeeman effect begins to produce measurable signatures in Stokes $V$. An early application of the Hanle effect to filaments on the disk was presented by Lin, Penn, & Kuhn (1998), but it was demonstrated subsequently that his approximate classical approach failed to account for important quantum effects (see the last paragraph of Casini *et al.* 2002). To date it appears that there have been no applications of the Hanle effect to infer the magnetic field in quiescent filaments, however recent studies of the magnetic structure of active region filaments, measured simultaneously in the chromosphere using He I 10830 Å and the photosphere using the nearby Si I line at 10827Å have demonstrated the presence of rather strong fields ($\sim$ 600 Gauss) at both heights with a vertical shear of the field direction (Kuckein *et al.* 2009; Kuckein, Martínez Pillet, & Centeno 2012). Those authors concluded that the field strengths within the filament were strong enough that the polarization may be described by the Zeeman effect alone after obtaining similar results for the field using analysis that incorporates both the Zeeman effect and scattering polarization. Trujillo Bueno & Asensio Ramos (2007) find that polarization from the scattering process in He I 10830 Å in low-lying prominences and filaments may be significant for field strengths up to 1000 Gauss, however they reason that the anisotropy of the radiation field within the filament could be reduced considerably due to the optical thickness of the He I 10830 Å line, thereby drastically reducing the scattering polarization.

## 3. The Photospheric Magnetic Field Under Prominences

Observations of the photospheric magnetic field shed some light on evolution of prominences and their associated large-scale magnetic structures. Studies of the photospheric vector magnetic field under active region filaments (Lites 2005, Okamoto *et al.* 2009, Lites *et al.* 2010) reveal features of the photospheric vector magnetic field that evolve in a way that is consistent with the emergence of a flux rope into the atmosphere, rather than formation within the atmosphere itself. Figure 2, from Lites *et al.* (2010), documents one well-observed active region magnetic field below a filament. The orientation of the spatial fine structure in the transverse apparent flux density (left panel) coincides with the orientation of the inferred magnetic field vector (arrows in right panel), revealing the inverse configuration of the field at the photosphere. Of particular interest is the upper part of the filament channel that not only shows a pronounced inverse magnetic configuration, but also this segment of the filament is not bordered on either side by strong photospheric plage. This filament presents a clear imprint upon the photospheric magnetic field, but its magnetic buoyancy does not appear to be constrained from above by magnetic loops of the immediately surrounding fields. The mass of the filament itself may be the dominant counterbalance to its magnetic buoyancy – a circumstance that would be common in the buoyant rise of a massive flux rope from below the photosphere.

There are very few studies of the vector magnetic field in the photosphere below quiescent filaments because they have weak magnetic fields and often reside relatively high in

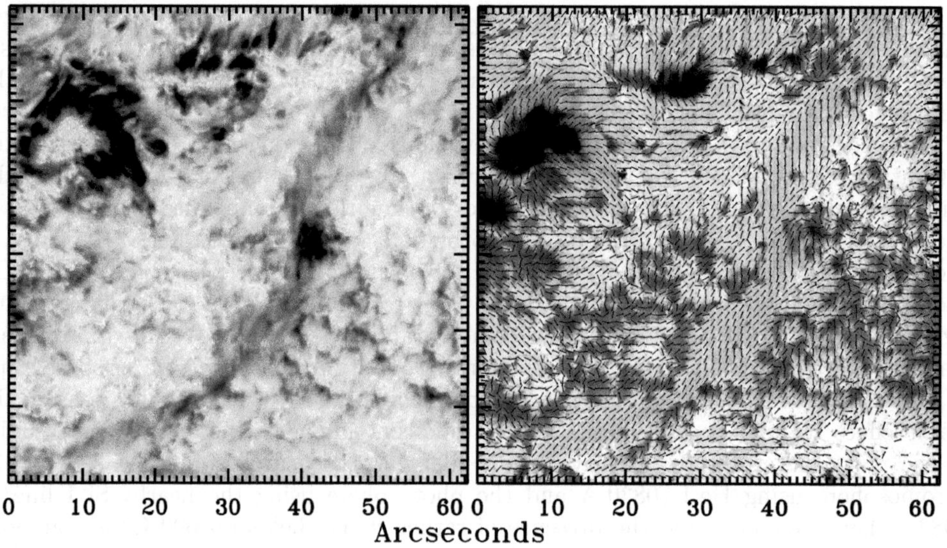


**Figure 2.** A filament channel whose evolution was documented by *Hinode*/SP observations reveals aspects suggesting that a flux rope emerged from below the photosphere. Left panel: the transverse apparent flux density is shown as a gray scale image with darker shades corresponding to higher values. The filament channel running from lower left to upper right is clearly distinguished by higher values of horizontal magnetic field strength. Right panel: intrinsic field strength is displayed as a gray scale with larger values corresponding to darker shades. The orientation of the horizontal component of the field is shown by arrows. The filament channel has rather uniform field strengths of $\sim$ 500 Gauss. Many areas of the channel have inverse polarity suggesting a flux rope configuration. See Lites *et al.* (2010) for further details.

the corona. Furthermore, fields in the quiet photosphere that define the magnetic structure in the corona originate in the stronger photospheric flux tubes which, by nature, are strongly buoyant and therefore nearly vertical there. Hence, little can be learned about the topology of the field at much greater heights through measurement of the field vector at the photosphere. Nonetheless, López Ariste *et al.* (2006) did encounter a situation where an inverse magnetic configuration (a "bald patch") was present under a filament "barbs". Filament barbs are known to be extensions of cool prominence material downward toward the photosphere from the main body of the prominence. It is likely that the filament magnetic field has a flux rope topology that dips low enough at the barbs to impose its presence at the photospheric level.

## 4. Prospects for the Future

Our field is poised for rapid advancement of our understanding of prominence magnetic fields. Parallel developments in comprehensive understanding of the theory polarization via the scattering process in stellar atmospheres and advanced inversion methods based on pattern recognition techniques now provide us with tools for inference of magnetic fields in prominences, both above the limb and seen as filaments on the solar disk.

Spectro-polarimeters are now capable of routinely producing maps of active regions with full spectral and spatial coverage and techniques for observing in the near IR region open the possibility of detailed observations in He I 10830 Å. These developments notwithstanding, new observational facilities are needed totake full advantage of advanced analysis techniques. At present, long integration times are required to achieve the S/N needed for quantitative analysis of the weak polarization signals from prominences. In practice, observations with current facilities require both spatial averaging and long integration times, with the result that the fine structure of prominences and filaments is essentially not observable when quantitative information on the polarization is required. This drawback may be addressed, of course, with larger aperture telescopes and more efficient polarimeters. We look to the future to the ATST and EST large solar telescopes to open the way to comprehensive, high-resolution observations from the ground. Another potentially dramatic advancement in observational capability would be the Japan/US/Europe *Solar-C* space mission that will be optimized for polarimetry of spectral features forming in the chromosphere. In the near future, we look to more modest advances from the ground-based Prominence Magnetometer (ProMag) instrument (http://www.hao.ucar.edu/research/stsw/science/promag.php) that will do spectro-polarimetry simultaneously in He I $D_3$, He I 10830 Å, and H$\alpha$; and the Japan-led Chromospheric Lyman-Alpha SpectroPolarimeter (CLASP) rocket program (Kano *et /al.* 2012) that will observe scattering polarization in Ly$\alpha$.

Because filaments on the disk hold such great promise for understanding the detailed structure of prominence magnetic fields, observations of filament magnetic fields will become increasingly important. In order to interpret those disk observations it will be necessary for analysis tools to embrace the significant optical thickness of the spectral features that provide good magnetic field diagnostics (i.e., He I 10830 Å). Of course, CLASP observations will also require analyses that account for the significant optical depth in the Ly$\alpha$ line. The few studies of the evolution of the photospheric vector magnetic field under filaments suggest the emergence of a flux rope, but it is unclear if this is the usual situation for active region filaments. It is now possible to use the Solar Dynamics Observatory/HMI space instrument to obtain a continuous record of the vector magnetic field for *every* active region filament on the disk since early 2010. The results of such a study could be very revealing.

## Acknowledgements

I thank the SOC of IAUS300 for the invitation to present this review, and R. Centeno for helpful suggestions on the manuscript. Work presented herein was supported in part under NASA contract NNM07AA01C for the *Hinode* program at LMSAL and HAO. The National Center for Atmospheric Research is sponsored by the National Science Foundation.

## References

Asensio Ramos, A., Trujillo Bueno, J., & Landi Degl'Innocenti, E. 2008, *ApJ*, 683, 542

Bommier, V. 1980, *A&A*, 87, 109

Bommier, V. & Leroy, J. L. 1998, in: Webb, D. F., Schmieder, B., & Rust, D. M. (eds.) *IAU Colloq. 167: New Perspectives on Solar Prominences*, (Astronomical Society of the Pacific Conference Series) 150, 434

Bommier, V. & Sahal-Brechot, S. 1978, *A&A*, 69, 57

Bommier, V., Sahal-Brechot, S., & Leroy, J. L. 1981, *A&A*, 100, 231

Bommier, V., Sahal-Brechot, S., & Leroy, J. L. 1986, *A&A*, 156, 79

Bommier, V., Landi Degl'Innocenti, E., Leroy, J.-L., & Sahal-Brechot, S. 1994, *Solar Phys.*, 154, 231

Brown, A., López Ariste, A., & Casini, R. 2003, *Solar Phys.*, 215, 295

Casini, R. 2002, *ApJ*, 568, 1056

Casini, R., López Ariste, A., Tomczyk, S., & Lites, B. W. 2003, *ApJ*, 598, L67

Casini, R., López Ariste, A., Paletou, F., & Léger, L. 2009, *ApJ*, 703, 114

Casini, R., Asensio Ramos, A., Lites, B. W., & López Ariste, A. 2013, *ApJ*, 773, 180

Collados, M., Lagg, A., Díaz Garcí A. J. J., Hernández Suárez, E., López López, R., Páez Mañá, E., & Solanki, S. K. 2007, in: Heinzel, P., Dorotovič, I., & Rutten, R. J. (eds.) *The Physics of Chromospheric Plasmas*, (Astronomical Society of the Pacific Conference Series) 368, 611

Harvey, J. W. 1969, *Magnetic Fields Associated with Solar Active-Region Prominences* PhD thesis, (University of Colorado at Boulder)

House, L. L. & Smartt, R. N. 1982, *Solar Phys.*, 80, 53

Hyder, C. L. 1965, *ApJ*, 141, 1374

Hyder, C. L. 1966, *ZfA*, 63, 78

Kano, R. *et al.* 2012, in: *Society of Photo-Optical Instrumentation Engineers (SPIE) Conference Series*, (Society of Photo-Optical Instrumentation Engineers (SPIE) Conference Series) 8443

Kuckein, C., Martínez Pillet, V., & Centeno, R. 2012, *A&A*, 539, A131

Kuckein, C., Centeno, R., Martínez Pillet, V., Casini, R., Manso Sainz, R., & Shimizu, T. 2009, *A&A*, 501, 1113

Landi Degl'Innocenti, E. 1982, *Solar Phys.*, 79, 291

Landi Degl'Innocenti, E., Bommier, V., & Sahal-Brechot, S. 1987, *A&A*, 186, 335

Leroy, J. L. 1977, *A&A*, 60, 79

Leroy, J. L. 1978, *A&A*, 64, 247

Leroy, J. L., Bommier, V., & Sahal-Brechot, S. 198, *Solar Phys.*, 83, 135

Leroy, J. L., Bommier, V., & Sahal-Brechot, S. 1984, *A&A*, 131, 33

Leroy, J. L., Ratier, G., & Bommier, V. 1977, *A&A*, 54, 811

Lin, H., Penn, M. J., & Kuhn, J. R. 1998, *ApJ*, 493, 978

Lin, Y., Wiik, J. E., Engvold, O., Rouppe van der Voort, L., & Frank, Z. A. (2005) *Solar Phys.*, 227, 283

Lites, B. W. 2005, *ApJ*, 622, 1275

Lites, B. W., Kubo, M., Berger, T., Frank, Z., Shine, R., Tarbell, T., Title, A., Okamoto, T. J., & Otsuji, K. 2010, *ApJ* 718, 474

López Ariste, A., Casini, R., Paletou, F., Tomczyk, S., Lites, B. W., Semel, M., Landi Degl'Innocenti, E., Trujillo Bueno, J., & Balasubramaniam, K. S. 2005, *ApJ*. 621, L145

López Ariste, A., Aulanier, G., Schmieder, B., & Sainz Dalda, A. 2006, *A&A*, 456, 725

López Ariste, A. & Aulanier, G. 2007, in: Heinzel, P., Dorotovič, I., Rutten, R. J. (eds.) *The Physics of Chromospheric Plasmas*, (Astronomical Society of the Pacific Conference Series) 368, 291

López Ariste, A. & Casini, R. 2002, *ApJ*, 575, 529

López Ariste, A. & Casini, R. 2003, *ApJ*, 582, L51

Mackay, D. H., Karpen, J. T., Ballester, J. L., Schmieder, B., & Aulanier, G. 2010, *Space Sci. Revs*, 151, 333

Okamoto, T. J., Tsuneta, S., Lites, B. W., Kubo, M., Yokoyama, T., Berger, T. E., Ichimoto, K., Katsukawa, Y., Nagata, S., Shibata, K., Shimizu, T., Shine, R. A., Suematsu, Y., Tarbell, T. D., & Title, A. M. 2009, *ApJ*, 697, 913

Merenda, L., Trujillo Bueno, J., Landi Degl'Innocenti, E. , & Collados, M. 2006, *ApJ*, 642, 554

Orozco Suárez, D., Asensio Ramos, A., & Trujillo Bueno, J. 2013, in:*Highlights of Spanish Astrophysics VII*, 786

Paletou, F. 2008, in: Charbonnel, C., Combes, F., & Samadi, R. (eds.) *SF2A-2008*, 559

Paletou, F. & Aulanier, G. 2003, in: Trujillo-Bueno, J. & Sanchez Almeida, J. (eds.) *Solar Polarization*, (Astronomical Society of the Pacific Conference Series) 307, 458

Rees, D. E., López Ariste, A., Thatcher, J., & Semel, M. 2000, *A&A*, 355, 759

Rust, D. M. 1967, *ApJ*, 150, 313

Sahal-Brechot, S., Bommier, V., & Leroy, J. L. 1977, *A&A*, 59, 223

Schmieder, B., Gunár, S., Heinzel, P., & Anzer, U. 2007, *Solar Phys.*, 241, 53

Schmieder, B., Kucera, T. A., Knizhnik, K., Luna, M., Lopez-Ariste, A., & Toot, D. 2013, *ApJ*, 777, 108

Tandberg-Hanssen, E. 1970, *Solar Phys.*, 15, 359

Trujillo Bueno, J. & Asensio Ramos, A. 2007, *ApJ*, 655, 642

Trujillo Bueno, J., Landi Degl'Innocenti, E., Collados, M., Merenda, L., & Manso Sainz, R. 2002, *Nature*, 415, 403

Zirin, H. & Severny, A. 1961, *The Observatory*, 81, 155

*Nature of Prominences and their role in Space Weather*
*Proceedings IAU Symposium No. 300, 2013*
*B. Schmieder, J.-M. Malherbe & S. T. Wu, eds.*

doi:10.1017/S174392131301082X

# A first look into the magnetic field configuration of prominence threads using spectropolarimetric data

**D. Orozco Suárez[1,2], A. Asensio Ramos[1,2], and J. Trujillo Bueno[1,2,3]**

[1]Instituto de Astrofísica de Canarias, E-38205 La Laguna, Tenerife, Spain,
[2]Departamento de Astrofísica, Universidad de La Laguna, E-38206 La Laguna, Tenerife, Spain
[3]Consejo Superior de Investigaciones Científicas, Spain

**Abstract.** We show preliminary results of an ongoing investigation aimed at determining the configuration of the magnetic field vector in the threads of a quiescent hedgerow solar prominence using high-spatial resolution spectropolarimetric observations taken in the He I 1083.0 nm multiplet. The data consist of a two-dimensional map of a quiescent hedgerow prominence showing vertical threads. The observations were obtained with the Tenerife Infrared Polarimeter attached to the German Vacuum Tower Telescope at the Observatorio del Teide (Spain). The He I 1083.0 nm Stokes signals are interpreted with an inversion code, which takes into account the key physical processes that generate and/or modify circular and linear polarization signals in the He I 1083.0 nm triplet: the Zeeman effect, anisotropic radiation pumping, and the Hanle effect. We present initial results of the inversions, i.e, the strength and orientation of the magnetic field vector along the prominence and in prominence threads.

**Keywords.** Sun: chromosphere, Sun: prominences, Sun: magnetic fields

## 1. Introduction

Quiescent hedgerow prominences are sheets of plasma characterized by the presence of thin, vertically oriented threads that show them as highly dynamic structures (e.g., Berger *et al.* 2008, Chae *et al.* 2008, and Berger *et al.* 2010). However, the observed threads evolution is at odds with what current theoretical prominence models predict: magneto-static structures where, as for instance in the classical Kippenhahn & Schlüter (1957) and Kuperus & Raadu (1974) models, the magnetic pressure exerted by horizontal, bowed field lines acts as the plasma supporting mechanism (for reviews on prominence models see Mackay *et al.* 2010 and Labrosse *et al.* 2010). Several authors have provided plausible physical mechanisms for the transport of plasma through prominence sheets. For instance, the Rayleigh-Taylor instability explains how hot plasma (plumes) and magnetic flux can be transported upwards from prominence bubbles (Hillier *et al.* 2012a, 2012b) or how the condensation of hot coronal plasma into the prominence can produce falling plasma droplets (Haerendel & Berger 2011).

If the highly dynamic plasma is coupled to the prominence magnetic field, we somehow expect the magnetic field to show local variations, at scales comparable to or smaller than the typical sizes of prominence threads. However, prominence fine scale structures (such as threads) have been elusive to magnetic field measurements mainly because of the low spatial resolution achieved in full spectropolarimetric ground-based observations. Thus, we have information about the global magnetic structuring only (Tandberg-Hanssen & Anzer 1970, Leroy *et al.* 1983, Bommier *et al.* 1994, Casini *et al.* 2003, Athay *et al.* 1983, Leroy *et al.* 1983, Casini *et al.* 2003, 2005, Merenda *et al.* 2006, 2007, Orozco Suárez *et al.*

2013). Knowing the magnetic field configuration in prominence threads is important in order to test the different proposed physical scenarios for the threads dynamic behaviour. Here, we present preliminary results of an ongoing investigation aimed at determining the configuration of the magnetic field vector in the threads of a quiescent hedgerow prominence.

## 2. Observations and data analysis strategy

To infer the magnetic field configuration in prominence threads we used spectropolarimetric observations taken in the 1083 nm spectral range. This spectral range contains the He I 1083.0 nm triplet which is sensitive to the joint action of atomic level polarization (i.e., population imbalances and quantum coherences among the levels sublevels, generated by anisotropic radiation pumping) and the Hanle (modification of the atomic level polarization due to the presence of a magnetic field) and Zeeman effects. In this triplet, the Stokes Q and U signals detected in prominences are dominated by atomic level polarization and the Hanle effect while Stokes V is mostly dominated by the longitudinal Zeeman effect. This makes the He I 1083.0 nm triplet to be sensitive to a wide range of field strengths, from dG to kG. Moreover, we know the physics of the Hanle and Zeeman effects in this triplet, which is described in detail in Trujillo Bueno *et al.* (2002), Socas-Navarro *et al.* (2004), Trujillo Bueno & Asensio Ramos (2007), and in the book of quantum theory of polarization by Landi and Landolfi (2004). The He I 1083.0 nm triplet has already provided remarkable results regarding the magnetic field structure of quiescent and active region filaments and prominences (e.g., Lin *et al.* 1998; Trujillo Bueno *et al.* 2002; Merenda *et al.* 2006, Orozco Suarez *et al.* 2012, 2013, Kuckein *et al.* 2009, 2012a,b, Xu, Z. *et al.* 2012).

The data we present here correspond to a two-dimensional map of a quiescent hedgerow prominence. They were taken at the German Vacuum Tower Telescope at the Observatorio del Teide (Spain) with the Tenerife Infrared Polarimeter (Collados *et al.* 2007). The pixel sampling is 0.5 arcseconds and the spectral sampling 1.1 picometers. The exposure time per slit position was about 30 seconds that allowed us to achieve high signal-to-the-noise ratios in linear and circular polarization. It took 1.5 hours to scan the whole prominence during stable seeing conditions. Figure 1 (top left panel) shows the Stokes I peak intensity corresponding to the 2D map. The field of view is about $60'' \times 80''$. We estimate a spatial resolution of about $1''$-$1''\!.5$, which is rather at the limit to detect prominence fine scale structures. In the map, vertical threads, a prominence foot, and a cavity are clearly distinguishable. The integrated Stokes Q, U, and V absolute value signals are also represented in Fig. 1. The Stokes Q and U signals are prominent at the prominence foot (right part). In the prominence body we mostly detect Stokes Q signals. Interestingly, there is a remarkable amount of Stokes V signals in the upper part of the prominence. Note that, close to the solar limb, there is a considerable amount of Stokes signals corresponding to solar spicules.

For modelling and interpreting the He I 1083.0 nm triplet polarization signals we have applied an user-friendly diagnostic tool called "HAZEL" (from HAnle and ZEeman Light, Asensio Ramos *et al.* 2008). This code allows us to determine the strength, inclination and azimuth of the magnetic field vector in all pixels with measurable linear polarization signals. It takes into account the key physical processes that generate and/or modify circular and linear polarization signals in the He I 1083.0nm triplet: the Zeeman effect, anisotropic radiation pumping, and the Hanle effect. The radiative transfer is taken into account assuming a suitable slab model.

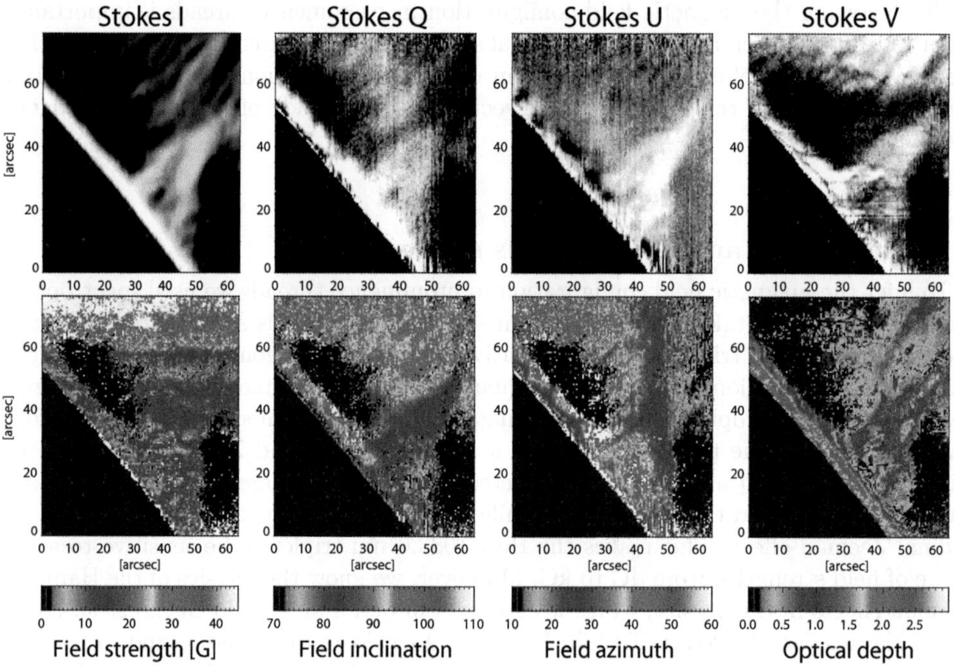


**Figure 1.** Top panels display the Stokes I peak intensity and the integrated Stokes Q, U, and V absolute value signals corresponding to the observed quiescent prominence. Bottom panels show the inferred field strength, inclination and azimuth values corresponding to the prominence as well as the optical depth. Black (purple) colour in the top (bottom) panels represents the solar disk as well as pixels with little signal (and therefore excluded from the analysis). Note that the inclination is measured with respect to the local vertical and the azimuth is measured counterclockwise with respect to the line of sight direction.

## 3. Preliminary results and future work

Figure 1 (bottom panels) shows the strength, inclination, and azimuth of the field vector inferred with the HAZEL inversion code. We display only the horizontal solution† since it is in line with most model predictions, i.e., with the fact that the prominence material should be sitting in dipped magnetic field lines. The field strength is about 10-15 Gauss and increases up to 40 Gauss in the upper part of the prominence. The field vector is about 90 degree inclined with respect to the local vertical in the prominence body. It becomes slightly more vertical at the foot. Finally, the field vector azimuth is about 30 degree. This result roughly agrees with previous measurements (e.g., Casini *et al.* 2003). In the same figure we show the optical depth of the He I 1083.0 nm triplet. We obtain optical depths of the order of 0.5 in the prominence body and of about 1.5 in the prominence foot.

In Figure 2 we show a zoom corresponding to a central part of the prominence. Note that the image has been rotated in order to show the threads vertically aligned. At least, two threads can be outlined in the peak intensity plot. However, at a glance we do not

† When interpreting the He I 1083.0 nm multiplet, there are several magnetic field vector configurations compatible with the observed polarization signals. These result from the 180° azimuth ambiguity and from the so-called 90° ambiguity of the Hanle effect. The latter provides two solutions which differ by about 90 degree in the field inclination, leaving us with a horizontal and vertical solutions. A detailed explanation about the existing ambiguities in the He I 1083.0 nm triplet can be found in Asensio Ramos *et al.* (2008).

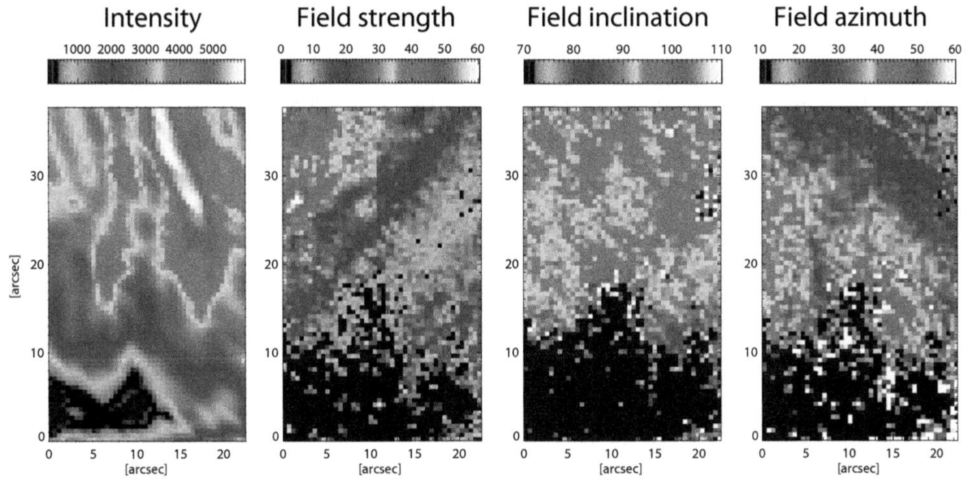


**Figure 2.** Intensity map showing a zoom to vertical prominence threads and the inferred magnetic field strength, inclination, and azimuth. Note that the map has been rotated to put the threads vertically. In the field strength, inclination, and azimuth maps the purple areas correspond to pixels with no emission in the He I 1083.0 nm triplet.

appreciate neither local variations in the field vector or correlations between the thread pattern in the intensity images and the magnetic field vector. Interestingly, there is a field strength gradient (the field strength decreases from left to right). This gradient in the field strength is uncorrelated with the peak intensity or with the inclination and azimuth maps. The inclination of the field is about constant. The azimuth also shows a slight variation from the left (30°–40°) to the right (15°–20°). In summary, we can conclude that the magnetic field along the prominence body where there are vertical threads is rather homogeneous. This suggests that the plasma concentrations that give rise to the threads are detached from the magnetic field. This would be in line with the magneto-convection scenario proposed by Berger *et al.* (2008). Since this is the first measurement of the field vector in quiescent hedgerow threads we wonder whether long integration times and/or lack of spatial resolution may be hampering the analysis. For instance, highly dynamic structures (such as threads) may washout the magnetic field vector information encoded in the measured Stokes parameters. This possibility seems not very plausible because we spatially resolve the prominence threads. These data are being analysed in very much details and we hope to publish the final results very soon.

As a concluding comment, we highlight that presently it is possible to determine the strength and orientation of the magnetic field in solar prominences by interpreting the Stokes I, Q, U, and V profiles of the He I 1083 nm triplet. In this case, we have analyzed data corresponding to a quiescent hedgerow prominence. The data allowed us to look at the magnetic configuration of prominence threads. We found that the orientation of the magnetic field vector is constant and does not show small-scale spatial variations at a resolution of 1″–1.″5. Certainly, full spectropolarimetric observations in the He I 1083.0 nm multiplet at very high spatial resolutions such as those achievable with GREGOR at present time or EST, ATST, and SOLAR-C in the near future, will help us to determine the local variations (if any) of the magnetic field vector in solar prominences.

## References

Asensio Ramos, A., Trujillo Bueno, J., & Landi Degl'Innocenti, E. 2008, *ApJ*, 683, 542

Athay, R. G., Querfeld, C. W., Smartt, R. N., Landi Degl'Innocenti, E., & Bommier, V. 1983, *Solar Phys.*, 89, 3

Berger, T. E., Shine, R. A., Slater, G. L., *et al.* 2008, *ApJl*, 676, L89

Berger, T. E., Slater, G., Hurlburt, N., *et al.* 2010, *ApJ*, 716, 1288

Bommier, V., Landi Degl'Innocenti, E., Leroy, J.-L., & Sahal-Brechot, S. 1994, *Solar Phys.*, 154, 231

Casini, R., Bevilacqua, R., & López Ariste, A. 2005, *ApJ*, 622, 1265

Casini, R., López Ariste, A., Tomczyk, S., & Lites, B. W. 2003, *ApJl*, 598, L67

Chae, J., Ahn, K., Lim, E.-K., Choe, G. S., & Sakurai, T. 2008, *ApJl*, 689, L73

Collados, M., Lagg, A., Díaz Garcí A, J. J., *et al.* 2007, The Physics of Chromospheric Plasmas, 368, 611

Kippenhahn, R. & Schlüter, A. 1957, *ZAp*, 43, 36

Kuckein, C., Centeno, R., Martínez Pillet, V., *et al.* 2009, *A&A*, 501, 1113

Kuckein, C., Martínez Pillet, V., & Centeno, R. 2012a, *A&A*, 542, A112

Kuckein, C., Martínez Pillet, V., & Centeno, R. 2012b, *A&A*, 539, A131

Kuperus, M. & Raadu, M. A. 1974, *A&A*, 31, 189

Haerendel, G. & Berger, T. 2011, *ApJ*, 731, 82

Hillier, A., Berger, T., Isobe, H., & Shibata, K. 2012, *ApJ*, 746, 120

Hillier, A., Hillier, R., & Tripathi, D. 2012, *ApJ*, 761, 106

Labrosse, N., Heinzel, P., Vial, J.-C., *et al.* 2010, *Space Sci. Revs*, 151, 243

Landi Degl'Innocenti, E. & Landolfi, M. 2004, Astrophysics and Space Science Library, 307,

Leroy, J. L., Bommier, V., & Sahal-Brechot, S. 1983, *Solar Phys.*, 83, 135

Lin, H., Penn, M. J., & Kuhn, J. R. 1998, *ApJ*, 493, 978

Mackay, D. H., Karpen, J. T., Ballester, J. L., Schmieder, B., & Aulanier, G. 2010, *Space Sci. Revs*, 151, 333

Merenda, L., Trujillo Bueno, J., Landi Degl'Innocenti, E., & Collados, M. 2006, *ApJ*, 642, 554

Merenda, L., Trujillo Bueno, J., & Collados, M. 2007, The Physics of Chromospheric Plasmas, 368, 347

Orozco Suárez, D., Asensio Ramos, A., & Trujillo Bueno, J. 2012, *ApJl*, 761, L25

Orozco Suárez, D., Asensio Ramos, A., & Trujillo Bueno, J. 2013, Highlights of Spanish Astrophysics VII, 786

Socas-Navarro, H., Trujillo Bueno, J., & Landi Degl'Innocenti, E. 2004, *ApJ*, 612, 1175

Trujillo Bueno, J. & Asensio Ramos, A. 2007, *ApJ*, 655, 642

Trujillo Bueno, J., Landi Degl'Innocenti, E., Collados, M., Merenda, L., & Manso Sainz, R. 2002, *Nature*, 415, 403

Xu, Z., Lagg, A., Solanki, S., & Liu, Y. 2012, *ApJ*, 749, 138

*Nature of prominences and their role in Space weather*
*Proceedings IAU Symposium No. 300, 2013*
*B. Schmieder, J.-M. Malherbe & S. T. Wu, eds.*

doi:10.1017/S1743921313010831

# Primary observations of solar filaments using the multi-channel imaging system of the New Vacuum Solar Telescope

**Zhi Xu, Zhen Y. Jin, Fang Y. Xu and Zhong Liu and the NVST team**

Fuxian Solar Observatory, Yunnan Observatory,
Chinese Academy of Science, Kunming, 650011
email: xuzhi@ynao.ac.cn

**Abstract.** The New Vacuum Solar Telescope (NVST) is a new generation ground-based solar facility of China. One of the post-focus instruments is the multi-channel high-resolution imaging system, which is designed to simultaneously observe the dynamic gas motion in the solar photosphere and chromosphere. Since October of 2010 it has been operational in the NVST and some necessary updates were performed in past 2 years. Here we first give a general introduction of this system, and then we exhibit one near-limb observation of solar filaments obtained using this system. By this communication, we would like to show the potential ability to perform the high resolution observation of solar filaments (prominences) using the multi-channel imaging system in the NVST.

**Keywords.** ground-based solar facilities, the NVST, solar prominences

## 1. Introduction

The New Vacuum Solar Telescope (NVST) belongs to a new generation of large and high-technology solar facilities of China. It is designed to observe the Sun in the range of 0.3 - 2.5 micron using high resolution imaging systems and spectrometers with a polarization analyzer. The NVST is a 1-m diameter telescope (a pure aperture of 980 millimeter) on an alt-azimuthal mount with an effective focal length of 45 meter (Liu and Xu, 2011). The NVST is located at the northeast bank of Fuxian Lake of Yunnan province. It is the most important facility of the Fuxian Solar Observatory (FSO). See the website *http : //fso.ynao.ac.cn* for more information about the FSO.

The first-light of the NVST came in October 2010. Two instruments are operational. One is the multi-channel high-resolution imaging system, the other one is the multi-wavelength spectrometer in visible lines (Wang *et al.*, 2013). These two systems are designed and arranged to be perpendicular to each other. Fig. 1 shows the present layout of the instruments in the focus plane. In this way, the imaging system can simultaneously serve as the slit-jaw recorder of the spectrometers.

The imaging system comprises 5 channels, the work wavelengthes and the main properties of each channel are summarized as follows:

• H$\alpha$ line, to monitor magnetic structures in the chromosphere. The central wavelength can be tunable in the range of 6562.8 $\pm$ 4 Å and the full bandpass width is 0.25 Å (e.g., giving a FWHM of an order of 11km/s). The time cadence of Level-1 data (post-processed based on a luck image selection algorithm ) is about 12 seconds.

• G band, to indicate small magnetic structures in the deep photosphere (Sütterlin *et al.*, 2001). The central wavelength is at 4300 Å and the full bandpass width is 10 Å.

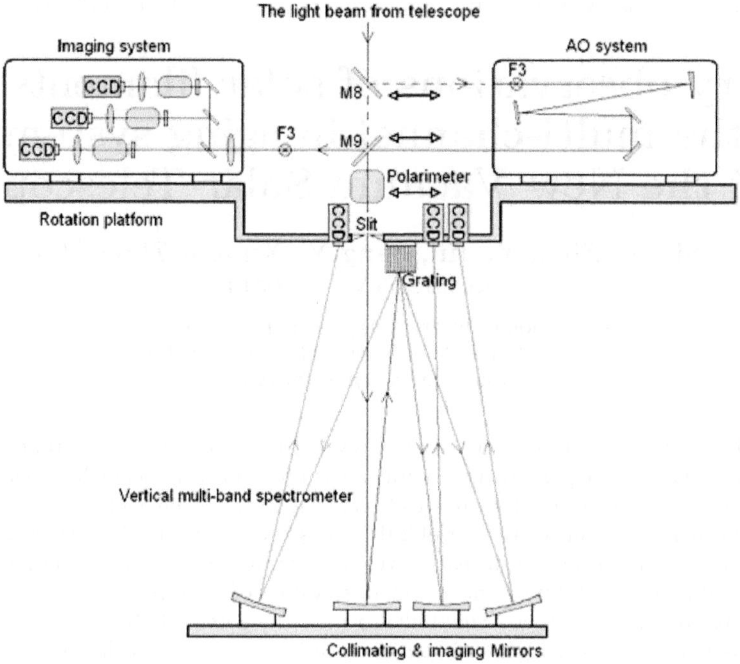


**Figure 1.** Layout of the instruments in the focal plane of the NVST. The main instruments include a multi-channel high-resolution imaging system, a vertical multi-wavelength spectrometer and a low-level AO system.

- TiO band, sensitive to the temperature, to allow easy detect bright points in granular lanes or umbral dots. It is centered at 7058 Å and the full bandpass width is 10 Å. The time cadence of Level-2 data (post-processed based on the speckle reconstruction) is about 40 seconds.
- Ca II 8542 Å, to monitor magnetic structures in the upper chromosphere. The central wavelength can be tunable in the range of 8542 $\pm$ 10 Å and the full bandpass width is 0.2 Å (an alternative width is 0.4 Å).
- He I 10830 Å, a near infrared proxy of coronal holes, to explore magnetic fields in the low corona. The parameters need to be determined.

Although up to now only three channels including the H$\alpha$, TiO and G-band are being used, the observations impressively demonstrate the high-resolution capability of the NVST (as shown in Fig.2). In particular, the post-processing with the speckle masking method produces the results with the resolution close to the diffraction limit. We use the speckle masking method of Weigelt (1977). It is adapted by Liu *et al.* (1998). To do so, we have to take a large number of short-exposure images ($t_{H\alpha} \leqslant 20$ ms, $t_{Tio} \sim 1$ ms). Using the large number of frames (200 frames) and a statistical algorithm, we are able to reconstruct one frame with the whole field of view ($\approx$ 3 arcmin). There are normally three processive levels of the data reduction. Level-0 represents the raw status. Level-1 is achieved based on a luck image selection algorithm, followed by the dark current and flat field modification. Level-2 is reconstructed by a speckle masking method. Quick look of the daily observation in the H$\alpha$ channel (Level-1) can be viewed from the website of $http://fso.ynao.ac.cn/dataarchive\_ql.aspx$.

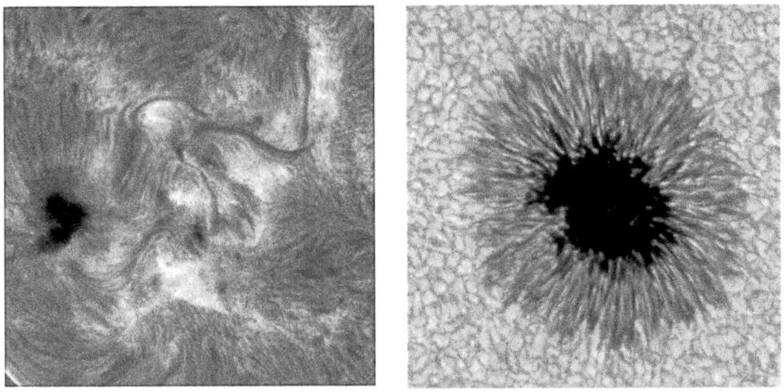

**Figure 2.** Observations (Level-1 data) in the H$\alpha$ line center (left) and TiO band (right) channels. The spatial resolution are about 0.165 and 0.04 arcsec/pixel, respectively.

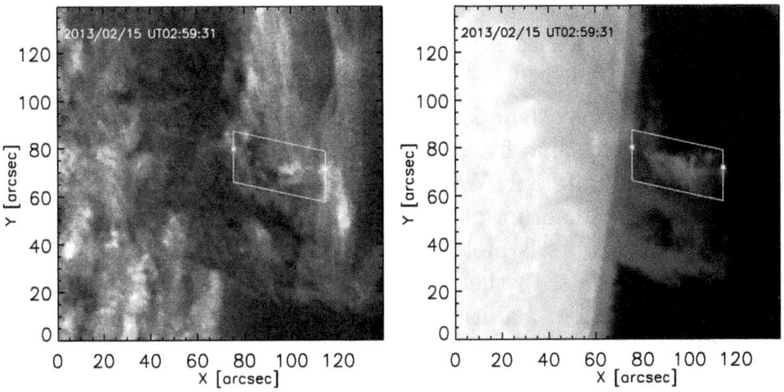


**Figure 3.** Simultaneous observations of a near-limb filament taken by the AIA/*SDO* at the 304 Å channel (left) and the NVST at the H$\alpha$ channel (right) on Feb. 15 2013. The spatial resolution of the NVST is reduced to be the same as the AIA image. A parallelogram delimitates a dynamic structure where successive outflows are observed.

## 2. Filaments observations in the H$\alpha$ channel

Using this imaging system, we carried out several time observations about solar filaments or prominences in both the H$\alpha$ and TiO channels. The H$\alpha$ Lyot filter was updated in 2013 April. The transmission rate is quite increased but the main properties, such as the tunable central wavelength and the bandpass width are consistent. As we know, in order to extract some physical information, e.g., the Doppler velocity, from such narrow-band H$\alpha$ filtergrams, one needs to obtain a quasi-profile by scanning through this line. Considering the requirement of the data post-processing, we needs to take at least 200 frames at one wavelength scan position. Therefore,we normally use 3 samplings across the profile to obtain the cadence below 1 minute. The default sampled wavelengths are at $-0.7$ Å, 0 Å and $+0.7$ Å off-band positions, which can be certainly changed by observers.

In this letter, we would like to present one high resolution observation of a near-limb filament obtained on Feb. 15 2013. It is observed in the H$\alpha$ line center without wavelength scan to achieve a temporal resolution of 12 seconds. We took observations for more than one and a half hours in the NVST. In Fig. 3, we exhibit one H$\alpha$ frame and compare with the simultaneous observation taken by the AIA/*SDO* (Lemen *et al.*, 2012) in the 304 Å channel (where emission is predominantly from He II). The 304 Å channel is used to monitor the chromosphere and transition region (see the review of Labrosse *et al.*, 2010). The resolution in the H$\alpha$ is about 0.165 arcsec/pixel, which is 3.6 times as high as that of the AIA 304 Å image. The main body and the barbs/legs of the filament are stable. By

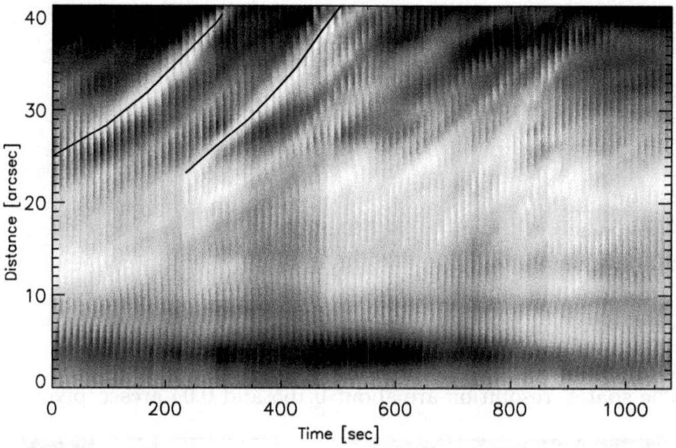


**Figure 4.** Time-space plot of a slice inside the parallelogram shown in Fig. 3. At each time, we take a slice of 1 arcsec wide and 40 arcsec long. Two black curves indicate polynomial fit results.

contrast, an adjacent structure delimitated by a parallelogram is very dynamic. It is seen that in the H$\alpha$ line center, some fragments are going outwards along the local vertical direction for more than a half hour. We investigate the proper motion of these fragments by showing the space-time diagram of a slice of 1 arcsec wide and 40 arcsec long inside this parallelogram. The results is plotted in Fig. 4. We deduce that the proper motion can reach up to 50 $-70$ km/s. In the AIA 304 Å movie, it is difficult to recognize this motion due to a low resolution. Additional observations in the 193 Å channel imply that this structure exhibits a rotating feature. Combining the NVST and SDO observations, we find that this outflow motion is approximately along a spiral axis of this rotating feature. More detailed analysis is beyond the content of this paper.

In conclusion, by the communication, we would like to demonstrate the potential capability of the multi-channel imaging system installed in the NVST to observe filaments or prominences with a high resolution. In addition, the imaging system of the Ca II 8542 Å line is now on trail. The remaining He I 10830 Å channel will successively follow in 2014. We expect fruitful observations in future.

**Acknowledgement**

We thank the NVST instrument team for their technique support. *SDO* is a project of NASA. Xu Z. gratefully acknowledges the anonymous referee and the support of K.C.Wong Education Foundation, Hong Kong and the National Natural Science Foundations of China (grants No. 11103075).

## References

Labrosse, N., Heinzel, P., Vial, J.-C. *et al.* 2010, Space Science Review, 151, 243

Lemen, J. R., Title, A., Akin, D. J. *et al.* 2012, *Sol. Phys.*, 275, 17

Liu, Z., Qiu, Y., & Lu, R. 1998, *SPIE*, 3561, 326

Liu, Z. & Xu, J. 2011, in First Asia-Pacific Solar Physics Meeting, ASI Conference Series, edited by Arnab Rai Choudhuri & Dipankar Banerjee, 2, 9

Weigelt, G. P. 1977, Optics Communications, 21, 55

Wang, R., Xu, Z., Jin, Z. Y. *et al.* 2013, Research in Astronomy and Astrophysics, 13, 1240

Sütterlin, P., Hammerschlag, R. H., Bettonvil, F. C. M. & Rutten, R. J. 2001, in 20th International Sacramento Peak Summer Workshop, *ASP Condference Series*, edited by Sigawrth, M., 236, 431

*Nature of Prominences and their role in Space Weather*
*Proceedings IAU Symposium No. 300, 2013*
*B. Schmieder, J.-M. Malherbe & S. T. Wu, eds.*

doi:10.1017/S1743921313010843

# Modeling Magnetic Flux Ropes

**Chun Xia and Rony Keppens**

Centre for mathematical Plasma Astrophysics, KU Leuven,
email: chun.xia@wis.kuleuven.be

**Abstract.** The magnetic configuration hosting prominences can be a large-scale helical magnetic flux rope. As a necessary step towards future prominence formation studies, we report on a stepwise approach to study flux rope formation. We start with summarizing our recent three-dimensional (3D) isothermal magnetohydrodynamic (MHD) simulation where a flux rope is formed, including gas pressure and gravity. This starts from a static corona with a linear force-free bipolar magnetic field, altered by lower boundary vortex flows around the main polarities and converging flows towards the polarity inversion. The latter flows induce magnetic reconnection and this forms successive new helical loops so that a complete flux rope grows and ascends. After stopping the driving flows, the system relaxes to a stable helical magnetic flux rope configuration embedded in an overlying arcade. Starting from this relaxed isothermal end-state, we next perform a thermodynamic MHD simulation with a chromospheric layer inserted at the bottom. As a result of a properly parametrized coronal heating, and due to radiative cooling and anisotropic thermal conduction, the system further relaxes to an equilibrium where the flux rope and the arcade develop a fully realistic thermal structure. This paves the way to future simulations for 3D prominence formation.

**Keywords.** Sun: prominences, filaments — MHD

---

Helical magnetic flux ropes are the most promising magnetic structures for solar prominences and their surrounding coronal cavities. More and more evidence for their existence is found in various observations (Gibson *et al.* 2010), with especially the latest observations using coronal magnetometry on cavities (Bak-Stęślicka *et al.* 2013) showing twist or shear of magnetic field extending up into the cavity and a pattern of concentric rings in line-of-sight velocity. Analytical models (Low & Zhang 2004) and numerical models (Fan 2010) of magnetic flux ropes have been proposed to explain observations on prominences and cavities. The physical processes that generate magnetic flux ropes in the corona are studied by many models based on magnetic flux cancellation in the photosphere. Many of these models focus on the magnetic structure alone and realistic temperature and density structures are missing.

To understand the formation process of a flux rope and to ultimately build a realistic model to explain prominences and cavities, we decouple the huge challenge of prominence formation modeling into two stages: (1) creating a magnetic flux rope from physically meaningful processes including gas pressure and gravity effects; (2) simulating the prominence plasma formation inside the pre-existing flux rope by including thermodynamics and chromospheric evaporation. This two-step approach is justified since the time scales for radiative processes that trigger condensations forming prominence plasma are much shorter than those of the large-scale evolution of the magnetic field.

## 1. First stage

We here summarize our recent results, discussed more extensively in Xia *et al.* 2013. In a 3D Cartesian box with extension of $-120 < x < 120$ Mm, $-90 < y < 90$ Mm, and $3 < z < 123$ Mm, an initial magnetic field adopts a linear force-free field extrapolation from an analytic bipolar magnetogram $B_z^0(x,y)$. In this first stage, we use an isothermal assumption with temperature $T_0 = 1$ MK throughout the domain representing the solar corona and solve the following isothermal MHD equations:

$$\frac{\partial \rho}{\partial t} + \nabla \cdot (\rho \mathbf{v}) = 0, \tag{1.1}$$

$$\frac{\partial (\rho \mathbf{v})}{\partial t} + \nabla \cdot \left( \rho \mathbf{v}\mathbf{v} + p_{\text{tot}}\mathbf{I} - \frac{\mathbf{B}\mathbf{B}}{\mu_0} \right) = \rho \mathbf{g} - \nabla \cdot \Pi, \tag{1.2}$$

$$\frac{\partial \mathbf{B}}{\partial t} + \nabla \cdot (\mathbf{v}\mathbf{B} - \mathbf{B}\mathbf{v}) = 0, \tag{1.3}$$

where $\rho$, $\mathbf{v}$, $\mathbf{B}$, and $\mathbf{I}$ are the plasma density, velocity, magnetic field, and unit tensor, respectively, while the total pressure is $p_{\text{tot}} \equiv p + \frac{B^2}{2\mu_0}$ and $\mathbf{g} = -g_\odot r_\odot^2/(r_\odot + z)^2\hat{\mathbf{z}}$ is the solar gravitational acceleration. A stress tensor $\Pi$ incorporates viscous effects. The code we use is the parallelized Adaptive Mesh Refinement Versatile Advection Code (MPI-AMRVAC, Keppens *et al.* 2012). A third-order accurate, shock-capturing scheme is used. The usage of a three-level AMR grid gives an effective resolution of $512 \times 384 \times 256$ and has a spatial resolving ability of 469 km. The initial density is derived from hydrostatic equilibrium under gravity.

To adjust the initial magnetic arcade to become more realistic, we impose a twisting velocity field, which is composed of two large-scale vortices rotating around the two main polarities, in the bottom plane. The normal velocity at the bottom is kept zero, and also on the other faces of the box, we keep zero velocity during this phase. The magnetic field on the boundaries is extrapolated by one-sided finite differencing. We use zeroth order extrapolation for the density on side boundaries, fixed density at the bottom, and adopt a gravitationally stratified density profile at the top. After the bottom driving and a relaxation, the lower smaller loops are further sheared while the overlying larger loops become less sheared and closer to potential field (see Fig. 1(a)(b)).

After this adjustment, we impose a converging velocity field toward the polarity inversion line (PIL) on the bottom boundary with the horizontal velocity formulated as

$$v_x^b = -f(t)v_1 \frac{\partial |B_z^0(x,y)|}{\partial x} \exp(-y^2/y_d^2),\ v_y^b = -f(t)v_1 \frac{\partial |B_z^0(x,y)|}{\partial y} \exp(-y^2/y_d^2), \tag{1.4}$$

where $y_d = 50$ Mm, $f(t)$ is a linear function to switch the driving on and off smoothly and the amplitude $v_1$ keeps the driving speed below 2% of the initial local Alfvén speed. We keep zero velocity on the four side boundaries and adopt limited open boundary conditions at the bottom and the top boundaries by extrapolations ensuring zero normal-gradient velocity and by setting an upper limit of 10% of the local Alfvén speed on the normal velocity. Similarly, the speed of downflows through the bottom is limited to be smaller than 1% of the local Alfvén speed. The density on the bottom boundary is extrapolated by gravitational stratification and kept no less than its initial value. Other boundary conditions are the same as the previous phase. The converging flows drive the system for 50 minutes before they cease. The system relaxes to a stable state over 60 minutes with zero horizontal velocity at the bottom.

The converging flows drive the footpoints of magnetic loops to approach the polarity inversion line (PIL), and the loops get sheared even further. In locations near the PIL, magnetic loops are brought closer to be parallel to the PIL. Flux elements of opposite polarities, where the loops are rooted in, are forced to collide at the PIL, and magnetic reconnection occurs there. As a result, pairs of arched magnetic loops, originally separate in $x$-extension, now become linked together in a head-tail style. The end result produces long helical flux tubes which have a concave magnetic dip at their middle portion and

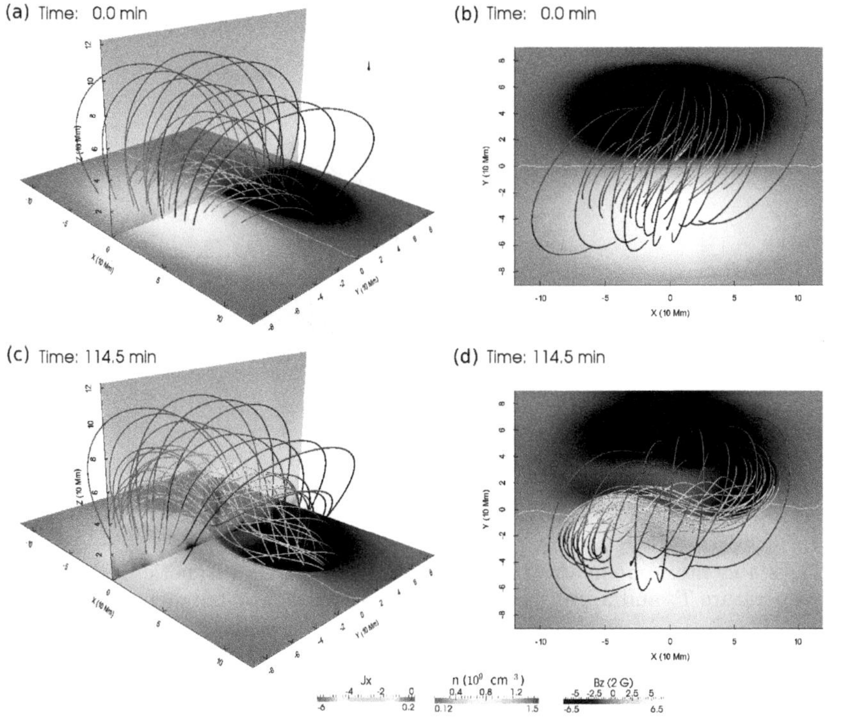


**Figure 1.** Adjusted sheared magnetic arcade (upper panels) and the stable magnetic flux rope (lower panels). The bottom magnetograms are shown in gray with the PIL plotted in white. Magnetic field lines are colored by $x$-component of the local current density $J_x$ in the rainbow color table. The vertical planes are colored by number density in blue-red colors. Side views and top views are shown in the left and right column respectively. Adopted from Xia *et al.* 2013.

hence ascend due to magnetic tension. The shorter reconnected loops which form below the reconnection sites, sink down through the bottom, causing magnetic flux cancellation. As the driving continues, new helical flux tubes form and wrap around prior-formed ones and together they assemble into a large scale helical flux rope. This is then found to rise, expand, stretch overlying loops, and relax to a stable state (see Fig. 1(c)(d)).

## 2. Second stage

To use an evaporation-condensation model to produce prominences (as in previous 2.5D work in Xia *et al.* 2012), a chromospheric layer must be included with temperature and density stratification, and thermodynamics must be considered. In this stage, we extend the isothermal MHD to thermodynamic MHD by adding the following energy equation:

$$\frac{\partial E}{\partial t} + \nabla \cdot \left( E\mathbf{v} + p_{\rm tot}\mathbf{v} - \frac{\mathbf{BB}}{\mu_0} \cdot \mathbf{v} \right) = \rho \mathbf{g} \cdot \mathbf{v} + \nabla \cdot (\boldsymbol{\kappa} \cdot \nabla T) - Q + H, \qquad (2.1)$$

where $p_{\rm tot} \equiv p + B^2/(2\mu_0)$ is the total pressure, composed of gas pressure $p$ and magnetic pressure $B^2/(2\mu_0)$; $E = p/(\gamma - 1) + \rho v^2/2 + B^2/(2\mu_0)$ is the total energy density and adiabatic index $\gamma = 5/3$. The last three terms stand for thermal conduction, radiative cooling, and coronal heating respectively.

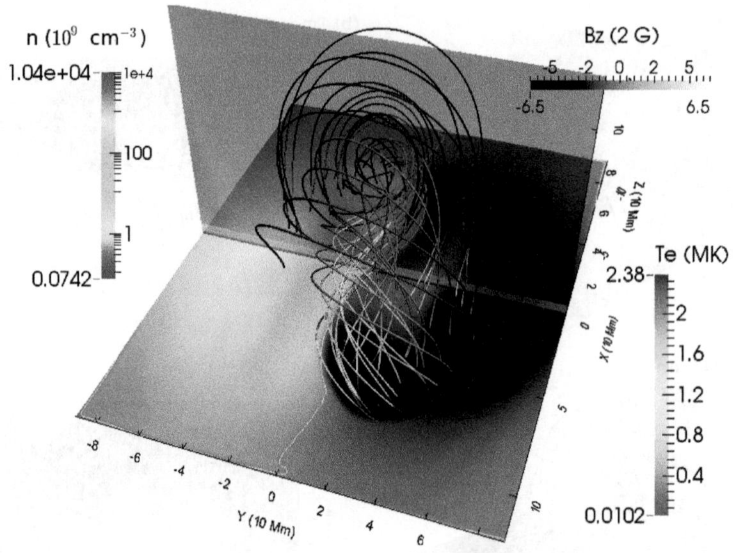


**Figure 2.** The stable flux rope in stratified solar atmosphere with magnetic field lines colored by temperature $T_e$ and a translucent vertical cut colored by number density $n$.

To start a full MHD simulation from the isothermal MHD data at the end of the first stage with a stable flux rope in the corona, we calculate gas pressure from the density distribution and constant 1 MK temperature and hence obtain the total energy $E$. In the bottom layer from 3 Mm to 7 Mm, the constant temperature is replaced by a hyperbolic tangent profile, where a region at 10000 K in the chromosphere (with thickness of 3 Mm) connects to the 1 MK in the corona. The number density in this layer is then determined by assuming a hydrostatic atmosphere with the bottom value of $10^{13}$ cm$^{-3}$. We restart the simulation from this modified end state of the first stage using MPI-AMRVAC, and relax the system towards an equilibrium with a time-dependent coronal heating. The latter contains a cosine function of time during the first 57 minutes, tuning the heating from $2 \times 10^{-4}$ to $6 \times 10^{-5}$ erg cm$^{-3}$ s$^{-1}$ and maintained constant since then. After a relaxation of about 100 min, this full thermodynamic MHD system reaches a quasi-equilibrium with residual velocity smaller than 10 km s$^{-1}$. Representative field lines colored by temperature and a central vertical slice colored by density are shown in Fig. 2. The dipped parts of the helical loops in the lower part of the flux rope are cooler, at around 0.8 MK. The top of the flux rope and the large arcade loops are hotter, about 2.4 MK.

Summarizing, we presented successive stages of simulations with isothermal MHD followed by thermodynamic MHD approaches, to finally get a realistic filament channel with a magnetic flux rope from chromosphere up to corona. From this state, a prominence formation inside the flux rope can be simulated in the future.

## References

Bak-Stęślicka, U. *et al.* 2013, *ApJ*, 770, L28
Fan, Y. 2010, *ApJ*, 719, 728
Gibson, S. E. *et al.* 2010, *ApJ*, 724, 1133
Keppens, R. *et al.* 2012, *J. Comput. Phys.*, 231, 718
Low, B. C. & Zhang, M. 2004, *ApJ*, 609, 1098
Xia, C., Chen, P. F., & Keppens, R. 2012, *ApJ*, 748, L26
Xia, C., Keppens, R., & Guo Y. 2013, *ApJ*, submitted

# Session I - 1.4

## FILAMENT ENVIRONMENT

*Nature of Prominences and their role in Space Weather*
*Proceedings IAU Symposium No. 300, 2013*
*B. Schmieder, J.-M. Malherbe & S. T. Wu, eds.*

doi:10.1017/S1743921313010855

# Structure and Topology of Magnetic Fields in Solar Prominences

**Adriaan A. van Ballegooijen**[1] **and Yingna Su**[1]

[1]Harvard-Smithsonian Center for Astrophysics,
60 Garden Street MS-15, Cambridge, MA 02138, USA
email: vanballe@cfa.harvard.edu

**Abstract.** Recent observations and models of solar prominences are reviewed. The observations suggest that prominences are located in or below magnetic flux ropes that lie horizontally above the PIL. However, the details of the magnetic structure are not yet fully understood. Gravity likely plays an important role in shaping the vertical structures observed in quiescent prominences. Preliminary results from a time-dependent model describing the interaction of a magnetic flux rope with photospheric magnetic elements are presented.

**Keywords.** Sun: prominences, Sun: filaments, Sun: magnetic fields

## 1. Observations

In this section we present an overview of various observations relevant to the magnetic structure of solar prominences. For more detailed reviews, see Tandberg-Hanssen (1995), Labrosse *et al.* (2010) and Mackay *et al.* (2010).

**Filament channels**: It is well known that filaments are located above polarity inversion lines (PIL) in the photospheric magnetic field (e.g., Babcock & Babcock 1955; Howard 1959). Most filaments are located within channels, traditionally defined as corridors surrounding PILs where the chromospheric H$\alpha$ fibrils are aligned with the PIL (e.g., Martres *et al.* 1966; Foukal 1971; Gaizauskas *et al.* 1997; Gaizauskas *et al.* 1998; Martin 1998). This fibril alignment implies that at chromospheric heights the magnetic field is dominated by its component along the PIL, and suggests the presence of a horizontal magnetic flux bundle in the low corona above the channel. The magnetic field in this flux bundle is highly sheared in the direction along the PIL, but may have some twist consistent with the direction of the overlying arcade field. The base of the (weakly twisted) flux bundle lies on the chromosphere, so the direction of its magnetic field is imprinted on the chromospheric fibrils. Filament channels can be classified as either *dextral* or *sinistral* depending on the direction of the magnetic field along the PIL as seen by an observer standing on the positive-polarity side of the channel (Martin *et al.* 1992). The chirality of filament channels can also be inferred from the plume-like tails of coronal emission cells observed with instruments on the Solar Dynamics Observatory (SDO) and the Solar Terrestrial Relations Observatory (STEREO) spacecraft (Sheeley *et al.* 2013). Filament channels exhibit a strong hemispheric pattern: those in the north are dominantly dextral, and those in the south are dominantly sinistral (Martin *et al.* 1994; Zirker *et al.* 1997).

The above view of a filament channel as a weakly twisted flux bundle lying on the chromosphere may apply to most low- and mid-latitude filaments, but may not be correct for polar crown prominences. At the polar crown PILs the chromospheric fibrils are nearly randomly oriented, even when prominences are present at larger heights. Therefore, the alignment of chromospheric fibrils is apparently not a necessary requirement for prominence formation. We suggest that the flux bundles supporting polar crown prominences may be somewhat elevated above the chromosphere, so that they have little effect on the chromospheric fibrils. Therefore, we will use the term "filament channel" also for polar crown prominences, even though alignments of chromospheric fibril along the PIL are not observed in this case.

**Filament environment**: In disk observations the filament channels appear as voids in coronal soft X-ray emission (e.g., Engvold 1989). EUV channels are darker than their surroundings, which may be due to a combination of lower emission in the channel and absorption of EUV line radiation in the H I Lyman continuum (Heinzel *et al.* 2001; Schmieder *et al.* 2003). In off-limb observations the channels appear as relatively dark coronal cavities surrounding the prominence. These cavities can be seen in white light coronagraph observations (Gibson *et al.* 2006), and also in the EUV and soft X-rays (e.g., Hudson *et al.* 1999). The cavity is located in the lower part of a helmet streamer that extends far from the Sun. Régnier *et al.* (2011) observed coronal plasma on U-shaped magnetic field lines sitting at the bottom of a cavity located above the polar crown. This U-shape of the field lines is consistent with the presence of a large flux rope supporting the prominence. Coronal cavities can support complex internal flows (Schmit *et al.* 2009; Wang & Stenborg 2010), indicating that the flux rope is not in a steady state. Rotational motions are also observed during eruptions (e.g., Panasenco *et al.* 2012). Immediately outside the cavity is an arcade of coronal loops that prevent the filament flux rope from expanding into the heliosphere, hence these loops play an important role in the overall force balance of the flux rope.

Su *et al.* (2010) presented observations of filament channels on the quiet Sun, and found a distinct asymmetry in the morphology of the bright structures on the two sides of the channel (also see Sheeley *et al.* 2013). As shown in Figure 1, one side of the channel has curved features, while the other side has straight features. This asymmetry is thought to be due to a variation in axial flux along the channel, which causes the field lines from one polarity to turn into the flux rope (curved features) while the field lines from the other polarity are connected to distant sources (straight features).

**Spectro-polarimetry**: The most direct way of studying the prominence magnetic fields is by inverting spectro-polarimetric data from Zeeman-sensitive spectral lines (e.g., Paletou & Aulanier 2003; López Ariste & Aulanier 2007). For example, the He I 1083.0 nm triplet is sensitive to the joint action of atomic level polarization (generated by anisotropic radiation pumping), the Hanle effect (modification of atomic level polarization due to a magnetic field) and the Zeeman effect (Lin *et al.* 1998; Trujillo Bueno *et al.* 2002; Orozco Suárez *et al.* 2013). Measurements using the He I 587.6 nm (D3) line above the solar limb have shown that the magnetic field strengths in quiescent prominences are in the range 3 - 15 G (Leroy 1989). The field is mostly horizontal and makes an angle of about 40° with respect to the long axis of the prominence (Bommier & Leroy 1998). The field strength increases slightly with height, indicating the presence of dips in the field lines. Most prominences have inverse polarity, i.e., the component of magnetic field perpendicular to the prominence long axis has a direction opposite to that of the potential field. The component along the prominence axis exhibits the hemispheric pattern discussed above (Leroy *et al.* 1983). Using full-Stokes polarimetry, Casini *et al.* (2003) were the first to make a map of the vector field in a prominence. Their results are consistent with the earlier measurements by Leroy and collaborators, but they also found localized regions with strong magnetic field (up to 80 G) within the prominence. Merenda *et al.* (2006) observed a polar crown prominence above the limb and found evidence for a magnetic field oriented only 25° from the vertical.

Spectro-polarimetry can also be applied on the solar disk. For example, Kuckein *et al.* (2012) observed an active region filament simultaneously in the photosphere and chromosphere. They found that the inferred vector fields are consistent with the presence

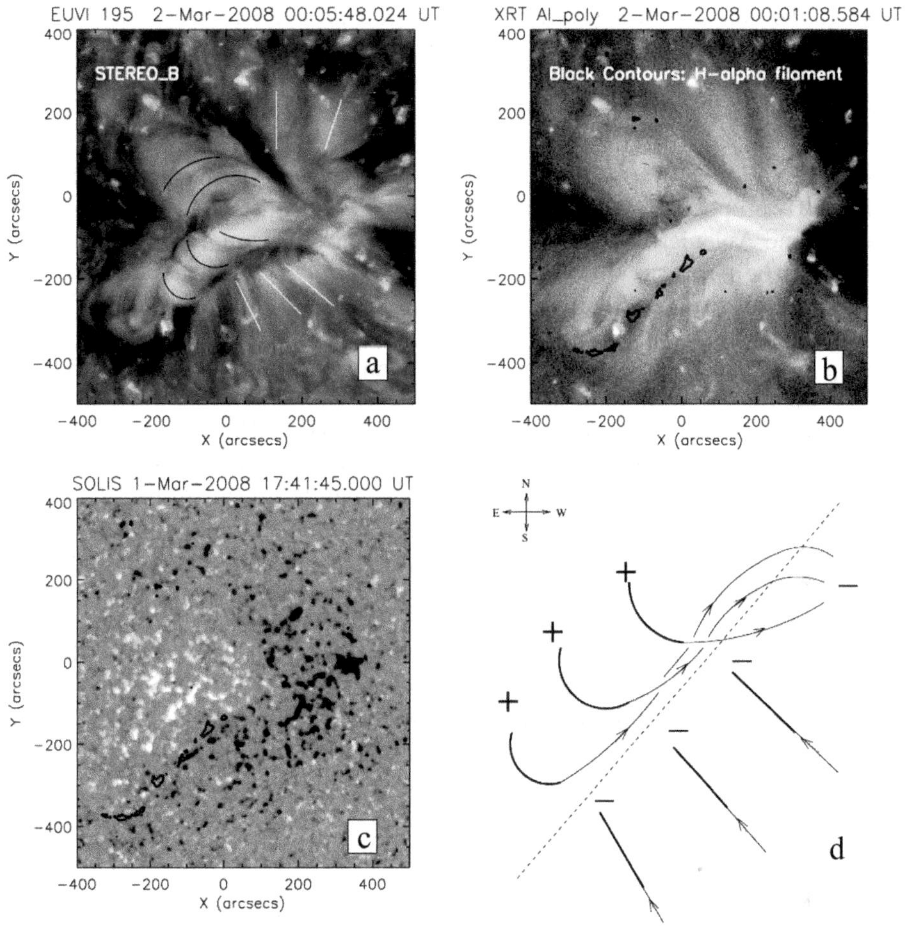


**Figure 1.** Asymmetry of coronal structures near filament channels. (a) STEREO/EUVI image of a channel observed on 2008 March 2 in the 195 Å passband. The bright features on the east side (black) are curved into the channel, while those on the west side (white) are more straight and point away from the channel. (b) Hinode/XRT image shows sheared loops overlying the channel; black contours indicate location of H$\alpha$ filament. (c) Photospheric magnetogram taken by NSO/SOLIS on previous day. (d) Interpretation of curved structures in terms a model in which the axial magnetic flux increases in NW direction along the PIL (dotted line). Adapted from Su *et al.* (2010).

of a magnetic flux rope that lies at low heights in the solar atmosphere. Yelles Chaouche *et al.* (2012) used these observations to construct a non-linear force-free field (NLFFF) model for the filament.

**Filament fine structures**: When observed on the solar disk with high spatial resolution, filaments show thin thread-like structures that continually evolve (Lin *et al.* 2008). A typical solar filament is composed of a spine, barbs, and two extreme ends. The spine defines the upper horizontal part of a filament, and the barbs are lateral extensions that extend from the spine to the chromosphere below (e.g., Martin 1998; Lin 2011). Active region filaments consist of narrow spines without barbs, intermediate filaments have a combination of both a spine and barbs, and quiescent filaments are dominated by barbs and vertical threads. When viewed from above, the barbs are either right-bearing or left-bearing, like ramps off a highway (Martin *et al.* 1992). There is a strong correlation between the orientation of the barbs and the chirality of the filament channel: filaments in dextral channels usually have right-bearing barbs, and those in sinistral channels usually have left-bearing barbs. At present the cause of this relationship is not well understood. Recently, Liu *et al.* (2010) found some cases where the barbs change their orientation within a period of a few hours, indicating there are exceptions to the above rule. Martin & Echols (1994) proposed that the barbs are rooted in parasitic polarity elements in the photosphere, and Aulanier *et al.* (1998) showed that the barbs move in accordance with the changes in the parasitic polarities. Others have argued that the ends of the barbs are located near small-scale PILs between majority and minority polarities (Wang 1999; Chae *et al.* 2005; Lin *et al.* 2005). Flux cancellation between parasitic polarities and the neighboring dominant polarity is believed to play an important role in the formation of the barbs (Wang 2001; Wang & Muglach 2007).

Recent observations with the Solar Optical Telescope on the Hinode satellite have revolutionized our understanding of quiescent and intermediate prominences. When observed above the solar limb, such prominences always show many thin thread-like structures. In some cases the threads are mainly horizontal (e.g., Okamoto *et al.* 2007), in other cases they are mainly vertical (e.g., Berger *et al.* 2008; Berger *et al.* 2010; Chae *et al.* 2008; Chae 2010). The horizontal threads may be understood as plasma being supported at the dips of weakly twisted, nearly horizontal field lines, but the structure of the vertical threads is not yet fully understood. The vertical threads are often not clearly visible (or not recognized) in on-disk H$\alpha$ observations; to see them on the disk requires high-resolution observations in other passbands, such as SDO/AIA 171 Å. Hedgerow prominences consist of many thin vertical threads organized in a vertical sheet or curtain. Upward-moving plumes and bubbles have been observed in between the denser, downflowing threads (e.g., Berger *et al.* 2008). Other prominences consist of isolated dark columns standing vertically above the PIL, and such prominences often exhibit rotational motions reminicent of "tornados" in the Earth's atmosphere (e.g., Su *et al.* 2012; Li *et al.* 2012; Panesar *et al.* 2013). The rotational motions have been confirmed using Doppler shift measurements (e.g., Liggett & Zirin 1984; Orozco Suárez *et al.* 2012). Some quiescent prominences have horn-like extensions that protrude from the top of the spine into the cavity above (e.g., Berger 2012; Schmit & Gibson 2013). These horns may outline a flux rope located above the prominence (Berger 2012).

## 2. Prominence Models

The cool prominence plasma must somehow be supported against gravity because without such support the plasma would fall to the chromosphere on a time scale of about 10 minutes. It has long been assumed that prominences are threaded by horizontal magnetic fields and electric currents, which provide the upward Lorentz force needed to counter gravity (e.g., Kippenhahn & Schlüter 1957; Kuperus & Raadu 1974; Pneuman 1983). However, the global topology of the magnetic field threading the prominence is not yet understood. For example, for tornado-like prominences it is unclear how horizontal fields can survive in the presence of rotational motions of the vertical structures (Liggett & Zirin 1984).

To understand the global structure of the filament magnetic field, Martin & Echols (1994) and Lin *et al.* (2008) proposed a conceptual model in which the filament plasma is located on arched field lines that extend along the length of the filament. The longest field lines represent the spine, and shorter ones splay sideways on both sides of the filament to form the barbs. In this model the barbs are located on inclined field lines, so this model does not explain how the plasma is supported against gravity. Aulanier *et al.* (1998) argued that filaments are located in weakly twisted flux ropes overlying the PIL. The filament plasma is located at dips in the field lines, and the barbs represent local distortions of the flux rope, causing the dips to be displaced from the main filament path. These distortions are due to parasitic polarities or bipoles in the photosphere below the filament channel (also see Dudik *et al.* 2012). Lionello *et al.* (2002) presented a data-driven MHD model describing the formation of an observed active region filament. They showed that the magnetic flux changes in the photosphere can produce a flux rope at the appropriate location, and that plasma condensations can form at the dips of the field lines. Su & van Ballegooijen (2012) developed a NLFFF model of a quiescent prominence observed with SDO and STEREO, and found that the flux rope has magnetic connections with the surrounding quiet Sun. Other variants of the flux rope model are the quadrupolar model (Magara *et al.* 2011) and the double-decker filament model (Liu *et al.* 2012).

It seems likely that gravity plays an important role in shaping the observed vertical structures. Berger *et al.* (2008) and Hillier *et al.* (2012) argued that the observed structures must be the result of a Rayleigh-Taylor (RT) instability. Van Ballegooijen & Cranmer (2010) proposed that hedgerow prominences are located in vertical current sheets, and that the plasma is supported by small-scale "tangled" magnetic fields within the sheet. Berger (2012) proposed the current sheet is located below an elevated flux rope (also see Fan 2012). One problem with these ideas is that small-scale tangled fields are not consistent with the apparent smoothness of the magnetic fields derived from Hanle measurements (e.g., Orozco Suárez *et al.* 2013). Haerendel & Berger (2011) suggested that the observed downflows may consist of plasma packets that squeeze themselves through the dominantly horizontal field under the action of gravity (also see Low *et al.* 2012b). Low *et al.* (2012a) investigated the coupling between the force balance and energy transport in quiescent prominences, and they suggest that the vertical threads may be falling across magnetic fields, with optically thick cores much denser and less ionized than conventionally assumed. Hillier & van Ballegooijen (2013) developed a 2.5D model for the support of prominence material by the dips of a coronal flux rope, and found that the magnetic field is significantly distorted by the weight of the prominence plasma. For some magnetic configurations force balance may not be possible unless the ratio of gas- and magnetic pressures is quite small, $\beta < 0.1$. This is opposite to Hillier *et al.* (2012), who need $\beta > 0.1$ to obtain RT instability.

Luna *et al.* (2012) simulated the formation and evolution of a multi-threaded solar prominence inside a flux rope. The process is governed by thermal nonequilibrium of the coronal plasma. They find that the condensations in the corona can be divided into two groups: threads and blobs. Threads are massive condensations that linger in the dips of the field lines, while blobs are small condensations that rapidly fall to the chromosphere. This model holds great promise for understanding how mass is supplied to the prominence. Models for the radiation emitted by prominence plasmas supported in magnetic dips have been developed by many authors (see Berlicki *et al.* 2011, Gunár *et al.* 2013, and references therein).

The observed hemispheric pattern of chirality (Zirker *et al.* 1997) has been investigated using models for the coupled evolution of the photospheric and coronal magnetic fields (see Mackay *et al.* 2010, and references therein). Recently, Yeates & Mackay (2012) simulated the evolution of the magnetic field over a full solar cycle. This model includes the effects of solar differential rotation, meridional flows and supergranular diffusion on the surface magnetic field, and new active regions are injected into the model in accordance with observed magnetograms (synoptic maps from NSO Kitt Peak). The modeled corona evolves through a series of nearly force-free equilibrium states, and coronal flux

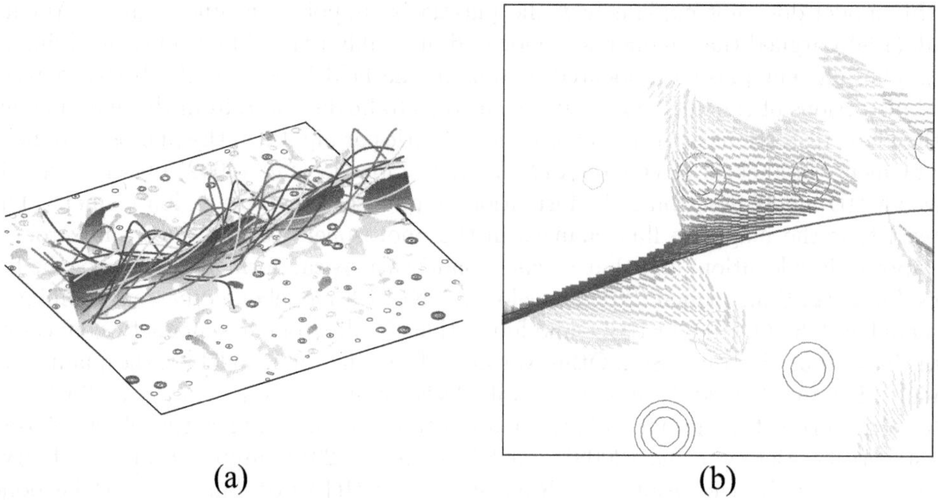


**Figure 2.** Model for the interaction of magnetic elements with a twisted flux rope. (a) Magnetic field at time $t = 200$ hrs. The red (green) features are positive (negative) magnetic flux elements on the photosphere. The colored curves are selected field lines within the flux rope (overlying field not shown). The blue feature are dips in the field lines (darkness increases with height). The front side is dominantly positive polarity (red) but has many parasitic (green) elemnts. (b) Close-up of several right-bearing "barbs" that formed when magnetic elements crossed the PIL.

ropes develop over the PILs. The predicted chirality of these flux ropes is compared with the observed hemispheric pattern. The authors find that differential rotation can lead to opposite chirality at high latitudes, but only for about 5 years of the solar cycle following the polar field reversal. At other times the transport of magnetic helicity from lower latitudes overcomes the effect of *in situ* differential rotation, producing the majority chirality even on the polar crowns (Yeates & Mackay 2012). These results indicate that helicity transport from lower latitudes is important for understanding the hemispheric pattern of filament chirality, and that the corona has long-term memory of its helicity.

Recently, we developed a model describing the interaction of a coronal flux rope with magnetic elements in the photosphere. The evolution of the coronal magnetic field is simulated using the magneto-frictional method (e.g., van Ballegooijen *et al.* 2000). The time-dependent boundary conditions include the effects of emerging bipoles, as well as random motion, splitting, merging and cancellation of magnetic flux elements. The initial field contains a dextral flux rope with left-helical twist. Figure 2a shows a perspective view of the flux rope after 200 hours solar time. The blue features are the field-line dips in the lower half of the twisted flux rope. The curtain of dips is highly distorted by parasitic polarities below the flux rope, resulting in barb-like extensions. A close-up view from above (Figure 2b) shows that the barbs are right-bearing for this dextral channel. Therefore, the model reproduces the observed relationship between dextral (sinistral) channels and right (left) bearing barbs (Martin *et al.* 1992), consistent with earlier results from LFFF models (Aulanier *et al.* 1998).

In summary, various observations suggest that solar prominences are located in magnetic flux ropes that lie horizontally above the PIL. For active region filaments these flux ropes are thin and almost untwisted; for intermediate and quiescent filaments some twist seems to be required by the spectro-polarimetric measurements. The plasma may be supported by dips in the field lines. Models for the 3D magnetic fields in and around prominences have been developed using various techniques (e.g., NLFFF and MHD models). Including the weight of the prominence plasma into such models is difficult because the pressure scale height of the prominence plasma ($H_p \approx 300$ km) is much smaller than the typical size of the flux rope. Therefore, modeling quiescent prominences represent a challenging computational problem.

Acknowledgements: This project is supported by NASA grant NNX12AI30G as well as NASA contract SP02H1701R from LMSAL to SAO.

## References


Aulanier, G., Démoulin, P., van Driel-Gesztelyi, L., Mein, P., & Deforest, C. 1998, *A&A*, 335, 309

Babcock, H. W. & Babcock, H. D. 1955, *ApJ*, 121, 349

Berger, T. E. 2012, in Second ATST-EAST Meeting: Magnetic Fields from the Photosphere to the Corona, eds. T. R. Rimmele, M. Collados, *et al.*, ASP Conf. Series, Vol. 463 (Astron. Soc. of the Pacific, San Francisco), p.147

Berger, T. E., Shine, R. A., Slater, G. L., *et al.* 2008, *ApJ*, 676, L89

Berger, T. E., Slater, G., Hurlburt, N., *et al.* 2010, *ApJ*, 716, 1288

Berlicki, A., Gunár, S., Heinzel, P., Schmieder, B., & Schwartz, P. 2011, *A&A*, 530, A143

Bommier, V. & Leroy, J.-L. 1998, in New Perspectives on Solar Prominences, IAU Colloq. 167, eds. D. Webb, D. Rust, B. Schmieder, ASP Conf. Series, Vol. 150 (Astron. Soc.of the Pacific, San Francisco), p. 434

Casini, R., Löpez Ariste, A., Tomczyk, S., & Lites, B. W. 2003, *ApJ*, 598, 67

Chae, J., Moon, Y.-J., & Park, Y.-D. 2005, *ApJ*, 626, 574

Chae, J., Ahn, K., Lim, E.-K., Choe, G. S., & Sakurai, T. 2008, *ApJ*, 689, L73

Chae, J. 2010, *ApJ*, 714, 618

Engvold, O. 1989, in Dynamics and Structure of Quiescent Solar Prominences, ed. E. R. Priest (Kluwer, Dordrecht), p.47

Fan, Y. 2012, *ApJ*, 758, 60

Foukal, P. 1971, *Sol. Phys.*, 19, 59

Gunar, S., Mackay, D. H., Anzer, U., & Heinzel, P. 2013, *A&A*, 551, id.A3

Gaizauskas, V., Zirker, J. B., Sweetland, C., & Kovacs, A. 1997, *ApJ*, 479, 448

Gaizauskas, V. 1998, in New Perspectives on Solar Prominences (ASP Conference Series, Vol. 150, IAU Colloquium 167), eds. D. Webb, D. M. Rust, & B. Schmieder, p.257

Gibson, S. E., Foster, D., Burkepile, J., de Toma, G., & Stanger, A. 2006, *ApJ*, 641, 590

Haerendel, G. & Berger, T. 2011, *ApJ*, 731, 82

Heinzel, P., Schmieder, B., & Tziotziou, K. 2001, *ApJ*, 561, L223

Hillier, A., Berger, T., Isobe, H., & Shibata, K. 2012, *ApJ*, 746, 120

Hillier, A. & van Ballegooijen, A. A. 2013, *ApJ*, 766, 126

Howard, R. 1959, *ApJ*, 130, 193

Hudson, H. S., Acton, L. W., Harvey, K. L., & McKenzie, D. E. 1999, *ApJ*, 513, L83

Kippenhahn, R. & Schlüter, A. 1957, *Z. Astrophys.*, 43, 36

Kuckein, C., Martinez Pillet, V., & Centeno, R. 2012, *A&A*, 539, A131

Kuperus, M. & Raadu, M. A. 1974, *A&A*, 31, 189

Labrosse, N., Heinzel, P., Vial, J.-C., Kucera, T., Parenti, S., Gunár, S., Schmieder, B., & Kilper, G. 2010, *Space Sci Rev*, 151, 243

Leroy, J.-L. 1989, in Dynamics and Structure of Quiescent Solar Prominences, ed. E. R. Priest (Kluwer, Dordrecht), p.77

Leroy, J.-L., Bommier, V., & Sahal-Brechot, S. 1983, *Sol. Phys.*, 83, 135

Li, X., Morgan, H., Leonard, D., & Jeska, L. 2012, *ApJ*, 752, L22

Liggett, M. & Zirin, H. 1984, *Sol. Phys.*, 91, 259

Lionello, R., Mikić, Z., Linker, J. A., & Amari, T. 2002, *ApJ*, 581, 718

Lin, H., Penn, M. J., & Kuhn, J. R. 1998, *ApJ*, 493, 978

Lin, Y., Engvold, O., Rouppe van der Voort, L., Wiik, J. E., & Berger, T. E. 2005, *Sol. Phys.*, 226, 239

Lin, Y., Martin, S. F., & Engvold, O. 2008, in Subsurface and Atmospheric Influences of Solar Activity, ed. R. Howe, R. W. Komm, K. S. Balasubramaniam, & G. J. D. Petrie, ASP Conf. Series, Vol. 333 (Astron. Soc. of the Pacific), p. 235
Lin, Y. 2011, *Space Sci. Rev.*, 158, 237
Liu, R., Xu, Y., & Wang, H. 2010, *Mem. S. A. It.*, Vol. 81, 796
López Ariste, A. & Aulanier, G. 2007, in The Physics of the Chromospheric Plasmas, ed. P. Heinzel, I. Dorotovic, R. J. Rutten, ASP Conf. Series, Vol. 368 (Astron. Soc. of the Pacific, San Francisco), p. 291
Low, B. C., Berger, T., Casini, R., & Liu, W. 2012a, *ApJ*, 755, 34
Low, B. C., Liu, W., Berger, T., & Casini, R. 2012b, *ApJ*, 757, 21
Luna, M., Karpen, J. T., & DeVore, C. R. 2012, *ApJ*, 746, 30
Magara, T., An, J.-M. Lee, H., & Kang, J. 2011, *J. Korean Astron. Soc.*, 44, 143
Mackay, D. H., Karpen, J. T., Ballester, J. L., Schmieder, B., & Aulanier, G. 2010, *Space Sci Rev*, 151, 333
Martin, S. F. 1998, *Sol. Phys.*, 182, 107
Martin, S. F., Billamoria, R., & Tracadas, P. W. 1994, in Solar Surface Magnetism, eds. R. J. Rutten, & C. J. Schrijver, Kluwer Academic Pubishers, Dordrecht, Holland, 303
Martin, S. F. & Echols, Ch.R. 1994, in Solar Surface Magnetism, eds. R. J. Rutten & C. J. Schrijver, Kluwer Academic Pubishers, Dordrecht, Holland, 339
Martin, S. F., Marquette, W., & Billamoria, R. 1992, in The Solar Cycle, ed. K. Harvey, Proceedings of the 12th Summer Workshop, National Solar Observatory, 53
Martres, M.-J., Michard, R., & Soru-Iscovici, I. 1966, *A&A*, 29, 249
Merenda, L., Trujillo Bueno, J., Landi Degl'Innocenti, E., & Collados, M. 2006, *ApJ*, 642, 554
Okamoto, T. J., Tsuneta, S., Berger, T. E., *et al.* 2007, *Science*, 318, 1577
Orozco Suárez, D., Asensio Ramos, A., & Trujillo Bueno, J. 2012, *ApJ*, 761, L25
Orozco Suárez, D., Asensio Ramos, A., & Trujillo Bueno, J. 2013, *A&A*, in preparation
Paletou, F. & Aulanier, G. 2003, in Solar Polarization Workshop 3, ed. J. Trujillo Bueno & J. Sánchez Almeida, ASP Conf. Series, Vol. 236 (Astron. Soc. of the Pacific, San Francisco), p. 458
Panasenco, O., Martin, S. F., Velli, M., & Vourlidas, A. 2012, in Solar Dynamics and Magnetism from the Interior to the Atmosphere, eds. R. Komm, A. Kosovichev, D. Longcope, & N. Mansour (Springer, Dordrecht)
Panesar, N. K., Innes, D. E., Tiwari, S. K., & Low, B. C. 2013, *A&A*, 549, A105
Pneuman, G. W. 1983, *Sol. Phys.*, 88, 219
Régnier, S., Walsh, R. W., & Alexander, C. E. 2011, *A&A*, 533, L1
Schmieder, B., Tziotziou, K., & Heinzel, P. 2003, *A&A*, 401, 361
Schmit, D. J., Gibson, S. E., Tomczyk, S., *et al.* 2009, *ApJ*, 700, L96
Schmit, D. J. & Gibson, S. E. 2013, *ApJ*, 770, id. 35
Sheeley, N. R., Jr., Martin, S. F., Panasenco, O., & Warren, H. P. 2013, *ApJ*, 772, article id. 88
Su, Y., Wang, T., Veronig, A., Temmer, M., & Gan, W. 2012, *ApJ*, 756, L41
Su, Y., van Ballegooijen, A. A., & Golub, L. 2010, *ApJ*, 721, 901
Tandberg-Hanssen, E. 1995, The nature of solar prominences, Astrophys. Space Sci. Lib., Vol. 199 (Dordrecht: Kluwer Academic Publishers)
Trujillo Bueno, J., Landi Degl'Innocenti, E., Collados, M., Merenda, L., & Manso Sainz, R. 2002, *Nature*, 415, 403
van Ballegooijen, A. A. Priest, E. R., & Mackay, D. H. 2000, *ApJ*, 539, 983
van Ballegooijen, A. A. & Cranmer, S. R. 2010, *ApJ*, 711, 164
Wang, Y.-M. 1999, *ApJ*, 520, L71
Wang, Y.-M. 2001, *ApJ*, 560, 456
Wang, Y.-M. & Muglach, K. 2007, *ApJ*, 666, 1284
Wang, Y.-M. & Stenborg, G. 2010, *ApJ*, 719, L181
Yeates, A. R. & Mackay, D. H. 2012, *ApJ*, 753, L34
Yelles Chaouche, L., Kuckein, C., Martinez Pillet, V., & Moreno-Insertis, F. 2012, *ApJ*, 748, 23
Zirker, J. B., Martin, S. F., Harvey, K., & Gaizauskas, V. 1997, *Sol. Phys.*, 175, 27

*Nature of Prominences and their role in Space Weather*
*Proceedings IAU Symposium No. 3000, 2013*
*B. Schmieder, J.-M. Malherbe & S. T. Wu, eds.*

doi:10.1017/S1743921313010867

# Hemispheric Patterns in Filament Chirality and Sigmoid Shape over the Solar Cycle

**Petrus C. Martens**[1], **Anthony R. Yeates**[2], **and Karthik G. Pillai**[3]

[1]Physics Department, Montana State University,
Bozeman, MT 59717, USA,
email: `martens@physics.montana.edu`

[2]Department of Mathematical Sciences, Durham University,
Science Laboratories, South Road, Durham DH1 3LE, UK,
email:`anthony.yeates@durham.ac.uk`

[3]Computer Science Department, Montana State University,
Bozeman, MT 59717, USA,
email: `karthikgp@gmail.com`

**Abstract.** The motivation for our research was to study the correlation between the chirality of filaments and the handedness (S- or Z-shape) of sigmoids. It was assumed that sigmoids would mostly coincide with filaments and that the S-shaped sigmoids would correlate well with filaments of sinistral chirality, which we found that to be at best a very weak relation. Since we had a full solar cycle of filament metadata at hand it was easy to verify the supposedly known hemispheric preference of filament chirality. We discovered that the hemispheric chirality rule was confirmed for the epoch where a thorough manual study had been performed, but that at other phases of the solar cycle the rule seems to disappear and sometimes even reverse.

**Keywords.** Solar Filaments, Solar Sigmoids, Hemispheric Chirality Rule

---

## 1. Introduction

The overwhelming amount of data coming from the Solar Dynamics Observatory (SDO, $\approx 1.5$ TB/day), and soon from the Advanced Technology Solar Telescope (ATST) forces the solar physics community to handle observations in a different way from what we are used to. One method of dealing with the data deluge is to develop automated feature recognition codes that produce metadata on solar events and phenomena, catalogs in fact, that until now have been painstakingly put together manually by solar scientists, often graduate students. NASA had foreseen this and funded two teams to develop automated feature recognition codes from prior to the SDO launch in February 2010.

One of these teams is headed by the first author of this paper and has produced 16 modules for the detection and analysis of different solar features (Martens *et al.* 2012), see also the more up-to-date website†. The analysis of the metadata from two of these modules, the "Advanced Automated Filament Detection and Characterization Code (AAFDCC)", see Bernasconi, Rust & Hakim (2005), and the "Sigmoid Sniffer" (Martens *et al.* 2012) will be presented in this paper, and the results contain surprises.

Apart from the scientific conclusions, one important lesson we have learned from using the metadata from automated feature detection modules is that it enables us solar physicists to analyze very large datasets in a much more efficient and economical way than we did before. Hence we have the capability now to move away from the analysis of single or

† solar.physics.montana.edu/sol_phys/fft/

small sets of events to large and complete sets of data for a given phenomenon. This may lead to the detection of hitherto unknown correlations, and thereby raise now challenging questions and the opportunity to develop new physical insights. The scientific conclusions from this paper are a case in point. The repository of metadata for Solar Physics is the Heliophysics Events Knowledgebase (HEK‡), and the metadata are accessible from there as well as from the Virtual Solar Observatory¶.

## 2. Filament Chirality

Sarah Martin and co-workers introduced the concept of chirality for filaments in a remarkable series of papers reviewed by Martin (1998). A key result of their work is that the chirality of filaments – dextral or sinistral – can be determined from H$\alpha$ observations alone, no magnetic observations are needed. The bearing of the barbs – little sideways protrusions emanating from filaments – can be either left or right (imagine the on- or off-ramps from highways). The dominant direction of the bearing of the barbs determines the chirality of the filament.

As Martin shows filament chirality in essence is the sense of winding of the magnetic field around the axis of a filament (left-handed or right-handed), and hence is related to the force-fee $\alpha$ and the sign of the helicity of filaments.

Bernasconi's AAFDC code detects the barbs of filaments and their bearing, and hence can assign chirality to a filament. The usual attribution criterion is that two more barbs have to be of one bearing versus the opposite one to assign filament chirality. Georgoulis' "Sigmoid Sniffer" (see Martens *et al.* 2012) likewise detects whether the contour of a sigmoid has an "S" (forward) or a "Z"-shape (reverse).

## 3. Filament-Sigmoid Correlation

We used AIA 94 Å data from October 11 2010 until March 27 2011, analyzed by the "Sigmoid Sniffer", and H$\alpha$ data from the same period obtained by Big Bear Solar Observatory, at a rate of about four per day, analyzed by the AAFDC. The correlation matrices, shown in Fig. 1 show that only a small fraction of filaments have an accompanying sigmoid, no surprise since filaments are much more prevalent than sigmoids. More of a surprise is that 2/3 of sigmoids do not have an accompanying filament. Perhaps the reason is that cool filament material has evaporated by the time a sigmoid forms, although there are clear examples of the opposite (e.g. Fig. 2 in Martens & Zwaan, 2001). The low cadence of the H$\alpha$ observations may have something to do with that as well; future versions of AAFDC will analyze images from ground stations worldwide.

The correlation between dextral filament chirality and inverse S-shape for sigmoids (and v.v.) seems clear, but the accuracy of prediction is only 64.4%, while the $\phi$-coefficient (see Cramer, 1946, p. 282, second paragraph) of the matrix is only $-0.25$ indicating a statistically very weak to non-existent negative correlation.

## 4. Hemispheric Rule for Chirality

Fully aware of the hemispheric preferences for filament chirality published by Pevtsov, Balasubramaniam & Rogers (2003) for the period July 15, 2000 to June 22, 2002, and closely reproduced for the same dataset by the AAFDC (Bernasconi *et al.* 2005) we used AAFDC metadata to verify the hemispheric rule for the dataset described above. To our

‡ http://www.lmsal.com/hek/
¶ http://sdac.virtualsolar.org

## Experimental Results

| Colocation Rule | Conditional Probability |
|---|---|
| *Sigmoid ⇒ Filament* | 32.85% |
| *Filament ⇒ Sigmoid* | 7.98% |

| Shape-Chirality | Dextral | Sinistral |
|---|---|---|
| Forward | 16.75% | 41.36% |
| Inverse | 23.04% | 18.85% |

**Figure 1.** Correlation between filaments and sigmoids.

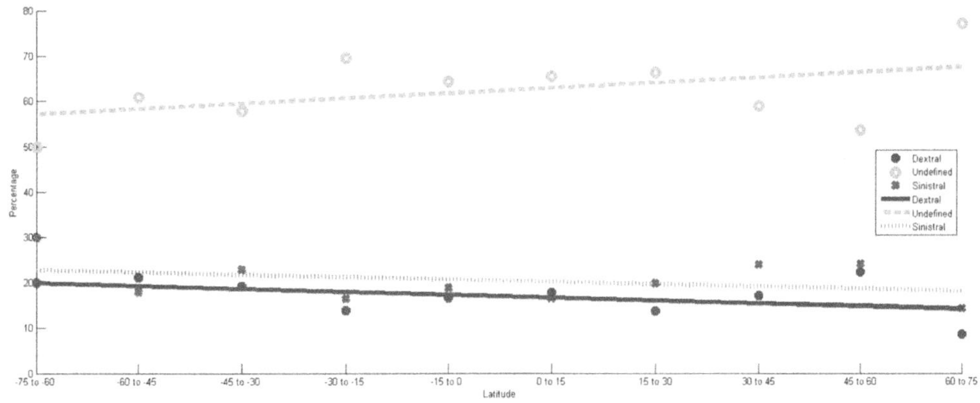


**Figure 2.** Linear fit of percentage of chirality in 15 degree latitude bins.

great surprise we discovered that there was absolutely no dependence of filament chirality on latitude for the period October 11 2010 to March 27 2011. In Fig. 2 our counts are presented in 15 degree latitude bins, and the result speaks for itself. A large percentage of filaments has undetermined chirality which is normal for solar minimum when filaments typically carry only few barbs. We found the same latitude independent distribution for sigmoid shape, not shown here, with no undetermined values in that sample.

This result motivated us to study the hemispheric chirality preference for all the AAFFDC data available for solar cycle 23 and the beginning of cycle 24, the period from the beginning of 2001 to the spring of 2012. The results are shown in Fig. 3. Our concern about the validity of the result in Fig. 2 was alleviated by that result because the analysis of the AAFDC metadata again reproduced the results from the study by Pevtsov *et al.*, as well as the earlier results from the AAFDC. Clearly there is a variation of the hemispheric rule with the phase of the solar cycle 23. During the extended solar minimum AAFDC had a hard time detecting the barb bearing of the limited number of filaments at all. For several months (around the 2007 tick mark in Fig. 3) during the two years prior to the minimum, 2006 and 2007, it seems that the hemispheric preference was reversed, a very puzzling result, and at the onset of cycle 24 the hemispheric pattern had difficulty reestablishing itself, a considerable amount of negative chirality remains in the northern hemisphere. It is important to keep in mind here that the hemispheric chirality preference does not reverse from cycle to cycle, as was established by Martin, Billimoria & Tracadas (1994).

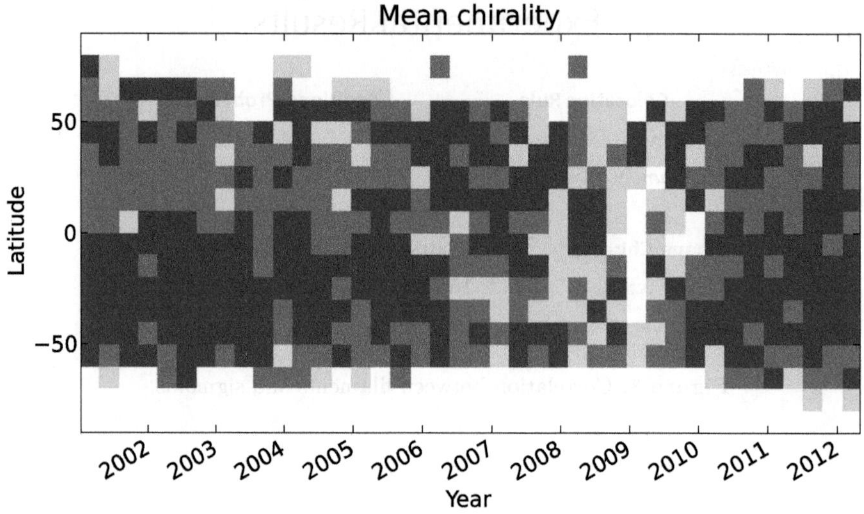


**Figure 3.** Mean chirality per time-latitude box for 2001-2012. Red (black) indicates dextral chirality, blue (dark grey) sinistral, and yellow (light grey) denotes undetermined chirality.

## 5. Conclusions

We have demonstrated, from a limited three month data set, that there is a weak correlation between the shape of sigmoids ("S" or "Z") and the chirality of the corresponding filaments. We also found from the same dataset that for 2/3 of the detected sigmoids no corresponding filament was detected, and that there was no hemispheric preference law for filament chirality and sigmoid shape in the period of observation.

The most surprising and intriguing result of our analysis is that in the epoch from early 2001 to early 2012 the hemispheric chirality preference law for filaments waxes and wanes; sometimes it is strongly present, at other times wholly absent, and sometimes even reversed. These results are preliminary; we intend to verify the AAFDC produced metadata with manual studies for appropriate time intervals, and we will carefully consider the statistical underpinnings of our results.

## References

Bernasconi, P. N., Rust, D. M., & Hakim, D. 2005, *Sol. Phys.*, 228, 97

Cramer, H. 1946, *Mathematical Methods of Statistics* (Princeton: Princeton University Press,)

Martens, P. C. & Zwaan, C. 2001, *ApJ*, 558, 872

Martens, P. C. H., Attrill, G. D. R., Davey, A. R., Engell, A., Farid, S., Grigis, P. C., Kasper, J., Korreck, K., Saar, S. H., Savcheva, A., Su, Y., Testa, P., Wills-Davey, M., Bernasconi, P. N., Raouafi, N.-E., Delouille, V. A., Hochedez, J. F., Cirtain, J. W., Deforest, C. E., Angryk, R. A., de Moortel, I., Wiegelmann, T., Georgoulis, M. K., McAteer, R. T. J., & Timmons, R. P. 2012, *Sol. Phys.*, 275, 79

Martin, S. F. 1998, *Sol. Phys.*, 182, 107

Martin, S. F., Billimoria, R., & Tracadas, P. W. 1994, in *Solar Surface Magnetism*, 303–338

Pevtsov, A. A., Balasubramaniam, K. S., & Rogers, J. W. 2003, *ApJ*, 595, 500

*Nature of Prominences and their role in Space Weather*
*Proceedings IAU Symposium No. 300, 2013*
*B. Schmieder, J.-M. Malherbe & S. T. Wu, eds.*

doi:10.1017/S1743921313010879

# Magnetism and the Invisible Man: The mysteries of coronal cavities

**Sarah Gibson**†

High Altitude Observatory/National Center for Atmospheric Research
3080 Center Green Dr. Boulder, CO, 80027, USA
email: sgibson@ucar.edu

**Abstract.** Magnetism defines the complex and dynamic solar corona. Twists and tangles in coronal magnetic fields build up energy and ultimately erupt, hurling plasma into interplanetary space. These coronal mass ejections (CMEs) are transient riders on the ever-outflowing solar wind, which itself possesses a three-dimensional morphology shaped by the global coronal magnetic field. Coronal magnetism is thus at the heart of any understanding of the origins of space weather at the Earth. However, we have historically been limited by the difficulty of directly measuring the magnetic fields of the corona, and have turned to observations of coronal plasma to trace out magnetic structure. This approach is complicated by the fact that plasma temperatures and densities vary among coronal magnetic structures, so that looking at any one wavelength of light only shows part of the picture. In fact, in some regimes it is the lack of plasma that is a significant indicator of the magnetic field. Such a case is the coronal cavity: a dark, elliptical region in which strong and twisted magnetism dwells. I will elucidate these enigmatic features by presenting observations of coronal cavities in multiple wavelengths and from a variety of observing vantages, including unprecedented coronal magnetic field measurements now being obtained by the Coronal Multichannel Polarimeter (CoMP). These observations demonstrate the presence of twisted magnetic fields within cavities, and also provide clues to how and why cavities ultimately erupt as CMEs.

**Keywords.** Sun: prominences, Sun: corona, Sun: magnetic fields.

---

*"No hand – just an empty sleeve... Then, I thought, there's something odd in that. What the devil keeps that sleeve up and open, if there's nothing in it?"* – The Invisible Man (Wells 1897)

## 1. Introduction

Coronal cavities are dark, elliptical structures that surround prominences (Figure 1). Like prominences, cavities are long-lived and may be stable for days or even weeks (Gibson *et al.* 2006). Also like prominences, cavities exhibit dynamic behavior even when not erupting, with swirling flows of coronal plasma within the cavity tracing out helical structure (Li *et al.* 2012). Cavities do eventually erupt along with their embedded prominences as coronal mass ejections (CMEs): roughly a third were observed to do so in a survey of over one hundred polar-crown-filament (PCF) cavities (Forland *et al.* 2013). Since a typical (median) length of time these cavities were visible at the limbs was about four and a half days, the time spent at the two limbs during the approximately 27-day solar rotation is about one-third; this implies that if one could observe all the way around the Sun, one would see all the cavities erupt eventually.

† NCAR is supported by the National Science Foundation

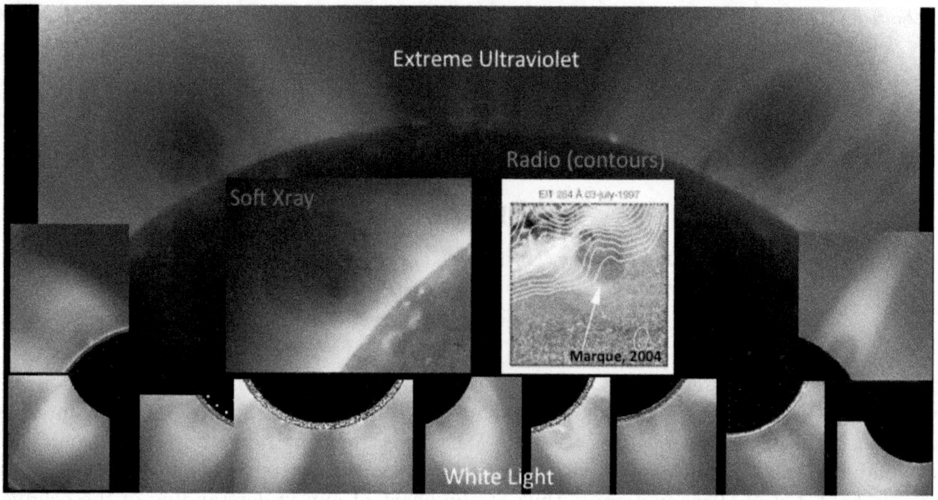


**Figure 1.** Cavities are visible in a broad range of wavelengths. Extreme ultraviolet (EUV) observations from Solar Dynamics Observatory/Atmospheric Imaging Assembly (SDO/AIA); Soft-Xray (SXR) from Hinode X-ray Telescope (XRT); white light images from Mauna Loa Solar Observatory Mk4 K-coronameter (MLSO/Mk4); Radio contours (Nancay) overlaid on Solar and Heliospheric Observatory EUV Imaging Telescope (SOHO/EIT) observations (Marque 2004).

Since cavities represent the bulk of the combined erupting prominence-cavity volume, it is their magnetic structure that maps to the magnetic cloud passing the Earth. If cavities are the "Invisible Man" of solar physics, prominences are his footprints: more visible perhaps, but only representative of a fraction of the magnetic structure that erupts. Cavities thus are key to understanding the nature of pre-CME equilibria and the mechanisms that trigger their loss.

But how does one measure the invisible? Luckily, as I will describe below, cavities are not truly empty. Their detection is subject to stringent line-of-sight constraints, and it is likely that many remain unobserved because of obscuration by surrounding bright structures in the optically-thin corona. However, recent work modeling the 3D geometry of PCF cavities, combined with new observations, has enabled a detailed analysis of cavity physical properties. A self-consistent picture has emerged explaining cavity morphology, sub-structure, and dynamic evolution of the cavity that is consistent with the theory that the cavity is a magnetic flux rope (e.g., Low & Hundhausen (1995)).

In Section 2, I will review observations of stable (non-erupting) cavity plasma and magnetic properties. In Section 3, I will present present observations of cavities in relation to CMEs. In Section 4, I will conclude by discussing how these observations map to magnetic flux ropes.

## 2. Coronal cavities: Observations

Cavities were first observed in white light in eclipses (see e.g. Waldmeier (1970), Tandberg-Hanssen (1974)). The advent of coronagraphs, radio, EUV and SXR telescopes have given us a means to observe them on a daily basis (Figure 1). In a six-year study of white light images, Gibson *et al.* (2006) found 98 distinct cavity systems, with cavities visible approximately one in ten days (Figure 2, left). However, these observations were taken with an occulted coronameter, so only cavities with heights > 1.15 solar radius could be identified. In a 19-month-long study of cavities at EUV wavelengths, Forland *et al.* (2013) found 129 distinct cavity systems, with cavities visible 78% of the days (Figure 2, right). This surveywas able to measure many smaller cavities than would have

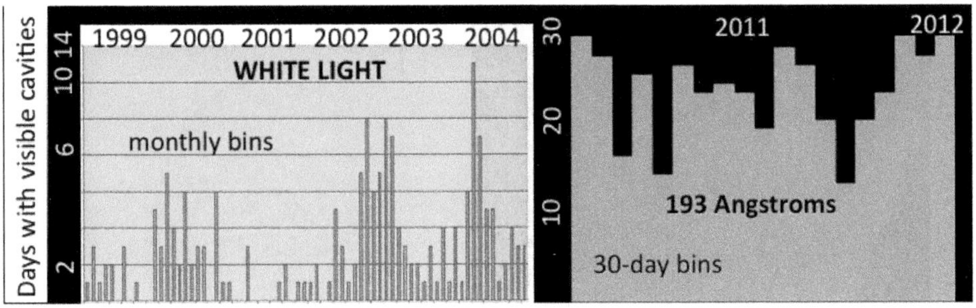


**Figure 2.** Cavities are ubiquitous. Left: white light survey of cavities from November 1998-September 2004 using MLSO/MK4 coronameter observations (Gibson *et al.* 2006; figure courtesy Joan Burkepile). Right: EUV (193 Å ) survey of cavities from June 1, 2010 - Dec 31, 2012 using SDO/AIA images (Forland *et al.* 2013).

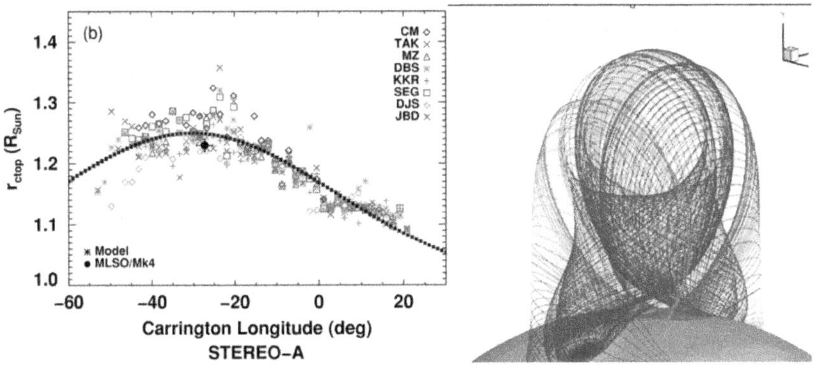


**Figure 3.** Cavities have arched, tunnel-like morphologies with elliptical cross-sections. Left: from Gibson *et al.* (2010); cavity ellipse height vs. longitude/date. Right: flux surfaces of Gibson & Fan (2006) simulation of flux rope.

been occulted in the white light survey. Another difference between the surveys was that, while the white light survey encompassed years of solar maximum, when a complexity of bright structures along the line of sight may well have obscured cavities, the EUV survey took place during the ascending phase of the solar cycle, a time when PCFs were common. PCFs are large, longitudinally-extended, quiescent filaments at high latitudes, and as such present near-ideal viewing conditions for cavities.

Gibson *et al.* (2010) studied the 3D morphology of a cavity using observations at multiple wavelengths, vantage points, and covering multiple days. The cavity was modeled as a tunnel-like structure, with a Gaussian height (Figure 3, left) and elliptical cross-section. Forland *et al.* (2013) fit ellipses to all the EUV cavities in the survey, and found a strong tendency (93%) for cavity ellipses to be taller than they were wide.

Building on the 3D morphology found by Gibson *et al.* (2010), Schmit & Gibson (2011) extracted density of a coronal cavity from multiwavelength observations. The density was found to be approximately 30% depleted at the center of the cavity relative to the surrounding streamer at the same height. This was consistent with prior analyses which found that cavities were, in general, significantly more dense than coronal holes (thus, not truly "invisible") and possessed on average 25% depletion and maximum 60% depletion relative to the surrounding streamer (Fuller & Gibson 2009). Building on both the morphology and the density analyses, Kucera *et al.* (2012)) found that the average temperature in the cavity was similar to that of the surrounding streamer (about $1.5MK$);

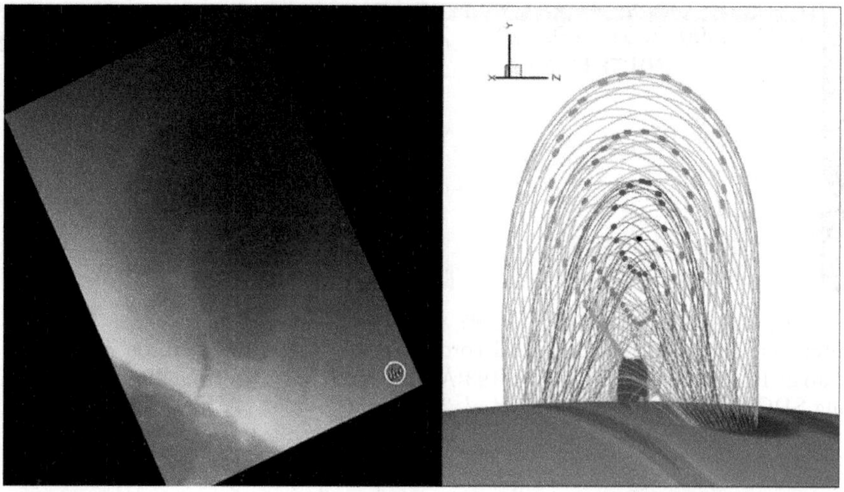


**Figure 4.** Cavities have substructure. Left: July 14, 2013 observations of a cavity from SDO/AIA (193 Å ), showing prominence and horn within larger-scale cavity. Right: magnetic flux surfaces (colored lines, with dots indicating intersection with plane of sky) and dips in field lines (lower sheet of brown dots) in magnetic flux rope simulation (Gibson & Fan 2006).

however, the cavity exhibited more thermal variability, indicating multiple temperatures were present at a given height.

The thermal variability within cavities is likely related to their often dynamic substructures. In particular, disk or ring-like structures lying at the center of the cavity (sometimes referred to as "chewy nougats") are commonly observed in soft X-ray, indicating regions of elevated temperature (Hudson *et al.* 1999; Reeves *et al.* 2012). These nougats sometimes appear immediately above the prominence, like a lollypop on a stick. In EUV, flows trace out horn-like structure in a similar central location above the prominence (Figure 4; left), and are temporally and spatially linked to flows in the cooler prominence plasma (Schmit & Gibson 2013). Flows along the line of sight have also been measured (Schmit *et al.* 2009). These flows are of order $5 - 10$ $km/sec$, have length scales of tens of megameters, and persist for at least one hour. They typically have outer boundaries corresponding to that of the cavity or its central substructure, and occasionally exhibit nested ring-like structure (Bąk-Stęślicka *et al.* 2013) (Figure 5; top).

Recently, a new means of directly probing the magnetic structure of cavities has become available through the Coronal Multichannel Polarimeter (CoMP): a coronagraph that measures Stokes polarimetry vectors and line-of-sight velocities using optically-thin coronal emission lines (Tomczyk *et al.* 2008). Linear polarization ($\sqrt{Q^2 + U^2}$, where $Q$ and $U$ are Stokes vectors) has turned out to be a particularly useful diagnostic for coronal cavities (Dove *et al.* 2011). Over the past few years, CoMP has shown that the linear polarization of PCF cavities systematically exhibit a structure akin to that of a rabbit's head ("lagomorph")(Figure 7; Bąk-Stęślicka *et al.* (2013)). As is evident by comparing Figures 6 and 7, linear-polarization lagomorphs generally scale with cavity size (Bąk-Stęślicka *et al.* 2014 (this issue)).

## 3. Cavites and CMEs

The properties of cavities prior to eruptions and their evolution leading up to CMEs may provide clues to the mechanisms that trigger them. Gibson *et al.* (2006) found an upper limit to cavity height of approximately 1.5 solar radii (Figure 8). This may imply a

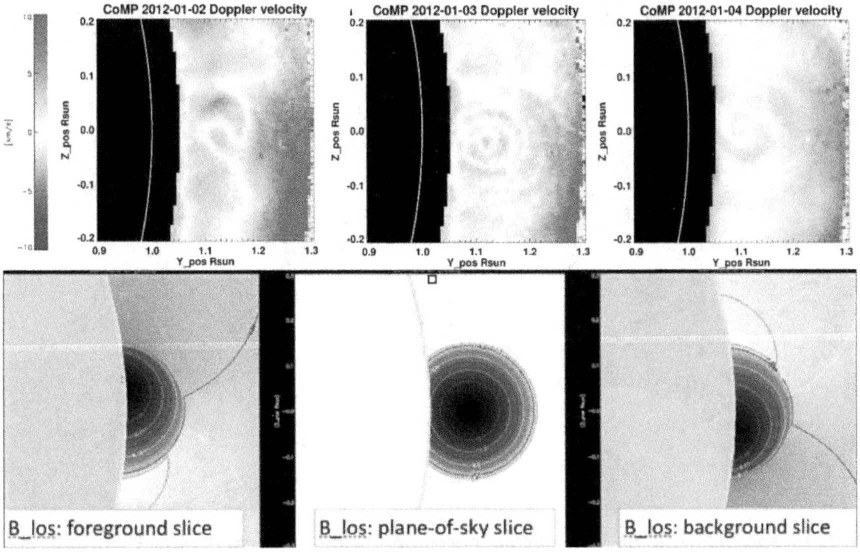


**Figure 5.** Cavities contain line-of-sight flows, sometimes with a bulls-eye pattern. Top: MLSO Coronal Multichannel Polarimeter (CoMP) Doppler velocity observations for a cavity seen over three days (from Bąk-Stęślicka *et al.* (2013)). Bottom: line-of-sight magnetic field in the plane of sky for flux rope model of Low & Hundhausen (1995).

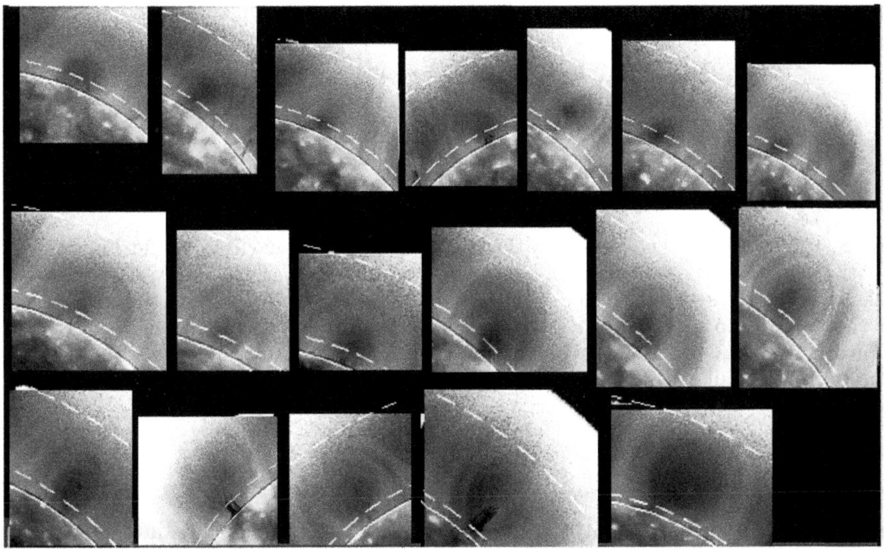

**Figure 6.** Cavities of a variety of sizes and shapes as seen in SDO/AIA 193 Å. Solid white line is at solar photosphere; dashed white line is at 1.05 solar radii (location of occulting disk for MLSO/CoMP telescope). Dates, starting at upper left: 5/25/11; 6/14/11; 6/24/11; 7/9/11; 7/14/11; 7/26/11; 7/27/11. Next row: 7/28/11; 7/29/11; 8/1/11; 8/10/11; 8/11/11; 8/12/11; Next row: 8/14/11; 8/24/11; 8/30/11; 11/11/11; 1/2/12.

global limit beyond which cavities are unstable. The EUV cavities of the survey of Forland *et al.* (2013) lie well below this height, but Figure 8 (right) indicates a slight tendency for higher cavities to be eruptive (red diamonds) rather than not (green triangles).

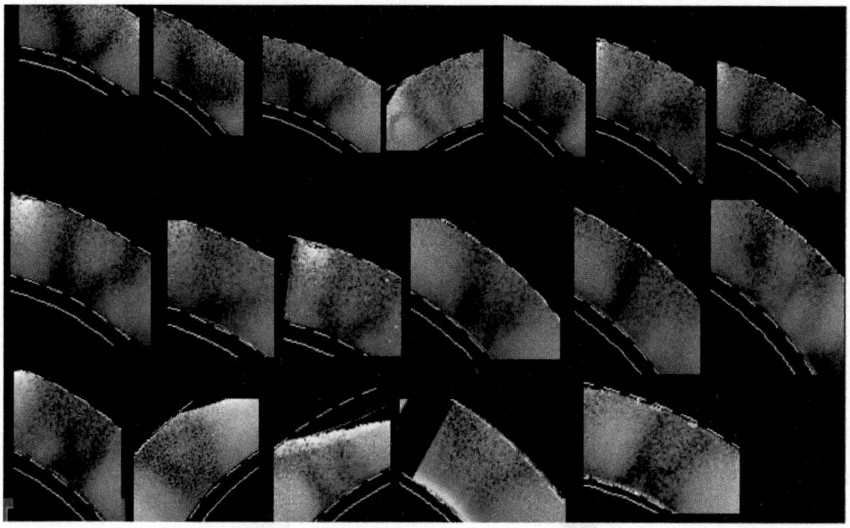

**Figure 7.** Linear polarization lagomorphs corresponding to the cavities of Figure 6, as seen by MLSO/CoMP.

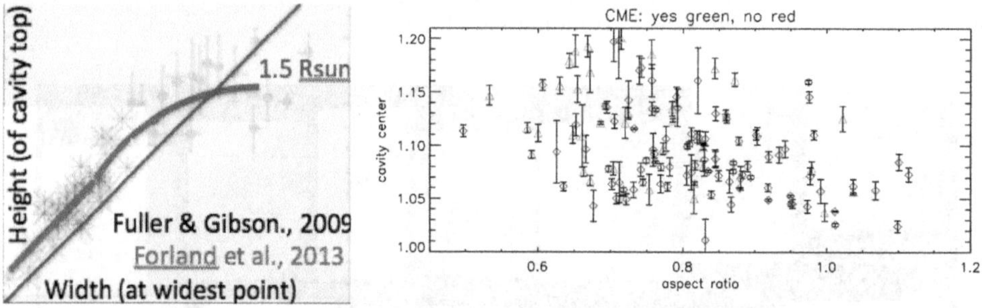


**Figure 8.** Cavity height and aspect ratio vs CME. Left: the cavities of the Fuller & Gibson (2009) white-light cavity study (error bars) and Forland *et al.* (2013) EUV cavity survey (red asterisks). Most cavities have aspect ratio less than one; cavity heights do not in general reach higher than about 1.5 solar radii. Right: cavity center heights tend to be higher for those that erupt (green triangles) than those that don't (red diamonds), and cavity morphology tends to be more narrow (small aspect ratio) (Forland *et al.* 2013).

Perhaps the strongest indicator of an impending eruption is the shape of the cavity. As seen in Figure 8, cavities tend to have aspect ratios less than one (i.e., their width is smaller than their height). Moreover, eruptive cavities in general have smaller aspect ratios than non-eruptive cavities. These aspect ratios are based on fitting the cavities with elliptical shapes. Forland *et al.* (2013) noted that in some cases cavities were better characterized as teardrop-shaped, and found that 68% of teardrop-shaped cavities erupted as CMEs as compared to 23% of elliptical cavities (and 10% of semicircular cavities). Gibson *et al.* (2006) noted similar behavior; due to the occulting disk, the full shape of the white-light cavities was not measured, but a quality referred to as "necking" was noted when cavities had narrower bases than tops. They found that 10/10 cases of cavities which erupted within 24 hours had necking, vs 25/99 of the entire sample.

## 4. Conclusions: cavities as magnetic flux ropes

Cavities are observed to be ubiquitous (Figures 1 and 2). If cavities are flux ropes, this is to be expected. A constant-alpha force-free state is the minimum energy configuration for a given boundary condition (Woltjer 1958): given sufficient helicity, this will be a flux rope. A large-scale force-free equilibrium of minimum energy conserving helicity can be reached through turbulent inverse cascade of helicity, from small scale to large (Taylor 1974). Since helicity is very nearly conserved even through magnetic reconnection (Berger & Field 1984), the free energy stored in the still-twisted large-scale magnetic fields represent "flare un-releasable" magnetic free energy (Zhang & Low 2005).

Cavities have arched, tunnel-like morphology with skinny-elliptical cross-section. Simulations have demonstrated that a flux rope expanding upwards into closed magnetic fields may find an equilibrium configuration as the forces causing the upward expansion are countered by confining magnetic tension forces. The equilibrium flux rope will then have an arched, tunnel-like morphology with narrow aspect ratio (Figure 3 (right)).

Cavities have low density, substructure, and are multithermal and dynamic (Figures 4-5). Schmit & Gibson (2014) (this issue) used hydrostatic models to argue that field lines at the center of the cavity, which are arched and non-dipped and relatively short (see blue lines in Figure 4), will have low density relative to surrounding winding/dipped field lines. The degree of depletion found was about 30% for a flux rope of aspect ratio (width, height, length) reasonable for a PCF. Schmit & Gibson (2013) argued that dynamic flow along dipped field lines driven, for example, by thermononequilibrium (Antiochos *et al.* 1999) provided an explanation for EUV horns above the cavity consistent with observations of cavities and prominences. Alternatively, Fan (2012) argued that heating and reconnection-driven flows at the top of the prominences could explain these structures, as well as the elevated temperatures of chewy nougats. In general, flows along magnetic flux surfaces, particularly those where dynamics might be expected such as at the interface of dipped and non-dipped field, may explain disk and ring-like structures and flows within cavities. Figure 5 (bottom) demonstrates this; if flows are field-aligned and assuming constant velocity, the line-of-sight component would peak at a flux rope's axis. Moreover, the shift of the flux rope in front or behind the plane of sky might introduce asymmetries in line-of-sight flow such as have been observed.

Polar crown filament cavities exhibit lagomorphic linear polarization signals (Figures 6-7). Bąk-Stęślicka *et al.* (2013) used forward modeling techniques to demonstrate that such lagomorphs are to be expected for a cylindrical flux rope extended along the line of sight. Rachmeler *et al.* (2013) discussed the often subtle differences between the flux-rope and the sheared-arcade model linear polarization signatures that might be expected in PCF cavities. Interestingly, Dove *et al.* (2011) found a different, ring-like linear polarization signal in a cavity that matched that predicted by a spheromak-type flux rope (e.g., Gibson & Low (1998)). This observation was taken in 2005 before CoMP was deployed at MLSO in Hawaii, and was for a large, but not PCF cavity.

The clear association of CMEs with high, narrow, and teardrop-shape cavities is an intriguing clue to why eruptions occur. Such a shape may occur if, for instance, a current sheet forms below a flux rope. If this is followed by reconnections at this current sheet and the slow rise of the flux rope, the increased height for cavities immediately prior to eruption may be explained. This is the picture painted by simulations which find such behavior leading up to the ultimate loss of stability through a "torus instability" (Aulanier *et al.* 2010, Savcheva *et al.* 2012, Fan 2012).

**Acknowledgements.** I thank and acknowledge the members of the International Space Science Institute (ISSI) working groups on coronal cavities (2008–2010) and coronal magnetism (2013–2014). I am particularly indebted to Urszula Bąk-Stęślicka, Giuliana de Toma, James Dove, Yuhong Fan, Blake Forland, Jim Fuller, Terry Kucera, B. C. Low, Laurel Rachmeler, Kathy Reeves, and Don Schmit.

## References

Antiochos, S. K., MacNeice, P. J., Spicer, D. S., & Klimchuk, J. A. 1999, Astrophys. J., 512, 985

Aulanier, G., Török, T., Démoulin, P., & DeLuca, E. E. 2010, Astrophys. J., 708, 314

Bąk-Stęślicka, U., Gibson, S. E., Fan, Y., Bethge, C., Forland, B., & Rachmeler, L. A. 2013, Astrophys. J., 770, L28

—. 2014, IAU S300 Proceedings

Berger, M. A. & Field, G. B. 1984, J. Fluid Mech., 147, 133

Dove, J., Gibson, S., Rachmeler, L. A., Tomczyk, S., & Judge, P. 2011, Astrophys. J., 731, 1

Fan, Y. 2012, Astrophys. J., 758, 60

Forland, B. F., Gibson, S. E., Dove, J. B., Rachmeler, L. A., & Fan, Y. 2013, Solar Phys., doi:10.1007/s11207-013-0361-1, *Online First*

Forland, B. F., Gibson, S. E., Dove, J. B., & Kucera, T., 2014, *IAU S300 Proceedings*

Fuller, J. & Gibson, S. E. 2009, Astrophys. J., 700, 1205

Gibson, S. E. & Fan, Y. 2006, J. Geophys. Res., 111

Gibson, S. E., Foster, D., Burkepile, J., & de Toma, G., A., S. 2006, Astrophys. J., 641, 590

Gibson, S. E., Kucera, T. A., Rastawicki, D., Dove, J., de Toma, G., Hao, J., Hill, S., Hudson, H. S., Marque, C., McIntosh, P. S., Rachmeler, L., Reeves, K. K., Schmieder, B., Schmit, D. J., Seaton, D. B., Sterling, A. C., Tripathi, D., Williams, D. R., & Zhang, M. 2010, Astrophys. J., 723, 1133

Gibson, S. E. & Low, B. C. 1998, Astrophys. J., 493, 460

Hudson, H. S., Acton, L. W., Harvey, K. A., & McKenzie, D. M. 1999, Astrophys. J., 513, 83

Kucera, T. A., Gibson, S. E., Schmit, D. J., Landi, E., & Tripathi, D. 2012, Astrophys. J., 757, 73

Li, X., Morgan, H., Leonard, D., & Jeska, L. 2012, Astrophys. J. Lett., 752, L22

Low, B. C. & Hundhausen, J. R. 1995, Astrophys. J., 443, 818

Marqué, C. 2004, Astrophys. J., 602, 1037

Rachmeler, L. A., Gibson, S. E., Dove, J. B., DeVore, C. R., & Fan, Y. 2013, Solar Phys., doi:10.1007/s11207-013-0325-5, *Online First*

Reeves, K. K., Gibson, S. E., Kucera, T. A., & Hudson, H. S. 2012, Astrophys. J., 746, 146

Savcheva, A., Green, L., van Ballegooijen, A., & DeLuca, E. 2012, Astrophys. J., 759, 105

Schmit, D. J. & Gibson, S. E. 2011, Astrophys. J., 733, 1

Schmit, D. J. & Gibson, S. E. 2013, Astrophys. J., 770, 35

Schmit, D. J. & Gibson, S. E. 2014, IAU S300 proceedings

Schmit, D. J., Gibson, S. E., Tomczyk, S., Reeves, K. K., Sterling, A. C., Brooks, D. H., Williams, D. R., & Tripathi, D. 2009, Astrophys. J. Lett., 700, 96

Tandberg-Hanssen, E. 1974, Geophysics and Astrophysics Monographs, 12

Taylor, J. B. 1974, Phys. Rev. Lett., 33, 19

Tomczyk, S., Card, G. L., Darnell, T., Elmore, D. F., Lull, R., Nelson, P. G., Streander, K. V., Burkepile, J., Casini, R., & Judge, P. G. 2008, Solar Phys., 247, 411

Waldmeier, M. 1970, Solar Phys., 15, 167

Wells, H., G. 1897, The invisible man, a grotesque romance (Bartleby.com, online edition, 2000), 84–85

Woltjer, L. 1958, Proceedings of the National Academy of Science, 44, 489

Zhang, M. & Low, B. C. 2005, Ann. Rev. of Astron. and Astrophys., 43, 103

*Nature of Prominences and their role in Space Weather*
*Proceedings IAU Symposium No. 300, 2013*
*B. Schmieder, J.-M. Malherbe, & S. T. Wu, eds.*

doi:10.1017/S1743921313010880

# The Formation of a Cavity in a 3D Flux Rope

**Donald Schmit[1] & Sarah Gibson[2]**

[1]Max Planck Institute for Solar System Research
Max Planck Str. 2
Katlenburg-Lindau, Germany
email: schmit@mps.mpg.de

[2]National Center for Atmospheric Research
Boulder, Colorado, USA

**Abstract.** There are currently no three dimensional numerical models which describe the magnetic and energetic formation of prominences self-consistently. Consequently, there has not been significant progress made in understanding the connection between the dense prominence plasma and the coronal cavity. We have taken an ad-hoc approach to understanding the energetic implications of the magnetic models of prominence structure. We extract one dimensional magnetic field lines from a 3D MHD model of a flux rope and solve for hydrostatic balance along these field lines incorporating field-aligned thermal conduction, uniform heating, and radiative losses. The 1D hydrostatic solutions for density and temperature are then mapped back into three dimensional space, which allows us to consider the projection of multiple structures. We find that the 3D flux rope is composed of several distinct field line types. A majority of the flux rope interior field lines are twisted but not dipped. These field lines are density-reduced relative to unsheared arcade field lines. We suggest the cavity may form along these short interior field lines which are surrounded by a sheath of dipped field lines. This geometric arrangement would create a cavity on top of a prominence, but the two structures would not share field lines or plasma.

## 1. Introduction

The energetic formation has a been central question for solar physics for several decades. We know that the solar atmosphere is heated, most likely by the dissipation of magnetic energy driven by convection. The energy input in the corona is partially dissipated by radiative losses primarily from optically thin UV/EUV emission lines of highly ionized elements. The primary mechanism for coronal energy loss is field-aligned thermal conduction. The energy that is input into the corona that cannot be effectively radiated is redirected through thermal conduction to the transition region and the chromosphere. These foundational ideas were brought together by Rosner *et al.* (1978, hereafter referred to as RTV). One of basic consequences of the equations for hydrostatic balance in the corona is that the magnetic loop geometry has a determining effect on the properties of the plasma embedded along that loop. In the constant pressure approximation imposed in RTV scaling, the loop length enters the equations in relation to the scale of thermal conduction and heat input. In a gravitationally stratified loop (Serio *et al.* 1981), the pressure gradient throughout the loop varies based on the gravity component parallel to the field orientation. We extend the work of these authors by studying how variations in the geometry of flux rope loops affect the three dimensional density structure of the flux rope. An example of a hydrostatic solutions are given in Figure 1.

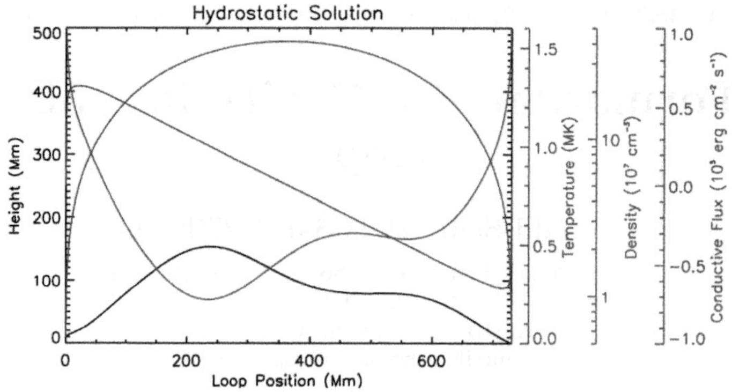


**Figure 1.** Solutions for density (blue), temperature (red), and conductive flux (green) along an asymmetric flux rope field line geometry (black).

## 2. Flux Rope Model

Fan & Gibson (2006) emerge a flux rope through a line-tied lower boundary into a low-beta, isothermal corona. The model evolves through quasi-static equilibrium as the flux rope is kinematically emerged. We extract magnetic field lines from a partially emerged equilibrium state (identical dataset as in Gibson & Fan 2006). Field lines from within the flux rope exhibit several distinct geometries. We find only a small fraction of the flux rope volume contains field lines with magnetic dips. Of the 1300 field lines which were extracted in uniform spacing in the r = 1.02 plane, 270 were found to contain dips. Magnetic dips are important as they are able to stably hold prominence plasma from falling into the chromosphere. Dipped field lines form a sheath-like surface that surrounds the axis of the flux rope. Arcade loops that surround the emerging rope have expanded to accommodate the additional flux.

While these loops have geometric differences discussed above, they also differ strongly in length. Figure 2 shows how field line length varies through cross sections of the model volume. The interior of the flux rope is composed of short field lines. The field line geometry changes as a function of radial distance from the axis (height of 1.4 $R_s$); there is a gradual lengthening of the loops that is related to the increasing winding number. There is a sharp boundary at the outer edge of the flux rope. The outer flux rope field lines are 20% longer than the neighboring external arcade field lines. Both length and geometry affect the energy balance of the corona. We will now discuss how these quantities enter into the equations, and how we solve for the plasma properties within these disparate structures.

## 3. Hydrostatic Calculation

The equations for hydrostatic balance within a coronal flux tube can be written as a three first-order ordinary differential equations as described in Vesecky *et al.* (1979). The three equations are:

$$\frac{dn}{ds} = \frac{-mg_s - 2kF_c}{2kT} n, \quad \frac{dF_c}{ds} = E - n^2 \Lambda(T), \quad \frac{dT}{ds} = F_c T^{-5/2},$$

where $n$ is the density, $T$ is the temperature, $F_c$ is the conductive flux, $m$ is the mass of the proton, $g_s$ is the component of gravity parallel to $\hat{s}$, $E$ is the heat input, and $\Lambda$ is the temperature-dependent radiative loss function. In this experiment, we use an identical $\Lambda(T)$ to Vesecky *et al.* (1979). $E$ is set to $2\times10^{-6}$ erg cm$^{-3}$ s$^{-1}$. Unlike Vesecky *et al.*

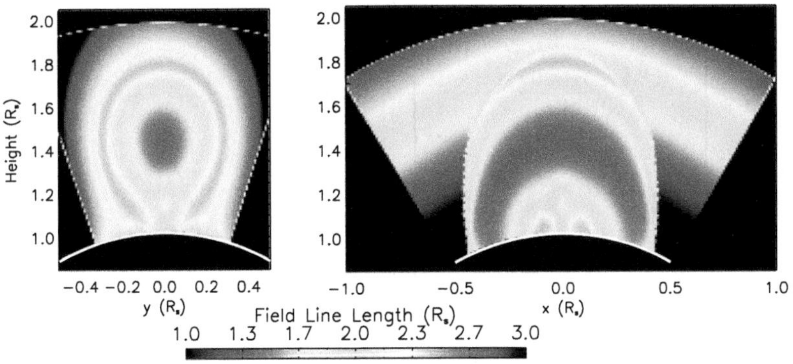


**Figure 2.** Variation of field line length within and surrounding the flux rope. The left panel displays a cross-section (x = 0) of the model volume viewed along the axis of the rope. The right panel displays the cross section (y = 0) across the axis.

(1979), we integrate the hydrodynamic equations foot point to foot point. This presents a problem as we have two foot points and three boundary conditions. We assume that the temperature and the conductive flux are fixed at the initial foot point ($3\times10^4$K and $10^{-3}$ erg cm$^{-2}$ s$^{-1}$, respectively) and that the temperature at far-side foot point is also $3\times10^4$K. We numerically integrate the hydrostatic equations using the Adams-Bashford-Moulton scheme (Shampine & Gordon 1975). The density is varied at the near-side foot point until a solution matching the far-side temperature boundary condition is met within a threshold.

We have applied this method to the 1300 field lines which were extracted from the MHD model. The density solutions from these field lines are compiled into an irregular three dimensional grid. Figure 3 shows the density for two different lines of sight through the 3D grid: one in the flux rope (blue diamonds) and one in the arcade (red diamonds) at the same projected height (1.4 $R_S$). There are several important elements to understand in Figure 3. The flux rope line of sight maintains a density depletion between $\Delta x = 0$ and $\Delta x = 00.32$ $R_s$. The peak depletion is around 35% at $\delta x = 0$. The flux rope interior and the arcade are characterized by different field line geometries, which are signified by colored arrows in Figure 3a which match with the representative field lines in Figure 3b and 3c. The pink field line in the flux rope interior reaches a maximum height of 1.4 $R_s$ while the green field line in the arcade achieves a maximum height of 1.7 $R_s$. In hydrostatic equilibrium with a fixed foot point temperature, the density at the foot points must increase as a function of maximum loop height; higher pressure is needed to support the additional mass in taller loops. For the arcade line of sight, there is a decrease in density as a function of $\Delta x$. This is caused the the arched shape of the partially emerged flux rope. For the flux rope line of sight, there is a increase in density as a function $\Delta x$. Our model suggests that this is related to the transition from short axial field lines to long dipped field lines (similar to light blue line in Figure 3b and 3c). These field lines wrap around the density depleted interior field lines and are significantly higher density.

## 4. Conclusions

This experiment presents us with evidence for a simple interpretation of the contrast between the low density cavity and the high density streamer. The cavity is composed of short field lines while the streamer is composed of taller, longer field lines. This interpretation is completely consistent with RTV scaling as well our current models magnetic models for prominence structure. This model makes many assumptions: constant

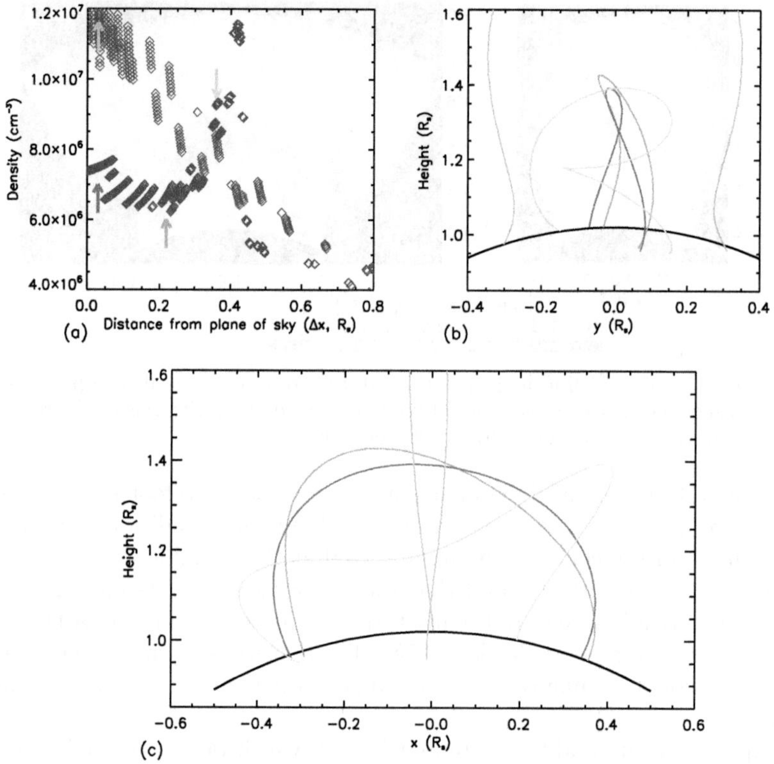


**Figure 3.** Line of sight projection of density structure. (a) Density as a function of x-position looking through the cavity (z = 1.4, y = 0) is denoted by blue diamonds. A line of sight through the arcade (z = 1.4, y = 0.3) is denoted by red diamonds. Example field lines which intersect the lines of sight viewed along the axis (b) and across the axis (c). The color of the field lines match the colored arrows in Figure 3a showing the density of the line.

area flux tubes, uniform heating, a highly idealized magnetic structure. However, we believe this simple model illustrates the important role that field line geometry may play in the morphology of prominence-cavity structure.

*The National Center for Atmospheric Research is funded by the National Science Foundation. This work was aided by discussions at the International Space Science Institute in Bern.*

## References

Fan, Y. & Gibson, S. E. 2006, Astrophys. J. Lett., 641, 149

Gibson, S. E. & Fan, Y. 2006, J. Geophys. Res.

Rosner, R., Tucker, W. H., & Vaiana, G. S. 1978, Astrophys. J., 220, 643

Serio, S., Peres, G., Vaiana, G. S., Golub, L., & Rosner, R. 1981, Astrophys. J., 243, 288

Shampine, L. & Gordon, M. 1975, Computer Solution of Ordinary Differential Equations: the Initial Value Problem (W. H. Freeman and Co.)

Vesecky, J. F., Antiochos, S. K., & Underwood, J. H. 1979, Astrophys. J., 233, 987

*Nature of Prominences and their role in Space Weather*
*Proceedings IAU Symposium No. 300, 2013*
*B. Schmieder, J.-M. Malherbe & S. T. Wu, eds.*

doi:10.1017/S1743921313010892

# Observation of the prominence cavity region using slitless eclipse flash spectra and space borne EUV filtergrams

**Cyrille Bazin[1,2], Serge Koutchmy[1], Philippe Lamy[2], and Ehsan Tavabi[3]**

[1]IAP, CNRS-UMR 7095 -UPMC, 98 bis Bd Arago 75014 Paris, France
email: bazin@iap.fr; koutchmy@iap.fr
[2]LAM, CNRS-UMR 6110 -AMU, Pole de l'Etoile, 38 Rue Joliot Curie 13388 Marseille, France
email: philippe.lamy@oamp.fr [3]Payame Noor University of Tehran, 14155-6466, I.R. of Iran
email: etavabi@gmail.com

**Abstract.** We used total solar eclipse free of parasitic light for studying the prominence to corona interface, and the corresponding cavity in the context of the coronal physics. We analysed the visible continuum between the prominences to directly look at the electron density. We demonstrate some enhanced heating in the cavity region. Some similarities with the interface regions are shown: the photosphere to the chromosphere and the prominence to the corona interface. The optically thin neutral Helium at 4713 Å and the singly ionized Helium 4686 Å Paschen $\alpha$ lines are considered. We summed 80 slitless visible eclipse flash spectra that we compare with simultaneously obtained EUV SWAP/Proba2 174 Å images of ESA and AIA/SDO 171Å 193 Å 304 Å and 131 Å filtergrams. Intensity profiles in a radial direction are studied. We deduce the variation of the intensity ratio $I$(He I 4713) / $I$(He II 4686). Discussion: the temperature rises at the edge of the prominences. We evaluate for the first time with spectrophotometric accuracy the continuum modulations in prominence spectra. W-L intensity deficits are observed near the prominence boundaries in both eclipse spectra and in EUV images, confirming that the prominence -cavity regions correspond to a relative depression of plasma density of the surrounding corona. Conclusion: we demonstrate some enhanced heating occurring in these regions assuming hydrostatic equilibrium.

**Keywords.** flash spectra, corona, prominence, cavity, EUV lines, He lines

## 1. Introduction

We take advantage of the occultation of the Sun by the Moon to analyse the low layers of the solar atmosphere at different heights without parasitic light (since the occultation takes place in space). Eclipse slitless flash spectra were obtained during the second contact, just before the totality of the solar eclipse of 11 July 2010 observed in French Polynesia. We used the grating objective technique and fast CCD imaging (see Bazin, Koutchmy and Tavabi 2012). The motivations are to try to explain some similarities between the high First Ionisation Potential- FIP of helium lines intensity ratio in the Chromosphere-Corona Interface CCI and the Prominence to Corona Interface PCI. The low FIP of the faint metallic emission lines and the helium shells above the solar limb are considered in Bazin and Koutchmy (2013). We discuss the importance of the prominence continuum. The optically thin helium line emissions (He I 4713 Å and He II 4686 Å) occurring at total solar eclipses were already analysed using slit spectrograph by Hirayama and Irie (1984). Simultaneous space borne SDO/AIA images (free of seeing effects), described in Cheimoto *et al.*(2009) are now additionally used.

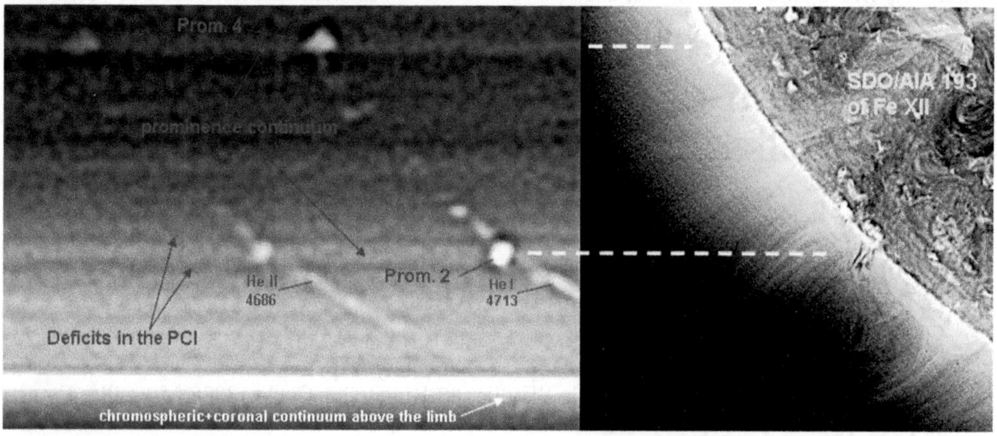


**Figure 1.** Simultaneous observations of the prominences using the 2010 eclipse flash spectra (left panel) and an image of 12 stacked EUV filtergrams (right panel). The left panel presents the sum of 80 individual flash spectra. The spatial dispersion runs along the horizontal direction and extends over $\approx$ 65 ÅThe field -of-view along the vertical direction extends over 300 Mm. The prominence studied here labeled 2 is highlighted by its main helium 4713 Å and 4686 Å emission lines and its continuum as well as the deficit at the interface between the prominence and the corona. The bright horizontal band near the bottom of the figure is not an artefact but corresponds to the intense chromospheric and coronal continuum above the limb. The right panel presents the summed image at 193 Å (Fe XII emission at a temperature of 1.5 MK) showing the same prominence 2 in absorption (note also spicules seen in absorption).

## 2. Observation and data analysis

We summed twelve Fe XII 193 Å images taken around the time of the eclipse for improving the signal to noise ratio. Figure 1 shows the comparison between the prominences 2 and 4 seen in He I 4713 Å and He II 4686 Å lines from the 80 summed flash spectra and the prominence 2 seen in absorption with the surrounding cavity using the SDO/AIA Fe XII 193 Å 1.5 MK temperature coronal line.

### 2.1. *Helium flash spectra*

We then removed the coronal background continuum to obtain the true intensity ratio of He II 4686 Å / He I 4713 Å with the radial distance. Figure 2 shows the resulting values. A comparable analysis was recently done in Labrosse *et al.*(2011) for the He II 256 Å EUV line which is unfortunatly blended like the He II 304 Å line is.

The intensity ratio of He II/He I at heights from 0 to 3 Mm shows a more than 2 times increase (see Figure 2). The same behaviour is seen in the CCI as in the PCI at heights starting from 4 Mm to 22 Mm. In both cases, a dynamic state is observed at small scale, often attributed in the past, to a turbulent effect. These results allow to diagnose the helium shell structuration and the contribution of high FIP elements in these regions.

### 2.2. *EUV filtergrams*

The image in the Fe XII 193 Å line suggests some twisted channels seen in absorption corresponding to the core of the prominence 2. The surrounding cavity shows plasma deficits, see Figure 1 at right. Intensity radial cuts taken along and outside the cavity region in 193 Å are plotted in Figure 3 for evaluating the scale height using second order and first order exponential decays that fitted the radial profiles along and outside the cavity, respectively. The plasma deficit in the PCI could be associated with some higher temperature (assuming hydrostatic equilibrium), because of the rather low radial gradient

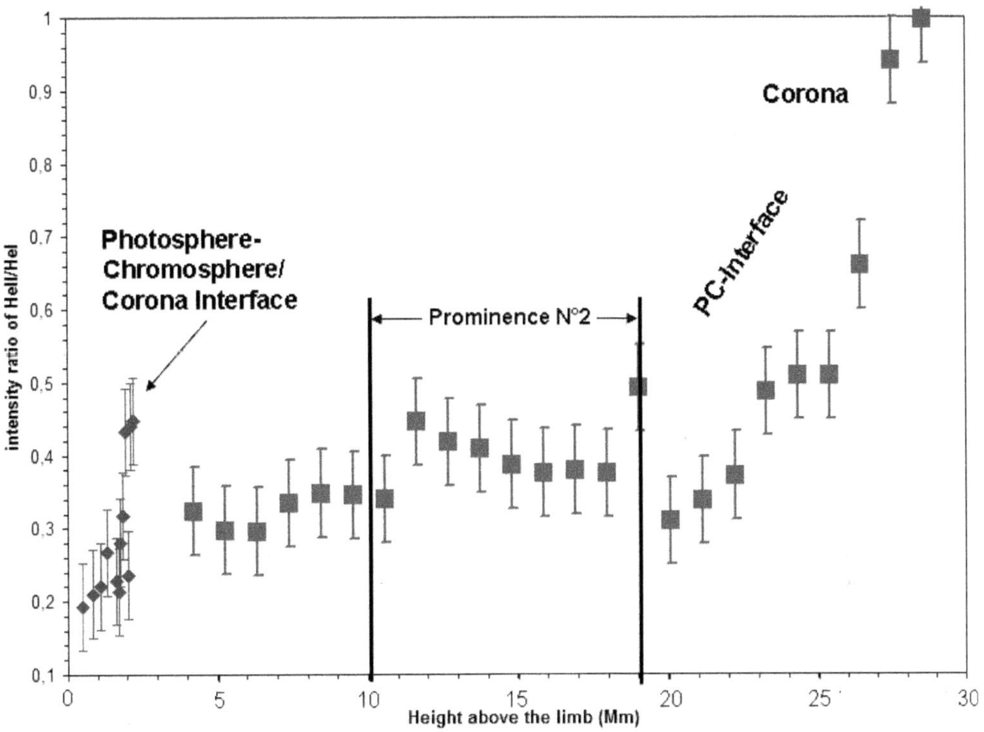


**Figure 2.** Intensity ratio measurements of the He II 4686 Å / He I 4713 Å as function of the radial distance starting above the limb and upwards to a prominence from eclipse flash spectra. The error bars correspond to one sigma standard deviation, after the coronal continuum background was removed. The losanges correspond to lower altitudes where the crescents of helium lines emissions are recorded, and squares indicate the higher altitudes prominence values.

of densities around and above this prominence. We found $I(h) = 1224*\exp(-h/18.7) + 454*\exp(-h/227.4)$, where $h$ is in Mm, along the cavity and $I'(h) = 1724*\exp(-h/36.1) + 23.5$ outside the cavity. The intensity distribution expressed by the Differential Emission Measure will be more extensively discussed in a forthcoming paper.

## 3. Discussion

The plots in figure 2 show approximately the same intensity ratio values around 0.4 for both the solar CCI in the lower altitudes, where the gravity could still dominate, and in the PCI at higher altitudes, where the more extended magnetic field dominates. The $\beta$ of the plasma is low. The continuum in the prominence spectra indicates Thomson scattering from electrons, and possibly a Paschen continuum contribution in denser parts, see Vial *et al.*(1992). The "cold" plasma of the prominence core where the gravity is negligible is surrounded by the hot corona seen in emission of EUV lines. One proposed mechanism responsible of the ionisation of the optically thin He II 4686 Å P$\alpha$ line is the photo-ionisation process by photons coming from the corona like the EUV lines of Fe XII and hotter ones. Because of the rather low radial gradients of densities around the prominence suggesting high temperature when hydrostatic equilibrium is assumed, the plasma deficit could be associated with some enhanced heating. Figure 3 points the slopes taken on radial cuts along and outside the cavity region.

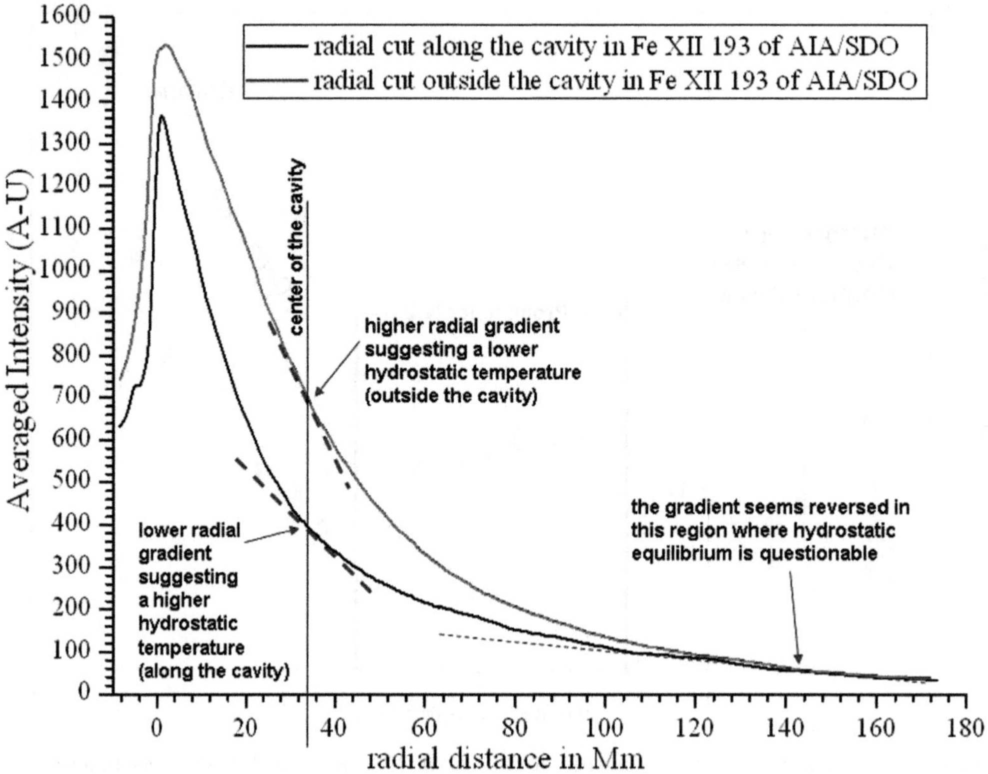


**Figure 3.** Averaged intensity radial cuts along and outside the south- east cavity at 138° in the 193 Å Fe XII SDO-AIA deduced from the image shown in figure 1

## 4. Conclusion

Eclipse flash spectra permit to probe the deep solar interfaces without any parasitic light for evaluating their extension, including the temperature gradients. Eclipse observation gives snapshots and no temporal coverage. Unfortunatly, the P$\alpha$ He II line cannot be observed near the limb outside of eclipse conditions, see Worden Beckers and Hirayama (1973). The enhancement of heating in the Fe XII 193 Å cavity compared to the surrounding corona is in agreement with the results of Habbal *et al.*(2010) coming from the analysis of eclipse filtergrams. A better spatial resolution is needed to definitely disentangle the LOS integration effect related to the density irregularities. Visible and EUV results seem compatible but their combined analysis to study the temperature gradients is beyond the scope of this contribution.

### References

Bazin, C. Koutchmy, S., & Tavabi, E. 2012, *Solar Physics*, 286, 255

Bazin, C. & Koutchmy, S. 2013, *Journal of Advanced Research*, 4, 307

Cheimets, P. Caldwell, D. C. Chou, C. *et al.* 2009, *Solar Physics and Space Instrumentation III*, Proc. SPIE 7438, 14

Habbal, S. R. Druckmuller, M. Morgan *et al.* 2010, *Astrophysical Journal*, 719, 1362

Hirayama, T. & Irie, M. 1984, *Solar Physics*, 90, 291

Labrosse, N. Schmieder, B. Heinzel, P. *et al.* 2011, *Astronomy and Astrophysics*, 531, 69

Vial, J.-C. Koutchmy, S. *et al.* 1992, *ESA SP* 344, 87

Worden, S. P. Beckers, J. M., & Hirayama, T. 1973, *Solar Physics*, 28, 73

*Nature of Prominences and their role in Space Weather*
*Proceedings IAU Symposium No. 300, 2013*
*B. Schmieder, J.-M. Malherbe & S. T. Wu, eds.*

doi:10.1017/S1743921313010909

# Large-amplitude longitudinal oscillations in solar prominences

**Manuel Luna[1,2], Judith Karpen[3], Antonio Díaz[1,2], Kalman Knizhnik[4,3], Karin Muglach[5,3], Holly Gilbert[3], and Therese Kucera[3]**

[1]Instituto de Astrofísica de Canarias,
E-38200 La Laguna, Tenerife, Spain
[2]Universidad de La Laguna, Dept. Astrofísica,
E-38206 La Laguna, Tenerife, Spain
email: mluna@iac.es

[3]NASA/GSFC, Greenbelt, MD 20771, USA
[4]Johns Hopkins University, Baltimore, MD USA
[5]ARTEP, Inc., Maryland, USA

**Abstract.** Large-amplitude longitudinal (LAL) prominence oscillations consist of periodic mass motions along a filament axis. The oscillations appear to be triggered by an energetic event, such as a microflare, subflare, or small C-class flare, close to one end of the filament. Observations reveal speeds of several tens to 100 km/s, periods of order 1 hr, damping times of a few periods, and displacements equal to a significant fraction of the prominence length. We have developed a theoretical model to explain the restoring force and the damping mechanism. Our model demonstrates that the main restoring force is the projected gravity in the flux tube dips where the threads oscillate. Although the period is independent of the tube length and the constantly growing mass, the motions are strongly damped by the steady accretion of mass onto the threads. We conclude that the LAL movements represent a collective oscillation of a large number of cool, dense threads moving along dipped flux tubes, triggered by a nearby energetic event. Our model yields a powerful seismological method for constraining the coronal magnetic field strength and radius of curvature at the thread locations.

**Keywords.** Solar prominence, Oscillations, Magnetic structure.

---

## 1. Observational LAL Oscillations Features

Large-amplitude longitudinal (LAL) oscillations in prominences were first reported by Jing *et al.* (2003); few additional observations have been reported which are summarized in Table 1 (Jing *et al.* 2006, Vršnak *et al.* 2007, and Zhang *et al.* 2012). The oscillations are longitudinal because the displacements are mainly along the axis of the filament or slightly skewed. According to the standard filament models and sparse observations of filament magnetic fields (see, Trujillo-Bueno *et al.* 2002 & Casini *et al.* 2003) the LAL displacements are along the field. These oscillations have long periods of 50–160 minutes, and are damped in 2.3–6.2 cycles, with high velocity amplitudes in the range 30-100 $\mathrm{km\,s^{-1}}$. LAL oscillations are apparently triggered by an energetic event: a subflare, a microflare, or a flare close to the filament in all reported events.

## 2. Theoretical Interpretation

LAL oscillation events are impressive due to the high speeds and large displacements reached by the massive prominences. Although the energy and momentum of the prominence are huge, the acceleration reached in the oscillatory motion can be as large as

**Table 1.** Some observational features of LAL oscillations reported so far.

| Observation | Amp. ($km\ s^{-1}$) | Period (min) | Damping Time (min) | Trigger |
|---|---|---|---|---|
| Jing *et al.* (2003) | 92 | 80 | 210 (2.6 periods) | Nearby sub-flare |
| Jing *et al.* (2006) | 50 | 160 | 600 (3.8 periods) | Nearby C-Class flare |
| Jing *et al.* (2006) | 30 | 150 | — | Nearby microflare |
| Jing *et al.* (2006) | 100 | 100 | — | — |
| Vršnak *et al.* (2007) | 51 | 50 | 115 (2.3 periods) | Nearby sub-flare |
| Zhang *et al.* (2012) | 40 | 52 | 133 (2.6 periods) | Nearby C-Class flare |
| Li & Zhang (2012) | 30-60 | 44-67 | — | Nearby flare activity |

100 m s$^{-2}$, comparable to the solar gravity acceleration ($g_0 = 274$ m s$^{-2}$). This motion is rapidly damped in only a few periods, demonstrating that a very efficient damping mechanism is operating. Several mechanisms have been proposed to explain the restoring force and damping of the LAL oscillations, but most of them do not successfully describe the thread motions (see review by Tripathi *et al.* 2009). Some authors claim that the restoring force is associated with the magnetic tension; in that case, the associated displacements would be perpendicular to the local magnetic field, contrary to what the observations show. Another possible restoring force is the gas pressure (slow MHD mode), but the temperatures necessary to accelerate the prominence threads to observed values are on the order of 10 MK. Although these temperatures are not observed in the oscillations, they could be present in the initiation event, e.g. a flare.

We have studied the oscillations of threads forming the basic components of a prominence (Luna & Karpen 2012) in a 3D sheared arcade (DeVore *et al.* 2005, Luna *et al.* 2012a). According to the thermal nonequilibrium model (Antiochos *et al.* 1999), the cool prominence threads form as the result of heating at the base of long coronal loops in filament channels, which produces chromospheric evaporation and subsequent dynamic radiative collapse. Inspired by the transient oscillations seen in our earlier simulations of this process, we developed the first self-consistent model for the observed large-amplitude oscillations along filament axes that explains both the restoring force and the damping mechanism. Because the corona is low-$\beta$, it is reasonable to assume that the threads are ducted along concave-upward (dipped) magnetic flux tubes as a solid body. We also assume that the density of the threads is much larger than the ambient coronal density. The simulations reveal that the pressure difference force between both ends of the thread is too small to produce the acceleration associated with the oscillations (see details in Luna *et al.* 2012a). Thus, the pressure force is not the restoring mechanism in the LAL oscillations, and the corona has little influence on the thread dynamics.

As long as the footpoint heating is steady, the mass of the threads continuously increases with time ($t$) by accretion of evaporated chromospheric plasma. Our simulations show that $m(t) = m_0 + \alpha(t - t_0)$ with a typical accretion rate $\alpha = 20 \times 10^6\ kg/hr$. The center of mass position of each thread, $s$, is described by a zero-order Bessel equation (see Eq. 2 in Luna & Karpen 2012). The solutions are the zero-order Bessel functions that are intrinsically damped. However, it is necessary to introduce an additional damping term, $e^{-(t-t_0)/\tau_w}$, in order to account for additional weak damping found in the simulations. With these considerations the motion of the threads is described by

$$s(t) = s_0 + A\, J_0[\omega(t - t_0) + \omega m_0/\alpha]\; e^{-(t-t_0)/\tau_w}\ . \qquad (2.1)$$

The angular frequency of the oscillations is

$$\omega = \sqrt{\frac{g_0}{R}} , \tag{2.2}$$

where $R$ is the radius of curvature of the dipped portion of the flux tubes supporting the threads. This oscillation resembles the motion of a gravity-driven pendulum. In Luna *et al.* (2012b) we studied the normal modes of oscillation of threads in curved field lines, and found that there are two contributions to the oscillation frequency: the gravity-driven term (Eq. 2.2) and the pressure-driven slow mode. For typical prominence parameters the pressure-driven term is small, however, confirming that the restoring force is the projected gravity. Equation (2.2) allows us to directly calculate the radius of curvature of the field lines that support the prominence.

Two mechanisms damp the thread motion (Eq. 2.1). The most important is associated with the mass accretion and related to the phase of the Bessel function $\omega m_0/\alpha$. For a large accretion rate $\alpha$, the phase is close to zero and the Bessel function is strongly damped. In contrast, for small values of $\alpha$ the phase is large and the Bessel function is essentially undamped. The second mechanism is the weak damping characterized by $\tau_{\mathrm{w}}$ . The simulations show that this is unimportant during the first oscillations of the threads. We speculate that the weak damping is associated with a non-adiabatic effect of energy dissipation.

The observed LAL oscillations appear to be triggered by small flares close to the filaments. However, the precise mechanism is not known. In Luna & Karpen (2012) we proposed that the energetic event increases the heating impulsively at one end of the filament, increasing the evaporation of chromospheric plasma into the affected flux tubes. We speculate that extra heat is deposited only on those footpoints located closest to the energetic event, so only parts of the filament are excited simultaneously and oscillate collectively in phase. Zhang *et al.* (2013) studied numerically the triggering by impulsive microflare heating of one leg of a prominence, and found that threads start to oscillate. This study revealed that condensations are remarkably robust; increased heating at one footpoint produces a flow of hot evaporated plasma that pushes a thread to a new location, rather than destroying it in situ. Note that this result is consistent with our earlier studies of thermal nonequilibrium (*e.g.*, Karpen & Antiochos 2008).

## 3. Prominence Seismology

It is generally agreed that the dense prominence threads must be magnetically supported. Thus, the magnetic tension in the dipped part of the tubes must be larger than the weight of the threads. With this consideration and Equation (2.2), we derived the following equation that constrains the minimum field strength as a function of the thread oscillation period (see Eq. 7 of Luna & Karpen 2012):

$$B[\mathrm{G}] \geqslant 26 \left(\frac{n}{10^{11}\ \mathrm{cm}^{-3}}\right)^{1/2} P[\mathrm{hours}] . \tag{3.1}$$

For typical electron densities ($n \sim 10^{11}$ cm$^{-3}$) and LAL oscillation periods ($\sim 1$ hour) the estimated magnetic field is consistent with the sparse observations of intermediate-type prominence magnetic fields of around 20 Gauss (see review by Mackay *et al.* 2010 & T. Berger paper in this volume).

In addition to providing valuable information on the difficult-to-measure coronal magnetic field in filament channels, our model can also determine the radius of curvature of the prominence flux-tube dips from the observed oscillation properties. We have used the

Jing *et al.* (2003) and Vršnak *et al.* (2007) observations to solve Equation (2.2) for $R$. The observed periods are 80 min and 50 min respectively, yielding corresponding curvature radii of 152 Mm and 62 Mm. Assuming a typical prominence density of $n = 10^{11}$ cm$^{-3}$, we estimate the minimum field strength in the two prominences as $\sim$ 35 G and 22 G.

The capabilities of recent instruments such as AIA/SDO (see Lemen *et al.* 2012) and NASA's Interface Region Imaging Spectrograph (IRIS) allow more LAL events to be discovered and characterized. We have identified several LAL events in the AIA data, and have analysed the oscillations of one such filament observed on August 20, 2010. The analysis and results will be described in a forthcoming publication (Luna *et al.* 2013); preliminary results are presented in the Knizhnik *et al.* paper in this volume.

## 4. Conclusions

We conclude that the observed LAL motions represent a collective oscillation of many cool, dense threads moving along the magnetic field, triggered by a nearby energy release event (e.g. a small flare). The main restoring force is the projected gravity in the flux tube dips where the threads oscillate, even in threads with large radii of curvature. The period is independent of the tube length and of the constantly growing mass, but the oscillations are strongly damped by the mass accretion caused by persistent footpoint heating. After several oscillations the excited threads lose their coherence due to slight differences in the oscillation periods, causing different parts of the filament to oscillate increasingly out of phase as observed. The mechanism that initiates the collective oscillation is not determined in the present investigation. We speculate that the impulsive energetic event temporarily increases the evaporation at the closest filament flux-tube footpoints. This raises the plasma pressure only at one end of the threads, initiating the oscillations. More observations of oscillating prominences are needed to further refine and test our model.

## References


Antiochos, S. K., MacNeice, P. J., Spicer, D. S., & Klimchuk, J. A. 1999, *ApJ*, 512, 985

Casini, R., López Ariste, A., Tomczyk, S., & Lites, B. W. 2003, *ApJL*, 598, L67

DeVore, C. R., Antiochos, S. K., & Aulanier, G. 2005, *ApJ*, 629, 1122

Jing, J., Lee, J., Spirock, T. J., *et al.* 2003, *ApJL*, 584, L103

Jing, J., Lee, J., Spirock, T. J., & Wang, H. 2006, *Sol. Phys.*, 236, 97

Karpen, J. T. & Antiochos, S. K. 2008, *ApJ*, 676, 658

Lemen, J. R., Title, A. M., Akin, D. J., *et al.* 2012, *Sol. Phys.*, 275, 17

Luna, M., Karpen, J. T., & Devore, C. R. 2012a, *ApJ*, 746, 30

Luna, M. & Karpen, J. 2012, *ApJ*, 750, L1

Luna, M., Díaz, A. J., & Karpen, J. 2012b, *ApJL*, 757, L98

Luna, M., Knizhnik, K., Muglach, K., Gilbert, H, Kucera, T., & Karpen, 2013, *in preparation*

Mackay, D. H., Karpen, J. T., Ballester, J. L., Schmieder, B., & Aulanier, G. 2010, *Space Sci. Rev.*, 151, 333

Tripathi, D., Isobe, H., & Jain, R. 2009, *Space Sci. Rev.*, 149, 283

Trujillo Bueno, J., Landi Degl'Innocenti, E., Collados, M., Merenda, L., & Manso Sainz, R. 2002, *Nature*, 415, 403

Vršnak, B., Veronig, A. M., Thalmann, J. K., & Žic, T. 2007, *Astron. Astrophys.*, 471, 295

Zhang, Q. M., Chen, P. F., Xia, C., & Keppens, R. 2012, *Astron. Astrophy.*, 542, A52

Zhang, Q. M., Chen, P. F., Xia, C., Keppens, R., & Ji, H. S. 2013, *Astron. Astrophys.*, 554, A124

# Session I - 1.5

## SOLAR CYCLE EVOLUTION OF PROMINENCES AND ERUPTIONS

*Nature of Prominences and their role in Space Weather*
*Proceedings IAU Symposium No. 300, 2013*
*B. Schmieder, J. Malherbe & S. T. Wu, eds.*

doi:10.1017/S1743921313010910

# Unusual migration of the prominence activities in recent solar cycles

**Masumi Shimojo**

National Astronomical Observatory of Japan
2-21-1, Osawa, Mitaka, Tokyo, 181-8588, Japan
email: masumi.shimojo@nao.ac.jp

**Abstract.** We investigated the prominence eruptions and disappearances observed with the Nobeyama Radioheliograph during over 20 years for studying the anomaly of the recent solar cycle. Although the sunspot number of Cycle 24 is smaller than the previous one dramatically, the occurrence rate, size and radial velocity of the prominence activities are not changed significantly. We also found that the occurrence of the prominence activities in the northern hemisphere is normal from the duration of the cycle and the migration of the producing region of the prominence activities. On the other hand, the migration in the southern hemisphere significantly differs from that in the northern hemisphere and the previous cycles. Our results suggest that the anomalies of the global magnetic field distribution started at the solar maximum of Cycle 23.

**Keywords.** Prominence, solar cycle, microwave

## 1. Introduction

The solar activity of the recent solar cycles 23–24, shows significant differences from the previous cycles that were observed over the course of half of a century with modern instruments such as a magnetograph. The noticeable differences are the long cycle duration of Cycle 23 and the low activity of Cycle 24. Many authors have already reported other differences (Livingston *et al.* 2012, Petrie 2012, Wang *et al.* 2009, Gopalswamy *et al.* 2012, Shiota *et al.* 2012, Svalgaard & Kamide 2013, Lee *et al.* 2009, Wang *et al.* 2009, Thompson *et al.* 2011). To understand the anomalies of the solar activity in the recent solar cycles, we need to investigate the global distribution and evolution of the photospheric magnetic field. One way is to investigate directly the magnetic field distribution using the magnetograms. However, it is relatively difficult to understand an outline of the global magnetic field distribution because the magnetograms include the numerous fine structures. Dark filaments were used for understanding the global magnetic field distribution because they always lie on a magnetic neutral line. The Solar-Geophysical Data (SGD) published by NOAA National Geophysical Data Center had carried the Carrington maps that indicate the locations of dark filaments until the 1990s for this purpose.

The Nobeyama Radioheliograph (NoRH) is an interferometer in the microwave range, and can observe the thermal emission from a prominence well. Long-term studies were performed using this detection technique to understand polarity reversal (Copalswarmy *et al.* 2013a) and the relation between prominence activity and coronal mass ejections (Copalswarmy *et al.* 2013b) Based on the advantage of the NoRH for prominence observations, Shimojo *et al.* (2006) developed a semi-automatic detection method of prominence eruptions and disappearances for the NoRH data, and made a butterfly diagram of the prominence eruptions and disappearances. Considering that all dark filaments disappear

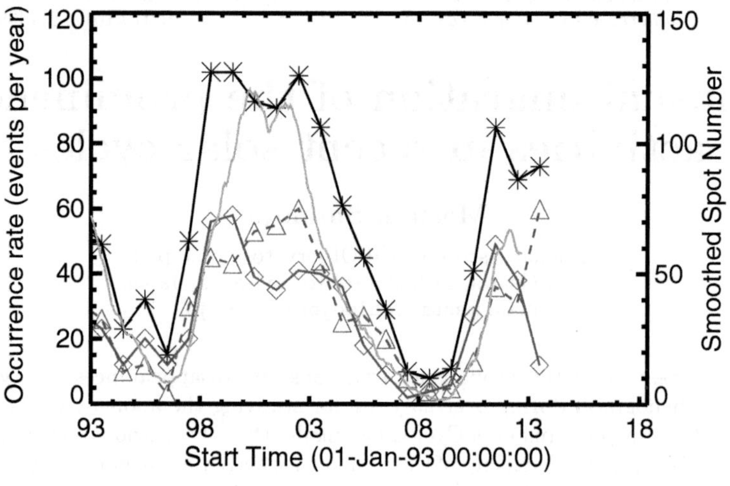


**Figure 1.** The occurrence rate of the prominence activities and sunspots from 1993 to 2013. The dark-gray line & diamond and the dark-gray dashed line & triangles show the occurrence rate in the northern and southern hemisphere. The black line is the total. The light gray line indicates the smoothed sunspot number.

by erupting and/or falling down, the distribution of the prominence eruptions and disappearances indicate that of the dark filaments. For this reason, the prominence eruption and disappearance are not distinguished and they are called "prominence activity" in this paper. To understand the global distribution and evolution of magnetic field of the Sun, we applied the prominence activity detection method developed by Shimojo *et al.* (2006) to over 20 years of NoRH data (July 1992 – March 2013) and detected 1131 events. Based on the events, we made the prominence activity database that includes the time, size, position, radial velocity and MPEG movie of them. The database is available at the website of the Nobeyama Solar Radio Observatory of the National Astronomical Observatory of Japan (http://solar.nro.nao.ac.jp/norh/html/prom_html_db/ see. Shimojo 2013).

From the next section, we show the solar cycle dependence of the occurrence rate, size, radial velocity of the prominence activities, and describe the migration of the producing region.

## 2. Solar cycle dependence of the occurrence, size & radial velocity

### 2.1. *Occurrence rate of prominence activities*

Figure 1 shows the time variations of the occurrence rate of the prominence activities and sunspots from 1993 to 2013. The value of 2013 hasa large uncertainty because the database includes only three months' data. The sunspot number in the figure is the six-month running mean calculated from the data that are released by the Solar Influence Data Center of the Royal Observatory of Belgium (SIDC-team 1992–2013).

Basically, the number variation of the the prominence activities is similar to that of the sunspot, but there are significant differences in the peak time of the values. The first peak of the prominence activities is faster than that of the sunspot number. It maybe caused by the high-latitude prominence activities that has no connection to sunspots as discuss in Copalswarmy *et al.* 2013a. The second peak of the prominence is delayed from the second peak of the sunspot number. We do not know what does caused the difference of the second peak. Although the sunspot number of Cycle 24 is smaller than that of

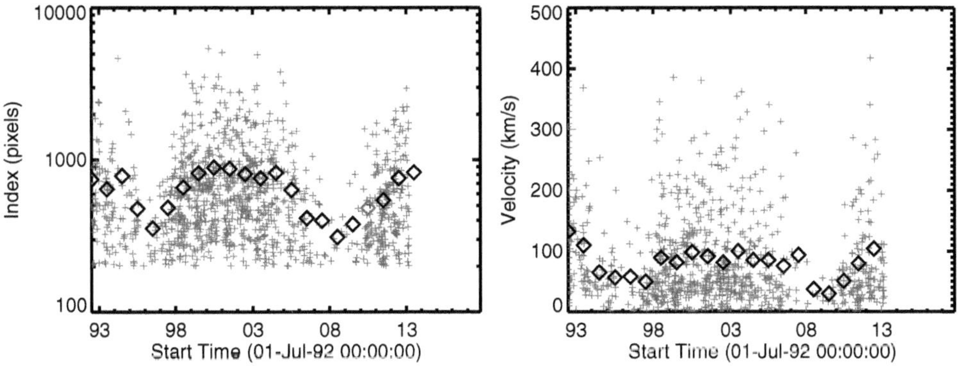


**Figure 2.** The time variation of the size (left panel) and the radial velocity (right panel).

the previous one significantly (about 50% down), the occurrence rate of the prominence activities is decreased only by 10%.

When we pay attention to the timing of the minimum values between Cycles 22 and 23, the figure shows that the number of the prominence activities (in the northern/southern hemisphere) and the sunspots became the minimum in 1996. On the other hand, between Cycles 23 and 24, the timings of the minimum values are different. Although the number of prominence activities occurring in the northern hemisphere became the minimum in 2007, the data of the sunspots and the prominence activities in the southern hemisphere indicates that the minimum year is 2008. The number of prominence activities suggests that the solar cycle period of the southern hemisphere in Cycle 23 is longer than the period of the northern hemisphere, 11 years.

### 2.2. *Size and radial velocity of prominence activities*

In this study, to evaluate of the size of the prominence, the number of the pixels that are brighter than 6 times of the average brightness temperature of the day at the pixel is used. We name the number of the bright pixels "Index". Since the pixel size of NoRH is 2.5 arcsec, the Index "1000" is correspond with about $6 \times 10^4$ km.

The left panel of Figure 2 shows the solar cycle dependence of the size. The small gray plus indicates the size of each prominence activity and the big diamonds indicate the yearly averages. Based on the yearly averages, the size of the prominence activities is only decreased around the solar minimum. However it is not correct understanding because the size distributions of prominence activities shows the power-law distribution (Shimojo *et al.* 2006). Figure 3 shows the size distribution of each phase of the solar cycles. Except the solar minimum, the distributions of the phases show the power-law and the index is 2.5∼2.9. Considering the precision of the fitting, we conclude that there is no difference between the values. In the solar minimum, the prominence activities that are larger than 1000 Index do not happen.

The right panel of Figure 2 shows the solar cycle dependence of the radial velocity. The radial velocity of the prominence activities is derived from the center of gravity of the brightness temperature. Hence, the radial velocities of our database are slower than the other results (Copalswarmy *et al.* 2013b). Only the events that have the upward velocities are plotted in the figure, because there are the events that have only the downward velocities (for example, the observation was started at the decay phase of the eruption). The small gray plus symbol in the figure indicates the radial velocity of each prominence activity and the big diamonds indicate the yearly averages. The figure of

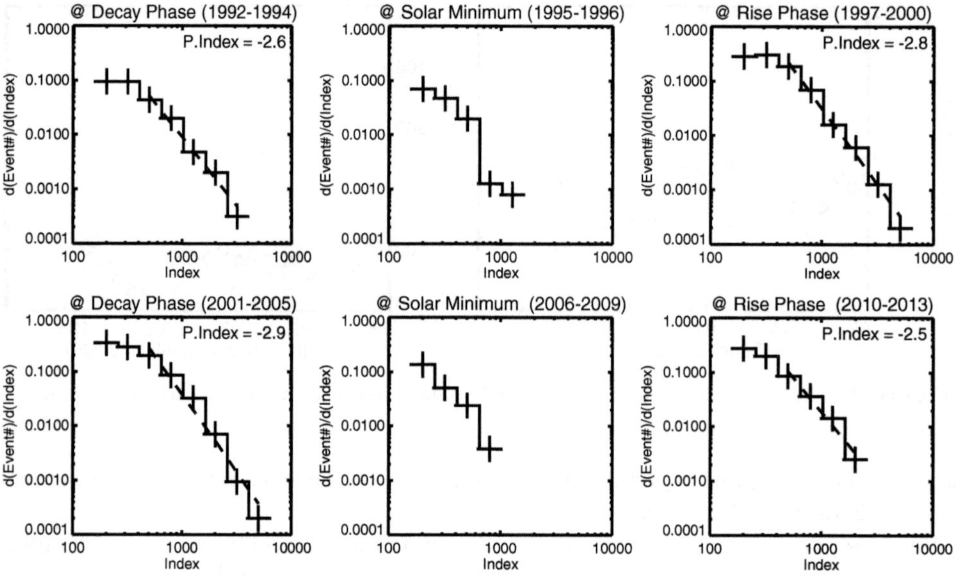


**Figure 3.** The size distribution of each phase.

the radial velocity is similar to that of the size, the yearly average of the radial velocity is about 100 km/s except the solar minimum. The difference might be caused by the producing region of the prominence activities. Except the solar minimum , most of the prominence activities occurred near the active regions, or the quiet regions with relative strong magnetic field that is caused by "the rush to the pole" magnetic field motion. Hence, the magnetic field strength around the prominence activities is larger than that in the solar minimum and the upward velocity becomes high.

## 3. Migration of the producing region of the prominence activities

### 3.1. *Northern hemisphere*

Figure 4 is the butterfly diagram of the prominence activities, and covers over the 20 years that correspond to the period from the decay phase of Cycle 22 to the rising phase of Cycle 24. We concentrate on the northern hemisphere at this subsection. The production region of the prominence activities is expanding toward the pole in the rising phase of the solar cycle. When the polemost prominence activity occurs near the pole around the solar maximum, the polarity reversal has taken place. The appearance region of dark filaments observed in H-alpha line also shows the similar poleward expansion and the relation with the polarity reversal (Waldmeier 1973, Topka *et al.* 1982, Mouradiam & Soru-Escaut 1994, Makarov *et al.* 2001, Li *et al.* 2009), and Gopalswamy *et al.* (2003b) also reported the relation between the prominence activities and the polarity reversal. We derive the expansion velocity from the polemost prominence activities in the butterfly diagram. The expansion velocities in the northern hemisphere are 5.9 m/sec at Cycle 23 and 5.2 m/sec at Cycle 24. Hathaway & Rightmire (2010) derived the latitudinal profile of the meridional flow speed from the magnetic features observed by the Michelson Doppler Imager (MDI) aboard the Solar and Heliospheric Observatory (SOHO) from May 1996 to June 2009. The expansion velocity that is derived from the prominence activities corresponds to the meridional flow speed around 60 degrees latitude. The coincidence suggests that the poleward migration of the polemost prominence activities

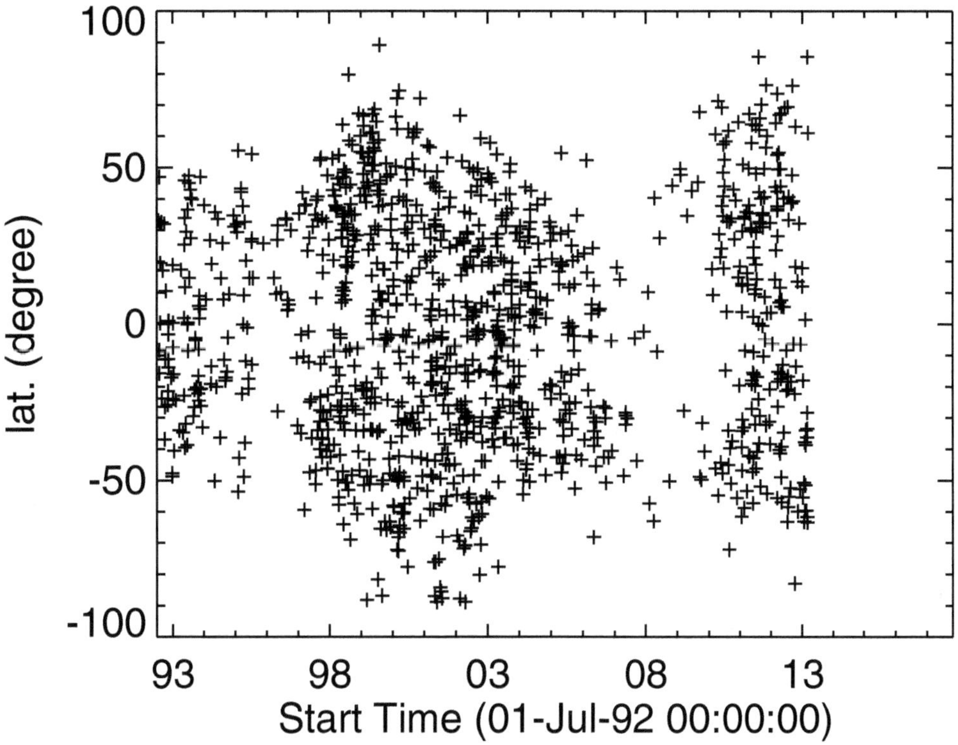


**Figure 4.** The butterfly diagram of the prominence activities.

is an indicator of the meridional flow. Our result is slower than the expansion velocity (~ 10 m/sec) derived from the dark filaments observed in Cycle 19 (Topka *et al.* 1982). The disagreement might show that the meridional flow speed in Cycle 23–24 is slower than that in the previous cycles.

After the solar maximum, the latitude of the prominence activity region is shrinking toward to the equator. The variation of the source latitude agrees with the variation of the active region filaments rather than all dark filaments (Mouradiam & Soru-Escaut 1994). This suggests that most of the prominence activities in the decay phase of a solar cycle are produced by the prominences in active regions. Li (2010) investigated the appearance region of the dark filaments in Cycle 16–21, and suggests that a third-order polynomial curve could give a satisfactory fit to the monthly mean latitude of dark filaments. We applied the fitting to our database, and found that the mean latitude in the northern hemisphere at Cycle 23 is similar to that at the previous cycles given in Figure 4 of Li (2010).

According to the observing facts, we can say that the migration of the source region of the prominence activities in the northern hemisphere at Cycle 23–24 is similar to the previous cycles. When the butterfly diagram of prominence activities that is shifted 11 years is overlaid on Figure 4 in the northern hemisphere (the overlaid figure is not included in the paper), the prominence activities that occurred after the solar maximum of Cycle 23 trace the butterfly diagram of the prominence activities from the decay phase of Cycle 22 to the rising phase of Cycle 23. The fact also suggests that the northern hemisphere at Cycle 23–24 is normal from the point of view of prominence activities.

### 3.2. *Southern hemisphere*

At the rising phase of Cycle 23, the producing region of the prominence activities is expanding toward the pole in the southern hemisphere. The expansion velocity is 7.3 m/sec and is faster than that in the northern hemisphere. The same trend of the hemispheric asymmetry is reported by Hathaway & Rightmire (2010), and it also suggests that the poleward migration of polemost prominence activities indicates the meridional flow.

After the solar maximum of Cycle 23, the migration of the source region of the prominence activities became unusual. Although the difference of the mean latitude migration in the northern and southern hemisphere at the previous cycles is small (Li 2010), the migration of the southern hemisphere is significantly different from that of the northern hemisphere at Cycle 23 and the previous cycles. The monthly mean latitude in the southern hemisphere did not decrease quickly after the solar maximum as in the northern hemisphere, and it stayed over −30 degree latitude after 2006. Figure 4 shows that the prominence activities occurred at over −50 degree latitude after 2006. Surprisingly, some prominence activities occurred at over −60 degrees latitude even in 2008, the solar minimum year. So, such high latitude prominence activities caused the unusual migration in the decay phase at the southern hemisphere.

The rising phase of Cycle 24 began in 2009. Although we can see the prominence activities associated with the active regions from late 2009, the expansion of the source region in the southern hemisphere is not seen in the butterfly diagram of prominence activities. Only the two prominence activities occurred at over −65 degrees latitude after 2009.

## 4. Summary

In the previous sections, we showed that the occurrence of the prominence activities in the northern hemisphere is normal from the duration of the cycle and the migration of the producing region of the prominence activities. On the other hand, the migration in the southern hemisphere significantly differs from that in the northern hemisphere and the previous cycles. The unusual migration in the southern hemisphere started from the solar maximum of Cycle 23. It is clear that the unusual migration was caused by the anomalous prominence activities in the high-latitude region (over −50 degrees). However, the origin of the anomalies is hidden under the photoshere, and we need the progress of solar dynamo studies to fully understand the anomalies. A polemost prominence activity is a good indicator of the meridional flow. The tracing of the polemost filament or prominence in the previous cycles might be one of the keys to understand the meridional flow and the solar activities.

## References

Gopalswamy, N., Lara, A., Yashiro, S., & Howard, R. A. 2003 *ApJ*, 598, L63

Gopalswamy, N., Shimojo, M., Lu, W., Yashiro, S., Shibasaki, K., & Howard, R. A. 2003 *ApJ*, 586, 562

Gopalswamy, N., Yashiro, S., Mäkelä, P., Michalek, G., Shibasaki, K., & Hathaway, D. H. 2012 *ApJ*(Letter), 750, L42

Hathaway, D. H. & Rightmire, L. 2010 *Science*, 327, 1350

Lee, C. O., Luhmann, J. G., Zhao, X. P., Liu, Y., Riley, P., Arge, C. N., Russell, C. T., & de Pater, I. 2009 *Solar Phys.*, 256, 345

Li, K. J., Li, Q. X., Gao, P. X., & Shi, X. J. 2008 *JGR*, 113, A11108

Li, K. J., 2010, *MNRAS*, 405, 1040

Livingston, W., Penn, M. J., & Svalgaard, L. 2012 *ApJ*(Letter), 757, L8

Makarov, V. I., Tlatov, A. G., & Sivaraman, K. R. 2001 *Solar Phys.*, 202, 11
Mouradiam, I. Z. & Soru-Escaut, I. 1994 *A&A*, 290, 279
Nakajima, H., *et al.* 1994 *Proceeding of the IEEE*, 82, 5, 705
Petrie, G. J. D. 2012 *Solar Phys.* 281, 577
Shimojo, M.,Yokoyama, T., Asai, A., Nakajima, H., & Shibasaki, K. 2006 *PASJ*, 58, 1, 85
Shimojo, M., 2013 *PASJ*, 65, S1, in press
Shiota, D., Tsuneta, S., Shimojo, M., Sako, N., Orozco Suárez, D., & Ishikawa, R. 2012 *ApJ*, 753, 157
SIDC-team 1992–2013 World Data Center for the Sunspot Index, Royal Observatory of Belgium, Monthly Report on the International Sunspot Number, online catalogue of the sunspot index: http://sidc.oma.be/sunspot-data/
Svalgaard, L. & Kamide, Y., 2013, *ApJ*, 763, 23
Thompson, B. J., *et al.* 2011 *Solar Phys.*, 274, 29
Topka, K., Moore, R., Labonte, B. J., & Howard, R. 1982 *Solar Phys.*, 79, 231
Waldmeier, M. 1973 *Solar Phys*, 28, 389
Wang, Y.-M., Robbrecht, E., & Sheeley, N. R. Jr. 2009 *ApJ*, 707, 1372

*Nature of Prominences and their role in Space Weather*
*Proceedings IAU Symposium No. 300, 2013*
*B. Schmieder, J.-M. Malherbe & S. T. Wu, eds.*

doi:10.1017/S1743921313010922

# Global magnetic field cycle evolution and prominence eruptions

**Irina A. Bilenko**

Moscow M. V. Lomonosov State Univercity, Sternberg Astronomiczl Institute, Moscow, Russia
email: bilenko@sai.msu.ru

**Abstract.** A comparison of changes in the structure of the global solar magnetic field and that in the prominence parameters, in solar cycles 21–23, are presented. It is proposed that the observed global magnetic field structure changes and periodicities in the mean solar magnetic field are the result of the excitation of large-scale Rossby waves. The changes in the prominence parameters are assumed to be the result of the global magnetic field structure changes, which may be triggered or modulated quasi-periodically by large-scale Rossby waves.

**Keywords.** magnetic fields, prominences, oscillations.

## 1. Introduction

Prominences are large cool plasma condensations in the low solar corona, typically 2 orders of magnitude denser and cooler than the million-degree coronal plasma. They appear only above long-lived magnetic channels in the chromosphere, trapped by the underlying magnetic fields (Tandberg-Hanssen 1995; Martin 1998). Although solar filaments are observed at all latitudes on the Sun, they always form above polarity inversion lines, which divide regions of positive and negative magnetic fields. Solar prominences are associated with a wide variety of solar activity phenomena. A close relationship between prominence eruptions and coronal mass ejections (CMEs), and flares were established (Munro *et al.* 1979; Hori & Culhane 2002; Gopalswamy *et al.* 2003). Prominences were one of the first activity phenomena associated with mass ejections from the Sun (Tandberg-Hanssen 1995). Prominences/filaments are also known to be the good traces of the evolutionary changes of the global solar magnetic field during solar cycles (Hyder 1965; Makarov & Sivaraman 1989).

The aim of this paper is to investigate the dependence of prominence parameters on global magnetic field structure (GMFS) cycle evolution.

## 2. Data

The mean magnetic field data and source surface synoptic maps from the Wilcox Solar Observatory were used. The coronal magnetic field is calculated from photospheric fields with a potential field model with the source surface location at 2.5 solar radii (Altschuler & Newkirk 1969; Schatten, Wilcox & Ness 1969; Hoeksema, Wilcox & Scherrer 1983).

Data on prominence parameters were taken from the Kislovodsk Mountain Astronomical Station of the Pulkovo Observatory. The daily observations on solar prominences and their parameters, such as area, height, and length were carried out on the station in line $H\alpha$ ($\lambda = 6563$ Å) since 1957 (Gnevyshev *et al.* 1963; Guseva *et al.* 2006).

## 3. Results

The longitudinal distribution of positive-polarity and negative-polarity magnetic fields, resulting from the PFSS extrapolation at 2.5 solar radii, is displayed in Fig. 1a. In Fig. 1b the longitudinal diagram of the mean solar magnetic field is presented. The brightness at a certain point is proportional to the magnetic field strength, averaged over latitude,

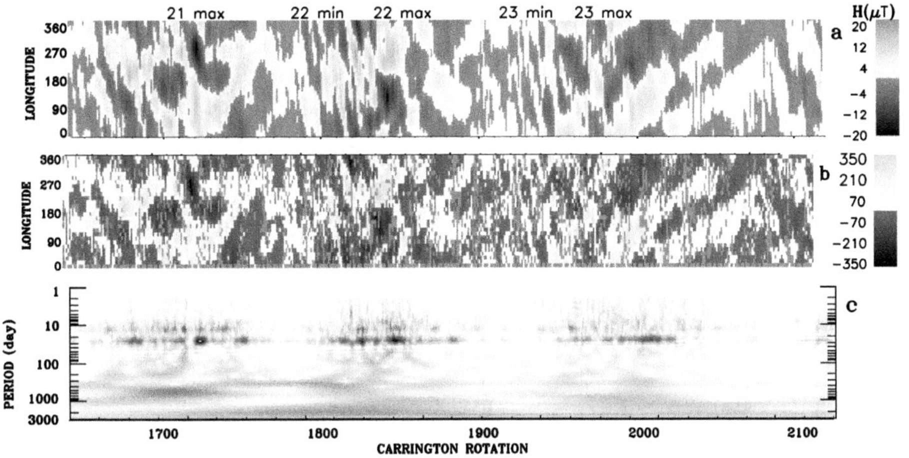


**Figure 1.** The longitudinal diagrams of (a) coronal magnetic field; (b) the mean solar magnetic field. Light denotes positive-polarity magnetic fields and dark - negative-polarity ones. (c) Wavelet power spectra of the mean solar magnetic field.

for each Carrington rotation (CR). The coronal magnetic field is anchored in the subphotospheric magnetic fields and is forced to evolve in accordance with the changing photospheric magnetic fields. Magnetic fields form a multi-scale GMFS, depending on the phase of a cycle. The largest structures, with the life-time ~1 to 5 years, were observed during the maxima and declining phases. There were three fast (~1 to 3 solar rotations) redistributions of the GMFS, covering a considerable part of the solar surface, during the maxima and declining phases. Magnetic structures during the minima and rising phases were smaller compared to those of the maxima and declining phases.

Gilman (1969) proposed, that observed solar magnetic fields can be the result of Rossby waves generated in the Sun's convection zone. In Tikhomolov (1995, 1996) the Rossby vortices, excited within a thin layer beneath the convection zone, were considered to explain the observed GMFS. The Rossby vortices are the result of heating from the solar interior and the deformation of the convection zone lower boundary. According to Zaqarashvili *et al.* (2010a, 2010b), the periodicity of 155–160 days and ~2 years, observed in different solar activity indices, can be connected to the dynamics of magnetic Rossby waves in the solar tachocline, since in the layer they are unstable due to the joint effect of the toroidal magnetic field strength and latitudinal differential rotation.

The wavelet power spectra of the mean solar magnetic field is shown in Fig. 1c. It is proposed, that the GMFS in Fig. 1a and 1b is a consequence of the excitation of Rossby waves of different periods during different solar cycle phases. Noteworthy is the fact that the wave period do not remains constant, but undergoes abrupt changes. There were several "switches" in wave periods during each maxima of cycles 21 – 23: from 300d–400d to 40d–50d and then to 300d–400d again. The intensity of the waves was different in different cycles. These changes in wave periods are reflected in the reorganizations of the GMFS. So, the GMFS could be a consequence of the exitation of Rossby waves of different periods.

Fig. 2 shows the evolution of prominence parameters such as (a) daily counts of prominence events (N), (b) prominence heights, (c) length, and (d) areas. Dots represent data for each prominence, thin solid lines represent CR averaged data (the scales are shown on the right y-axis). The latitudinal distributions of prominences depending on their area are shown in Fig. 2(e-h). It is known that the parameters of prominences varies with solar cycles. From Fig. 2(a-d) it is seen that in addition to the general cycle variation in the prominence parameters, the local changes are also observed. The comparison of Fig. 1

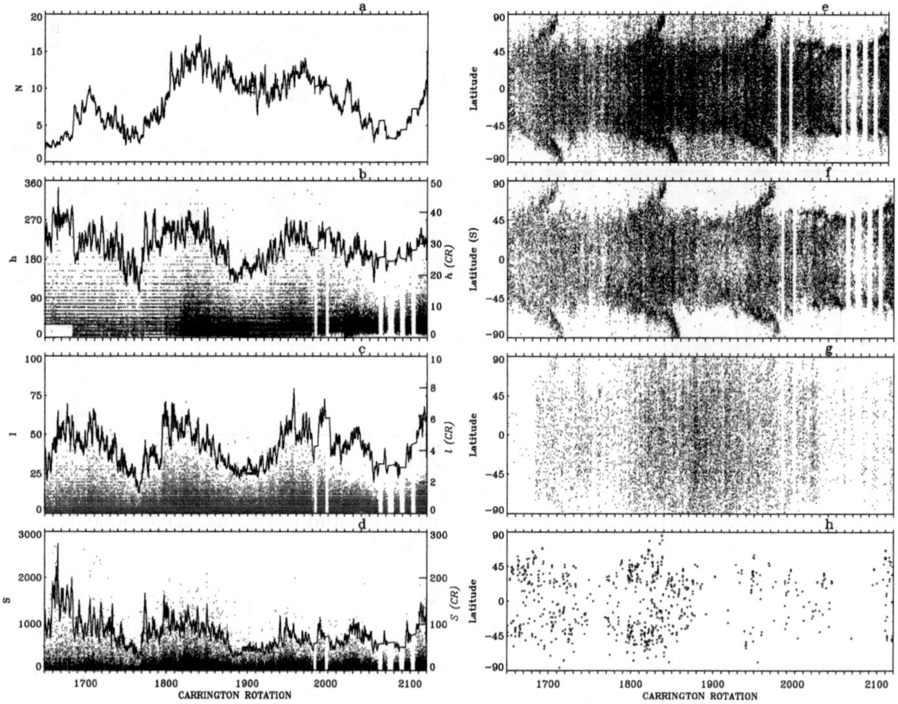


**Figure 2.** Prominence parameters.

and 2 shows that the number of prominences was higher in times of the GMFS reorganization. Prominences had, on average, lower height, length, and area. Thus, changes in the GMFS determine the number and parameters of prominences. The reorganization of the GMFS result in the growth of weak CME events (Bilenko 2012), that can be associated with some prominence eruptions. This may be caused by the changes in Rossby wave regime generation and a new magnetic flux emergence, resulting in the GMFS reorganization, which in turn leads to an increase in the number of prominence eruptions and associated CMEs. The coronal magnetic field strength follows the GMFS evolution and undergoes abrupt changes, reflecting changes in activity within the large-scale magnetic patterns. The field strength decreases during the times of the reorganization of the GMFS. The strength of magnetic fields surrounding prominences play an important role in the filaments/prominences evolution (Svetska 1986). Decrease of magnetic field strength could be responsible for some prominence/filament eruption (Schmieder *et al.* 2008). The prominence eruptions can be also the result of the removal of the restraining coronal magnetic field. When the coronal magnetic structure is destroyed and the field diminishes, the force, which prevent a filament/prominence from eruption, decreases and a prominence can erupt. During the reorganization of the structure of the magnetic field in the solar corona the conditions are not stable. Therefore, large, long, prominenves/filamens do not have time to form, and, hence, they are smaller in size and length. Moreover, since external magnetic field, preventing an eruption, decreases, the critical height (Filippov *et al.* 2006) for prominences to erupt will be lower.

It is well known that the latitude distribution of prominences and their activity changes during solar cycles (Gnevyshev & Makarov 1985; Lorenc, Pastorek, & Rybanský 2003; Shimojo *et al.* 2006). The latitudinal interrelation of prominence eruptions and CMEs depends on a solar-cycle phase (Gopalswamy *et al.* 2012). From Fig. 2(e-h) it is seen that the behaviour of prominences with different parameters is different. Prominences with an area more that 700 (height > 100", length > 25°) change their position in latitude during cycles (Fig. 2h), whereas prominences with an area less than 15 (7", 2°) are distributed uniformly (Fig. 2g). For these prominences, the concentration to the moments of the GMFS reorganization is observed. The drift of prominence location to the north and the south poles is most pronounced for that in size from 40 to 600 (80"–100", 10°–15°) (Fig. 2f). The drift occurred when the large-scale GMFS formed.

In summary, from the above, we can conclude that the changes in the prominence parameters, during solar cycles, could be the result of GMFS changes, which may be triggered or modulated quasi-periodically by large-scale Rossby waves.

## 4. Acknowledgements

I would like to thank the Local Organizing Committee and IAU for the financial support and the chance to present at the conference. Wilcox Solar Observatory data used in this study was obtained via the web site *http : //wso.stanford.eduat*2011 : 03 : 22_05 : 18 : 09 PDT courtesy of J.T. Hoeksema. The Wilcox Solar Observatory is currently supported by NASA. Kislovodsk Mountain Astronomical Station of the Pulkovo Observatory data was obtained via the web site *//www.solarstation.ru.*

### References

Altschuler, M. D. & Newkirk, G. 1969, *Solar Phys.*, 9, 131

Bilenko, I. A. 2012, *Geomagnetism and Aeronomy*, 52, 1005

Filippov, B. P., Zagnetko, A. M., Ajabshirizadeh, A., & Den, O. G. 2006, *Sol. Syst. Res.*, 40, 319

Gilman, P. A. 1969, *Solar Phys.*, 8, 316

Gnevyshev, M. N. 1963, *AZh*, 40, 401

Gnevyshev, M. N. & Makarov, V. I. 1985, *Solar Phys.*, 95, 189

Gopalswamy, N., Shimojo, M., Lu, W., Yashiro, S., Shibasaki, K., & Howard, R. A. 2003, *ApJ*, 586, 562

Gopalswamy, N., Yashiro, S., Mäkelä, P., Michalek, G., Shibasaki, K., & Hathaway, D. H. 2012, *ApJ*, 750, L42

Guseva, S. A., Kim,G.-D. & Tlatov, A. G. 2007, *in proc. "Multiwavelength investigations of the Sun and the problems in solar activity" SAO RAN, publ. Sankt-Peterburg*, p. 269

Hoeksema, J. T., Wilcox, J. M., & Scherrer, P. H. 1983, *JGR*, 88, 9910

Hori, K. & Culhane, J. L. 2002, *A&A*, 382, 666

Hyder, C. L. 1965, *ApJ*, 141, 272

Lorenc, M., Pastorek, L., & Rybanský, M. 2003, *in proc. ISCS 2003 Symposium "Solar Varuability as an Input to the Earth's Environment", Tatranská Lomnica, Slovakia*, p. 129

Makarov, V. I. & Sivaraman, K. R. 1989, *Solar Phys.*, 123, 367

Martin, S. F. 1998, *Solar Phys.*, 182, 107

Munro, R. H., Gosling, J. T., Hildner, E., MacQueen, R. M., Poland, A. I., & Ross, C. L. 1979, *Solar Phys.*, 61, 201

Schatten, K. N., Wilcox, J. M., & Ness, N. F. 1969, *Solar Phys.*, 6, 442

Schmieder, B., Bommier, V., Kitai, R., Matsumoto, T., Ishii, T. T., Hagino, M., Li, H., & Golub, L. 2008, *Solar Phys.*, 247, 321

Shimojo, M., Yokoyama, T., Asai, A., Nakajima, H., & Shibasaki, K. 2006, *PASJ*, 58, 85

Svetska, Z. 1986, *The Lower Atmosphere of Solar Flares, NSO/Sac Peak Pub.*, 332.

Tandberg-Hanssen, E. 1995 *The Nature of Solar Prominences, Kluwer Acad., Norwell, Mass.*

Tikhomolov, E. 1995, *Solar Phys.*, 156, 205

Tikhomolov, E. & Mordvinov, V. I. 1996, *ApJ*, 472, 389

Zaqarashvili, T. V., Carbonell, M., Oliver, R., & Ballester, J. L. 2010a, *ApJ*, 709, 749

Zaqarashvili, T. V., Carbonell, M., Oliver, R., & Ballester, J. L. 2010b, *Ap. Lett.*, 724, L95

*Nature of Prominences and their role in Space Weather*
*Proceedings IAU Symposium No. 300, 2013*
*B. Schmieder, J.-M. Malherbe & S. T. Wu, eds.*

doi:10.1017/S1743921313010934

# Explaining the Hemispheric Pattern of Filament Chirality

**Duncan H. Mackay[1] and Anthony R. Yeates[2]**

[1]School of Mathematics and Statistics, University of St Andrews, St Andrews, KY16 8HB, UK
email: duncan@mcs.st-and.ac.uk

[2]Department of Mathematical Sciences, Durham University, Durham, UK
email: anthony.yeates@durham.ac.uk

**Abstract.** Solar filaments are known to exhibit a hemispheric pattern in their chirality, where dextral/sinistral filaments dominate in the northern/southern hemisphere. We show that this pattern may be explained through data driven 3D global magnetic field simulations of the Sun's large-scale magnetic field. Through a detailed comparison with 109 filaments over a 6 month period, the model correctly reproduces the filament chirality and helicity with a 96% agreement. The data driven simulation is extended to run over a full solar cycle, where predictions are made for the spatial and temporal dependence of the hemispheric pattern over the solar cycle.

**Keywords.** Sun: filaments, Sun: magnetic fields, Sun: prominences

## 1. Introduction

In recent years, solar filaments have been classified by the orientation of their main axial magnetic field. This orientation, named the filament chirality (Martin *et al.*, 1994) may take one of two forms: dextral or sinistral. Dextral/sinistral filaments have an axial magnetic field that points to the right/left when the main axis of the filament is viewed from the positive polarity side of the PIL. A dextral filament is expected to contain dominantly negative helicity, a sinistral filament positive helicity. As filaments and their channels form over a wide range of latitudes on the Sun, they may be regarded as useful indicators of sheared non-potential fields and magnetic helicity.

A surprising feature of the chirality of filaments is that it displays a large-scale hemispheric pattern: dextral/sinistral filaments dominate in the northern/southern hemispheres, respectively (Martin *et al.*, 1994; Zirker *et al.*, 1997; Yeates *et al.*, 2007). Although dextral/sinistral filaments dominate in the northern/southern hemisphere, observations show that exceptions to this pattern do occur. Any model which tries to explain the formation of filaments must explain not only the origin of this hemispheric pattern but also why exceptions to it arise.

## 2. Global Non-linear Force-Free Field Model

To explain the hemispheric pattern of solar filaments we will apply a global non-linear force-free magnetic field model. The model developed by van Ballegooijen *et al.* (2000) and Mackay & van Ballegooijen (2006) considers the long-term evolution of coronal magnetic fields. The technique describes the build-up of free magnetic energy and electric currents in the corona by coupling together two distinct models. The first is a data driven surface flux transport model (Yeates *et al.*, 2007). This models the large-scale surface motions of differential rotation, meridional flow and surface diffusion. In addition

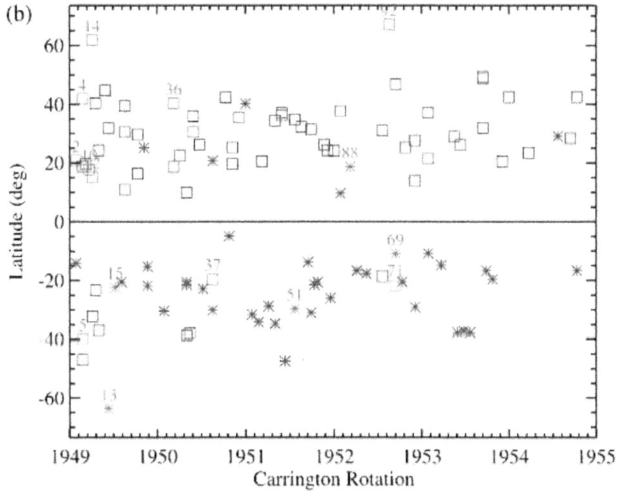


**Figure 1.** Results of the comparison between theory and observations. The symbols denote the observed chirality of each filament (squares ~ dextral, stars ~ sinistral). The color denotes agreement between the simulation and observations (blue ~ correct chirality in simulation, red ~ incorrect chirality in simulation). A color version is available online.

to this observations of newly emerging magnetic bipoles are included to produce a continuous evolution of the observed photospheric magnetic flux over long periods of time. Coupled to this is a quasi-static coronal evolution model (Mackay & van Ballegooijen, 2006; Yeates *et al.*, 2008) which evolves the coronal magnetic field through a sequence of nonlinear force-free fields in response to the observed photospheric evolution and flux emergence.

## 3. Six Month Comparison Between Theory and Observations

To determine the origin of the hemispheric pattern of filaments we carry out a direct comparison between theory and observations (see Yeates *et al.*, 2007; Yeates *et al.*, 2008). First, H$\alpha$ observations from BBSO over a 6 month period are studied to determine the location and chirality of 109 filaments (Yeates *et al.*, 2007) relative to the underlying magnetic flux. In Figure 1 the symbols denote the observed chirality of each filament (squares ~ dextral, stars ~ sinistral). It is clear that the dominant chirality pattern occurs in each hemisphere, but also a number of exceptions exist. Next the combined magnetic flux transport and magneto-frictional simulations (Section 2), based on actual photospheric magnetic distributions found on the Sun is run. The run is carried out for the whole six month period without ever resetting the surface field back to that found in observations or the coronal field to potential. To maintain accuracy over this time, 119 bipoles are emerged within the simulations where their properties (location, size, flux, tilt angle) are determined from observations. The simulations describe the long term helicity transport across the solar surface from low to high latitudes.

A direct one-to-one comparison of the chirality produced by the model and the observed chirality of the filaments is carried out (Yeates *et al.*, 2008). Through varying the sign and amount of helicity emerging within the bipoles and by emerging dominantly negative helicity in the northern hemisphere and positive in the southern, a 96% agreement can be found between the observations and simulations. The locations where the simulations produce the same chirality as the observations are shown in blue. Where they disagree and the simulations produce the incorrect chirality are shown in red. An important feature is that the agreement is equally good for minority chirality filaments as well as for dominant chirality filaments. A key feature of the simulations is that a better agreement between the observations and simulations is found the longer the simulations are run. The majority of the disagreements arise within the first six months. This indicates that the Sun has a long term memory of the transport of helicity from low to high latitudes. The reason for this high agreement is described in the paper of Yeates & Mackay (2009).

The results demonstrate that the combined effects of differential rotation, meridional flow, supergranular diffusion along with the emergence of non-potential active regions are sufficient to produce the observed hemispheric pattern of filaments. While the model obtained an excellent agreement with the observed chirality of filaments, the 6 month simulation was unable to reproduce dextral/sinistral chirality along the polar crown PILs in the northern/southern hemispheres. Note that filament 92 that lies at 68° latitude between CR1952 and CR1953 is incorrect. This is due to the fact that the simulation was not run for long enough to transport helicity from low to high latitudes which occurs over the meridional flow timescale of 2 year. The topic of the polar crown chirality will be considered in the next section.

## 4. Full Solar Cycle Predictions

To consider the chirality at high latitudes a longer simulation is run for a full solar cycle (Yeates & Mackay, 2012). The simulation is initialised with a potential field extrapolation based on a Kitt Peak synoptic magnetogram for Carrington Rotation CR1910, corrected for differential rotation to represent 1996 May 15. The coronal magnetic field is then continuously evolved. Over the 15 year period 1838 active regions are emerged where their flux varies from $2 \times 10^{20}$ Mx to $5.3 \times 10^{22}$ Mx.

Figure 2(a) shows a longitudinal average of the photospheric radial magnetic field $B_{0r}$ in the simulation which reproduces well the observed field. Figure 2(b) shows the longitude-averaged skew angle along PILs where red denotes dextral skew and blue sinistral skew. Several features are apparent: (a) Below $\pm 50°$ latitude the majority pattern of chirality predominates, i.e., dextral in the northern hemisphere and sinistral in the southern hemisphere, although the overall pattern has significant fluctuations (minority chirality exists at all latitudes in each hemisphere); (b) During the period of few active regions from 2007 to 2010, there is a more mixed chirality at lower latitudes; (c) Until 1998 and during the declining phase from 2001 to 2006, there is a tendency for minority chirality on the high-latitude PILs (sinistral in the north, dextral in the south); (d) During the rush-to-the-poles between 1998 and mid-1999, the polar crowns exhibit the majority chirality pattern; (e) From 2006 onward, the majority chirality dominates at high latitudes once more, continuing into Cycle 24.

It should be noted that the minority chirality at high-latitudes in either hemisphere until 1998 is due the initial condition which is a potential field. Once this is removed by the transport of helicity poleward which occurs over a 2 year time period the correct chirality is found at high latitudes in the rising phase. Note that due to the continuous nature of the simulation, from 2006 onwards the correct chirality is found at high latitudes in the rising phase of the present cycle. This indicates that the incorrect results for filament 92, which lies at high latitudes (Figure 1), was a result of the simulation not being run for long enough.

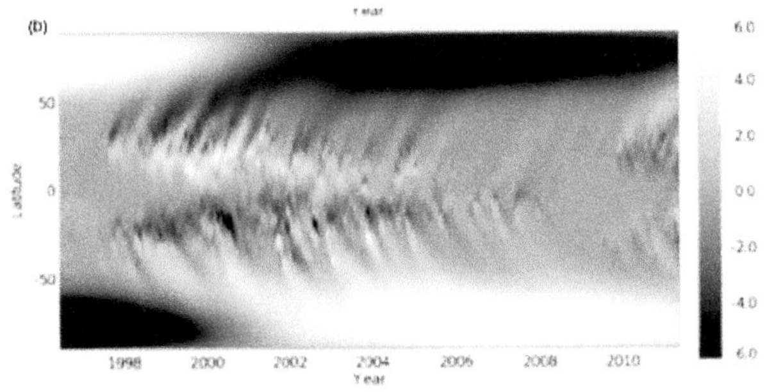


(a)

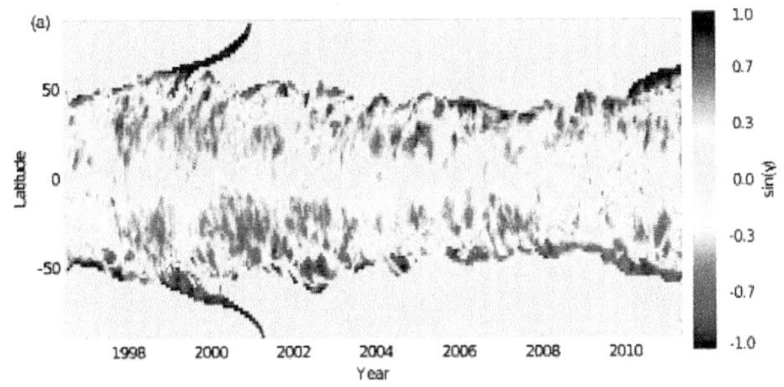


(b)

**Figure 2.** Butterfly diagrams showing longitude-averaged over Cycle 23. (a) Radial magnetic field $B_{0r}$ in the simulation and (b) skew $\sin\gamma$ measured at height $r = 1.033R_\odot$. A color version is available online.

## 5. Conclusions

Within the present study we have shown that the hemispheric pattern of filaments, including exceptions, may be explained through the combined effects of differential rotation, meridional flow and surface diffusion, where these effect act on new magnetic bipoles, which emerge in each hemisphere containing the dominant sign of helicity for that hemisphere. Full solar cycle simulations show that the correct chirality may be found along the polar crown PIL as long as the simulations are run for long enough to remove the initial condition. Predictions for the hemispheric pattern have been made for the declining phase of the cycle which may now be tested with observations in Solar Cycle 24.

## References

Mackay, D. H. & van Ballegooijen, A. A. 2006, *ApJ*, 641, 577
Martin, S. F., Bilimoria, R., & Tracadas, P. W. 1994, *Solar Surface Magnetism*, 303
van Ballegooijen, A. A., Priest, E. R., & Mackay, D. H. 2000, *ApJ*, 539, 983
Yang, W. H., Sturrock, P. A., & Antiochos, S. K. 1986, *ApJ*, 309, 383
Yeates, A. R., Mackay, D. H., & van Ballegooijen, A. A. 2007, *Solar Phys.*, 245, 87
Yeates, A. R., Mackay, D. H., & van Ballegooijen, A. A. 2008, *Solar Phys.*, 247, 103
Yeates, A. R. & Mackay, D. H. 2009, *Solar Phys.*, 254, 77
Yeates, A. R. & Mackay, D. H. 2012, *ApJL*, 753, L34
Zirker, J. B., Leroy, J.-L., & Gaizauskas, V. 1997, *Solar Phys.*, 176, 279

# Session II - 2.1

PROMINENCE DESTABILIZATION

CORONAL MASS EJECTIONS

3D RECONSTRUCTIONS

*Nature of Prominences and their role in Space Weather*
*Proceedings IAU Symposium No. 300, 2013*
*B. Schmieder, J.-M. Malherbe & S. T. Wu, eds.*

doi:10.1017/S1743921313010946

# On the dynamics of eruptive prominences

**Laura A. Balmaceda[1,6], Hebe Cremades[2], Guillermo Stenborg[3], Carlos Francile[4], Leonardo Di Lorenzo[5] and Fernando López[1]**

[1]ICATE/CONICET-UNSJ,
CC 49, 5400 San Juan, Argentina
email: lbalmaceda@icate-conicet.gob.ar

[2]UTN - Facultad Regional Mendoza and CONICET, Mendoza, Argentina
[3]George Mason University, Fairfax, VA 22030, USA
[4]OAFA - Universidad Nacional de San Juan, San Juan, Argentina
[5]Universidad Nacional de San Luis, San Luis, Argentina
[6]Instituto Nacional de Pesquisas Espaciais, São José dos Campos, Brazil

**Abstract.** To contribute to the understanding of the physical mechanisms at work during the initial phase and early evolution of erupting prominences, we analyze combined observations from ground-based and space-borne instruments. We present two case studies, which occurred at two different phases of the solar cycle, namely on March 2, 2002 and on April 16, 2012. In particular, we show the results of a morphological and kinematical analysis and interpret them in terms of available theoretical models.

**Keywords.** Prominence dynamics, prominence eruption.

## 1. Introduction

Solar prominences have been studied for many decades using both ground-based and space-borne data. They usually show up in white-light coronagraph observations as a bright feature behind the leading edge (LE) of the associated CME, i.e., as the main part of the CME inner core (Webb & Hundhausen, 1987). They are also imaged onto the solar disk (as well as above the limb up to some fractions of a solar radii) at visible wavelengths in H$\alpha$, or in the extreme ultraviolet range as in HeII or highly-ionized Fe lines. Combined observations from such instruments provide a great opportunity to get insight on their early development. During the SOHO/LASCO (Brueckner *et al.*, 1995) era, it was well established that CMEs and filament/prominence eruptions are strongly related (Srivastava *et al.*, 1999, Gopalswamy *et al.*, 2003; Jing *et al.*, 2004, and references therein).

Prominences are believed to be related to the formation of a magnetic flux rope, which eventually erupts to form the CME (Low, 2001). Thus, the study of their kinematic profile during their onset and early rise in the low corona at high cadences provides crucial information on the forces/physical mechanisms that trigger/drive the eruption.

On a theoretical aspect, many models have been developed to describe the CME initiation. In particular, Chen (1989) and Chen (1996) considered a line-tied current-carrying loop, e.g., a flux rope, holding a prominence at its bottom, and derived the total force exerted on its apex. Given that a prominence is generally stable for weeks before erupting, they proposed that a flux rope can be accelerated and erupt when new poloidal flux is injected into the system. An instability, named *torus instability* from laboratory plasma experiments (Kliem & Török, 2006), would take place in the flux rope due to the "toroidal force" and eventually act as the trigger mechanism for the eruption.

In this work, we use data from two ground-based solar instruments in combination with satellite images of two particular events, which we further contrast with the theoretical model developed by Chen (1996) to shed light into the dynamics of prominence eruptions during their early stages.

## 2. The observations

The two ground-based instruments are operating in San Juan, Argentina (31.8 S, 69.3 W, at 2400 m of altitude) since 1999. The observing time window runs between roughly 12 UT and 20 UT. MICA (Mirror Coronagraph for Argentina, Stenborg *et al.* 1999) is an internally-occulted mirror coronagraph with a plate scale of 3.6"/pixel. Two auxiliary devices allow the constant monitoring of the atmospheric conditions and hence its automatic operation. It images the solar corona at a nominal cadence of $\sim$ 1 min between 1.05 and 2.0 $R_\odot$ in the Fe XIV line (at 530.3 nm, hereafter green line) and in the continuum emission (at 526.0 nm) using interference filters that have a full width at half maximum (FWHM) passband of 0.13 and 1.1 nm, respectively. On the other hand, HASTA† (H-alpha Solar Telescope for Argentina, Bagalá *et al.* 1999) is a full-disk imager, with a plate scale of $\sim$2"/pixel operating in two cadence modes: 90 seconds in patrol mode and 3 seconds in high speed mode. It images the solar disk in the H$\alpha$ emission line at 656.27 nm with a filter of 0.03 nm (FWHM).

Unfortunately, the detection of faint dynamic events in MICA observations is not a straightforward task due to the highly varying atmospheric conditions at the observing site. Therefore, in order to help reveal the events of interest we first process the images with a customized version of the wavelet-based technique developed by Stenborg & Cobelli (2003) to clean the images and thus increase the relative intensity contrast of the coronal features in the images (the detailed description of the devised technique will appear elsewhere).

### 2.1. *Event A: March 2, 2002*

On March 2, 2002, the onset phase of an erupting prominence was identified in both HASTA H$\alpha$ and MICA green-line images. The prominence initial height ($Z_{p_0}$) in HASTA observations, as measured from the disk center at 14:24 UT, was 1.05 $R_\odot$, the separation between the footpoints being $\sim$0.3 $R_\odot$. These measurements were confirmed with observations from the Extreme Ultraviolet Imaging Telescope (EIT, Delaboudinière *et al.*, 1995) onboard SOHO. The prominence was a long-lived one, which had been observed in HASTA images since some weeks prior to the eruption. The eruption gave rise to a well structured three-part CME observed later in the LASCO-C2 and -C3 coronagraphs (Bruecker *et al.*, 1995) as seen in the composite of MICA and LASCO-C2 images (Fig. 1, left panel). From MICA images we could determine the main acceleration phase of the prominence eruption. The prominence shows a gradual increase during its early phase (open circles in Fig. 2, top left panel). At distances beyond $\sim$3.5 $R_\odot$, the height–time profile is approximately linear with a constant velocity of $\sim$1000 km $s^{-1}$.

### 2.2. *Event B: April 16, 2012*

At the time of the event, the coronagraph's observing mode was set to take images only in the green-line continuum, allowing for a cadence of 40 seconds. An erupting prominence was first observed above the eastern limb at $\sim$17:30 UT. We identified this eruption as the initial phase of a CME observed later in LASCO-C2 and C3 data (see in Fig.1, central

† http://www.oafa.fcefn.unsj-cuim.edu.ar/Hasta/

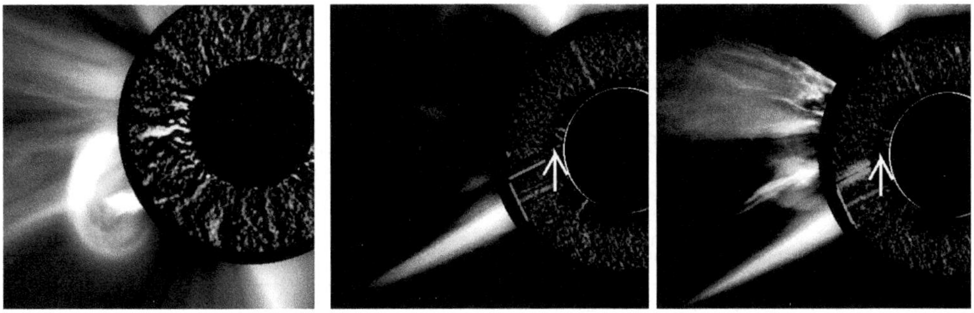

**Figure 1.** Composite of quasi-simultaneous MICA and LASCO-C2 images. *Left:* Event A, at ~15:30 UT. *Center:* Event B, early development of the prominence eruption (~17:36 UT). *Right:* Event B at ~18:24 UT. The arrows pinpoint the prominence legs.

and right panels the composite of MICA and LASCO-C2 FOVs). The eruption develops gradually in the lower corona. A large amount of the prominence material is observed to drain back to the Sun in HASTA images. At higher distances from the Sun, the CME LE moves at ~1000 km $s^{-1}$, with an acceleration of $-30$ m $s^{-2}$ (as determined from the corresponding height–time plot for heights $> 2R_\odot$, at ~71° position angle. See Fig. 2, bottom left panel, where the LASCO measurements are indicated by crosses).

## 3. Comparison with a theoretical model

In the integrated MHD approach developed by Chen (1996), the modeled flux rope is described by a non-axisymmetric geometry, with an average major radius of curvature during expansion $R$, stationary footpoints with separation $S_f$, and a non-uniform minor radius $a$. The initial structure is assumed to be in equilibrium. The conditions for the ambient coronal field are specified by an independent model. The eruption is triggered by the injection of poloidal magnetic flux whose profile is imposed (a generic pulse whose variation is described by Eq. 9 in Chen & Kunkel, 2010). The flux rope is driven mainly by the Lorentz force arising from the toroidal currents and is assumed to remain connected to the solar surface. The model solves two coupled differential equations (Eq. 2 and 9 in Chen, 1996) that describe the forces acting on the major and minor radii and are evaluated in the apex of the structure. These are: Lorentz force arising from the poloidal component of the magnetic field and the downward tension due to the toroidal component, the gravitational force, the drag force, and the pressure gradient.

The geometric characteristics of the initial flux rope, as well as the initial mass derived from observational data are used to constrain the model following the methodology described in Chen *et al.* (2006). The height of the CME LE is given by $Z_{LE} = Z + 2a_a$, while the prominence height can be obtained from: $Z_p = Z - a$ and the prominence footpoint separation is related to the flux rope footpoint separation by $S_p = S_f - 2a_f$. Here, $Z$ denotes the height of the apex and the subindices $f$ and $a$ indicate the flux rope and apex parameters.

From solar images we determine the observed quantities: $S_p$ and $Z_p$. We estimate the CME mass using LASCO data and following the Vourlidas *et al.* (2010) technique. We deduce $Z_0$=$1.6\times10^5$ km, $S_0$=$1.2\times10^5$ km and $M_0$= $10^{16}$ g for Event A, and $Z_0$=$8\times10^4$ km, $S_0$=$1\times10^4$ km, $M_0$= $10^{15}$ g for Event B. We consider a maximum flux injection rate of $d\phi/dt_{max}$=$10^{19}$ Mx $s^{-1}$ in both cases. In order to find the time constants that define the flux injection profile (set as free parameters in the model), the genetic algorithm PIKAIA

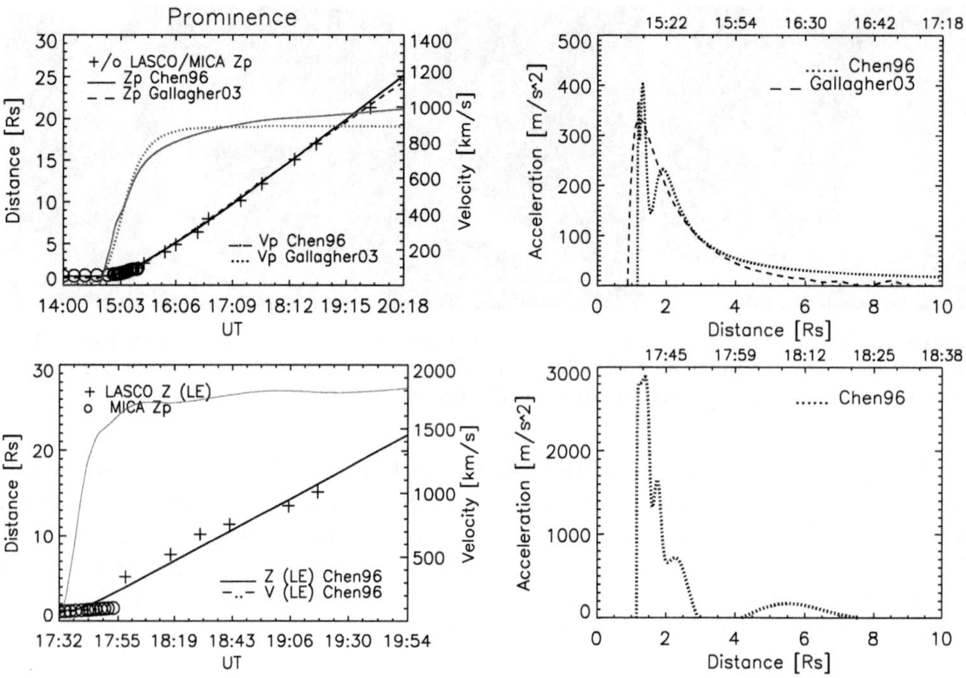


**Figure 2.** Height-time plot for the prominence and for the CME leading edge: Event A (top left), Event B (bottom left). The solid curves correspond to the fitting of the observations by the given models. Right panels: Acceleration profile derived from the models for Event A (top) and Event B (bottom).

by Charbonneau (1995) is used. The model is set to fit the observations up to $\sim$20 $R_\odot$ by minimizing a $\chi^2$-function.

The comparison between the observed heights and the results of the model is shown in Fig.2 (left panels) for both events. For Event A (top panel), the model is evaluated by comparing the measured prominence heights through MICA and LASCO FOVs. The results of the model describe fairly well the observations (solid line). The velocity exhibits a rapid increase at heights $< 2$ $R_\odot$ (dash-dotted line). This is also in good agreement with the velocity profile obtained assuming the hypothesis of a exponential varying acceleration suggested by Gallagher *et al.* (2003) (dashed line for the H-T and dotted line for the velocity-time curve in Fig. 2 top left panel). The acceleration reaches a maximum of $\sim$300–400 m s$^{-2}$ when the prominence top is at $\sim$1.5 $R_\odot$ ($\sim$15:05 UT, Fig. 2 top right panel). For Event B (Fig. 2, bottom panel) the prominence front is only well detected in MICA FOV (open circles). It is seen to expand gradually and erupt after reaching $\sim$1.7 $R_\odot$. Its top is not clearly discernible in LASCO-C2 FOV. The model is then evaluated by comparing only the CME LE in LASCO FOV (crosses). MICA observations are used only to constraint the initial morphological parameters of the flux rope. The modeled acceleration profile shows a main acceleration phase also at low heights with a peak at $\sim$ 1.7 $R_\odot$ and a residual acceleration phase at distances beyond 4 $R_\odot$ (bottom right panel). These results, however, should be further contrasted using complementary data from current missions. A complete analysis of these events will be published in an upcoming paper.

## 4. Concluding remarks

In this work, we have studied the evolution of two erupting prominences associated to CMEs using combined data from ground and space. We have also compared these observations with the model by Chen (1996) of the eruption of a flux rope and found good agreement.

Despite the limitations imposed by the observing time window and weather conditions, the ground-based data used in this work proved to be a suitable complement to existent space-based EUV and white-light coronagraph data, in particular to SOHO/EIT observations (the EIT FOV was limited to 1.4 $R_\odot$ with a nominal temporal cadence of only 12 min), and to LASCO observations before the advent of STEREO (Kaiser *et al.*, 2008) and SDO (Pesnell *et al.*, 2012) missions (the extended field of view of the MICA coronagraph fills the gap between the outer edge of AIA FOV and the inner edge of LASCO-C2). We plan to perform a morphological and kinematical analysis of a selection of events that occurred at different phases of solar cycle 23, and check the results against theoretical models, to shed light on the dynamics of eruptive phenomena, in particular during their early stages.

## Acknowledgements

L.B. and H.C. are members of the Carrera del Investigador Científico of CONICET. L.B. thanks the organizers and the IAU for the financial support to attend the Symposium.

## References

Bagalá, L. G., Bauer, O. H., Fernández Borda, R., *et al. Magnetic Fields and Solar Processes*, 1999, ESA Special Publication, 448, 469
Brueckner, G. E., Howard, R. A., Koomen, M. J., *et al.* 1995, *Sol. Phys.* 162, 357–402
Charbonneau, P. 1995, *ApJS*, 101, 309
Chen, J. 1989, *ApJ* 338, 453–470
Chen, J. 1996, *JGR*, 101, 27499–27520
Chen, J., Marqué, C., Vourlidas, A., Krall, J., & Schuck, P. W. 2006, *ApJ*, 649, 452–463
Chen, J. & Kunkel, V. 2010, *ApJ*, 717, 1105–1122
Delaboudinière, J.-P., Artzner, G. E., Brunaud, J., *et al.* 1995, *Sol. Phys.* 162, 291–312
Gallagher, P. T., Lawrence, G. R., & Dennis, B. R. 2003, *ApJL*, 588, L53–L56
Gopalswamy, N., Shimojo, M., Lu, W., *et al.* 2003, *ApJ*, 586, 562–578
Jing, J., Yurchyshyn, V. B., Yang, G., Xu, Y., & Wang, H. 2004, *ApJ*, 614, 1054–1062
Kaiser, M. L., Kucera, T. A., Davila, J. M., *et al.* 2008, *Sp. Sci. Rev.*, 136, 5–16
Kliem, B. & Török, T. 2006, *Phys. Rev. Let.*, 96, 25, 255002
Low, B. C. 2001, *JGR*, 106, 25141–25164
Pesnell, W. D., Thompson, B. J., & Chamberlin, P. C. 2012, *Sol. Phys.*, 275, 3–15
Srivastava, N., Schwenn, R. & Stenborg, G. 1999, 8th SOHO Workshop: Plasma Dynamics and Diagnostics in the Solar Transition Region and Corona, ESA Special Publication, 446, 621
Stenborg, G., Schwenn, R., Srivastava, N., *et al.* 1999, *Sp. Sci. Rev.*, 87, 307–310
Stenborg, G. & Cobelli, P. J. 2003, *Astron. Astroph.*, 398, 1185–1193
Vourlidas, A., Howard, R. A., Esfandiari, E., *et al.* 2010, *ApJ*, 722, 1522–1538
Webb, D. F. & Hundhausen, A. J. 1987, *Sol. Phys.*, 108, 383–401

*Nature of Prominences and their role in Space Weather*
*Proceedings IAU Symposium No. 300, 2013*
*B. Schmieder, J.-M. Malherbe & S. T. Wu, eds.*

doi:10.1017/S1743921313010958

# The physical mechanisms that initiate and drive solar eruptions

**Guillaume Aulanier**
Observatoire de Paris, LESIA, CNRS, UPMC, Univ. Paris Diderot,
5 place Jules Janssen, 92190 Meudon, France
email: guillaume.aulanier@obspm.fr

**Abstract.** Solar eruptions are due to a sudden destabilization of force-free coronal magnetic fields. But the detailed mechanisms which can bring the corona towards an eruptive stage, then trigger and drive the eruption, and finally make it explosive, are not fully understood. A large variety of storage-and-release models have been developed and opposed to each other since 40 years. For example, photospheric flux emergence vs. flux cancellation, localized coronal reconnection vs. large-scale ideal instabilities and loss of equilibria, tether-cutting vs. breakout reconnection, and so on. The competition between all these approaches has led to a tremendous drive in developing and testing all these concepts, by coupling state-of-the-art models and observations. Thanks to these developments, it now becomes possible to compare all these models with one another, and to revisit their interpretation in light of their common and their different behaviors. This approach leads me to argue that no more than two distinct physical mechanisms can actually initiate and drive prominence eruptions: the *magnetic breakout* and the *torus instability*. In this view, all other processes (including flux emergence, flux cancellation, flare reconnection and long-range couplings) should be considered as various ways that lead to, or that strengthen, one of the aforementioned driving mechanisms.

**Keywords.** Solar corona, prominences, coronal mass ejections, MHD

## 1. Introduction

Eruptive prominences are large clouds of magnetized plasma, which are ejected from the low solar corona into interplanetary space, in the form of Coronal Mass Ejections (CMEs). They can erupt either from within active regions, or from long filament channels. During the eruption, the system accelerates up to typical velocities of $100 - 1000$ km/s (although slower and faster CMEs also exist) while flare loops always form in the wake of the eruption (even though they can be hard to see in weak events).

Since the low corona is a sufficiently collisional plasma, its evolution can be studied in the frame of MHD. Also, the ratio between thermal and magnetic pressure is there very small, i.e. $\beta \ll 1$. Therefore, the magnetic energy dominates all other forms of energy in the source regions of solar eruptions (see Forbes 2000, Table 2). Current-free (potential) magnetic fields correspond to the minimum magnetic energy for a given distribution of magnetic flux through the dense photosphere. Since the photospheric flux distribution does not significantly change during the time-scales of eruptions, and since the powering of eruptions requires the magnetic energy to decrease, the coronal magnetic field must therefore be highly non-potential prior to eruption onset, i.e. it must contain strong electric currents. Due to the slow evolution of the photospheric magnetic field (as compared to typical coronal velocities), currents which are injected into the corona must accumulate slowly, such that the coronal field evolves quasi-statically, as a sequence of force-free equilibria. The triggering of CMEs therefore requires the coronal field to reach some threshold above which the balance between magnetic pressure (which points upward) and magnetic tension (which points downward) is broken. When the system suddenly enters a regime in which the pressure dominates, it can erupt in a catastrophic way, leading to a CME. The resulting ideal expansion of the magnetic field, as well as the resistively driven magnetic reconnection in the current layer that forms in the wake of the expanding system, both contribute to decrease the magnetic energy. These arguments are the root of the "storage-and-release" MHD models for solar eruptions.

Even though it is now widely accepted that solar eruptions are due to such a violent destabilization of previously energized coronal magnetic fields, the detailed mechanisms which bring a system into an eruptive stage, and which eventually drive the eruption, are not yet fully understood. A large variety of storage-and-release models has been put forward in the past decades (see Forbes *et al.* 2006; Schmieder *et al.* 2013, for two extensive reviews that also describe observations). Firstly, most of these models nicely describe many observed aspects of solar eruptions. Therefore it is difficult to estimate their respective merits solely based on observational criteria. Secondly, the models qualitatively share many common physical ingredients. So they may be difficult to distinguish from one another. So, to date, two questions remain open: Which *physical mechanisms* drive prominence eruptions? Which *solar drivers* can gradually bring stable prominences to eruptive states?

This paper aims at reviewing the existing storage-and-release models that are realistic enough, in terms of the solar physical conditions, and at considering them all together in a common frame, so as to bring some up-to-date answers to the two aforementioned questions. So this review focuses only on the *onset and driving mechanisms* of eruptions, not on their ensuing development in the large-scale corona and in the heliosphere.

## 2. Non-equilibrium and instability electric-wire models

The oldest prominence eruption model that remains considered to date is the loss-of-equilibrium model, that was initially put forward in the physical paradigm of electric-wires, and that was further proven to occur in fully 3D MHD simulations.

### 2.1. *Straight-wire geometry*

The original model was developed in 2D, in cartesian geometry (van Tend & Kuperus 1978; van Tend 1979; Molodenskii & Filippov 1987; Filippov & Den 2001). The set-up consists of a line current $I$ that is inserted at some height $z = h$ above the photospheric plane, $z = 0$, an ambient coronal field $B_{\rm ex}$, and a so-called "image current" $-I$ is added at $z = -h$ to emulate one effect of photospheric line tying, i.e. so that the photospheric magnetic field does not change when $h$ changes. The resulting coronal magnetic field consists of a detached plasmoid (or flux rope) that mimics a prominence that is embedded in a coronal arcade, and whose apex is located at $z = h$.

In the "electric paradigm", the equilibrium of the system results from the competition between two Laplace forces, namely the downward force that $B_{\rm ex}$ excerts on the coronal line current, and the upward force generated by the repulsion of the two line currents. In the "MHD paradigm", the former corresponds to restraining magnetic tension of the potential field overlying the flux rope, and the latter to magnetic pressure that results from the increase of the magnetic field strength below the coronal line current induced by the photospheric boundary.

With these settings, the equilibrium curve $h(I)$ has a critical point $(I_c; h_c)$, beyond which the line-current $I \geqslant I_c$ cannot stay in equilibrium and must move to infinite $z$. The altitude $z = h_c$ of this critical point is given by the height at which $B_{\rm ex}(z)$ starts to drop faster than $z^{-1}$.

The cartesian model has been refined several times, e.g. by giving a finite width to the coronal current, by taking into account the conservation of magnetic flux during the eruption, and by treatig the line-tying at the photospheric part of the flux rope. The latter yields the formation of a vertical current sheet below the flux rope during its eruption (Martens & Kuin 1989; Amari & Aly 1990). This current sheet exerts an extra restraining force on the line current, such that the flux rope cannot move to infinity, but finds a new equilibrium position at finite $z$ (Forbes & Isenberg 1991). In 2D, a full eruption requires the dissipation of this current sheet by sufficiently fast magnetic reconnection (Lin & Forbes 2000). But this may not be required in 3D. This is in line with analytical MHD considerations on the energy of fully open (so unreconnected) magnetic fields. Indeed this energy is infinite in 2D cartesian geometry, while it remains finite in 2D axisymmetric spherical systems and in all 3D geometries (Aly 1984, 1991; Sturrock 1991).

### 2.2. *Curved-wire geometry*

The model has also been investigated in 2.5D axisymmetric (toroidal) geometry. In a first approach, the coronal line-current is replaced by a detached ring-current at some height above the photospheric spherical surface (Lin *et al.* 1998). If an image current is added below the photosphere so that the coronal arcades surrounding the flux rope are line-tied, the same repulsive and restraining forces as discussed above contribute to the force balance. However, a new repulsive force (which the current exerts on itself due to its bending) comes into play. This curvature (or "hoop") force is radially outward directed and can be balanced by an external magnetic field, $B_{\rm ex}$ (Shafranov 1966; Chen 1989; Titov & Démoulin 1999).

In these spherical models, the requirement for magnetic reconnection below the rope as identified in cartesian geometry (Lin & Forbes 2000) still holds, but it is less important because the rope can rise ideally to tens of solar radii before the Laplace force of the vertical current sheet can halt the eruption.

In a second approach, half of the ring-current of radius $R$ is emerged above a planar photosphere, and the other half located below the photosphere somehow plays the role of the image current. With these settings, the untied ring-current can freely expand radially, as a result of a so-called "torus instability". This instability occurs when the restoring force due to the external field drops faster with the altitude than the hoop force. For external poloidal fields (i.e. perpendicular to the current) with $B_{\rm ex} \sim R^{-n}$, the instability threshold is given by $n_c \sim 3/2$ (Bateman 1978; Kliem & Török 2006).

Qualitatively similar instability thresholds have been identified when the line-tying of the ring-current is treated, through the addition of multiple image current segments in the model (Isenberg & Forbes 2007; Olmedo *et al.* 2013b).

### 2.3. *Discussion on electric-wire models*

The cartesian and the axisymmetric models had initially been developed separately. The former studied the conditions for "loss-of-equilibria", and the latter calculated onset criteria for "instabilities". Both approaches were recently revisited by Démoulin & Aulanier (2010). Non-circular current paths were later considered (Olmedo & Zhang 2010; Olmedo *et al.* 2013a,b). All geometries were shown to share almost the same analytical equations, and therefore the same physics. It was then proposed to join both approaches in a single "torus instability" mechanism.

This model has been criticized by several MHD physicists. Indeed the physical simplifcations of the electric-wire paradigm, and the qualitative nature of their link with the (correct) MHD paradigm, are a priori quite disputable. Nevertheless, the analytical elecric-wire predictions for eruptive thesholds have been found to match the onset of eruptions in some line-tied MHD simulations. Those include suspended flux ropes in 2.5D (Forbes 1990) and in 3D (Inoue & Kusano 2006; Nishida *et al.* 2013), and fully 3D line-tied flux ropes (Roussev *et al.* 2003; Török & Kliem 2005, 2007; Schrijver *et al.* 2008; Török *et al.* 2010; Jiang *et al.* 2013). So the electric-wire model was found to be consistent with its correct MHD treatment. But even then, some questions were left open. Indeed, even if all the aforementioned simulations correctly prescribed force-free flux ropes as initial conditions, firstly all but one used analytical flux rope solutions from Titov & Démoulin (1999) that contain very idealized current distributions (much simpler than those produced by solar MHD processes [see e.g. Aulanier *et al.* 2005], which are themselves more compatible with photospheric observations [see e.g. Schmieder & Aulanier 2012; Georgoulis *et al.* 2012]), and secondly these ropes were already unstable, so that their pre-eruptive evolution was not self-consistently treated.

In spite of all these issues, the torus instability was found to occur in some recent MHD simulations in which 3D flux ropes were gradually formed by photospheric drivers that mimic solar processes (see Aulanier *et al.* 2010; Fan 2010, as described further below). So the torus instability appears as a robust process to initiate and drive solar eruptions.

## 3. MHD models based on increasing manetic pressure

Any realistic eruption model must involve non-potential pre-eruptive coronal fields. There are various ways to generate them, as listed below. Some models investigated the role of the increasing magnetic pressure alone to drive an eruption.

### 3.1. *Axial flux increase*

The first models that were developed in the correct MHD paradigm were analytical and two-dimensional. There the prominence axis was oriented perpendicularly to the 2D plane of the models, and the magnetic shear was substituted for the electric current $I$ as a primary variable. But in the absence of a self-consistent way to prescribe increasing magnetic shear along the prominence, these models rather prescribed the prominence axial magnetic field (or flux) as a free parameter. The stability properties of the modeled systems were analyzed, in pretty much the same way as in the electric-wire models. Equilibrium curves were identified, and the lack of existing solutions were found for specific parameters, in particular for strong axial fields and when thermal pressure was taken into account (Low 1977; Birn *et al.* 1978; Heyvaerts *et al.* 1982; Zwingmann 1987).

These models remained theoretical, until the development of the flux-insertion method through magneto-frictional numerical relaxations in 3D (van Ballegooijen 2004). This novel approach allowed to model observed prominences and to find some eruptions, by "manually" inserting axial fluxes of different prescribed magnitudes (Su *et al.* 2011).

The early 2D models and the recent 3D ones qualitatively interpreted their modeled eruptions as evidences for losses of equilibria that could occur when the ratio $\mathfrak{R}$ of the axial prominence flux to the overlying arcade flux exceeds some unidentified threshold (as discussed by Heyvaerts *et al.* 1982; Green *et al.* 2011).

It is only very recently that Kliem *et al.* (2013) performed new detailed analysis of the 3D models. They found that the eruption onset condition matches the threshold as predicted by the electric-wire models, namely the torus instability. This result is important in two ways. Fistly it shows that, even though the ratio $\mathfrak{R}$ is defined from the right MHD paradigm, its unclear condition for eruptiveness has to be substituted by the clearer criterion for torus instability, even if that one comes from the disputable electric-wire paradigm. Secondly, this result provides one more case of torus instability in numerical simulations.

### 3.2. *Line-tied shearing and twisting motions*

The development of 3D line-tied MHD simulations showed that, when the system is driven by horizontal photosphetic motions, the axial flux cannot increase arbitrarily. Two situations were identified.

Firstly, if shearing or twisting flux tubes are restrained by strong non-moving overlying arcades in 3D, the axial flux eventually saturates. Then the system can either remain stable (Antiochos *et al.* 1994; DeVore & Antiochos 2000; Aulanier *et al.* 2002) or eventually develop a kink instability that subsequently disrupts the whole configuration (Amari & Luciani 1999). Secondly, if the overlying arcades are either too weak or also sheared or twisted, the whole system starts to expand. This bulging increases the length of the field lines, which in turn reduces the electric currents that have been induced by the photospheric motions. Analytical arguments (Aly 1985; Klimchuk & Sturrock 1989; Sturrock *et al.* 1995) and numerical simulations (Mikic & Linker 1994; Roumeliotis *et al.* 1994; Amari *et al.* 1996a,b; Aulanier *et al.* 2005) have shown that, in ideal MHD, the expansion-driven current decrease eventually dominates the shear-driven current increase. This effect prevents the magnetic field from reaching any loss of equilibrium.

In all 3D cases, no undriven expansion and therefore no eruption occurs. There are two counter-examples only, in 3D (Török & Kliem 2003; Rachmeler *et al.* 2009). But those may not be applicable to prominence eruptions. Indeed the related loss of equilibrium there develops when the flux rope has strongly expanded, long before the eruption. In 2D, shearing motions can easily produce eruptions if reconnection is allowed (see Mikic & Linker 1994; Amari *et al.* 1996a; Jacobs *et al.* 2006, that are further discussed below). But this behavior has never been reproduced in 3D, except maybe in Archontis & Hood (2008). All these results suggest that, in general, in 3D, simple line-tied shearing/twisting motions alone are not sufficient to drive an eruption. Nevertheless, line-tied motions provide a natural process to enhance the departure from non-potentiality that is required to power prominence eruptions.

### 3.3. *Twisted flux emergence*

Electric current and magnetic pressure can also be directly injected into the corona by the emergence through the photosphere of twisted flux ropes that rose through the convection zone (Emonet & Moreno-Insertis 1998; Jouve & Brun 2009) .

Some "kinematic flux emergence" simulations do achieve this. There the emergence is prescribed as time-dependent boundary conditions for the magnetic field in a line-tied photospheric boundary, and the whole flux rope can be allowed to emerge from the photospheric boundary into the corona. Such simulations indeed lead to eruptions (Fan & Gibson 2004; Amari *et al.* 2004, 2005; Fan & Gibson 2007; Fan 2010). A clear result came from the careful analysis of some of those. There, eruptions have been unambigously shown to be attributed to the torus instability, as shown by Fan & Gibson (2007) and later by Fan (2010). The former and latter constitute the first and third report, respectively, of a simulation that involved a torus-unstable flux rope that was gradually formed in the corona, and not prescribed as initial conditions as in the first MHD simulations of the torus instability.

Unfortunately, simulations of twisted flux emergence through a stratisfied medium (hence, non-kinematic emergence) show that, due to the weight of photospheric plasma which is trapped in its lower windings the flux rope hardly emerges as a whole (Fan 2001; Magara & Longcope 2001; Archontis *et al.* 2004, 2009) as it does in the kinematic models. Unless the flux rope is not strongly curved (e.g. as in MacTaggart & Hood 2009c), the only way for the lower part of the flux rope to emerge is to dispose of the dense plasma trapped in the photospheric dipped portions of the field. According to the "resistive flux emergence model" this may take place through magnetic reconnection photospheric U-loops (e.g. Pariat *et al.* 2004; Isobe *et al.* 2007). This difficulty still raises questions about the results of the kinematic simulations.

Nevertheless, a few non-kinematic simulations of flux emergence have successfully produced eruptions, using different codes and initial conditions (Manchester *et al.* 2004; Archontis & Hood 2008; Archontis & Török 2008; MacTaggart & Hood 2009b; Archontis & Hood 2012). But the physical mechanism that drive eruptions in these simulations remains unclear. Some self-induced shear flows in the photosphere may cause eruptions (Manchester *et al.* 2004). Magnetic reconnection with an ambient horizontal coronal field seems to trigger eruptions (Archontis & Török 2008; MacTaggart & Hood 2009b), like in the breakout model (see Sect. 4.2). But eruptions are not always successful with this process (MacTaggart & Hood 2009a; Leake *et al.* 2010). The development of low-altitude magnetic reconnection within the emerging fields could also cause, or at least contribute to, the eruption of a newly-formed flux rope (Manchester *et al.* 2004; Archontis & Hood 2008), like in the tether-cutting model (see Sect. 4.1). Finally, the relative strength of the overlying confining arcades as compared to that the emerging rope appears determining (Archontis & Hood 2012), maybe like in the axial-flux increase models (see Sect. 3.1). So more investigation is required in terms of physical analysis. One other issue concerns the too small sizes of the modeled flux ropes in these simulations, relative to the thickness of the modeled photosphere.

## 4. MHD models based on decreasing magnetic tension

Instead of increasing the current to a value $I \geqslant I_c$ an alternative approach is to reduce the restraining tension of coronal arcades which overlie initially stable current-carrying magnetic fields. Most eruption models actually fall into this class.

### 4.1. *Tether-cutting*

Magnetic tension can decrease due to the breakdown of ideal MHD in the vertical current-sheet that forms within a shearing arcade (Amari & Aly 1990; Forbes & Isenberg 1991), resulting in magnetic reconnection that eventually forms flare loops and ribbons in the wake of the CME, i.e. *below* the current-carrying field lines. This non-ideal effect creates and feeds a twisted envelope around the initial current-carrying fields, from the flux of the overlying arcades. So the flare reconnection is an efficient process for reducing the downward tension of the arcades: it can "cut the tethers" (Sturrock 1989). This process is self-sustaining, since the more the flux rope rises during the eruption, and the more reconnection happens, the weaker is the restraining tension, so the more the flux rope can rise (Moore & Roumeliotis 1992; Moore *et al.* 2001; Nishida *et al.* 2013). Thus, in principle, it can become explosive.

The tether-cutting effect alone has indeed been shown to trigger and to drive eruptions in 2.5D cartesian (Amari *et al.* 1996a) and axisymmetric (Mikic & Linker 1994; Jacobs *et al.* 2006) MHD simulations. Early tether-cutting reconnection has also been found to sustain the formation of twisted envelopes in 3D MHD simulations of sheared arcades (DeVore & Antiochos 2000), of flux cancellation (Aulanier *et al.* 2010) and of kinematic flux emergence (Fan 2010). But it did not cause the eruption, when there was one, in any of these simulations. Also this reconnection there tends to stall when the photospheric driving is supressed during non-eruptive stages. Still, the late onset of this reconnection clearly accelerates eruptions in some 2.5D and 3D MHD simulations. But these eruptions were previously initiated by another mechanism, such as the magnetic breakout process (Lynch *et al.* 2008; Karpen *et al.* 2012, as described hereafter) and ideal instabilities

(Nishida *et al.* 2013). Some 3D flux emergence simulations did report a qualitative role for tether-cutting reconnection in their eruptions (Manchester *et al.* 2004; Archontis & Hood 2008). But they did not show that it was explicitly driving the eruptions.

So the tether-cutting has never been proven to initiate, alone, an eruption in any 3D simulations. This negative result was found (but rarely written) by independent groups using different codes. This raises strong doubts about the validity of the tether-cutting as an eruption driving mechanism. However, this reconnection is obviously an important aspect of every solar eruption. Indeed it provides an extra-acceleration to the erupting prominence and, of course, it releases a lot of magnetic energy and it produces the most energetic particles in the flare that develops in the wake of the CME (Masson *et al.* 2013).

### 4.2. *Magnetic breakout*

A new idea was proposed by Antiochos *et al.* (1999), for lowering the flux and the tension of the overlying arcades, by invoking magnetic reconnection occurring at a magnetic null point, being located at high altitude *above* the current-carrying field lines.

Observationally, this model requires a quadrupolar topology for the photospheric magnetic field. This condition can be satisfied in many active regions, especially young ones (see e.g. Ugarte-Urra *et al.* 2007). But is not guaranteed for older decaying active regions that look bipolar (see e.g. van Driel-Gesztelyi *et al.* 2003), although large remote connections may still be invoked. Theoretically, Antiochos *et al.* (1999) proved for axisymmetric systems that this "magnetic breakout" alone can drive eruptions, provided that the onset of null point reconnection was delayed during the slow energy build-up phase, and that the rate of reconnection was slow enough during the fast eruptive phase. As for the tether-cutting reconnection, the breakout reconnection could create a feedback-loop, leading to an explosive behavior, hence to an eruption. In addition, DeVore & Antiochos (2005) found that the efficiency of this mechanism also depends on the ratio of the magnetic fluxes located above and below the null point: if the flux of the largest overlying arcades is too weak (resp. too strong), there is not enough (resp. too much) flux to reconnect for the breakout mechanism to be sustained long enough for a full eruption.

Full simulations of the breakout were first achieved in 2.5D MHD simulations (MacNeice *et al.* 2004), including with very high spatial resolutions (Karpen *et al.* 2012; Lynch & Edmondson 2013) and with the solar wind (van der Holst *et al.* 2007; Masson *et al.* 2013). A key difference with the tether-cutting model, though, is that the breakout was also found to occur in a 3D line-tied simulation (Lynch *et al.* 2008), and very probably in a 3D flux emergence simulation (Archontis & Török 2008). Also, this original MHD model found unambiguous support in several observational analyses (e.g. Aulanier *et al.* 2000; Sterling & Moore 2001; Gary & Moore 2004; Ugarte-Urra *et al.* 2007).

So the breakout mechanism appears as a robust process to initiate and drive solar eruptions, although it has several requirements that prevents it from being general (DeVore & Antiochos 2005), and it may require a relatively strong flare reconnection to produce a fast eruption (Karpen *et al.* 2012; Masson *et al.* 2013).

### 4.3. *Side-reconnections and remote couplings*

Some other models also explain eruptions through coronal reconnection, which occurs aside of the prominence instead of below or above it. In several cases this reconnection can increase the length of the overlying arcades, and lower their tension.

Eruptions driven by this process were modeled in the context of small-scale flux emergence in the vicinity of the flux rope, in the 2D electric-wire paradigm (Lin *et al.* 2001), in 2.5D MHD simulations (Chen & Shibata 2000) and recently in 3D simulations (Kusano *et al.* 2012; Toriumi *et al.* 2013). It was also found to operate in 3D MHD models of interacting active regions (Jacobs *et al.* 2009), and possibly to trigger sympathetic eruptions in models where several current-carrying flux tubes are included (Török *et al.* 2011; Lynch & Edmondson 2013), in line with the concept proposed by Schrijver & Title (2011) and further developed by Schrijver *et al.* (2013).

The dominant mechanism that drive eruptions in these side-reconnection models is still uncertain. Chen & Shibata (2000) attributes the eruption to the tether-cutting reconnection triggered by the side-reconnection. The broad coverage of the paramater space achieved by Kusano *et al.* (2012) shows that the eruptivity strongly depends on the magnetic field configuration. Schrijver *et al.* (2013) and Lynch & Edmondson (2013) argue that eruptions are triggered sympathetically because the corona is constantly reconfiguring from the previous eruption. Lin *et al.* (2001) show that an ideal loss of equilibrium is triggered in the new system that results from the appearance of a new bipole. And finally Török *et al.* (2011) show that the coronal reconfiguration that results from high-altitude reconnection, actually leads pre-eruptive flux ropes that are almost torus-unstable to enter the instability regime and then erupt one after the other.

So, like in flux emergence models, a loss of equilibrium /torus instability can be triggered by remote reconnections that result in small-scale or large-scale couplings. But other interpretations for the cause of the eruptions have also been proposed.

### 4.4. *Converging motions*

Models driven by photospheric motions that converge toward polarity inversion lines, above which prominences are located, have also been considered. Quasi-static theory shows that reducing the length-scale of the photospheric magnetic field also reduces the magnitude of the coronal field at large heights, and makes the field drop faster with height. A priori, both can facilitate the torus instability.

This is strongly suggested by the landmark electric-wire model by Forbes & Priest (1995) and by the MHD simulations of Török & Kliem (2007). They explored eruptive behaviors, by making several independent calculations for different ratios between the current or the height of the pre-eruptive flux rope, and the horizontal extent of the surrounding photospheric bipolar field.

Eruptions of current-carrying fields subject to dynamically-treated converging motions, have also been found in MHD simulations, both in 2.5D (Inhester *et al.* 1992) and in 3D (Amari *et al.* 2003a). Recently, a direct MHD simulation of an observed event, forced by ideal converging motions, also produced an eruption (Zuccarello *et al.* 2012).

Such motions are frequently observed at the Sun's surface. This makes the model appealing. Some questions remain open, though: very extended motions as used in the models are rarely observed; and the physical mechanism that actually drives the eruptions has not yet been firmly identified in the MHD models.

### 4.5. *Decreasing photospheric magnetic field*

This class of models can somehow be viewed as the exact opposite as the axial flux increase models. They rely on a homogeneous magnetic field decrease in an extended section of the photosphere around the flux rope. This decrease is imposed, either by reducing the magnetic momentum of the external subphotospheric magnetic field sources (see e.g. Lin *et al.* 1998), or by prescribing adequate horizontal electric fields in the photosphere (see e.g. Amari *et al.* 2000). In axisymmetric geometry, this process produces eruptions of detached flux ropes, that are either pre-existing (Lin *et al.* 1998), or slowly formed during the magnetic field decrease (Linker *et al.* 2003; Reeves *et al.* 2010). It can also form and trigger the eruption of line-tied flux ropes of various sizes in 3D (Amari *et al.* 2000, Lin *et al.* 2002; Linker *et al.* 2003).

Qualitatively, the origin of the eruptions can be directly attributed to the diminishing of the coronal restraining tension, that naturally results from the gradual disappearance of the photospheric magnetic flux. Quantitatively, the eruptions occur when the diminishing magnetic energy of the fully open field reaches down to a value that is equal to that of the current-carrying fields, as identified by Amari *et al.* (2000). This interpretation is particularly interesting because it provides a very clear eruption threshold in the correct MHD paradigm (like the torus instability does in the electric-wire paradigm).

The physical validity of these models is still debated, as it is difficult to find a self-consistent MHD process that diminishes the photospheric magnetic field over large areas. Amari *et al.* (2000) qualitatively noted that flux rope emergence can actually lead to an apparent flux decrease on the side rope, after the emergence of the rope axis. But it is unclear whether this process produces the magnetic field decrease as required for an eruption. Amari *et al.* (2010) quantitatively calculated that photospheric flows that mimic flux dispersal in decaying active regions (as described below) can account for the prescribed flux decrease. But this interpretation requires flows that accelerate to infinite speeds towards the polarity inversion line, which may be problematic.

### 4.6. *Flux dispersal and cancellation*

The "flux-cancellation" model is based on the observed long-term evolution of magnetic flux concentrations in the photosphere, within or between bipolar active regions (e.g. Wang *et al.* 1989; Démoulin *et al.* 2002; van Driel-Gesztelyi *et al.* 2003; Schmieder *et al.* 2008; Green *et al.* 2011). Over time-periods of days to months, depending on their sizes, flux concentrations disperse and spread in all directions. Their apparent diffusion leads their peak and mean magnetic field magnitude to decrease, while their total magnetic flux slowly and weakly decreases through local flux convergence and cancellation at polarity inversion lines, right below prominences.

The landmark references for this model are van Ballegooijen & Martens (1989) and Forbes & Isenberg (1991). They showed that converging motions and flux cancellation combined all together (with no flux dispersal or decrease) yield the gradual formation of a flux rope through a tether-cutting-like photospheric reconnection, that involves coronal arcades that have previously been sheared in a 2.5D geometry. On the long run, the rope grows in size and in altitude until it erupts, as calculated in 2.5D electric-wire models (Forbes & Isenberg 1991; Isenberg *et al.* 1993).

By treating the large-scale decay of the photospheric magnetic field with an extra photospheric diffusion term in the induction equation (as introduced by Wang *et al.* 1989), 3D flux ropes were also found to form and erupt, firstly by Amari *et al.* (2003b) with MHD simulations, and later by Mackay & van Ballegooijen (2006) and Yeates & Mackay (2009) with magneto-frictional simulations. These results were also found in non-symmetric MHD models (Aulanier *et al.* 2010, 2012; Pagano *et al.* 2013), and in symmetric models in which the flux dispersal was instead treated by line-tied flows diverging from the center each flux concentration (Amari *et al.* 2011).

The magnetic flux decrease model (see Sect. 4.5) is often regarded as a flux cancellation model. But both are physically very different. Firstly, in the flux cancellation model the magnetic flux decreases locally because of magnetic field annihilation at the inversion line (Wang *et al.* 1989). That is different than a flux decrease induced by a diminishing magnetic field over a large area. Secondly, the flux cancellation model does not involve a dimishing of the open field energy down to the magnetic field energy of the pre-eruptive field (see Amari *et al.* 2003b).

Detailed analysis of one MHD simulation, and its comparison with electric-wire models, showed that "photospheric flux-cancellation and tether-cutting coronal reconnection do not trigger CMEs in bipolar magnetic fields, but are key pre-eruptive mechanisms for flux ropes to build up and to rise to the critical height above the photosphere at which the torus instability causes the eruption" (Aulanier *et al.* 2010).

## 5. Discussion

A large variety of storage-and-release eruption models have been developed during the last forty years. At first sight they look similar to each other. Indeed, they predict similar observable features, and they share common physical ingredients. But they also contain important differences, either in their equations, their geometries, and their prescriptions. So they have often been opposed to each other. This emulation fostered fine-tuned developments and analyses, up to a stage at which they can now be classified and compared with one another, so that *the physical mechanisms* that initiate, drive, and contribute to prominence eruptions may now be identified independently of *the models* themselves.

When acknowledging that prominence eruptions occur once the magnetic pressure explosively wins over the magnetic tension (exerted on the system by the overlying coronal arcades), then the present review along with that of Schmieder *et al.* (2013) suggest that, to date, no more than two physical mechanisms can initiate and sustain this explosive loss of force balance.

The ideal loss of equilibrium of a flux rope is the first mechanism. 3D models have shown that the prominence flux rope does not actually need to be very twisted: the mechanism works with ropes that have less than one turn. The eruption there occurs once the rope axis has reached an altitude, above which all stable equilibria cease to exist. The threshold is reached when the magnitude of magnetic fields of the overlying arcades decrease faster with height than the magnetic pressure which pushes the flux rope upwards, which also decreases with time during the rise of the flux rope. The process was first proposed by van Tend & Kuperus (1978), and it was shown by Démoulin & Aulanier (2010) to correspond to the "torus instability" first proposed by Bateman (1978) in tokamaks, and first revisited for solar eruptions by Kliem & Török (2006).

The removal of the arcades that overlay and confine the prominence by means of high-altitude magnetic reconnection is the second mechanism. It has initially been proposed to occur at null points, but it may also operate at separators and quasi-separatrix layers. Once the reconnection has begun, it transfers overlying arcades into connectivity domains that are located aside of the prominence. So the amount of magnetic flux that overlays the prominence is reduced. The associated diminishing of the confinement makes the prominence rise to larger altitudes. This provides a loop-feedback on the high-altitude reconnection, so that an eruption can occur. This process, called the "magnetic breakout", was first proposed by Antiochos *et al.* (1999). The efficiency of the breakout requires the magnetic fluxes located above and below the reconnection region be comparable in magnitude (DeVore & Antiochos 2005).

To date, only the torus instability and the magnetic breakout were found to occur in many different 3D MHD simulations. The torus instability has been identified to cause eruptions with prescribed unstable flux ropes (Roussev *et al.* 2003; Török & Kliem 2005, 2007; Schrijver *et al.* 2008; Török *et al.* 2010), with kinematic flux emergence (Fan & Gibson 2007; Fan 2010), with flux cancellation (Aulanier *et al.* 2010, 2012), with sympathetically erupting flux ropes (Török *et al.* 2011), and with non-linear force-free relaxations (Kliem *et al.* 2013; Jiang *et al.* 2013). The magnetic breakout mechanism has been shown to operate with shearing bipoles in multipolar geometry (Lynch *et al.* 2008), with twisted flux tubes emerging through a stratified medium into a pre-existing horizontal field (Archontis & Török 2008; MacTaggart & Hood 2009b), and with sympathetically erupting sheared loops (Lynch & Edmondson 2013). So it can be conjectured that eruptions can only be initiated and driven by one of these two mechanisms, or their combination.

It follows that, depending on the solar conditions, all the other processes may be considered as different ways to either bring the system to the threshold of one of these two mechanisms, or to help making the resulting eruption faster. For example, flux emergence (Fan 2010) or flux cancellation (Aulanier *et al.* 2010) can initiate a torus instability. Also, reconnection-driven long-range couplings around flux ropes (Török *et al.* 2011) or sheared arcades (Lynch & Edmondson 2013) can initiate sympathetic torus instabilities and magnetic breakouts. And flare reconnection can accelerate eruptions initiated by a torus instability (Nishida *et al.* 2013) and a magnetic breakout (Karpen *et al.* 2012; Masson *et al.* 2013).

In this line each and every solar process that can contribute to solar eruptions should be taken into account, all together with the few physical mechanisms that initiate and drive eruptions, so as to reach a comprehensive understanding of observed events, and so as to predict the occurrence of future events.

## References


Aly, J. J. 1984, *Astrophys. J.*, 283, 349

—. 1985, *Astron. Astrophys.*, 143, 19

—. 1991, *Astrophys. J. Lett.*, 375, L61

Amari, T. & Aly, J. J. 1990, *Astron. Astrophys.*, 227, 628

Amari, T., Aly, J.-J., Luciani, J.-F., Mikic, Z., & Linker, J. 2011, *Astrophys. J. Lett.*, 742, L27

Amari, T., Aly, J.-J., Mikic, Z., & Linker, J. 2010, *Astrophys. J. Lett.*, 717, L26

Amari, T. & Luciani, J. F. 1999, *Astrophys. J. Lett.*, 515, L81

Amari, T., Luciani, J. F., & Aly, J. J. 2004, *Astrophys. J. Lett.*, 615, L165

—. 2005, *Astrophys. J. Lett.*, 629, L37

Amari, T., Luciani, J. F., Aly, J. J., Mikic, Z., & Linker, J. 2003a, *Astrophys. J.*, 585, 1073

—. 2003b, *Astrophys. J.*, 595, 1231

Amari, T., Luciani, J. F., Aly, J. J., & Tagger, M. 1996a, *Astron. Astrophys.*, 306, 913

—. 1996b, *Astrophys. J. Lett.*, 466, L39

Amari, T., Luciani, J. F., Mikic, Z., & Linker, J. 2000, *Astrophys. J. Lett.*, 529, L49

Antiochos, S. K., Dahlburg, R. B., & Klimchuk, J. A. 1994, *Astrophys. J. Lett.*, 420, L41

Antiochos, S. K., DeVore, C. R., & Klimchuk, J. A. 1999, *Astrophys. J.*, 510, 485

Archontis, V. & Hood, A. W. 2008, *Astrophys. J. Lett.*, 674, L113

—. 2012, *Astron. Astrophys.*, 537, A62

Archontis, V., Hood, A. W., Savcheva, A., Golub, L., & Deluca, E. 2009, *Astrophys. J.*, 691, 1276

Archontis, V., Moreno-Insertis, F., Galsgaard, K., Hood, A., & O'Shea, E. 2004, *Astron. Astrophys.*, 426, 1047

Archontis, V. & Török, T. 2008, *Astron. Astrophys.*, 492, L35

Aulanier, G., DeLuca, E. E., Antiochos, S. K., McMullen, R. A., & Golub, L. 2000, *Astrophys. J.*, 540, 1126

Aulanier, G., Démoulin, P., & Grappin, R. 2005, *Astron. Astrophys.*, 430, 1067

Aulanier, G., DeVore, C. R., & Antiochos, S. K. 2002, *Astrophys. J. Lett.*, 567, L97

Aulanier, G., Janvier, M., & Schmieder, B. 2012, *Astron. Astrophys.*, 543, A110

Aulanier, G., Török, T., Démoulin, P., & DeLuca, E. E. 2010, *Astrophys. J.*, 708, 314

Bateman, G. 1978, *MHD instabilities*, ed. G. Bateman

Birn, J., Goldstein, H., & Schindler, K. 1978, *Solar Phys.*, 57, 81

Chen, J. 1989, *Astrophys. J.*, 338, 453

Chen, P. F. & Shibata, K. 2000, *Astrophys. J.*, 545, 524

Démoulin, P. & Aulanier, G. 2010, *Astrophys. J.*, 718, 1388

Démoulin, P., Mandrini, C. H., van Driel-Gesztelyi, L., Thompson, B. J., Plunkett, S., Kovári, Z., Aulanier, G., & Young, A. 2002, *Astron. Astrophys.*, 382, 650

DeVore, C. R. & Antiochos, S. K. 2000, *Astrophys. J.*, 539, 954
—. 2005, *Astrophys. J.*, 628, 1031
Emonet, T. & Moreno-Insertis, F. 1998, *Astrophys. J.*, 492, 804
Fan, Y. 2001, *Astrophys. J. Lett.*, 554, L111
—. 2010, *Astrophys. J.*, 719, 728
Fan, Y. & Gibson, S. E. 2004, *Astrophys. J.*, 609, 1123
—. 2007, *Astrophys. J.*, 668, 1232
Filippov, B. P. & Den, O. G. 2001, *Journal of Geophysical Research*, 106, 25177
Forbes, T. G. 1990, *Journal of Geophysical Research*, 95, 11919
—. 2000, *Journal of Geophysical Research*, 105, 23153
Forbes, T. G. & Isenberg, P. A. 1991, *Astrophys. J.*, 373, 294
Forbes, T. G., Linker, J. A., Chen, J., Cid, C., Kóta, J., Lee, M. A., Mann, G., Mikić, Z., Potgieter, M. S., Schmidt, J. M., Siscoe, G. L., Vainio, R., Antiochos, S. K., & Riley, P. 2006, *Space Science Reviews*, 123, 251
Forbes, T. G. & Priest, E. R. 1995, *Astrophys. J.*, 446, 377
Gary, G. A. & Moore, R. L. 2004, *Astrophys. J.*, 611, 545
Georgoulis, M. K., Titov, V. S., & Mikić, Z. 2012, *Astrophys. J.*, 761, 61
Green, L. M., Kliem, B., & Wallace, A. J. 2011, *Astron. Astrophys.*, 526, A2
Heyvaerts, J., Lasry, J. M., Schatzmann, M., & Witomsky, P. 1982, *Astron. Astrophys.*, 111, 104
Inhester, B., Birn, J., & Hesse, M. 1992, *Solar Phys.*, 138, 257
Inoue, S. & Kusano, K. 2006, *Astrophys. J.*, 645, 742
Isenberg, P. A. & Forbes, T. G. 2007, *Astrophys. J.*, 670, 1453
Isenberg, P. A., Forbes, T. G., & Démoulin, P. 1993, *Astrophys. J.*, 417, 368
Isobe, H., Tripathi, D., & Archontis, V. 2007, *Astrophys. J. Lett.*, 657, L53
Jacobs, C., Poedts, S., & van der Holst, B. 2006, *Astron. Astrophys.*, 450, 793
Jacobs, C., Roussev, I. I., Lugaz, N., & Poedts, S. 2009, *Astrophys. J. Lett.*, 695, L171
Jiang, C., Feng, X., Wu, S. T., & Hu, Q. 2013, *Astrophys. J. Lett.*, 771, L30
Jouve, L. & Brun, A. S. 2009, *Astrophys. J.*, 701, 1300
Karpen, J. T., Antiochos, S. K., & DeVore, C. R. 2012, *Astrophys. J.*, 760, 81
Kliem, B., Su, Y., van Ballegooijen, A., & DeLuca, E. 2013, ArXiv e-prints
Kliem, B. & Török, T. 2006, *Physical Review Letters*, 96, 255002
Klimchuk, J. A. & Sturrock, P. A. 1989, *Astrophys. J.*, 345, 1034
Kusano, K., Bamba, Y., Yamamoto, T. T., Iida, Y., Toriumi, S., & Asai, A. 2012, *Astrophys. J.*, 760, 31
Leake, J. E., Linton, M. G., & Antiochos, S. K. 2010, *Astrophys. J.*, 722, 550
Lin, J. & Forbes, T. G. 2000, *Journal of Geophysical Research*, 105, 2375
Lin, J., Forbes, T. G., & Isenberg, P. A. 2001, *Journal of Geophysical Research*, 106, 25053
Lin, J., Forbes, T. G., Isenberg, P. A., & Démoulin, P. 1998, *Astrophys. J.*, 504, 1006
Lin, J., van Ballegooijen, A. A., & Forbes, T. G. 2002, *Journal of Geophysical Research (Space Physics)*, 107, 1438
Linker, J. A., Mikić, Z., Lionello, R., Riley, P., Amari, T., & Odstrcil, D. 2003, *Physics of Plasmas*, 10, 1971
Low, B. C. 1977, *Astrophys. J.*, 212, 234
Lynch, B. J., Antiochos, S. K., DeVore, C. R., Luhmann, J. G., & Zurbuchen, T. H. 2008, *Astrophys. J.*, 683, 1192
Lynch, B. J. & Edmondson, J. K. 2013, *Astrophys. J.*, 764, 87
Mackay, D. H. & van Ballegooijen, A. A. 2006, *Astrophys. J.*, 641, 577
MacNeice, P., Antiochos, S. K., Phillips, A., Spicer, D. S., DeVore, C. R., & Olson, K. 2004, *Astrophys. J.*, 614, 1028
MacTaggart, D. & Hood, A. W. 2009a, *Astron. Astrophys.*, 501, 761
—. 2009b, *Astron. Astrophys.*, 508, 445
—. 2009c, *Astron. Astrophys.*, 507, 995
Magara, T. & Longcope, D. W. 2001, *Astrophys. J. Lett.*, 559, L55
Manchester, IV, W., Gombosi, T., DeZeeuw, D., & Fan, Y. 2004, *Astrophys. J.*, 610, 588

Martens, P. C. H. & Kuin, N. P. M. 1989, *Solar Phys.*, 122, 263
Masson, S., Antiochos, S. K., & DeVore, C. R. 2013, *Astrophys. J.*, 771, 82
Mikic, Z. & Linker, J. A. 1994, *Astrophys. J.*, 430, 898
Molodenskii, M. M. & Filippov, B. P. 1987, *Soviet Astronomy*, 31, 564
Moore, R. L. & Roumeliotis, G. 1992, *Lecture Notes in Physics*, Springer Verlag, IAU Colloq. 133: Eruptive Solar Flares, 399, 69
Moore, R. L., Sterling, A. C., Hudson, H. S., & Lemen, J. R. 2001, *Astrophys. J.*, 552, 833
Nishida, K., Nishizuka, N., & Shibata, K. 2013, *Astrophys. J. Lett.*, 775, L39
Olmedo, O. & Zhang, J. 2010, *Astrophys. J.*, 718, 433
Olmedo, O., Zhang, J., & Kunkel, V. 2013a, *Astrophys. J.*, 771, 125
Olmedo, O., Zhang, J., & Chen, J. 2013b, *Astrophys. J.*, submitted
Pagano, P., Mackay, D. H., & Poedts, S. 2013, *Astron. Astrophys.*, 554, A77
Pariat, E., Aulanier, G., Schmieder, B., Georgoulis, M. K., Rust, D. M., & Bernasconi, P. N. 2004, *Astrophys. J.*, 614, 1099
Rachmeler, L. A., DeForest, C. E., & Kankelborg, C. C. 2009, *Astrophys. J.*, 693, 1431
Reeves, K. K., Linker, J. A., Mikić, Z., & Forbes, T. G. 2010, *Astrophys. J.*, 721, 1547
Roumeliotis, G., Sturrock, P. A., & Antiochos, S. K. 1994, *Astrophys. J.*, 423, 847
Roussev, I. I., Forbes, T. G., Gombosi, T. I., Sokolov, I. V., DeZeeuw, D. L., & Birn, J. 2003, *Astrophys. J. Lett.*, 588, L45
Schmieder, B., Bommier, V., Kitai, R., Matsumoto, T., Ishii, T. T., Hagino, M., Li, H., & Golub, L. 2008, *Solar Phys.*, 247, 321
Schmieder, B. & Aulanier, G. 2012, *Advances in Space Research*, 49, 1598
Schmieder, B., Démoulin, P., & Aulanier, G. 2013, *Advances in Space Research*, 51, 1967
Schrijver, C. J., Elmore, C., Kliem, B., Török, T., & Title, A. M. 2008, *Astrophys. J.*, 674, 586
Schrijver, C. J. & Title, A. M. 2011, *Journal of Geophysical Research (Space Physics)*, 116, 4108
Schrijver, C. J., Title, A. M., Yeates, A. R., & DeRosa, M. L. 2013, *Astrophys. J.*, 773, 93
Shafranov, V. D. 1966, Reviews of Plasma Physics, 2, 103
Sterling, A. C. & Moore, R. L. 2001, *Astrophys. J.*, 560, 1045
Sturrock, P. A. 1989, *Solar Phys.*, 121, 387
—. 1991, *Astrophys. J.*, 380, 655
Sturrock, P. A., Antiochos, S. K., & Roumeliotis, G. 1995, *Astrophys. J.*, 443, 804
Su, Y., Surges, V., van Ballegooijen, A., DeLuca, E., & Golub, L. 2011, *Astrophys. J.*, 734, 53
Titov, V. S. & Démoulin, P. 1999, *Astron. Astrophys.*, 351, 707
Toriumi, S., Iida, Y., Bamba, Y., Kusano, K., Imada, S., & Inoue, S. 2013, *Astrophys. J.*, 773, 128
Török, T., Berger, M. A., & Kliem, B. 2010, *Astron. Astrophys.*, 516, A49
Török, T. & Kliem, B. 2003, *Astron. Astrophys.*, 406, 1043
—. 2005, *Astrophys. J. Lett.*, 630, L97
—. 2007, *Astronomische Nachrichten*, 328, 743
Török, T., Panasenco, O., Titov, V. S., Mikić, Z., Reeves, K. K., Velli, M., Linker, J. A., & De Toma, G. 2011, *Astrophys. J. Lett.*, 739, L63
Ugarte-Urra, I., Warren, H. P., & Winebarger, A. R. 2007, *Astrophys. J.*, 662, 1293
van Ballegooijen, A. A. 2004, *Astrophys. J.*, 612, 519
van Ballegooijen, A. A. & Martens, P. C. H. 1989, *Astrophys. J.*, 343, 971
van der Holst, B., Jacobs, C., & Poedts, S. 2007, *Astrophys. J. Lett.*, 671, L77
van Driel-Gesztelyi, L., Démoulin, P., Mandrini, C. H., Harra, L., & Klimchuk, J. A. 2003, *Astrophys. J.*, 586, 579
van Tend, W. 1979, *Solar Phys.*, 61, 89
van Tend, W. & Kuperus, M. 1978, *Solar Phys.*, 59, 115
Wang, Y.-M., Nash, A. G., & Sheeley, Jr., N. R. 1989, *Science*, 245, 712
Yeates, A. R. & Mackay, D. H. 2009, *Astrophys. J.*, 699, 1024
Zuccarello, F. P., Meliani, Z., & Poedts, S. 2012, *Astrophys. J.*, 758, 117
Zwingmann, W. 1987, *Solar Phys.*, 111, 309

*Nature of Prominences and their role in Space Weather*
*Proceedings IAU Symposium No. 300, 2013*
*B. Schmieder, J.-M. Malherbe & S. T. Wu, eds.*

doi:10.1017/S174392131301096X

# Magnetohydrodynamic study on the effect of the gravity stratification on flux rope ejections

**Paolo Pagano[1], Duncan H. Mackay[1], and Stefaan Poedts[2]**

[1]School of Mathematics and Statistics, University of St Andrews,
North Haugh, KY16 9SS, St Andrews, Scotland
email: ppagano@mcs.st-andrews.ac.uk

[2]Centre for mathematical Plasma-Astrophysics, K. U. Leuven ,
Celestijnenlaan 200B, 3001, Leuven, Belgium

**Abstract.** Coronal Mass Ejections (CMEs) are one of the most violent phenomena found on the Sun. One model to explain their occurrence is the flux rope ejection model where these magnetic structures firt form in the solar corona then are ejected to produce a CME. We run simulations coupling two models. The Global Non-Linear Force-Free Field (GNLFFF) evolution model to follow the quasi-static formation of a flux rope and MHD simulations for the production of a CME through the loss of equilibrium and ejection of this flux rope in presence of solar gravity and density stratification. Our realistic multi-beta simulations describe the CME following the flux rope ejection and highlight the decisive role played by the gravity stratification on the CME propagation speed.

**Keywords.** Sun, CME, Solar Corona, MHD

## 1. Introduction

Coronal Mass Ejections (CME's) are the main drivers of Space Weather, a term used to describe the effect of plasmas and magnetic fields on the near Earth environment.

Although not all the aspects of the CME origin have been understood, the ejection of a magnetic flux rope from the solar corona successfully describes many of the general features of CMEs, and in some of these models the flux rope naturally undergoes a loss of equilibrium Forbes & Isenberg (1991). CMEs propagate in the solar corona and interact with it. In the solar corona, plasma motions are primary driven by the Lorentz force. However, under equilibrium conditions the plasma distribution in the corona can be described as being stratified by the effect of gravity.

Here we specifically address the role of gravitational stratification on the CME propagation. In our framework, the ejection is caused by a non-equilibrium magnetic field configuration. The non-equilibrium magnetic field is added to the stratified solar corona which is initially in equilibrium and decoupled from the magnetic field. In particular, we start from the model of Pagano *et al.* (2013) where an eruptive magnetic configuration is used and the full life span of a magnetic flux rope is described by coupling the global non linear force free model of Mackay & van Ballegooijen (2006) with MHD simulations run with the AMRVAC code Keppens *et al.* (2012). We explore how the parameter space affects the eruption characteristics by tuning both the temperature of the solar corona and the intensity of the magnetic field.

## 2. The Model

In order to study the effect of gravitational stratification on the propagation of a CME we start from the work of Pagano *et al.* (2013) where a magnetic configuration has been shown to be suitable for the ejection of a magnetic flux rope. We carry out a set of MHD simulations that consider variations in the temperature of the background coronal atmosphere and the magnetic field intensity.

We use the AMRVAC code developed at the KU Leuven to run the simulations Keppens *et al.* (2012). The code solves the ideal MHD equations and the terms that account for gravity are included and the expression for the solar gravitational acceleration is

$$\vec{g} = -\frac{GM_\odot}{r^2}\hat{r}, \tag{2.1}$$

where $G$ is the Gravitational constant and $M_\odot$ denotes the mass of the Sun. The configuration of the magnetic field is taken from Day 19 in the simulation of Mackay & van Ballegooijen (2006). Pagano *et al.* (2013) in Sec.2.2 explain in detail how the magnetic field distribution is imported. We assume an initial atmosphere of a constant temperature corona stratified by solar gravity, where we also allow for a small background pressure and density.

$$\rho(r) = \rho_0 e^{\frac{M_\odot G \mu m_p}{T k_b r}} + \rho_{bg}, \tag{2.2}$$

$$p(r) = \frac{k_b T}{\mu m_p}\rho_0 e^{\frac{M_\odot G \mu m_p}{T k_b r}} + p_{bg} \tag{2.3}$$

where we use $\rho_0$ to tune the density at the lower boundary, $\mu = 1.31$ is the average particle mass in the solar corona, $m_p$ is the proton mass, $k_b$ is Boltzmann constant and $T$ is the temperature of the corona. We tune $\rho_0$ depending on the temperature $T$ in order to always have the same density at the bottom of our computational domain. Finally, we set $\vec{v} = 0$, as initial condition for the velocity. In the present work we use the maximum value of the magnetic field intensity of the initial magnetic field configuration, $B_{max}$, as a simulation parameter. The simulation domain extends over 3 $R_\odot$ in the radial dimension starting from $r = R_\odot$. The colatitude, $\theta$, spans from $\theta = 30°$ to $\theta = 100°$ and the longitude, $\phi$, spans over $90°$. To define the MHD quantities in the portion of the domain from 2.5 $R_\odot$ to 4 $R_\odot$ we use Eq.2.2 and Eq.2.3 for density and thermal pressure and the magnetic field for $r > 2.5\ R_\odot$ is assumed purely radial ($B_\theta = B_\phi = 0$) where the magnetic flux is assumed to be conserved. The boundary conditions are treated with a system of ghost cells. Open boundary conditions are imposed at the outer boundary, reflective boundary conditions are set at the $\theta$ boundaries and the $\phi$ boundaries are periodic. At the lower boundary we impose constant boundary conditions taken from the first four $\theta$-$\phi$ planes of cells derived from the GNLFFF model. The computational domain is composed of $256 \times 128 \times 128$ cells distributed in a uniform grid. Full details of the grid can be found in Sec.2.3 of Pagano *et al.* (2013).

In order to analyse the role of the background stratified corona we run a set of 9 simulations using 3 different temperatures ($T = 1.5,2,3\ MK$) of the corona and 3 different maximum intensity of the magnetic field ($B_{max} = 7,21,42\ G$).

In our model, a higher corona temperature implies a more uniform density and pressure gradient and a higher amount of mass that constitutes the solar corona. At the same time, the higher temperature leads to remarkably higher $\beta$ in the outer corona, while in the lower corona ($r < 1.2R_\odot$) $\beta$ is clearly below $\sim 10^{-1}$ regardless of the temperature. The outer corona switches from a low to high $\beta$ regime when the temperature increases from 1.5 to 3 $MK$.

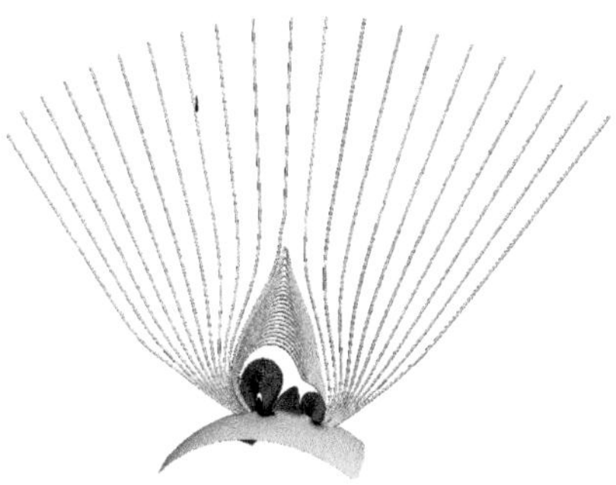

**Figure 1.** Magnetic field configuration used as the initial condition in all the MHD simulations. Red lines represent the flux rope, blue lines the arcades, green lines the external magnetic field. The lower boundary is coloured according to the polarity of the magnetic field from blue (negative) to red (positive) in arbitrary units.

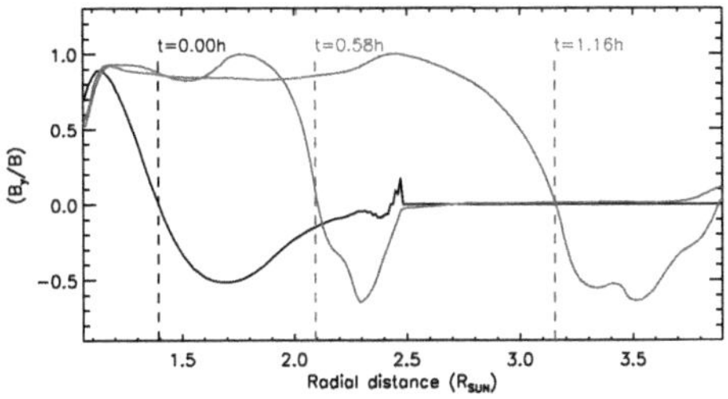


**Figure 2.** Profile of $B_\theta/|B|$ above the centre of the LHS bipole, at different times (different colours). Dashed lines of different colours indicate where $B_\theta/|B| = 0$.

Simulations with different $B_{max}$ basically differ in the plasma $\beta$ which uniformly decreases as $B_{max}$ increases. Thus, by changing the parameter $B_{max}$ we globally modify the capacity of the solar corona to react to the flux rope ejection.

## 3. Results

The initial magnetic field configuration is identical in all of the simulations and it produces an ejection of the flux rope due to an initial excess of the Lorentz force. Fig.1 shows the initial magnetic configuration. The flux rope (red lines) that is about to erupt lies under the arcade system (blue lines), and external magnetic field lines (green lines) are shown above.

In the simulations, the flux rope (lying in the $\theta$ direction) is ejected outwards and it leads to an increase in density at larger radii and the propagation and expansion of the region where the magnetic field is mostly axial, i.e. $B_\theta/|B| \sim 1$. This suggests that the magnetic flux rope is propagating upwards, perpendicular to its axial magnetic field lines, lifting coronal plasma.

Since the region where $B_\theta/|B| > 0$ reproduces the CME, we use this quantity to track the ejection. Fig.2 shows the profile of $B_\theta/|B|$ radially from the centre of the LHS bipole at different times for the simulation with $T = 2\ MK$, and $B_{max} = 21\ G$. The radial position where $B_\theta/|B| = 0$ along the radial direction vertically from the centre of the LHS bipole is defined to be the top of the magnetic flux rope. In the plot in Fig.2 the top of the flux rope is at 1.4 $R_\odot$ at $t = 0$h (black line) and it reaches 3.15 $R_\odot$ after 1.16h (red line). Some of the simulations show an ejection of the flux rope, while others show a quenched ejection. We here do not present a detailed analysis for each simulation, but plot the position of the top of the flux rope as a function of time. In Fig.3a we show the

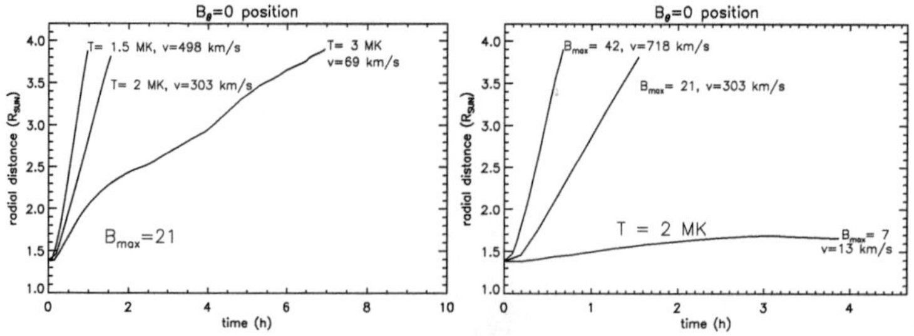


**Figure 3.** (a) Position of the top of the flux rope as a function of time in the three simulations with $B_{max} = 21\ G$. (b) Position of the top of the flux rope as function of time in the three simulations with $T = 2\ MK$.

position of the top of the flux rope for the simulations where $B_{max} = 21\ G$. The speed quoted in Fig.3 is the average speed of propagation. The average speed of the CME spans a wide range from 69 $km/s$ to 498 $km/s$. The lower is the temperature, the faster the CME. Note also that the CMEs propagating in a 2 $MK$ or 1.5 $MK$ corona are at near constant speed. Fig.3b shows the height of the top of the flux rope as a function of time for the three simulations with $T = 2\ MK$. The higher is the magnetic field ($B_{max}$) the larger the initial force due to the unbalanced Lorentz force under the magnetic flux rope and the CME travels faster. With $B_{\max} = 7\ G$ we have a quenched ejection, while we have a 718 $km/s$ fast CME with $B_{\max} = 42\ G$.

## 4. Conclusions

The present study aims at understanding the role of gravitational stratification in CMEs, and to do so we run several simulations varying the stratification temperature ($T$) and plasma $\beta$. In all simulations an identical magnetic field configuration is used.

This study shows that gravitational stratification has an effect on the propagation of CMEs in the solar corona through the way it specifies how large the plasma $\beta$ becomes. We also find that the plasma $\beta$ distribution is a crucial parameter determining whether a flux rope ejection escapes the solar corona, turning into a CME, or if it just makes the flux rope find a new equilibrium at a higher height. Similarly we find that a cooler solar corona ($T \sim 1.5\ MK$) can help the escape of the CME and make it travel faster. The gravitational stratification turns to play a role because the coronal stratified plasma reacts to the perturbation due to the flux rope ejection with thermal pressure gradients contrasting the motion. However, these gradients have a significant effect (either in quenching the ejection or braking it), only when we depart from the low $\beta$ regime in which the Lorentz force completely governs the dynamics.

However, the domain of application of the current work is limited to the coronal events where we find similar values of plasma parameters and magnetic field intensity. Therefore, it can be certainly applied to large quiescent and intermediate prominences that have moderate magnetic fields, but in active region filaments the magnetic field is significantly more intense and thus the plasma $\beta$ significanly low in spite of the coronal temperature. In such circumstances, the gravtitional stratification is likely to play no role.

## References

Pagano, P. and Mackay, D. H., & Poedts, S. 2013, *A&A*, 554, A77

Forbes, T. G. & Isenberg, P. A. 1991, *ApJ*, 373, 294

Mackay, D. H. & van Ballegooijen, A. A. 2006, *ApJ*, 641, 577–589

Keppens, R., Meliani, Z., van Marle, A. J., Delmont, P., Vlasis, A., & van der Holst, B. 2012, *Journal of Computational Physics*, 231, 718–744

*Nature of Prominences and their role in Space Weather*
*Proceedings IAU Symposium No. 300, 2013*
*B. Schmieder, J.-M. Malherbe & S. T. Wu, eds.*

doi:10.1017/S1743921313010971

# Initiation of Coronal Mass Ejections by Sunspot Rotation

**G. Valori[1], T. Török[2], M. Temmer[3,4], A. M. Veronig[3], L. van Driel-Gesztelyi[1,5,6], and B. Vršnak[7]**

[1]LESIA, Observatoire de Paris, CNRS, UPMC, Université Paris Diderot, 5 place Jules Janssen, 92190 Meudon, France; email: gherardo.valori@obspm.fr
[2]Predictive Science Inc., 9990 Mesa Rim Rd., Suite 170, San Diego, CA 92121, USA; email: tibor@predsci.com
[3]IGAM/Kanzelhöhe Observatory, Institute of Physics, Universität Graz, Universitätsplatz 5, A-8010 Graz, Austria
[4]Space Research Institute, Austrian Academy of Sciences, Schmiedlstrasse 6, A-8042 Graz, Austria
[5]University College London, Mullard Space Science Laboratory, Holmbury St. Mary, Dorking, Surrey RH5 6NT, UK
[6]Konkoly Observatory, Hungarian Academy of Sciences, Budapest, Hungary
[7]Hvar Observatory, Faculty of Geodesy, University of Zagreb, Kačićeva 26, HR-10000 Zagreb, Croatia

**Abstract.** We report observations of a filament eruption, two-ribbon flare, and coronal mass ejection (CME) that occurred in Active Region NOAA 10898 on 6 July 2006. The filament was located South of a strong sunspot that dominated the region. In the evolution leading up to the eruption, and for some time after it, a counter-clockwise rotation of the sunspot of about 30 degrees was observed. We suggest that the rotation triggered the eruption by progressively expanding the magnetic field above the filament. To test this scenario, we study the effect of twisting the initially potential field overlying a pre-existing flux rope, using three-dimensional zero-$\beta$ MHD simulations. We consider a magnetic configuration whose photospheric flux distribution and coronal structure is guided by the observations and a potential field extrapolation. We find that the twisting leads to the expansion of the overlying field. As a consequence of the progressively reduced magnetic tension, the flux rope quasi-statically adapts to the changed environmental field, rising slowly. Once the tension is sufficiently reduced, a distinct second phase of evolution occurs where the flux rope enters an unstable regime characterized by a strong acceleration. Our simulation thus suggests a new mechanism for the triggering of eruptions in the vicinity of rotating sunspots.

**Keywords.** MHD, instabilities, Sun: activity, magnetic fields, sunspots, coronal mass ejections (CMEs), filaments, methods: numerical, data analysis

---

## 1. Introduction

The eruption on 6 July 2006 in active region NOAA 10898 was a two-ribbon flare accompanied by a filament eruption and a halo CME, the latter being most prominent in the southwest quadrant and reaching a linear plane-of-sky velocity of $\approx$900 km s$^{-1}$ (Temmer *et al.* 2008). The event was associated with an EIT wave, a type II burst, and very distinct coronal dimming regions. The flare was of class M2.5/2N, located at the heliographic position S9°, W34°. It was observed in soft X-rays (SXR) by GOES (peak time at $\approx$08:37 UT) as well as in hard-X rays (HXR) with RHESSI, with the two highest peaks of nonthermal HXR emission occurring during 08:20 – 08:24 UT.

The morphology and evolution of the bipolar active region in the days preceding the eruption were studied using photospheric line-of-sight magnetograms obtained by the

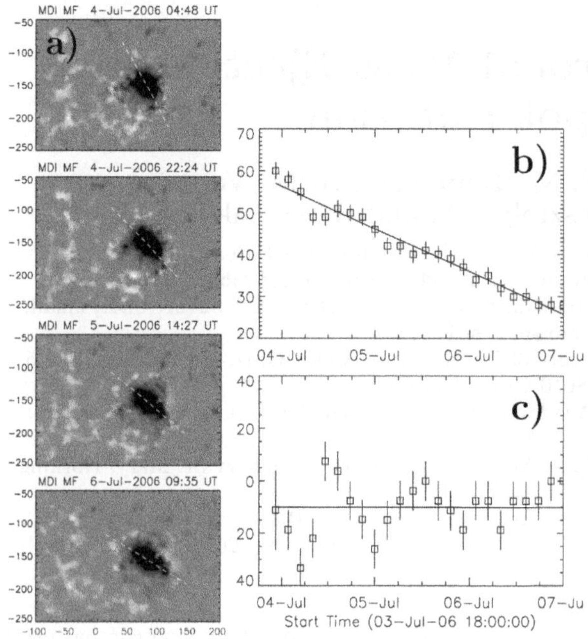


**Figure 1.** (**a**) Representative MDI longitudinal magnetic-field maps of the sunspot evolution during 4–6 July 2006: The dashed yellow line outlines the major axis of the sunspot that was used to measure the sunspot rotation. The images are all differentially rotated to the first image of the series, when the sunspot was closer to disk centre. (**b**) Sunspot rotation determined from the MDI magnetic-field maps over the period 3 July 2006, 22:00 UT, to 7 July 2006, 8:00 UT, showing the orientation of the sunspot's major axis, measured clockwise from solar East. (**c**) Sunspot rotation rate in degrees per day, determined as the temporal derivative of the rotation measurements.

MDI instrument (Scherrer *et al.* 1995). The region consisted of a compact negative polarity (the sunspot) surrounded by a dispersed positive polarity, most of which was extending eastwards (Fig. 1a). The two polarities were surrounded by a large, "inverse C-shaped" area of dispersed negative flux to the west of the region (Fig. 2a). The magnetic-flux measurements indicate a mere 5% negative surplus flux in this major bipolar active region of $2.1 \times 10^{22}$ Mx total flux and maximum-field strengths (negative:positive) in a roughly 10:1 ratio. The sequence in Fig. 1a shows that the sunspot is rotating counter-clockwise during the considered period. The total rotation observed over the three days preceding the eruption is about 30°, with sunspot's rotation rate of about 10° $\mathrm{day}^{-1}$ (Fig. 1b,c).

The flare and the filament eruption were observed in full-disk H$\alpha$ filtergrams by the Kanzelhöhe Observatory and, over a smaller field-of-view around the active region, by the Hvar Observatory. These observations reveal that the filament consisted of a double structure before and during the eruption. Significant rising motions of the filament could be seen from about 08:23 UT. The H$\alpha$ flare started by the appearance of very weak double-footpoint brightening at 08:15 UT. We also estimated the kinematics of the filament and the CME front from a time sequence of running-difference images obtained from TRACE, EIT, and LASCO C2/C3 observations. We obtain that the coronal loops overlying the filament started their slow rising phase at 08:15 UT, i.e., about five–ten minutes before the filament. Similarly, the CME front reached its final, almost constant velocity a few minutes before the filament. More details about the methods employed to obtain the above results can be found in Török *et al.* (2013), hereafter Paper I. We refer to that article and to the references therein for further details on the eruption.

Guo *et al.* (2010) suggested that the eruption was triggered by recurrent chromospheric mass injection in the form of surges or jets into the filament channel. Here we propose a different mechanism: Assuming that the filament was suspended in the corona by a magnetic flux rope, we suggest that the continuous rotation of the sunspot led to a slow expansion of the arcade-like magnetic field overlying the filament, i.e., to a continuous weakening of its stabilizing tension, until a critical point was reached at which equilibrium

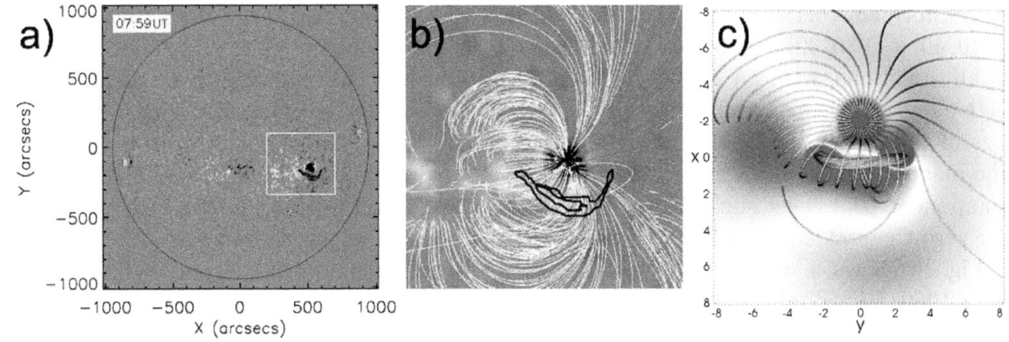


**Figure 2.** **(a)** Full-disk MDI magnetogram on 6 July 2006, 07:59 UT. AR10898 is marked by the white box. **(b)** PFSS magnetic field lines in the AR calculated for 6 July 2006, 06:04 UT, overlaid on a synoptic MDI magnetogram. Pink (white) field lines depict open (closed) fields. The outer contours of the filament, based on H$\alpha$ data taken at 07:59 UT on 6 July 2006, are outlined with black lines. **(c)** Magnetic configuration used in the simulation, after the initial relaxation of the system, showing the core of the TD flux rope (orange field lines) and the ambient potential field (green field lines). $B_z(z = 0)$ is shown, with red (blue) corresponding to positive (negative) values, and is saturated at 4% $\max(B_z)$ to depict weaker flux distributions.

could not be maintained and the flux rope erupted. We note that we do not claim that the eruption was triggered *exclusively* by this mechanism. Filaments are often observed to spiral into the periphery of sunspots (see,e.g., Green *et al.* 2007), and also in our case an inspection of the TRACE and H$\alpha$ images during the early phase of the eruption suggests a possible magnetic connection between the western extension of the filament-carrying core field and the sunspot area. Thus, the sunspot rotation may have added stress to this field, thereby possibly contributing to drive it towards eruption. On the other hand, for an injection of twist to occur, the core field must be rooted in the centre of the sunspot, not just in its periphery, which is difficult to establish from observations. It appears reasonable to assume that a clear connection between core field and sunspot centre is not always present, and that the stressing of the overlying ambient field by sunspot rotation may be more relevant for the destabilization of the system in such cases. In order to test this scenario, we perform a three-dimensional (3D) MHD simulation in which we twist the stabilizing potential field overlying a stable coronal flux rope. Differently from previous works (e.g., Amari *et al.* 1996), the photospheric vortex motions we use do not directly affect the flux rope, but solely the field surrounding it.

## 2. Numerical simulation setup

As in previously published simulations (e.g., Török, Kliem, and Titov 2004; Kliem, Titov, and Török 2004), we integrate the $\beta = 0$ compressible ideal MHD equations, ignoring the effects of thermal pressure and gravity, and we employ the coronal flux rope model of Titov and Démoulin (1999), hereafter TD, to construct the initial magnetic field. The main ingredient of the TD model is a current ring of major radius $[R]$ and minor radius $[a]$ that is placed such that its symmetry axis is located at a depth $[d]$ below a photospheric plane. The outwardly directed Lorentz self-force (or "hoop force") of the ring is balanced by a potential field created by a pair of sub-photospheric point sources $\pm q$ that are placed at the symmetry axis, at distances $\pm L$ from the ring centre. The resulting coronal field consists of an arched and line-tied flux rope embedded in an arcade-like potential field.

We normalize lengths by $l = R - d$ and use a Cartesian grid discretizing the volume $[-40, 40] \times [-30, 30] \times [0, 60]$, resolved by $307 \times 257 \times 156$ points. The grid is nonuniform in all directions, with an almost uniform resolution $\Delta = 0.05$ in the area mimicking the active region. The top and lateral boundaries are closed, which is justified given the large size of the simulation box. Below the photospheric plane, the tangential components of the magnetic field $[B_{x,y}]$ are extrapolated from the integration domain, and the normal component $[B_z]$ is set such that $\nabla \cdot \mathbf{B} = \mathbf{0}$ in $z = 0$ at all times. The vertical velocities are zero there at all times, and the mass density is fixed at its initial values.

Fig. 2b shows a coronal potential-field source-surface model (Schatten, Wilcox, and Ness 1969, PFSS), obtained from a synoptic MDI magnetogram for Carrington Rotation 2045. It can be seen that the field lines rooted in the main polarity (the sunspot) form a fan-like structure, which partly overlies the pre-eruption filament. In order to build an initial magnetic configuration that resembles this coronal field and the underlying highly asymmetric magnetic flux distribution (Sect. 1), we modify the standard TD model by replacing the pair of sub-photospheric point charges by an ensemble of ten sub-photospheric sources. These are adjusted in order to mimic: the approximate flux balance between the concentrated leading negative polarity and the dispersed following positive polarity; the ratio of approximately 10:1 between the peak field strengths in the leading polarity and the following polarity; the size ratio between these polarities; the presence of an "inverse C-shaped" area of dispersed negative flux to the West of the leading polarity; the fan-like shape of the coronal field rooted in the leading polarity. Since the model is still relatively idealized, all these features can be matched only approximately. We then add a TD flux rope, setting $R = 2.75$, $a = 0.8$, and $d = 1.75$. The position of the rope within the ambient field is guided by the observed location of the filament (Fig. 2b), and its magnetic field strength is chosen such that it is in approximate equilibrium with the ambient potential field. We use an initial density distribution $\rho_0(\mathbf{x}) = |\mathbf{B_0}(\mathbf{x})|^2$, corresponding to a uniform initial Alfvén velocity. In order to obtain a numerical equilibrium as a starting point, we first perform a numerical relaxation for $75\,\tau_a$, after which the time is reset to zero.

In order to mimic the observed sunspot rotation, we then twist the main negative flux concentration by imposing tangential velocities at the bottom boundary. They produce a horizontal counterclockwise rotation, chosen such that the velocity vectors always point along the contours of $B_z(x, y, 0, t = 0)$, which assures that the distribution of $B_z(x, y, 0, t)$ is conserved to a very good approximation. The velocities are zero at the polarity centre, located at $(x, y, z) = (-2, 0, 0)$, and decrease towards its edge from their maximum value, equal to 0.005 times the initial Alfvén velocity $[v_{a0}]$. The equations and parameters used to compute the tangential velocities at each time are given in Paper I. The twist injected by such motions is nearly uniform close to the polarity centre and decreases monotonically towards its edge, such that it does not directly affect the flux rope field.

## 3. Results

The magnetic configuration resulting after the initial numerical relaxation is shown in Fig. 2c and in Fig. 4a. The fan-structure inferred from the PFSS extrapolation is qualitatively well reproduced. The TD flux rope is stabilized by flux rooted towards the southern edge of the main polarity, and the rope is inclined with respect to the vertical, which is due to the asymmetry of the potential field surrounding it.

Figure 4a shows that electric currents are present in the ambient field volume. The strongest current concentrations are located in the front of the flux rope and exhibit an X-shaped pattern in the vertical cut shown. This pattern outlines the locations of quasi-separatrix layers (QSL, see, e.g., Démoulin *et al.* 1996) that separate different connectivity

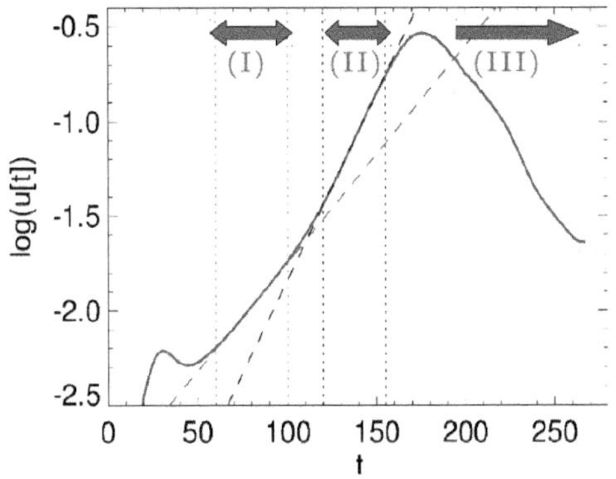


**Figure 3.** Logarithmic presentation of the velocity of the axis apex of the TD flux rope during the twisting phase, as a function of time. The dashed lines show linear fits, obtained within the time periods marked by the vertical dotted lines of the same color. Thick arrows mark different evolution phases described in the text.

domains. The QSLs are present in the configuration from the very beginning and arise from the complexity of the potential field. Their presence is evident also in the left panel of Fig. 4a: the green field lines show strong connectivity gradients in the northern part of the main polarity and in the vicinity of the western flux rope footpoint. It has been demonstrated that current concentrations form preferably at the locations of QSLs as a system containing such structures is dynamically perturbed (see, e.g., Aulanier, Pariat, and Démoulin 2005). In our case the perturbation results from the – relatively modest – dynamics during the initial relaxation of the system.

After the relaxation, at $t = 0$, we start twisting the main negative polarity, and we quantify the evolution of the TD flux rope by monitoring the velocity at the axis apex of the rope (Fig. 3). Due to the pronounced fan-structure of the field rooted in the main polarity, the photospheric twisting does not lead to the formation of a single twisted flux tube that rises exactly in vertical direction above the TD rope, as it was the case earlier studies (Amari *et al.* 1996; Török and Kliem 2003; Aulanier, Démoulin, and Grappin 2005). Rather, the twisting leads to a slow, global expansion of the fan-shaped field lines, as shown in Fig. 4. Since we are mainly interested in the destabilization of the flux rope, we did not study the detailed evolution of the large-scale field. We expect it to be very similar to the one described in Santos, Büchner, and Otto (2011), since the active region those authors simulated was also dominated by one main polarity, and the field rooted therein had a very similar fan-shaped structure (cf. our Fig. 4 with their Fig.1).

Important for our purpose is the evolution of the arcade-like part of the initial potential field that directly overlies the TD flux rope. Those field lines are directly affected only by a fraction of the boundary flows and therefore get merely sheared (rather than twisted), which still leads to their slow expansion. As a result, the TD rope starts to rise, adapting to the successively decreasing magnetic tension of the overlying field (phase I in Fig. 3). This initial phase of the evolution is depicted in Fig. 4b. Note that some of the flux at the front of the expanding arcade reconnects at the QSL current layer, which can be expected to aid the arcade expansion to some degree. As can be seen in Fig. 3, the TD rope rises, after some initial adjustment, exponentially during this slow initial phase.

As the twisting continues, a transition to a rapid acceleration takes place after $t \approx 100\,\tau_a$, when the rise curve leaves the quasi-static regime. After the transition phase, the TD rope again rises exponentially, but now with a significantly larger growth rate than during the slow rise phase (phase II in Fig. 3). Such a slow (quasi-static) rise phase, followed by a rapid acceleration, is a well-observed property of many filament eruptions in the early evolution of CMEs (see, e.g., Schrijver *et al.* 2008, and references therein), and is also seen for the event studied here (see also Paper I). The evolution of the TD rope after $t \approx 100\,\tau_a$ can be associated with the development of the torus instability

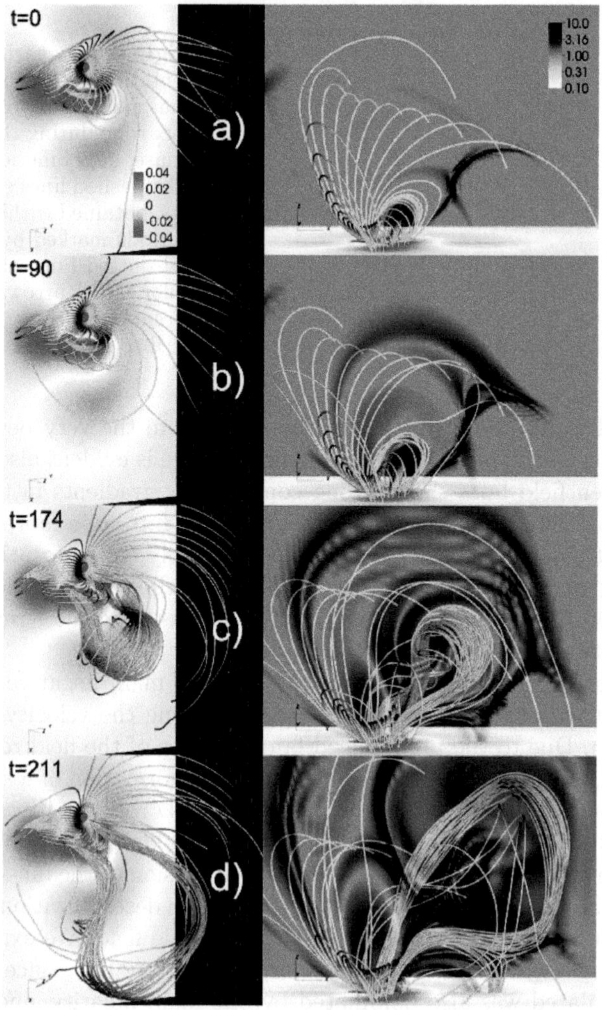


**Figure 4.** Magnetic configuration after initial relaxation (a), during slow rise phase (b), at time of peak flux rope velocity (c), and during flux rope deceleration (d). The flux rope core is depicted by orange field lines; ambient field lines are green. $B_z(z = 0)$ is shown, with red (blue) corresponding to positive (negative) values. Left panels use a view similar to the observations (see paper I); right panels show a side view. The transparent grey-scales show a logarithmic distribution of $|\mathbf{j}|/|\mathbf{B}|$ in the plane $x = 0$, outlining the locations of strongest currents. The sub-volume $[-10, 16] \times [-11, 11] \times [0, 18]$ is used for all panels.

(Kliem and Török 2006; Démoulin and Aulanier 2010), as has been shown under similar conditions in various simulations of erupting flux ropes (Török and Kliem 2007; Fan and Gibson 2007; Schrijver *et al.* 2008; Aulanier *et al.* 2010; Török *et al.* 2011). The right panels in Fig. 4 show that the trajectory of the flux rope is far from being vertical. Such lateral eruptions have been reported frequently in both observations and simulations (see, e.g., Williams *et al.* 2005; Panasenco *et al.* 2011; Yang *et al.* 2012), and are usually attributed to an asymmetric structure of the field overlying the erupting core flux. We believe that this causes the lateral rise also in our case.

As the eruption continues, the trajectory of the flux rope becomes increasingly horizontal, resembling the so-called "roll effect" (Panasenco *et al.* 2011) and indicating that the rope cannot overcome the tension of the large-scale overlying field. Moreover, as a consequence of its increasing expansion, the flux rope strongly pushes against the QSL current layer, which results in reconnection between the front of the rope and the ambient field. Eventually, the rope splits into two parts, similar to what has been found in simulations of confined eruptions (Amari and Luciani 1999; Török and Kliem 2005). These two effects slow down the rise of the rope after $t \approx 175\,\tau_a$ and inhibit its full eruption, i.e., the development of a CME in our simulation (phase III in Fig. 3).

Since QSLs can affect the evolution of an eruption, but are not expected to play a significant role for its initiation, we did not investigate in detail whether or not QSLs were present in the pre-eruption configuration of the 6 July 2006 event. The PFSS extrapolation indicates the presence of a QSL to the North and the West of the main polarity (see the field-line connectivities in Fig. 2b), but less clearly so to its South. Since we merely aim to model the initiation of the eruption rather than its full evolution into a CME, we refrained from further improving our model to obtain a configuration without a strong QSL in front of the flux rope.

## 4. Summary and Conclusions

We presented a 3D MHD simulation that was designed to test a possible scenario for the initiation and early evolution of the filament eruption and CME that occurred on 6 July 2006 in active region NOAA 10898. Our conjecture was that the slow rotation of the sunspot that dominated the active region progressively reduced the tension of the magnetic field overlying the pre-eruption filament, until the latter could not be stabilized anymore and erupted, resulting in the CME. Using the TD coronal flux rope model as a starting point, we constructed an initial magnetic field that resembles the photospheric flux distribution and coronal magnetic field structure of the active region around the time of the event. In particular, the highly asymmetric flux density and the resulting overall fan-shape of the coronal magnetic field are well captured by the model, while the approximative flux balance of the region is kept. We then mimicked the observed sunspot rotation by imposing photospheric vortex flows localized at the main magnetic polarity of the model. The flows were chosen such that they do not directly affect the flux rope.

As a result of this twisting, the field lines overlying the flux rope start to expand and the rope undergoes a quasi-static adaptation to the changing surrounding field in the simulation, which manifests in a slow rise phase. As the weakening of the overlying field reaches an appropriate level, the torus instability sets in and rapidly accelerates the rope upwards, leading to a second, fast rise phase and eruption. The asymmetry of the ambient field leads to a markedly lateral eruption. This evolution in two phases resembles the often observed slow rise phase and subsequent strong acceleration of filaments in the course of their eruption. However, the presence of a QSL-related current layer in the front of the erupting flux rope in the simulation results in magnetic reconnection which eventually splits the rope before it can evolve into a CME, in contrast to the observations. Although we are not able to follow the expansion of the flux rope beyond this phase, the simulation successfully models the early phases of the eruption (the slow rise and the initial rapid acceleration of the flux rope) in a setting that is qualitatively similar to the observed magnetic configuration around the time of the eruption.

Our simulation thus demonstrates that the continuous expansion due to sunspot rotation of the magnetic field that stabilizes the current-carrying core flux, i.e., the progressive decrease of magnetic tension, can lead to filament eruptions and CMEs.

**Acknowledgements**

We acknowledge the use of data provided by the SOHO/MDI consortium. SOHO/EIT was funded by CNES, NASA, and the Belgian SPPS. The SOHO/LASCO data used here are produced by a consortium of the Naval Research Laboratory(USA), Max–Planck–Institut für Aeronomie (Germany), Laboratoire d'Astrophysique de Marseille (France), and the University of Birmingham (UK). SOHO is a mission of international cooperation between ESA and NASA. The *Transition Region and Coronal Explorer* (TRACE) is a mission of the Stanford–Lockheed Institute for Space Research, and part of the NASA Small Explorer program. Hα data were provided by the Kanzelhöhe Observatory, University of Graz, Austria, and by the Hvar Observatory, University of Zagreb, Croatia. The research leading to these results has received funding from the European Commission's Seventh Framework Programme (FP7/2007-2013) under the grant agreements nn° 218816 (SOTERIA project, www.soteria-space.eu) and n° 284461 (eHEROES, http://soteria-space.eu/eheroes/html). TT was partially supported by NASA's HTP, LWS, and SR&T programs. LvDG acknowledges funding through the Hungarian Science Foundation grant OTKA K81421.

## References

Amari, T. & Luciani, J. F. 1999, *Astrophys. J. Lett.* 515, L81.

Amari, T., Luciani, J. F., Aly, J. J., & Tagger, M. 1996, *Astrophys. J. Lett.* 466, L39.

Aulanier, G., Démoulin, P., & Grappin, R. 2005, *Astron. Astrophys.* 430, 1067.

Aulanier, G., Pariat, E., & Démoulin, P. 2005, *Astron. Astrophys.* 444, 961.

Aulanier, G., Török, T., Démoulin, P., & DeLuca, E. E. 2010, *Astrophys. J.* 708, 314.

Démoulin, P. & Aulanier, G. 2010, *Astrophys. J.* 718, 1388.

Démoulin, P., Henoux, J. C., Priest, E. R., & Mandrini, C. H. 1996, *Astron. Astrophys.* 308, 643.

Fan, Y. & Gibson, S. E. 2007, *Astrophys. J.* 668, 1232.

Gopalswamy, N., Yashiro, S., Kaiser, M. L., Howard, R. A., & Bougeret, J. L. 2001, *Astrophys. J. Lett.* 548, L91.

Green, L. M., Kliem, B., Török, T., van Driel-Gesztelyi, L., & Attrill, G. D. R. 2007, *Solar Phys.* 246, 365.

Guo, J., Liu, Y., Zhang, H., Deng, Y., Lin, J., & Su, J. 2010, *Astrophys. J.* 711, 1057.

Kliem, B. & Török, T. 2006, *Phys. Rev. Lett.* 96(25), 255002.

Kliem, B., Titov, V. S., & Török, T. 2004, *Astron. Astrophys.* 413, L23.

Panasenco, O., Martin, S., Joshi, A. D., & Srivastava, N. 2011, J. Atmos. Solar-Terr. Phys. 73, 1129.

Santos, J. C., Büchner, J., & Otto, A. 2011, *Astron. Astrophys.* 535, A111.

Schatten, K. H., Wilcox, J. M., & Ness, N. F. 1969, *Solar Phys.* 6, 442.

Scherrer, P. H., Bogart, R. S., Bush, R. I., Hoeksema, J. T., Kosovichev, A. G., Schou, J., Rosenberg, W., Springer, L., Tarbell, T. D., Title, A., Wolfson, C. J., & Zayer, I., MDI Engineering Team 1995, *Solar Phys.* 162, 129.

Schrijver, C. J., Elmore, C., Kliem, B., Török, T., & Title, A. M. 2008, *Astrophys. J.* 674, 586.

Temmer, M., Veronig, A. M., Vršnak, B., Rybák, J., Gömöry, P., Stoiser, S., & Maričić, D. 2008, *Astrophys. J. Lett.* 673, L95.

Titov, V. S. & Démoulin, P. 1999, *Astron. Astrophys.* 351, 707.

Török, T. & Kliem, B. 2003, *Astron. Astrophys.* 406, 1043.

Török, T. & Kliem, B. 2005, *Astrophys. J. Lett.* 630, L97.

Török, T. & Kliem, B. 2007, *Astronom. Nachr.* 328, 743.

Török, T., Kliem, B., & Titov, V. S. 2004, *Astron. Astrophys.* 413, L27.

Török, T., Panasenco, O., Titov, V. S., Mikić, Z., Reeves, K. K., Velli, M., Linker, J. A., & De Toma, G. 2011, *Astrophys. J. Lett.* 739, L63.

Török, T., Temmer, M., Valori, G., Veronig, A. M., van Driel-Gesztelyi, L., & Vršnak, B. 2013, *Solar Phys.* 286, 453. referred to as **Paper I**

Williams, D. R., Török, T., Démoulin, P., van Driel-Gesztelyi, L., & Kliem, B. 2005, *Astrophys. J. Lett.* 628, L163.

Yang, J., Jiang, Y., Yang, B., Zheng, R., Yang, D., Hong, J., Li, H., & Bi, Y. 2012, *Solar Phys.* 279, 115.

*Nature of Prominences and their role in Space Weather*
*Proceedings IAU Symposium No. 300, 2013*
*B. Schmieder, J.-M. Malherbe & S. T. Wu, eds.*

doi:10.1017/S1743921313010983

# Observations of flux rope formation prior to coronal mass ejections

**Lucie M. Green[1] and Bernhard Kliem[2]**

[1]Mullard Space Science Laboratory, UCL, Holmbury St. Mary, Dorking, Surrey, UK [2] University of Potsdam, Institute of Physics & Astronomy, Potsdam, Germany

**Abstract.** Understanding the magnetic configuration of the source regions of coronal mass ejections (CMEs) is vital in order to determine the trigger and driver of these events. Observations of four CME productive active regions are presented here, which indicate that the pre-eruption magnetic configuration is that of a magnetic flux rope. The flux ropes are formed in the solar atmosphere by the process known as flux cancellation and are stable for several hours before the eruption. The observations also indicate that the magnetic structure that erupts is not the entire flux rope as initially formed, raising the question of whether the flux rope is able to undergo a partial eruption or whether it undergoes a transition in specific flux rope configuration shortly before the CME.

**Keywords.** Sun: coronal mass ejections (CMEs), Sun: activity

## 1. Introduction

Twisted bundles of magnetic field lines known as flux ropes are central to all models of coronal mass ejections. They are invoked as being present either before the eruption onset (e.g. Török & Kliem 2005) or being formed during the eruption itself (e.g. Antiochos *et al.* 1999). In light of this, one way to discriminate between these two sets of CME models is to determine the pre-eruption magnetic configuration. However, there is an inherent difficulty in the confident identification of flux ropes as it is not currently possible to directly measure the magnetic field above the photosphere/chromosphere. Instead, proxies for the presence of flux ropes need to be used, either from solar observations or from reconstructions of the coronal magnetic field using the photospheric field as the boundary condition.

One well developed observational approach to investigate the pre-eruption magnetic configuration is to study so-called sigmoidal active regions (Rust & Kumar 1996). Sigmoidal regions contain S shaped EUV and X-ray emission structures and are regions that have a very high likelihood of producing a coronal mass ejection. Sigmoids are seen as predictors of an eruption (Canfield *et al.* 1999). The sigmoid can be S or reverse S shaped, depending on the chirality of the magnetic field in which it forms (Pevtsov *et al.* 1997). Rust & Kumar (1996) suggested a link between sigmoids and kinking flux ropes, which has been a lively area of research ever since. Theoretical expectations from modelled flux ropes show that layers of enhanced current, and presumably heating, are located at the interface between a flux rope and its surrounding magnetic arcade (Titov & Démoulin 1999). These layers should appear S shaped when viewed from above, building the case that sigmoids represent flux ropes in the solar atmosphere.

Recent investigations into the magnetic configuration of sigmoids have shown support that, at least a sub-set of sigmoids, do indeed indicate the presence of a magnetic flux rope. Green & Kliem (2009) showed that continuous S shaped threads that make an inverse crossing of the photospheric polarity inversion line (PIL) in their centre strongly suggest field lines that spiral around the flux rope. These threads additionally cross the PIL in the normal direction in the sigmoid's elbows. They trace field lines at the periphery of the rope and are likely to extend down to the lower atmosphere, where they form bald patches.

As well as forming in active regions, sigmoids have been observed on a smaller scale in X-ray bright points (Mandrini *et al.* 2005) and on a larger scale in the quiet Sun (Jiang *et al.* 2007). It has been suggested that there are different types of sigmoid relating to their lifetime (transient and long-lived) or detailed observational appearance (continuous S threads or double J's that overall look S shaped) (Pevtsov 2002a).

In summary, flux ropes can be investigated in the solar atmosphere prior to a coronal mass ejection by using observations of sigmoids. These features provide an opportunity to follow the evolution of the magnetic configuration and investigate how the flux rope forms, the magnetic flux content of the rope and the specific magnetic configuration. In this paper we focus on flux rope formation in a small sample of sigmoidal active regions in the days leading up to a coronal mass ejection.

## 2. Observations of flux rope formation

Here we present the evolution of four active regions; (1) NOAA region 10930 that was observed on the Sun during December 2006, (2) NOAA region 8005 that was seen during December 1996, (3) an un-numbered region that was seen in February 2007 and (4) NOAA active region 10977 that was on the disk during December 2007. This sample includes one region which forms a sigmoid during a flux emergence event and three regions which show sigmoidal structure forming during the decay phase of the region.

### 2.1. *NOAA active region 10930*

Active region 10930 rotated over the eastern limb of the Sun on 5 December 2006. The region contained a large negative polarity sunspot with a corresponding dispersed and spotless positive magnetic field to the east. Immediately to the south of the negative polarity sunspot was a smaller magnetic bipole. On 10 December 2006, new flux began to emerge in the location of the smaller bipole. The positive polarity of the emerging flux underwent an eastward motion and a strong counter clockwise rotation of up to 540 degrees between 10 and 14 December (Min & Chae 2009). The positive polarity of the emerging flux was directly next to the pre-existing negative spot meaning that flux cancellation was likely to be taking place.

The XRT/Hinode (Golub *et al.* 2007) data show that as soon as the flux emergence begins, the magnetic field of the positive sunspot formed connections with the pre-existing negative sunspot producing a magnetic arcade between them. The shear in this arcade field increased rapidly, most likely due to the rapid eastward motion of the emerging flux and its strong counter-clockwise rotation. By 23:15 UT on 11 December the arcade field had taken on the appearance of a 'double J' configuration and by 19:00 UT on 12 December a continuous S shaped structure was seen. See the top row of Fig. 1 for the coronal evolution of this region. On 13 December 2006 a GOES X3.4 class flare occurred in the sigmoidal region at around 02:15 UT. A coronal mass ejection was associated with this flare.

### 2.2. *NOAA active region 8005*

NOAA active region 8005 had a bipolar magnetic configuration. The region emerged on the far side of the Sun and as it rotated over the limb it was already in its decay phase with dispersed magnetic polarities and no sunspots. The photospheric field exhibited dispersal of the magnetic polarities and no episodes of flux emergence. On many occasions opposite polarity fragments approached the active region's polarity inversion line and cancelled. See Green & Kliem (2009) and Mackay *et al.* (2011) for details of the evolution of the photospheric magnetic field in this region. In the days leading up to the eruption the active region loops as seen in the soft X-ray data had an overall S shape which evolved from a sheared arcade, to a double J shape (by 19 December 07:24 UT) to a sigmoid with continuous S shaped threads being observed by 10:37 UT on 19 December. See the second from top row of Fig. 1 for the coronal evolution of this region. The active region produced an eruption and a GOES C2.3 class flare on 19 December 1996 at 15:21 UT.

### 2.3. *February 2007 region*

In February 2007 a sigmoid was observed in a very dispersed bipolar active region that had no NOAA number assigned to it. The region produced a coronal mass ejection on 12 February 2007 beginning around 07:00 UT. There was no flare emission associated with this eruption. In the days leading up to the eruption the active region evolved from a double J configuration to having continuous S shaped loops by end of the day on 11 February 2007. See the second row from the bottom of Fig. 1 for the coronal evolution of this region. The evolution of the photospheric magnetic field showed ongoing dispersal and episodes of flux cancellation at the polarity inversion line. For more details on the evolution of this active region see Savcheva *et al.* (2012).

### 2.4. *NOAA active region 10977*

NOAA active region 10977 had a bipolar configuration and the whole lifetime of the active region was observed from emergence to dispersal into the surrounding quiet sun. The emergence began on 3 December 2007 and the flux concentrations began to disperse on 4 December 2007. During the decay phase of the active region the photospheric field was dominated by ongoing dispersal, an elongation of the polarities in the north-south direction and cancellation of flux along the polarity inversion line.

Hinode/XRT data show that the active region loops appeared to have relatively little shear during the emergence phase and into 5 December after the emergence had ceased. The shear began to build up during the decay phase when the loops became much more aligned to the polarity inversion line. Early on 6 December the region appeared as a region of double J shaped loops with some remnant arcade field in the south. Then, by 6 December 15:50 UT continuous S shaped sigmoidal threads were seen in the region. See the bottom row of Fig. 1 for the coronal evolution of this region. On 7 December at around 04:20 UT the region produced a coronal mass ejection. For a more detailed description of the evolution of this active region see Green *et al.* (2011).

## 3. Filament formation

All of the above active regions exhibited the formation of a filament along the polarity inversion line where the sigmoid formed. Big Bear Solar Observatory H-alpha data show that a filament was present in NOAA active region 10930 by 11 December 18:00 UT. In NOAA region 8005 the filament had formed by 18 December 17:40 UT. In the unnamed region seen on the disk in February 2007, Kanzelhöhe Observatory data show that there was a filament present by 10 February 2007 12:00 UT. In NOAA region 10977 Hinode/SOT H-alpha data show that the filament was forming by 5 December around 20:00 UT. The filaments in all these regions formed during the phase where the coronal arcade was becoming more sheared. The presence of a filament indicates low lying loops that have dips, or twisted field lines which can support the dense plasma against gravity.

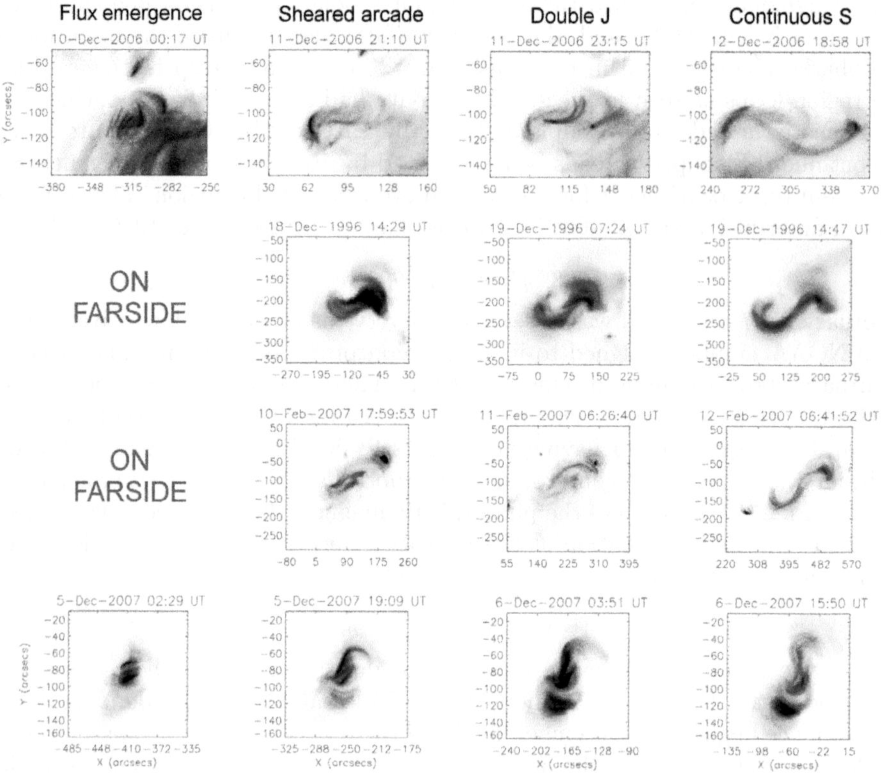


**Figure 1.** Evolution of sigmoidal active regions showing the three phases of evolution from flux emergence, to sheared arcade, to 'double J' shaped loops and then finally the continuous S shaped threads of the sigmoid. Top row: NOAA region 10930. Second row from the top: NOAA region 8005. Second row from the bottom: un-numbered region observed on the disk during February 2007. Bottom row: NOAA region 10977.

## 4. Coronal mass ejections from the sigmoidal regions

The observation of the sigmoid allows the magnetic configuration to be probed and a time for the formation of the flux rope to be identified. The S shaped field lines can indicate the presence of magnetic field lines with dips which suggests the presence of a flux rope, even if it is not fully formed. The time between sigmoid (and flux rope) formation and the onset of the coronal mass ejection is given in Fig. 2. To increase the study size this figure also includes the sigmoid and coronal mass ejection that occurred in the bright point study of Mandrini *et al.* (2005) and the sigmoidal region NOAA 11047 that produced a coronal mass ejection and that was studied in Savcheva *et al.* (2012). The colours indicate whether the region was observed from the emergence phase, and hence whether all activity is seen, or whether the region emerged on the far-side so that aspects of the photospheric and coronal evolution may have been missed. In all cases the flux rope exhibits a stable phase of several hours between its formation and eruption.

In all four regions of this study the coronal mass ejection does not involve the filament material, as has been previously shown when sigmoidal active regions produce coronal mass ejections (Pevtsov 2002b). The events presented here show that the S shaped loops also do not rise during the eruption. Instead, the erupting structure in three out of the

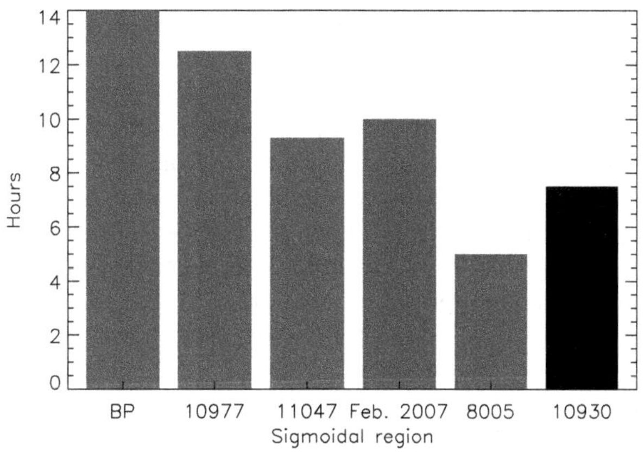


**Figure 2.** Bar chart showing the time between first observation of continuous S shaped loops in the regions and the time of the coronal mass ejection. The colours indicate different region characteristics. Red: isolated bipolar regions that have been followed from emergence to eruption and which produce the sigmoid and the eruption in their decay phase. Blue: isolated bipolar regions that emerged on the far side of the Sun and which produced the sigmoid and eruption during their decay phase. Black: a multipolar active region tracked from a flux emergence episode to the sigmoid formation and eruption and which formed the sigmoid during the flux emergence phase.

four sigmoidal regions is a faint linear or loop-like feature that rises up as the flare loops brighten underneath. See Fig. 3 in Su *et al.* (2007) for region 10930, McKenzie & Canfield (2008) for the case in February 2007 and Green *et al.* (2011) for region 10977.

## 5. Discussion

There is growing observational support that a flux rope is present before the onset of a coronal mass ejection in some cases (see Aulanier *et al.* 2010, Green *et al.* 2011 and references therein). Active regions that show a sigmoidal structure have been fruitful in revealing these flux ropes and their formation mechanism. The observations presented here strongly suggest that flux ropes in the solar atmosphere are built by the flux cancellation mechanism proposed by van Ballegooijen & Martens (1989). Flux cancellation involves reconnection low down in the solar atmosphere between converging opposite polarity fragments in a sheared arcade. The evolution of the coronal configuration is driven by the motions of the photospheric plasma and in this study is seen to pass through three stages as arcade field evolves into a flux rope. During the first stage shear in the coronal arcade field increases due to photospheric motions associated with flux emergence or flux dispersal and flux cancellation. Filaments start to form during this stage. During stage two there is an accumulation of a significant amount of axial flux running along the polarity inversion line as flux cancellation, further shearing and/or rotation of the magnetic polarities takes place. Remnant arcade field takes on the appearance of two J's either side of this axial flux. In stage three, flux cancellation produces field lines that are twisted around the axial flux and which contribute poloidal flux to the rope. This flux cancellation scenario appears to be relevant in dispersed and isolated bipolar active regions in their decay phase, where the flux rope forms along the polarity inversion line of the bipole, or during the flux emergence phase of a multipolar active region where the flux rope forms along the polarity inversion line between neighbouring bipoles which are butted up against each other. The observations of the evolutionary phases suggest that sigmoids do not exhibit different types, rather their appearance evolves as the magnetic configuration changes.

The details of the eruptions from these regions are also important for understanding the pre-eruption magnetic configuration and the aspects of this configuration that are involved in the coronal mass ejection. The eruptions from these regions do not involve either the continuous S shaped threads or the filament. Since both are likely to involve field lines that are located in, or extend down to, the dense plasma of the lower atmosphere, it is not surprising that they are immobile. The flux rope cannot erupt in its entirety in this situation. The observations show that the structure which does erupt is instead a collection of loops that connect between the elbows of the S shaped threads and which can be seen before the onset of the eruption. These loops have been called a linear feature and tentatively been associated with the core of the erupting flux rope in Green *et al.* (2011). However, in Aulanier *et al.* (2010) the same feature is interpreted as being a consequence of heating in a current shell above the rope and called an erupting loop-like feature. Such an erupting structure was also seen by Moore *et al.* (2001) but has become more frequently observed with Hinode/XRT due to its large dynamic range and also with SDO due to the increased plasma temperature coverage. See, for example, Liu *et al.* (2010) and Zharkov *et al.* (2011).

The observations presented here suggest that either the flux rope is able to partially erupt, allowing the accumulated axial flux to escape and become the erupting linear feature, whilst leaving the flux associated to the S shaped threads and filament behind. In other cases the flux rope may evolve prior to the eruption so that its underside is detached from the lower atmosphere, allowing it to erupt fully, as demonstrated in the simulations of Aulanier *et al.* (2010).

## References

Antiochos, S. K., DeVore, C. R., & Klimchuk, J. A. 1999, *ApJ*, 510, 484
Aulanier, G., Török, T., Démoulin, P., & DeLuca, E. 2010, *ApJ*, 708, 314
Canfield, R. C., Hudson, H. S., & McKenzie, D. E. 1999, *GeoRL*, 26, 627
Golub, L., Deluca, E., Austin, G., *et al.* 2007, *Solar Phys.*, 243, 63
Green, L. M. & Kliem, B. 2009, *ApJ*, 700, L83
Green, L. M., Kliem, B., & Wallace, A. J. 2011, *A&A*, 526, 2
Jiang, Y., Chen, H., Shen, Y., Yang, L., & Li, K. 2007, *Solar Phys.*, 240, 77
Liu, R., Liu, C., Wang, S., Deng, N., & Wang, H. 2010, *ApJ*, 725, 84
Mackay, D. H., Green, L. M., & van Ballegooijen, A. A. 2011, *ApJ*, 729, 97
Mandrini, C. H., Pohjolainen, S., Dasso, S., *et al.* 2005, *A&A*, 434, 725
McKenzie, D. E. & Canfield, R. C. 2008, *A&A*, 481, 65
Min, S. & Chae, J. 2009, *Solar Phys.*, 258, 203
Moore, R. L., Sterling, A. C., Hudson, H. S., & Lemen, J. R. 2001, *ApJ*, 552, 833
Pevtsov, A. A. 2002a, in P.C.H. Martens and D. Cauffman (eds.), "Multi-Wavelength Observations of Coronal Structure and Dynamics – Yohkoh 10th Anniversary Meeting", p.125
Pevtsov, A. A. 2002b, *Solar Phys.*, 207, 111
Pevtsov, A. A., Canfield, R. C., & McClymont, A. N. 1997, *ApJ*, 481, 973
Rust, D. M. & Kumar, A. 1996, *ApJ*, 464, 199
Savcheva, A. S., Green, L. M., van Ballegooijen, A. A., & DeLuca, E. 2012, *ApJ*, 759, 105
Su, Y., Golub, L., & van Ballegooijen, A. A., *et al.* 2007, *PASJ*, 59, 785
Titov, V. S. & Démoulin, P. 1999, *A&A*, 351, 707
Török, T. & Kliem, B. 2005, *ApJ*, 630, L97
van Ballegooijen, A. A. & Martens, P. C. H.. 1989, *ApJ*, 343, 971
Zharkov, S., Green, L. M., Matthews, S. A., & Zharkova, V. V. 2011, *ApJ*, 741, L35

*Nature of Prominences and their role in Space Weather*
*Proceedings IAU Symposium No. 300, 2013*
*B. Schmieder, J.-M. Malherbe & S. T. Wu, eds.*

doi:10.1017/S1743921313010995

# 3D reconstruction of erupting filaments with STEREO data

**Ting Li[1] and Jun Zhang[1]**

[1]Key Laboratory of Solar Activity, National Astronomical Observatories, Chinese Academy of Sciences, Beijing 100012, China; [liting;zjun]@nao.cas.cn

**Abstract.** The launch of *STEREO* spacecraft in October 2006 provided an opportunity to view filament eruptions from two viewpoints giving new insights into their three-dimensional (3D) geometry and true trajectory. The kinematical parameters (velocity and acceleration vectors), the rotation motion, non-radial motion and the helical twist around the filament axis in 3D space have been obtained for the first time. All these properties of erupting filaments are very important to our understanding of the physical phenomena triggering the eruption and their early evolution. In the present paper we review different reconstruction techniques of erupting filaments and the main results obtained using the *STEREO* observations.

**Keywords.** filaments, coronal mass ejections (CMEs), flares, etc.

## 1. Introduction

The study of filament eruptions is of great significance because of their close association with coronal mass ejections (CMEs). In the outer corona, CMEs are generally observed to have a three-part structure: the bright core, the dark cavity, and the leading edge (see, e.g., Illing & Hundhausen 1986). The bright core is thought to be composed of filament material. In the inner corona, the filament is located within a flux rope which is the pre-eruption structure of a CME (Li & Zhang 2013).

The eruption mechanism is still unclear although a large number of studies about the nature of filament eruptions and CMEs have been carried out (Filippov 2013; Jiang *et al.* 2011; Liu *et al.* 2012; Shen *et al.* 2012). According to these studies, ideal MHD instabilities such as the kink instability (Fan 2005; Török & Kliem 2005) and the torus instability (Kliem & Török 2006) are thought as one type of eruption mechanism. Other eruption mechanisms include "tether cutting" model (Sturrock 1989; Moore *et al.* 2001) and "break out" model (Antiochos *et al.* 1999; Gilbert *et al.* 2007), both of which focus on magnetic reconnection as the driver of eruption. The kinematic properties of erupting filaments are very important to our understanding of the associated physical mechanisms.

According to previous studies, filaments often exhibit a relatively slow rise ("slow-rise phase") prior to eruption (e.g., Tandberg-Hanssen *et al.* 1980; Kahler *et al.* 1988). For quiet-region events, the pre-eruption slow-rise phase can last several hours (Sterling *et al.* 2007), but it can be much shorter, ~10 min for active-region eruptions (Schrijver *et al.* 2008). The slow-rise phase is followed by a rapid-acceleration phase ("fast-rise phase") during which velocities increase to a range of 100 to over 1000 km $s^{-1}$. The rapid-acceleration phase finally transitions into a phase with a nearly constant velocity or even a deceleration into the heliosphere.

However, previous studies are based on a single spacecraft which are influenced by projection effect. The launch of the twin *Solar Terrestrial Relations Observatory* (*STEREO*; Kaiser *et al.* 2008; Howard *et al.* 2008) spacecraft in October 2006 provided us an opportunity to view filament eruptions from two viewpoints giving new insights into their three-dimensional (3D) geometry and true trajectory. Each spacecraft of the *STEREO* mission carries four remote sensing and in situ instrument suites. The Sun Earth Connection Coronal and Heliospheric Investigation (SECCHI; Howard *et al.* 2008) imaging package on each spacecraft consists of the following five telescopes: the EUVI imager, inner (COR1) and outer (COR2) coronagraphs, and inner (HI1) and outer (HI2) heliospheric imagers. EUVI images are taken at four wavelengths centered at 304 Å (6-8 $\times 10^4$ K, primarily the He II line), 171 Å ($10^6$ K, primarily Fe IX/X), 195 Å ($1.4 \times 10^6$ K, primarily Fe XII line) and 284 Å ($2.2 \times 10^6$ K, Fe XV line). The standard SECCHI synoptic program provides simultaneous A-B image pairs at a ten min cadence for 304 Å and 195 Å and 2.5 min cadence for 171 Å. COR1 has a field of view (FOV) from 1.4 to 4 $R_{sun}$ and COR2 from 2.5 to 15 $R_{sun}$.

In this paper, we mainly review the studies on filament reconstructions in the *STEREO* era. The reconstruction techniques are described in Section 2, the main results obtained from the analysis of *STEREO* EUVI data are described in Section 3 and the summary is presented in Section 4.

## 2. Reconstruction techniques

Since the launch of the *STEREO*, a few works have been published so far on the 3D reconstruction of filaments (Gissot *et al.* 2008; Liewer *et al.* 2009, 2013; Gosain *et al.* 2009; Li *et al.* 2010, 2011; Bemporad 2009, 2011). The main reconstruction techniques include the following three methods.

### 2.1. *Optical-flow algorithm*

The method of optical-flow algorithm was presented by Gissot *et al.* (2008) to reconstruct an erupting filament. This method is based on a novel algorithm which was developed to estimate displacement and brightness variation maps with *SOHO*/EIT data. Gissot *et al.* (2008) applied their algorithm to obtain the maps of the estimated apparent displacement of different parts of a filament due to different viewpoints in the *STEREO* A and B (304 Å) images. From the displacement maps, the radial distance from the solar center is derived pixel by pixel in EUVI images.

### 2.2. *Tie-pointing and triangulation*

The method of tie-pointing and triangulation is widely used by many authors to reconstruct erupting filaments (Liewer *et al.* 2009, 2013; Gosain *et al.* 2009; Li *et al.* 2010, 2011; Bemporad 2009, 2011; Joshi & Srivastava 2011). With this method, the same feature is identified in both images seen from different viewpoints, then the position of the feature in 3D coordinate system is determined by triangulation (Thompson 2006). This technique makes use of the epipolar geometry (Inhester 2006). The two observer positions and any object point to be reconstructed exactly define a plane, which is known as an epipolar plane. By definition, epipolar planes are projected on both observer's images as lines known as the epipolar lines. Once we identify a feature in one image, it is possible to determine the projection of the epipolar plane (i.e., epipolar line) passing the same feature in the second image, thus reducing the problem of placing the second tie-point from a 2D to 1D problem.

Many works mentioned above make use of the software tools for stereoscopy developed within the Solar Software library, in particular the scc_measure.pro routine (developed by W. Thompson). The routine uses triangulation to determine the 3D coordinate of the featured tie-pointed. It is a widget based application that allows the user to locate (and select with a cursor) the same feature in EUVI A and B images. Once the user selects a

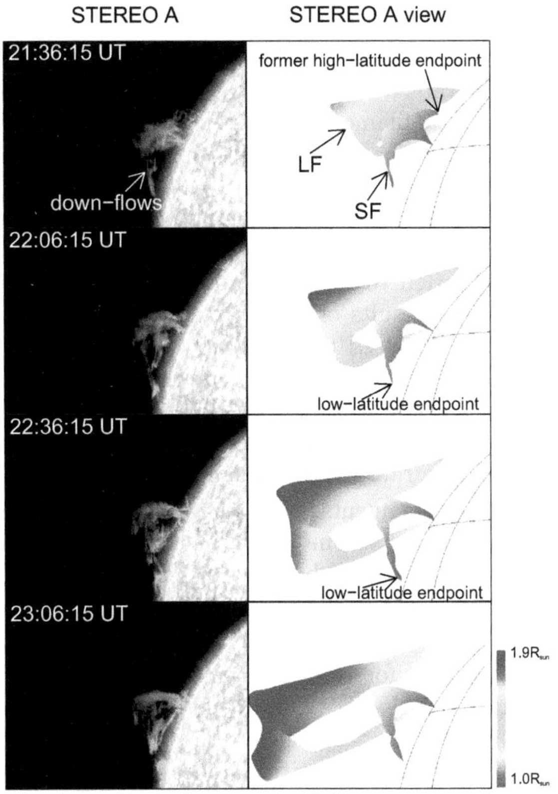


**Figure 1.** Comparison of the observed filaments and reconstructed filaments seen from STEREO A view. Left: a series of STEREO A/EUVI 304 Å observed images showing the evolution of the filaments on 2009 Sep 26. Right: a series of reconstructed images seen from STEREO A showing the eruption process of the filaments. (Li *et al.* 2010)

feature in one image, immediately the epipolar line is displayed and the placement of the tiepoint in the second image is constrained to this line. After selecting the same feature in both images, the 3D coordinates of the feature are then determined as longitude, latitude, and radial distance from the center of the Sun.

### 2.3. *Tomographic method*

Gosain *et al.* (2012) recently developed a new tomographic method for the 3D reconstruction. The basic principle of this method is as follows. For the intensity image of the Sun such as in He II 304 Å, the same feature of a filament in *STEREO* A and B can be projected into heliographic coordinates. This projection is known as Marinus projection. If the heliographic projection is attempted (the Sun as a sphere with a larger radius than one solar radius) and a correct radius of the sphere is assumed, then the latitude and longitude of the common feature are the same. They generated the generalized Carrington maps for different assumed radius of the spherical grid, using a 5 Mm step from R = 700 Mm to R=1500 Mm, and compared the latitude-longitude positions of the same feature in three Carrington maps until they are the same. Then the 3D coordinate of the feature is obtained.

## 3. Results

From *STEREO* 3D reconstructions, the kinematic property and the twist or writhe of the filament could be obtained. Gosain *et al.* (2009) used mainly the tie-pointing technique and reconstructed the filament observed on 2008 May 22. They found that the filament sheet was not normal to the solar surface but highly inclined, with an inclination angle of about 47°. In Gosain & Schmieder (2010), they developed a new method for estimating the width and inclination of the filament by assuming that the filament is a rectangular sheet. They applied this method on the same filament analyzed by Gosain *et al.* (2009), and the inclination to the solar normal was estimated to be 54°, which is in good agreement with the results of their previous study by using the tie-pointing technique. Bemporad (2009) analyzed the 3D shape and orientation of the filament on 2007 May 9 and found that the filament can be approximated mainly as a 2D ribbon-like feature during the early expansion. In this work, the non-isotropic expansion was found because the expansion rate observed in the direction parallel to the filament plane was much larger than the expansion rate perpendicular to that plane.

The kinematical properties of filaments in 3D space could be obtained by reconstructing the featured points of the filaments. The height-time plot of the 3D trajectory of the filament material was initially presented by Liewer *et al.* (2009) which revealed evidence of a slow rise before the main eruption. The rapid rise from one to two solar radius immediately follows the flare peak.

A similar 3D height-time plot for the evolution of the filament was reported by Li *et al.* (2010). In this work, the velocities and accelerations in 3D space of different features along the filament axis were analyzed for the first time. It was found that the velocity and acceleration of features vary with the measured location. The highest points are not always the fastest as the points on the low-latitude leg later become the fastest. A new visualization method was applied after selecting a series of common features in EUVI A and B images by using the scc_measure.pro routine. This method is as follows. The top edge, the main-body line and the bottom edge are chosen as three baselines of the filaments. About 20 pairs of points are placed along each baseline. 500 points are interpolated by cubic spline along each baseline among the selected points to smooth the baseline. Then the enclosed region is simply filled in between any two baselines with 1000 triangular elements to produce the extended regions. Using this visualization method, the reconstructed filaments seen for any viewing angle can be displayed. The two reconstructed filaments seen from the *STEREO A* viewpoint on 2009 Sep 26 are shown in Figure 1. Colors denote different altitudes from the solar center. The two filaments almost simultaneously erupted and were thought to lie in the same filament channel. The latitudinal and longitudinal variations of the large filament are quite different, with the latitudinal motion faster than the longitudinal, which implies the non-isotropic eruption.

It is of great importance to apply different reconstruction techniques to the same event and compare the results. Gosain *et al.* (2012) recently applied their new tomographic method to the same filament as Li *et al.* (2010). The reconstruction from two independent methods agree quite well, considering the general scatter in the reconstructed coordinates. By fitting different functional forms to the true height-time profile of the filament apex during the rapid acceleration phase, they found that an exponential function fits the rise phase slightly better than parabolic or cubic functions.

With the increase of the separation angle between the two *STEREO* spacecraft, it is disadvantageous to reconstruct a 3D configuration of solar features by only using STEREO data. With the launch of the Solar Dynamics Observatory (SDO; Pesnell *et al.*

2012) in February 2010, 3D reconstruction would be improved by using observations of SDO and STEREO. Li *et al.* (2011) made use of observations from the three different viewpoints and reconstructed the 3D evolution process of a polar crown filament on 2010 August 01. It was found the filament moved towards the low-latitude region with a change in inclination by $\sim 48°$ as the main body of the filament rose up. The filament expanded only in altitude and in the latitudinal direction, as also shown by Bemporad (2009). The feature at the highest location had the largest value of acceleration during this eruption process by investigating the true velocities and accelerations of different locations along the filament. Seen from the north pole, the velocity of the filament had a large Earth–directed component. This was consistent with the Earth–directed CME accompanying the eruptive filament.

The same event was also analyzed by Joshi & Srivastava (2011) and Chifu *et al.* (2012). In the work of Joshi & Srivastava (2011), they observed the variations in true longitude and latitude of the reconstructed features in the two legs of the filaments and attributed the variations to an interplay of two motions: the overall non-radial motion of the filament toward the equator and the helical twist in the filament spine. By studying the kinematics of the filament in the slow-rise and fast-eruptive phases, they found a constant acceleration for each reconstructed feature in each phase and the acceleration was found different for different features along the filaments. Chifu *et al.* (2012) found that the filament and CME core material did not evolve in a self-similar way.

Several authors analyzed the rotation motion of erupting filaments by reconstructing the 3D trajectory of the whole filament. Panasenco *et al.* (2010) analyzed three erupting filaments with sideways roll effect and found that all the erupting filaments and associated CMEs were non-radial and occurred near large coronal holes. The roll effect and non-radial motion were away from the closest coronal hole. Thompson (2011) derived the 3D structure of an erupting filament on 2007 June 5-6 and found that the filament underwent substantial rotation of at least 90° along the radial axis as it rose up. It was interpreted that the helical kink instability initiated the eruption and the counter-clockwise rotation of the filament.

Bemporad *et al.* (2011) investigated the filament on 2007 August 31 and showed evidence for a progressive clockwise rotation by about 90° during the filament eruption. Interestingly, the filament rotated by about 40° counter-clockwise in the week before the eruption and the overlying extrapolated potential field lines before the eruption rotated in the same direction as the rotation of the erupting filament. This suggested that the magnetic helicity storage occurred not only in the filament itself, but in the global magnetic field configuration of the surrounding corona. The same filament was also analyze by Liewer *et al.* (2013), who drew similar conclusions as Bemporad *et al.* (2011). They found that the filament rotation began in the slow rise phase and most of the rotation occurred in the fast rise phase, after the CME started to appear.

A larger rotation angle of the erupting filament was reported by Thompson *et al.* (2012), who revealed that the dextral filament rotated counter-clockwise by about 115° up to a height of 2.5 $R_\odot$ where the rotation leveled off. Zhu & Alexander (2013) reported the eruption of a bifurcated filament and found that the upper branch of the filament rotated approximately 120° in a counter-clockwise direction. Bi *et al.* (2013) analyzed the simultaneous non-radial and rotation motion of the filament axis by reconstructing the 3D geometry of the filament and suggested that the non-radial motion was influenced by the overlying pseudostreamer and the rotation motion was a representation of the asymmetric deflection between the eastern and western segments of the filament.

## 4. Summary

By reconstructing the 3D eruption process of filaments with *STEREO* data, the kinematical parameters (velocity and acceleration vectors), rotation motion, non-radial motion and helical twist around the filament axis have been studied in the past several years. However, the full potential of *STEREO* data in addressing many open questions, such as the eruption mechanism of filaments and CMEs, has not been utilized so far. It is therefore necessary to combine *STEREO* observations with data from other ground-based and space-based solar observatories for exploring the physical nature of filament eruptions.

**Acknowledgements**

This work is supported by the National Basic Research Program of China under grant 2011CB811403, the National Natural Science Foundations of China (11303050,11025315, 11221063, 10921303 and 11003026) and the CAS Project KJCX2-EW-T07.

## References

Antiochos, S. K., DeVore, C. R., & Klimchuk, J. A. 1999, *ApJ*, 510, 485
Bemporad, A. 2009, *ApJ*, 701, 298
Bemporad, A. 2011, *Journal of Atmospheric and Solar-Terrestrial Physics*, 73, 1117
Bemporad, A., Mierla, M., & Tripathi, D. 2011, *A&A*, 531, A147
Bi, Y., Jiang, Y., Yang, J., *et al.* 2013, *ApJ*, 773, 162
Chifu, I., Inhester, B., Mierla, M., Chifu, V., & Wiegelmann, T. 2012, *Sol. Phys.*, 281, 121
Fan, Y. 2005, *ApJ*, 630, 543
Filippov, B. 2013, *ApJ*, 773, 10
Gilbert, H. R., Alexander, D., & Liu, R. 2007, *Sol. Phys.*, 245, 287
Gissot, S. F., Hochedez, J.-F., Chainais, P., & Antoine, J.-P. 2008, *Sol. Phys.*, 252, 397
Gosain, S. & Schmieder, B. 2010, *Annales Geophysicae*, 28, 149
Gosain, S., Schmieder, B., Artzner, G., Bogachev, S., & Török, T. 2012, *ApJ*, 761, 25
Gosain, S., Schmieder, B., Venkatakrishnan, P., Chandra, R., & Artzner, G. 2009, *Sol. Phys.*, 259, 13
Howard, R. A., Moses, J. D., Vourlidas, A., *et al.* 2008, *Space Sci. Riv.*, 136, 67
Illing, R. M. E. & Hundhausen, A. J. 1986, *JGR*, 91, 1095
Inhester, B. 2006, arXiv:astro-ph/0612649
Jiang, Y., Yang, J., Hong, J., Bi, Y., & Zheng, R. 2011, *ApJ*, 738, 179
Joshi, A. D. & Srivastava, N. 2011, *ApJ*, 730, 104
Kahler, S. W., Moore, R. L., Kane, S. R., & Zirin, H. 1988, *ApJ*, 328, 824
Kaiser, M. L., Kucera, T. A., Davila, J. M., *et al.* 2008, *Space Sci. Riv.*, 136, 5
Kliem, B. & Török, T. 2006, *Physical Review Letters*, 96, 255002
Li, T. & Zhang, J. 2013, *ApJL*, 770, L25
Li, T., Zhang, J., Zhang, Y., & Yang, S. 2011, *ApJ*, 739, 43
Li, T., Zhang, J., Zhao, H., & Yang, S. 2010, *ApJ*, 720, 144
Liewer, P. C., de Jong, E. M., Hall, J. R., Howard, R. A., Thompson, W. T., Culhane, J. L., Bone, L., & van Driel-Gesztelyi, L. 2009, *Sol. Phys.*, 256, 57
Liewer, P. C., Panasenco, O., & Hall, J. R. 2013, *Sol. Phys.*, 282, 201
Liu, R., Kliem, B., Török, T., *et al.* 2012, *ApJ*, 756, 59
Moore, R. L., Sterling, A. C., Hudson, H. S., & Lemen, J. R. 2001, *ApJ*, 552, 833
Panasenco, O., Martin, S., Joshi, A. D., & Srivastava, N. 2011, *Journal of Atmospheric and Solar-Terrestrial Physics*, 73, 1129
Pesnell, W. D., Thompson, B. J., & Chamberlin, P. C. 2012, *Sol. Phys.*, 275, 3
Schrijver, C. J., Elmore, C., Kliem, B., *et al.* 2008, *ApJ*, 674, 586
Shen, Y., Liu, Y., & Su, J. 2012, *ApJ*, 750, 12
Sterling, A. C., Harra, L. K., & Moore, R. L. 2007, *ApJ*, 669, 1359

Sturrock, P. A. 1989, *Sol. Phys.*, 121, 387
Tandberg-Hanssen, E., Martin, S. F., & Hansen, R. T. 1980, *Sol. Phys.*, 65, 357
Thompson, W. T. 2006, *A&A*, 449, 791
Thompson, W. T. 2011, *Journal of Atmospheric and Solar-Terrestrial Physics*, 73, 1138
Thompson, W. T., Kliem, B., & Török, T. 2012, *Sol. Phys.*, 276, 241
Török, T. & Kliem, B. 2005, *ApJL*, 630, L97
Zhu, C. & Alexander, D. 2013, *Sol. Phys.*, 200

*Nature of Prominences and their role in Space Weather*
*Proceedings IAU Symposium No. 300, 2013*
*B. Schmieder, J.-M. Malherbe & S. T. Wu, eds.*

doi:10.1017/S1743921313011009

# FIP bias in a sigmoidal active region

**D. Baker[1], D. H. Brooks[2], P. Démoulin[3], Lidia van Driel-Gesztelyi[1,3,4], L. M. Green[1], K. Steed[5], J. Carlyle[1]**

[1]University College London, Mullard Space Science Laboratory, Holmbury St Mary, Dorking, Surrey, RH5 6NT, UK
email: deborah.baker@ucl.ac.uk

[2]College of Science, George Mason University, 4400 University Drive, Fairfax, VA 22030, U.S.A.

[3]LESIA, Observatoire de Paris, CNRS, UPMC Université Paris-Diderot, 92195, Meudon, France

[4]Konkoly Observatory, Research Centre for Astronomy and Earth Sciences, Hungarian Academy of Sciences, PO Box 67, 1525, Budapest, Hungary

[5]Centre for Mathematical Plasma Astrophysics, KU Leuven, Celestijnenlaan 200B, 3001 Leuven, Belgium

**Abstract.** We investigate first ionization potential (FIP) bias levels in an anemone active region (AR) - coronal hole (CH) complex using an abundance map derived from *Hinode*/EIS spectra. The detailed, spatially resolved abundance map has a large field of view covering 359″ × 485″. Plasma with high FIP bias, or coronal abundances, is concentrated at the footpoints of the AR loops whereas the surrounding CH has a low FIP bias, ~1, i.e. photospheric abundances. A channel of low FIP bias is located along the AR's main polarity inversion line containing a filament where ongoing flux cancellation is observed, indicating a bald patch magnetic topology characteristic of a sigmoid/flux rope configuration.

**Keywords.** Sun: abundances, Sun: active region, Sun: coronal hole, solar wind.

## 1. Introduction

Elemental abundances and their variations in space and time are all-important to our understanding of the physical processes fundamental to space weather. Seemingly, the most abundant elements of the Sun can be divided into two distinct groups according to their first ionization potential (FIP): low-FIP (<10 eV) and high-FIP (>10 eV) elements. In the corona, low-FIP elements are enhanced over photospheric abundance levels by a factor of 3–4, in contrast with high-FIP elements which maintain the elemental distribution of the photosphere. It is convenient to express the so-called FIP effect in terms of FIP bias, which is the ratio of the elemental abundance in the corona to the elemental abundance in the photosphere.

Plasma with photospheric abundances can be found in erupting prominences (Widing *et al.* 1986), newly emerging active regions (ARs; Sheeley 1995, 1996; Widing 1997), and in coronal holes (CHs; Brooks & Warren 2011; Feldman *et al.* 1998; Feldman & Widing 1993). Observational studies indicate that coronal abundances vary substantially from structure to structure and with time. FIP bias levels in established ARs are ~4–6 and increase to 8–16 in older ARs (Feldman & Widing 2003; Widing & Feldman 1995). In situ measurements of the solar wind (SW) have established that low-FIP elements are enhanced by a factor of 3–4 in the slow wind, comparable to FIP bias levels observed in ARs, whereas composition of the fast wind is ~1, similar to levels found in CHs (Gloeckler

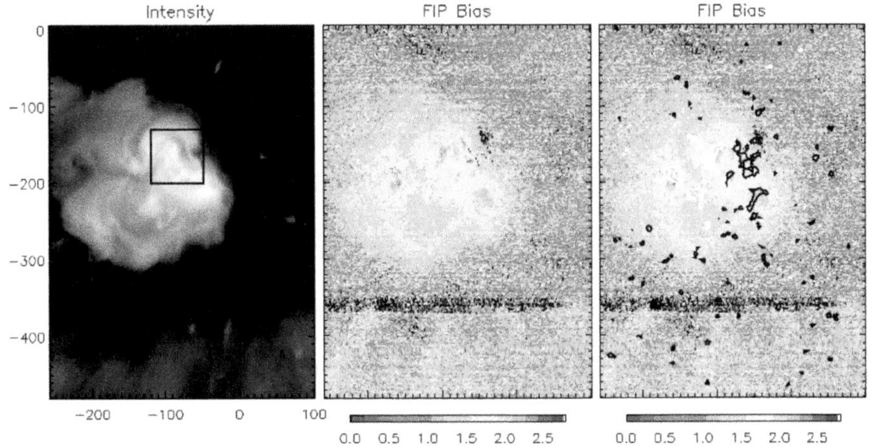


**Figure 1.** *Hinode*/EIS maps at 02:47 UT on October 17, 2007. Left: Fe XII 195.119 Å intensity map. The black box shows the field of view of Figure 2. Middle and right: S X 264.223 Å – Si X 258.375 Å abundance map (middle) overplotted with SOHO/MDI contours of ±100 G (right).

& Geiss 1989). Hence, elemental abundances have been used to connect SW plasmas to their source regions.

## 2. Observations

A small AR was observed inside an on-disk CH on October 17, 2007. During emergence of the bipole inside the CH, the AR interacts with the surrounding CH field and its magnetic connectivities get reorganized by way of interchange reconnection. As a result, new loops extend radially from the location of the included AR polarity, creating the characteristic anemone AR configuration. At the time of the *Hinode*/EIS raster, a filament is present in the AR along the main polarity inversion line (PIL). By 18:00 UT on the 17th, a sigmoid/flux rope is forming via flux cancellation. This mechanism was proposed by van Ballegooijen & Martens (1989). Finally, a coronal mass ejection (CME) erupts at approximately 07:30 UT on the 18th. The *Hinode*/EIS Fe XII 195.119 Å intensity map (left panel) and S X 264.223 Å – Si X 258.375 Å (middle and right panels) abundance map are shown in Figure 1. See Baker *et al.* (2012) for a complete account of the AR-CH complex observations and Baker *et al.* (2013) for a description of the method used to construct the abundance map derived from *Hinode*/EIS spectra.

## 3. Results

Large-scale structures are evident in the abundance map (middle panel of Figure 1). The composition of the CH surrounding the AR is photospheric with a FIP bias of ~1. Levels in the anemone AR range between ~2 and 3. Although the general FIP bias in the AR is somewhat low compared with previous compositional studies of ARs (e.g. Widing & Feldman 1995), it is clearly above the level of FIP bias in the CH, therefore, the AR has a coronal composition.

Detailed structure is discernible within the AR. Patches of high-FIP bias are located close to concentrations of relatively strong magnetic flux density at AR loop footpoints. Figure 1, right panel, shows the abundance map overplotted with SOHO/MDI contours of ±100 G. The contours are cospatial with areas of high-FIP bias. Furthermore, pathways

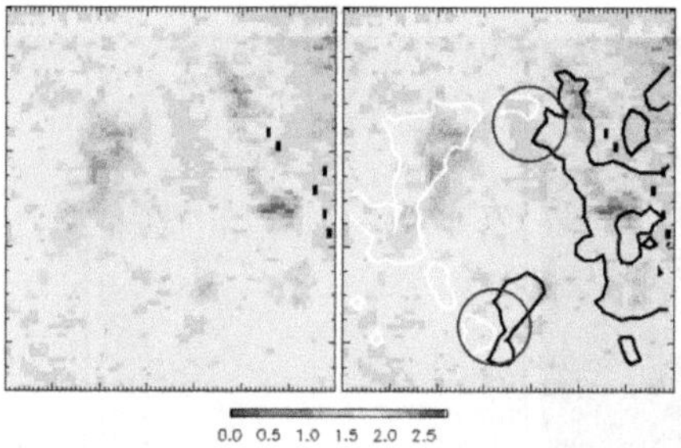


**Figure 2.** Left: Zoomed abundance map showing the sigmoid region indicated by the black box in the left panel of Figure 1. Right: Zoomed abundance map overplotted with SOHO/MDI contours of ±100 G. Red circles indicate locations of principal flux cancellation along the main PIL.

of slightly enhanced FIP bias (~2–2.5) appear to trace loops connecting opposite polarity magnetic flux within the AR.

In Figure 2 (left panel), the abundance map is zoomed to highlight the sigmoid region within the box overplotted on the *Hinode*/EIS Fe XII 195.119 Å intensity map in Figure 1. The zoomed map shows an inverse S-channel of low-FIP bias along the main PIL which hosts a filament within the AR where a sigmoid/flux rope is forming and will erupt soon thereafter (Baker *et al.* 2012). The two principal sites of flux cancellation along the PIL, indicated by the red circles in Figure 2, are located in low-FIP bias areas.

## 4. Discussion

FIP bias in established ARs can reach values significantly greater than 4 which is higher than the levels of 2–3 found in this anemone AR. Large variation in AR plasma composition is related to the variation in the average age of ARs (McKenzie & Feldman 1992; Widing & Feldman 2001). New flux emergence has photospheric composition (Sheeley 1995, 1996; Widing 1997; Widing & Feldman 2001), however, the FIP bias of evolving ARs progresses at approximately a constant rate toward coronal levels within days of emergence (Widing & Feldman 2001). In this case, the low level of FIP bias may be due to the fact that the anemone AR is mainly comprised of recently formed loops instead of the older loop structures of mature ARs (Young & Mason 1997). Although the AR is visible behind the eastern limb in STEREO/EUVI images one week prior to the *Hinode*/EIS observation, the AR's age should not be greater than ≈10 days as the average lifetime for such a small AR is measured in days (Schrijver & Zwaan 2008).

The low-FIP bias plasma of the surrounding CH may also contribute to the low levels in the AR. Since CH plasma undergoes very little modification upon emerging from the photosphere, its composition remains at levels close to 1, as was confirmed by this study. The anemone AR is entirely surrounded by low-FIP CH plasma which can readily mix with the high-FIP AR plasma via interchange reconnection. It is plausible that anemone ARs would have lower FIP bias levels compared with ARs surrounded by mixed magnetic polarities.

It is generally accepted that elemental fractionation takes place mainly in the chromosphere, where low-FIP elements are mostly ionized and high-FIP elements are partially neutral. Meyer (1996) and Widing & Feldman (2001) concluded that the footpoints and legs of loop-like structures are the likely sites for elemental fractionation and uplift to occur. We find strong evidence in support of their conclusions in the anemone AR. High-FIP bias is concentrated at the AR's loop footpoints close to where fractionation is believed to take place. In the young AR, there is insufficient time for high-FIP plasma to fill the coronal loops so we would expect to observe a concentration of high-FIP bias at the footpoints combined with the start of plasma mixing in some of the coronal loops. Loop traces of slightly enhanced FIP bias is suggestive of partial plasma mixing within AR loops.

Finally, the sigmoidal filament channel of low-FIP bias in Figure 2 is atypical of the global pattern of the AR FIP bias. We propose that this channel could be field lines in a flux rope formed by flux cancellation along the main PIL. This is highly suggestive of a bald patch (BP) topology where field lines are tangent to the photosphere. Converging/shearing motions induce the formation of a current sheet along the BP separatrix (Aulanier *et al.* 2010). Reconnection is expected to take place low down along the separatrix, lifting up photospheric plasma in the magnetic dips, implying a low-FIP bias when mixed with the coronal plasma of the reconnecting loops. *Hinode*/EIS abundance maps have a potential role to play in space weather forecasting by enhancing our ability to identify the early formation of flux ropes, which are possible sites of CMEs.

## Acknowledgements

Hinode is a Japanese mission developed and launched by ISAS/JAXA, collaborating with NAOJ as a domestic partner, NASA and STFC (UK) as international partners. It is operated by these agencies in co-operation with ESA and NSC (Norway). LvDG and KS acknowledge the European Community FP7/2007-2013 programme through the eHEROES Network (EU FP7 Space Science Project Nos. 284461). KS also acknowledges the SWIFF Network (EU FP7 Space Science Project No. 263340). LvDG acknowledges the Hungarian government for grant OTKA K-081421. The work of DHB was performed under contract with the Naval Research Laboratory and was funded by the NASA Hinode program. JC thanks UCL and the Max Planck Institute for an Impact Studentship award. DB's work was supported by STFC.

## References

Aulanier, G., Török, T., Démoulin, P., & DeLuca, E. E. 2010, *ApJ*, 708, 314

Baker, D., van Driel-Gesztelyi, L., & Green, L. M. 2012, *Solar Phys.*, 276, 219

Baker, D., Brooks, D. H., Démoulin, P., van Driel-Gesztelyi, L., Green, L. M., Steed, K., & Carlyle, J. 2013, *ApJ*, 778, 69B

Brooks, D. H. & Warren, H. P. 2011, *ApJ* (Letters), 727, L13

Feldman, U., Schühle, U., Widing, K. G., & Laming, J. M. 1998, *ApJ*, 505, 999

Feldman, U. & Widing, K. G. 1993, *ApJ*, 414, 381

—. 2003, *Space Sci. Revs*, 107, 665

Gloeckler, G. & Geiss, J. 1989, in American Institute of Physics Conference Series, Vol. 183, Cosmic Abundances of Matter, ed. C. J. Waddington, 49

McKenzie, D. L. & Feldman, U. 1992, *ApJ*, 389, 764

Meyer, J.-P. 1996, in Astronomical Society of the Pacific Conference Series, Vol. 89, Coronal Abundances, ed. Holt, S. S. and Sonneborn, G., 127

Schrijver, C. J. & Zwaan, C. 2008, Solar and Stellar Magnetic Activity (Cambridge, UK: Cambridge University Press)

Sheeley, Jr., N. R. 1995, *ApJ*, 440, 884

—. 1996, *ApJ*, 469, 423

van Ballegooijen, A. A. & Martens, P. C. H.. 1989, *ApJ*, 343, 971V

Widing, K. G. 1997, *ApJ*, 480, 400
Widing, K. G. & Feldman, U. 1995, *ApJ*, 442, 446
—. 2001, *ApJ*, 555, 426
Widing, K. G., Feldman, U., & Bhatia, A. K. 1986, *ApJ*, 308, 982
Young, P. R. & Mason, H. E. 1997, *Solar Phys.*, 175, 523

*Nature of Prominences and their role in Space Weather*
*Proceedings IAU Symposium No. 300, 2013*
*B. Schmieder, J.-M. Malherbe & S. T. Wu, eds.*

doi:10.1017/S1743921313011010

# A confined flare above filaments

**K. Dalmasse[1], R. Chandra[2], B. Schmieder[1], and G. Aulanier[1]**

[1]LESIA, Observatoire de Paris, CNRS, UMPC, Univ. Paris Diderot,
5 place Jules Janssen, 92190 Meudon, France
email: kevin.dalmasse@obspm.fr

[2]Dept. of Physics, DSB Campus, Kumaun University, Nainital- 263 002, India

**Abstract.** We present the dynamics of two filaments and a C-class flare observed in NOAA 11589 on 2012 October 16. We used the multi-wavelength high-resolution data from SDO, as well as THEMIS and ARIES ground-based observations. The observations show that the filaments are progressively converging towards each other without merging. We find that the filaments have opposite chirality which may have prevented them from merging. On October 16, a C3.3 class flare occurred without the eruption of the filaments. According to the standard solar flare model, after the reconnection, post-flare loops form *below* the erupting filaments whether the eruption fails or not. However, the observations show the formation of post-flare loops *above* the filaments, which is not consistent with the standard flare model. We analyze the topology of the active region's magnetic field by computing the quasi-separatrix layers (QSLs) using a linear force-free field extrapolation. We find a good agreement between the photospheric footprints of the QSLs and the flare ribbons. We discuss how slipping or slip-running reconnection at the QSLs may explain the observed dynamics.

**Keywords.** Filaments, flare, MHD

## 1. Introduction

Filaments are dark, elongated structures consisting of chromospheric plasma embedded in the much hotter corona (van Ballegooijen & Martens 1989; Chae *et al.* 2001). They are cool ($\approx$ 8000 K), dense material confined in highly stressed magnetic fields overlying polarity inversion lines (PILs; Aulanier & Démoulin 1998; Schmieder *et al.* 2006). In the standard picture, the magnetic structure of filaments is formed through shearing motions along PILs and/or, through magnetic flux cancellation due to converging motions of opposite magnetic polarities towards the PILs (van Ballegooijen & Martens 1989; Antiochos *et al.* 1994).

Eventually, filaments may become unstable (see Moore *et al.* 2001; Martens & Zwaan 2001). According to the standard solar flare model (hereafter, CSHKP model), the instability of the filament may lead to two different types of flares, namely, eruptive or confined flares (see review by Shibata & Magara 2011). Eruptive flares correspond to cases for which the filament erupts, leading to the formation of a coronal mass ejection (CME). On the other hand, confined flares are sometimes associated with cases for which the eruption of the filament fails (*e.g.*, Török & Kliem 2005). Confined flares also comprise flares induced by magnetic reconnection of different magnetic flux tubes, or magnetic coronal loops, for which no filament is present (*e.g.*, Berlicki *et al.* 2004; Chandra *et al.* 2006). In the context of flares involving the presence of a filament, the CSHKP model predicts for both eruptive and confined flares, that the flare will be associated with two flare ribbons, and with the formation of hot post-flare loops *below* the erupting filament, regardless of whether it is a successful or failed eruption.

In this study, we present the evolution of two filaments and a confined flare observed in NOAA 11589, which cannot be explained by the CSHKP model. We propose an alternative flare scenario which accounts for the observed flare signatures and filaments evolution during the flare.

## 2. Observations

Our study was performed by combining observations from the Solar Dynamic Observatory (SDO) satellite, the french Télescope Héliographique pour l'Etude du Magnétisme et des Instabilités Solaires (THEMIS), and an indian telescope of the Aryabhatta Research Institute of observational Sciences (ARIES).

NOAA 11589 appeared on 2012 October 10 at the heliographic coordinates N13 E61. The AR quickly developed into two decaying magnetic polarities (see Fig. 1a). During its on-disk passage, the AR was associated with large-scale magnetic flux cancellation, and a few localized magnetic flux emergence events.

The flux cancellation in the internal part of the AR led to the formation of two filaments of opposite chirality which eventually converged. However, the filaments did not merge probably due to their axial field being oriented in opposite direction along the PIL (*e.g.*, Schmieder *et al.* 2004; DeVore *et al.* 2005).

The AR also presented some recurring and localized magnetic flux emergence associated with Ellerman bombs (EBs) in its northern part (as in Pariat *et al.* 2004).

On October 16, the AR produced a confined C3.3 class flare which started around 16:00 UT and ended around 16:39 UT. A first analysis of the flare signatures with AIA 1600 Å and AIA 171 Å channels seem to be in agreement with the CSHKP model: apparently presenting two flare ribbons, and the formation of hot post-flare loops.

However, a careful analysis of the EUV data from the AIA 171 Å channel reveals that the flare did not lead to the eruption of any of the filaments. It also shows a striking result: the post-flare loops were formed *above* the filaments contrary to what is expected from the CSHKP model, and the filaments were not disturbed by the flare.

## 3. Analysis

### 3.1. *Magnetic field extrapolation*

To understand and explain the evolution of the filaments during the flare, we study the magnetic topology of the AR by means of an LFFF extrapolation ($\vec{\nabla} \times \vec{B} = \alpha \vec{B}$, with $\alpha$ being the force-free parameter) to identify the key sites for the development of magnetic reconnection that led to the flare.

We only considered the global magnetic field of the AR because (i) the filaments were in plage regions where the magnetic field is weak, and thus, the currents are not well measured, and (ii) the filaments did not seem to play any role in the flare.

The extrapolations were performed using the fast Fourier transform method (Alissandrakis 1981) with a non-uniform grid of $1024^2 \times 351$ points covering a domain of $700^2 \times 2000$ Mm$^3$. Within the set of performed extrapolations, we kept the solution $\alpha = 7 \times 10^{-3}$ Mm$^{-1}$ because it gave the best match with the northern loops of the AR (Fig. 1a), *i.e.*, the region where the flare was initiated.

### 3.2. *Topological analysis*

The topology is then analyzed by computing the quasi-separatrix layers (QSLs; *e.g.*, Démoulin *et al.* 1996). QSLs are thin 3D volumes of very sharp gradients of the magnetic field connectivity. QSLs are preferential sites for the build-up of strong and thin

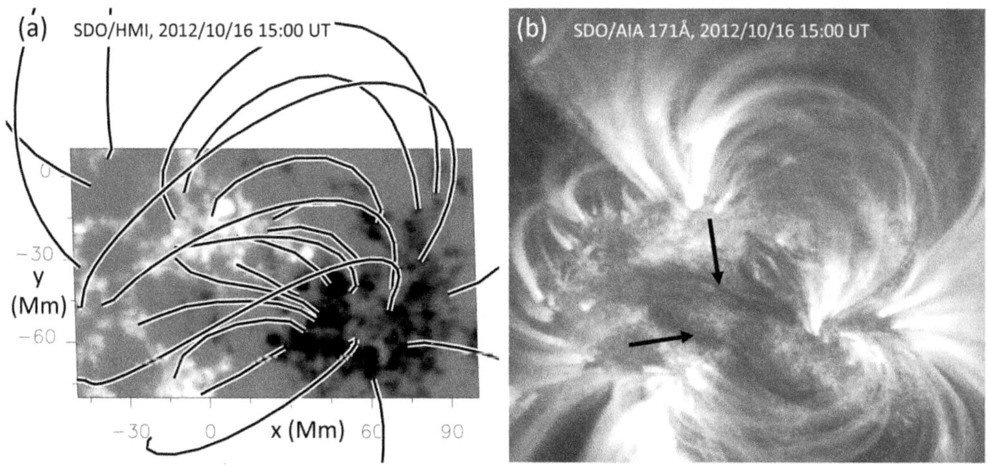


**Figure 1.** Central part of NOAA 11589. (a) Photospheric vertical magnetic field, $B_z$, in greyscale overplotted with selected magnetic field lines from the extrapolation (black lines). (b) AIA171 image showing some of the AR loops and the two observed filaments highlighted by black arrows.

current layers, and for the development of magnetic reconnection at these current layers (see review by Démoulin 2006). As separatrices, QSLs are preferential sites for particle acceleration (Aulanier *et al.* 2006). Many observational studies have thus successfully compared and associated the photospheric footprints of QSLs to flare ribbons providing indirect evidence of magnetic reconnection as the triggering mechanism of solar eruptive events (*e.g.*, Démoulin *et al.* 1996; Mandrini *et al.* 1997; Schmieder *et al.* 1997). The photospheric mapping of QSLs can be obtained by computing the squashing degree, $Q$ (Titov *et al.* 2002). QSLs are thus identified as 3D regions of strong $Q$-values ($Q \gg 2$).

We computed the squashing degree for our LFFF extrapolation using "method 3" of Pariat & Démoulin (2012). Fig. 2(a) displays the photospheric mapping of QSLs for the same field of view as Fig. 1(a) by representing $\log_{10} Q$ at the photosphere. By plotting magnetic field lines over the photospheric $Q$-map, we identified two double-C shape QSLs, $Q_{1,2}$, similar to Aulanier *et al.* (2005), and a circular-like one (overlaid with a white circle), $Q_3$, similar to Masson *et al.* (2009). We find a few discrepancies between the QSLs footprints, $Q_i$, and the three flare ribbons of Fig. 2(b), $R_i$. This is due to the assumptions made by extrapolating the AR's magnetic field in LFFF, which do not model the highly-stressed filament magnetic fields, and which results in local modifications of the magnetic connectivity that slightly modifies the location and shape of the QSLs in our extrapolation. Nevertheless, there is a good qualitative agreement between the QSLs footprints and the flare ribbons (Fig. 2).

## 4. Conclusion

From the previous analysis, it is clear that the magnetic field of AR 11589 presents a complex topology formed by three entangled QSLs. Such a complex topology was favorable to the build-up of electric current layers and to the development of magnetic reconnection at any of these QSLs. The flare might thus have been the result of the stress of, at least, one of the QSLs eventually triggering magnetic reconnection at all QSLs.

Analyzing the AIA and HMI data prior to, and after the flare, we found signatures of localized, recurring magnetic flux emergence in the northern part of the AR — in the region below $Q_1$, *i.e.*, between the western part of $Q_{1,curv}$ and the southern part of $Q_{1,arc}$.

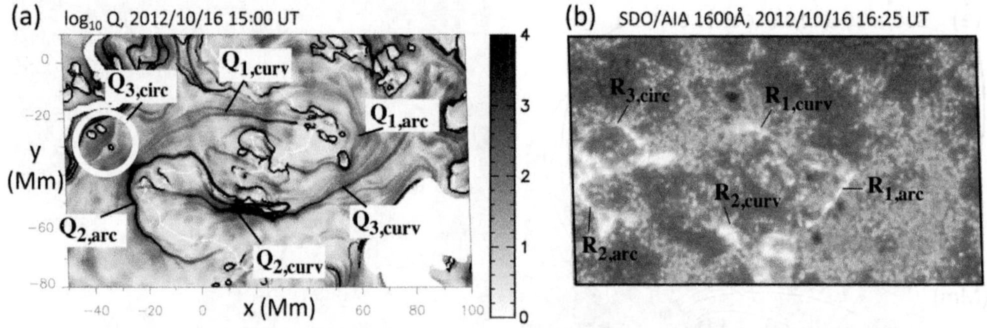


**Figure 2.** Central part of NOAA 11589. (a) Photospheric mapping of $\log_{10} Q$ displaying the photospheric footprints of QSLs at 15:00 UT. (b) Flare ribbons at 16:25 UT. The footprints of three QSLs, labelled $Q_i$, are identified with the three flare ribbons labelled $R_i$.

Consequently, we propose that this episodic flux emergence was the driver of the C3.3 class flare: this continuous magnetic flux emergence may have stressed the magnetic field of $Q_1$, resulting in the formation of a strong thin current layer, at least, within this QSL. Eventually, this can trigger slipping or slip-running reconnection at $Q_1$ (see Aulanier *et al.* 2006), which, in turn, can trigger magnetic reconnection at all the other intersecting QSLs, $Q_2$ and $Q_3$. This would have induced particle acceleration at all QSLs (*e.g.*, Masson *et al.* 2009), and hence, the formation of a complex distribution of flare ribbons (such as shown by Fig. 2b). Since both filaments were located below the QSLs involved in the flare mechanism, our scenario naturally accounts for the development of post-flare loops *above* these non-erupting filaments.

## References


Alissandrakis, C. E. 1981, *A&A*, 100, 197
Antiochos, S. K., Dahlburg, R. B., & Klimchuk, J. A. 1994, *ApJ*, 420, L41
Aulanier, G. & Démoulin, P. 1998, *A&A*, 329, 1125
Aulanier, G., Pariat, E., & Démoulin, P. 2005, *A&A*, 444, 961
Aulanier, G., Pariat, E., Démoulin, P., & DeVore, C. R. 2006, *Solar Phys.*, 238, 347
Berlicki, A., Schmieder, B., Vilmer, N., Aulanier, G., & Del Zanna, G. 2004, *A&A*, 423, 1119
Chae, J., Wang, H., Qiu, J., Goode, P. R., Strous, L., & Yun, H. S. 2001, *ApJ*, 560, 476
Chandra, R. *et al.* 2006, *Solar Phys.*, 239, 239
Démoulin, P. 2006, *Adv. Sp. Res.*, 37, 1269
Démoulin, P., Henoux, J. C., Priest, E. R., & Mandrini, C. H. 1996, *A&A*, 308, 643
DeVore, C. R., Antiochos, S. K., & Aulanier, G. 2005, *ApJ*, 629, 1122
Mandrini, C. H. *et al.* 1997, *Solar Phys.*, 174, 229
Martens, P. C. & Zwaan, C. 2001, *ApJ*, 558, 872
Masson, S., Pariat, E., Aulanier, G., & Schrijver, C. J. 2009, *ApJ*, 700, 559
Moore, R. L., Sterling, A. C., Hudson, H. S., & Lemen, J. R. 2001, *ApJ*, 552, 833
Pariat, E., Aulanier, G. *et al.* 2004, *ApJ*, 614, 1099
Pariat, E. & Démoulin, P. 2012, *A&A*, 541, A78
Schmieder, B., Aulanier, G., Démoulin, P. *et al.* 1997, *A&A*, 325, 1213
Schmieder, B., Aulanier, G., Mein, P., & López Ariste, A. 2006, *Solar Phys.*, 238, 245
Schmieder, B., Mein, N. *et al.* 2004, *Solar Phys.*, 223, 119
Shibata, K. & Magara, T. 2011, *Living Reviews in Solar Phys.*, 8, 6
Titov, V. S., Hornig, G., & Démoulin, P. 2002, *Journal of Geophysical Research*, 107, 1164
Török, T. & Kliem, B. 2005, *ApJ*, 630, L97
van Ballegooijen, A. A. & Martens, P. C. H.. 1989, *ApJ*, 343, 971

*Nature of Prominences and their role in Space Weather*
*Proceedings IAU Symposium No. 300, 2013*
*B. Schmieder, J.-M. Malherbe & S. T. Wu, eds.*

doi:10.1017/S1743921313011022

# Successive filament eruptions within one solar breakout event

**Yuandeng Shen**[1,2]

[1]Yunnan Astronomical Observatory, Chinese Academy of Sciences, Kunming 650011, China
[2]Kwasan and Hida Observatories, Kyoto University, Kyoto 6078471, Japan
email: ydshen@ynao.ac.cn

**Abstract.** The magnetic breakout model has been widely used to explain solar eruptive activities. Here, we apply it to explain successive filament eruptions occurred in a quadrupolar magnetic source region. Based on the high temporal and spatial resolution, multi-wavelengths observations taken by the Atmospheric Imaging Assembly (AIA) on board the *Solar Dynamic Observatory* (*SDO*), we find some signatures that support the occurrence of breakout-like external reconnection just before the start of the successive filament eruptions. Furthermore, the extrapolated three-dimensional coronal field also reveals that the magnetic topology above the quadrupolar source region resembles that of the breakout model. We propose a possible mechanism within the framework of the breakout model to interpret the successive filament eruptions, in which the so-called magnetic implosion mechanism is firstly introduced to be the physical linkage of successive filament eruptions. We conclude that the structural properties of coronal fields are important for producing successive filament eruptions.

**Keywords.** Activity, flares, filaments, magnetic fields, Coronal mass ejections.

## 1. Introduction

Filament (prominence) eruption is one of the most spectacular, large-scale activity on the Sun, which often associates with solar flare and coronal mass ejection (CME). The eruption of a filament can severely impact the solar-terrestrial environment and human activities; and the study of these phenomena has developed into a new discipline dubbed space weather. However, the physical mechanism of filament eruption is still not well understood, even though extensive observational and theoretical works have been made in recent decades.

Generally speaking, a filament eruption always starts from a closed magnetic system in quasi-static equilibrium, in which the upward magnetic pressure force of the low-lying sheared field is balanced by the downward tension force of the overlying field. When the eruption begins, the equilibrium is destroyed catastrophically, and part of the non-potential magnetic flux and the plasma are expelled violently from the Sun. Given different magnetic environments, the eruption of a filament can be failed, partial, and complete eruptions (e.g., Gilbert *et al.* (2001), Liu *et al.* (2009), Shen *et al.* (2011)). Sometimes, several filaments far from each other or resided in one complex active region can erupt successively within a short time period. A key question of successive eruptions is whether they are physically connected or not. It seems that the answer is positive, and the connection is often of a magnetic nature (e.g., Jiang *et al.* (2008), Jiang *et al.* (2011), Török *et al.* (2011), Shen, Liu & Su (2012), Titov *et al.* (2012), Lynch & Edmondson (2013), Schrijver & Title (2011), Schrijver *et al.* (2013)).

Currently, solar physicists have developed a number of models for interpreting filament/CME eruptions. Among various models, the magnetic breakout model assumes

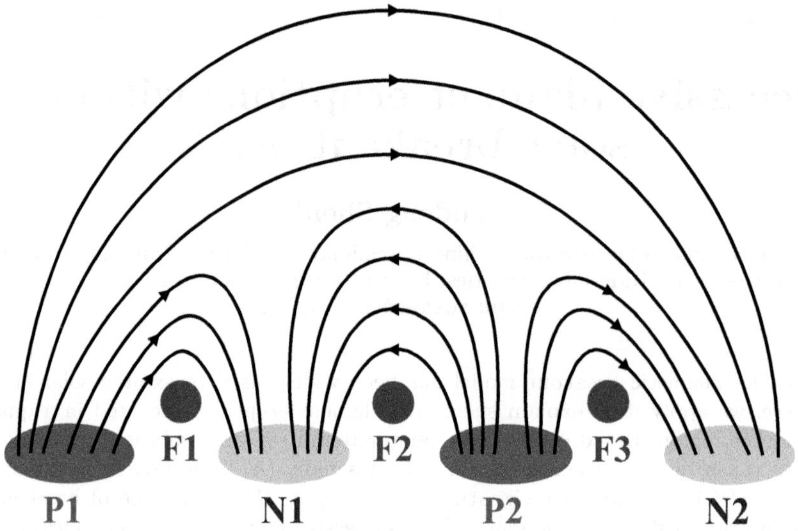


**Figure 1.** Topology structure of the breakout mode. The red (P1 and P2) and blue (N1 and N2) patches indicate the positive and negative polarities of the magnetic field, respectively. The field direction is indicated by a series of arrows. The gray patches represent the filaments (F1, F2, and F3) confined by the three low-lying lobes.

a large-scale quadrupolar field configuration; the core field is increasingly sheared by photospheric motions, which is surrounded by an overlying antiparallel loop system (Antiochos (1998), Antiochos *et al.* (1999)). Naturally, a null point is formed between the core and the overlying loop system. This model can be used to interpret many eruptions which occur in complex multipolar active regions (e.g., Aulanier *et al.* (2000), Maia *et al.* (2003), Shen, Liu & Su (2012)). Here, we apply the magnetic breakout model to explain successive filament eruptions which occurred in a quadrupolar magnetic source region, and propose a possible physical linkage between the filament eruptions.

## 2. Results & Interpretation

The basic magnetic topology of the breakout model is shown in Figure. 1. It can be seen that the four magnetic poles (P1, N1, P2, and N2) are connected by three low-lying lobes and one overlying loop system; and a coronal null resides inbetween the middle lobe and the overlying antiparallel loop system. In addition, one can assume the existence of a filament under each lobe (F1, F2, and F3). In such a configuration, a small disturbance to the system could lead to the eruption of the whole system. Typically, there are two types of disturbance to the system. The first type is that the disturbance acts on F2, which will lead to the rising of F2 and the middle lobe, and further results in the external reconnection around the null point, which will removes the confining field of F2 to the lateral lobes and thereby reduce (increase) the confining capacity of the middle (lateral) lobe. Therefore, this type of disturbance often lead to failed eruptions of F1 and F3, while the eruption of F2 should be a successful one. The second type is that the disturbance acts on F1 or F2, which will not lead to any reconnection, and therefore no filament eruption occurs. Here, we present another type of filament eruption in a solar breakout event, in which successive partial and full filament eruptions are involved.

The detailed analysis of the event can be found in Shen, Liu & Su (2012). Here we just give a brief summery of the results. As shown in Fig. 2(a), we find that the extrapolated

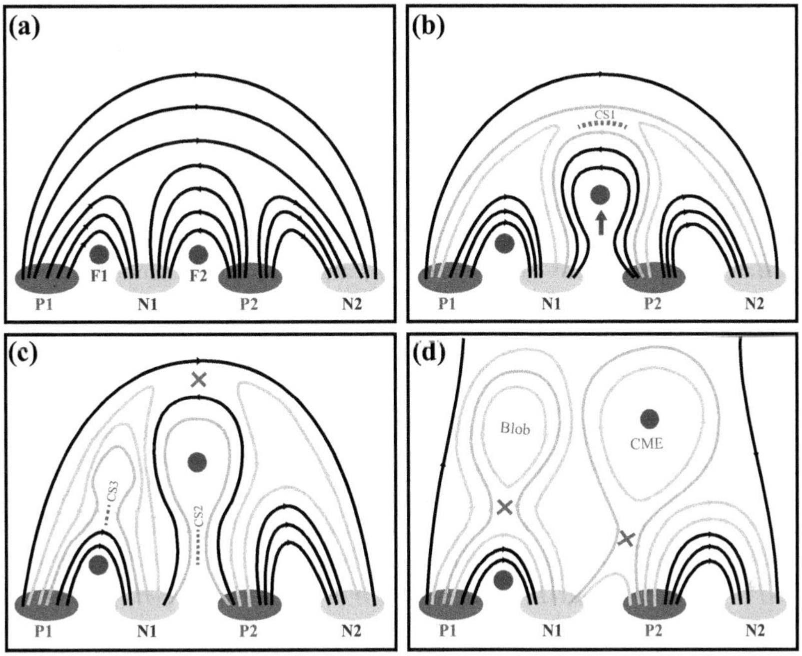


**Figure 2.** Schematic demonstrating the successive filament eruptions. (a) The initial magnetic configuration. (b) The rising of F2, formation of CS1 and the external reconnection. (c) Formation of CS2 (CS3) underneath (above) F2 (CS3). (d) Reconnections in CS2 and CS3, and the production of the nearly simultaneous CMEs. The arrow pointing to F2 represent the disturbance. The red dotted lines indicate locations of the current sheets, while the reconnection sites are labeled by red "X" symbols. The yellow lines represent the field lines to be reconnected, while the green lines are reconnected ones.

coronal field above the magnetic source region is of the topology of the breakout model, and the two filaments are located below the middle and the left lobes respectively. The initiation of the successive filament eruptions started from a small mass ejection, which directly interacted with the southern part of F2 and thus resulted in the slow rise of this filament. The slow rise of F2 lasted for about 23 minutes and a speed of 8 km/s. During this period, some signatures for breakout-like external reconnection were observed. For example, two brightening patches at both sides of F1, the appearance of bright loops and a weak hard X-ray source above F1. After the slow rising phase, F2 was quickly accelerated to 102 km/s, and finally, it erupted successfully and caused a CME. The activation of F1 started around the end of F2's slow rising phase, which erupted with strong writhing motions. When F1 reached its maximum height, the eruption of a blob-like structure was observed above the filament. In the meantime, F1 began to fall back to the solar surface. These results indicate that the eruption of F1 should be a typical partial flux rope eruption. According to the model proposed by Gilbert *et al.* (2001), the reconnection site should be located above the filament.

We interpret the observations using the breakout model as shown in Fig. 2. Panel (a) presents basic magnetic topology. Due to the disturbance introduced by a small plasma ejection, F2 slowly rises, expanding the middle lobe, which will result in the external magnetic reconnection within the current sheet formed around the coronal null point (see CS1 in panel (b)). According to the magnetic implosion mechanism proposed by Hudson (2000), the magnetic pressure around the reconnection site will decrease due to energy released during the energy conversion process in coronal transients such as flares. The reduction of magnetic pressure will lead to the contraction of the overlying loop system and the expansion of the low-lying lobes. In addition, the strong writhing of F1 indicates that the eruption of this filament was driven by the kink instability. The reduction of the magnetic tension force of the left lobe facilitates triggering the kink instability within F1. Hence, the magnetic implosion could be a possible physical linkage between the successive filament eruptions within the framework of the breakout model. As the rising of F1 and F2, new current sheet CS2 (CS3) will form underneath (above) F2 (F1). The reconnection within CS2 (CS3) will lead to the successful (partial) eruption of F2 (F1), and the CME (blob) (see panels (c) and (d)). In this model, we can expect two simultaneous CMEs.

## 3. Summary

Based on multi-wavelengths observations, we propose an interpretation for the successive eruptions of two filaments in a solar breakout event. We first introduce the magnetic implosion mechanism to be the physical linkage of the successive filament eruptions. The observations of both the pre-eruptive signatures and the extrapolated three-dimensional coronal fields are in good agreement with the breakout model. Our scenario presented in this article implies the occurrence of nearly simultaneous CMEs. Therefore, this interpretation is important for the forecast of space weather. It should be noted that the breakout scenario is a possible explanation for the observations. We do not intend to exclude other possibilities. In any case, the structural properties of coronal fields are important for producing successive filament eruptions.

**Acknowledgements** This work is supported by the Western Light Youth Project of Chinese Academy of Sciences (CAS), the CAS open research programs (KLSA201204, DMS2012KT008). Y. Shen thank the financial support for young researches to participate the 300th symposium (IAUS300: Nature of Prominences and their role in Space Weather) of the International Astronomical Union.

### References

Antiochos, S. K. 1998, *ApJ*, 502, L181
Antiochos, S. K., DeVore, C. R., & Klimchuk, J. A. 1999, *ApJ*, 510, 485
Aulanier, G., DeLuca, E. E., Antiochos, K. S., McMullen, R. A., & Golub, L. 2000, *ApJ*, 540, 1126
Gilbert, H. R., Holzer, T. E., Low, B. C., & Burkepile, J. T. 2001, *ApJ*, 549, 1221
Hudson, H. S. 2000, *ApJ*, 531, L75
Jiang, Y., Shen, Y., Yi, B., Yang, J., & Wang, J. 2008, *ApJ*, 677, 699
Jiang, Y., Yang, J., Hong, J., Bi, Y., & Zheng, R., 2011, *ApJ*, 738, 179
Liu, Y., Su, J., Xu, Z., Lin, H., Shibata, K. *et al.* 2009, *ApJ*, 696, L70
Lynch, B. J. & Edmondson, J. K. 2013, *ApJ*, 764, 87
Maia, D., Aulanier, G., Wang, S. J., *et al.* 2003, *A&A*, 405, 313
Plunkett, S. P., Vourlidas, A., Šimberová, S. *et al.* 2000, *Sol. Phys.*, 194, 371
Schrijver, C. J., Title, A. M., Yeates, A. R., & DeRosa, M. L. 2013, *ApJ*, 773, 93
Schrijver, C. J. & Title, A. M. 2011, *J.* Geophys. Res., 116, A04108
Shen, Y., Liu, Y., & Liu, R. 2011, *Res. Astron. Astrophys.*, 11, 594
Shen, Y., Liu, Y., & Su, J. T. 2012, *ApJ*, 750, 12
Titov, V. S., Mikic, Z., Török, T., Linker, J. A., & Panasenco. O. 2012, *ApJ*, 759, 70
Török, T., Panasenco, O., Titov, V. S., *et al.* 2011, *ApJ*, 739, L63

*Nature of Prominences and their role in Space Weather*
*Proceedings IAU Symposium No. 300, 2013*
*B. Schmieder, J.-M. Malherbe & S. T. Wu, eds.*

doi:10.1017/S1743921313011034

# A solar tornado caused by flares

**N. K. Panesar,[1,3] D. E. Innes,[1] S. K. Tiwari,[1] and B. C. Low[2]**

[1]Max-Planck Institut für Sonnensystemforschung, Max-Planck-Str. 2, 37191, Katlenburg-Lindau
email: panesar,innes,tiwari@mps.mpg.de

[2]High Altitude Observatory, National Center for Atmospheric Research, P.O. Box 3000, Boulder, CO 80307, USA
email: low@ucar.edu

[3]Institut für Astrophysik, Georg-August-Universität Göttingen, Friedrich-Hund-Platz 1, D-37077 Göttingen

**Abstract.** An enormous solar tornado was observed by SDO/AIA on 25 September 2011. It was mainly associated with a quiescent prominence with an overlying coronal cavity. We investigate the triggering mechanism of the solar tornado by using the data from two instruments: SDO/AIA and STEREO-A/EUVI, covering the Sun from two directions. The tornado appeared near to the active region NOAA 11303 that produced three flares. The flares directly influenced the prominence-cavity system. The release of free magnetic energy from the active region by flares resulted in the contraction of the active region field. The cavity, owing to its superior magnetic pressure, expanded to fill this vacated space in the corona. We propose that the tornado developed on the top of the prominence due to the expansion of the prominence-cavity system.

**Keywords.** Sun, prominences, cavity, corona and flares

## 1. Introduction

Solar prominences are composed of cool plasma, embedded in the hotter solar corona (Tandberg-Hanssen 1995). They typically form above the magnetic polarity inversion line (Martin 1973). They are known as prominences when seen above the solar limb and filaments, when observed on the solar surface (Zirker 1989; Mackay *et al.* 2010). Quiescent prominences are often seen with large-scale coronal cavities (Hudson *et al.* 1999; Gibson & Fan 2006). Prominences sometimes appear to have rotating flows that resemble a tornado. Such prominences are known as prominence tornadoes (Pettit 1925, 1932). According to Pettit (1932) tornadoes appear like tightly wound flux ropes or vertical open spirals.

A huge quiescent-prominence tornado was observed by the Atmospheric Imaging Assembly (AIA) onboard the Solar Dynamic observatory (SDO) (Lemen *et al.* 2012) on 25 September 2011 (Li *et al.* 2012; Panesar *et al.* 2013). Observations showed the flow of plasma along a helical structure and tornado-like rotations that lasted for more than 3 hours. We have analysed the tornado-like prominence and found that the acceleration of the tornado was correlated in time with the flares taking place in a neighbouring active region (Panesar *et al.* 2013).

In this paper, we show the relationship between the flares and the prominence. For this work, we have used the data from two different spacecraft - SDO/AIA and Solar TErrestrial RElations Observatory (STEREO)/ Extreme UltraViolet Imager (EUVI). We combine the observations from the two different directions and reconstruct the 3D geometry to study the tornado event. In the next section, we describe the observations and then discuss the results.

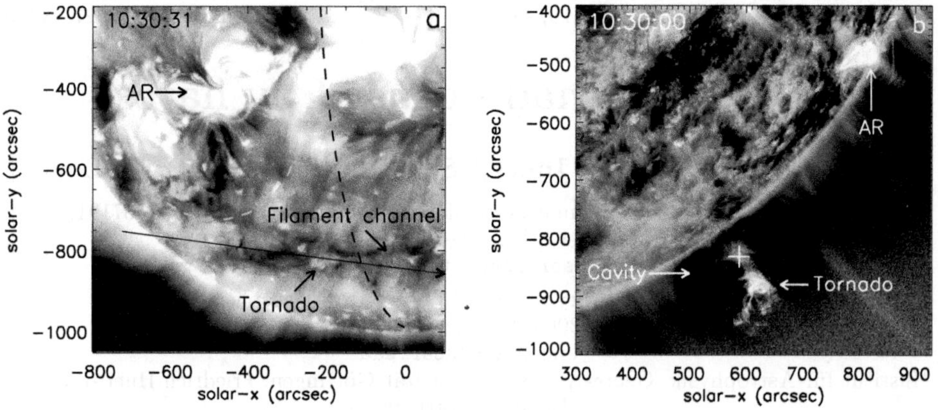


**Figure 1.** Prominence tornado on 25 September 2011: a) EUVI-A 195Å b) AIA 171Å images. In (a) the black dashed line shows the position of SDO limb line on the EUVI-A image. The orange dashed line highlights the southern edge of the active region corona and the black long arrow is the epipolar line for the prominence position which is marked as a '+' sign on AIA image. The active region (11303) is indicated with an arrow on both images. (Image adapted from Panesar *et al.*, A&A, 549, A105, 2013, reproduced with permission © ESO).

## 2. Observations

The prominence was observed by the SDO/AIA on the south-west solar limb. It appeared near the disk center in STEREO/EUVI-A. The separation angle was 103° between both the spacecrafts. For studying the limb perspective, we have used the SDO/AIA 171Å full-disk images of 12 s cadence. STEREO/EUVI-A 195Å images with a time cadence of 5 min have been used for the study of the filament on the solar disk.

Fig. 1 shows the overview of the prominence/filament system from two different angles - STEREO/EUVI-A and SDO/AIA. The active region (AR) and tornado are indicated with an arrows. We confirm the coordinates of the filament channel on disk by plotting the SDO limb line on STEREO images (black dashed line in Fig. 1a). The + sign on the prominence stem represents the position of selected coordinates along the long black arrow (epipolar line) on the EUVI image. The bright head of the tornado coincides with the epipolar line in the filament channel. The orange dashed line highlights the southern edge of the active region, which is only at a distance of ~50 arcsec from the filament channel.

There were three M-class flares from the active region at 02:45, 07:00 and 09:40 UT. All three flares were associated with CMEs and EUV waves. The third flare was the strongest and produced the most visible EUV wave front. Fig. 2 shows the 195Å running ratio image during the third flare. The faint EUV wave front swept over the filament channel about half an hour after the flare. After each flare, we noticed increased activity in the prominence. After the third flare there was accelerated tornado-like motion on the top of the prominence.

To investigate the relation between the flares and prominence dynamics, we used SDO limb images. We observed that the wave buffeted the prominence cavity system for all the flares. After the first flare, clear oscillations have been seen in the prominence system for about one hour (see Figure 7a of Panesar *et al.* (2013)). The second flare was at 06:45 UT, at the time of an SDO eclipse. Since this was an important time for the tornado activity, we used SWAP data to see the relation between the second flare and the tornado activation. The SWAP observations clearly show that the increase in prominence activity

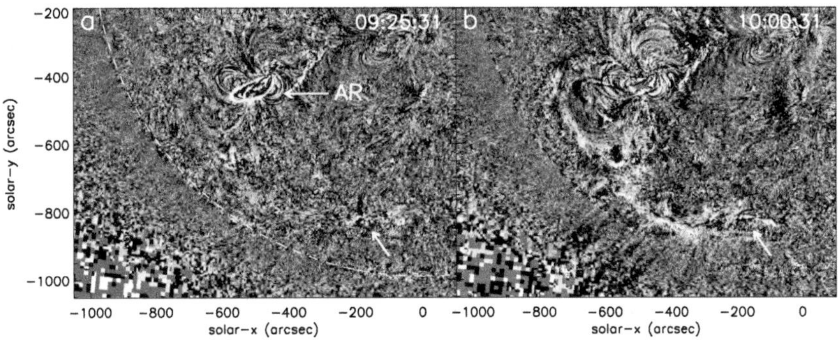


**Figure 2.** STEREO/EUVI-A 195Å running ratio images. a) The onset of the third flare b) is associated with an EUV wave. The dashed line is the solar limb line and the small arrows point to the tornado site.

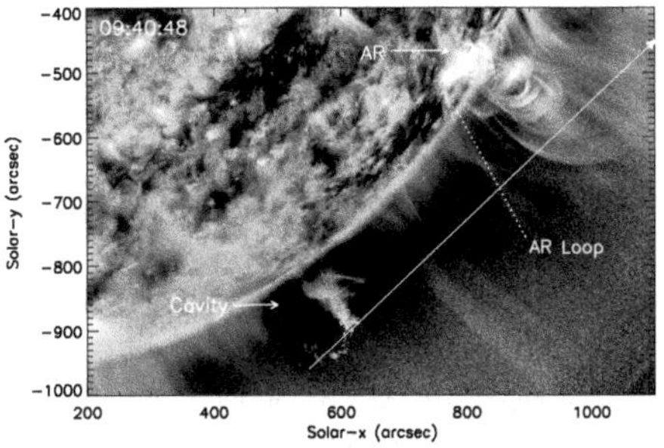


**Figure 3.** SDO/AIA 171Å intensity image. The long arrow shows the position of time-series in Fig. 4 taken through the prominence, cavity, active region loop (dotted line) and active region.

after the second flare. We noticed from the SDO images, that there was an extension of the prominence towards the active region after the second flare.

The main change occurred in the cavity after the third flare. Fig. 4 shows the time-series along the diagonal arrow in Fig. 3. The flare started at 09:20 UT and the active region loop started to move southward just after that. This resulted in a sharp contraction in the cavity boundary (indicated by an arrow in Fig. 4). The cavity contracted about 10 Mm in width. After that the active region loop swayed back towards the active region. Then the cavity started to expand in the direction of the neighbouring active region. The cavity grew in size from 140 Mm to 167 Mm in 3.5 hours. At the same time, the prominence also grew in size (see Fig. 4) and developed the tornado at the top.

## 3. Discussion

A huge solar tornado was observed by SDO/AIA on the SW solar limb and by STEREO/EUVI-A at disk center, on 25 September 2011. We noticed from the SDO/AIA images, that there was an active region near to the prominence-cavity system, which produced three flares on 25 September. We suspected that the flares influenced the prominence-cavity system. A careful analysis of STEREO/EUVI-A images showed that all three flares were associated with CMEs and EUV waves. EUV waves swept over the filament channel and increased the prominence activity. A tornado developed after the third flare.

After the first flare, there were oscillations in the prominence main body. The second flare caused an arm-like extension of the prominence towards the active region. The biggest change occurred after the third flare. It influenced the surrounding corona and the neighbouring cavity. During the third flare, the active region loop started to expand

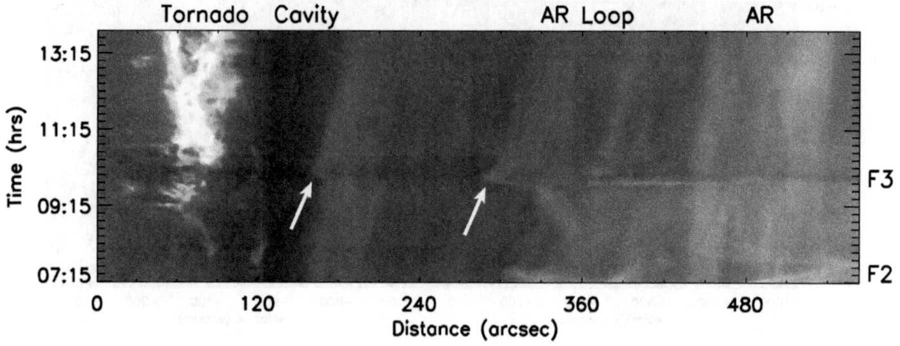


**Figure 4.** SDO/AIA 171Å intensity time-series image along the long diagonal arrow in Fig. 3. F2 and F3 show the time of the second and third flare. The small white arrows point to the contraction in the cavity edge (~09:40 UT) and expanding of the active region loop.

and move towards the edge of the cavity, which resulted in a sharp contraction in the cavity boundary, followed by an outward expansion in the cavity after the flare. The cavity continued visibly expanding for 3.5 hours. During this time the prominence increased in height and formed a tornado-like structure. It is likely that plasma moving along the helical field lines produced the tornado appearance.

We speculate that the expansion of the prominence was a result of the cavity expansion. We know that flares and CMEs are a means to release free magnetic energy (Zhang & Low 2005), which can result in a contraction of the field lines at the flare site (Hudson 2000; Zhang & Low 2003; Janse & Low 2007). In the present scenario, the neighbouring cavity expanded towards the active region to fill the surrounding corona, that had contracted after the flare. The cavity expanded because it had a higher magnetic pressure than that on the flare site. Expansion of the cavity led to a decrease of the magnetic pressure inside the cavity which resulted in the expansion of the prominence field, and this led to the development of the tornado-like activity. A detailed mathematical model of this observation is given in Panesar et al. (2013).

**Acknowledgements**

We are obliged to the SDO and STEREO teams. The presentation of this paper in IAU symposium 300 was possible due to the partial support by IAU and DAAD.

**References**

Gibson, S. E. & Fan, Y. 2006, *Journal of Geophysical Research (Space Physics)*, 111, 12103
Hudson, H. S. 2000, ApJ, 531, L75
Hudson, H. S., Acton, L. W., Harvey, K. L., & McKenzie, D. E. 1999, ApJ, 513, L83
Janse, Å. M. & Low, B. C. 2007, A&A, 472, 957
Kitiashvili, I. N., Kosovichev, A. G., *et al.* A. A. 2013, ApJ, 770, 37
Lemen, J. R., Title, A. M., Akin, D. J., *et al.* 2012, Solar Phys., 275, 17
Li, X., Morgan, H., Leonard, D., & Jeska, L. 2012, ApJ, 752, L22
Mackay, D. H., Karpen, J. T., Ballester, J. L., *et al.* 2010, Space Sci. Rev., 151, 333
Martin, S. F. 1973, 31, 3
Panesar, N. K., Innes, D. E., Tiwari, S. K., & Low, B. C. 2013, A&A, 549, A105
Pettit, E. 1925, *Publications of the Yerkes Observatory*, 3, 4
Pettit, E. 1932, ApJ, 76, 9
Tandberg-Hanssen, E., ed. 1995, *ASSL*, Vol. 199, The nature of solar prominences
Zhang, M. & Low, B. C. 2003, ApJ, 584, 479
Zhang, M. & Low, B. C. 2005, ARA&A, 43, 103
Zirker, J. B. 1989, Solar Phys., 119, 341

*Nature of Prominences and their role in Space Weather*
*Proceedings IAU Symposium No. 300, 2013*
*B. Schmieder, J.-M. Malherbe & S. T. Wu, eds.*

doi:10.1017/S1743921313011046

# The contribution of X-ray polar blowout jets to the solar wind mass and energy

**Giannina Poletto[1], Alphonse C. Sterling[2], Stefano Pucci[3] and Marco Romoli[3]**

[1]INAF - Arcetri Astrophysical Observatory,
Largo Fermi, 5, 50125, Firenze, Italy
email: poletto@arcetri.astro.it

[2]Space Science office, VP 62, MSSC,
Huntsville, AL 35812, USA
email: alphonse.sterling@nasa.gov

[3]University of Firenze, Firenze, Italy
email: stpucci@arcetri.astro.it

[4]University of Firenze, Firenze, Italy
email: mromoli@unifi.it

**Abstract.** Blowout jets constitute about 50% of the total number of X-ray jets observed in polar coronal holes. In these events, the base magnetic loop is supposed to blow open in what is a scaled-down representation of two-ribbon flares that accompany major coronal mass ejections (CMEs): indeed, miniature CMEs resulting from blowout jets have been observed. This raises the question of the possible contribution of this class of events to the solar wind mass and energy flux. Here we make a first crude evaluation of the mass contributed to the wind and of the energy budget of the jets and related miniature CMEs, under the assumption that small-scale events behave as their large-scale analogs. This hypothesis allows us to adopt the same relationship between jets and miniature-CME parameters that have been shown to hold in the larger-scale events, thus inferring the values of the mass and kinetic energy of the miniature CMEs, currently not available from observations. We conclude our work estimating the mass flux and the energy budget of a blowout jet, and giving a crude evaluation of the role possibly played by these events in supplying the mass and energy that feeds the solar wind.

**Keywords.** Sun: activity, Sun: solar wind, Sun: Coronal Mass Ejections.

## 1. Introduction

As a result of the analysis of HINODE data, the estimated number of X-ray polar jets has been soaring from a few per day to an average of 10 per hour (Cirtain *et al.* 2007). Successively, it became clear that jets may be classified either as "standard", or as "blowout" events, depending on their characteristics (Moore *et al.* 2010). Standard jets are suspected to arise from reconnection between an emerging bipolar arch and the open unipolar ambient field of the polar coronal hole regions. In blowout jets, it is suggested that an emerged arch, similar to that of standard jets, becomes unstable, erupts, and blows open as it reconnects with the open ambient field. That erupting arch can contain a cool filament that becomes a cool jet mimicking on a miniature scale what happens in large-scale coronal mass ejections (CMEs). Fig. 1 shows a typical polar coronal hole, as seen in the HINODE/XRT Al/Poly filter on March 7, 2008, where a narrow pencil-like jet is clearly seen together with other smaller structures.

After Moore *et al.* (2010) paper, a number of authors presented further evidence of the occurrence of two classes of jets and even detected the mini CMEs associated with

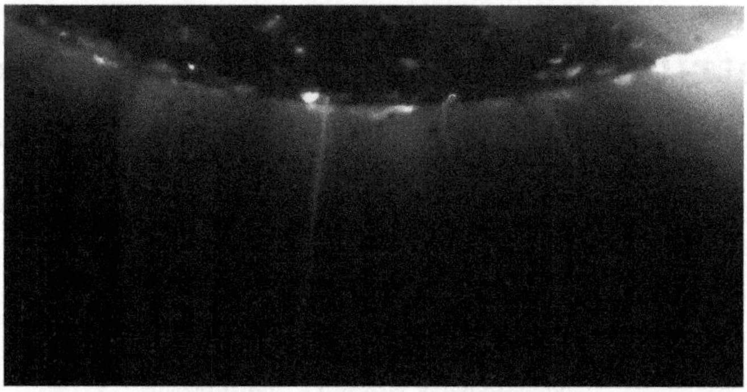

**Figure 1.** HINODE XRT image of the southern polar region acquired in the Al/Poly filter on March 7, 2008, at 08:43:48 UT, showing a pencil like jet and other smaller features rooted in the coronal hole area [Courtesy of P. Grigis].

blowouts (Hong *et al.* 2011, Shen *et al.* 2012), although the CME masses could not be measured. Analogous to the situation of confined vs. eruptive flares, blowout jets have a longer duration and are more energetic as compared to the standard jets Pucci *et al.* (2013). Observations in polar coronal holes reveal about an equal number of standard and blowouts (Moore *et al.* 2013).

Nowadays the properties of standard and blowout jets seem to be, morphologically, thoroughly defined, but their physical characteristics have been scarcely explored. Knowledge of the density, temperature, outflow speed of the jets is relevant *per se*, to understand the behavior of this miniature class of events with respect to their large-scale counterparts, and because it allows us to check whether jets represent the long-searched source of solar wind mass and energy. Here we adopt the physical parameters that have been inferred, from an analysis of typical ejections, by Pucci *et al.* (2013) and, on this basis, we calculate the energy budget of jets, focusing on a blowout jet. Blowout jets are more energetic than standard jets, and so if they turn out to be inadequate to supply the solar wind mass and energy, then standard jets will be ineffective in this regard as well. In the next section, we summarize the jet parameters and then proceed to calculate their energy budget. In Section 3, we evaluate the energy associated with the cool component of blowouts and give an estimate of the mini-CME mass. We conclude discussing the jet mass flux and energy in the context of the solar wind mass and energy requirements.

## 2. The physical parameters of jets and their energy budget

Spectroscopic techniques allow us to infer the physical parameters of jets seen in different X-ray bands by different experiments/spacecraft. Pucci *et al.* (2013) used data from HINODE/XRT and STEREO/SECCHI experiments to evaluate temperature, density and outflow speed of standard and blowouts. They found that, for a standard jet, the electron temperatures ($T_e = 1.6 \times 10^6\ K$) was lower by $\approx 15\%$, densities ($6. \times 10^8\ cm^{-3}$) were about the same and outflow speeds ($\approx 250\ kms^{-1}$) were lower by $\approx 40\%$ than in the blowout jet. Because blowouts, which are composed of multiple structures, last longer than standard jets, the role of the latter, if any, in contributing to the wind mass and energy is minor, and justifies our previous statement that blowouts are the most energetic representatives of the jets family. Hence, from here on, we focus on this class of jets.

As mentioned earlier, reconnection is crucial in the origin and development of jets. The magnetic energy flux delivered in reconnection episodes is partitioned among enthalpy, wave, kinetic, potential and radiative energy fluxes of the ensuing jets. Because temperature was approximately constant along the axes of the jets studied in Pucci *et al.* (2013), we ignore the conductive energy flux and we write the above terms as

$$F_{enth} = \frac{\gamma}{\gamma - 1} pv \tag{2.1}$$

$$F_w = \sqrt{(\frac{\rho}{4\pi})}\xi^2 B \tag{2.2}$$

$$F_{kin} = 1/2\rho v^3 \tag{2.3}$$

$$F_{pot} = \rho g L v \tag{2.4}$$

$$F_{rad} = n_e n_H \chi T^{\alpha} L \tag{2.5}$$

where $p, \rho, g, \gamma$ are, respectively, the plasma pressure, mass density, gravity, $\gamma$ is the ratio of the specific heats ($\gamma = 5/3$), $n_e$ is the plasma number density ($n_H = n_e$), $\xi$ is the amplitude of unresolved non-thermal plasma motion, B is the field strength and L is the jet length. The parameters $\chi$ and $\alpha$ appear in the analytical approximation to $F_{rad}$ given by Rosner *et al.* (1978). The most uncertain energy flux is $F_w$: lacking direct evidence of the amplitude of Alfvénic waves excited by reconnection, we assume that waves show up only as unresolved plasma motions and adopt a wave amplitude of 100 $kms^{-1}$ in agreement with values found by Kim *et al.* (2007) from EIS observations of line broadening in jets. The field strength B has been derived from the assumption that outflows occur at the Alfvén speed. An evaluation of the above stated energy fluxes leads to a blowout total energy of $\approx 2 \times 10^{27}$ $erg$. Hence blowouts are typically $10^{-5} - 10^{-6}$ times less energetic than the largest flares associated with CMEs. Pucci *et al.* (2013) provide more details of these calculations.

## 3. The cool component of blowouts: the associated CME and its energy

When calculating the energy budget of blowouts, we should include also the cool component associated with the event: in their large-scale counterparts the kinetic/potential energy associated with the ejecta plays a major role (Emslie *et al.* 2012). Although cool-material outflows have been detected above the limb by white light coronagraphs, the amount is usually too tiny to be estimated via conventional methods. On the other hand, knowledge of the CME mass is a prerequisite for calculating the CME's energy; we resorted to an indirect technique for estimating the mass, and subsequently the energy, by assuming blowout jets behave as their large-scale counterparts.

Yashiro & Gopalswamy (2009) have shown that there is a relationship between the 1-8 Å X-ray flare fluence and the kinetic energy of the corresponding CME. Analogously Aarnio *et al.* (2011) have shown that the flare X-ray flux is correlated with the mass of the associated CME. Hence, we estimated the X-ray emission of the blowout from its density and temperature (which can be easily done either analytically or via standard codes like CHIANTI) and inferred the CME mass and its kinetic energy, via the relationships given by the above authors. It turns out that the expected CME kinetic energy is of the order of a few times $10^{26}$ erg and its mass of the order of $10^{12}$ g. Although these estimates are quite crude, they are based on extrapolations to small-scale events of well established relations for larger-scale events. We conclude that the energy residing in the cool component of blowouts is of the same order as that of the hot component.

## 4. Jets and the solar wind

We now discuss our results in terms of the relevance of jets as contributors to the solar wind mass and energy flux. The mass flux $nv$ originating from the ejections can be estimated from the density and outflow speed values given above. We limit our analysis to blowouts whose occurrence rate is still not precisely known, but we assume here 50 events per day. Assuming a base radius of the order of 1000 km, for the bright components of the blowout and an event duration of the order of 20 min, it turns out that blowouts are the source of a mass flux of $5 \times 10^5\ cm^{-2}s^{-1}$, which is smaller than typical wind mass flux ($\approx 2 \times\ 10^8\ cm^{-2}s^{-1}$) by about a factor of $10^3$. This estimate is affected by large uncertainties: however, even if the total number of jets, their outflows, their base area, were each underestimated by a factor 3 (which is unlikely) jets could supply only a few hundredths of the wind mass flux.

We point out that we neglected the contributions that mini CMEs might supply to the wind. If it would be proven that each blowout is accompanied by a mini-CME, we might hypothesize that this is a further source of mass flux. However, a quick order-of-magnitude estimate shows that even if this were the case, a rate of 50 small CMEs per day cannot contribute appreciably to the wind mass flux.

The wind energy flux $F_{wind}$ is of the order of $10^5\ erg\ cm^{-2}s^{-1}$ (see, e.g. Le Chat, 2012 and Schwenn, 2006). We have shown that the total energy released by blowouts is on the order of a few units times $10^{27}$ erg per event. Taking into account their total number per day and their duration, we end up with an energy flux of $\approx 10^{-4}\ F_{wind}$, much too small to be accounted for by uncertainties in our calculations. If we underestimated the value of B - which is likely, as bulk flows may not be as fast as small-scale reconnection flows - the wave flux $F_w$ will also be underestimated. Analogously, outflow speeds may have been underestimated, leading to low values of the kinetic flux $F_{kin}$. Nevertheless, our estimates cannot be off by more than a factor of 100, which is insufficient to make blowouts contribute appreciably to the wind. We conclude that blowouts cannot provide for the mass and energy of solar wind. However, we did not include other features, possibly associated with blowouts, nor we counted type II spicules as blowouts. If the latter turn out to be part of the same class of events, our conclusion may need to be revised.

GP acknowledges support from ASI I/015/07/0. ACS was supported by funding from NASA's Office of Space Science through the Living with a Star Targeted Research and Technology Programs.

### References

Aarnio, A. N., Stassun, K. G., Hughes, W. J., & McGregor, S. L. 2011, *SP*, 268, 195
Cirtain, J. W., Golub, L., Lundquist, A., van Ballegoijen, Savcheva, A., *et al.* 2007, *Science*, 318, 1580
Emslie, A. G., Dennis, B. R., Shin, A. Y., *et al.* 2012, *ApJ*, 759, 71
Hong, J., Jiang, Y., Zheng, R., Yang, J., Bi, Y., Zinner, E., & Yang, B. 2011, *ApJ* (Letters), 738, L20
Kim, Y. H., Moon, Y.-J., Park, Y.-D., Sakurai, T., Chae, J. *et al.* 2007, *PASJ*, 59, S763
Le Chat, G., Issautier, K., & Meyer-Vernet, N. 2012, *SP*, 279, 197
Moore, R. L., Cirtain, J. W., Sterling, A. C., & Falconer, D. A. 2010, *ApJ*, 720, 757
Moore, R. L., Cirtain, J. W., Sterling, A. C., Falconer, D. A., & Robe, D. 2013, *ApJ*, 769, 134
Pucci, S., Poletto, G., Sterling, A. C., & Romoli, M. 2013, *ApJ*, 776, 16
Rosner, R., Tucker, W. H., & Vaiana, G. S. 1978, *ApJ*, 220, 643
Schwenn, R. 2006, *SSR*, 124, 51
Shen, Y., Liu, Y., Su, J., & Deng, J. 2012, *ApJ*, 745, 164
Yashiro, S. & Gopalswamy, N. 2009, *IAU Symp.*, 17, 233

# Section II - 2.2

CORONAL MASS EJECTIONS

MAGNETIC CLOUDS IN THE HELIOSPHERE

IMPACTS ON EARTH'S ENVIRONMENT

*Nature of Prominences and their role in Space Weather*
*Proceedings IAU Symposium No. 300, 2013*
*B. Schmieder, J.-M. Malherbe & S. T. Wu, eds.*

doi:10.1017/S1743921313011058

# Evolution of interplanetary coronal mass ejections and magnetic clouds in the heliosphere

**Pascal Démoulin**

Observatoire de Paris, LESIA, UMR 8109 (CNRS), F-92195 Meudon Principal Cedex, France
email: Pascal.Demoulin@obspm.fr

**Abstract.** Interplanetary Coronal Mass Ejections (ICMEs), and more specifically Magnetic Clouds (MCs), are detected with in situ plasma and magnetic measurements. They are the continuation of the CMEs observed with imagers closer to the Sun. A review of their properties is presented with a focus on their magnetic configuration and its evolution. Many recent observations, both in situ and with imagers, point to a key role of flux ropes, a conclusion which is also supported by present coronal eruptive models. Then, is a flux rope generically present in an ICME? How to quantify its 3D physical properties when it is detected locally as a MC? Is it a simple flux rope? How does it evolve in the solar wind? This paper reviews our present answers and limited understanding to these questions.

**Keywords.** magnetic fields, (magnetohydrodynamics:) MHD, Sun: coronal mass ejections (CMEs), Sun: magnetic fields, interplanetary medium

## 1. Main characteristics of ICMEs and magnetic clouds

Interplanetary coronal mass ejections (ICMEs) are detected by in situ measurements having unusual characteristics of the solar wind. Depending on authors and available instruments, ICMEs have been defined by a broad set of signatures of the plasma, energetic particles and magnetic field (e.g., see the reviews of Zurbuchen & Richardson 2006; Wimmer-Schweingruber *et al.* 2006). The main ICME characteristics are summarized in Fig. 1. In the solar wind, the proton temperature is well correlated to its bulk velocity (e.g., Elliott *et al.* 2005; Démoulin 2009). Then, departure from this law, more precisely where the proton temperature is below half the proton temperature found in the solar wind with the same speed, is a typical definition of an ICME (Liu *et al.* 2005; Richardson & Cane 2010). ICMEs also have typically enhanced ion charge states which are formed in the corona below the height where the collision rate becomes negligible (see Lepri 2014). ICMEs are also characterized by enhanced abundances of ions with a low first ionisation potential (FIP) compared to those with a high FIP (Fig. 1). This separation of the elements occurs in the low solar atmosphere (e.g., see Baker *et al.* 2014). All these ICME signatures are typically not present in the same interval of time, they are frequently variable in intensity during an ICME crossing, and even they are not always all present (Richardson & Cane 2010).

Magnetic clouds (MCs) are defined by an enhanced magnetic field strength, a smooth rotation of the magnetic field direction through a large angle and a low proton temperature compare to the expected one in the solar wind (see the reviews of Gosling *et al.* 1995; Dasso *et al.* 2005). These local measurements are typically interpreted with a flux rope model (i.e., a twisted magnetic configuration, e.g., Lepping *et al.* 1990; Lynch *et al.* 2003; Dasso *et al.* 2006; Leitner *et al.* 2007). MCs are present within ICMEs, e.g., Fig. 1. An ICME can coincide with the full time range of the associated MC, be more extended

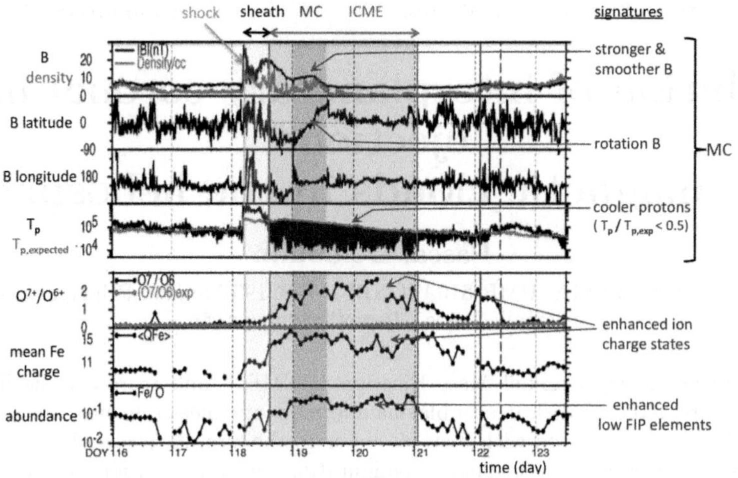


**Figure 1.** Main characteristics of MCs and ICMEs from in situ measurements. This event was observed from 28 April to 1 May 2001 at 1 AU by ACE spacecraft. A MC is present inside the ICME as shown by the four upper panels. The ICME is defined by the region $T_{\rm p} < T_{\rm p,expected}/2$ (fourth panel, the region between $T_{\rm p}$ and $T_{\rm p,expected}$ is set in black inside the ICME). It corresponds approximatively to the regions of enhanced ion charge states and abundance anomaly (lower panels). Adapted from Richardson & Cane (2010).

or contain no MC. On average a MC is detected in about 30% of ICMEs (Wu & Lepping 2011). This ratio evolves with the solar cycle from $\approx 15\%$ at solar maximum to $\approx 100\%$ at solar minimum (Richardson & Cane 2010; Kilpua *et al.* 2012).

ICMEs are the counterpart of coronal mass ejections (CMEs) observed with coronagraphs (e.g., Howard 2011; Lugaz & Roussev 2011). The link between coronal and in situ observations were done by using as many as possible constraint derived by both type of data: the relative location of the solar source and the spacecraft detecting the ICME, the transit time (duration of displacement from the Sun to in situ location), and by the estimation of the same physical parameters (orientation of the magnetic configuration, magnetic fluxes and helicity, see e.g., the review of Démoulin 2007). With STEREO twin spacecraft having both in situ and imager instruments, this link is presently well established (e.g., Harrison *et al.* 2009; Kilpua *et al.* 2011; Rouillard 2011; Lugaz *et al.* 2012, and references therein). An example is shown in Fig. 2 where a CME is followed from its launch in the corona to 1 AU where the ICME/MC was observed by ACE and Wind spacecraft (see Fig. 4 of DeForest *et al.* 2013). The build up of mass in the sheath by a snowplow effect, as well as the kinetic energy budget were also followed over such a distance.

The solar origin of a CME is the progressive buildup of a flux rope (FR) by magnetic reconnection at a photospheric magnetic inversion line in and also outside active regions, typically where a filament is present. When the magnetic stress is too large, an instability occurs, launching the flux rope (e.g., Török & Kliem 2007; Aulanier *et al.* 2010). A large amount of the surrounding magnetic arcade is reconnecting below the erupting flux rope, leading to a flare with its two ribbons separating and with the formation of flare loops (see e.g., the review of Schmieder *et al.* 2013). This process further builds up a very hot FR ($\approx 10$ MK) as imaged with AIA EUV observations (e.g., Reeves & Golub 2011; Cheng *et al.* 2011, 2013). Such extreme temperatures are expected to be at the origin of the higher ion charge states in ICMEs (e.g., Gopalswamy *et al.* 2013). These observations also confirm the global scenario of an eruptive flare leading to a CME (see the reviews of Aulanier 2014; Shibata 2014). Moreover, some specificities of 3D reconnection without magnetic null points, like the continuous slippage of field lines, have been found in AIA

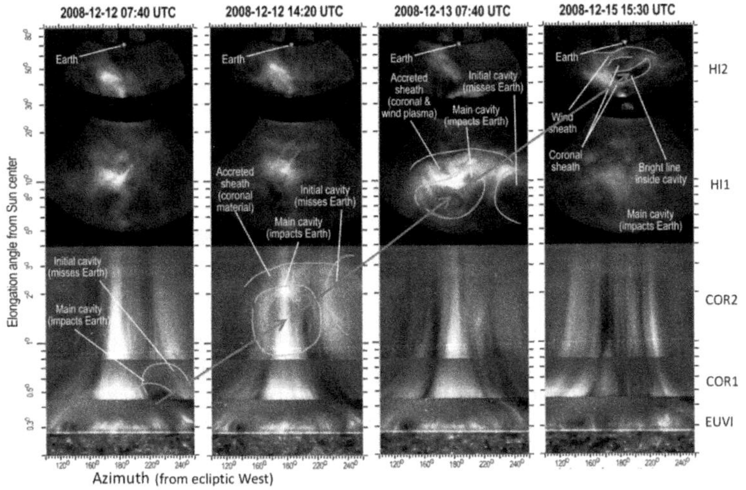


**Figure 2.** Launch of a CME on December 12, 2008 from the corona and evolution of the associated ICME as observed by the different instruments of STEREO-A (EUVI,COR1,COR2,HI1,HI2). The images have been transformed to the same cylindrical coordinate system (elongation, azimuth). The CME starts by the destabilization of a coronal cavity (plausibly the signature of a flux rope). This cavity is followed all the way to Earth where it is identified as a MC by in situ measurements. Adapted from DeForest *et al.* (2013).

observations and a 3D MHD simulation (Aulanier *et al.* 2012; Janvier *et al.* 2013a; Dudik *et al.* 2014b).

With SDO/AIA observations many clear examples of the formation, then eruption of twisted structures are available (e.g., Title 2014). However, this does not imply that FR signatures are easily observable in all solar events. First, the twist is only just above one turn at the start of an eruption. This weakly twisted FR is even more difficult to visualize when it is bent in a 3D configuration, and even more with typically only very few field lines partially outlined by a sufficiently dense plasma to be emissive enough (it is also very difficult to visualize, from an arbitrary point of view, the 3D configuration involved in MHD simulations with few and partially drawn field lines). More over, background/foreground emissions are typically present, and projection effects should be favorable (e.g., along the FR axis). During the eruption, further twist is built up by reconnection, below the flux rope, of the surrounding arcades. The more potential the arcade is, typically higher up, the more twisted is the resulting wrapping field around the erupting FR (while preserving magnetic helicity). This builds the higher twist observed in situ at the periphery of MCs, but this is typically not observable in EUV images. At the CME stage, it is even more difficult to find evidences of a FR as the scattered light comes mainly from the sheath built up in front of the CME (Fig. 2). This difficulty further increases in the heliospheric imagers as the signal/noise ratio decreases and the interpretation of the images becomes more model dependent (Lugaz 2014). Finally, the in situ observations only provide a 1D cut through a 3D evolving magnetic configuration, so modeling and interpretation are required. In brief, in order to detect the FR within an erupting configuration, a CME or an ICME, the Sun must "cooperate" with us!

## 2. Structure and evolution of magnetic clouds

Since in situ data are typically available only along the 1D crossing of some MCs by one spacecraft, a magnetic model is needed to derive, at least around the trajectory, the magnetic configuration. A FR model is typically due to the smooth rotation of the mag-

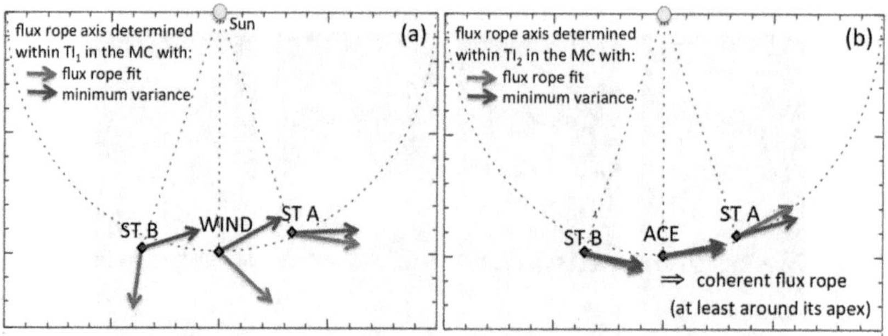


**Figure 3.** A MC crossed by three spacecraft (20 Nov. 2007 at $\approx$ 1 AU). The local axis is determined from a minimum variance (MV) or a fit of a Lundquist's model to the in situ $\vec{B}$ data. The methods are applied to two different Time Intervals (TI) within the MC. While the MV and the fit techniques significantly disagree in (a), they give comparable directions in (b) and moreover a coherent global shape of the MC axis. Adapted respectively from Farrugia *et al.* (2011) and Ruffenach *et al.* (2012).

netic field across a MC, as well as the relationship with erupting twisted configurations on the Sun. Its free parameters are determined by a least square fit to the in situ data. So far, no model has been proven to best represent a large set of MCs. Many models consider a straight FR as a local approximation. The simplest and most used model is a linear force-free field model, called Lundquist's model (see e.g. Lepping *et al.* 1990; Leitner *et al.* 2007). Extensions to non circular cross-section (e.g. Vandas & Romashets 2003; Démoulin & Dasso 2009b), or non force-free models (e.g. Mulligan & Russell 2001; Hidalgo 2011) have been developed. An alternative is to solve magneto-hydrostatic equations, which reduce to the Grad-Shafranov equation with an axis of invariance, together with the observed boundary conditions for the integration procedure (e.g., Isavnin *et al.* 2011; Hu & Qiu 2014).

An appealing approach is to extend the above models to toroidal geometry in order to include the curvature of the FR axis as in the schema of Fig. 5a (e.g. Marubashi & Lepping 2007; Romashets & Vandas 2009). This is especially needed when the spacecraft is crossing one MC leg as the spacecraft explores only a fraction of the FR (e.g. Marubashi *et al.* 2012; Owens *et al.* 2012). However, toroidal models imply a larger number of free parameters and it is not yet clear if they can all be constrained by the data of a single spacecraft, even for a MC crossing with a low impact parameter (distance of the spacecraft to the MC axis normalized by its radius). Two well separated spacecraft provide more constraints to the toroidal model (Nakagawa & Matsuoka 2010). However, the number of MCs observed is very limited in this configuration as it requires a FR close to the ecliptic where spacecraft are typically located (see the review of Kilpua *et al.* 2011).

The simplest way for estimating the FR orientation is to apply the Minimum Variance (MV) method to the normalized series of magnetic field measurements within the MC (e.g., Lepping *et al.* 1990). It provides a good orientation (with a bias lower than 20°) if the MC is not too asymmetric and the boundaries are well defined (i.e., includes only the flux rope, Gulisano *et al.* 2007). A fit of a model, like Lundquist's, to the data also provides an estimation of the local axis direction. Both methods, and others, can provide very different orientations if they are applied to different ICMEs boundaries while a better agreement is found with the same boundaries (Al-Haddad et al. 2013). For example, MV and FR fit methods give significant different orientations, and moreover without coherence along the MC when the time intervals $TI_1$ are used at the three spacecraft (Fig. 3a, Farrugia *et al.* 2011), while they give consistent results when the time intervals $TI_2$ are rather used (Fig. 3b, Ruffenach *et al.* 2012).

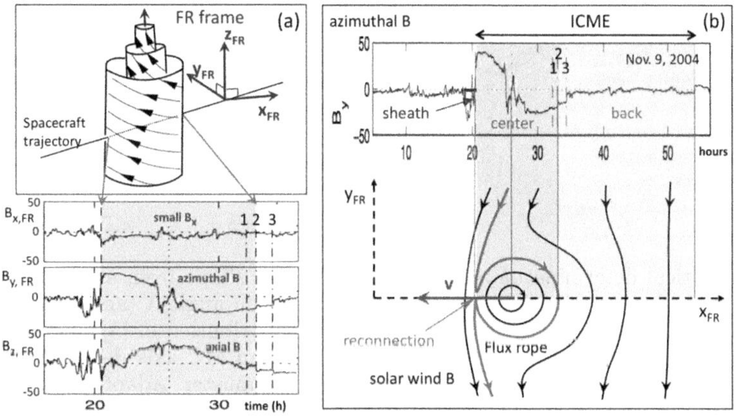


**Figure 4.** Definition of the flux rope (FR) frame and MC back region. **(a)** Schema of a FR and its associated local frame. Below, an example of observed magnetic field (9 Nov. 2004 at 1 AU) in the FR frame. This MC was crossed by a spacecraft close to the FR axis (small $< B_x > / < B >$ value). The numbers 1,2,3 show possible rear boundaries associated to discontinuities of $\vec{B}$. The azimuthal flux balance identifies the rear boundary as the number 2. **(b)** Extended time interval of the same MC which extends up to $t \approx 55$ h. At the time of the spacecraft crossing, the FR is limited to the grey region, while a long back region (green region) is present with a mix of MC and solar wind characteristics. This back region was part of the FR closer to the Sun before reconnection arround the FR front occurred (bottom schema and Fig. 5). Adapted from Dasso *et al.* (2007).

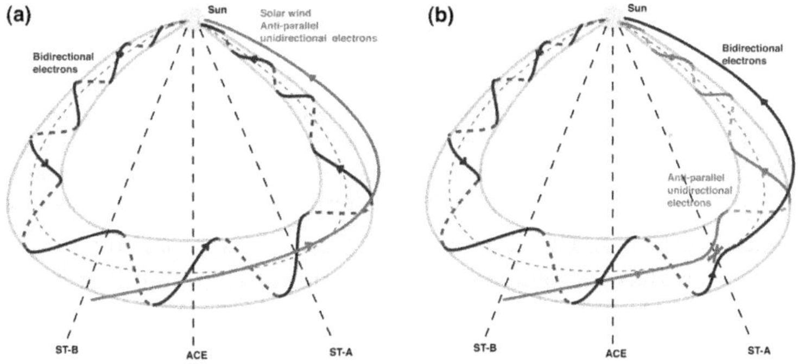


**Figure 5.** Example of reconnection between a MC, still attached to the Sun to simplify, and an inward sector magnetic field. **(a)** Before reconnection. **(b)** Part of the magnetic flux is removed from the MC front around the reconnection region (by shifting on the FR sides) creating a back region (Fig. 4). However, the in situ detection of this process depends on the location of the spacecraft crossing (here a back is present at ST-A, but not at ST-B and ACE). Adapted from Ruffenach *et al.* (2012).

A MC travels typically at a different speed than the surrounding solar wind, then their respective magnetic fields are pushed together, generally implying magnetic reconnection. This leads to a FR progressively peeling off layers until only the central region remains as a coherent FR when crossed by a spacecraft (Dasso *et al.* 2006). Figures 4 and 5 illustrate two MCs where reconnection happened in the front leaving a back region behind the remaining FR. This region has mixed properties between MC and solar wind ones. The leading edge of the back region was determined by the azimuthal flux balance. This was confirmed by the presence of a magnetic discontinuity (Dasso *et al.* 2007; Nakwacki *et al.* 2011) as well as by in situ reconnection signatures (Ruffenach *et al.* 2012). Finally, the FR orientation, determined, e.g., by an MV or a FR fit, can be significantly biased if the

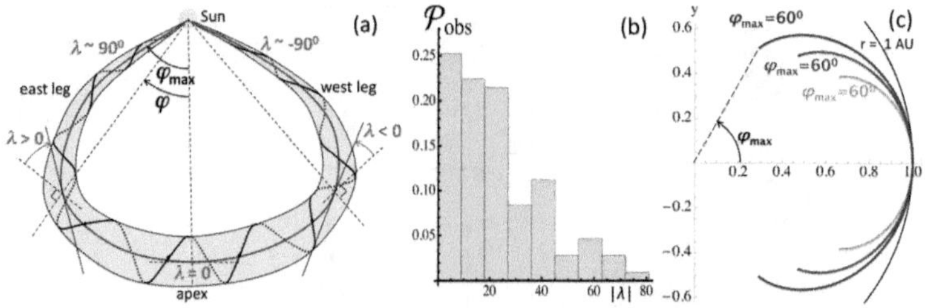


**Figure 6.** Statistical determination of the mean MC axis shape from a set of MCs observed at 1 AU. **(a)** Schema of a FR and definition of the location angle $\lambda$ (angle between the local axis direction and the ortho-radial from the Sun), **(b)** Distribution of $|\lambda|$ as observed at 1 AU for 107 MCs (Lepping & Wu 2010). **(c)** Shape of the mean MC axis deduced from (b), with the maximum angular extension, $\varphi_{\rm max}$, as the only free parameter. Adapted from Janvier *et al.* (2013b).

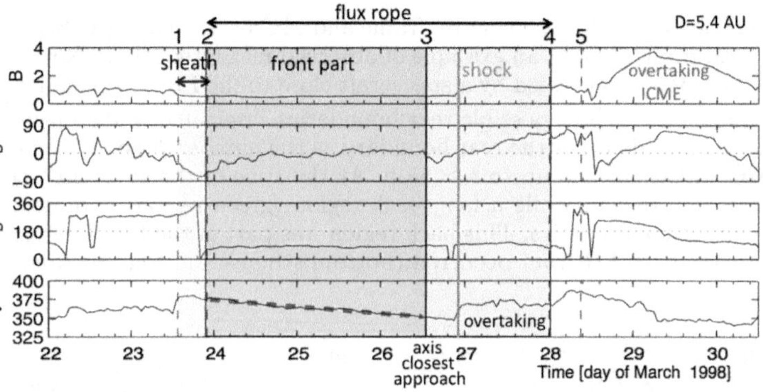


**Figure 7.** Example of a MC strongly overtaken by an ICME when observed by Ulysses at 5.4 AU. Only the first part, before the FR axis, is unperturbed (dark grey region) while the second part is strongly compressed and deformed (light grey region). The same MC was observed at 1 AU by ACE when it was not yet overtaken. Its magnetic field was similar to the MC in Fig. 4, with a smaller back region ($\approx 1/5$ of the FR extension). Adapted from Nakwacki *et al.* (2011).

method is applied to the full MC time interval. Then, it is important to define the MC boundaries in order to have a balance of the azimuthal magnetic flux before and after the closest approach to the FR axis (Fig. 4b). This is achieved in the FR frame (Fig. 4a), and it removes the back region from the time interval selected to apply an MV or a FR fit (Dasso *et al.* 2006, 2007).

While few MCs have been observed by at least two spacecraft with a low impact parameter, more than one hundred MCs have been observed by Wind spacecraft and fitted by the Lundquist model (Lepping & Wu 2010). The inclination of the FR axis on the ortho-radial direction defines the angle $\lambda$ which is related to the location of the spacecraft crossing along the FR axis (Fig. 6a). From the distribution of $\lambda$ (Fig. 6b), Janvier *et al.* (2013b) showed that one can derive a mean axis shape (Fig. 6c). It depends only on one free parameter, $\varphi_{\rm max}$, which can be constrained from heliospheric imagers (as on STEREO) in favorable cases (Janvier *et al.* 2014). The mean derived axis shape (Fig. 6c) is compatible with the MC observed by three spacecraft (Fig. 3b).

Apart from reconnection, MCs are also evolving as they move away from the Sun. In particular they are expanding since the surrounding total pressure decreases rapidly with the distance (as $\approx D_{\odot}^{-2.9\pm0.3}$ where $D_{\odot}$ is the distance to the Sun, Démoulin & Dasso 2009a; Gulisano *et al.* 2010). This expansion is detected in situ by a faster velocity in front than at the rear of the MC, with typically a linear profile (see the V panel of Fig. 7). Démoulin *et al.* (2008) quantifies this expansion with an nondimensional parameter called $\zeta$ which can be derived from in situ measurements. When $\zeta$ is constant, the FR radius evolves with $D_\odot$ as $R \propto D_\odot^\zeta$. From observations of different MCs at significantly different helio-distances (Wang *et al.* 2005; Leitner *et al.* 2007) and from observations of the velocity profile slope from single spacecraft observations (Démoulin *et al.* 2008; Gulisano *et al.* 2010, 2012), from the inner heliosphere to 5 AU, it has been found that $\zeta \approx 0.9 \pm 0.3$ for MCs which are not perturbed (e.g., not overtaken by a fast stream or ICME). In particular, one MC was observed both at 1 AU and 5.4 AU by ACE and Ulysses spacecraft, and indeed $\zeta \approx 0.7$ at both distances (only the front part is used at 5.4 AU because the part of the MC after its axis was strongly distorted by the overtaken ICME, Fig. 7). It implies an expansion by a factor $\approx 3.5$. From a fit of the Lundquist model and also from a direct integration of the data of the front part, assuming cylindrical symmetry, Nakwacki *et al.* (2011) estimated the magnetic fluxes, helicity and energy. They showed that fluxes and helicity were well conserved while magnetic energy decayed almost as expected (as $\approx D_\odot^{-\zeta}$).

## 3. Are MC and non-MC ICMEs different types of events?

The observed magnetic field in MCs is typically modeled by FRs (Sect. 2), but only about one third of ICMEs have a detected MC inside (Sect. 1). Then, are non-MC ICMEs formed by the simple blowout of magnetic arcades from the Sun, so without FR (Gosling 1990), or rather do the spacecraft miss it? Even more, does a MC necessarily include a FR? Al-Haddad *et al.* (2011) rather proposed the eruption of a strongly sheared (not writhed) arcade to explain the coherent rotation of the magnetic field in MCs. They further argue that magnetic reconstruction methods, such as the Grad-Shafranov one, are biased to recover a FR even when the analyzed field is only a sheared arcade. However, they numerically simulate only very broad cases (almost as broad as the solar distance) and it is difficult to imagine how such model can explain typical MCs, with a radius of 0.1 AU at 1 AU, and even more the smaller ones. Moreover, there is presently no model which blows up an arcade from the Sun without non-physical numerical forcing (Aulanier 2014).

Could spacecraft miss the FR in many events as argued by Jian *et al.* (2006) from an analysis of the total (almost) pressure profile? Indeed, some MCs have a modest rotation of the magnetic field (down to $\approx 40°$, Lepping & Wu 2010). This is more an observing bias than an intrinsic difference of twist amount, as they are almost equally well fitted by a Lundquist model. Although MCs are expected to be crossed at random positions by a spacecraft, leading to a flat distribution of the impact parameter, the observed probability distribution of the impact parameter actually shows a strongly decreasing function (Lepping & Wu 2010). Does this imply that many MCs, with a moderate or large impact parameter, are not recognized as MCs? This would explain why only one third of ICMEs have a MC inside. However, a quantitative analysis shows that this bias has a small effect; rather the impact parameter distribution is naturally explained by an oblateness of the MC cross section which is elongated by a factor 2 to 3 in the direction perpendicular to the radial from the Sun (Démoulin *et al.* 2013), in global agreement with other estimations (e.g., Antoniadou *et al.* 2008; Savani *et al.* 2010) and magnetohydrodynamics simulations (e.g., Lugaz *et al.* 2005; Xiong *et al.* 2006).

Two workshops were organized in order to further investigate the possible difference between MC and non-MC ICMEs. Shock-driven ICMEs observed at 1 AU were selected in the time period 1996-2006. Only the ICMEs with an identified solar source within a longitude range $\pm 15^0$ were selected which reduced the number of ICMEs to 54 events. If the CMEs are FRs launched radially from the Sun, a large proportion of MCs is expected to be detected near Earth. However, only 23 MCs were detected. Does it imply that the 31 non-MC ICMEs have no FR, so that a different erupting configuration is present on the Sun? In fact no significant difference was found between the flare temperature, the

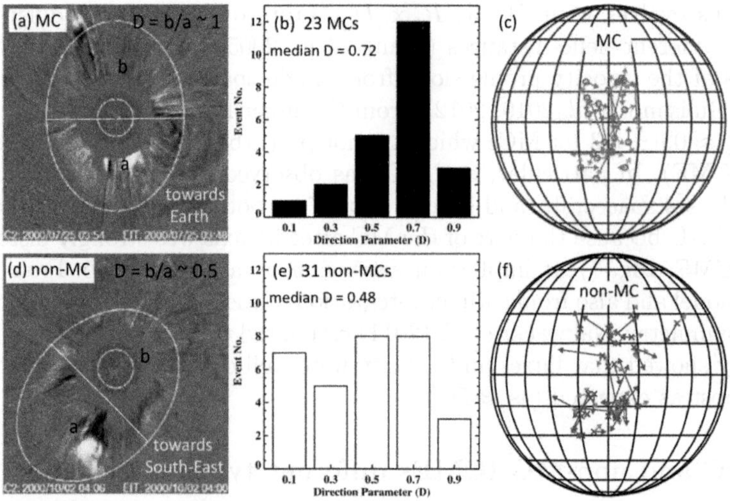


**Figure 8.** Deflection of CMEs by coronal holes. All CMEs start close to the central meridian (right panels). CMEs are detected close to Earth as MCs (a-c) and non-MC ICMEs (d-f). **(a,d)** The CME direction is estimated by the ratio $D = b/a$ of an ellipse fitted by eyes to the CME front observed by SoHO/LASCO. **(b,e)** Distributions of $D$ showing that CMEs associated to a MC have larger $D$ values, so are more Earth directed, than non-MC CMEs. **(c,f)** Amount of CME deflection from the CME source location towards its propagation direction in the outer coronagraph (red and blue arrows). The green arrows are for a model of coronal hole deflection. Adapted from Yashiro *et al.* (2013), Kim *et al.* (2013), and Mäkelä *et al.* (2013).

flare loop properties (size, duration, tilt) as well as the latitude and longitude of the source regions between the two groups (MC and non-MC ICMEs, Gopalswamy *et al.* 2013; Yashiro *et al.* 2013). The main difference was the flare magnitude, the median being M1.5 and C4.1 for MC and non-MC ICMEs, respectively, while the distributions were broad and largely overlapped.

The next step was to compare the CME properties. Again no significant difference between the two groups was found for the CME velocity, acceleration and angular width (all projected in the plan of sky, Gopalswamy *et al.* 2013; Xie *et al.* 2013). However, a significant difference was found in the CME propagation direction estimated by fitting visually the CME front with an ellipse (Kim *et al.* 2013; Zhang *et al.* 2013). More precisely, they analyzed the direction parameter, $D$ (Fig. 8a,d), and found that the CMEs associated to MCs are more directed towards the observing spacecraft than the ones associated to non-MC ICMEs (Fig. 8b,e). This result is confirmed by fitting visually the CME front with a flux-rope like model (Xie *et al.* 2013). Moreover, Mäkelä *et al.* (2013) confirmed that CMEs are deflected by the magnetic field of coronal holes as found in MHD simulations (e.g. Lugaz *et al.* 2011). The deflection is stronger with a closer coronal hole with larger magnetic flux, as expected. The key point is that CMEs associated to MCs are deviated towards the spacecraft direction, while the ones associated to non-MC ICMEs are deflected away (Fig. 8d,f). The main conclusion of all these papers is that the difference between MC and non-MC ICMEs is a selection effect, and not an intrinsic property of the solar source or of the ejected configuration, in agreement with Jian *et al.* (2006).

## 4. Overview

There is a broad range of observations and models which are pointing towards a key role of flux ropes (FRs) in the corona, CMEs and ICMEs as follows. Well before an eruption, the presence of magnetic dips, then of a twisted field in bipolar configurations, is needed to support the dense plasma of filaments (Dudik *et al.* 2014a). The structure of coronal cavities is also naturally explained by the presence of a FR (Gibson 2014). Next, an evolution from an arcade to a sigmoid is typically observed in EUV and X-rays before and during eruptions (Green 2014). The main mechanism of eruption is thought to be the loss of equilibrium / torus instability which involves a FR (Aulanier 2014; Olmedo & Zhang 2014). In some eruption, the kink instability of a FR could also play a role. Finally, in situ data of MCs are compatible with a FR configuration (Sect. 2) and all evidences point toward the conclusion that ICMEs should all contain FRs (Sect. 3).

All these FR evidences are remarkable taking into account the difficulties in observing 3D FRs with both imagers and in situ data. Further, the FRs are initially embedded in complex 3D coronal magnetic configurations and they reconnect with the surrounding encountered fields as they propagate, creating a back region (e.g., Ruffenach *et al.* 2012). It is important to well define the FR boundaries before fitting a model to the in situ data in order to derive more global quantities. To further progress on the magnetic configuration involved, we need to combine the analyze of some clear events with multi-instruments and spacecraft (e.g. Möstl *et al.* 2009), with statistical studies (e.g. Janvier *et al.* 2013b) and with numerical simulations (as reviewed by Lugaz 2014).

**Acknowledgements.** I thank Sergio Dasso, Miho Janvier, Alexis Ruffenach, and David Webb for useful comments.

## References

Al-Haddad, N., Nieves-Chinchilla, T., Savani, N., *et al.* 2013, *Solar Phys.*, 284, 129
Al-Haddad, N., Roussev, I. I., Möstl, C., *et al.* 2011, *ApJL*, 738, L18
Antoniadou, I., Geranios, A., Vandas, M., *et al.* 2008, *Planet. Spa. Sci.*, 56, 492
Aulanier, G. 2014, IAUS 300, this issue
Aulanier, G., Janvier, M., & Schmieder, B. 2012, *A&A*, 543, A110
Aulanier, G., Török, T., Démoulin, P., & DeLuca, E. E. 2010, *ApJ*, 708, 314
Baker, D., Brooks, D., Démoulin, P., *et al.* 2014, IAUS 300, this issue
Cheng, X., Zhang, J., Ding, M. D., *et al.* 2013, *ApJL*, 769, L25
Cheng, X., Zhang, J., Liu, Y., & Ding, M. D. 2011, *ApJL*, 732, L25
Dasso, S., Mandrini, C. H., Démoulin, P., & Luoni, M. L. 2006, *A&A*, 455, 349
Dasso, S., Mandrini, C. H., Démoulin, P., Luoni, M. L., & Gulisano, A. M. 2005, *Adv. Spa. Res.*, 35, 711
Dasso, S., Nakwacki, M. S., Démoulin, P., & Mandrini, C. H. 2007, *Solar Phys.*, 244, 115
DeForest, C. E., Howard, T. A., & McComas, D. J. 2013, *ApJ*, 769, 43
Démoulin, P. 2007, *Ann. Geophys.*, 26, 3113
Démoulin, P. 2009, *Solar Phys.*, 257, 169
Démoulin, P. & Dasso, S. 2009a, *A&A*, 498, 551
Démoulin, P. & Dasso, S. 2009b, *A&A*, 507, 969
Démoulin, P., Dasso, S., & Janvier, M. 2013, *A&A*, 550, A3
Démoulin, P., Nakwacki, M. S., Dasso, S., & Mandrini, C. H. 2008, *Solar Phys.*, 250, 347
Dudik, J., Aulanier, G., Schmieder, B., Zapior, M., & Heinzel, P. 2014a, IAUS 300, this issue
Dudik, J., Janvier, M., Del Zanna, G., *et al.* 2014b, IAUS 300, this issue
Elliott, H. A., McComas, D. J., Schwadron, N. A., *et al.* 2005, *JGR*, 110, A04103
Farrugia, C. J., Berdichevsky, D. B., Möstl, C., *et al.* 2011, *JASTP*, 73, 1254
Gibson, S. 2014, IAUS 300, this issue
Gopalswamy, N., Mäkelä, P., Akiyama, S., *et al.* 2013, *Solar Phys.*, 284, 17
Gosling, J. T. 1990, In Physics of magnetic flux ropes, American Geophysical Union, 58, 343
Gosling, J. T., Bame, S. J., McComas, D. J., *et al.* 1995, *Space Sci. Revs*, 72, 133
Green, L. 2014, IAUS 300, this issue
Gulisano, A. M., Dasso, S., Mandrini, C. H., & Démoulin, P. 2007, *Adv. Spa. Res.*, 40, 1881
Gulisano, A. M., Démoulin, P., Dasso, S., & Rodriguez, L. 2012, *A&A*, 543, A107

Gulisano, A. M., Démoulin, P., Dasso, S., Ruiz, M. E., & Marsch, E. 2010, *A&A*, 509, A39
Harrison, R. A., Davies, J. A., Rouillard, A. P., *et al.* 2009, *Solar Phys.*, 256, 219
Hidalgo, M. A. 2011, *JGR*, 116, 2101
Howard, T. A. 2011, *JASTP*, 73, 1242
Hu, Q. & Qiu, J. 2014, IAUS 300, this issue
Isavnin, A., Kilpua, E. K. J., & Koskinen, H. E. J. 2011, *Solar Phys.*, 273, 205
Janvier, M., Aulanier, G., Pariat, E., & Démoulin, P. 2013a, *A&A*, 555, A77
Janvier, M., Démoulin, P., & Dasso, S. 2013b, *A&A*, 556, A50
Janvier, M., Démoulin, P., & Dasso, S. 2014, IAUS 300, this issue
Jian, L., Russell, C. T., Luhmann, J. G., & Skoug, R. M. 2006, *Solar Phys.*, 239, 393
Kilpua, E. K. J., Jian, L. K., Li, Y., & et al. 2011, *JASTP*, 73, 1228
Kilpua, E. K. J., Jian, L. K., Li, Y., Luhmann, J. G., & Russell, C. T. 2012, *Solar Phys.*, 56
Kim, R.-S., Gopalswamy, N., Cho, K.-S., Moon, Y.-J., & Yashiro, S. 2013, *Solar Phys.*, 284, 77
Leitner, M., Farrugia, C. J., Möstl, C., *et al.* 2007, *JGR*, 112, A06113
Lepping, R. P., Burlaga, L. F., & Jones, J. A. 1990, *JGR*, 95, 11957
Lepping, R. P. & Wu, C. C. 2010, *Ann. Geophys.*, 28, 1539
Lepri, S. 2014, IAUS 300, this issue
Liu, Y., Richardson, J. D., & Belcher, J. W. 2005, *Planet. Spa. Sci.*, 53, 3
Lugaz, N. 2014, IAUS 300, this issue
Lugaz, N., Downs, C., Shibata, K., *et al.* 2011, *ApJ*, 738, 127
Lugaz, N., Kintner, P., Möstl, C., *et al.* 2012, *Solar Phys.*, 279, 497
Lugaz, N., Manchester, IV, W. B., & Gombosi, T. I. 2005, *ApJ*, 627, 1019
Lugaz, N. & Roussev, I. 2011, *JASTP*, 73, 1187
Lynch, B. J., Zurbuchen, T. H., Fisk, L. A., & Antiochos, S. K. 2003, *JGR*, 108, A01239
Mäkelä, P., Gopalswamy, N., Xie, H., *et al.* 2013, *Solar Phys.*, 284, 59
Marubashi, K., Cho, K.-S., Kim, Y.-H., Park, Y.-D., & Park, S.-H. 2012, *JGR*, 117, 1101
Marubashi, K. & Lepping, R. P. 2007, *Ann. Geophys.*, 25, 2453
Möstl, C., Farrugia, C. J., Temmer, M., *et al.* 2009, *ApJL*, 705, L180
Mulligan, T. & Russell, C. T. 2001, *JGR*, 106, 10581
Nakagawa, T. & Matsuoka, A. 2010, *JGR*, 115, 10113
Nakwacki, M., Dasso, S., Démoulin, P., & et al. 2011, *A&A*, 535, A52
Olmedo, O. & Zhang, J. 2014, IAUS 300, this issue
Owens, M. J., Démoulin, P., Savani, N. P., Lavraud, B., & Ruffenach, A. 2012, *Solar Phys.*, 278, 435
Reeves, K. K. & Golub, L. 2011, *ApJL*, 727, L52
Richardson, I. G. & Cane, H. V. 2010, *Solar Phys.*, 264, 189
Romashets, E. & Vandas, M. 2009, *A&A*, 499, 17
Rouillard, A. P. 2011, *JASTP*, 73, 1201
Ruffenach, A., Lavraud, B., Owens, M. J., *et al.* 2012, *JGR*, 117, A09101
Savani, N. P., Owens, M. J., Rouillard, A. P., & et al. 2010, *ApJL*, 714, L128
Schmieder, B., Démoulin, P., & Aulanier, G. 2013, *Adv. Spa. Res.*, 51, 1967
Shibata, S. 2014, IAUS 300, this issue
Title, A. 2014, IAUS 300, this issue
Török, T. & Kliem, B. 2007, Astronomische Nachrichten, 328, 743
Vandas, M. & Romashets, E. P. 2003, *A&A*, 398, 801
Wang, C., Du, D., & Richardson, J. D. 2005, *JGR*, 110, A10107
Wimmer-Schweingruber, R. F., Crooker, N. U., Balogh, A., *et al.* 2006, *Space Sci. Revs*, 123, 177
Wu, C.-C. & Lepping, R. P. 2011, *Solar Phys.*, 269, 141
Xie, H., Gopalswamy, N., & St. Cyr, O. C. 2013, *Solar Phys.*, 284, 47
Xiong, M., Zheng, H., Wang, Y., & Wang, S. 2006, *JGR*, 111, A08105
Yashiro, S., Gopalswamy, N., Mäkelä, P., & Akiyama, S. 2013, *Solar Phys.*, 284, 5
Zhang, J., Hess, P., & Poomvises, W. 2013, *Solar Phys.*, 284, 89
Zurbuchen, T. H. & Richardson, I. G. 2006, *Space Sci. Revs*, 123, 31

*Nature of Prominences and their role in Space Weather*
*Proceedings IAU Symposium No. 300, 2013*
*B. Schmieder, J.-M. Malherbe & S. T. Wu, eds.*

doi:10.1017/S174392131301106X

# Complex Evolution of Coronal Mass Ejections in the Inner Heliosphere as Revealed by Numerical Simulations and STEREO Observations: A Review

**Noé Lugaz[1], Charles J. Farrugia[1] and Nada Al-Haddad[2]**

[1]Space Science Center and Department of Physics, University of New Hampshire, Morse Hall, 8 College Rd, Durham, NH, 03824, USA
email: noe.lugaz@unh.edu

[2]Center for Plasma Astrophysics, KU Leuven, Leuven, BE

**Abstract.** The transit of coronal mass ejections (CMEs) from the Sun to 1 AU lasts on average one to five days. As they propagate, CMEs interact with the solar wind and preceding eruptions, which modify their properties. In the past ten years, the evolution of CMEs in the inner heliosphere has been investigated with the help of numerical simulations, through the analysis of remote-sensing heliospheric observations, especially with the SECCHI suite onboard STEREO, and through the analysis of multi-spacecraft *in situ* measurements. Most studies have focused on understanding the characteristics of the magnetic flux rope thought to form the core of the CME. Here, we first review recent work related to CME propagation in the heliosphere, which point towards the need to develop more complex models to analyze CME observations. In the second part of this article, we review some recent studies of CME-CME interaction, which also illustrate the complexity of phenomena occurring in the inner heliosphere.

**Keywords.** Sun: coronal mass ejections (CMEs), solar-terrestrial relations, methods: numerical, methods: data analysis

## 1. Introduction

The heliospheric propagation of coronal mass ejections (CMEs) has historically been one of the "poor children" of space physics research as it appears to lack the fundamental physical questions at the core of the research about CME initiation, coronal heating, solar wind acceleration or geomagnetic storms and substorms. Early work in the 1970s and 1980s took advantage of the presence of multiple spacecraft measuring the solar wind at different heliocentric distances in the inner heliosphere. However, after the mid-1980s and the end of the Helios missions, most *in situ* measurements were made from 1 AU or beyond. In the meantime, coronagraphic observations by SMM and SOHO have been providing remote-sensing views of CMEs up to distances of at most 0.15 AU, leaving the large majority of the inner heliosphere without direct remote-sensing observations or *in situ* measurements. In the past ten years, CMEs have been remotely imaged in the inner heliosphere by SMEI and the Heliospheric Imagers (HIs) onboard STEREO (Davies *et al.* 2009; Howard *et al.* 2013). In addition, multiple, simultaneous *in situ* measurements of CMEs at the same heliocentric distance but at different longitudes have been made possible by the presence of the two STEREO spacecraft and ACE and Wind at the L1 point. Other observations (interplanetary scintillation –IPS–, type II and type III radio bursts, etc.) can also be useful to obtain additional information about CMEs and CME-driven shocks.

Most analyses of CMEs in the heliosphere rely on the concept of twisted magnetic flux ropes (TMFRs) to understand and describe their properties. In this way, one often refers to the CME width or radius, the orientation of its axis, etc. In fact, TMFR models of CMEs have been extremely successful to study CME properties in the corona (e.g., Chen 1996; Thernisien *et al.* 2006), and in the heliosphere (e.g., Lepping *et al.* 1990; Wood *et al.* 2009). They are also the foundation of many numerical and theoretical studies, where the TMFR is either initiated out of equilibrium (Manchester *et al.* 2004b), emerged from below the photosphere (Archontis *et al.* 2004), perturbed (Titov & Démoulin 1999; Amari *et al.* 1999) or formed and destabilized (Linker & Mikic 1995; Antiochos *et al.* 1999).

In addition to isolated CMEs, further complexity of heliospheric transients comes from the interaction of successive CMEs (Burlaga *et al.* 2003; Lugaz *et al.* 2012). In fact, there are a number of fundamental processes which need to be better understood: the deflection of a CME by another, the momentum exchange between colliding CMEs and the fate of shock waves as they propagate inside magnetic ejecta. In this article, we first review observations and modeling of CMEs in the heliosphere, with an emphasis on the need for improved models of TMFRs. Next, we summarize recent progresses in our understanding of CME-CME interaction obtained by combining numerical simulations and the analysis of heliospheric observations and measurements. We conclude with a short discussion of further progresses expected from future missions.

## 2. What we really know about CMEs and how we know it

### 2.1. *In Situ Measurements*

From the analyses of *in situ* measurements of CMEs at multiple points came the definition of magnetic clouds, magnetically dominated structures characterized by an enhanced and smoothly varying magnetic field, low proton $\beta$ and temperature as a subset of magnetic ejecta observed at 1 AU (Burlaga *et al.* 1981). To explain the smooth rotation of the magnetic field, Goldstein (1983) and Marubashi (1986) proposed that magnetic clouds can be described as twisted magnetic flux ropes satisfying the force-free condition. Burlaga (1988) showed that a linear force-free model such as that described by Lundquist (1950) was a good approximation to the *in situ* data. This resulted in the development of force-free fitting techniques with a twisted flux rope (Lepping *et al.* 1990), which is still one of the most commonly used techniques to analyze *in situ* observations of CMEs to date. Improvements upon this technique comprise the inclusion of the expansion of the magnetic cloud (Farrugia *et al.* 1993; Shimazu & Vandas 2002) and the possibility of taking into consideration a curved axis (the torus models of Romashets & Vandas 2003; Marubashi & Lepping 2007; Janvier et al. 2013; Janvier 2014). Some techniques have also been developed to incorporate the effect of reconnection of the magnetic cloud on its way to Earth (Dasso et al. 2006; Ruffenach *et al.* 2012).

Other models that dispense with the force-free approximation are the Grad-Shafranov reconstruction (Hu & Sonnerup 2001) based on the magneto-hydrostatic approximation, elliptical non-force-free models (e.g., see Hidalgo *et al.* 2002) and other non-cylindrical models (Mulligan & Russell 2001; Owens *et al.* 2012, among others). Common to all these methods is the assumption of near-invariance along the reconstructed magnetic cloud axis, (2.5-D approximation). Recently, using a numerical simulation of a CME with a field rotation dominated by writhe, Al-Haddad *et al.* (2011) showed that this approximation, which is central to all frequently used models to reconstruct CMEs, could predict a twisted flux rope even when it is not present. Only spheroidal models (Gosling 1990; Vandas *et al.* 1993), proposed in the early 1990s, dispense with the assumption; however, they have been all but abandoned by the large majority of researchers. It should be kept

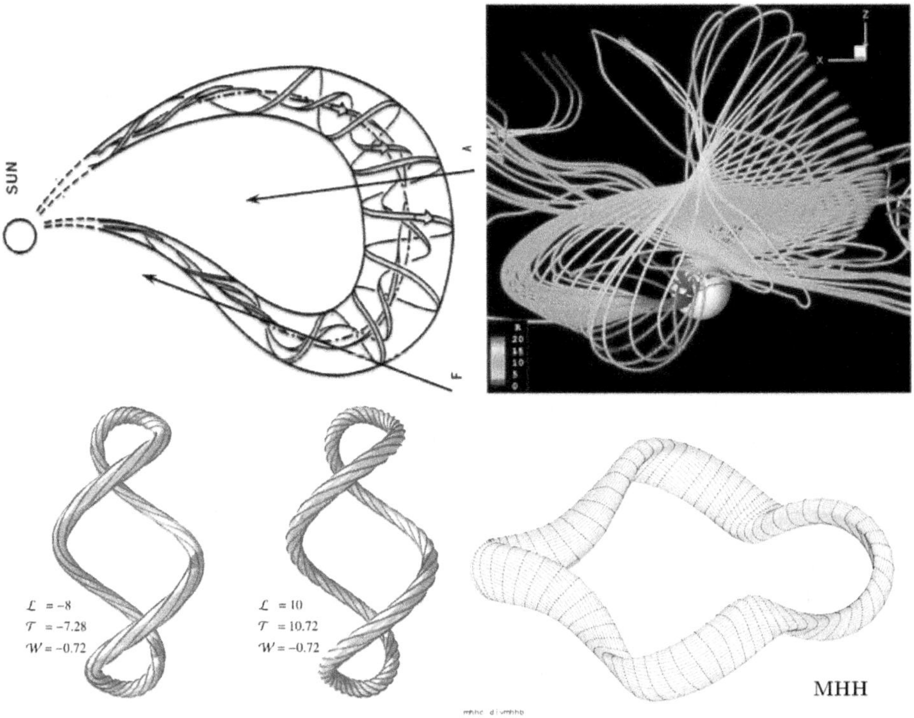


**Figure 1.** Four different "models" of CMEs. *Top left*: TMFR (classical picture with a bent axis). *Top right*: 3-D simulated rotating magnetic field without significant twist. *Bottom*: Writhed and twisted fields, which cannot yet be analyzed by current reconstruction and fitting models (right from stellarator).

in mind that reconstruction of CMEs from *in situ* measurements at 1 AU is intrinsically an ill-posed problem (there is no unique solution) even under the assumption that CMEs do not evolve as they pass over the spacecraft. This is because it is necessary to solve a set of 3-D differential equations with only a 1-D boundary condition (obtained from the time series at a single point).

Thanks to the launch of the STEREO spacecraft in 2006, it is, for the first time, possible to routinely and simultaneously probe the same CME at different locations but at almost the same heliocentric distance (within ± 10% of 1 AU). Farrugia *et al.* (2011) analyzed one such event observed by *Wind* and the two STEREO spacecraft in November 2007 and found very different orientations of the CME axis depending on which measurements were used. Other multi-spacecraft measurements also resulted in different reconstructed orientations from different spacecraft, but they may be reconciled by considering that one of the two spacecraft measured the "leg" of a flux-rope CME (Möstl *et al.* 2012; Nieves-Chinchilla *et al.* 2012), that the global morphology of CMEs is not cylindrically symmetric (Mulligan & Russell 2001), or by using different techniques (see Démoulin 2014, for another interpretation of the November 2007 CME). Lastly, there has been a number of studies that found that the amount of twist is nearly uniform throughout the CME, in contrast to the Lundquist model (Hu & Sonnerup 2001; Kahler *et al.* 2011).

Figure 1 shows the classical picture of a CME used for *in situ* reconstruction (and also for fitting remote-sensing observations, see next section) as drawn by Marubashi & Lepping (2007) and inspired by previous works by Burlaga *et al.*, the numerical simulation of Jacobs *et al.* (2009) without an axial invariance and two examples of twisted and writhed field from Berger & Prior (2006) and from the stellarator in nuclear fusion (Lyon *et al.* 1997), which might be more appropriate models for CMEs measured *in situ*. Burlaga *et al.* (2002), when analyzing a complex ejecta resulting from the merging of multiple CMEs drew an intricate picture of the magnetic field inside CMEs, and drew a parallel to a DNA molecule (their Figures 7 and 8), an example of a structure incorporating twist and writhe.

### 2.2. *Remote-sensing observations*

Coronagraphic observations of CMEs were first made in the 1970s (MacQueen *et al.* 1974; Gosling *et al.* 1975); following comparisons with *in-situ* measurements, the concept of CMEs as flux ropes was developed and it is currently the accepted paradigm. TMFRs are also a result of flare reconnection during all proposed CME initiation models.

Chen (1996) developed an analytical model of a CME as a flux rope driven by the $\mathbf{j}\times\mathbf{B}$ force. It has been used to understand their evolution in the corona and, recently extended into the heliosphere (Kunkel & Chen 2010). In addition to this model, visual fitting of CMEs in coronagraph and heliospheric imagers images with a flux rope shape has been performed by a number of researchers (Thernisien *et al.* 2006; Wood *et al.* 2009). Other models to analyze remote-sensing observations of CMEs, which do not assume a flux rope shape, include direct triangulation (Liu *et al.* 2010) and methods based on the idea of a circular or elliptical cross-section of the CME front (Byrne *et al.* 2009; Lugaz *et al.* 2009a; Davies *et al.* 2012). The most advanced of these models fit observations using concave or convex segments of circles (Tappin & Howard 2009). As for *in situ* measurements, these reconstructions are not unique, even for a given model. This is because a visual fitting of 6–8 parameters representing the CME is attempted using only 2 or 3 images. For example, CME rotation and over-expansion may be mistaken for one another (as pointed out by Poomvises *et al.* 2010, for the 2008 March 25 CME). Differences between different analysis techniques can be even more significant.

While not providing heliospheric observations, SDO has been able to reveal the formation of the possible TMFR in the low corona (e.g., Cheng *et al.* 2011; Su *et al.* 2012). In SDO images (see for example Zhang *et al.* 2012), the TMFR may already be strongly distorted by 1.3 $R_\odot$ to a point which is captured neither by models used to analyze coronagraphic images or *in situ* measurements, nor by most numerical simulations.

Recent progress has been made thanks to SECCHI observations and, particularly with a new post-processing developed by DeForest *et al.* (2011). There has been a few observations of CME distortion (Savani *et al.* 2010), which is expected to be due to the interaction with the bimodal solar wind and has been inferred from *in situ* measurements (Owens *et al.* 2006) and predicted from numerical simulations (Riley *et al.* 2002; Manchester *et al.* 2004b). CMEs are often observed in heliospheric imagers bounded by two density enhancements, probably associated with their expansion. Taking advantage of these observations, it is possible to determine the radial and temporal evolution of the CME width, its expansion and aspect ratio (Savani *et al.* 2009; Nieves-Chinchilla *et al.* 2012; Lugaz *et al.* 2012). The CME expansion has been found to be proportional to $\mathrm{r}^{0.5-0.9}$, in agreement with statistical studies based on measurements of different CMEs at different distances (Bothmer & Schwenn 1998; Liu *et al.* 2005) and also with theoretical studies (Démoulinet al. 2008; Gulisano *et al.* 2010). In a series of recent works, Howard & DeForest (2012) and DeForest *et al.* (2013) have been able to directly link remote-sensing and *in situ* measurements, to confirm the identification of the magnetic ejecta as the dark cavity in corona graphic and HI images, to determine the CME mass increase during its heliospheric propagation (found to be by a factor of $\sim 2$ as found in numerical simulations Lugaz *et al.* 2005b) and to identify the different sources of material for the CME sheath region, in part made of coronal material and in part of swept-up

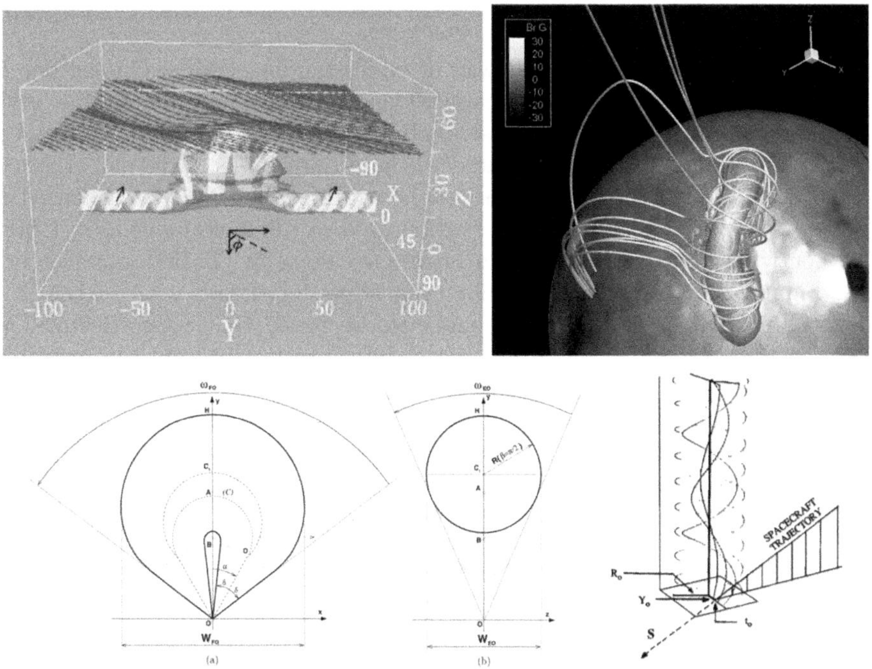


**Figure 2.** *Top*: Distorted TMFRs from numerical simulations from Galsgaard *et al.* (2005) and Lugaz *et al.* (2011). *Bottom*: TMFR models used to analyze remote-sensing and *in-situ* measurements from Thernisien *et al.* (2006) and Lepping *et al.* (1990).

solar wind material. Lastly, an area of active research, due in part to its importance for space weather forecasting, is the deceleration of CMEs as they propagate due to their interaction with the solar wind (e.g., see Tappin 2006; Vršnak *et al.* 2010).

### 2.3. *Numerical modeling*

Flux ropes initiated at the Sun in numerical models are able to "reproduce" typical *in situ* measurements of magnetic clouds (Riley *et al.* 2002; Manchester *et al.* 2004b; Chané *et al.* 2006; Shen *et al.* 2011). These models used an out-of-equilibrium flux rope (Titov & Démoulin 1999; Roussev *et al.* 2003) or flux-rope like structures (Gibson & Low 2000; Manchester *et al.* 2004b). Numerical simulations have been primarily used to study the initiation of CMEs. In this case, the models include, at least, the study of the loss of equilibrium of the TMFR. Some models start from a TMFR initially in equilibrium and destabilize it with boundary motions, which represent either flux cancellation or flux emergence (Linker & Mikic 1995; Török *et al.* 2011), while in other models it emerge from below the photosphere (Archontis *et al.* 2004; Manchester *et al.* 2004a; Fan & Gibson 2004; Roussev *et al.* 2012). In all these models, the TMFR is initiated with a non-distorted axis, although it may kink or get distorted during the eruption phase (Lugaz *et al.* 2011; Török *et al.* 2004) (see top panels of Figure 2). It is one important reason to initiate simulations in the lower corona or even lower down to reproduce as accurately as possible the early phase of eruption and the associated distortion. Other models attempt to create the TMFR from boundary motions (Amari *et al.* 1996; Lynch *et al.* 2008; Aulanier *et al.* 2010; Zuccarello *et al.* 2012) and study the further destabilization and eruption of the TMFR. In Lugaz & Roussev (2011), we gave an overview of comparison of numerical simulations with heliospheric images and the interested reader is invited to refer to that review for additional information.

### 2.4. *Summary*

There is an obvious dichotomy in our way of understanding CMEs. On the one hand, numerical simulations and the latest observations by SDO, clearly show that TMFRs, when they are present, may be already strongly distorted and kinked in the low corona (see top panels of Figure 2) to a point where it is hard to relate local and global properties such as the orientation of the CME "axis" or the width of the CME. On the other hand, most models to analyze heliospheric observations and *in situ* measurements of CMEs are based on non-writhed structures with an invariance along the TMFR axis (see bottom panels of Figure 2). Most models do not even allow for deformation of the TMFRs beyond an elliptical cross-section. It should also be noted that properties of TMFRs reconstructed from coronal observations often differ drastically from the properties of flux ropes reconstructed from *in situ* measurements (Yurchyshyn *et al.* 2001; Isavnin *et al.* 2013). A better understanding of CMEs in the heliosphere may require a shift in paradigm by acknowledging the complexity of their evolution and the development of models combining remote-sensing observations and *in situ* measurements, with input from modeling, in order to reconcile local and global properties of CMEs.

## 3. CME-CME Interaction

Another source of complexity in the heliosphere is the interaction of successive CMEs. With an average of more than three coronal mass ejections (CMEs) per day near solar maximum, CME-CME interaction should occur regularly in the inner heliosphere. Some of the earliest reports of likely CME-CME interaction were associated with the series of events in early August 1972 (Ivanov 1982), as well as multi-sapcecraft measurements of series of CMEs during the Helios era (Burlaga *et al.* 1987). With the improvement of coronagraph observations, and specifically the large field-of-view of LASCO/C3 covering distances up to 32 $R_\odot$, the 1990s witnessed the first direct observations of CME-CME interaction (Gopalswamy *et al.* 2001) and further confirmations that they are associated with complex ejecta or compound streams at 1 AU (Burlaga *et al.* 2003; Wang *et al.* 2002). It has also been proposed that some seemingly isolated CMEs measured *in situ* may in fact result from the interaction of multiple CMEs on their way to Earth (Dasso *et al.* 2009). The past six years have seen a similar increase in detection of CME-CME interaction thanks to the wide field-of-view of the HIs onboard STEREO (e.g., see Rouillard 2010). Recent HI observations of CME-CME interaction include the 2008 November CMEs (Shen *et al.* 2012), the 2010 May CMEs (Lugaz *et al.* 2012) and the 2010 August CMEs (Harrison *et al.* 2012; Temmer *et al.* 2012). Analysis of these events often combine remote-sensing observations with *in situ* measurements and sometimes numerical simulations (Lugaz *et al.* 2009b; Webb *et al.* 2009, 2013).

Combining numerical simulations, theoretical works and the analysis of *in situ* and remote-sensing observations, a number of studies have investigated the complex physical processes occurring during instances of CME-CME interaction. The change in speed during the collisions of two CMEs had been previously investigated via numerical simulations (Lugaz *et al.* 2005a; Xiong *et al.* 2006). Recently, it has also been investigated by analyzing SECCHI observations, which revealed unexpected behavior of either deceleration of the overtaking CME below the speed predicted from a perfectly inelastic collision (Temmer *et al.* 2012) or acceleration of the overtaken CME beyond the speed predicted from a perfectly elastic collision (super-elastic, Lugaz *et al.* 2009a; Shen *et al.* 2012). This latter phenomenon was then further identified in another numerical simulation (Shen *et al.* 2013). Recent observations (Lugaz *et al.* 2012) have also confirmed the numerical results of the deflection of one CME by another as they collide (Schmidt &

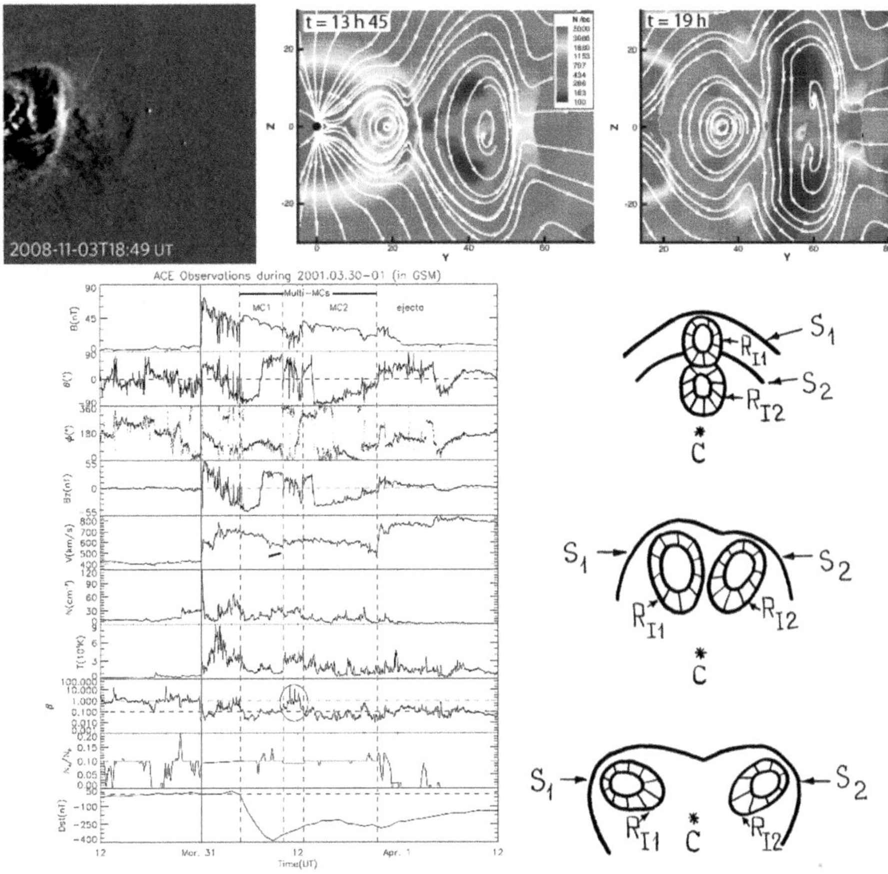


**Figure 3.** Different manifestations or studies of CME-CME interaction. From top left to bottom right: HI observations (Shen *et al.* 2012), MHD simulation (Lugaz *et al.* 2005a), in-situ measurements (Wang *et al.* 2003) and cartoon representation from the early 1980s (Ivanov 1982).

Cargill 2004; Xiong *et al.* 2009) and the potential over-expansion of the overtaken CME after its compression (Gulisano *et al.* 2010, 2012). The potential merging of successive shock waves was already recognized by Ivanov (1982) (see Figure 3) and has been detected from multiple *in situ* measurements at different heliocentric distances (Farrugia & Berdichevsky 2004) and studied via numerical simulations (Lugaz *et al.* 2005a). The effect of successive and interacting CMEs on the acceleration of particles (e.g., see Li *et al.* 2012) and the potential for sympathetic eruptions (Török *et al.* 2011; Schrijver & Title 2011) are also two active areas of research associated with CME-CME interaction but go beyond the scope of this review. Figure 3 illustrates some manifestations and studies of CME-CME interaction.

Overall, a complex but consistent picture of the interaction of two CMEs is starting to emerge. Analyses must move beyond the simple kinematic model to incorporate the CME expansion, the effect of the shock waves as well as the internal magnetic pressure and magnetic tension in the CMEs. Interaction between two CMEs typically takes 6-12 hours and can result in strong compression of the overtaken CME and an increase in its internal magnetic field. If the collision is head-on and the relative orientation of the two CMEs do not allow for full reconnection, this instance of CME-CME interaction is likely to result in a multiple-magnetic cloud event (Wang *et al.* 2002). However, if the collision further results in the deflection of one CME or in a strong reconnection between the two CMEs (akin to full "cannibalism"), the first CME, in a state of enhanced magnetic pressure, is bound to over-expand. We speculate that the enhanced magnetic pressure inside the CME may be converted into kinetic energy through this over-expansion, resulting in a higher speed than expected. Such an event may also be mistaken for an isolated CME at 1 AU (Dasso *et al.* 2009; Lugaz *et al.* 2012). Lastly, the interaction of more than two CMEs in a short span can destroy the regularity of the magnetic field in magnetic clouds resulting in complex ejecta (Burlaga *et al.* 2003). This general picture must be confirmed by performing more dedicated studies of CME-CME interaction as well as series of parametric numerical investigation to evaluate the effect of the CME relative speed, direction and orientation on the interaction process.

## 4. Conclusion

We have reviewed some recent works related to the heliospheric propagation of CMEs, focusing first on what they teach us about the flux rope nature of CMEs. We have emphasized that TMFRs are a sufficient model to understand CMEs, but there is no indication that it is necessary. In addition, and more importantly, most TMFR models used to analyze observations appear overly simplistic as compared to observations, which often reveal bends and writhe as well as a varying cross-section shape along the CME "axis". SECCHI observations are able to reveal the density structure of the magnetic ejecta and the CME sheath, and this global view of the CME should be incorporated into models in order to analyze the local properties of CMEs as provided by *in situ* measurements. We have also summarized some recent investigations of CME-CME interaction. For the first time, remote-sensing observations can complement *in situ* measurements and numerical simulations to shed light on the acceleration and deceleration of CMEs following their collision and the changes in the expansion of CMEs. With planned missions such as Solar Probe+ and Solar Orbiter (Müller *et al.* 2013) providing *in situ* measurements in the inner heliosphere and upper corona, the next decade should allow us to better study the evolution of CMEs as they propagate, including their interaction, especially if heliospheric remote-sensing imaging is available. Other future missions should include remote-sensing observations from out of the ecliptic plane (as will be provided, partially, by Solar Orbiter). Coronagraphs and heliospheric images from a solar polar orbit may allow us to better understand the longitudinal properties of CMEs: their extent, deflection, the extent of the CME-driven shocks and the shock stand-off distance among others. By combining these observations and measurements and numerical simulations, we should be able to more fully understand the richness and complexity of the heliospheric evolution of CMEs.

### Acknowledgements

We would like to thank S. Poedts, K. Galsgaard and P. Démoulin for useful discussions regarding the nature of CMEs. N. L. would like to thank B. Schmieder and E. Pariat for the great symposium in Paris. The work for this manuscript was supported by the grants AGS1239699, AGS1239704 and NNX13AH94G, as well as the STEREO grant to UNH.

### References

Al-Haddad, N., Roussev, I. I., Möstl, C., *et al.* 2011, *Astrophys. Journ. Lett.*, 738, L18
Amari, T., Luciani, J. F., Aly, J. J., & Tagger, M. 1996, *Astron. Astrophys.*, 306, 913
Amari, T., Luciani, J. F., Mikic, Z., & Linker, J. 1999, *Astrophys. Journ. Lett.*, 518, L57
Antiochos, S. K., DeVore, C. R., & Klimchuk, J. A. 1999, *Astrophys. J.*, 510, 485

Archontis, V., *et al.* 2004, *Astron. Astrophys.*, 426, 1047
Aulanier, G., Török, T., Démoulin, P., & DeLuca, E. E. 2010, *Astrophys. J.*, 708, 314
Berger, M. A. & Prior, C. 2006, *Journal of Physics A Mathematical General*, 39, 8321
Bothmer, V. & Schwenn, R. 1998, *Annales Geophysicae*, 16, 1
Burlaga, L. & Berdichevsky, D., *et al.* 2003, *J. Geophys. Res.*, 108, 2
Burlaga, L., Sittler, E., Mariani, F., & Schwenn, R. 1981, *J. Geophys. Res.*, 86, 6673
Burlaga, L. F. 1988, *J. Geophys. Res.*, 93, 7217
Burlaga, L. F., Behannon, K. W., & Klein, L. W. 1987, *J. Geophys. Res.*, 92, 5725
Burlaga, L. F., & Plunkett, S. P., St. Cyr, O. C. 2002, *J. Geophys. Res.*, 107, 1
Byrne, J. P. & Gallagher, P. T., *et al.* 2009, *Astron. Astrophys.*, 495, 325
Chané, E. & van der Holst, B., *et al.* 2006, *Astron. Astrophys.*, 447, 727
Chen, J. 1996, *J. Geophys. Res.*, 101, 27499
Cheng, X., Zhang, J., Liu, Y., & Ding, M. D. 2011, *Astrophys. Journ. Lett.*, 732, L25
Dasso, S., Mandrini, C. H., Démoulin, P., & Luoni, M. L. 2006, *Astron. Astrophys.*, 455, 349
Dasso, S., Mandrini, C. H., Schmieder, B., *et al.* 2009, *J. Geophys. Res.*, 114, 2109
Davies, J. A., Harrison, R. A., Rouillard, A. P., *et al.* 2009, *Geophys. Res. Lett.*, 36, L02102
Davies, J. A., Harrison, R. A., Perry, C. H., *et al.* 2012, *Astrophys. J.*, 750, 23
Démoulin, P., Nakwacki, M. S., Dasso, S., & Mandrini, C. H. 2008, *Solar Phys.*, 250, 347
Démoulin, P. *this issue*
DeForest, C. E., Howard, T. A., & McComas, D. J. 2013, *Astrophys. J.*, 769, 43
DeForest, C. E., Howard, T. A., & Tappin, S. J. 2011, *Astrophys. J.*, 738, 103
Fan, Y. & Gibson, S. E. 2004, *Astrophys. J.*, 609, 1123
Farrugia, C. & Berdichevsky, D. 2004, *Annales Geophysicae*, 22, 3679
Farrugia, C. J., Burlaga, L. F., Osherovich, V. A., *et al.* 1993, *J. Geophys. Res.*, 98, 7621
Farrugia, C. J., Berdichevsky, D. B., Möstl, C., *et al.* 2011, *J. Atmos. Solar-Terr. Phys.*, 73, 1254
Galsgaard, K., Moreno-Insertis, F., *et al.* 2005, *Astrophys. Journ. Lett.*, 618, L153
Gibson, S. E. & Low, B. C. 2000, *J. Geophys. Res.*, 105, 18187
Goldstein, H. 1983, *in NASA Conference Publication*, Vol. 228, 731–733
Gopalswamy, N., Yashiro, S., *et al.* 2001, *Astrophys. Journ. Lett.*, 548, L91
Gosling, J. T. 1990, *AGU Geophysical Monograph Series*, 58, 343
Gosling, J. T., Hildner, E., MacQueen, R. M., *et al.* 1975, *Solar Phys.*, 40, 439
Gulisano, A. M., Démoulin, P., Dasso, S., & Rodriguez, L. 2012, *Astron. Astrophys.*, 543, A107
Gulisano, A. M., Démoulin, P., *et al.* 2010, *Astron. Astrophys.*, 509, A39
Harrison, R. A., Davies, J. A., Möstl, C., *et al.* 2012, *Astrophys. J.*, 750, 45
Hidalgo, M. A., Nieves-Chinchilla, T., & Cid, C. 2002, *Geophys. Res. Lett.*, 29, 130000
Howard, T. A. & DeForest, C. E. 2012, *Astrophys. J.*, 746, 64
Howard, T. A., Bisi, M. M., Buffington, A., *et al.* 2013, Space Sci. Rev.
Hu, Q. & Sonnerup, B. U. Ö. 2001, *Geophys. Res. Lett.*, 28, 467
Isavnin, A., Vourlidas, A., & Kilpua, E. K. J. 2013, Solar Phys.
Ivanov, K. G. 1982, *Space Sci. Rev.*, 32, 49
Jacobs, C., Roussev, I. I., Lugaz, N., & Poedts, S. 2009, *Astrophys. Journ. Lett.*, 695, L171
Janvier, M., Démoulin, P., & Dasso, S. 2013, *Astron. Astrophys.*, 556, A50
Janvier, M. *this issue*
Kahler, S. W., Krucker, S., & Szabo, A. 2011, *J. Geophys. Res.*, 116, 1104
Kunkel, V. & Chen, J. 2010, *Astrophys. Journ. Lett.*, 715, L80
Lepping, R. P., Burlaga, L. F., & Jones, J. A. 1990, *J. Geophys. Res.*, 95, 11957
Li, G., Moore, R., Mewaldt, R. A., Zhao, L., & Labrador, A. W. 2012, *Space Sci. Rev.*, 171, 141
Linker, J. A. & Mikic, Z. 1995, *Astrophys. Journ. Lett.*, 438, L45
Liu, Y., Davies, J. A., Luhmann, J. G., *et al.* 2010, *Astrophys. Journ. Lett.*, 710, L82
Liu, Y., Richardson, J. D., & Belcher, J. W. 2005, *Planet. Space Sci.*, 53, 3
Lugaz, N., Downs, C., Shibata, K., *et al.* 2011, *Astrophys. J.*, 738, 127
Lugaz, N., Farrugia, C. J., Davies, J. A., *et al.* 2012, *Astrophys. J.*, 759, 68
Lugaz, N., Manchester, W. B., & Gombosi, T. I. 2005a, *Astrophys. J.*, 634, 651

—. 2005b, *Astrophys. J.*, 627, 1019
Lugaz, N. & Roussev, I. I. 2011, *J. Atmos. Solar-Terr. Phys.*, 73, 1187
Lugaz, N., Vourlidas, A., & Roussev, I. I. 2009a, *Annales Geophysicae*, 27, 3479
Lugaz, N., Vourlidas, A., Roussev, I. I., & Morgan, H. 2009b, *Solar Phys.*, 256, 269
Lundquist, S. 1950, *Ark. Fys.*, 2, 361
Lynch, B. J., Antiochos, S. K., *et al.* 2008, *Astrophys. J.*, 683, 1192
Lyon, J. F., Rome, J. A., *et al.* 1997, Chapter 2: Final Report of Stellarator Power Plant Study
MacQueen, R. M., Eddy, J. A., Gosling, J. T., *et al.* 1974, *Astrophys. Journ. Lett.*, 187, L85+
Manchester, W. B., Gombosi, T., DeZeeuw, D., & Fan, Y. 2004a, *Astrophys. J.*, 610, 588
Manchester, W. B., Gombosi, T. I., Roussev, I., *et al.* 2004b, *J. Geophys. Res.*, 109, 2107
Marubashi, K. 1986, *Adv. Space Res.*, 6, 335
Marubashi, K. & Lepping, R. P. 2007, *Annales Geophysicae*, 25, 2453
Möstl, C., Farrugia, C. J., Kilpua, E. K. J., *et al.* 2012, *Astrophys. J.*, 758, 10
Müller, D., Marsden, R. G., St. Cyr, O. C., & Gilbert, H. R. 2013, *Solar Phys.*, 285, 25
Mulligan, T. & Russell, C. T. 2001, *J. Geophys. Res.*, 106, 10581
Nieves-Chinchilla, T., Colaninno, R., Vourlidas, A., *et al.* 2012, *J. Geophys. Res.*, 117, 6106
Owens, M. J., Démoulin, P., *et al.* 2012, *Solar Phys.*, 278, 435
Owens, M. J., Merkin, V. G., & Riley, P. 2006, *J. Geophys. Res.*, 111, 3104
Poomvises, W., Zhang, J., & Olmedo, O. 2010, *Astrophys. Journ. Lett.*, 717, L159
Riley, P., Linker, J. A., Mikić, Z., *et al.* 2002, *Astrophys. J.*, 578, 972
Romashets, E. P. & Vandas, M. 2003, *Geophys. Res. Lett.*, 30, 200000
Rouillard, A. P. 2010, *J. Atmos. Solar-Terr. Phys.*
Roussev, I. I., Forbes, T. G., Gombosi, T. I., *et al.* 2003, *Astrophys. Journ. Lett.*, 588, L45
Roussev, I. I., Galsgaard, K., Downs, C., *et al.* 2012, *Nature Physics*, 8, 845
Ruffenach, A., Lavraud, B., Owens, M. J., *et al.* 2012, *J. Geophys. Res.*, 117, 9101
Savani, N. P., Owens, M. J., *et al.* 2010, *Astrophys. Journ. Lett.*, 714, L128
Savani, N. P., Rouillard, A. P., Davies, J. A., *et al.* 2009, *Annales Geophysicae*, 27, 4349
Schmidt, J. & Cargill, P. 2004, *Annales Geophysicae*, 22, 2245
Schrijver, C. J. & Title, A. M. 2011, *J. Geophys. Res.*, 116, 4108
Shen, C., Wang, Y., Wang, S., *et al.* 2012, Nature Physics, 8, 923
Shen, F., Feng, X. S., *et al.* 2011, *J. Geophys. Res.*, 116, A04102
Shen, F., Shen, C., Wang, Y., Feng, X., & Xiang, C. 2013, *Geophys. Res. Lett.*, 40, 1457
Shimazu, H. & Vandas, M. 2002, *Earth, Planets, and Space*, 54, 783
Su, Y., Dennis, B. R., Holman, G. D., *et al.* 2012, *Astrophys. Journ. Lett.*, 746, L5
Tappin, S. J. 2006, *Solar Phys.*, 233, 233-248
Tappin, S. J. & Howard, T. A. 2009, *Space Sci. Rev.*, 147, 55
Temmer, M., Vršnak, B., Rollett, T., *et al.* 2012, *Astrophys. J.*, 749, 57
Thernisien, A. F. R., Howard, R. A., & Vourlidas, A. 2006, *Astrophys. J.*, 652, 763
Titov, V. S. & Démoulin, P. 1999, *Astron. Astrophys.*, 351, 707
Török, T., Kliem, B., & Titov, V. S. 2004, *Astron. Astrophys.*, 413, L27
Török, T., Panasenco, O., Titov, V. S., *et al.* 2011, *Astrophys. Journ. Lett.*, 739, L63
Vršnak, B., Žic, T., *et al.* , *Astron. Astrophys.*, 512, A43
Vandas, M., Fischer, S., Pelant, P., & Geranios, A. 1993, *J. Geophys. Res.*, 98, 11467
Wang, Y. M., Wang, S., & Ye, P. Z. 2002, *Solar Phys.*, 211, 333
Wang, Y. M., Ye, P. Z., & Wang, S. 2003, *J. Geophys. Res.*, 108, A10, 1370
Webb, D. F., Howard, T. A., Fry, C. D., *et al.* 2009, *Solar Phys.*, 256, 239
Webb, D. F., Möstl, C., Jackson, B. V., *et al.* 2013, *Solar Phys.*, 285, 317
Wood, B. E., Howard, R. A., Plunkett, S. P., & Socker, D. G. 2009, *Astrophys. J.*, 694, 707
Xiong, M., Zheng, H., & Wang, S. 2009, *J. Geophys. Res.*, 114, A11101
Xiong, M., Zheng, H., Wang, Y., & Wang, S. 2006, *J. Geophys. Res.*, 111, A08105
Yurchyshyn, V. B., Wang, H., Goode, P. R., & Deng, Y. 2001, *Astrophys. J.*, 563, 381
Zhang, J., Cheng, X., & Ding, M.-D. 2012, *Nature Communications*, 3
Zuccarello, F. P., Meliani, Z., & Poedts, S. 2012, *Astrophys. J.*, 758, 117

*Nature of Prominences and their role in Space Weather*
*Proceedings IAU Symposium No. 300, 2013*
*B. Schmieder, J.-M. Malherbe & S. T. Wu, eds.*

doi:10.1017/S1743921313011071

# Flux rope axis geometry of magnetic clouds deduced from in situ data

**Miho Janvier**[1], **Pascal Démoulin**[1] **and Sergio Dasso**[2]

[1]Observatoire de Paris, LESIA, UMR 8109 (CNRS), F-92195 Meudon Principal Cedex, France
email: miho.janvier@obspm.fr, pascal.demoulin@obspm.fr

[2]Departamento de Física e Instituto de Astronomía y Física del Espacio (UBA-CONICET), Facultad de Ciencias Exactas y Naturales, Universidad de Buenos Aires, 1428 Buenos Aires, Argentina
email: dasso@df.uba.ar

**Abstract.** Magnetic clouds (MCs) consist of flux ropes that are ejected from the low solar corona during eruptive flares. Following their ejection, they propagate in the interplanetary medium where they can be detected by in situ instruments and heliospheric imagers onboard spacecraft. Although in situ measurements give a wide range of data, these only depict the nature of the MC along the unidirectional trajectory crossing of a spacecraft. As such, direct 3D measurements of MC characteristics are impossible. From a statistical analysis of a wide range of MCs detected at 1 AU by the *Wind* spacecraft, we propose different methods to deduce the most probable magnetic cloud axis shape. These methods include the comparison of synthetic distributions with observed distributions of the axis orientation, as well as the direct integration of observed probability distribution to deduce the global MC axis shape. The overall shape given by those two methods is then compared with 2D heliospheric images of a propagating MC and we find similar geometrical features.

**Keywords.** Sun: coronal mass ejections (CMEs), Sun: magnetic fields, interplanetary medium

## 1. Introduction

Magnetic clouds (MCs) form a subclass of Interplanetary Coronal Mass Ejections (ICMEs), which are released in the interplanetary medium during eruptive flares (Gosling *et al.* 1990). As they transport a large amount of plasma material and magnetic flux, they are amongst the main drivers of space weather (Gosling 1993; Marubashi 2000). Although an ICME definition depends on measurements and authors (see review of Démoulin 2014), MCs are strictly defined, for example, by an enhanced magnetic field strength and a smooth rotation of the magnetic field direction through a large angle (Dasso *et al.* 2005). This magnetic field rotation indicates the existence of a flux rope (FR) (e.g., Lepping *et al.* 1990; Dasso *et al.* 2006) that is well correlated with observations of eruptive flare configurations (e.g., Zhang *et al.* 2012; Title 2014), as well as with theoretical models reproducing the underlying processes of eruptive flares (e.g., the tether cutting model of Moore *et al.* 1997 or the torus-unstable model of Aulanier *et al.* 2010).

Understanding the structure of MCs is important for several reasons. For example, knowing the characteristics of the magnetic FR can help us to understand the role of the field line length in the time delay of energetic particles detection (see Larson *et al.* 1997 and Masson *et al.* 2012). Similarly, it can help us to link MC structures with the 3D configuration of the associated solar source (see Nakwacki *et al.* 2011), as well as to calculate the budget for the magnetic helicity, magnetic energy and flux (e.g., Dasso *et al.* 2005 and Démoulin *et al.* 2002).

However, deriving the 3D MC structure is not at all straightforward from one sample of data; as the spacecraft is crossing the MC along a unidirectional trajectory, it can

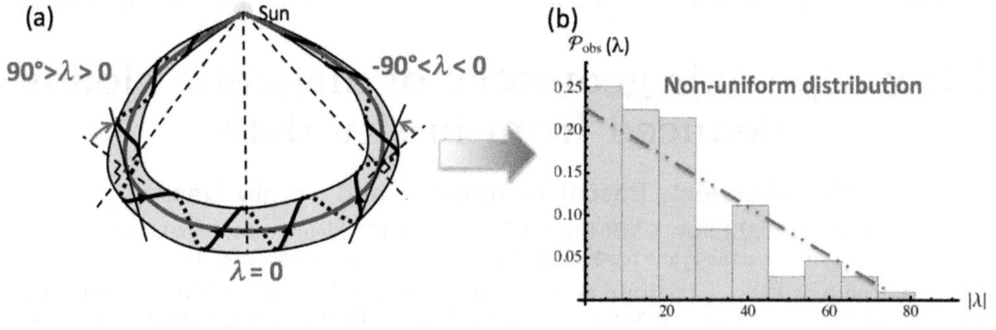


**Figure 1.** (a) Definition of the location angle, $\lambda$, and (b) the related non-uniform distribution from observations of a set of 107 MCs observed by *Wind.*

only locally measure MC characteristics. Then, from 1D data, several different fitting procedures can be applied (see review of Démoulin 2014) but all of them necessitate more or less drastic hypotheses (e.g., Sonnerup *et al.* 2006). All in all, information of one event extrapolated from 1D measured parameters to get the 3D characteristics of a MC can lead to a large error when estimating its properties.

One solution to this problem is to have multiple spacecraft crossings (e.g., Kilpua *et al.* 2011). However, the occurence of such an event is too rare to properly investigate the general characteristics of MCs. In the following, we propose a new study based on statistical analyses of a sample of several MCs crossed by the same spacecraft and fitted by the same analytical model. In particular, we are interested in the most probable shape of the FR axis of observed MCs. The methodology, results and discussion are summarized in the present paper, but more information are found in Janvier *et al.* (2013).

## 2. Location angle, $\lambda$, distribution

In the present work, MCs properties are investigated via a statistical analysis of MC measurements so as to deduce the general FR axis shape. The set of MCs that is chosen is a sample of 107 MCs detected by the *Wind* spacecraft (located at 1 AU) during 15 years (Lepping & Wu 2010). This list also gives their physical characteristics following the Lundquist fitting model (see Lundquist 1950), such as their sizes and their orientations. In addition, we introduce two new orientation angles, namely the inclination angle, $i$, which measures the angle between the plane of the MC axis and the ecliptic plane, and the location angle, $\lambda$, that measures the angle between the local direction of the MC axis and the orthoradial (see Fig. 1a, where $\lambda$ can be understood as the location of the spacecraft regarding the MC). These two parameters are more adapted to the study of the FR axis shape than the standard direction parameters (longitude and latitude).

From the list of MC parameters, a problem can be raised: since all MCs have different characteristics (in terms of speed, size, ...), can a sample of MCs be considered as a whole so as to statistically analyze the behavior of one parameter? Would it not be necessary to categorize MCs in sub-classes so as to properly investigate this parameter? In other words, if the FR axis shape is analyzed via the location angle $\lambda$, does not this shape also depend on other intrinsic properties of MCs?

To answer this question, the distribution of the location angle, $\lambda$, (Fig.1b) was thoroughly investigated as a function of all the other MCs parameters. Especially, sub-groups of MCs ordered in function of their characteristics (say, the radius) were made, so as to investigate the changes in the $\lambda$-distribution. Since no changes can be reported, and as

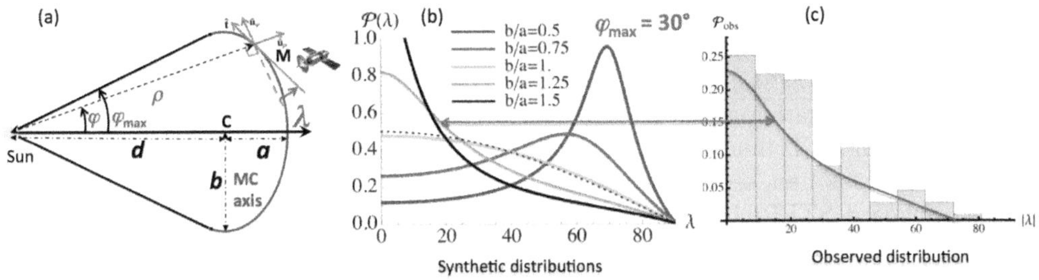

**Figure 2.** **(a)** Diagram representing the analytical ellipse shape given to the FR axis, with parameters $\rho$ and $\varphi$ in cylindrical coordinates. **(b)** Synthetic distributions deduced from the analytical shapes when varying the aspect ratio, at fixed $\varphi_{\max} = 30°$. **(c)** Matching of the synthetic distribution for aspect ratio = 1.25 with the observed probability distribution.

there is only a very weak correlation between $\lambda$ and all other MCs intrinsic parameters, we verified that the *Wind* sample of 107 MCs can be analyzed together as a whole, so as to deduce the most probable axis shape from this set of data.

Let us now focus on the properties of the $\lambda$ distribution function. Contrary to the distribution of the inclination angle, $i$, which is flat (see Janvier *et al.* 2013), the distribution of the location angle is non-uniform. A flat distribution for $i$ implies that there is no privileged inclination for the detection of a MC by a spacecraft: North/South or East/West MCs are detected the same. However, a non-uniform decreasing distribution of $|\lambda|$, $\mathcal{P}_{\rm obs}(\lambda)$, implies that MCs are more often detected at the apex (see Fig. 1a) than in the legs. This is an interesting property that has implications on the axis shape of the FR.

To investigate the characteristics of $\mathcal{P}_{\rm obs}(\lambda)$, we propose two joint methods. First, synthetic distributions are derived from an analytical MC model, and are then compared with $\mathcal{P}_{\rm obs}(\lambda)$. As a second step, $\mathcal{P}_{\rm obs}(\lambda)$ is directly integrated so as to derive the most probable FR axis shape.

## 3. Flux rope axis shape deduced from two statistical methods

*Synthetic distributions method.* This first method implies creating synthetic distributions from an analytical model of a FR axis shape and comparing them with $\mathcal{P}_{\rm obs}(\lambda)$. We analyze the shape of the FR axis as an ellipse joined at two ends to the Sun. The ellipse is parametrized in cylindrical coordinates by the radius, $\rho$, and the rotation angle, $\varphi$, that are themselves expressed as a function of the ellipse parameters (Fig.2a). The full extension of the ellipse is given as $\varphi_{\max}$. From this analytical shape, we can derive an expression of a probability distribution for $\lambda$: $\mathcal{P}_{\rm synth}(\lambda) = \mathcal{P}(\varphi)|d\varphi/d\lambda|$ where $\mathcal{P}(\varphi) = 1/(2\varphi_{\max})$. Different $\mathcal{P}_{\rm synth}(\lambda)$ are obtained by varying the axis shape. Then, when compared with the observed probability distribution (Fig.2c), we found a very good correlation for an aspect ratio of the ellipse, $b/a = 1.25$. Note that those results depend on one free parameter, $\varphi_{\max}$. However, by changing its values, we checked that $\varphi_{\max}$ has a small effect on the shapes taken by $\mathcal{P}_{\rm synth}(\lambda)$ contrary to the aspect ratio.

*Direct derivation of the FR axis shape from $\mathcal{P}_{\rm obs}(\lambda)$.* As a second method, we directly use the observed distribution. By integrating $\mathcal{P}_{\rm obs}(\lambda)$, we express the parameters $\rho$ and $\varphi$ of the FR axis shape in cylindrical coordinates (without assuming any preconceived ellipse shape). Note here that, similarly with the first method, there is one free parameter, $\varphi_{\max}$, that cannot be constrained. As such, different shapes are derived, depending on the values of the maximum elongation, but all are similar to that found with the first method, making both methods consistent.

*Comparison with heliospheric imagers.* Since heliospheric imagers give information on the shape of propagating structures from the Sun, we use such data to compare the FR axis shapes seen in 2D images with that found by statistical methods. For that, we chose an event that was best seen in terms of FR detection. This event was recorded by HI imagers onboard STEREO-A (see Möstl *et al.* 2009). We then repeated several times manual pointing of the FR axis in different images, and tracked its propagation. We found that although the FR grows larger with time, there is a self-similarity in the shape of the axis that can be directly compared with the shapes obtained with the previous methods. Furthermore, using the heliospheric images allowed us to constrain the free parameter, $\varphi_{\max}$, to 30°.

## 4. Conclusion

The present paper summarizes different methods used to derive the most probable FR axis shape. For that, we used in situ data from a sample of MCs detected by *Wind* spacecraft over 15 years. In particular, we studied the characteristics of the non-uniform distribution of the location angle, $\lambda$, a parameter that is directly related to the location of the spacecraft along the FR axis, and therefore to its shape. After verifying that $\lambda$ was strictly uncorrelated with all other MC parameters, to ensure the consistency of the full set of data, we compared synthetic distributions obtained from analytical FR axis shape with the observed distribution. Similar results with this method and with the direction integration of the observed distribution were found. Then, for completeness of the study, we finally used heliospheric images to compare the shape observed in a propagating MC with shapes determined with in situ data. All those methods prove to be consistent and we were able to find the most probable shape of FR axis (see Figs 10 and 12 in Janvier *et al.* 2013).

### References

Aulanier, G., Török, T., Démoulin, P., & DeLuca, E. E. 2010, *Astrophysical Journal*, 708, 314
Dasso, S., Mandrini, C. H., Démoulin, *et al.* 2005, *Adv. Spa. Res.*, 35, 711
Dasso, S., Mandrini, C. H., Démoulin, P., & Luoni, M. L. 2006, *A&A*, 455, 349
Démoulin, P., Mandrini, C. H., van Driel-Gesztelyi, L., *et al.* 2002, *A&A*, 382, 650
Démoulin, P. 2014, *Solar Physics*, this issue
Gosling, J. T., Bame, S. J., McComas, D. J., & Phillips, J. L. 1990, *Geo. Res. Let.*, 17, 901
Gosling, J. T. 1993, *Physics of Fluids B*, 5, 2638
Janvier, M., Démoulin, P., & Dasso, S. 2013, *A&A*, 556, A50
Kilpua, E. K. J., Jian, L. K., Li, *et al.* 2011, *Jour. Atmos. Sol.-Ter. Phys.*, 73, 1228
Larson, D. E., Lin, R. P., McTiernan, J. M. *et al.* 1997, *Geophysical Research Letters*, 24, 1911
Lepping, R. P., Burlaga, L. F., & Jones, J. A. 1990, *J. Geophys. Res.*, 95, 11957
Lepping, R. P. & Wu, C. C. 2010, *Ann. Geophys.*, 28, 1539
Lundquist, S. 1950, *Ark. Fys.*, 2, 361
Marubashi, K. 2000, *Adv. Spa. Res.*, 26, 55
Masson, S., Démoulin, P., Dasso, S., & Klein, K.-L. 2012, *A&A*,, 538, A32
Moore, R. L., Schmieder, B., Hathaway, D. H., & Tarbel, T. D. 1997, *Solar Physics*, 176, 153
Möstl, C., Farrugia, C. J., Temmer, M., *et al.* 2009, *Astrophysical Journal*, 705, L180
Nakwacki, M., Dasso, S., & Démoulin, P. 2011, *A&A*,, 535, A52
Sonnerup, B. U., Hasegawa, H., Teh, W.-L., & Hau, L.-N. 2006, *Jour. Geophys. Res.*, 111, A09204
Title, A. 2014, *Solar Physics*, this issue
Zhang, J., Cheng, X., & Ding, M. 2012, *Nature communications*, 3, 747

*Nature of Prominences and their role in Space Weather*
*Proceedings IAU Symposium No. 300, 2013*
*B. Schmieder, J.-M. Malherbe & S. T. Wu, eds.*

doi:10.1017/S1743921313011083

# Reconstruction of magnetic clouds from in-situ spacecraft measurements and intercomparison with their solar sources

**Qiang Hu**[1] **and Jiong Qiu**[2]

[1]Dept. of Space Science/CSPAR, University of Alabama in Huntsville,
Huntsville, AL 35805, USA
email: qh0001@uah.edu

[2]Physics Department, Montana State University,
Bozeman, MT 59717-3840, USA
email: qiuj@mithra.physics.montana.edu

**Abstract.** Coronal Mass Ejections (CMEs) are eruptive events that originate, propagate away from the Sun, and carry along solar material with embedded solar magnetic field. Some are accompanied by prominence eruptions. A subset of the interplanetary counterparts of CMEs (ICMEs), so-called Magnetic Clouds (MCs) can be characterized by magnetic flux-rope structures. We apply the Grad-Shafranov (GS) reconstruction technique to examine the configuration of MCs and to derive relevant physical quantities, such as magnetic flux content, relative magnetic helicity, and the field-line twist, etc. Both observational analyses of solar source region characteristics including flaring and associated magnetic reconnection process, and the corresponding MC structures were carried out. We summarize the main properties of selected events with and without associated prominence eruptions. In particular, we show the field-line twist distribution and the intercomparison of magnetic flux for these flux-rope structures.

**Keywords.** Magnetic Clouds, Magnetic Flux Rope, CME/ICME, In Situ Measurements, Prominence

## 1. Introduction

This report addresses the magnetic field topology of Magnetic Clouds (MCs) and the connection to their solar sources. We focus on the quantitative characterization of MC/flux-rope structures at 1 AU, and illustrate our approach to inter-connecting the solar and interplanetary analysis. We treat the magnetic structure of prominence as an integral part of the MC observed in-situ (Schmieder *et al.* 2013; van Ballegooijen & Martens 1989). We note that all events examined here were associated with solar flares. Some were accompanied by prominence eruption (PE) and others were not (non-PE), according to Li (2012) and our own assessment.

## 2. Grad-Shafranov Reconstruction of Magnetic Clouds

Magnetic Clouds are a subset of ICMEs that possess the following characteristics based on in-situ spacecraft measurements (e.g., Burlaga 1995): 1) relatively strong magnetic field, 2) smooth rotation of the magnetic field direction, and 3) relatively low proton temperature or proton $\beta$ value. This is a traditional definition and has enabled the modeling of such structures via traditional force-free approaches (Burlaga 1995). An alternative and now widely used approach, beyond the force-free assumption, is based on the plane Grad-Shafranov equation that allows the analysis of events that are not strictly

MCs. For example, when electron temperature is included, the total plasma pressure can be significantly increased, resulting in total plasma $\beta$ values $\geqslant 1$ where the GS method is still applicable (Hu *et al.* 2013). Overall all approaches have yielded flux-rope solutions for the structures embedded within MC or ICME intervals. A typical output of a GS reconstruction result is shown in Fig. 1.

Here a typical cylindrical flux-rope configuration is illustrated by three selected thick spiral field lines, lying on nested isosurfaces of the magnetic flux function $A$ whose projection onto the $x-y$ plane is shown in the back as equi-value contours of $A$. The flux rope axis along $z$ represents the invariance direction (i.e., $\partial/\partial z = 0$, and $B_z = B_z(A)$). Various physical quantities can be derived based on the GS reconstruction result. For example, the total toroidal (axial) magnetic flux $\Phi_t = \int B_z dxdy$ and the total poloidal magnetic flux $\Phi_p = |A_0 - A_b| \cdot L$, i.e., the flux across the shaded area in Fig. 1, for certain effective length $L$ along $z$. Correspondingly, the relative magnetic helicity within certain volume can be calculated (Webb *et al.* 2010). Additionally the axial magnetic field and electric current density distributions, the accumulative and total current etc. can also be obtained. In particular, we derive the field-line twist as a function of $A$ to examine its variation within the flux rope. A highly relevant study was carried out by Dasso *et al.* (2006) where four different analytic flux-rope models were employed to derive various physical quantities for one event.

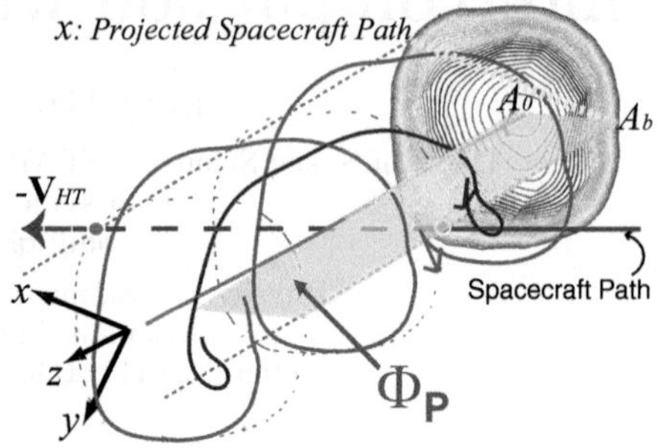


**Figure 1.** A typical GS reconstruction result of a cylindrical magnetic flux rope. See text for details.

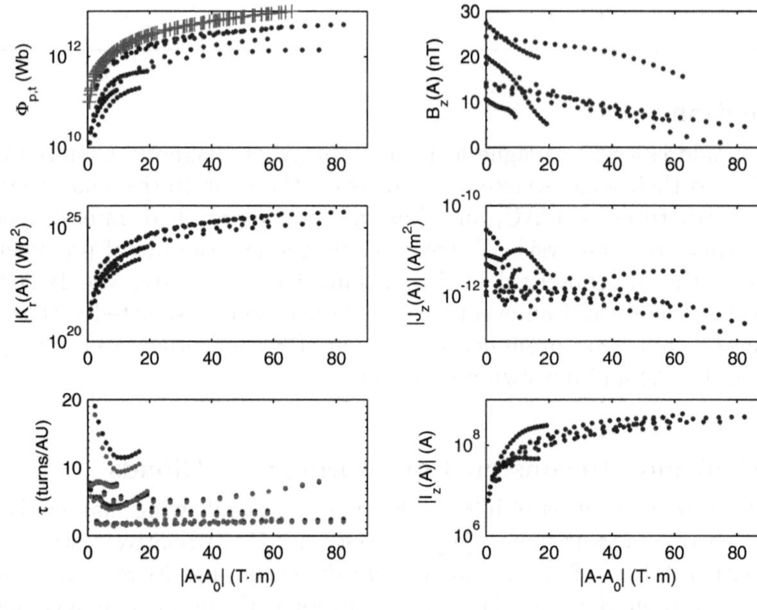


**Figure 2.** Summary plot of various quantities as functions of $A$ for selected events (Li 2012). (*from top to bottom*) Left column: toroidal and poloidal ('+') flux, relative magnetic helicity, and field-line twist; right column: axial magnetic field, axial current density, and accumulative axial current.

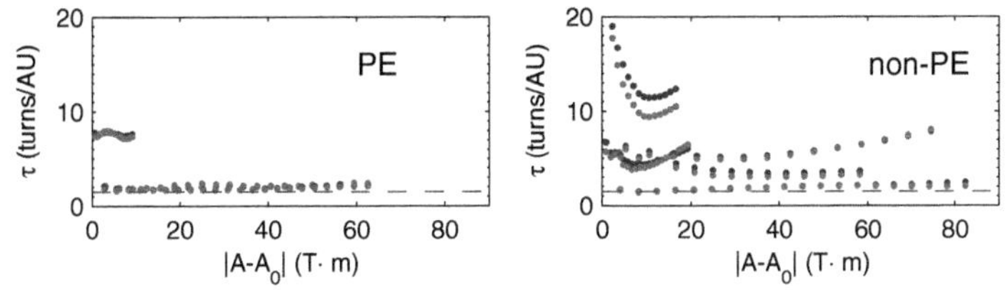


**Figure 3.** Field-line twist for PE (*left*) and non-PE (*right*) events. The horizontal dashed line is of value 1.5.

Fig. 2 shows the summary plot of various quantities for a handful of events, especially as organized by the shifted flux function $A$, which represents different cylindrical shells of varying radial distance away from the center ($A \equiv A_0$) of the flux rope, the larger the shifted $A$ values, the farther the distances away from the center. Generally speaking, the maximum $A$ value indicates the transverse size of each flux rope. The distributions of $B_z(A)$ and $|J_z(A)|$ show greater ranges of variation, while the integral quantities of magnetic flux, electric current and relative magnetic helicity do not. They all increase monotonically, and they do not appear to have clear distinctions between PE and non-PE events.

The average magnetic field line twist is approximated by $\tau(A) = K_r(A)/\Phi_t^2(A)$ (Berger & Field 1984), and similarly $\tau(A) = \Phi_p(A)/\Phi_t(A)$, based on the assumption of a constant twist. Both are functions of $A$, representing an average twist, in terms of number of turns per AU, within the volume enclosed by each $A$ shell. The bottom left panel of Fig. 2 shows such twist distributions of the two numbers (dots of blue and red colors, respectively) for each event. Each set of blue and red dots overlaps very well, except for the one of the largest values. The general trend is that the smaller the flux rope, the larger the twist becomes. However, for events of large sizes, the twist remains fairly constant (e.g., $\sim 2$ turns/AU for the largest event) throughout the $A$ shells.

We further separate the events into two categories of PE and non-PE events, and show their twist distributions, respectively, in Fig. 3. The main distinction between the two sets is that for the PE events, the twist remains largely constant within the flux rope, while for the non-PE events, the twist shows significant variations. Some exhibit declining gradient outward from the center. The implications of such behaviors are discussed in Section 4.

## 3. Intercomparison of Magnetic Flux

An important approach to utilize the GS reconstruction results outlined above is to make quantitative comparison with their solar sources. Following the original study of Qiu *et al.* (2007), we augmented the original list of events and show the magnetic flux comparison among $\Phi_p$, $\Phi_t$, and the corresponding flare-associated magnetic reconnection flux $\Phi_r$, in Fig. 4. The comparison indicates that $\Phi_p > \Phi_t$ and $\Phi_p \approx \Phi_r$ for $L = 1$ AU with uncertainty range $L \in [0.5, 2]$ AU, which conforms to prior result (Qiu *et al.* 2007). The caveat associated with the few low points in the right panel is that the poloidal flux was significantly underestimated due to selection of a rather short interval for the GS reconstruction ($\sim 2$ hours as opposed to normally $\sim 20$ hours) in some cases. There are generally no clear distinctions between PE and non-PE events, except for the two PE events of significantly greater poloidal flux than the corresponding reconnection flux. Whether or not that indicates the significant contribution from the pre-existing flux rope (prominence) is worth pursuing.

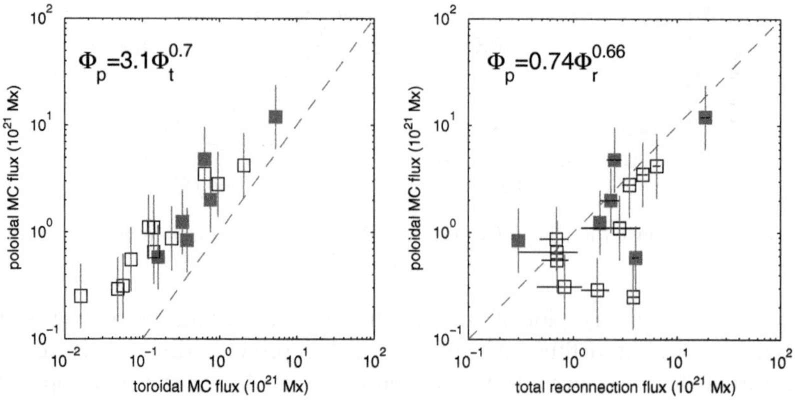


**Figure 4.** Magnetic flux comparison of $\Phi_p$ vs. $\Phi_t$ (*left*) and $\Phi_p$ vs. $\Phi_r$ (*right*). The events associated with PE are marked by filled squares. The least-squares fit to each data set is given and the dashed line indicates the one-to-one line.

## 4. Summary and Discussion

In summary, the GS reconstruction method has matured and been widely used in analyzing in-situ measurements of magnetic flux ropes. A software package has been developed and distributed world-wide for interested users. We presented the summary of the GS reconstruction results in a congregated form for a number of events, in terms of the distributions of various quantities along the $A$ shells. Among them are the total magnetic flux ranging between $10^{11}$-$10^{13}$ Wb, the total relative helicity ranging $10^{23}$-$10^{26}$ $\mathrm{Wb}^2/\mathrm{AU}$, and the total axial current ranging $10^7$-$10^9$ A. The poloidal MC flux compares well with the magnetic reconnection flux accumulated during flare in solar source region. In addition, the non-PE (flare dominant) events showed greater gradients in field-line twist variation, especially near the flux-rope center, corresponding to the formation of flux-rope core primarily via magnetic reconnection. As a distinction for PE events (lack of strong flares), the twist distribution remains constant, which might indicate a fundamentally different formation process of the core. We plan to further elaborate on this issue and present detailed case studies elsewhere.

### Acknowledgements

We acknowledge NASA grants NNG06GD41G, NNX12AF97G, and NNX12AH50G; NSF grants AGS-1062050 and ATM-0748428 for support. HQ acknowledges consultations with B. Dasgupta, A. Khare, and G.M. Webb.

### References

Berger, M. A., & Field, G. B. 1984, *Journal of Fluid Mechanics*, 147, 133

Burlaga, L. 1995, *Interplanetary Magnetohydrodynamics* (New York: Oxford Univ. Press), 89

Dasso, S., Mandrini, C. H., Démoulin, P., & Luoni, M. L. 2006, *Astron. Astrophys.*, 455, 349

Hu, Q., Farrugia, C. J., Osherovich, V. A., Möstl, C., Szabo, A., Ogilvie, K. W., & Lepping, R. P. 2013, *Solar Phys.*, 284, 275

Li, Y. 2012, *private communication*

Qiu, J., Hu, Q., Howard, T. A., & Yurchyshyn, V. B. 2007, *Astrophys. J.*, 659, 758

Schmieder, B., Démoulin, P., & Aulanier, G. 2013, *Adv. Space Res.*, 51, 1967. 1212.4014

van Ballegooijen, A. A., & Martens, P. C. H. 1989, *Astrophys. J.*, 343, 971

Webb, G. M., Hu, Q., Dasgupta, B., & Zank, G. P. 2010, *J. Geophys. Res.*, 115, A10112

*Nature of Prominences and their role in Space Weather*
*Proceedings IAU Symposium No. 300, 2013*
*B. Schmieder, J.-M. Malherbe & S. T. Wu, eds.*

doi:10.1017/S1743921313011095

# Properties and processes that influence CME geo-effectiveness

**Benoit Lavraud**[1,2] **and Alexis Rouillard**[1,2]

[1]Institut de Recherche en Astrophysique et Planétologie, Université de Toulouse (UPS), Toulouse, France

[2]UMR 5277, Centre National de la Recherche Scientifique, Toulouse, France
email: `benoit.lavraud@irap.omp.eu`

**Abstract.** The geo-effectiveness of coronal mass ejections (CME) is determined by a complex chain of processes. This paper highlights this fact by first discussing the importance of CMEs intrinsic properties set at the Sun (e.g., trajectory, eruption process, orientation, etc.). We then review other key processes that may occur during propagation (e.g., shocks, compressions, magnetic flux erosion) and in the specific interaction with Earth's magnetosphere (e.g., magnetic properties, preconditioning mechanisms). These processes sequentially have a significant influence on the final geo-effectiveness of CMEs. Their relative importance is discussed. While the CME's trajectory, magnetic field orientation, velocity and their duration as set at the Sun certainly are key ingredients to geo-effectiveness, other processes and properties, that at first appear secondary, often may be as important.

**Keywords.** Plasmas - Sun: coronal mass ejections (CMEs) - Sun: solar-terrestrial relations - Earth

## 1. Introduction

The impacts of solar phenomena, and in particular that of coronal mass ejections, in the heliosphere has been the focus of much attention in recent years. This theme is known under the general terminology of "Space Weather". It has been driven by the accumulation of evidences for significant societal and economical implications such as radiation hazard to astronauts, partial to full spacecraft power failures, GPS signal alterations, or ground power grid and transformer failures. It has led to numerous dedicated developments, including databases, analysis tools, virtual observatories, instruments and mission concepts. The purpose of the present paper is not to focus on these impacts, but rather to review the complex chain of physical processes that come into play for determining whether and how much a CME eruption may affect Earth environment.

Of particular importance are the solar wind velocity and the direction and strength of the interplanetary magnetic field, which combine into the Dawn-Dusk component of the solar wind electric field $\mathbf{E} = -\mathbf{V} \times \mathbf{B}$ (e.g., Gonzales *et al.* 1974, Siscoe & Crooker 1974, Burton *et al.* 1975, Perreault *et al.* 1978). This is because the solar wind electric field relates to the global dayside magnetic reconnection rate. The role of these parameters was already made evident in early works that attempted to derive driving function of the magnetospheric response to solar wind conditions. In his review, Akasofu (1981) already noted 13 such coupling functions. Many more complex coupling functions have been devised since then. Recent ones can be found in Newell et al. (2007), Borovsky (2008), and Tenfjord & Østgaard (2013). In addition to the electric field driver (often referred to as the VBz parameter in coupling functions), the duration of the driver is of course of paramount importance for cumulative energy input reasons. This cumulative effect needs to be accounted for when one implements coupling functions, e.g., such as the semi-empirical model of the ring current Dst index by O'Brien & McPherron (2000) (based on Burton *et al.* 1975).

The central role of Coronal Mass Ejections (CMEs) in driving geomagnetic activity was early recognized, and well put into context in the work by Gosling (1993). Magnetic Clouds (MCs) are a subset of CMEs. These are defined by a flux rope structure with enhanced magnetic field and a clear and smooth rotation of its components, and a lower-than-usual proton temperature (Burlaga *et al.* 1982). The fact that not all CMEs are observed as clear flux ropes (i.e., MCs) is generally attributed to the sampling geometry through the magnetic structure (e.g., Jian *et al.* 2006). For more details on the basic CME and MC properties, the reader is referred to the review by Démoulin (this issue; and references therein).

## 2. Properties set at the Sun

### 2.1. *Eruption location, trajectory and deflections*

The first key ingredient to assess the potential impact of a CME at Earth is of course its eruption location on the Sun and ensuing trajectory. The huge 23rd July 2012 CME recorded at STEREO-A is a good example with, despite the potential for a record geomagnetic storm, a trajectory at $\sim 120°$ from the Sun-Earth line (Russell *et al.* 2013).

On the other hand, processes in the corona during and after ejection may alter the trajectory. First, significant latitudinal trajectory deflection may occur through interaction of the CME with the strong magnetic fields from the adjacent polar coronal holes. This was recently highlighted through global magnetohydrodynamic (MHD) modeling by Zuccarello *et al.* (2012), as is illustrated in Figure 1. An important implication of this process is that even CMEs which originate at high latitudes will tend to be channeled towards the ecliptic plane (along the helmet streamers and heliospheric current sheet; HCS). This phenomenon will thus statistically tend to focus CMEs toward the Earth's latitude. It should be noted, however, that this effect plays a role mainly during non-maximum solar cycle periods when CMEs are generally isolated and the dipolar structure of the solar magnetic field is well in place. During solar maximum period, the much more complex structure of the streamers and active regions will tend to homogenize this effect in latitude.

In addition to latitudinal deflections, longitudinal deflections also occur for similar reasons; i.e., through interaction with low-latitude coronal holes in the presence of a warped

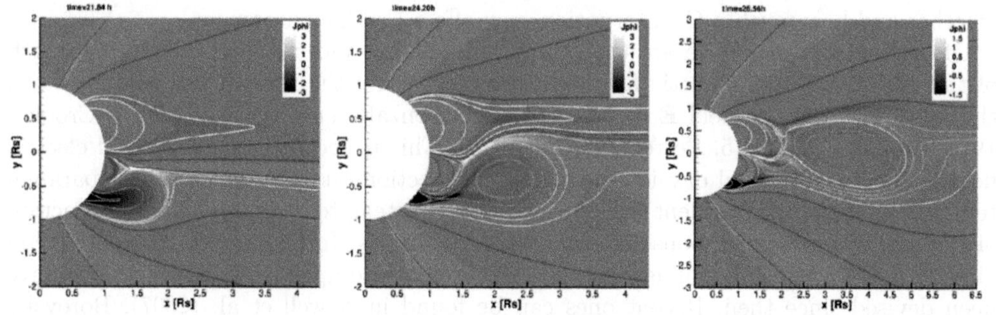


**Figure 1.** Three temporal snapshots of the evolution of a CME (orange and pink lines) erupting at low southern latitudes, from global MHD simulations. The CME is channeled towards lower latitudes and eventually propagates in the ecliptic plane. Adapted from Zuccarello *et al.* (2012).

helmet streamer/HCS. As a result CMEs tend to be channeled along the Parker spiral direction during their ejection and further propagation. This is the explanation typically put forth to explain the trend for the origin of Earth-impacting CMEs to be statistically skewed towards the West side of the solar disk (cf. Zhang *et al.* 2003; but see also Cane *et al.* 2000 for earlier, contradicting results).

### 2.2. *Ejection processes and models: a spectrum of CME strengths and sizes*

The geo-effectiveness of CMEs directed towards Earth of course largely relies on the energetics of the eruptive phenomenon at the Sun. In other words, the potential geo-effectiveness of a CME directly depends upon three key parameters: the size of the CME, its initial magnetic field strength and its bulk speed. The two latter set the VBz parameter (cf. introduction), while the former controls its duration. These directly impact the ability of the CME to drive strong coupling at Earth's magnetopause and in turn large-scale magnetospheric convection and a geomagnetic storm.

There exist various models of CME eruption, and inferred processes. The point worth conveying here is that there should exist a whole spectrum of processes involved, and thus also a whole spectrum of resulting CME strengths and sizes. We highlight this fact in Figure 2 by presenting what may be two extremes cases of this spectrum. Figures 2a and 2b display solar disk images during a flare and CME event in November 1997, with Figure 2b showing in addition reconstructed magnetic fields (using a coronal model). This work by Delannée & Aulanier (1999) investigated an eruption model, called transequatorial filament eruption based on large-scale loops threading both the southern and northern hemispheres. This model, and inferred processes, thus involves very large-scale structures by definition. Resulting CMEs are expected large in size, and may possess strong internal energy (e.g., magnetic field and speed).

Figure 2c shows a schematic describing the release of plasmoid-type structures through reconnection at the tip of a helmet streamer (Wang *et al.* 2000). This type of CME release mechanism is unlikely to produce strong and fast CMEs. Studies have actually linked these plasmoids with the in situ observation of small-scale flux ropes with low magnetic field and speed entrained by the solar wind at 1 AU (e.g., Rouillard *et al.* 2009 and Rouillard *et al.* 2011). In conclusion, the transequatorial filament eruption model is more likely to produce geo-effective CMEs than the streamer blowout mechanism, and a whole range of models and associated processes exists in between.

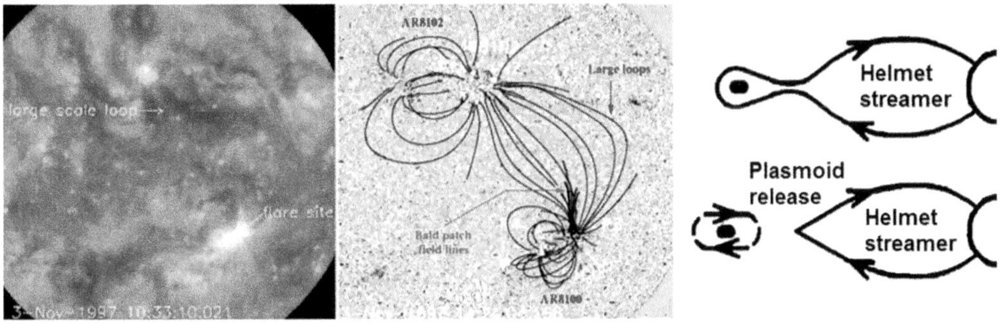


**Figure 2.** Illustrations of two drastically different CME eruption models: (a and b) the transequatorial filament eruption model (Figure courtesy of G. Aulanier; based on Delannée & Aulanier (1999)) and c) the plasmoid release model (Figure adapted from Wang *et al.* (2000)). The former involves large regions of the Sun and large-scale processes, while the latter proposes the release of small-scale structures through reconnection at the tip of the helmet streamer.

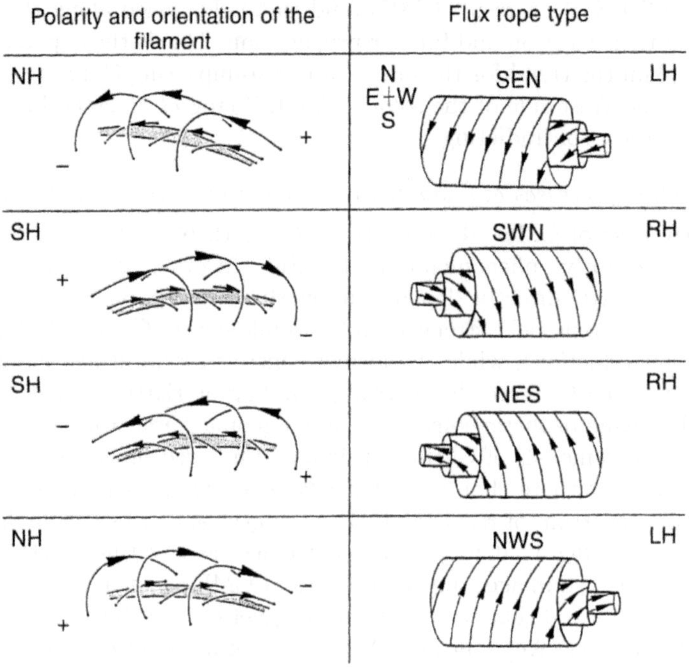


**Figure 3.** Relationship between the polarity and orientation of solar filaments and the associated CME flux rope structure, as deduced from a statistical study. From Bothmer & Schwenn (1998).

### 2.3. *CME acceleration and deceleration in the corona*

It is known that CME speeds as observed near the Sun (e.g., from coronograph) hardly correspond with CME speeds measured in situ at 1 AU. This is particularly true for halo CMEs (directed towards Earth), where projection effects further render the CME speed hard to determine. This has direct implications for our ability to estimate CME transit times between the Sun and Earth (e.g., Gopalswamy *et al.* 2001 and references therein). Studies of CME acceleration using coronograph images alone show that strong CME acceleration or deceleration occur in the corona, with initially slow (fast) CMEs accelerating (decelerating) (e.g. Gopalswamy *et al.* 2000). CMEs interact with the ambient solar wind, so that the dichotomy between acceleration and deceleration is centered for CME speeds of order of that of the slow solar wind: 400 $km/s$ (Gopalswamy *et al.* 2000). This process is akin to general drag (e.g., Cargill *et al.* 1995). These models predict that most of the drag occurs close to the Sun, soon after eruption, and that CME speed is more constant past few tens of solar radii. Of course, these processes are very important since the final CME speed is a key parameter for geo-effectiveness.

### 2.4. *Magnetic flux rope orientation and polarity*

Because the sign and magnitude of the IMF Bz component is critical for the triggering of geomagnetic storms, as mediated by magnetic reconnection at the dayside magnetopause, the orientation of the CME (or MC) plays a key role. For CMEs that do not show a flux rope structure (cf. review by Démoulin (this issue)), or for sheaths ahead of CMEs, the magnetic field strength and direction is rather unordered and thus the geo-effectiveness hard to anticipate. However, in the case of well-formed flux rope structures such as in MCs, the orientation and magnetic polarity of active regions and filaments at the origin of CME eruption at the Sun directly impact the timing and intensity of the ensuing geomagnetic storm. This is because the polarity at the Sun generally determines (statistically) whether it is the leading or trailing portion of the CME that contains the southward-oriented magnetic field (e.g., Bothmer & Schwenn 1998, cf. Figure 3). On the other hand, the tilt of the active region or filament, and other deflection/rotation processes in the corona, determine the MC main axis orientation in interplanetary space. The resulting latitudinal angle of the main flux rope axis orientation is particularly important because it can lead to a reduced southward Bz component at 1 AU. Often MCs at 1 AU have such a large latitudinal angle that the Bz component is very weak and no geomagnetic storm is triggered despite the MC being intense and fast. Finally, it should be noted that whether MCs have a South-North or North-South magnetic polarity is solar cycle dependent (Mulligan *et al.* 1998). This may have an impact when propagation and compression effects come in, as discussed next in Section 3.

## 3. Processes during propagation

### 3.1. *CME expansion, drag, shocks and sheath*

Following eruption near Sun where most of the solar wind and CME acceleration occurs, numerous additional processes come into play for determining geo-effectiveness. The first to mention is the global expansion of the CME. While small-scale CMEs likely have insufficient internal magnetic fields for significant expansion, large-scale CMEs with intense magnetic fields are know to expand significantly in both the radial and longitudinal/latitudinal directions. The expansion has been shown to be often self-similar (Démoulin and Dasso 2009), and models that represent such global expansion have been devised (Owens 2006; Owens *et al.* 2006).

While CMEs that significantly expand in the interplanetary medium will see their internal magnetic field intensity decrease significantly, the decrease is less for CMEs that strongly interact with the ambient solar wind. CMEs that expand the most have lower fields but longer duration. In terms of geo-effectiveness, the two effects are competing but overall a large magnetic field remains the critical component to produce large storms. Statistics show that compressions and the formation of shocks, ahead of CMEs that strongly interact with the ambient solar wind (and thus expand less), are the source of major storms. This was already found in early studies (Gosling *et al.* 1990). While a separate review would be required to cover geo-effectiveness aspects related to energetic particles, it should be noted here that CME-driven shocks are a major source of such particles, in particular for the largest and fastest CMEs.

While the formation of shocks and sheaths produces enhanced magnetic fields, CMEs are subject to a drag force that slows them down as they expand into a slower solar wind. The drag is primarily dependent upon the speed difference between the CME and solar wind into which it expands. For fast CMEs modelling efforts typically show strong deceleration close to the Sun, then weaker drag and more constant speed further out in the heliosphere, although unusual CME speed, size or density can lead to unusual behaviours (e.g., Vrsnak *et al.* 2010; Subramanian *et al.* 2012). The impact of drag on CME speed at 1 AU is very important, in particular in extreme cases, as speed strongly influences CME geo-effectiveness.

### 3.2. *HSS-CME and CME-CME interactions*

Interaction between CMEs and the ambient solar wind is not limited to the formation of a shock and sheath at its front. Interaction has often been reported at the rear of CMEs, for instance in the presence of a trailing high speed stream (HSS) emanating from a nearby coronal hole. The impact of such a configuration on geo-effectiveness was put forth in

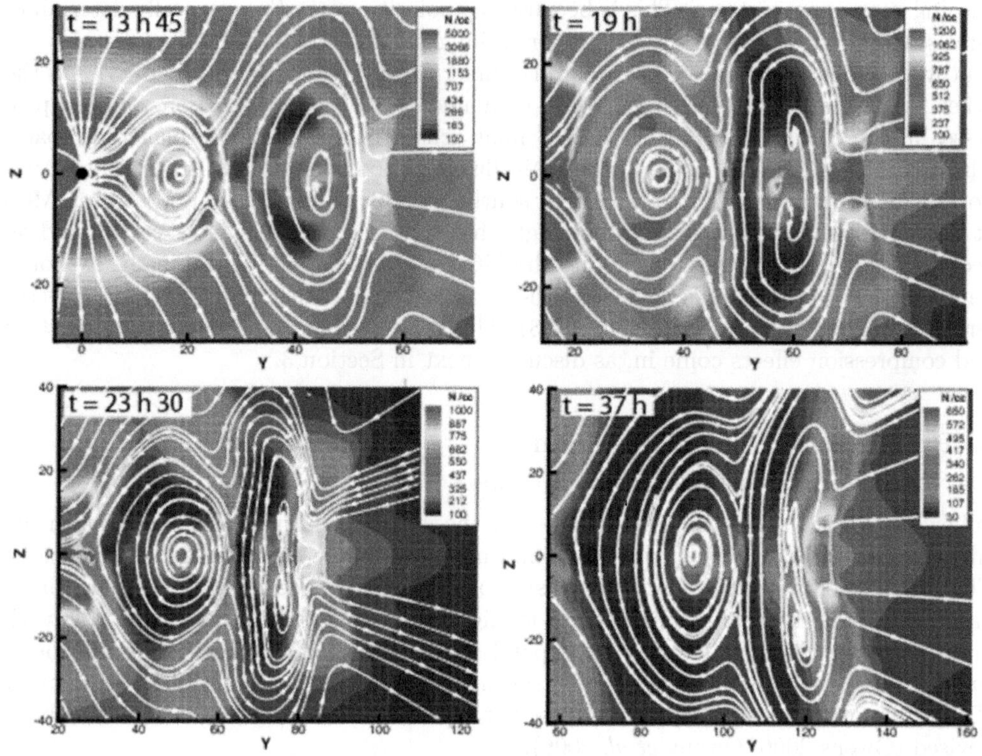


**Figure 4.** Snapshots of a 3D global MHD simulation for the case of CME-CME interaction at four times during the run by Lugaz *et al.* (2005). The faster CME (to the left) overcomes and compresses the first, slower CME. Magnetic reconnection and magnetic flux erosion are also likely to occur during such interaction.

the work by Fenrich & Luhmann (1998), who showed that in the case of a North-South polarity MC (i.e., with a trailing southward Bz) the compression exerted by the HSS led to increased magnetic fields at the rear of the MC and enhanced geo-effectiveness.

Such rear compression may also occur in the context of direct CME-CME interaction, i.e., when a CME is overcome by a faster one launched only slightly later from the Sun. Such cases are not rare, in particular during solar cycle maximum. Such a complex interaction was studied by Lugaz *et al.* (2005), and is depicted in Figure 4. Strong compression regions are formed both in front of the first CME and at the interface between the two CMEs. The resulting compressed fields may enhance ensuing storm strength depending on the orientation and magnetic polarity of each MC (cf. Section 2.4). In addition to compression, CME-CME interaction may induce magnetic reconnection and magnetic flux erosion, as further discussed in Section 3.3, with implications for geo-effectiveness.

With the advent of the two STEREO spacecraft, orbiting the Sun ahead and behind the Earth on similar orbits, it has been possible to track CMEs all the way from the Sun to the Earth. This is thanks to observation locations away from the Sun-Earth line and the use of dedicated new instruments - the Heliospheric Imagers - which observe heliospheric electron density fluctuations from Thomson scattering in white light (these are essentially coronographs with wider fields-of-view). Because they capture density fluctuations, these instruments have been widely used to track the propagation of CMEs by observing the compression regions either in front or trailing CMEs (e.g., Rouillard *et al.* 2010). These novel capabilities have permitted to better constrain the origin of CMEs,

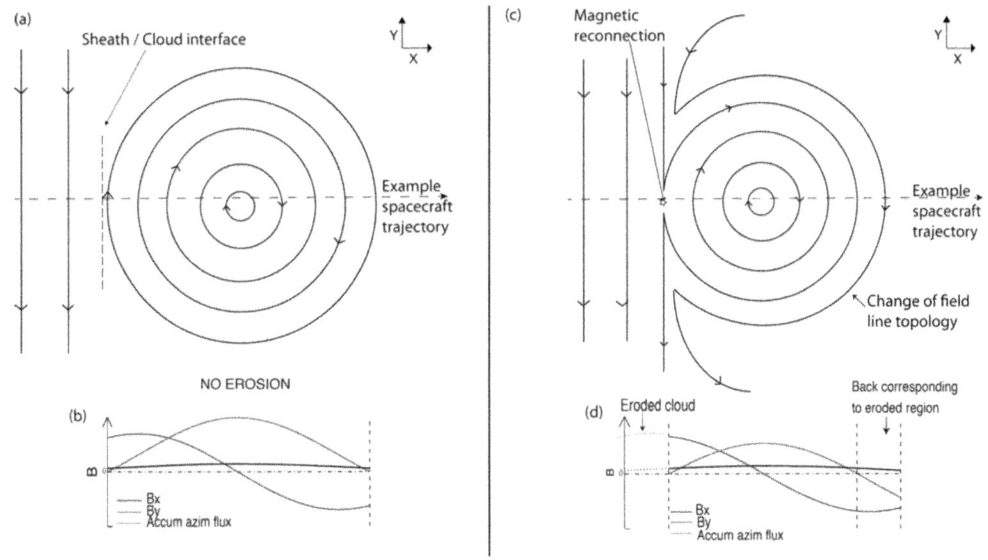


**Figure 5.** Schematic illustration of the impact of flux erosion by magnetic reconnection at the front of MCs. The magnetic structure and related magnetic field components in the MC frame for a non-eroded MC are shown in panels (a) and (b). Panels (c) and (d) show the structure that results from magnetic erosion at the front of a MC, with an imbalance in azimuthal magnetic flux. From Ruffenach *et al.* (2012)

and to determine their trajectory and arrival time at Earth (or other locations) with unprecedented advance timing and reliability.

### 3.3. *Magnetic flux erosion by magnetic reconnection*

Despite a reduced speed owing to drag, we noted that compressions during CME propagation into the ambient solar wind typically induce enhanced fields and geo-effectiveness. In this Section we highlight an other mechanism that occurs during propagation, and which may diminish CME geo-effectiveness: magnetic flux erosion by magnetic reconnection.

The possibility of magnetic flux erosion at the front of CMEs dates back to McComas *et al.* (1988). It was further studied and quantified by Dasso *et al.* (2007) and Ruffenach *et al.* (2012), in particular. A basic implication of this mechanism is the bulk removal i.e., the erosion of the CME magnetic flux, at its front. This is observed as an imbalance in azimuthal magnetic flux when the magnetic field of the CME is analysed in the proper frame, relative to the flux rope orientation. This analysis may thus only be applied to CMEs that have a flux rope structure, i.e., MCs. This is schematically presented in Figure 5, where the eroded MC on the right-hand side is characterized by an excess of magnetic flux at its rear. In addition to this signature in the magnetic field, Ruffenach *et al.* (2012) pointed out several other signatures of this process, including the observation of local magnetic reconnection at the MC front boundary and the observation of a distinct magnetic topology in the back region using suprathermal electrons.

Lavraud *et al.* (2013) recently used simple modelling to investigate the potential geo-effectiveness of this process. They suggested that the reconnection rates observed in the solar wind at 1 AU, despite appearing low at first glance, are overall consistent with the average reconnection rates required to produce the erosion calculated from the imbalance in azimuthal magnetic field (cf. also Ruffenach *et al.* 2012). For a MC with convenient

South-North polarity and geometry, their analysis suggested an ensuing storm strength of order 30% lower in terms of the Dst index (using a simple semi-empirical model) as a consequence of magnetic flux erosion at its front.

## 4. Specifics of the interaction with Earth's magnetosphere

This is an extremely broad subject; we therefore focus only on few key aspects that highlight the importance of the basic characteristics of the object considered in determining the effectiveness of a CME to perturb its nearby environment. We mainly discuss the case of Earth, but comparisons with other bodies are given.

### 4.1. *Properties of the object impacted*

Firstly, the object size and the strength of its magnetic dipole are very important. This determines for instance whether the magnetopsheric dynamics is controlled by processes primarily internal (giant planets) or external (e.g., Mercury or mostly unmagnetized planets such as Mars) to the body. The case of Earth can be considered as driven both ways. A key parameter is then of course the magnetic dipole orientation. Because external driving depends upon the magnetic shear at the magnetopause, whether the dipole axis is directed northwards like at Earth or southwards like at Saturn plays a key role in assessing how a CME of given magnetic polarity (cf. Figure 3) will interact. The case of Uranus, with a dipole axis close to the ecliptic plane is in this context highly unusual.

The magnetic dipole strength, the presence of moons within the magnetosphere, the proximity to the Sun and the properties of the atmosphere-exosphere system are all very important in assessing the impact of internal process. This is true in particular with regards to ion escape/outflows (Mars, Earth, etc.) and the values of ionospheric conductivities (e.g., Earth versus Mercury). The latter for instance affects global magnetospheric convection, such as in the case of polar cap potential saturation at Earth (e.g., Shepherd *et al.* 2007).

### 4.2. *Basic paradigm for coupling at Earth*

The interaction with the interplanetary medium essentially occurs at the magnetopause for magnetized planets. Although other processes occur at this boundary, during strong coupling the key energy transfer mechanism is magnetic reconnection at the front (sunward-side) of the magnetosphere. At Earth the basic paradigm is that dayside reconnection leads to energy and momentum transport through the magnetopause, which in turn drives magnetospheric convection. This leads to energy loading in the magnetotail and ensuing release and transport towards the inner magnetosphere through yet other reconnection mechanisms (either distant or nearer to Earth). The magnetic flux is ultimately transported back towards the dayside into the reconnection region so as to close the magnetic cycle. This is known as the Dungey cycle (Dungey 1961).

Although complexities arise in the details of how the solar wind properties control the local and global rates of magnetic field reconnection at the dayside magnetosphere (e.g., Borovsky 2008), to first order the key parameters are the value of the magnetic field Bz component and the bulk speed (Gonzales *et al.* 1974; Siscoe & Crooker 1974; Perreault *et al.* 1978; Newell *et al.* 2007). These combine into the Dawn-Dusk electric field, which is related to reconnection rates at the magnetopause through mapping into the dayside magnetic reconnection line. The reader is referred to Lavraud & Borovsky (2008), Borovsky (2008) and Lopez *et al.* (2010) for further discussion on this topic.

As mentioned previously, the duration of the strong driver is a key parameter to assess global geo-effectiveness. Semi-empirical models are thus typically built to represent the

**Table 1.** List of low Mach number solar wind effects at the Earth's magnetosphere and the current status of observational evidence and physical understanding for them. Table adapted from Lavraud & Borovsky 2008.

| Low Mach number solar wind effect | Observational evidence | Physical understanding |
|---|---|---|
| Low $\beta$ magnetosheath | Confirmed | Yes |
| Flow enhancements in magnetosheath | Confirmed | Yes |
| Asymmetric magnetosheath flows | To confirm | Yes |
| Asymmetric magnetopause | To confirm | Yes |
| Faster onset of the KH instability | Suggestive | Poor |
| Spiral auroral features | To confirm | Poor |
| Changes in dayside reconnection rate | To confirm | Yes |
| Density effect on reconnection rate | Suggestive | Yes |
| Cross polar cap potential saturation | Confirmed | Poor |
| Global sawtooth oscillations | Suggestive | Poor |
| Alfvén wings | To confirm | Yes |
| Enhanced plasmasphere effect | To confirm | Yes |
| Changes to plasma depletion layer | To confirm | Yes |
| Sunward displacement of the cusps | To confirm | Yes |
| Ti/Te ratio lower than 6-7 | To confirm | Yes |
| Lower bow shock reflection/acceleration | To confirm | Poor |
| Lower turbulence in magnetosheath | To confirm | Poor |

cumulative effect of strong driving; cf. O'Brien & McPherron (2000) for a widely used ring current strength model.

### 4.3. *Impact of the solar wind Alfvén Mach number*

The prime role of the solar wind electric field must be weighted by other considerations. It was shown in particular that the solar wind Alfvén speed and Mach number can lead to very significant alterations of the solar wind-magnetosphere interaction (cf. Lavraud & Borovsky 2008; Lopez *et al.* 2010; Siscoe 2011). Therefore, in addition to a large Dawn-Dusk electric field, combined strong magnetic field and low densities (which lead to low Mach number values) have significant effects on how Earth's magnetosphere reacts to the driving, and such low Mach numbers typically occur during CMEs and MCs (Lavraud & Borovsky 2008).

First, the solar wind Alfvén speed and Mach number are important because the Alfvén speed at meso-scales in the inflow region controls reconnection rate (e.g., Swisdak & Drake 2007). The inclusion of this effect into a complex coupling function for the dayside magnetopause was performed by Borovsky (2008).

The Mach number also has important consequences on the global dynamics of Earth's magnetosheath, and in turn on the magnetopause shape and processes that occur there (Lavraud & Borovsky 2008; Erkaev *et al.* 2012). The complex flow distributions and the changes to current systems that result from it then also impact global magnetospheric convection; i.e., low solar wind Mach numbers typically result in polar cap potential saturation (Lavraud & Borovsky 2008; Lopez *et al.* 2010).

These and many other processes take place when low Mach number CMEs hit the Earth. A non-exhaustive list taken from Lavraud & Borovsky (2008) is given in Table

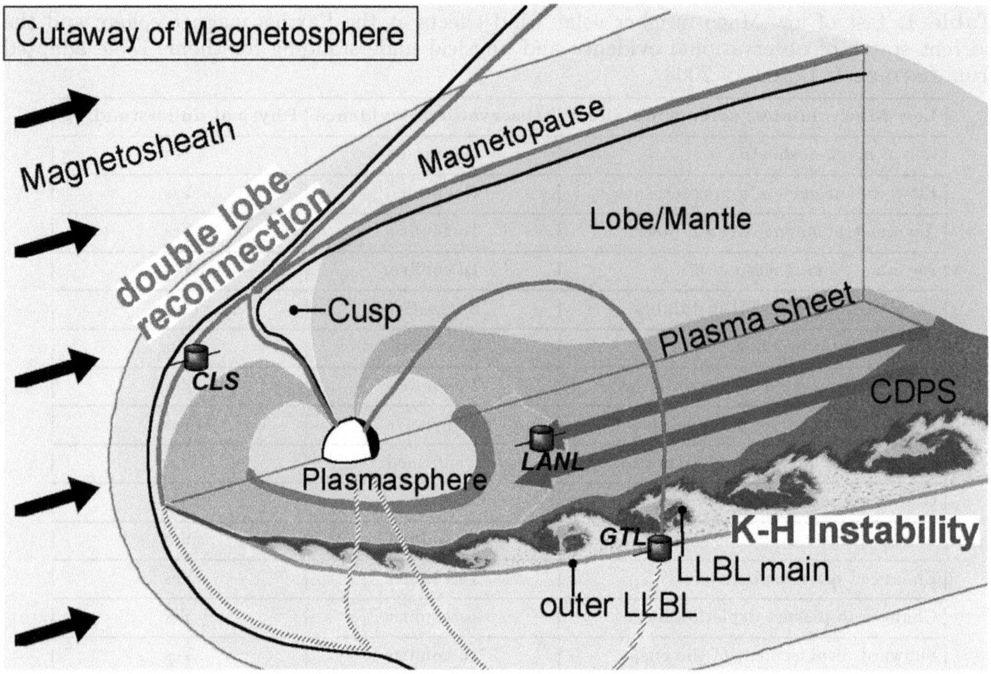


**Figure 6.** Structure of the magnetosphere with its main regions illustrated. Green field lines on the dayside depict the process known as double high-latitude reconnection, with a given northward-oriented IMF field line reconnecting in both the northern and southern hemisphere. This process creates newly closed flux tubes containing dense and cold plasma of solar wind origin, which may then be transported into the plasma sheet. An other CDPS formation mechanism is the Kelvin-Helmholtz (KH) instability. It is illustrated as waves on the flanks of the magnetopause. This mechanism may also allow direct entry of the cold and dense solar wind plasma into the magnetosphere. Figure courtesy of Kanako Seki (Nagoya University).

1. All these phenomena tend to occur simultaneously, so that solar wind-magnetosphere interaction is drastically altered as compared to the more common high Mach number interaction. Whether and how these effects come into play at other bodies is mostly unknown. It will be of particular interest to study Mach number effects at Mercury, with Messenger and Bepi-Colombo, since Mach numbers are typically lower closer to the Sun.

### 4.4. *Magnetospheric preconditioning mechanisms*

Specifics of the solar wind - magnetosphere interaction at Earth also include interesting processes known as pre-conditioning mechanisms. These arise from the fact that the state of the magnetosphere before strong driving changes how the system will react to that driving.

The first such process to mention is the formation of a cold and dense plasma sheet (CDPS) in the magnetotail. Such a CDPS may form during intervals of northward-oriented magnetic field. The processes responsible for the formation of the CDPS are depicted in Figure 6, and include primarily double high-latitude reconnection (Song *et al.* 2012) and the Kelvin-Helmhotz instability (e.g., Fujimoto & Terasawa 1995). The importance of the formation of a CDPS then comes from the fact that a CDPS is capable of producing a stronger ring current during the first hours of an ensuing storm (cf. Lavraud & Jordanova 2007 for details). The formation of a CDPS may easily occur just prior to CME arrival at Earth.

An other preconditioning mechanism is mediated by the properties of the plasmasphere (cf. Figure 6), in terms of its spatial extent and density. Indeed, a fully formed plasmasphere will extend all the way to the magnetopause (in the form of plumes) during enhanced convection in storm time. Because the plasmasphere is made of extremely dense and cold plasma, if this plasma is entrained into the reconnection region at the dayside magnetopause it will substantially lower the Alfvén speed locally. As previously discussed, this will affect the local reconnection rate in such a way that reconnection is chocked. This effect was demonstrated through statistical analysis of magnetospheric response during plasmaspheric plume observation (Borovsky & Denton 2006).

Other preconditioning mechanisms exist, for instance related to ring current and radiation belt, as well as to plasma transport in the magnetosphere or effects due to the composition of the plasma sheet as a function of ion outflows (e.g., with implications for reconnection rates in the magnetotail).

## 5. Conclusions

This review aimed to provide a discussion of our current knowledge of the processes that determine CME geo-effectiveness. There are many other additional properties and processes that have not been addressed due to space limitation. While key properties such as the CME trajectory, its magnetic field strength/orientation, its speed and their duration are first and largely set by the eruption processes at the Sun, highly influential processes also operate during propagation in the interplanetary medium and in the specific interaction with the object considered. Interplanetary processes in particular are capable of very significant alterations of CME geo-effectiveness (e.g., shock, compression, erosion). Those often are more than secondary in impact.

The impacts of each property and parameter still need to be fully understood and quantified. With regards to predictive capabilities, we note that among all the parameters that determine CME geo-effectiveness, predicting the North-South component of the magnetic field from solar data remains a key issue. As readers will notice, this review builds on the well-known work by Gosling (1993), the solar flare myth where he stated: "In this paper I outline a different paradigm of cause and effect that removes solar flares from their central position in the chain of events leading from the Sun to near-Earth space. Instead, this central role is given to events known as coronal mass ejections."

**Acknowledgments.** The authors wish to thank Brigitte Schmieder, Etienne Pariat, Jean-Marie Malherbe, and the other IAU300 conference organizers for the invitation to present at the conference and to write this review.

### References

Akasofu, S.-I. 1981, *Space Sci. Rev.*, 28, 2, 121
Borovsky, J. E. & Denton, M. H. 2006, *Geophys. Res. Lett.*, 33, L20101
Borovsky, J. E. 2008, *J. Geophys. Res.*, 113, A8, A08228
Bothmer, V. & Schwenn, R. 1998, *Ann. Geophys.*, 16, 1
Burlaga, L. F. *et al.* 1982, *Geophys. Res. Lett.*, 9, 1317
Burton, R. K., McPherron, R. L., & Russell, C. T. 1975, *J. Geophys. Res.*, 80, 4204
Cane, H. V., & Richardson, I. G., St. Cyr, O. C. 2000, *Geophys. Res. Lett.*, 27, 3591
Cargill, P. J., Chen, J., Spicer, D. S., & Zalesak, S. T. 1995, *Geophys. Res. Lett.*, 22, 647
Chen, J. 1996, *J. Geophys. Res.*, 101, 27 499
Dasso, S., Nakwacki, M. S., Demoulin, P., & Mandrini, C. H. 2007, *Sol. Phys.*, 244, 115
Delannée, C. & Aulanier, G. 1999, *Sol. Phys.*, 190, 107
Démoulin, P. & Dasso, S. 2009, *A&A*, 498, 551

Dungey, J. W. 1961, *Phys. Rev. Lett.*, 6, 47
Erkaev, N. V., *et al.* 2012, *Geophys. Res. Lett.*, 39, L01103
Fenrich, F. R. & Luhmann, J. G. 1998, *Geophys. Res. Lett.*, 25, 2999
Fujimoto, M. & Teresawa, T. 1995, *J. Geophys. Res.*, 100, 12025
Gonzalez, W. D. & Mozer, F. S. 1974, *J. Geophys. Res.*, 79, 4186
Gopalswamy, N., Lara, A., Lepping, R. P., Kaiser, M. L., & Berdichevsky, D., St. Cyr, O. C. 2000, *Geophys. Res. Lett.*, 27, 145
Gopalswamy, N., Lara, A., Yashiro, S., Kaiser, M. L., & Howard, R. A. 2001, *J. Geophys. Res.*, 106, 29 207
Gosling, J. T., Bame, S. J., McComas, D. J., & Phillips, J. L. 1990, *Geophys. Res. Lett.*, 17, 901
Gosling, J. T. 1993, *J. Geophys. Res.*, 98, A11, 18937
Jian, L., Russell, C. T., Luhmann, J. G., & Skoug, R. M. 2006, *Sol. Phys.*, 239, 393
Lavraud, B. & Jordanova, V. 2007, *Geophys. Res. Lett.*, 34, L02102
Lavraud, B. & Borovsky, J. E. 2008, *J. Geophys. Res.*, 113, A00B08
Lavraud, B., Ruffenach, A., Kajdic, P., Manchester, W. B., & Lugaz, N. 2013, *J. Geophys. Res.*, submitted
Lopez, R. E., Bruntz, R., Mitchell, E. J., Wiltberger, M., Lyon, J. G., & Merkin, V. G. 2010, *J. Geophys. Res.*, 115, A12, A12216
Lugaz, N., Manchester, W. B., & IV, Gombosi, T. I. 2005, *ApJ*, 634, 651
McComas, D., *et al.* 1988, *ApJ*, 93, 2519
Mulligan, T. & Russell, C. T., Luhmann J. G. 1988, *Geophys. Res. Lett.*, 25, 2959
Newell, P. T., Sotirelis, T., Liou, K., Meng, C.-I., & Rich, F. J. 2007, *J. Geophys. Res.*, 112, A01206
OBrien, T. P. & McPherron, R. L. 2000, *J. Geophys. Res.*, 105, 7707
Owens, M. J. 2006, *J. Geophys. Res.*, 111, A12109
Owens, M. J., Merkin, V. G., & Riley, P. 2006, *J. Geophys. Res.*, 111, A03104
Perreault, P. & Akasofu, S.-I. 1978, *Geophys. J. R. Astr. Soc.*, 54, 547
Rouillard, A. P., *et al.* 2011, *ApJ*, 734, 7
Rouillard, A. P., *et al.* 2010, *ApJ*, 719, 1385
Rouillard, A. P., *et al.* 2009, *Sol. Phys.*, 256, 307
Ruffenach, A., *et al.* 2012, *J. Geophys. Res.*, 117, A9, A09101
Russell, C. T., *et al.* 2013, *ApJ*, 770, 38
Shepherd, S. G. 2007, *J. Atmos. Solar-Terr. Phys.*, 69, 234
Siscoe, G. & Crooker, N. 1974, *Geophys. Res. Lett.*, 1, 17
Siscoe, G. 2011, *J. Atmos. Solar-Terr. Phys.*, 73, 402
Subramanian, P., Lara, A., & Borgazzi, A. 2012, *Geophys. Res. Lett.*, 39, L19107
Swisdak, M. & Drake, J. F. 2012, *Geophys. Res. Lett.*, 34, L11106
Tenfjord, P. & Østgaard, N. 2013, *J. Geophys., Res.*, in press
Vrsnak, B. *et al.* 2010, *A&A*, 512, A43
Wang, Y.-M. *et al.* 2000, *J. Geophys. Res.*, 105, 25133
Zhang, J., Dere, K. P., Howard, R. A., & Bothmer, V. 2003, *ApJ*, 582, 520
Zuccarello, F. P., Bemporad, A., Jacobs, C., Mierla, M., Poedts, S., & Zuccarello, F. 2012, *ApJ*, 744, 14

*Nature of Prominences and their role in Space Weather*
*Proceedings IAU Symposium No. 300, 2013*
*B. Schmieder, J.-M. Malherbe & S. T. Wu, eds.*

doi:10.1017/S1743921313011101

# Clarifying some issues on the geoeffectiveness of limb halo CMEs

**Consuelo Cid[1], Hebe Cremades[2], Angels Aran[3], Cristina Mandrini[4,5], Blai Sanahuja[3], Brigitte Schmieder[6], Michel Menvielle[7,8], Luciano Rodriguez[9], Elena Saiz[1], Yolanda Cerrato[1], Sergio Dasso[10,4], Carla Jacobs[11], Chantal Lathuillere[12], and Andrei Zhukov[9]**

[1]Departamento de Física y Matemáticas, Universidad de Alcalá,
Campus Universitario, A- II, km. 33,600, E-28871 Alcalá de Henares (Madrid), Spain
email: consuelo.cid@uah.es

[2]Universidad Tecnológica Nacional-Facultad Regional Mendoza /CONICET,
Rodriguez 243, M5502AJE, Mendoza, Argentina

[3]Departament d' Astronomia i Meteorologia and Institut de Ciències del Cosmos, Universitat de Barcelona, Martí i Franquès 1, E-08028 Barcelona, Spain

[4]Instituto de Astronomía y Física del Espacio, IAFE, CONICET-UBA,
CC 67 Suc. 28, 1428 Buenos Aires, Argentina

[5]Facultad de Ciencias Exactas y Naturales, UBA, Buenos Aires, Argentina

[6]Observatoire de Paris, LESIA, 92290, Meudon, France

[7]CNRS/INSU, Universitè Versailles St-Quentin,
LATMOS-IPSL, Guyancourt, France

[8]Univ Paris-Sud, Dèpartement des Sciences de la Terre, Orsay, France

[9]Solar-Terrestrial Center of Excellence SIDC,
Royal Observatory of Belgium, Av. Circulaire 3, 1180, Brussels, Belgium

[10]Departamento de Física, Facultad de Ciencias Exactas y Naturales,
UBA, Buenos Aires, Argentina

[11]Space Applications Services NV
Leuvensesteenweg 325, B-1932 Zaventem, Belgium

[12]UJF-Grenoble 1 / CNRS-INSU, Institut de Planètologie et d'Astrophysique de Grenoble (IPAG) UMR 5274, Grenoble, F-38041, France

**Abstract.** A recent study by Cid *et al.* (2012) showed that full halo coronal mass ejections (CMEs) coming from the limb can disturb the terrestrial environment. Although this result seems to rise some controversies with the well established theories, the fact is that the study encourages the scientific community to perform careful multidisciplinary analysis along the Sun-to-Earth chain to fully understand which are the solar triggers of terrestrial disturbances. This paper aims to clarify some of the polemical issues arisen by that paper.

**Keywords.** Sun: coronal mass ejections (CMEs), Sun: solar-terrestrial relations

## 1. Introduction

Since the beginning of the 80*s*, when Howard *et al.* (1982) reported the observation of a coronal transient directed at Earth as a bright halo around the coronagraphs of LASCO instrument on board SoHO, the term 'halo CME' has been associated with an ejection of material from the solar corona with a large probability to disturb the terrestrial environment, especially if it travels fast and if the magnetic field embebed on the interplanetary counterpart is properly oriented. However, both terms ('CME' and 'halo') are only related to the way something is observed, i.e., the term 'CME' is not related to material from the solar corona, but to the ejection recorded by an instrument designed to observe the solar corona, i.e., a coronagraph (Hudson *et al.* 2006). Also the term 'halo' does not mean material directed towards or away from the Earth, but just a bright halo that appears to surround the occulting disk of a coronagraph.

A very simple model for a full halo CME is that proposed by Howard *et al.* (1982), where the observed bright halo will consist on the 2D projection of solar material propagating approximately towards or away from the Earth, i.e., from the observer. In this model, a solar source close to the solar central meridian appears as a condition to the full halo signature. Otherwise, the solar source should be located at the back side and the CME will travel away from our planet. In this scenario, no full halo CME should arise close to the limb, and all full halo CMEs with a solar source close to the solar central meridian should reach the Earth. However, it is easy to find examples to prove that both statements are far to be right and that full halo CMEs close to the limb should not be discarded as potential hazards to the terrestrial environment.

An additional issue is to estimate the time of arrival of a solar ejection at the Earth. There are several CME and interplanetary shock propagation models, both physic-based and empirical (see for example Gopalswamy *et al.* 2001, Fry *et al.* 2003 and references therein). Checking the prediction capability of some of these models, Cho *et al.* (2003) obtained that there were 20% of cases where the predictions exceeded 24 hours in errors. Among others, they proposed as possible reasons for this large uncertainty a complex heliospheric environment involving interactions between shocks or interplanetary CMEs, or even some misidentifications, i.e., a wrong linkage between the solar source and the interplanetary consequence. In fact these two reasons might have also been involved in the different results obtained by analyzing the geoeffectiveness of halo CMEs on the basis of a large set of solar observations from statistical analysis. Just as an example, for a similar period of time, Gopalswamy *et al.* (2007) found that geoeffectiveness of frontside halo CMEs was 71% while Kim *et al.* (2005) obtained 40%.

## 2. Full halo CMEs with a solar source close to the solar limb

Figure 1 shows C2 and EIT images during four full halo CMEs that disturbed the Earth during solar cycle 23 (Cid *et al.* 2012). The solar sources of these events are listed in Table 1. In all cases their position is less than 10 degrees far from the solar limb. Three of them originate from the West limb and one is from the East limb, showing a western bias already found by Wang *et al.* (2002). These events not only prove the existence of full halo CMEs from the limb but all of them also present two common signatures:

(1) they are associated with flares of X (or almost) class (see Table 1) and

(2) they are highly asymmetric and they exhibit a bright feature close to the limb where the solar source is located.

The first signature, where three X-class X-ray flares and one M9.1 are related to the CMEs on Figure 1, indicates a huge amount of energy released. This suggests that the active region where both the flare and the CME originate is able to produce powerful events. It is so, that the large brightening coming from the limb is able to surround the whole coronagraph resulting in a full halo CME. This kind of events are far from the 2D projection model proposed by Howard *et al.* (1982).

The second signature indicates a large difference between these limb halo CMEs and the 'gradually expanding, Sun-centered excess brightness' originated by the sudden disappearance of a large filament at N05W03 and a relatively minor solar flare at N18E05, reported by Howard *et al.* (1982). This asymmetric excess brightness relative to solar

**Table 1.** Geoeffective LFH CMEs of solar cycle 23.

| CME date | DoY | CME time [UT] | CME speed [km/s] | Flare class | Flare max [UT] | Solar source | Flare location | $Dst_{min}$ [nT] | $Dst_{min}$ [DoY/hh] |
|---|---|---|---|---|---|---|---|---|---|
| 01 10 2001 | 274 | 05:30:05 | 1405 | M9.1 | 05:15 | AR9628 | S18W90 | -196 | 276 15 |
| 21 04 2002 | 111 | 01:27:20 | 2393 | X1.5 | 01:51 | AR9906 | S14W84 | -57 | 113 16 |
| 15 06 2003 | 166 | 23:54:05 | 2053 | X1.3 | 23:56 | AR10386 | S07E80 | -141 | 169 10 |
| 04 11 2003 | 308 | 19:54:05 | 2657 | X28 | 19:50 | AR10486 | S19W83 | -33 | 313 20 |

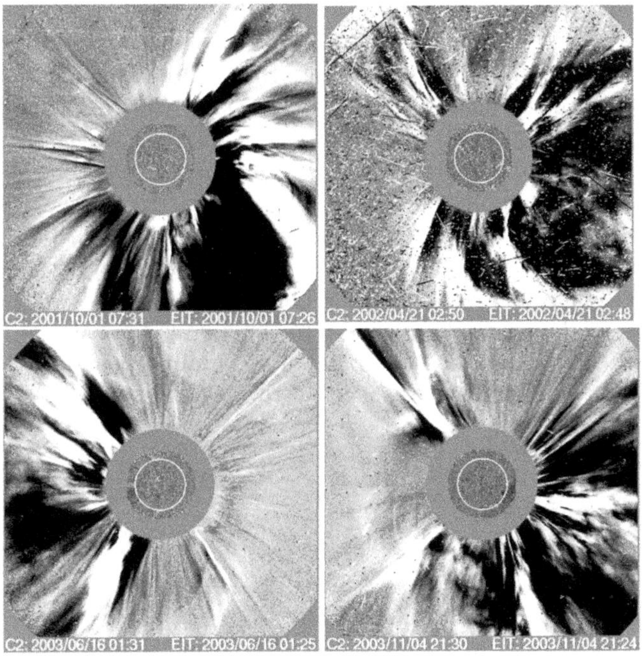


**Figure 1.** Four limb full halo CMEs that disturbed the terrestrial environment.

center, shown in the coronagraph images in Figure 1, may be used as an indicator of the location of the solar source. Moreover, the direction of maximum brightness might be related to the propagation direction, therefore, the direction of the interplanetary counterpart of the CME can be guessed (Cremades and Bothmer 2005).

## 3. Reaching the terrestrial environment from the solar limb

A controversial issue still remains in the scenario described above, regarding the fact that the interplanetary counterpart of a CME from the limb could reach the Earth and disturb it. The key point to solve the problem is to consider that disturbing the terrestrial environment does not mean that the ejected solar material during the CME reaches the Earth. But, how is it possible to disturb the Earth without reaching it? As indicated in Table 1, the speed of the four limb full halo CMEs of solar cycle 23 which are geoeffective is really large. As a result, all of them drive fast interplanetary shocks. Using Helios spacecraft data, de Lucas *et al.* (2011) found that 50% of CME driven shocks can be found generally to manifest themselves over ecliptic longitudinal distances of $90^0$. Considering the expansion in longitude of shock waves, de Lucas *et al.* (2011) concluded that one can expect with about a 50% chance that the shock driven by a limb CME will hit the Earth. Indeed, not only the shock driven by the CME but even the turbulent extended sheath behind the shock may disturb the terrestrial environment.This seems to be what happened in the case of the four full halo CMEs from Table 1, although only 16% of the limb halo CMEs analyzed by Cid *et al.* (2012) disturbed the Earth.

Moreover, Cid *et al.* (2012) showed that to assess the severity of a geomagnetic storm, not only the $Dst$ index needs to be checked but also other geomagnetic indices have to be analysed because the $Dst$ is only a proxy for the disturbances occurring at low latitudes. As an example, after the CME on November 4, 2003 the $Dst$ and $SYM - H$ indices did not fall below the –50 nT threshold. However, this CME is related to an active period (starting with a sharp increase in $SYM - H$ at around 19:40 UT on November 6) preceded and followed by periods of low magnetic activity in other indices as $ASY - H$, $AL$ or $am$, which clearly indicate a disturbed terrestrial environment.

As a final issue, we insist on the difficulty in tracking any event from its solar source up to its terrestrial consequences. The footprints at different stages are usually difficult to connect. Frequently, interplanetary Type II emissions and solar energetic particle data provide a way to glue some pieces but sometimes to choose among the different possibilities to fix the puzzle requires a thorough study to come up with a solution. Broadly speaking, the correct choice to build up the whole Sun-to-Earth chain may be quite different from that obtained by merely applying the time-window method commonly used in statistical studies. Detailed studies are the unique procedure to reach the right linkage among the solar source, the interplanetary counterpart and the terrestrial consequences, and even in this case evidences collected may not be enough to uniquely solve the problem.

## 4. Acknowledgements

The authors would like to thank to the International Space Science Institute (Bern, Switzerland) for supporting the project “From the Sun to the terrestrial surface: understanding the chain”, led by C. Cid and all data sources used in this work. C.C. thanks IAU for the support to participate in the IAUS300 Symposium.

## References

Cho, K.-S. *et al.* 2003, *J. Geophys.Res.*, 108, 1445, doi:10.1029/2003JA010029

Cid, C., *et al.* 2012, *J. Geophys. Res.*, 117, A11102, doi:10.1029/2012JA017536

Cremades, H. & Bothmer, V. 2005, *Coronal and Stellar Mass Ejections Proceedings IAU Symposium*, No. 226 , 2005, K. P. Dere, J. Wang & Y. Yan, eds., doi:10.1017/S174392130500013X

de Lucas, A., *et al.* 2011, *J. Atm. Sol. Terr. Phys.*, 73, 1281, doi:10.1016/j.jastp.2010.12.011

Fry, C. D., *et al.* 2003, *J. Geophys. Res.*, 108 , 1070, doi:10.1029/2002JA009474

Gopalswamy, N., Lara, A., Yashiro, S., Kaiser, M., & Howard R. A. 2001, *J. Geophys. Res.*, 106, 29,207, doi:10.1029/2001JA000177

Gopalswamy, N., Yashiro, S., & Akiyama, S. 2007, *J. Geophys. Res.*, 112, A06112, doi:10.1029/2006JA012149

Howard, R. A., Michels, D. J., Sheeley Jr., N. R., & Koomen, M. J. 1982, *Ap. J. Let.*, 263, 10.1

Hudson, H. S., Bougeret, J.-L., & Burkepile, J. 2006, *Space Sc. Rev.*, 123, 13

Kim, R.-S. *et al.* 2005, *J. Geophys. Res.*, 110, A11104, doi:10.1029/2005JA011218

Wang, Y. M., Ye, P. Z., Wang, S., Zhou, G. P., & Wang, J. X. 2002, *J. Geophys. Res.*, 107, 1340, doi:10.1029/2002JA009244

*Nature of Prominences and their role in Space Weather*
*Proceedings IAU Symposium No. 300, 2013*
*B. Schmieder, J.-M. Malherbe & S. T. Wu, eds.*

doi:10.1017/S1743921313011113

# The in-situ manifestation of solar prominence material

**Susan T. Lepri, Thomas H. Zurbuchen, Jacob R. Gruesbeck, and Jason A. Gilbert**

Dept. of Atmospheric, Oceanic and Space Sciences,
The University of Michigan, Ann Arbor, MI 48109-2143
email: slepri@umich.edu
thomasz@umich.edu
jagruesb@umich.edu
jagi@umich.edu

**Abstract.** Coronal mass ejections observed in the corona exhibit a three-part structure, with a leading bright front indicating dense plasma, a low density cavity thought to be a signature of the embedded magnetic flux rope, and the high density core likely containing cold, prominence material. When observed in-situ, as Interplanetary CMEs (or ICMEs), the presence of all three of these signatures remains elusive, with the prominence material rarely observed. We report on a comprehensive and long-term search for prominence material inside ICMEs as observed by the Solar Wind Ion Composition Spectrometer on the Advanced Composition Explorer. Using a novel data analysis process, we are able to identify traces of low charge state plasma created during prominence eruptions associated with ICMEs. We find that the likelihood of occurrence of cold material in the heliosphere is vastly lower than that observed in the corona but that conditions during the eruption do allow low charge ions to make it into the solar wind, preserving their expansion history. We discuss the implications of these findings.

**Keywords.** coronal mass ejections, prominence, solar wind, etc.

## 1. Introduction

Coronal Mass Ejections (CMEs) occur as a result of a rapid reconfiguration of the coronal magnetic field following destabilization of the magnetic field on the Sun. This reconfiguration via magnetic reconnection explosively ejects large quantities of material, magnetic field, and energy out of the corona and into the heliosphere (Klimchuk *et al.*, 2001). The in-situ manifestation of CMEs are called Interplanetary Coronal Mass Ejections (ICMEs), which strongly influence space weather and the interaction of the Sun with the near-Earth space environment (Gosling *et al.*, 1974; Webb and Howard, 1994; Howard *et al.*, 1997).

The physics responsible for CME eruption have yet to be fully resolved. However, the majority of CMEs observed remotely at the Sun are associated with filament eruptions (e.g., Gopalswamy *et al.*, 2003; Webb & Hundhausen 1987; Munro *et al.*, 1979), in which the filament is lifted out of the Sun's gravitational well and can become embedded in the CME (e.g., Webb & Hundhausen 1987, and references therein). Filaments are cool ($T \sim 10^4$ K) quasi-stationary structures in the low corona which sit over magnetic neutral lines. They form in locations with high plasma density, where radiative cooling strongly affects the plasma (Karpen & Antiochos 2008). In these regions, low-charge ions (($He^+$, $O^+$, $O^{2+}$, $C^{2+}$, etc.) and neutral atoms are likely to contribute to a significant portion of the local material.

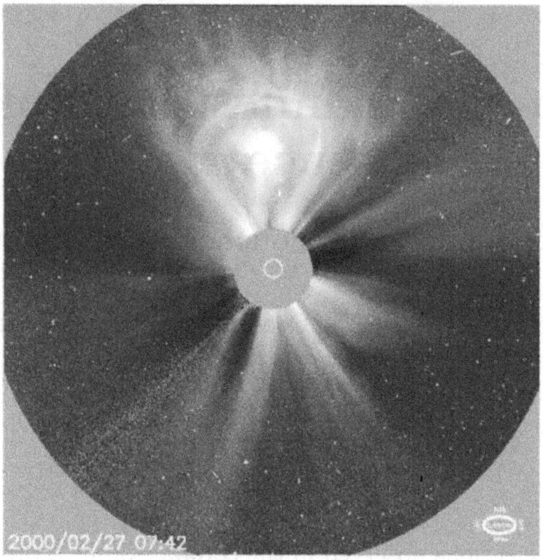


**Figure 1.** A CME with a clear three-part structure launches off the northern limb of the Sun and is observed by SOHO LASCO. The bright front rings the top of the exploding bubble, with a less dense cavity contained within. The bright central region is the core which is thought to contain the cold remnants of the erupted filament. Adapted from Riley *et al.* (2008).

As a CME erupts, it lifts filament and possibly flare material out of the low corona and carries it into the heliosphere along with the solar wind. The filament material plays a key role in forming the three-part structure commonly observed in coronagraph images as shown in Figure 1. The "bright front" is formed as the faster moving CME interacts with the ambient solar wind ahead of the ejecta. As the front forms, a low density region is left behind, forming the "cavity" which likely contains stronger magnetic fields ejected from low in the corona (Lynch *et al.*, 2004; Low 1994). The trailing feature of the three-part structure is the dense "core", consisting of cool filament plasma, which is thought to occupy a small fraction of the volume of the CME.

Based on the fact that ~70% of CMEs observed near the Sun exhibit this three-part structure (e.g., Gopalswamy *et al.* 2003; Webb & Hundhausen 1987; Munro *et al.* 1979), one might expect to find frequent evidence of the same three-part structure in ICMEs observed in the heliosphere. However, this structure is rarely observed in-situ and presents an observational puzzle (Zurbuchen & Richardson 2006, and references therein). In fact, ICMEs typically exhibit only two of the three structures, with the front appearing as a density pile up ahead of the low density, magnetically dominated cavity, often with flux-rope type topology. The core, containing filament material, is rarely reported to be present.

In-situ, one would expect the core to be identifiable based on the presence of usually low charge states, indicating cool electron temperatures associated with the filament, low in the corona where the charge states are frozen into the plasma. Observations by the SOHO Ultra-Violet Corona Spectrograph (UVCS; e.g., Raymond 2002) show these low charge states and temperatures in the low corona, however, in-situ measurements with ion spectrometers have not revealed the same ubiquitous signatures. In order to preserve the filament's low charge state ions during their expansion out of the corona, plasma heating from the eruption must be ineffective at further ionizing the filament material.

It has been established that the vast majority of ICMEs exhibit high charge states, reflecting elevated coronal electron temperatures in the source region of the CME (Lepri *et al.* 2001; Henke *et al.* 2001; Richardson & Cane 2004). It is thought that the heating responsible for these elevated charge states is driven by energy released during magnetic reconnection, possibly in flaring regions. (Lepri & Zurbuchen 2004; Reinard 2005; Rakowski *et al.* 2007). In fact, nearly all ICMEs exhibit bimodal charge state distributions with contributions from hot plasma and solar wind-like plasma (Gruesbeck *et al.*, 2011). This "mix" of plasma is also seen in remote-sensing observations from UVCS, although the cool contribution is often observed remotely to be far cooler than the typical solar wind type temperatures, and resembles the filament type cold plasma. Why then, do we not frequently see this cold plasma in the heliosphere?

Early observations with electrostatic analyzers did reveal the rare presence of cold ions in the solar wind associated with observations of ICMEs. Schwenn *et al.* (1980) and Zwickl *et al.* (1982) found 3 ICMEs over an 8 year period with enhanced He+. Zwickl *et al.* also found evidence for low charge states of heavy ions ($Fe^{5+}$, $O^{2+}$, and $C^{4+}$), although the measurement techniques at that time were not able to unambiguously determine the ions' identities. Until recently, only 2-3 more events were identified in in-situ observations (Burlaga *et al.* 1998; Skoug *et al.* 1999), with Gloeckler *et al.* (1998) presenting the first unambiguous determination of the presence of a range of cold ions, as measured with the Solar Wind Ion Composition Spectrometer (SWICS) on the Advanced Compsoition Explorer (ACE).

We review the first systematic search for cold ions originating in filament material embedded within ICMEs in this paper. A more in depth discussion of this study can be found in Lepri and Zurbuchen (2010). We utilize ACE SWICS data and capitalize on the ion retrieval algorithms that isolate and characterize charge states down to $C^{2+}$, $O^{2+}$ and $Fe^{4+}$ using observational techniques that significantly limit contributions from background noise sources. We report that the presence of low charge state material inside of ICMEs is more common than previously thought, but much more rare than expected based on remote observations.

## 2. Measurements

SWICS on ACE is a time-of-flight (TOF) mass spectrometer. Particles enter the instrument through an electrostatic analyzer (ESA) which selects ions based on their energy per charge (E/Q). After exiting the ESA, they pass through the TOF telescope and deposit their energy (E) in solid state detectors (SSDs). This technology can be used to uniquely identify a wide range of ions (from He to Fe) in the solar wind (Gloeckler *et al.*, 1998). Combining the ion's E/Q, E and TOF allows the the mass, charge and speed of an ion to be independently calculated. Integrating over one or more cycles in integer multiples of 12 minutes, one can obtain the velocity and charge state distribution of the thermal solar wind in the E/Q range of 0.5 - 60 keV/e as it passes the spacecraft. For this analysis we examine the charge state distributions calculated with 2-hour time resolution– long enough to provide sufficient counting statistics for reliable C, O and Fe charge states. More detail on the methods used to calculate the charge states can be found in Von Steiger *et al.* (2000), Lepri *et al.* (2001), Lepri and Zurbuchen (2004, 2010).

In general, the typical unperturbed solar wind exhibits charge states consistent with the 1MK corona, with the main ion contributors being $H^{+}$, $He^{2+}$, $C^{5+}$, $O^{6+}$ and $Fe^{10+}$ in the slower, equatorial solar wind as shown in the green histograms in Figure 2. Most ICMEs are characterized by elevated charge states, including significant contributions from $O^{7+}$ and $Fe^{16+}$ and higher as shown in the red histograms in Figure 2 (e.g. Lepri *et al.*, 2001, Lepri and Zurbuchen, 2004 and Zurbuchen and Richardson, 2006). These high charge states set ICME-associated solar wind apart from the nominal solar wind state, as the high charge states reflect temperatures ~4-5MK in the source regions. As a

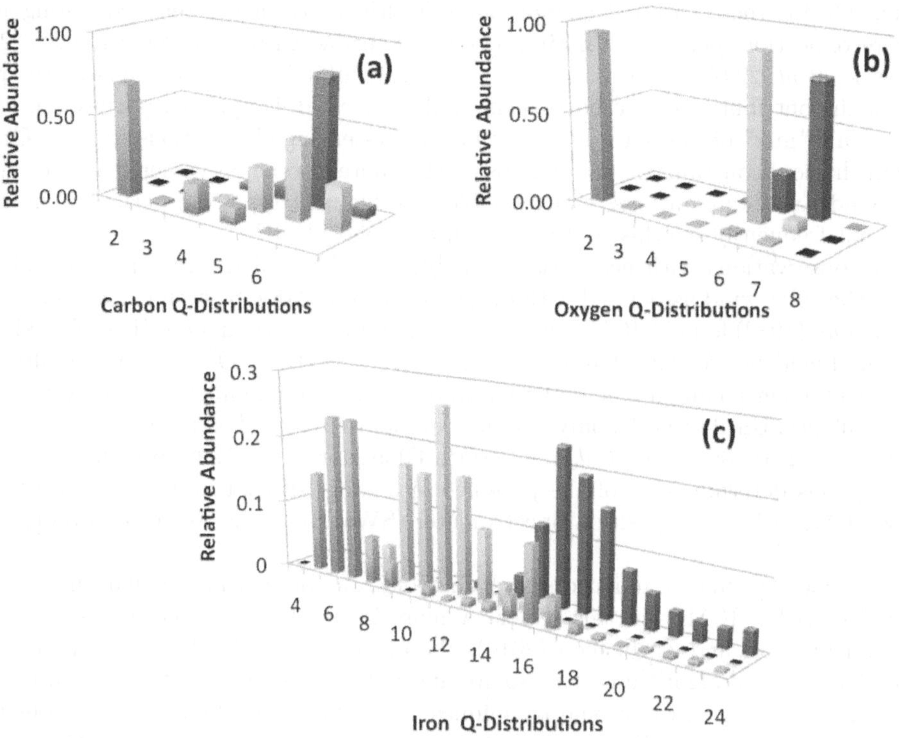


**Figure 2.** Charge state distributions measured by ACE/SWICS for C, O, and Fe for three different scenarios: a cold ICME (in blue), normal solar wind (in green), and a hot ICME (in red). Adapted from Lepri and Zurbuchen (2010).

result, these elevated charge states are excellent identifiers of ICME material in-situ. In the event that an ICME is characterized by unusually low charge states (shown in the blue histograms in Figure 2), as in the May 1-2, 1998 event first reported in Gloeckler *et al.* (1999), these low charge states can be observed simultaneously with contributions from the high charge states more typically associated with ICMEs.

## 3. Cold ICMEs

In order to mine a decade long range (1998-2008) of the ACE/SWICS data set for cold ICMEs, we compare periods with enhanced abundances of low charge state ions with known near-Earth ICME periods (as identified by Richardson and Cane (2010)). A stringent set of criteria for C, O, and Fe are established to identify periods with cold material and reduce the number of false positives due to local sources of low charge ions (e.g. comets, planetary sources). In order to identify intervals with enhanced contributions from low charge states, using 2-hour data from ACE/SWICS, we calculate relative abundances of $\frac{C^{2-3+}}{C_{Tot}}$ , $\frac{O^{4+}}{O_{Tot}}$ and $\frac{Fe^{4-7+}}{Fe_{Tot}}$. We calculate the mean value of these ratios across the mission as well as the standard deviation. We then search for periods where all three ion ratios are greater than one standard deviation above their mean values, thus ensuring we are finding significant enhancements of solar origin. We include an additional criteria to eliminate further confusion from false positives by requiring the ion speed to be within 15% of the solar wind speed, in line with expectations for heavy ions in the solar wind (see Hefti *et al.*, 1998 for more details). To summarize, the criteria used to identify low charge state intervals are as follows:

$$\frac{|\Delta V|}{V} < 0.15$$
$$\frac{C^{2-3+}}{C_{Tot}} > 0.034$$
$$\frac{O^{4+}}{O_{Tot}} > 0.009$$
$$\frac{Fe^{4-7+}}{Fe_{Tot}} > 0.121$$

Using these highly constrained criteria, we find 11 events with significant contributions from low charge state ions during the 10 years surveyed, all of which correlate with ICME periods. An example of one event, 20-22 May 2005, is shown in Figure 3. Panels (a), (b) and (c) show the proton density, temperature and velocity from ACE/SWEPAM, respectively. Panel (d) shows the magnetic field magnitude which becomes enhanced just inside the leading edge of the ICME, denoted by the solid vertical red line. Charge state distributions of Fe, O and C are shown in Panels (e), (f), and (g); the color bars represent the relative density of individual charge states in the given 2-hour time interval. The charge states deviate from their normal behavior inside the front boundary of the ejecta, shifting to higher charge states. They return to normal, lower charge states after the trailing boundary passes, as is denoted by the dashed red vertical line, although there are periods where the charge states appear nominal inside the ejecta as well. In fact, the second half of the ejecta is dominated by charge states of Fe similar to the nominal solar wind with another minor contribution from the high charge states during this period. The period dominated by enhanced low charge state ions is highlighted in yellow around noon on 20 May 2005.

Of the 284 ICMEs which were observed by ACE at L1 during the observation period, only 4%, or 11 events, contained cool material. In order to investigate whether these events were related with filament eruptions that were swept into the expanding CMEs, we reviewed EIT 195Å movies and examined observations from the Mauna Loa Solar Observatory, looking for erupting or disappearing filaments. 90% of the events that have LASCO CME counterparts were associated with erupting filaments, thus lending credibility to the idea that this cold material originated inside of ICMEs. The 11 low charge state events identified in the 10-year span of data indicate that these events are at least an order of magnitude more frequent than previously measured. With less restrictive criteria, we may find these numbers are much higher. An examination of the location of the low charge material within the 11 events does not reveal spatial ordering within the ICMEs.

Gilbert *et al.* (2012) went on to include the singly charged particles in his analysis of the same cold ICMEs and found that the 11 events identified in Lepri and Zurbuchen (2010) also included contributions from the lowest charge states, including C+ and O+. In general, when these singly charge ions were included in the analysis, the events tended to last on average 20% longer than reported using the more restrictive criteria laid out in Lepri and Zurbuchen (2010).

## 4. Ionization Models

Ionization models can be utilized to understand how cold material is preserved within some ICMEs. Gruesbeck *et al.* (2011) found success modeling the ionization and recombination of observed bimodal charge states in ICMEs assuming rapid heating of a dense plasma ($>10^9$ cm$^{-3}$) followed by adiabatic cooling during expansion out of the corona. This scenario is similar to what is expected during flares and successfully reproduced

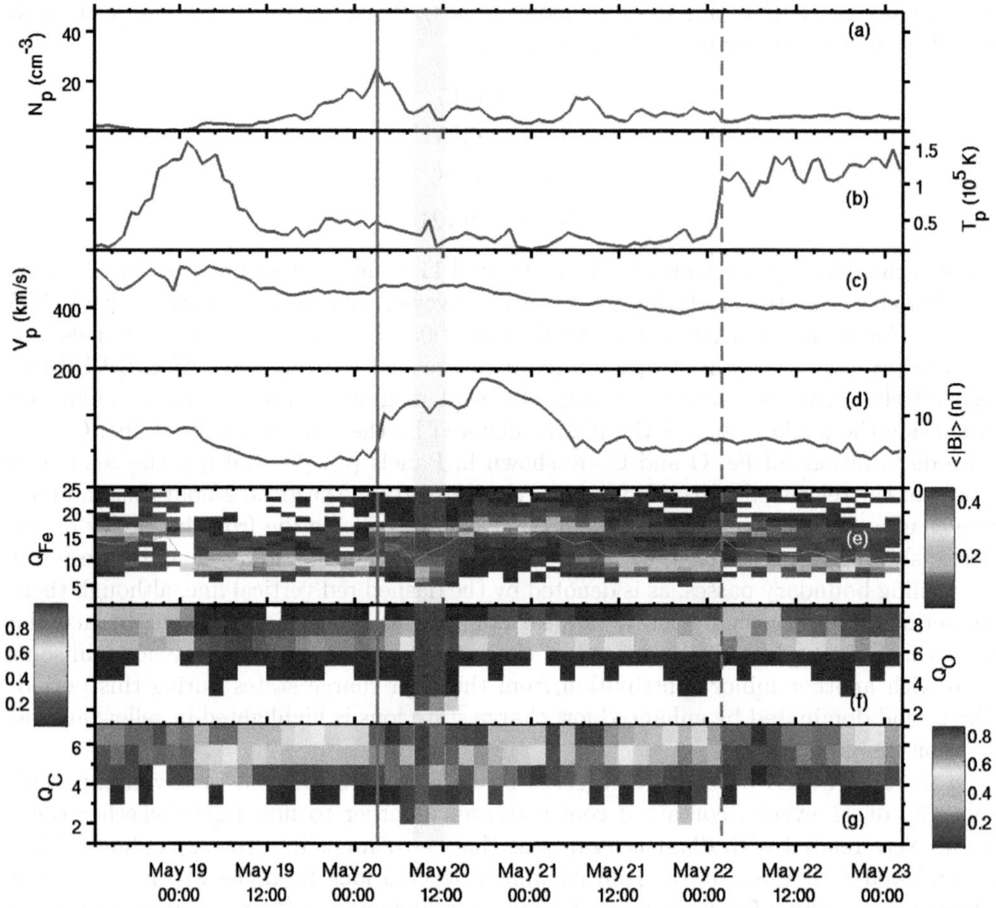


**Figure 3.** Solar wind plasma parameters and solar wind ion charge state distributions for an ICME associated with cold material (shaded in yellow in all panels). Panels are as follows: (a) Solar wind density, (b) Solar wind temperature, (c) Solar wind speed, (d) Solar magnetic field strength, (e) Fe charge distribution, (f) O charge distribution and (g) C charge distribution. Adapted from Lepri and Zurbuchen (2010).

Fe charge state distributions that had peaks at both $Fe^{10+}$ (nominal solar wind) and $Fe^{16+}$ (hot ICME solar wind). Gruesbeck *et al.* (2012) considers the challenges presented in explaining the simultaneous presence of both high charge state and low charge state material as is seen in the 11 events described above. Using the same ionization model as in the 2011 paper, they test two theories for the creation of the observed bimodal charge state distributions, with both very low and high charge states observed concurrently. They find that while a rapidly heated single dense plasma that subsequently cools adiabatically can produce bimodal charge state distributions for Fe as in the 2011 paper, they fail to produce the bimodal charge state distributions of C and O, especially failing to capture the low charge state peaks. However, by combining two different plasmas, which start off at the same temperature but which undergo different heating profiles, they were able to successful recover the observed bimodal charge state distribution. In this model, one plasma is rapidly heated to very high temperatures ∼3MK, similar to flare heated plasma, while the other plasma is an order of magnitude denser ($>10^{10}$ cm$^{-3}$) and is heated to much lower temperatures ∼100,000K representing the filament material. The key here is that the cooler plasma, due to its high density, has a chance to recombine to even lower charge states than the initial charge states before freezing into the solar wind. The best match of the observed charge state distributions is obtained with a mass-dependent mixing ratio model for the cold and hot plasma. This result is not unexpected as lower mass ions tend to freeze-in closer to the Sun while higher mass ions freeze farther out in the corona (out to $\sim 4R_s$). These results support the idea that disparate parts of the ICME can mix as magnetic reconnection rearranges the plasma during the eruption, allowing prominence plasma to mix with flare plasma during the eruption. Gruesbeck *et al.* (2012) suggests that this type of mixing supports the breakout model of CME eruption, in which reconnection allows the plasma to mix along field lines (Antiochos *et al.* 1999).

## 5. Summary

Despite the frequent association of filament eruptions with CMEs at the Sun, a dearth of in-situ evidence of such a linkage exists in the heliosphere. While the survey of Lepri and Zurbuchen (2010) increases the known number of ICMEs with prominence material embedded within by more than a factor of 4, questions still remain as to why these observations are so infrequent. Ionization model results suggest that in order to preserve the low charge state plasma a very dense cold plasma ($>10^{10}$ cm$^{-3}$) must initially be heated to temperatures ~100,000K and then given time to recombine further as it transitions toward the freeze-in point. It may be likely that the fraction of ICME mass filled with filament material is very low, thus lowering our probability of observing it with a single spacecraft. The low number of in-situ observations may also indicate that most filaments expand more rapidly or may be further heated during eruption, erasing the low charge state signatures of the filament by the time it freezes-into the solar wind.

## References

Antiochos, S. K., DeVore, C. R., & Klimchuk, J. A. 1999, *Astrophysical Journal*, 510, 485

Burlaga, L., *et al.* 1998, *Journal of Geophysical Research*, 103, 277

Gilbert, J. A., Lepri, S. T., Landi, E., & Zurbuchen, T. H. 2012, *Astrophysical Journal*, 751, 20

Gloeckler, G., *et al.* 1998, *Space Science Reviews*, 86, 497

Gloeckler, G., *et al.* 1999, *Geophysical Research Letters*, 26, 157

Gopalswamy, N., Shimojo, M., Lu, W., Yashiro, S., Shibasaki, K., & Howard, R. A. 2003, *Astrophysical Journal*, 586, 562

Gosling, J. T., Hildner, E., MacQueen, R. M., *et al.* 1974, *Journal of Geophysical Research*, 79, 4581

Gruesbeck, J. R., Lepri, S. T., & Zurbuchen, T. H. 2012, *Astrophysical Journal*, 760, 141

Gruesbeck, J. R., Lepri, S. T., Zurbuchen, T. H., & Antiochos, S. K. 2011, *Astrophysical Journal*, 730, 103

Hefti, S., *et al.* 1998, *Journal of Geophysical Research*, 103, 29697

Henke, T., Woch, J., Schwenn, R., Mall, U., Gloeckler, G., von Steiger, R., Forsyth, R. J., & Balogh, A. 2001, *Journal of Geophysical Research*, 106, 10597

Howard, R. A. *et al.*, Observations of CMEs from SOHO/LASCO, in *Coronal Mass Ejections*, edited by N. Crooker, J.-A. Joselyn, and Joan Feynman, pp. 17-26, Geophysical Monograph 99, Washington, DC., 1997

Karpen, J. T. & Antiochos, S. K. 2008, *Astrophysical Journal*, 676, 658

Klimchuk, J. A. 2001, *Space Weather (Geophysical Monograph 125)*, ed. P. Song, H. Singer, G. Siscoe (Washington: Am. Geophys. Un.), 143 (2001), 125, 143

Lepri, S. T., Zurbuchen, T. H., Fisk, L. A., Richardson, I. G., Cane, H. V., & Gloeckler, G. 2001, *Journal of Geophysical Research*, 106, 29231

Lepri, S. T. & Zurbuchen, T. H. 2004, *Journal of Geophysical Research (Space Physics)*, 109, 1112

Lepri, S. T. & Zurbuchen, T. H. 2010, Astrophysical Journall, 723, L22
Low, B. C. 1994, *Solar Dynamic Phenomena and Solar Wind Consequences*, the Third SOHO Workshop, 373, 123
Lynch, B. J., Antiochos, S. K., MacNeice, P. J., Zurbuchen, T. H., & Fisk, L. A. 2004, *Astrophysical Journal*, 617, 589
Munro, R. H., Gosling, J. T., Hildner, E., MacQueen, R. M., Poland, A. I., & Ross, C. L. 1979, *Solar Physics*, 61, 201
Rakowski, C. E., Laming, J. M., & Lepri, S. T. 2007, *Astrophysical Journal*, 667, 602
Raymond, J. C. 2002, *From Solar Min to Max: Half a Solar Cycle with SOHO*, 508, 421
Reinard, A. 2005, *Astrophysical Journal*, 620, 501
Richardson, I. G. & Cane, H. V. 2004, *Journal of Geophysical Research (Space Physics)*, 109, 9104
Richardson, I. G. & Cane, H. V. 2010, *American Institute of Physics Conference Series*, 1216, 683
Riley, P., Lionello, R., Mikić, Z., & Linker, J. 2008, *Astrophysical Journal*, 672, 1221
Schwenn, R., Rosenbauer, H., & Muehlhaeuser, K.-H. 1980, *Geophysical Research Letters*, 7, 201
Skoug, R. M., *et al.* 1999, *Geophysical Research Letters*, 26, 161
von Steiger, R., *et al.* 2000, *Journal of Geophysical Research*, 105, 27217.
Webb, D. F. & Howard, R. A. 1994, *Journal of Geophysical Research*, 99, 4201
Webb, D. F. & Hundhausen, A. J. 1987, *Solar Physics*, 108, 383
Zurbuchen, T. H. & Richardson, I. G. 2006, *Space Science Reviews*, 123, 31
Zwickl, R. D., Asbridge, J. R., Bame, S. J., Feldman, W. C., & Gosling, J. T. 1982, *Journal of Geophysical Research*, 87, 7379

*Nature of Prominences and their role in Space Weather*
*Proceedings IAU Symposium No. 300, 2013*
*B. Schmieder, J.-M. Malherbe & S. T. Wu, eds.*

doi:10.1017/S1743921313011125

# Interplanetary Disturbances Affecting Space Weather

**Robert F. Wimmer-Schweingruber**[1]

[1] Institute for Experimental and Applied Physics
Christian-Albrechts-University Kiel,
Kiel, Germany
email: wimmer@physik.uni-kiel.de

**Abstract.** The Sun somehow accelerates the solar wind, an incessant stream of plasma originating in coronal holes and some, as yet unidentified, regions. Occasionally, coronal, and possibly sub-photospheric structures, conspire to energize a spectacular eruption from the Sun which we call a coronal mass ejection (CME). These can leave the Sun at very high speeds and travel through the interplanetary medium, resulting in a large-scale disturbance of the ambient background plasma. These interplanetary CMEs (ICMEs) can drive shocks which in turn accelerate particles, but also have a distinct intrinsic magnetic structure which is capable of disturbing the Earth's magnetic field and causing significant geomagnetic effects. They also affect other planets, so they can and do contribute to space weather throughout the heliosphere. This paper presents a historical review of early space weather studies, a modern-day example, and discusses space weather throughout the heliosphere.

## 1. Introduction

On the tenth of September, 1580, the people of the town of Augsburg in Southern Germany saw "a large and very frightening magic sign in the sky after sunset" (Fig. 1). Its origins were unknown and even less understood, and hence the text in Fig. 1 continues in a religious manner. This heavenly appearance is likely to have been an aurora, a phenomenon usually not seen this far South in Europe. However, they were already known to the Romans, who termed them 'aurora', i.e., dawn, or the reddening of the morning sky. The description of the aurora in Fig. 1 is a witness that people were frightened by this unusual phenomenon which would take more than 300 years to be understood.

William Gilbert is traditionally credited with having initiated scientific studies of magnetism, especially the Earth's magnetism (Gilbert 1600). He correctly interpreted the deflection of a compass's needle as due to global terrestrial magnetism, stating that "Magnus magnes ipse est globus terrestris" (The Earth itself is a giant magnet). This then was an inadvertent first step towards understanding the origin of aurorae in a scientific way. Another ingredient is sunspots which as naked-eye phenomena were probably a puzzle already much earlier in history. For instance, Anaxagoras, a Greek philosopher, may have observed a sunspot in 467 BC (Bicknell 1968). Chinese astronomers reported sunspots in 164 B.C. (Clark & Stephenson 1978, Wittmann & Xu 1987), but we apparently have no records about Babylonian or Arab observations. It was Galileo who gave the correct interpretation despite the prevailing prejudice that sunspots must be transits of planets or moons (Galileo 2010). It is interesting to note that even in those times, priorities on discoveries were also bitterly disputed, in this case between Galileo and Scheiner (Galileo 2010). Sunspots were soon found to vanish nearly completely (during the so-called Maunder minimum (Eddy 1976)) and reappear around 1700. Schwabe (1844) showed clearly that the number of sunspots varies with a cycle (i.e., not an exact period) of roughly

**Figure 1.** Medieval print showing aurorae at the southern German city of Augsburg in 1580. Reprinted with permission by Zentralbibliothek Zürich, Graphische Sammlung und Fotoarchiv.

11 years. Around this time, Carl Friederich Gauß together with his assistant, Wilhelm Weber, was studying the Earth's magnetic field and variations thereof. They set up the "Magnetic Group" ("Magnetischer Verein") and with the aid of Alexander von Humboldt set up a world-wide network of magnetometers (Stern 2002). This allowed them to monitor rapid changes in the Earth's magnetic field and ultimately led to the discovery of the magnetosphere, ionosphere, and disturbances in their normal state. One famous such disturbance was - in retrospect - caused by the famous Carrington flare. Carrington had reported that "... when ... two patches of intensely bright and white light broke out in the positions ... A and B, in the forms of the spaces left white." The account continues and mentions that "Mr Carrington ... pointed out ... that toward four hours after midnight there commenced a great magnetic storm,... While the contemporary occurrence may deserve noting, he would not have it supposed that he even leans toward hastily connecting them. 'One swallow does not make a summer."' (Carrington 1859). Kristian Birkeland, a Norwegian researcher who studied the aurora in two expeditions in 1902 and 1903 (Birkeland 1908 and 1913), showed that magnetic storms and aurorae were intimately linked (Birkeland 1908), and that the Sun, solar (cathode) rays, and solar discharges cause them (Birkeland 1911).

Studies of ionizing radiation were much en vogue in the early twentieth century after the discovery of X-rays (Röntgen 1895) which earned him the Nobel proze in 1901. The penetrating nature of radiation was studied by Becquerel in 1896 (1896a, b) and by the Curies (1898a, b) which earned the three the Nobel prize in 1903 †, notably in the absence of the Curies. One of the researchers of these days was Victor Hess who investigated the ionization of air using an electrometer at different altitudes. He excluded the Sun as the ionizing source and concluded, after the final of a series of balloon flights, that there is a source of radiation which increases with height and enters the atmosphere from above (Hess 1912). He was not alone in investigating what Millikan later termed

† see `http://www.nobelprize.org/nobel_prizes/physics/laureates/1903/press.html`

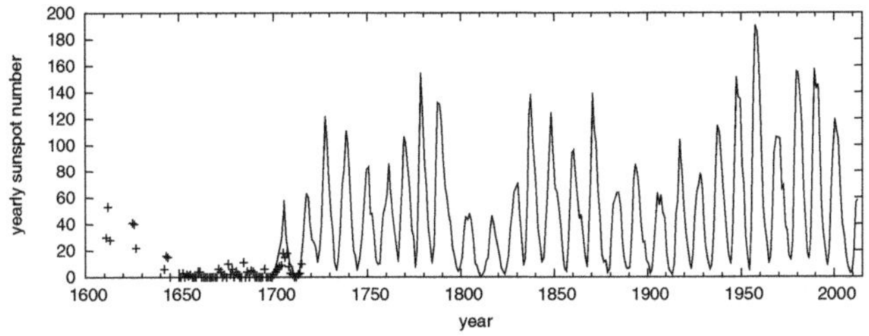


**Figure 2.** Yearly sunspot number plotted versus time. Early data are plotted as plus symbols, systematic data as a solid line. One easily recognizes the Maunder minimum in the second half of the seventeenth century. Data are from the SIDC-team (`http://www.sidc.be/sunspot-data/`).

'cosmic rays'. Pacini, an Italian researcher, performed similar experiments but also on the sea, on lakes, as well as under water. From these he also concluded that there must be radiation reaching the ground from above, i.e., through the atmosphere (Pacini 1912).

### 1.1. *Solar Activity*

The Sun is not constant, as can be seen in Fig. 2, which shows yearly sunspot numbers as obtained from `http://www.sidc.be/sunspot-data/` as a solid line and earlier data as plus symbols. The well-known Maunder minimum is clearly seen in the second half of the seventeenth century, a period of time when the Sun was virtually spotless. Later, solar activity waxes and wanes on a semi-regular basis which is termed the solar cycle (Schwabe 1844). There are historical reconstructions of solar activity beyond the sunspot record (e.g., Eddy 1976) using cosmogenic isotopes, notably $^{14}$C and $^{10}$Be which are produced by interactions of galactic cosmic rays (GCR) with the Earth's atmosphere and can then be measured in terrestrial archives. Possible linkages with terrestrial climate, as first discussed at length, e.g., in Eddy (1976) (but see Usoskin 2013 for a more recent account), are certainly interesting, but will not be addressed here, as it is not the intent of the paper. Nevertheless, reconstructions of historical solar activity may allow us to better understand space weather today because only they can put the observations into a larger context. We must not forget that we have only just begun to understand space weather during the space age which itself is very short compared to the age of the Sun, the sunspot record, or longer solar activity cycles such as the $\sim$ 80-year Gleissberg cycle (Wolf 1856, Gleissberg 1944, Eddy 1976). It is from such reconstructions that we understand that the Sun does not only vary with a 11 or 22-year cycle, but that there are longer cycles superimposed on this (e.g., Hathaway 2010), as can also be seen in Fig. 2. We can clearly see another minimum around 1800 (Dalton minimum) and another, less deep one around 1900. What is more conspicuous though, is the observation that the number of sunspots in the space age has been exceptionally high if viewed in a historical context. In fact, the sunspot number at maximum has been decreasing since the maximum in the late fifties, and there are indications that we may be leaving the 'Grand Solar Maximum' (e.g., Barnard *et al.* 2011) which has dominated the space age. This would entail an increase in the flux of cosmic rays, which indeed was observed during the last solar activity minimum in 2009/2010 when the intensities of heavy cosmic ray nuclei was $\sim$ 20% higher than in the previous minimum (Mewaldt *et al.* 2010).

I was asked to give a talk entitled "Interplanetary disturbances affecting space weather", and have just given a historical motivation for the study of space weather and some of the effects of space weather. It is now time to proceed with the actual topic - but what is it? What is a disturbance? What affects space weather? Where and what is space weather? This then is the plan of the paper: I will discuss the traditional aspects of space weather in the coming section (2), show that it matters at other locations than Earth in sections 3 and 3.2, and yes, even at other planets (section 3.3), and conclude in section 4.

## 2. What Affects Space Weather? – A Traditional Introduction

So what is space weather and what affects it? As already alluded to above, geomagnetists knew about the solar influence on the Earth's magnetic state (e.g., Birkeland 1908, 1911). More subtle variations than the spectacular aurorae were also uncovered to be induced by then Sun. Remarkably, even the virtually spotless Sun in the solar minimum around 1930 had a clear effect on the Earth's magnetism, as was recognized by Bartels (1934). The observation that there were steadily recurrent magnetic disturbances at Earth despite the fact that the Sun had hardly any spots (and thus was very inactive) had a profound implication. Looking at Figure 1 of Bartels (1934), one observes that the Sun has only few spots after about 1930 but that there is a strong pattern of recurrent geomagnetic disturbances which appeared to be stable for two or more years. Its frequency of recurrence was approximately once every 27 days, hinting at a link with the Sun and its 27-day rotation period. These recurrent magnetic disturbances were thought to be caused by so-called M-regions on the Sun (Bartels 1934) which remained, however, elusive, and could not be observed. Today we consider recurrent high-speed streams or rather their compressed leading edges (corotating interaction regions, CIRs) to be the origin of recurrent magnetic disturbances.

So it looks as if we have two ingredients of space weather at Earth, the leading edges of (not necessarily recurrent) high-speed streams or CMEs and/or particles from active regions on the Sun. Both can, but must not necessarily cause aurorae. Again, today we also understand the latter phenomenon a bit better. Active regions can somehow (we still disagree on how) be triggered (or trigger themselves) to rearrange magnetically, liberating a tremendous amount of energy in the form of radiation (photons and energetic particles) as well as (often, but not always) in the form of a coronal mass ejection (CME). Both can affect space weather, but in different ways. If you are interested in geomagnetic disturbances, then the CME is much more geoeffective than the energetic particles. If you are more interested in what will happen to astronauts in the International Space Station (ISS) or to electronics parts in satellites or other spacecraft, then the particles can be more important. So it looks very much as if space weather depends on your point of view! All three agents (high-speed streams (CIRs), flare particles, and CMEs) matter, but not at the same location and not for the same purpose.

CMEs or rather their interplanetary manifestations (ICMEs, see Wimmer-Schweingruber (2006), Wimmer-Schweingruber *et al.* (2006), Zurbuchen and Richardson (2006)) can strongly influence space weather. As CMEs leave the Sun, they carry with them an imprint of their solar origin in their magnetic structure (see, e.g., Crooker and Horbury 2006). Many ICMEs are magnetic clouds and can be recognized as a flux rope with a handedness (chirality) which is determined by Hale's law (Bothmer and Rust 1997). As they expand, their field becomes gradually more and more twisted in order to conserve both axial and azimuthal magnetic flux (e.g., Parker 1979). Thus, when a flux-rope ICME lies mainly in the ecliptic plane, it will have either at its leading or its trailing edge, a southward-pointing magnetic field component. A Southward-pointing leading edge, being compressed, is the more geoeffective because the Earth's magnetic field points roughly from the geographic South pole (approximately the magnetic North pole) to the geographic North pole (the magnetic South pole). Thus, the leading edge of the Earth's

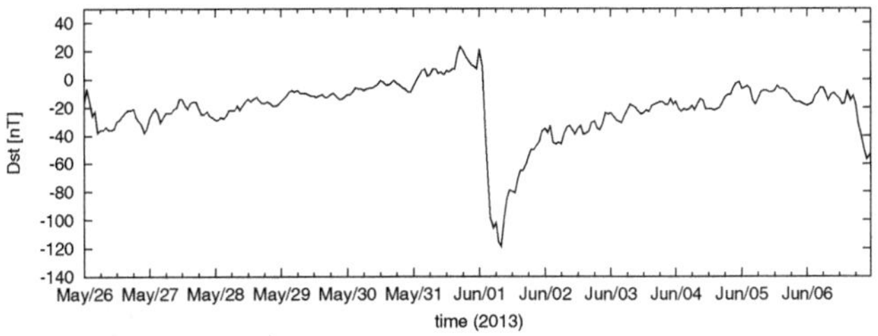


**Figure 3.** Major geomagnetic storm as observed on June 1, 2013. Data are from the Kyoto WDC (`http://wdc.kugi.kyoto-u.ac.jp/wdc/Sec3.html`).

magnetosphere has a northward-pointing magnetic field. As the southward interplanetary magnetic field (IMF) associated with an ICME hits the Earth's magnetic field, it can trigger magnetic reconnection (e.g., Parker 1979). In this case, the global field reconfigures itself, which induces a current, which in turn creates its own magnetic field. If the induced current returns in power lines or oil pipelines, this can adversely affect technological infrastructure. Today, such reconnection at the day side magnetosphere has been observed in situ, e.g., by the combined Cluster and Double Star missions (e.g., Dunlop *et al.* (2005)). Reconnection at the magnetotail and elsewhere can accelerate particles to sufficient energies that they can penetrate the uppermost layers of the Earth's atmosphere. After such a reconnection event, the polar regions are also open to the solar wind. Apart from the particles accelerated by the reconnection, the solar wind can als interact with upper atmospheric nitrogen and oxygen to form aurorae in the polar regions. The green and dark-reddish aurorae come from recombining oxygen, the blueish and light-red ones come from nitrogen.

The difficulty in predicting the geoeffectiveness of CMEs vs. CIRs was nicely illustrated in the week just before this workshop. A major geomagnetic storm with a Dst $< -100$ nT was observed at Earth on June 1, 2013, as shown in Fig. 3. What caused it? Initial inspections of images from the Solar Dynamics Observatory (SDO) revealed a pronounced equatorial coronal hole which was centered on the solar disk on May 29, 2013, strongly suggesting a CIR origin of this storm. In fact, Zhang *et al.* (2007) have studied 88 major storms (Dst $\leqslant -100$) between 1996 and 2005 and found that while 60% were due to ICMEs, 27% to complex solar wind flows, usually due to interacting CMEs, 11 storms out of 88 were due to CIRs. The June 1, 2013 geomagnetic storm was at the strong end of CIR-induced ones. The CIR and the driving high-speed stream arrived at Wind on June 1, 2013, a forward shock preceded it by nearly one day, as can be seen in the top panel of Fig. 4. The data look very much as if the storm occurred between two high-speed streams. Careful inspection of the SDO images reveals a very faint poleward CME leaving the Sun in the questionable time, and it was not clear whether it could have caused the storm. If one looks at the Wind in situ data more carefully however, one finds an indication that the storm was triggered by a CME. The preceding fast wind is likely to have been driven by the faint polar CME from the SDO images. The second panel from the top in Fig. 4 shows magnetic field strength, which remains very steady from approximately May 28, 03:00 until $\sim$ May 31, 15:00. This is one possible interplanetary signature of ICMEs (Wimmer-Schweingruber *et al.* (2006)). In the same period the magnetic field turned increasingly southward, as is shown in the next panel (second from the bottom). It shows the fraction of southward field, i.e., $B_z/B$. Again fluctuations are seen to be small in the time preceeding the storm, i.e., likely to be part of an ICME. The field turns northward in the compression region driven by the high-speed stream from the coronal hole. Thus it looks as if the storm of June 1 may have been

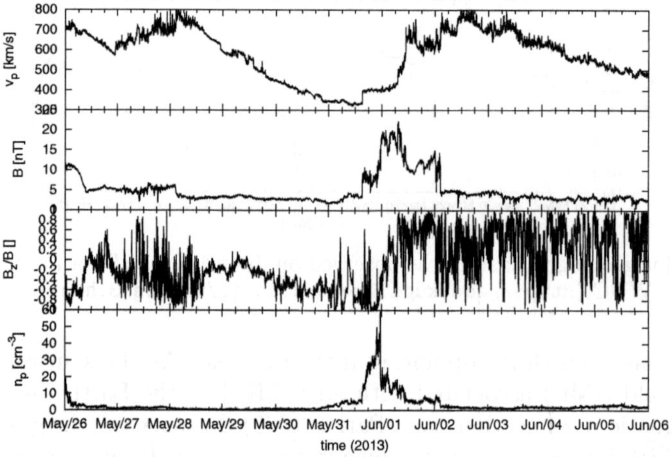


**Figure 4.** Solar wind proton and magnetic field data for the major geomagnetic storm as observed on June 1, 2013. Data are from the Wind SWE and MFI instrument and were acquired from the Omniweb data site (`http://ftpbrowser.gsfc.nasa.gov/wind_min_merge.html`)

caused by a complex interaction of a high-speed stream and its CIR interacting with the trailing edge of a preceeding ICME. Indeed if one also looks at STEREO (Kaiser *et al.* 2005) COR2 images, one sees clearly the polar CME which may be the driver of this storm. (Note also the large fluctuations in $B_z/B$ in the second high-speed stream, which are likely due to large-amplitude Alfvén waves. They too can be geoeffective.)

The message to take home from this discussion is that we still have difficulties identifying the source of a geomagnetic storm 'on the fly', especially when we have insufficient data. Having both STEREOs is very helpful in identifying Earth-directed CMEs!

## 3. Where is Space Weather?

Space weather as in "something that weathers something" is an important and ubiquitous process in the solar system and beyond. It results in spectral darkening, reddening and subdued absorption bands, which can greatly hamper the interpretation of remote sensing data of airless bodies. Sputtering by the solar wind and micrometeorite impact vaporization produce thin vapor-deposited films on regolith grains. These films contain metallic iron particles which change the optical properties of the grains (Hapke 2001). Space weather also produces the hermean and lunar exospheres (Killen and Ip 1999), and can be seen 'in action' in the sodium lines at Mercury (Killen *et al.* 2001). There solar wind and energetic particles can sputter the regolith, at least in the polar regions, thus liberating sodium atoms (among others) which can escape into the extended hermean exosphere. They are easily excited and their glow can then be seen (Killen *et al.*, 2001) and the ions also measured in situ (Zurbuchen *et al.* 2011). In fact, space weathering is likely to destroy grains in the interstellar medium and makes the interpretation of key abundances, such as that of life-enabling carbon, very difficult (Frisch & Slavin 2003) because it condenses in grains. Thus, it is clear that space weather also matters at other locations than only Earth.

### 3.1. *Space Weather and (Human) Exploration*

Not only is space weather a process which slowly alters grain reflective properties and composition, but it also matters because we have an increasing number of spacecraft at other planets. For instance, the Martian Radiation Environment Experiment (MARIE)

(Badhwar 2004), which was built to measure the radiation environment on the way to and at Mars, began malfunctioning probably due to several solar particle events and eventually ceased to operate in the fall of 2003, roughly two and a half years after launch. Incidentally, Mars Science Laboratory's (MSL) Radiation Assessment Detector (RAD) (Hassler *et al.* 2012), by pure serendipity began measuring the radiation environment on the surface of Mars on August 7, 2012, to the day exactly 100 years after Victor Hess' discovery of cosmic rays (Hess 1912). These measurements are partially also made in preparation of human exploration of the Moon, Mars, and beyond (Olson *et al.* 2011). Human solar system exploration will need to be accompanied by careful modeling of space weather and its effects on humans and life support systems. With increasing commercial access to space, including space tourism, this will become even more important. For instance, MarsOne plans to land humans on Mars in 2023 and 2024. In May 2013, with the application process still running until August 31, 2013, more than 78,000 people from over 120 countries had applied for selection (MarsOne 2013). Virgin Galactic, another commercial contestant for space tourism has recently added two pilots to its commercial flight team (Virgin Galactic 2013) and SpaceX, a commercial launcher company, successfully delivered cargo to the ISS (SpaceX 2013). These three examples show that commercial space utilization is increasing and larger-scale space tourism is not too distant in the future, underlining the importance of being able to predict and protect from space weather.

### 3.2. *An Example: Space Weather on the way to and at Mars*

Soon after MSL's launch on November 28, 2011, the RAD instrument was switched on on December 6 and started collecting data with a high duty cycle. Zeitlin *et al.* (2013) report the dose equivalent which would be experienced by a crew on its way to Mars and find a value of 465 mSv for the 253-day long cruise to Mars. If this duration could be shortened to 180 days, a crew member would thus experience $2 \times 465\text{mSv}/253 \times 180 = 662\text{mSv}$ on his/her way to Mars and back again (hence the factor 2). This is already uncomfortably close to NASA's career limit for astronauts, illustrating that space weather effects such as radiation are indeed a serious problem that requires careful study. The lion's share of the exposure was due to GCRs, with solar particle events only contributing approximately 5%. This fraction could have been quite different and higher at other times, as it depends strongly on solar activity.

On Mars itself, the radiation environment is complex because GCRs and solar particles interact with the Martian atmosphere and soil. Because of the thin atmosphere, the production of secondary particles peaks at or close to the surface. In other words, the Pfotzer maximum (Pfotzer 1936a, b), which at Earth is in a height of approximately 20 km, lies about at the surface of Mars. This has important consequences for possible life or biosignatures on Mars (Dartnell *et al.* 2007, Ehresmann *et al.* 2011).

Space weather has other, more severe consequences at Mars. Mars lost its global, dynamo-driven magnetic field (Schubert & Spohn 1990) a long time ago. Today it only has a remnant crustal field (Acuna *et al.* 1998 and 1999, Connerney *et al.* 2004). In fact, the field is so weak that the first probes to Mars only measured a field strength which was indistinguishable from the interplanetary field (Smith *et al.* 1965). Mariner 4 observed no bow shock up to its closest approach of four Mars radii. This missing magnetic field led to a gradual but massive loss of Mars' atmosphere, as recently confirmed by the Sample Analysis at Mars (SAM) instrument on MSL (Mahaffy *et al.* 2013). A Mars exospheric molecule, once ionized has a gyro radius which is larger than Mars' radius, and thus escapes from the planet.

### 3.3. *Space Weather at Other Planets*

Space weather can also be observed at other planets or solar system bodies. We have already discussed it for Mercury, the Moon, and asteroids at the beginning of this section 3. Venus, as the next logical planet, again has no intrinsic magnetic field (Russell 1993), and subsequently, is also loosing light exospheric ions through the same process as Mars, but with the difference that the stronger interplanetary field at Venus leads to particle gyro radii which are smaller than Venus' radius.

Farther out in the solar system we encounter the gas giants Jupiter and Saturn which have strong intrinsic magnetic fields and large magnetospheres. Because they themselves can accelerate particles to high energies, they generate their own space weather. Nevertheless, solar-induced aurorae have been observed at Jupiter (e.g., Caldwell *et al.* 1992) and at Saturn (Trauger *et al.* 1998) in truly beautiful Hubble images. That these planets would have aurorae was expected, as aurora-induced radio signals from Jupiter and Saturn were observed, as reviewed in Zarka (1998) and Zarka and Kurth (2005). Even farther out in the solar system the icy giants Uranus and Neptune are also exposed to space weather. Lamy *et al.* (2012) observed the chain of events from Sun to Uranus which lead up to the generation of aurorae at Uranus, as observed with the Hubble Space Telescope. Taking the next step, beyond the solar system, exoplanets are also expected to produce radio signals when their magnetospheres are excited sufficiently, although measurements have so far remained elusive (Bastian *et al.* 2000 and Lecavelier des Etangs *et al.* 2013).

## 4. Summary and Conclusions

Disturbances are often believed to be only due to coronal mass ejections (CMEs), but we have seen that even this term (disturbances) may not be as clear cut as sometimes thought. Recurrent CIRs do disturb geomagnetism, but don't necessarily lead to the more spectacular manifestations of space weather such as aurorae or intense geomagnetic storms. What exactly disturbs space weather also depends very much on your point of view. An astronaut on his way to Mars would probably welcome an ICME as a temporary magnetic shield against the GCR, whereas the same ICME may wreck havoc in the Earth's power grid. As we just saw, space weather effects are not limited to Earth. It has important effects on the regoliths of moons, asteroids, of Mercury, and possibly even of Mars. Thus, space weather is a ubiquitous phenomenon which has been tracked from the Sun all the way out to Uranus. Of course to us it matters most at Earth as our civilization is becoming more and more dependent on miniaturized technology which is susceptible to space weather effects. The good news may be that we appear to currently be leaving the Grand Solar Maximum which has dominated the space age so far. It will be interesting to experience what the Sun has in store for us in the future.

## References

Acuna, M. H., Connerney, J. E. P., Ness, N. F., Lin, R. P., Mitchell, D., Carlson, C. W., McFadden, J., Anderson, K. A., Reme, H., Mazelle, C., Vignes, D., Wasilewski, P., & Cloutier, P., 1999. *Science*, 284, 790.

Acuna, M. H., Connerney, J. E. P., Wasilewski, P., Lin, R. P., Anderson, K. A., Carlson, C. W., McFadden, J., Curtis, D. W., Mitchell, D., Reme, H., Mazelle, C., Sauvaud, J. A., D'Uston, C., Cros, A., Medale, J. L., Bauer, S. J., Cloutier, P., Mayhew, M., Winterhalter, D., & Ness, N. F., 1998. *Science*, 279, 1676.

Badhwar, G. D., 2004. *Space Sci. Revs.*, 110, 131.

Barnard, L., Lockwood, M., Hapgood, M. A., Owens, M. J., Davis, C. J., & Steinhilber, F., 2011. *Geophys. Res. Lett.*, 38, L16103.

Bartels, J., 1934. *Terrestrial Magnetism and Atmospheric Electricity (Journal of Geophysical Research)*, 39, 201.

Bastian, T. S., Dulk, G. A., & Leblanc, Y., 2000. *ApJ*, 545, 1058.

Bequerel, H., 1896a. *Comptes rendus*, 122, 420

Bequerel, H., 1896b. *Comptes rendus*, 122, 501

Bicknell, P. J., 1968. *Isis. Journal of the History of Science Society*, 59(1), 87

Birkeland, K., 1908a. *The Norwegian Aurora Polaris Expedition 1902–1903, vol. 1, On the Cause of Magnetic Storms and the Origin of Terrestrial Magnetism, first section.* H. Aschehoug and Co, Christiania

Birkeland, K., 1908b. *Comptes rendus*, 147, 539

Birkeland, K., 1911. *Comptes rendus*, 153, 513

Birkeland, K., 1913. *The Norwegian Aurora Polaris Expedition 1902–1903, vol. 1, On the Cause of Magnetic Storms and the Origin of Terrestrial Magnetism, second section.* H. Aschehoug and Co, Christiania

Bothmer, V. & Rust, D. M., 1997. In N. Crooker, J. A. Joselyn, & J. Feynman, eds., *Coronal Mass Ejections*, Geophysical Monograph Series. American Geophysical Union, Washington DC, USA, 139 – 146

Caldwell, J., Turgeon, B., & Hua, X. M., 1992. *Science*, 257, 1512.

Carrington, R. C., 1859. *MNRAS*, 20, 13

Clark, D. & Stephenson, F., 1978. *Quart. J. R. Astron. Soc.*, 19, 387

Connerney, J. E. P., Acuna, M. H., Ness, N. F., Spohn, T., & Schubert, G., 2004. *Space Sci. Revs.*, 111, 1.

Crooker, N. U. & Horbury, T. S., 2006. *Space Sci. Revs.*, 123, 93

Curie, P. & Curie, M., 1898a. *Comptes rendus*, 127, 1215

Curie, P. & Curie, M., 1898b. *Comptes rendus*, 127, 175

Dartnell, L. R., Desorgher, L., Ward, J. M., & Coates, A. J., 2007. *Biogeosciences*, 4, 545

Dunlop, M. W., Taylor, M. G. G. T., Davies, J. A., Owen, C. J., Pitout, F., Fazakerley, A. N., Pu, Z., Laakso, H., Bogdanova, Y. V., Zong, Q. G., Shen, C., Nykyri, K., Lavraud, B., Milan, S. E., Phan, T. D., Rème, H., Escoubet, C. P., Carr, C. M., Cargill, P., Lockwood, M., & Sonnerup, B., 2005. *Annales Geophysicae*, 23(8), 2867

Eddy, J. A., 1976. *Science*, 192, 1189.

Ehresman, B., Burmeister, S., Wimmer-Schweingruber, R. F., & Reitz, G., 2011. *J. Geophys. Res.*, 116, A10106 (1-9).

Frisch, P. C. & Slavin, J. D., 2003. *ApJ*, 594, 844.

Galactic, V., 2013. Virgin galactic adds two pilots to commercial flight team. `http://www.virgingalactic.com/news/item/virgin-galactic-adds-two-pilots-to-commercial-flight-team/`. [Online; accessed 2013-09-10]

Galileo, G. & Scheiner, C., 2010. *On sunspots.* University of Chicago Press. Translated and Introduced by Eileen Reeves and Albert Van Helden

Gilbert, W., 1600. *De Magnete*

Gleissberg, W., 1944. *Terrestrial Magnetism and Atmospheric Electricity*, 49, 243

Hapke, B., 2001. *J. Geophys. Res.*, 106, 10039.

Hassler, D. M., Zeitlin, C., Wimmer-Schweingruber, R. F., Böttcher, S., Martin, C., Andrews, J., Böhm, E., Brinza, D. E., Bullock, M. A., Burmeister, S., Ehresmann, B., Epperly, M., Grinspoon, D., Köhler, J., Kortmann, O., Neal, K., Peterson, J., Posner, A., Rafkin, S., Seimetz, L., Smith, K. D., Tyler, Y., Weigle, G., Reitz, G., & Cucinotta, F. A., 2012. *Space Sci. Revs.*, 170, 503.

Hathaway, D. H., 2010. *Living Reviews in Solar Physics*, 7(1) URL `http://www.livingreviews.org/lrsp-2010-1`. Accessed 2013-09-09

Hess, V. F., 1912. *Physikalische Zeitschrift*, 13, 1084

Kaiser, M. L., Kucera, T. A., Davila, J. M., St. Cyr, O. C., Guhathakurta, M., Christian, E., 2008. *Space Sci. Revs.*, 136, 5

Killen, R. M. & Ip, W. H., 1999. *Reviews of Geophysics*, 37, 361

Killen, R. M., Potter, A. E., Reiff, P., Sarantos, M., Jackson, B. V., Hick, P., & Giles, B., 2001. *J. Geophys. Res.*, 106, 20509.

Lamy, L., Prangé, R., Hansen, K. C., Clarke, J. T., Zarka, P., Cecconi, B., Aboudarham, J., André, N., Branduardi-Raymont, G., Gladstone, R., Barthélémy, M., Achilleos, N., Guio, P., Dougherty, M. K., Melin, H., Cowley, S. W. H., Stallard, T. S., Nichols, J. D., & Ballester, G., 2012. *Geophys. Res. Lett.*, 39, L07105.

Lecavelier des Etangs, A., Sirothia, S. K., Gopal-Krishna, & Zarka, P., 2013. *A&A*, 552, A65.

Mahaffy, P. R., Webster, C. R., Atreya, S. K., Franz, H., Wong, M., Conrad, P. G., Harpold, D., Jones, J. J., Leshin, L. A., Manning, H., Owen, T., Pepin, R. O., Squyres, S., Trainer, M., & Team, M. S., 2013. *Science*, 341(6143), 263

MarsOne, 2013. 78,000 sign up for one-way mission to Mars. `http://www.mars-one.com/en/mars-one-news/press-releases/11-news/437- 78000-sign-up-for-one-way-mission-to-mars/`. [Online; accessed 2013-09-10]

Mewaldt, R. A., Davis, A. J., Lave, K. A., Leske, R. A., Stone, E. C., Wiedenbeck, M. E., Binns, W. R., Christian, E. R., Cummings, A. C., de Nolfo, G. A., Israel, M. H., Labrador, A. W., & von Rosenvinge, T. T., 2010. *ApJ. Lett.*, 723, L1.

Olson, J., Craig, D., Maliga, K., Mullins, C., Hay, J., Graham, R., Graham, R., Smith, P., Johnson, S., & Simmons, A., 2011. Voyages - charting the course for sustainable human space exploration, brochure np-2011-06-395-larc. Technical report, NASA

Pacini, d., 1912. *Nuovo Cimento, Serie VI*, 3, 93

Parker, E. N., 1979. *Cosmical Magnetic Fields.* Oxford University Press

Pfotzer, G., 1936a. *Zeitschrift für Physik*, 102, 23.

Pfotzer, G., 1936b. *Zeitschrift für Physik*, 102, 41.

Röntgen, W. C., 1895. *Sitzungsberichte der Würzburger Physik.-medic. Gesellschaft*, 9, 132

Russell, C. T., 1993. *Reports on Progress in Physics*, 56, 687.

Schubert, G. & Spohn, T., 1990. *J. Geophys. Res.*, 95, 14095.

Schwabe, H., 1844. *Astron. Nachr.*, 21(495), 233.

SIDC-team, 1700-2013. *Monthly Report on the International Sunspot Number, online catalogue.* URL `http://www.sidc.be/sunspot-data/`. Accessed 2013-09-09

Smith, E. J., Davis, Jr., L., Coleman, Jr., P. J., & Jones, D. E., 1965. *Science*, 149, 1241.

SpaceX, 2013. Dragon delivers cargo to station. `http://www.spacex.com/`. [Online; accessed 2013-09-10]

Stern, D. P., 2002. *Rev. Geophys.*, 40, 1

Trauger, J. T., *et al.*, 1998. *J. Geophys. Res.*, 103, 237

Usoskin, I. G., 2013. *Living Reviews in Solar Physics*, 10(1) URL `http://www.livingreviews.org/lrsp-2013-1` Accessed 2013-09-08

Wimmer-Schweingruber, R. F., 2006. *Space Sci. Revs.*, 123, 471

Wimmer-Schweingruber, R. F., Crooker, N. U., Balogh, A., Bothmer, V., Forsyth, R. J., Gazis, P., Gosling, J. T., Horbury, T., Kilchenmann, A., Richardson, I., Richardson, J., Riley, P., Rodriguez, L., von Steiger, R., Wurz, P., & Zurbuchen, T. H., 2006. *Space Sci. Revs.*, 123, 177 .

Wittmann, A. & Xu, Z., 1987. *A&A Suppl.*, 70, 83

Wolf, R., 1856. *Viertelsjahrschr. Naturforsch. Ges. Zürich*, 1, 151

Zarka, P., 1998. *J. Geophys. Res.*, 103, 20159.

Zarka, P. & Kurth, W. S., 2005. *Space Sci. Revs.*, 116, 371.

Zeitlin, C., Hassler, D. M., Cucinotta, F. A., Ehresmann, B., Wimmer-Schweingruber, R. F., Brinza, D. E., Kang, S., Weigle, G., Bttcher, S., Bhm, E., Burmeister, S., Guo, J., Khler, J., Martin, C., Posner, A., Rafkin, S., & Reitz, G., 2013. *Science*, 340(6136), 1080

Zhang, J., Richardson, I. G., Webb, D. F., Gopalswamy, N., Huttunen, E., Kasper, J. C., Nitta, N. V., Poomvises, W., Thompson, B. J., Wu, C. C., Yashiro, S., & Zhukov, A. N., 2007. *Journal of Geophysical Research (Space Physics)*, 112, A10102.

Zurbuchen, T. H., Raines, J. M., Slavin, J. A., Gershman, D. J., Gilbert, J. A., Gloeckler, G., Anderson, B. J., Baker, D. N., Korth, H., Krimigis, S. M., Sarantos, M., Schriver, D., McNutt, R. L., & Solomon, S. C., 2011. *Science*, 333, 1862.

Zurbuchen, T. H. & Richardson, I. G., 2006. *Space Sci. Rev.*, 123, 31

# Session III

STELLAR EJECTA AND IMPACT ON EXOPLANETS

*Nature of Prominences and their role in Space Weather*
*Proceedings IAU Symposium No. 300, 2013*
*B. Schmieder, J.-M. Malherbe & S. T. Wu, eds.*

doi:10.1017/S1743921313011137

# Observations of stellar coronae and prominences

**Gaitee A. J. Hussain**

ESO, Karl-Schwarzschild-Strasse 2, D-85748, Garching bei München
Germany
email: ghussain@eso.org

**Abstract.** X-ray and EUV observations of young cool stars have shown that their coronae are extremely pressured environments with temperatures and densities that are up to two orders of magnitudes larger than those observed in the solar corona. At the same time rapidly transiting absorption features in optical and UV spectra reveal the presence of large cool, prominence-type complexes that can extend several stellar radii. I will give an overview of our current understanding of coronal structures in cool stars from multi-wavelength observations, detailing their properties and apparent dependence on spectral type. I will also outline future prospects in this field, particularly from observations of stellar coronal environments at radio and sub-mm wavelengths.

**Keywords.** stars: activity, stars: coronae, stars: circumstellar matter, stars: magnetic fields, X-rays: stars

## 1. Introduction

Thanks to almost fifteen years of observations from the great X-ray facilities, *Chandra* and XMM-*Newton*, we have gained numerous insights into the properties of stellar coronae. Grating X-ray spectra have enabled the measurement of coronal densities at a range of plasma temperatures while time-resolved studies allow us to place strong constraints on the extents and distribution of the hot, MK, plasma. Multiwavelength observations allow us to gain a more complete picture into the range of sizes of coronal structures, from compact X-ray and EUV flares to cool prominence-like structures that extend out to several stellar radii.

While it is now possible to follow the dynamic flows inside prominences on the Sun and categorise in detail the differences between different types of prominences, stellar coronal observations are of course necessarily limited in terms of spatial resolution. Stellar studies can only measure diagnostics that are integrated over the entire stellar disk. Despite this limitation some clear trends emerge with stars of different spectral types and activity levels. These observations combined with the modelling efforts currently underway to understand coronal environments, (e.g., Vidotto *et al.*, *this volume*) provide us with powerful tools to probe the early history of the Sun, the solar system and their place in a wider context.

## 2. X-ray/EUV spectra: coronal filling factors

X-ray grating spectra from *Chandra* and XMM-*Newton* encompass the He-like triplets of O, Ne, Mg and Si, which are useful diagnostics of plasma conditions (Gabriel & Jordan 1969). Fig. 1 shows the O VII triplet for an active K-type star. The ratio of the fluxes in the forbidden and intercombination lines are mainly density-sensitive in coronae, while all three lines are used to measure the temperature at which these lines have formed

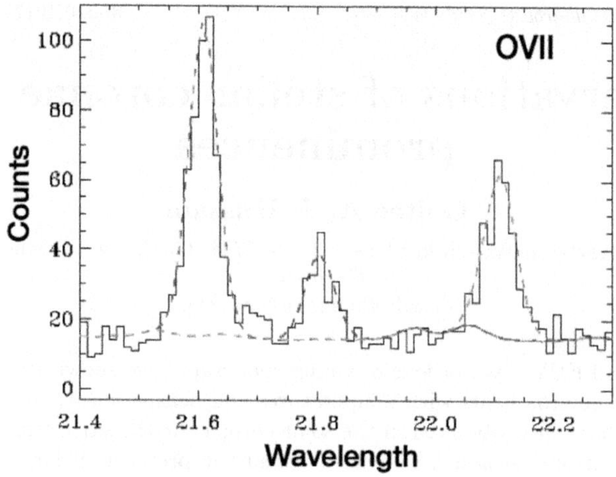


**Figure 1.** The dashed lines show the fits to the resonance ($r$: green), intercombination ($i$: red) and forbidden ($f$: orange) lines corresponding to the transitions between the n=2 shell and n=1 ground level in the O VII triplet. The flux ratio between the forbidden and intercombination lines, $f/i$, is predominantly density-sensitive while the ratio $(f+i)/r$ is temperature-sensitive. Data from Hussain *et al.* (2007).

(see Fig. 1). If we assume isothermal plasma and hydrostatic equilibrium it is possible to compute an upper limit for the filling factor of plasma using the emission measures, density and temperature measurements from these triplets (see Testa *et al.* 2004, Ness *et al.* 2004). While the emission measures and densities imply large coronal volume filling factors compared to the solar corona; the pressure scale heights must be small ($h \ll R_*$).

Coronal densities have been estimated for over 30 G-K-type stars encompassing a range of activity levels and X-ray luminosities ($\log L_X \sim$ 27-32 erg s$^-1$). Testa *et al.* (2004) focus on the O VII 22Å, and the Mg XI 9Å, He-like triplets that have temperatures of peak formation at a few MK and 10 MK respectively. The lower temperature diagnostic, O VII, indicates typical densities, $n_e \sim 10^{10} - 10^{11}\mathrm{cm}^{-3}$ in MK plasma, with filling factors ranging from 0.1 in the most inactive stars and increasing to 1.0 in the most rapidly rotating, active stars. This implies that the most active stars are completely covered in dense, MK plasma. The Mg XI diagnostic is sensitive to the hotter, 10 MK plasma, and is therefore associated with the more energetic flares. The observed emission measures and densities from Mg XI yield densities of $10^{12}$ cm$^{-3}$ and filling factors ranging from $10^{-4}$ up to 0.1 with increasing activity. This implies that with increasing activity levels, stellar coronae fill up with the type of MK plasma associated with active regions on the Sun until they are essentially completely covered. This rise is accompanied by a corresponding increase in 10 MK plasma at higher densities, with typical $\log n_e \sim 12\mathrm{cm}^{-3}$ though the filling factors remain much smaller.

## 3. X-ray rotational modulation & eclipses

Studies of X-ray rotational modulation in the most active, rapidly rotating stars indicate the presence of compact stable "quiescent" coronal active regions. If observations span several orbital periods, time intervals associated with large energetic flares can be identified and filtered from X-ray and EUV data and enable the study of the quiescent component of the corona. Long X-ray and EUV observations of the contact binaries, 44i Boo and VW Cep, revealed some of the strongest constraints on the locations of coronal active regions (Brickhouse, Dupree & Young 2001, Huenemoerder *et al.* 2003).

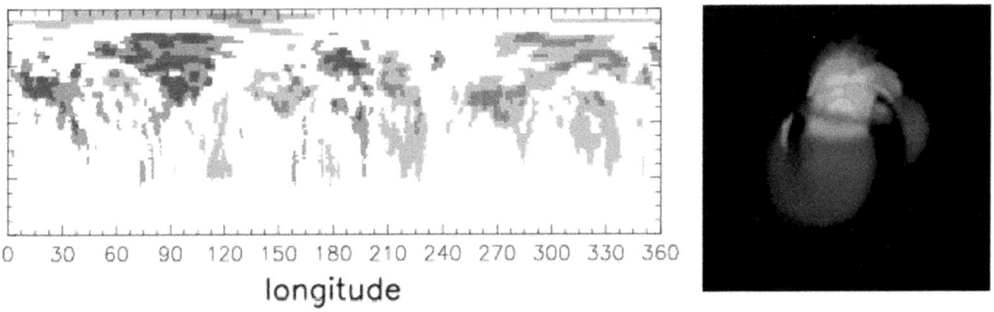


**Figure 2.** Surface magnetic field map of the K0V star, AB Dor obtained using Zeeman Doppler imaging. This map is a rectangular projection of the stellar surface showing only the radial vector, blue and red represent +/-1kG respectively. This surface map has been used to model the position of closed corona in the star and therefore to predict the X-ray modulation from the star's X-ray lightcurves from the quiescent active regions (Hussain *et al.* 2007).

Contact binaries are ideally suited to these types of studies as they have relatively short orbital periods (<< 1d), are X-ray luminous, and show large orbital velocity variations. This means that several orbits can be covered in a short period of time, with sufficient statistics in the X-ray lightcurves to disentangle flaring intervals from quiescent X-ray emission and search robustly for rotational modulation.

*Chandra* observations of the short period contact binary system 44i Boo (G0V+G0V, Prot=0.27d) spanning 2.56 orbits show clear modulation in both the X-ray lightcurves and spectra (Brickhouse, Dupree & Young 2001). Modelling of the observed rotational modulation places strong constraints on the extent and distribution of the X-ray emitting active regions. The strongest isolated emission lines (O VIII, Ne X, Fe XVI, Mg XII) show consistent velocity modulations, which closely trace the orbital motion of the primary star. These combined with the 20% X-ray lightcurve modulation indicate that a significant fraction of the coronal X-ray plasma is concentrated on the primary star at high latitudes with a smaller component of emission distributed over both component stars.

The rotation of single and binary stars can also help to measure the sizes and locations of loops associated with large energetic X-ray and EUV flares. In the Algol binary system (B8V+K2IV, $P_{\rm orb} = 2.86$d) a strong flare was observed to be completely eclipsed by the primary star, strongly constraining its location to $0.5R_*$ above the "south" pole of the K-star secondary (Schmitt & Favata 1999). This is also consistent with studies of flare heights made from the modelling of flaring loop lengths based on their decay timescales (see Section 4).

### 3.1. *Modelling X-ray coronae from surface magnetic field maps*

It is now possible to reconstruct the large scale magnetic field topologies at the surfaces of rapidly rotating cool stars using Zeeman Doppler imaging techniques (Semel 1989, Donati & Collier Cameron 1997, Hussain *et al.* 2000, Kochukhov *et al.* 2013; see Fig. 2). Zeeman Doppler imaging works by inverting a time-series of high resolution circularly polarised spectra obtained from large format optical echelle spectrographs such as CFHT/ESPADONS and the ESO 3.6-m/HARPS instrument in polarimetric mode. The circularly polarised signatures are combined from thousands of photospheric line profiles to detect polarisation signatures down to the 0.1% continuum intensity level. The velocity modulation and amplitude modulation of a time-series of polarisation signatures acquired over a full rotation cycle enable us to identify the position in latitude and longitude and the orientation of the large scale region in velocity and amplitude of

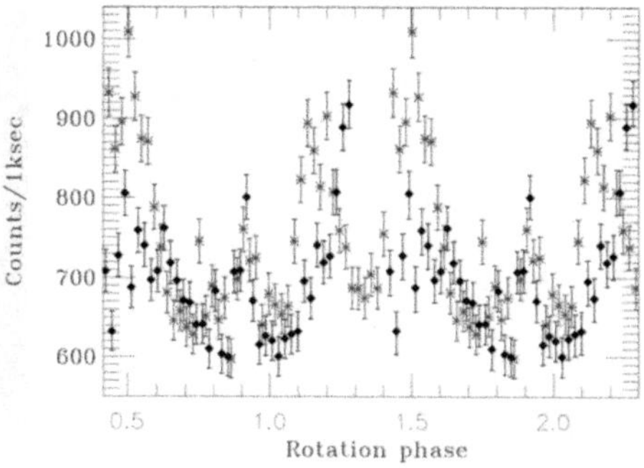


**Figure 3.** *Chandra* X-ray lightcurve of the active, K0 star, AB Dor covering two consecutive rotation periods with successive cycles marked by black diamonds and red asterisks (from Hussain *et al.* 2007). A period analysis shows significant rotational modulation in the X-ray lightcurve at the 12% level. X-ray models based on contemporaneous surface magnetic field maps of the star can reproduce X-ray emission measures and levels of rotational modulation that are consistent with the observed properties of the star.

these polarisation signatures enable us to identify the field orientation and position of large-scale magnetic field regions at the stellar photosphere.

The surface magnetic field maps from these techniques can be used to model the positions of footpoints of magnetic fields that extend into the corona. This coronal loop model can then be used to model the X-ray corona and predict the location and distribution of coronal active regions. The X-ray corona (e.g., Fig. 2) is modelled using several inputs: (a) the 3-D coronal field model extrapolated from the surface magnetic field map; assuming an isothermal corona with a temperature corresponding to the dominant temperature in its X-ray emission measure distribution (10 MK in the case of the K0 star, AB Dor shown above); (c) assume hydrostatic equilibrium. More details on how the X-ray models are constructed are available in Hussain *et al.* (2007) and references therein.

We find that in order to explain the emission measure, coronal densities and rotational modulation measured in the X-ray lightcurves and spectra (e.g., Fig. 3), and based on the complex multipolar large surface field, AB Dor's corona cannot extend beyond a scale height of 0.3-0.4$R_*$. However, the same dataset indicates that the compact X-ray emitting active regions co-exist with cooler and much larger structures, which subtend several stellar radii and appear to corotate with the star (Section 5).

## 4. Flaring statistics

Measurements of the flare emission measure and decay timescales can be used to estimate flaring properties including: the temperature, the density and the size of the loop length in which the flare originated. With the archived data from the EUVE satellite it is possible to analyse a large sample of EUV flares on a range of stars consistently (e.g., Mullan *et al.* 2006). They employ the Haisch Simplified Approach which assumes that radiative losses in the earliest decay phases are dominated by bremsstrahlung emission and requires few further assumptions to derive consistent properties over the entire sample. While there are clearly some limitations to this approach, by applying the same technique to a homogeneous dataset it is possible to make a comparative study of the flare properties across a wide range of stars.

From studies of over 100 flares on 33 stars Mullan *et al.* (2006) find clear trends with spectral type (see Fig. 4). There appears to be a transition in the flare loop lengths from compact heights of $L < 0.5R_*$ in the hotter stars, to a range of heights up to $1.5R_*$ in

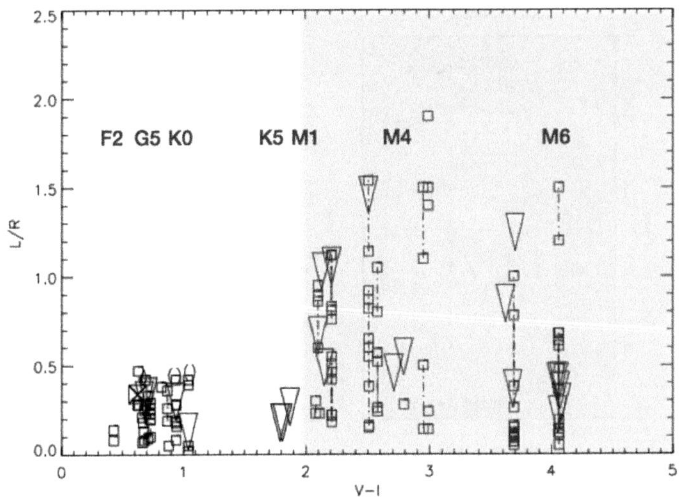


**Figure 4.** Flaring loop length in 33 main sequence stars derived from EUVE data. The loop length as a fraction of the radius is plotted against (V-I). The corresponding spectral types are denoted along the top. The point at which loop lengths appear to be systematically larger are marked by grey shaded area. Main sequence stars possess fully convective interiors beyond a spectral type of M4 (V-I$\geqslant$2.7) (Mullan *et al.* 2006, reproduced by permission of the AAS).

the cooler K2-M0 stars. It is unclear why the transition in loop sizes occurs and similar transitions are not clear in other diagnostics.

Lightcurves are now available that are more sensitive than ever before in "white-light" and cover much longer timescales thanks to high sensitivity photometry obtained from missions such as *Kepler* and COROT. Flares detected serendipitously in wide field studies indicate that energetic flares, hitherto supposed to be the domain of the most rapidly rotating cool stars, are also found in relatively slow rotators (Maehara *et al.* 2012, Shibata *et al. this volume*). Follow-up studies are being carried out that aim at characterising these stars in more detail; it remains to be seen whether similar trends with spectral type are found in these 'white-light' flares.

## 5. Stellar prominences

Several stars show evidence of extended structures; these were first detected as fast-moving absorption transient features that move through the H$\alpha$ line profile revealing the presence of cool ($10^4$K) material subtended out to 5 $R_*$, i.e. at and beyond the Keplerian co-rotation radius of the star (Collier Cameron & Robinson 1989). These have been found on almost all rapidly rotating cool stars where dense time-series spectra have been obtained, ranging from G to M-type systems (e.g., Jeffries 1993, Byrne ,Eibe & Rolleston 1996, Barnes *et al.* 1998). Up to 7 clouds can be found to exist at any one time, with the number changing from night to night. There is also a hint that some stars host much smaller prominences than others even with similar spectral types (e.g., Barnes *et al.* 1998) though this has not been examined in detail.

The masses of these prominences can be estimated if a number of different diagnostics are available (e.g., H$\alpha$, Ca II and Mg II), from estimates of a column density and the projected cross-sectional area typical masses of 2-6 $10^{14}$ kg are recovered (see Dunstone *et al.* 2006).

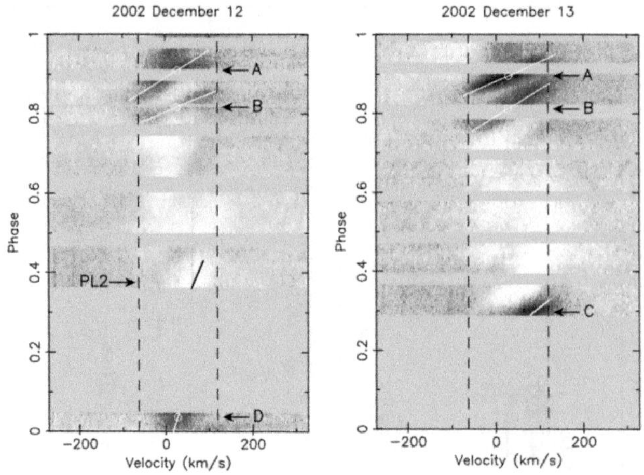


**Figure 5.** Time-series of Hα spectra on the K0 dwarf, AB Dor (from Hussain *et al.* 2007). Clear transient features are seen moving from $-v_e \sin i$ to $+v_e \sin i$ (marked by the dashed vertical lines). These data were taken on consecutive nights and two prominence complexes are recovered at the same phase though with changing velocities, implying the heights changed within 1 day ($2P_{\rm rot}$). The complex A moves out from 2.55 to $3.48R_*$ while the complex B moves in from 3.74 to 2.55 $R_*$. Given the stellar inclination angle of 60° these prominences must be at high latitudes in order to transit across the projected stellar disk.

Assuming the magnetic field confining the structures is rigid the rotation period of the cloud, and its rate of transit across the stellar disk indicates the distance of the "prominence" from the stellar rotation axis (also see Collier Cameron & Robinson 1989):

$$v_a(t) = d_a v_e \sin i cos l \sin \frac{2\pi t}{P_{\rm rot}} \tag{5.1}$$

where $v_a$ is the velocity of the absorption transient, $d_a$ is the height of the absorption transient in stellar radii, $l$ is the latitude and $P_{\rm rot}$ is the rotation period of the star. By measuring the slope of the transient it is therefore possible to derive the longitude and height of each cloud causing the transient. Occasionally there is structure in these transients so it is only possible to determine the mean velocity of the larger complex.

The exact nature of these prominence type complexes has been a subject of debate but there is evidence from other diagnostics that active cool stars can support stable structures several stellar radii above the stellar surface. UV spectra of the binary system, V471 Tau (DA+K2V, $P_{\rm rot}$ = 0.52d) indicate the presence of discrete structures above $2R_*$ as absorption in various chromospheric lines with temperatures between 8000 to $10^4$ K (Walter *et al.* 2004). Radio observations also imply the presence of extended coronae in M stars (Alef *et al.* 1997).

### 5.1. *Mass ejections*

Very few stars have been as intensely monitored with dense spectroscopic time-series as the K0 star, AB Dor. Over a re-examination of data acquired almost annually over a decade only three possible ejection events have been identified (Fig. 6).

Leitzinger *et al.* (2011) report evidence for an ejection event in the low mass star, AD Leo (M4.5V, $P_{\rm rot}$ =2.24 d) as indicated by enhanced blue-shifted emission in the transition line OVI 1032Å (Fig. 7). The velocity measured for this event would correspond to a distance of almost $28R_*$, i.e., well outside the $11.7R_*$ Keplerian corotation radius and suporting the interpretation of an ejection event. Studies are currently underway

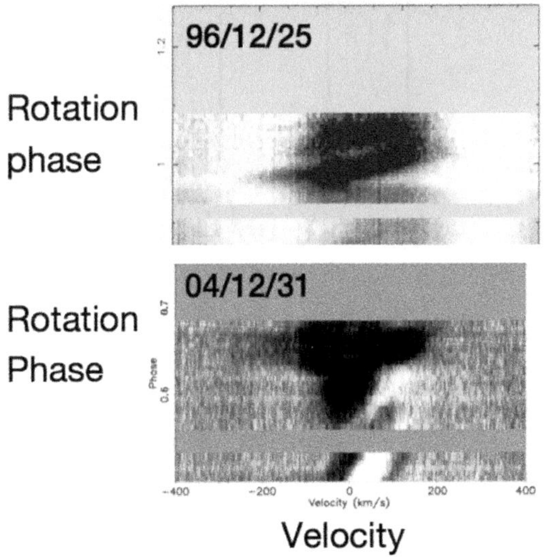


**Figure 6.** Observations of blue-shifts associated with absorption transients, indicating that part of the cool absorbing cloud is being ejected, reaching velocities of up to -180 km/s (Donati *et al.* 1997, Dunstone 2008). Only three such events have been found despite intense spectroscopic monitoring of this system, AB Dor, over 3-5 night observing runs every 1-2 years for 20 years. Ejection events such as these are difficult to observe without intense monitoring as they only last 15-20 minutes.

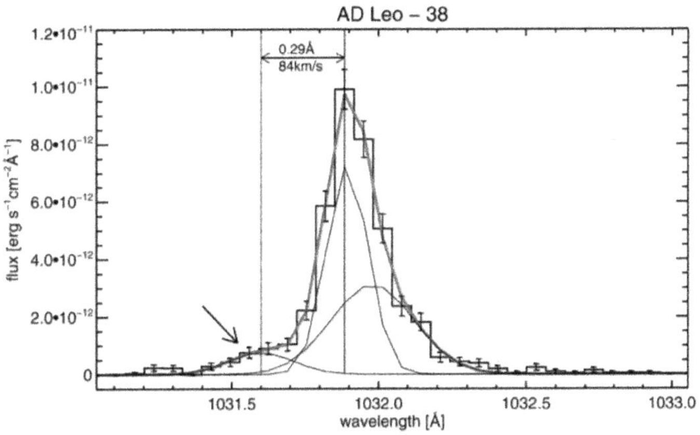


**Figure 7.** An enhancement in the blue-wing of the O VI 1032Å line of the M star AD Leo following a flare event (Leitzinger *et al.* 2011). This indicates a velocity shift of -84km s$^{-1}$ and is likely caused by an ejectione vent.

that are following this up ona larger sample of stars to get better statistics on how often these events occur.

### 5.2. *Coronal streamer model*

Any model to explain the nature of these large-scale structuers must explain how they can be stable out to several stellar radii. The evidence from X-ray and EUV data suggests that the closed corona must be relatively compact. Magnetic field maps of the surfaces of these active stars reveal complex, multipolar fields that must drop off quickly with height (as opposed to simple dipole fields) and are therefore unlikely to support a closed corona beyond $0.5R_{*}$. Jardine & van Ballegooijen (2005) propose that these stellar prominences

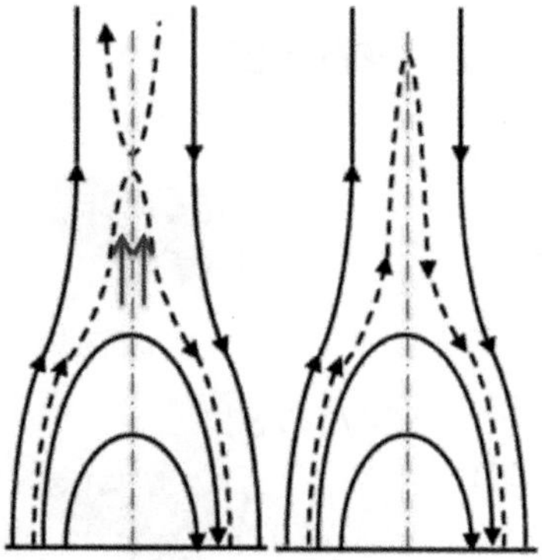

**Figure 8.** A schematic diagram showing how cool condensations may form above helmet streamers (adapted from Jardine & van Ballegooijen (2005). Reconnection within the helmet streamer causes a closed loop to form (left). The stellar wind continues to act (arrows), which increases the density at the top of this loop. Radiative losses cool the loop and the internal pressure causes a new equilibrium (right). This implies that for a particular stellar magnetic field distribution there will be preferred longitudes at which these condensations continuously form above the neutral polarity lines (dot-dashed blue vertical line).

could be supported in the open field. Cool static equilibria can be found above coronal helmet streamers (Fig. 8). A clear prediction from this model is that prominences should be found above neutral polarity lines. This can now be tested with spectro-polarimetric studies of cool stars as the same datasets can be used to reconstruct the surface magnetic field and ascertain the positions of the prominences. Prominence positions and heights can be measured and used to predict the locations of neutral polarity lines that can be tested against the magnetic field maps reconstructed from the photospheric circularly polarised profiles.

## 6. Summary & Conclusions

While hot plasma appears confined to well within 0.5R*, particularly in G and K stars there is evidence of extended cooler structures with temperature ranging from 1000 to $10^4$K from a variety of diagnostics, ranging from Balmer lines, to UV to radio wavelengths. There is also some evidence, though better statistics are needed, that the sizes of flaring loops change with spectral type, with increasing loop lengths found in late K and M-type stars. In contrast the prominence properties don't appear to change significantly with spectral type though stars with similar spectral types can host prominences that range from several stellar radii (up to $5R_*$) to very compact structures that appear much closer to the stellar surace. The model that can best explain the presence of these extended structrues demonstrates that static equilibria can exist above neutral polarity lines with prominences forming above coronal helmet streamers, i.e., in the open field. This model can explain why prominences are often found at similar phases but different heights, implying that they are constantly being formed and ejected at preferential points related to the positions of neutral polarity lines in the stellar magnetic field. Ejection events themselves are only rarely observed - they can last up to approximately 20 min and are therefore challenging without continuous observations. It is is possible to search for these events in both time-series of Balmer line profiles and as enhancements in the blue wings of transition and chromospheric line profiles. Future studies will enable us to gather better statistics on the frequency of these events across a range of spectral types.

## References

Alef, W., Benz, A. O., & Güdel, M. 1997, *A&A*, 317, 707

Barnes, J. R., Collier Cameron, A., Unruh, Y. C., Donati, J. F., & Hussain, G. A. J. 1998, *MNRAS*, 299, 904

Brickhouse, N. S., Dupree, A. K., & Young, P. R. 2001, *ApJ*, 562, L75

Byrne, P. B., Eibe, M. T., & Rolleston, W. R. J. 1996, *A&A*, 311, 651

Collier Cameron, A. & Robinson, R. D. 1989, *MNRAS*, 236, 57

Donati, J.-F. & Collier Cameron, A. 1997, *MNRAS*, 291, 658

Donati, J.-F., Collier Cameron, A., Hussain, G. A. J., & Semel, M. 1999, *MNRAS*, 302, 437

Dunstone, N. J. 2008, *PhD thesis*, University of St Andrews, UK

Dunstone, N. J., Collier Cameron, A., Barnes, J. R., & Jardine, M. 2006, *MNRAS*, 373, 1308

Gabriel, A. H. & Jordan, C. 1969, *MNRAS*, 145, 241

Huenemoerder, D. P., Testa, P., & Buzasi, D. L. 2006, 650, 1119

Hussain, G. A. J., Donati, J.-F., Collier Cameron, A., & Barnes, J. R. 2000, *MNRAS*, 318, 961

Hussain, G. A. J., Jardine, M., Donati, J.-F., Brickhouse, N. S., Dunstone, N. J., Wood, K., Dupree, A. K., Collier Cameron, A., & Favata, F. 2007, *MNRAS*, 377, 1488

Jardine, M. & van Ballegooijen, A. A. 2005, *MNRAS*, 361, 1173

Jeffries, R. D. 1993, *MNRAS*, 262, 369

Kochukhov, O., Mantere, M. J., Hackman, T., & Ilyin, I. 2013, *A&A*, 550, 84

Leitzinger, M., Odert, P., Ribas, I., Hanslmeier, A., Lammer, H., Khodachenko, M. L., Zaqarashvalli, T. V., & Rucker, H. O. 2011, *A&A*, 536, A62

Maehara, H., Shibayama, T., Notsu, S., Notsu, Y., Nagao, T., Kusaba, S., Honda, S., & Nogami, D., Shibata K. 2012, *Nature*, 485, 478

Mullan, D. J., Mathioudakis, M., Bloomfield, D. S., & Christian, D. J. 2006, *ApJS*, 164, 173

Ness, J.-U., Güdel, M., Schmitt, J. H. M. M., Audard, M., & Telleschi, A. 2004, *A&A*, 427, 667

Semel, M. 1989, *A&A*, 225, 456

Schmitt, J. H. M. M. & Favata, F. 1999, *Nature*, 401, 44

Testa, P., Drake, J. J., & Peres, G. 2004, *ApJ*, 617, 508

Walter, F. M. 2004, *AN*, 325, 241

*Nature of Prominences and their role in Space Weather*
*Proceedings IAU Symposium No. 300, 2013*
*B. Schmieder, J.-M. Malherbe & S. T. Wu, eds.*

doi:10.1017/S1743921313011149

# Coronal Mass Ejections and Angular Momentum Loss in Young Stars

**Alicia N. Aarnio**[1], **Keivan G. Stassun**[2,3] **and Sean P. Matt**[4]

[1]Dept. of Astronomy, University of Michigan, Ann Arbor, MI, 48109, USA
email: aarnio@umich.edu

[2]Dept. of Physics & Astronomy, Vanderbilt University, Nashville, TN, 37235, USA

[3]Dept. of Physics, Fisk University, Nashville, TN, 37208 USA

[4]School of Physics, University of Exeter,
Stocker Road, Exeter, EX4 4QL, UK

**Abstract.** In our own solar system, the necessity of understanding space weather is readily evident. Fortunately for Earth, our nearest stellar neighbor is relatively quiet, exhibiting activity levels several orders of magnitude lower than young, solar-type stars. In protoplanetary systems, stellar magnetic phenomena observed are analogous to the solar case, but dramatically enhanced on all physical scales: bigger, more energetic, more frequent. While coronal mass ejections (CMEs) could play a significant role in the evolution of protoplanets, they could also affect the evolution of the central star itself. To assess the consequences of prominence eruption/CMEs, we have invoked the solar-stellar connection to estimate, for young, solar-type stars, how frequently stellar CMEs may occur and their attendant mass and angular momentum loss rates. We will demonstrate the necessary conditions under which CMEs could slow stellar rotation.

**Keywords.** Sun: flares, Sun: coronal mass ejections, stars: activity, stars: flare, stars: evolution, stars: rotation, stars: planetary systems: protoplanetary disks

## 1. Introduction

On young stars, we observe flares hundreds to ten thousand times more energetic and frequent than solar flares. Along with energy scales greater by orders of magnitude, we also observe physical scales far greater than in the solar case: while solar prominences soar around 1 $R_\odot$ above the solar surface and CMEs launch from similar radii in the Sun's atmosphere, magnetic structures on T Tauri Stars (TTS, young solar analogs)–post-flare loops and prominences–can extend tens of stellar radii from the star's surface. The discovery of such large magnetic structures arose from solar-stellar analogy, applying solar flare models to the X-ray light curve data from young stars (e.g., Reale *et al.* 1998).

In characterizing the solar-stellar connection, overwhelming evidence has been found in support of the idea that the fundamental physics of magnetic reconnection is the same, despite differences in stellar parameters (e.g., mass, radius, $B$, age). As such, we approach analysis of young stars' flares and CMEs under this supposition, and aim to assess how the physical properties of these events–and their frequency–may scale accordingly with stellar parameters. Ultimately, we seek to understand the consequences of exoplanetary space weather on protostellar systems and their forming planets.

## 2. Estimating Angular Momentum Loss via Stellar CMEs

The inference of magnetic loops many stellar radii in extent (Favata *et al.* 2005; hereafter F05) inspired three questions: one, are these loops interacting with circumstellar disks? In Aarnio *et al.* (2010), we did not find evidence for this. Second, if there is not

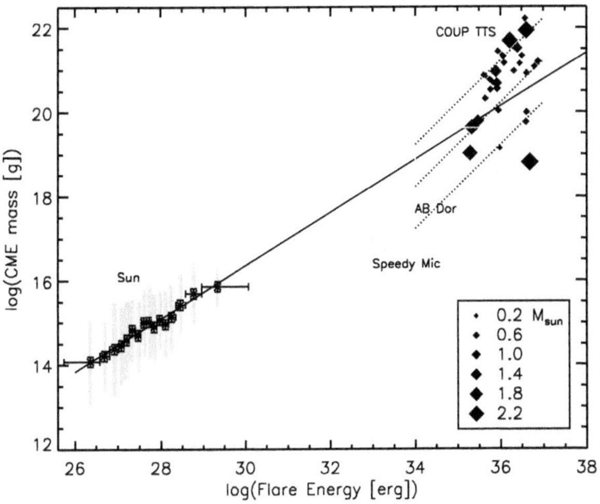


**Figure 1.** CME mass/flare energy relationship of AA11 shown with TTS post-flare loop masses (black diamonds) derived in AA12 plotted as a function of the flare's energy. Point size denotes the mass of the star on which the flare was observed. Black dotted lines show predicted post-flare stellar loop masses (Eqn. 2.1 and discussion in text) for a range of observed densities from $10^{10}$-$10^{12}$ $cm^{-3}$ and assuming a confining field strength of 50G. Gray, solid-lined boxes denote the range of observed X-ray flare energies (Maggio *et al.* 2000, Collier Cameron *et al.* 1988) and cool H$\alpha$ prominence mass estimates (Collier Cameron & Robinson, 1989a,b) for AB Dor and Speedy Mic, 20-50Myr K dwarfs.

a star-disk link, how do the loops remain stable for the multiple rotation periods over which the X-ray flares are observed to decay? We showed in Aarnio *et al.* (2012) that when modeled as hot prominences, the addition of a scaled-up wind consistent with TTS observations provided sufficient support for the loops to be stable. Finally, here (and in Aarnio, Matt, & Stassun, 2012; hereafter AA12) we address the third question of what happens when stability is lost: at many stellar radii, is the specific angular momentum shed significant enough to slow stellar rotation?

In order to estimate the effects of eruptive prominences and stellar CMEs on the rotation of young stars, we must procure two ingredients: the mass lost via these events, and their frequency of occurrence. Despite ongoing and historical efforts to observe stellar CMEs, we lack definitive detections and thus frequency distributions. It is known that at times, magnetic reconnection on the Sun will produce both a flare and an associated CME; for stars, the flare is the observable quantity, and so we characterize stellar CME frequency by using stellar flare frequency as a proxy.

In Fig. 1, we show our solar flare energy/CME mass relationship (Aarnio *et al.* 2011, hereafer AA11) extrapolated up to the energies of young stellar flares. We calculate loop masses for the 32 "superflaring" stars from the *Chandra* Orion Ultradeep Project (Getman *et al.* 2005) from the parameters reported by F05. Interestingly, these loop masses are close in parameter space to the extrapolated solar relationship. This is perhaps unsurprising, as the plasma confined in a post-flare loop has properties which relate to the energy of the flare. In AA11, we found that for associated flares and CMEs, that is to say, flares and CMEs which likely originated from a shared magnetic reconnection event, the CME mass and flare energy were related. As such, the post-flare loop mass and mass of an associated CME should then also be related.

From Shibata & Yokoyama (2002), we can estimate the total energy released by a flare is related to the total magnetic energy in a flare loop:

$$E_{mag} = \frac{B^2 L^3}{8\pi}. \tag{2.1}$$

If we substitute the loop volume expressed in terms of mass (i.e., $L^3 \sim V = m_{\rm loop}/\rho$), we find a relationship between the total flare energy and the mass confined in the magnetic loop (dashed, parallel lines in Fig. 1).

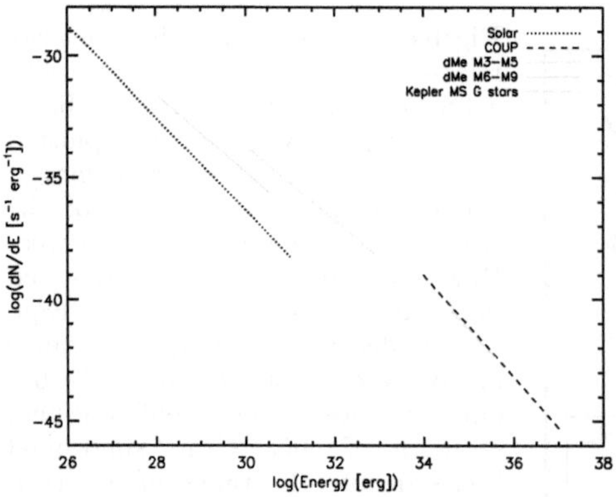


**Figure 2.** Flare frequencies for the Sun, M dwarfs (Hilton *et al.* 2011), TTS in the ONC (Albacete Colombo *et al.* 2007), and active, main sequence G stars from Kepler (Maehara *et al.* 2012). Interestingly, the active G stars are almost indistinguishable from the TTS. Note the solar and TTS frequencies are derived from X-ray flare data, while the M dwarf and G stars are optical flare frequencies.

Here, to be consistent with the analysis of F05, we assume an Euclidian loop filling but do note that recent solar X-ray flare imaging has indicated that a fractal scaling of $V(L) \propto L^{2.4}$ is likely a more accurate characterization (e.g., Aschwanden, Stern, & Güdel, 2008).

It is remarkable that the post-flare loop masses even lie near the extrapolated solar CME mass/flare energy relationship, several orders of magnitude away in parameter space. In Fig. 1, we have also shown representative ranges of X-ray flare energy and prominence mass for two K dwarfs intermediate in age to the TTS sample and the Sun; with an eruptive prominence thought to be the core of a CME, these mass ranges likely represent lower limits on the range of CME masses on these stars. In the following calculations, to represent a fiducial TTS case, we will simply extrapolate the solar relationship to generate a stellar CME mass distribution.

In Fig. 2, we show the frequency distributions for flares observed on the Sun, TTS, M dwarfs, and active, main sequence G stars. For this work, we use the TTS frequency distribution. Clearly, not all CMEs are flare-associated, nor are all flares CME-associated; AA11 found, however, that the association fraction increases with increasing flare energy, so for young stars for which we observe flares several orders of magnitude more energetic than in the solar case, we simply assume this association fraction to be of order unity.

### 2.1. *Angular momentum loss*

In AA12, we extrapolate the solar CME mass/flare energy relationship (Fig. 1) to TTS flare energies and frequencies (Fig. 2) and construct a CME frequency distribution as a function of CME mass. Given the observational completeness limits on the flare distributions that went into the CME distribution, we derive lower and upper limits on the mass loss rate by empirical (integrating the distribution) and analytical (integrating a fit to the distribution) means. The range of mass loss rates we estimate for the TTS case is $10^{-12}$-$10^{-9}$ $M_\odot$ yr$^{-1}$.

To assess the torque applied against stellar rotation by these CMEs, we apply stellar wind models with mass loss rates set as determined above. Given the episodic nature of CMEs, we included an efficiency parameter to account for the fact that steady-state winds are more efficient at removing angular momentum than "clumpy" winds (cf. AA12 and references therein). We adopt a dipolar field with strength 600G, consistent with observations of TTS fields, and allow the stellar radius to contract as stellar evolution models predict.

In a protostellar system, multiple torques act simultaneously to spin up and spin down the star. In this analysis, we compared spin up due to contraction and spin down due to mass loss from stellar CMEs to see if, at any point in the pre-main sequence our fiducial

TTS could have its rotation slowed due to CMEs. Comparing parameters of efficiency and the range of mass loss rates we calculated, it became clear that only towards the end of the pre-main sequence phase (ages $\gtrsim$6 Myr) could a very efficient, high CME mass loss rate begin to counteract spin up from contraction. We have left out factors such as spin up from accretion and mass loss via stellar wind; Matt & Pudritz (2005) explore these two torques in depth and the necessary conditions for an accretion powered stellar wind to slow stellar rotation.

## 3. Discussion

We have shown that for young, solar-type stars, spin down due to CMEs might play a significant role in stellar rotation evolution after the star has ceased accreting. Our Figs. 1 and 2 illustrate a critical selection effect in performing this kind of calculation: we only have data for the most active young stars, or the star conveniently located at 1 AU. There is a dearth of data for older, less active stars, and we suggest that filling in the gaps in flare X-ray energy could trace age evolution in these parameter spaces. The addition of data from the ~20-50 Myr old K dwarfs in Fig. 1 hints at this, but more data are needed to conclusively show age dependence. In both figures, we have taken care to specify the masses of the stars involved: how would evolution with stellar age look in these parameter spaces as a function of stellar mass? While the fundamental physics are the same, the scaling could change, and the ramifications certainly would. For low-mass stars in particular, high activity levels are observed for longer fractions of the stars' lives; this could have grave implications for exoplanets as these stars' habitable zones could be within range of extreme exo-weather.

### Acknowledgements

A. N. A. thanks K. Shibata for helpful discussion regarding stellar post-flare loops.

### References

Aarnio, A. N., Matt, S. P., & Stassun, K. G. 2012, *ApJ*, 760, 9
Aarnio, A. N., Llama, J., Jardine, M., & Gregory, S. G. 2012, *MNRAS*, 421, 1797
Aarnio, A. N., Stassun, K. G., Hughes, W. J., & McGregor, S. L. 2011, *Solar Phys.*, 268, 195
Aarnio, A. N., Stassun, K. G., & Matt, S. P. 2010, *ApJ*, 717, 93
Albacete Colombo, J. F., Flaccomio, E., Micela, G., Sciortino, S., & Damiani, F. 2007, *A&A*, 464, 211
Aschwanden, M. J., Stern, R. A., & Güdel, M. 2008, *ApJ* 672, 659
Collier Cameron, A. & Robinson, R. D. 1989, *MNRAS*, 238, 657
Collier Cameron, A. & Robinson, R. D. 1989, *MNRAS*, 236, 57
Collier Cameron, A., Bedford, D. K., Rucinski, S. M., Vilhu, O., & White, N. E. 1988, *MNRAS*, 231, 131
Dunstone, N. J., Collier Cameron, A., Barnes, J. R., & Jardine, M. 2006, *MNRAS*, 373, 1308
Favata, F., Flaccomio, E., Reale, F., Micela, G., Sciortino, S., Shang, H., Stassun, K. G., & Feigelson, E. D. 2005, *ApJS*, 160, 469
Getman, K. V. *et al.* 2005, *ApJS*, 160, 319
Hilton, E. J., Hawley, S. L., Kowalski, A. F., & Holtzman, J. 2011, *ASPC 16th Cambridge Workshop on Cool Stars, Stellar Systems, and the Sun*, 448, 197
Maehara, H., Shibayama, T., Notsu, S., Notsu, Y., Nagao, T., Kusaba, S., Honda, S., Nogami, D., & Shibata, K. 2012, *Nature*, 485, 478
Maggio, A., Pallavicini, R., Reale, F., & Tagliaferri, G. 2000, *A&A*, 356, 627
Matt, S. P. & Pudritz, R. E. 2005, *ApJL*, 632, L135
Reale, F. & Micela, G. 1998, *A&A*, 334, 1028
Shibata, K. & Yokoyama, T. 2002, *ApJ*, 577, 422

*Nature of Prominences and their role in Space Weather*
*Proceedings IAU Symposium No. 300, 2013*
*B. Schmieder, J.-M. Malherbe & S. T. Wu, eds.*

doi:10.1017/S1743921313011150

# Magnetised winds of low-mass stars and their impact on exoplanets

**A. A. Vidotto**

SUPA, University of St Andrews, North Haugh, KY16 9SS, UK
email: Aline.Vidotto@st-andrews.ac.uk

**Abstract.** The proper characterisation of stellar winds is crucial to constrain interactions between exoplanets and their surrounding environments and also essential for the study of space weather events on exoplanets. Although the great majority of exoplanets discovered so far are orbiting cool, low-mass stars with properties (mass, radius and effective temperatures) similar to solar, the stellar magnetism can be significantly different from the solar one, both in topology and intensity. Due to the current technology used in exoplanetary searches, most of the currently known exoplanets are found orbiting at extremely close distances to their host stars ($< 0.1$ au). The dramatic differences in stellar magnetism and orbital radius can make the interplanetary medium of exoplanetary systems remarkably distinct from the one present in the solar system. In addition, the interaction of the stellar winds with exoplanets can lead, among others, to observable signatures that are absent in our own solar system.

**Keywords.** MHD, stars: coronae, stars: late-type, stars: magnetic fields, stars: mass loss, stars: planetary systems, stars: winds, outflows

## 1. Introduction

Stars lose mass in the form of winds during their entire lives. For some stars, at certain evolutionary phases, the amount of mass lost through stellar winds can amount to a significant portion of its own mass. For cool stars at the main sequence phase, we believe their winds carry only a small amount of mass compared to the star total mass. Never the less, magnetised winds carry a large amount of angular momentum, which affect the stellar rotational evolution (e.g., Bouvier *et al.* 1997).

P-Cygni profiles, the traditional mass-loss signatures observed in the denser winds of giant and super-giant stars, are not formed in the rarefied winds of cool dwarf stars. To detect these winds, other indirect methods have been proposed (e.g., Lim and White 1996; Wood *et al.* 2001; Wargelin and Drake 2002). So far, the method developed by Wood *et al.* (2001) has been the most successful one, enabling estimates of mass-loss rates for about a dozen cool dwarf stars.

Due to the lack of observational constraints on winds of cool low-mass stars, many works assume these winds to be identical to (or a scaled version of) the solar wind. In the next Section, I highlight a few observed properties of cool dwarf stars that suggest that their winds can actually be significantly different from the solar wind.

### 1.1. *What can we infer from winds of cool stars?*

Temperatures: Parker (1958) developed the first hydrodynamical model of thermally-driven winds. In this simple, spherically symmetric, isothermal model, Parker (1958) showed that the stellar wind terminal velocity is very sensitive to the choice of temperature of the stellar wind. Although we believe the solar coronal heating is ultimately caused by the Sun's magnetism, it is still debatable which mechanism heats the solar corona. In the stellar case, coronal heating is much less understood. Stellar coronae can be much hotter than the solar corona. The temperature of the X-ray emitting coronae of solar analogs can exceed 10 MK (Guedel 2004), more than one order of magnitude larger than the solar coronal temperature of $\sim 1$ MK. If the temperature of the X-ray emitting (closed) corona is related to the temperature of the stellar wind (flowing along open field lines), as one would naively expect, then we may expect stellar winds of cool dwarf stars to have temperatures that could be much larger than the solar wind temperature. As a consequence, the winds of cool dwarf stars might have a variety of terminal velocities.

Magnetism: Due to our privileged position immersed in the solar wind, we have access to a great quantity of data that allow a detailed understanding of the physics that is operating in the Sun. Measurements of mass flows and magnetic field of the solar wind have revealed an asymmetric solar wind (McComas *et al.* 1995; Suess and Smith 1996; Jones *et al.* 1998; Wilhelm 2006). In particular, Ulysses measurements showed that the solar wind structure depends on the characteristics of the solar magnetic field (McComas *et al.* 2008). The solar wind characteristics change along the solar cycle, presenting a simple bimodal structure of fast and slow flows during solar minimum, when the geometry of the solar magnetic field is closest to that of an aligned dipole. At solar maximum, when the the axis of the large-scale dipole becomes nearly perpendicular to the solar axis of rotation, the solar wind shows a more complex structure.

Although the richness of details of the magnetic field configuration is only known for our closest star, modern techniques have made it possible to reconstruct the large-scale surface magnetic fields of other stars. The Zeeman-Doppler Imaging (ZDI) technique is a tomographic imaging technique (Donati and Brown 1997) that allows us to reconstruct the large-scale magnetic field (intensity and orientation) at the surface of the star from a series of circular polarisation spectra. This method has now been used to investigate the magnetic topology of stars of different spectral types (Donati and Landstreet 2009) and has revealed fascinating differences between the magnetic fields of different stars. For example, solar-type stars that rotate about two times faster than our Sun show the presence of substantial toroidal component of magnetic field, a component that is almost non-existent in the solar magnetic field (Petit *et al.* 2008). The magnetic topology of low-mass ($< 0.5\ M_\odot$) very active stars seem to be dictated by interior structure changes: while partly convective stars possess a weak non-axisymmetric field with a significant toroidal component, fully convective ones exhibit strong poloidal axisymmetric dipole-like topologies (Morin *et al.* 2008; Donati *et al.* 2008a).

Rotation: In addition to magnetic field characteristics, the symmetry of a stellar wind also depends on the rotation rate of the star. In the presence of fast rotation, a magnetised wind can lose spherical symmetry, as centrifugal forces become more important with the increase of rotation rate (Washimi and Shibata 1993). The distribution of rotation periods in cool main-sequence stars is very broad, ranging from stars rotating faster than once per day to indefinitely long periods.

*

The variety of observed rotation rates, intensities and topologies of the magnetic fields of cool, dwarf stars indicate that their winds might come in different flavours and might be significantly different from the solar one. One might also bear in mind that, similarly to the Sun, low-mass stars are also believed to host magnetic and activity cycles, which imply that the characteristics of their winds vary in a time scale of the cycle periods.

## 2. Models and simulations of winds of cool stars

There are two most commonly used ways to model stellar winds. One approach consists of computing the detailed energetics of the wind, starting from the photosphere, passing through the chromosphere, until it reaches the stellar corona (e.g., Hollweg 1973; Holzer *et al.* 1983; Hartmann and MacGregor 1980; Jatenco-Pereira and Opher 1989; Vidotto and Jatenco-Pereira 2006; Falceta-Gonçalves *et al.* 2006; Cranmer 2008; Cranmer and Saar 2011; Suzuki *et al.* 2012). Depending on the physics that is included in such models, this approach can become computationally expensive. As a result, it has been limited to analytical, one- and two-dimensional solutions. In addition, it is also usually focused in the inner most part of the corona and usually only adopts simplified magnetic field topologies. The second approach commonly used to model winds of cool stars consists of adopting a simplified energy equation, usually assuming the wind to be isothermal or with a polytropic equation of state (in which thermal pressure $p$ is related to density $\rho$ as $p \propto \rho^\gamma$). In this case, one is allowed to perform multi-dimensional numerical simulations of stellar winds and can incorporate more complex magnetic field topologies (e.g., Mestel 1968; Pneuman and Kopp 1971; Tsinganos and Low 1989; Washimi and Shibata 1993; Keppens and Goedbloed 2000; Lima *et al.* 2001; Vidotto *et al.* 2009b,a, 2010b; Pinto *et al.* 2011; Jardine *et al.* 2013)

Because of the simplified energetics that are considered in the second approach, its domain can extend considerably farther out than the first one. As a result, polytropic winds can be useful in the characterisation of the interplanetary medium and also to characterise interactions between exoplanets and the winds of their host-stars. In the next Section, I present how to make more realistic stellar wind simulations by incorporating recent insights acquired on the magnetic topology of different stars into stellar wind models.

### 2.1. *Data-driven wind simulations*

To illustrate how observationally reconstructed surface maps can be incorporated in the simulations of stellar winds, I will present the work done in Vidotto *et al.* (2012), where we performed numerical simulations of the stellar wind of the planet-hosting star $\tau$ Boo (spectral type F7V). $\tau$ Boo is a remarkable object, not only because it hosts a giant planet orbiting very close to the star (located at 0.046 au from the star), but also because it is the only star other than the Sun for which a full magnetic cycle has been reported in the literature (Donati *et al.* 2008b; Fares *et al.* 2009, 2013). These observations suggest that $\tau$ Boo undergoes magnetic cycles similar to the Sun, but with a cycle period that is about one order of magnitude smaller than the solar one (about 2 years as opposed to 22 years for the solar magnetic cycle).

The surface magnetic maps of $\tau$ Boo reconstructed by Catala *et al.* (2007), Donati *et al.* (2008b) and Fares *et al.* (2009) were used as boundary conditions for the stellar wind simulations. It was found that variations of the stellar magnetic field during the cycle directly influence the outflowing wind. Therefore, the rapid variation of the large-scale magnetic field of $\tau$ Boo implies that the environment surrounding the close-in planet should be varying quite rapidly. In addition, Vidotto *et al.* (2012) estimated the mass-loss rate ($\dot{M}$) of $\tau$ Boo, showing that this star seems to have a denser wind than that of the Sun, with $\dot{M}$ that are 2 orders of magnitude larger than the solar value $\dot{M}_\odot$ ($\dot{M} \approx 135\ \dot{M}_\odot$).

## 3. Interaction between stellar winds and exoplanets

When the wind outflows from the star, it permeates the entire extrasolar system, interacting with any body that it encounters on its way. The interaction between stellar winds and exoplanets can lead to observable signatures, some of which are absent in our own solar system.

### 3.1. *Planetary radio emission*

In the solar system, the giant planets and the Earth emit at radio wavelengths. Such planetary radio emission is due to the interaction between the magnetic planets and the solar wind, and its power is proportional to the stellar wind energy dissipated in the wind-planet interaction. By analogy to what is observed in the solar system, it is expected that exoplanets also interact with the winds of their host stars and should, therefore, also generate radio emission. Because the energy dissipated by the stellar wind is larger at closer distances to the star (because the wind density and magnetic fields are significantly larger than further out from the star), radio emission of close-in planets, such as $\tau$ Boo b, is expected to be several orders of magnitude larger than the emission from the planets in the solar system (Zarka 2007). In addition, the relatively dense wind of $\tau$ Boo estimated in Vidotto *et al.* (2012) implies that the energy dissipated in the stellar wind-planet interaction can be significantly higher than the values derived in the solar system. Combined with the close proximity of the system ($\sim$ 16 pc), the $\tau$ Boo system has been one of the strongest candidates to verify exoplanetary radio emission predictions.

Using the detailed stellar wind model developed for its host-star, Vidotto *et al.* (2012) estimated radio emission from $\tau$ Boo b, exploring different values for the *assumed* planetary magnetic field. For example, they showed that, for a planet with a magnetic field similar to Jupiter's ($\simeq$ 14 G), the radio flux is estimated to be $\simeq 0.5-1$ mJy, occurring at an emission frequency of $\simeq$ 34 MHz. Although small, this emission frequency lies in the observable range of current instruments, such as LOFAR. To observe such a small flux, an instrument with a sensitivity lying at a mJy level is required. The same estimate was done considering the planet has a magnetic field similar to the Earth ($\simeq$ 1 G). Although the radio flux does not present a significant difference to what was found for the previous case, the emission frequency ($\simeq$ 2 MHz) falls at a range below the ionospheric cut-off, preventing its possible detection from the ground. In fact, due to the ionospheric cutoff at $\sim$ 10 MHz, radio detection with ground-based observations from planets with magnetic field intensities $\lesssim$ 4 G should not be possible (Vidotto *et al.* 2012).

### 3.2. *Bow shock signatures in transit observations*

Despite many attempts, radio emission from exoplanets has not been detected so far (e.g., Smith *et al.* 2009; Lazio *et al.* 2010; Lecavelier des Etangs *et al.* 2013). Such a detection would not only constrain local characteristics of the stellar wind, but would also demonstrate that exoplanets are magnetised. Fortunately, there may be other ways to probe exoplanetary magnetic fields, in particular for transiting systems, through signatures of bow shocks during transit observations.

Based on Hubble Space Telescope/Cosmic Origins Spectrograph (HST/COS) observations using the narrow-band near-UV spectroscopy, Fossati *et al.* (2010b) showed that the transit lightcurve of the close-in giant planet WASP-12b presents both an early ingress when compared to its optical transit, as well as excess absorption during the transit, indicating the presence of an asymmetric distribution of material surrounding the planet. Motivated by these transit observations, Vidotto *et al.* (2010a) suggested that the

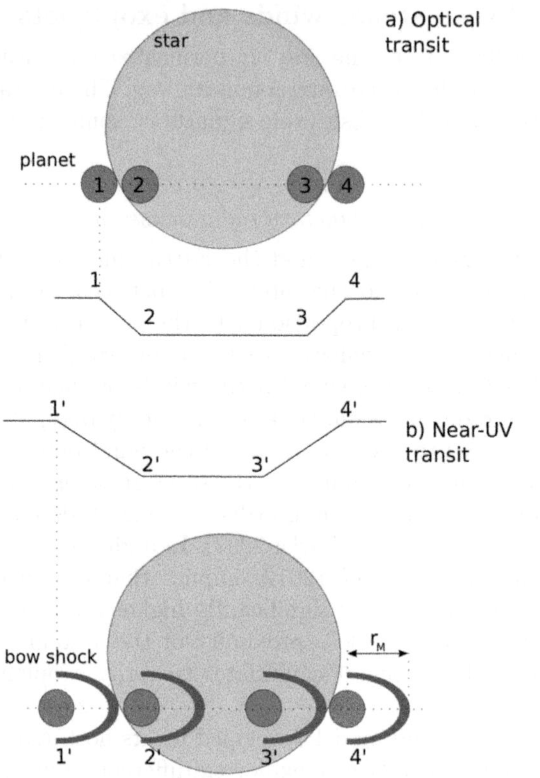


**Figure 1.** Sketches of the light curves obtained through observations in the a) optical and b) near-UV, where the bow shock surrounding the planet's magnetosphere is also able to absorb stellar radiation. Figure adapted from Vidotto *et al.* (2011b).

presence of bow shocks surrounding close-in planets might lead to transit asymmetries at certain wavelengths, such as the one observed in WASP-12b at near-UV wavelengths.

The main difference between bow shocks formed around exoplanets and the ones formed around planets in the solar system is the shock orientation, determined by the net velocity of the particles impacting on the planet's magnetosphere. In the case of the Earth, the solar wind has essentially only a radial component, which is much larger than the orbital velocity of the Earth. Because of that, the bow shock surrounding the Earth's magnetosphere forms facing the Sun (a dayside shock). However, for close-in exoplanets that possess high orbital velocities and are frequently located at regions where the host star's wind velocity is comparatively much smaller, a shock may develop ahead of the planet (ahead shock). In general, we expect that shocks are formed at intermediate angles.

Due to their high orbital velocities, close-in planets offer the best conditions for transit observations of bow shocks. If the compressed shocked material is able to absorb stellar radiation, then the signature of bow shocks may be observed through both a deeper transit and an early-ingress in some spectral lines with respect to the broadband optical ingress (Vidotto *et al.* 2010a). The sketches shown in Figure 1 illustrate this idea.

In the case of WASP-12b, Vidotto *et al.* (2010a) suggested that the shocked material, which is able to absorb enough stellar radiation in the near-UV, can cause an asymmetry in the lightcurve observed (see lightcurve sketches in Figure 1), where the presence of compressed material ahead of the planetary orbit causes an early ingress (compare points

1 and 1′ in Figure 1), while the lack of compressed material behind the planetary orbit causes simultaneous egresses both in the near-UV transit as well as in the optical one (compare points 4 and 4′ in Figure 1). This suggestion was verified by Llama *et al.* (2011), who performed Monte Carlo radiation transfer simulations of the near-UV transit of WASP-12b. They confirmed that the presence of a bow shock indeed breaks the symmetry of the transit lightcurve

### 3.3. *Planetary magnetic fields: a new detection method?*

An interesting outcome of the observations of bow shocks around exoplanets is that it permits one to infer the magnetic field intensity of the transiting planet. By measuring the phases at which the near-UV and the optical transits begin (phases given by points 1 and 1′ in the lightcurve sketches of Figure 1), one can derive the stand-off distance from the shock to the centre of the planet, which is assumed to trace the extent of the planetary magnetosphere $r_M$. At the magnetopause, pressure balance between the coronal total pressure and the planet total pressure requires that

$$\rho_c \Delta u^2 + \frac{[B_c(a)]^2}{8\pi} + p_c = \frac{[B_p(r_M)]^2}{8\pi} + p_p, \tag{3.1}$$

where $\rho_c$, $p_c$ and $B_c(a)$ are the local coronal mass density, thermal pressure, and magnetic field intensity at orbital radius $a$, and $p_p$ and $B_p(r_M)$ are the planet thermal pressure and magnetic field intensity at $r_M$. In the case of a magnetised planet with a magnetosphere of a few planetary radii, the planet total pressure is usually dominated by the contribution from the planetary magnetic pressure (i.e., $p_p \sim 0$). Vidotto *et al.* (2010a) showed that, in the case of WASP-12b, Eq. (3.1) reduces to $B_c(a) \simeq B_p(r_M)$. Further assuming that stellar and planetary magnetic fields are dipolar, we have

$$B_p = B_\star \left( \frac{R_\star / a}{R_p / r_M} \right)^3, \tag{3.2}$$

where $B_\star$ and $B_p$ are the magnetic field intensities at the stellar and planetary surfaces, respectively. Eq. (3.2) shows that the planetary magnetic field can be derived directly from *observed* quantities. For WASP-12, using the upper limit of $B_\star < 10$ G (Fossati *et al.* 2010a) and the stand-off distance obtained from the near-UV transit observation $r_M = 4.2\ R_p$ (Lai *et al.* 2010), we predicted an upper limit for WASP-12b's planetary magnetic field of $B_p < 24$ G.

### 3.4. *Searching for magnetic fields in other exoplanets*

In theory, the suggestion that through transit observations one can probe the planetary magnetic field is quite straightforward - all it requires is a measurement of the transit ingress phase in the near-UV. In practice, however, acquisition of near-UV transit data requires the use of space-borne facilities, making follow-ups and new target detections rather difficult.

In order to optimise target selection, Vidotto *et al.* (2011a) presented a classification of the known transiting systems according to their potential for producing shocks that could cause observable light curve asymmetries. The main considered assumption was that, once the conditions for shock formation are met, planetary shocks absorb in certain near-UV lines, in a similar way as WASP-12b. In addition, for it to be detected, the shock must compress the local plasma to a density sufficiently high to cause an observable level of optical depth. This last hypothesis requires the knowledge of the local ambient medium that surrounds the planet.

By adopting simplified hypotheses, namely that up to the planetary orbit the stellar corona can be treated as in hydrostatic equilibrium and isothermal, Vidotto *et al.* (2011a) predicted the characteristics of the ambient medium that surrounds the planet for a sample of 125 transiting systems, and discussed whether such characteristics present favourable conditions for the presence and detection of a bow shock. Excluding systems that are quite far ($\gtrsim 400$ pc), the planets that were top ranked are: WASP-19b, WASP-4b, WASP-18b, CoRoT-7b, HAT-P-7b, CoRoT-1b, TrES-3, and WASP-5b.

## 4. Conclusion

As the wind outflows from the star, it permeates the interplanetary medium, interacting with any planet encountered on its way. The proper characterisation of stellar winds is therefore crucial to constrain interactions between exoplanets and their surrounding environments and also essential for the study of space weather events on exoplanets. Stellar winds are affect by the stellar rotation, magnetism and coronal temperature, properties that can vary significantly from star to star. As a consequence, stellar winds of cool stars can actually be significantly different from the solar wind. In this paper, I illustrated how one can take an extra step towards more realistic models of stellar winds of low-mass stars by incorporating observationally reconstructed surface magnetic maps into simulations of stellar winds. I also showed that dramatic differences in stellar magnetism and orbital radius can make the interplanetary medium of exoplanetary systems remarkably distinct from the one present in the solar system. In addition, I showed that the interaction of the stellar winds with exoplanets can lead, among others, to observable signatures that are absent in our own solar system.

## Acknowledgements

AAV acknowledges support from a Royal Astronomical Society Fellowship and thanks the IAU for travel support to attend the Symposium.

## References

Bouvier, J., Forestini, M., & Allain, S., 1997, *A&A* 326, 1023

Catala, C., Donati, J.-F., Shkolnik, E., Bohlender, D., & Alecian, E., 2007, *MNRAS* 374, L42

Cranmer, S. R., 2008, *ApJ* 689, 316

Cranmer, S. R. & Saar, S. H., 2011, *ApJ* 741, 54

Donati, J. & Landstreet, J. D., 2009, *ARA&A* 47, 333

Donati, J., Morin, J., Petit, P., Delfosse, X., Forveille, T., Aurière, M., Cabanac, R., Dintrans, B., Fares, R., Gastine, T., Jardine, M. M., Lignières, F., Paletou, F., Velez, J. C. R., & Théado, S., 2008a, *MNRAS* 390, 545

Donati, J.-F. & Brown, S. F., 1997, *A&A* 326, 1135

Donati, J.-F., Moutou, C., Farès, R., Bohlender, D., Catala, C., Deleuil, M., Shkolnik, E., Collier Cameron, A., Jardine, M. M., & Walker, G. A. H., 2008b, *MNRAS* 385, 1179

Falceta-Gonçalves, D., Vidotto, A. A., & Jatenco-Pereira, V., 2006, *MNRAS* 368, 1145

Fares, R., Donati, J., Moutou, C., Bohlender, D., Catala, C., Deleuil, M., Shkolnik, E., Cameron, A. C., Jardine, M. M., & Walker, G. A. H., 2009, *MNRAS* 398, 1383

Fares, R., Moutou, C., Donati, J.-F., Catala, C., Shkolnik, E., Jardine, M., Cameron, A., & Deleuil, M., 2013, *ArXiv e-prints*

Fossati, L., Bagnulo, S., Elmasli, A., Haswell, C. A., Holmes, S., Kochukhov, O., Shkolnik, E. L., Shulyak, D. V., Bohlender, D., Albayrak, B., Froning, C., & Hebb, L., 2010a, *ApJ* 720, 872

Fossati, L., Haswell, C. A., Froning, C. S., Hebb, L., Holmes, S., Kolb, U., Helling, C., Carter, A., Wheatley, P., Cameron, A. C., Loeillet, B., Pollacco, D., Street, R., Stempels, H. C.,

Simpson, E., Udry, S., Joshi, Y. C., West, R. G., Skillen, I., & Wilson, D., 2010b, *ApJ* 714, L222
Guedel, M., 2004, *A&A Rev.* 12, 71
Hartmann, L. & MacGregor, K. B., 1980, *ApJ* 242, 260
Hollweg, J. V., 1973, *ApJ* 181, 547
Holzer, T. E., Fla, T., & Leer, E., 1983, *ApJ* 275, 808
Jardine, M., Vidotto, A. A., van Ballegooijen, A., Donati, J.-F., Morin, J., Fares, R., & Gombosi, T. I., 2013, *MNRAS* 431, 528
Jatenco-Pereira, V. & Opher, R., 1989, *A&A* 209, 327
Jones, G. H., Balogh, A., & Forsyth, R. J., 1998, *Geophys. Res. Lett.* 25, 3109
Keppens, R. & Goedbloed, J. P., 2000, *ApJ* 530, 1036
Lai, D., Helling, C., & van den Heuvel, E. P. J., 2010, *ApJ* 721, 923
Lazio, T. J. W., Carmichael, S., Clark, J., Elkins, E., Gudmundsen, P., Mott, Z., Szwajkowski, M., & Hennig, L. A., 2010, *AJ* 139, 96
Lecavelier des Etangs, A., Sirothia, S. K., & Gopal-Krishna, and Zarka, P., 2013, *A&A* 552, A65
Lim, J. & White, S. M., 1996, *ApJ* 462, L91+
Lima, J. J. G., Priest, E. R., & Tsinganos, K., 2001, *A&A* 371, 240
Llama, J., Wood, K., Jardine, M., Vidotto, A. A., Helling, C., Fossati, L., & Haswell, C. A., 2011, *MNRAS* 416, L41
McComas, D. J., Barraclough, B. L., Gosling, J. T., Hammond, C. M., Phillips, J. L., Neugebauer, M., Balogh, A., & Forsyth, R. J., 1995, *J. Geophys. Res.* 100, 19893
McComas, D. J., Ebert, R. W., Elliott, H. A., Goldstein, B. E., Gosling, J. T., Schwadron, N. A., & Skoug, R. M., 2008, *Geophys. Res. Lett.* 35, 18103
Mestel, L., 1968, *MNRAS* 138, 359
Morin, J., Donati, J., Petit, P., Delfosse, X., Forveille, T., Albert, L., Aurière, M., Cabanac, R., Dintrans, B., Fares, R., Gastine, T., Jardine, M. M., Lignières, F., Paletou, F., Ramirez Velez, J. C., & Théado, S., 2008, *MNRAS* 390, 567
Parker, E. N., 1958, *ApJ* 128, 664
Petit, P., Dintrans, B., Solanki, S. K., Donati, J.-F., Aurière, M., Lignières, F., Morin, J., Paletou, F., Ramirez Velez, J., Catala, C., & Fares, R., 2008, *MNRAS* 388, 80
Pinto, R. F., Brun, A. S., Jouve, L., & Grappin, R., 2011, *ApJ* 737, 72
Pneuman, G. W. & Kopp, R. A., 1971, *Sol. Phys.* 18, 258
Smith, A. M. S., Collier Cameron, A., Greaves, J., Jardine, M., Langston, G., & Backer, D., 2009, *MNRAS* 395, 335
Suess, S. T. and Smith, E. J., 1996, *Geophys. Res. Lett.* 23, 3267
Suzuki, T. K., Imada, S., Kataoka, R., Kato, Y., Matsumoto, T., Miyahara, H., & Tsuneta, S., 2012, *ArXiv e-prints*
Tsinganos, K. and Low, B. C., 1989, *ApJ* 342, 1028
Vidotto, A. A., Fares, R., Jardine, M., Donati, J.-F., Opher, M., Moutou, C., Catala, C., & Gombosi, T. I., 2012, *MNRAS* 423, 3285
Vidotto, A. A., Jardine, M., & Helling, C., 2010a, *ApJ* 722, L168
Vidotto, A. A., Jardine, M., & Helling, C., 2011a, *MNRAS* 411, L46
Vidotto, A. A., Jardine, M., & Helling, C., 2011b, *MNRAS* 414, 1573
Vidotto, A. A. and Jatenco-Pereira, V., 2006, *ApJ* 639, 416
Vidotto, A. A., Opher, M., Jatenco-Pereira, V., & Gombosi, T. I., 2009a, *ApJ* 703, 1734
Vidotto, A. A., Opher, M., Jatenco-Pereira, V., & Gombosi, T. I., 2009b, *ApJ* 699, 441
Vidotto, A. A., Opher, M., Jatenco-Pereira, V., & Gombosi, T. I., 2010b, *ApJ* 720, 1262
Wargelin, B. J. and Drake, J. J., 2002, *ApJ* 578, 503
Washimi, H. and Shibata, S., 1993, *MNRAS* 262, 936
Wilhelm, K., 2006, *A&A* 455, 697
Wood, B. E., Linsky, J. L., Müller, H., & Zank, G. P., 2001, *ApJ* 547, L49
Zarka, P., 2007, *Planet. Space Sci.* 55, 598

*Nature of Prominences and their role in Space Weather*
*Proceedings IAU Symposium No. 300, 2013*
*B. Schmieder, J.-M. Malherbe & S. T. Wu, eds.*

doi:10.1017/S1743921313011162

# Modeling magnetized star-planet interactions: boundary conditions effects

**Antoine Strugarek[1,2], Allan Sacha Brun[2], Sean P. Matt[3] and Victor Reville[2]**

[1]Département de physique, Université de Montréal,
C.P. 6128 Succ. Centre-Ville, Montréal, QC H3C-3J7, Canada
email: strugarek@astro.umontreal.ca

[2]Laboratoire AIM Paris-Saclay, CEA/Irfu Université Paris-Diderot CNRS/INSU,
F-91191 Gif-sur-Yvette

[3]Department of Physics & Astronomy, University of Exeter, Exeter EX2 4QL, UK

**Abstract.** We model the magnetized interaction between a star and a close-in planet (SPMIs), using global, magnetohydrodynamic numerical simulations. In this proceedings, we study the effects of the numerical boundary conditions at the stellar surface, where the stellar wind is driven, and in the planetary interior. We show that is it possible to design boundary conditions that are adequate to obtain physically realistic, steady-state solutions for cases with both magnetized and unmagnetized planets. This encourages further development of numerical studies, in order to better constrain and undersand SPMIs, as well as their effects on the star-planet rotational evolution.

**Keywords.** planet-star interactions; stars: winds, outflows; magnetohydrodynamics (MHD)

---

## 1. Introduction

The growing number of known exoplanet systems raises the question of how to properly define the habitability zone around a star (Kasting *et al.* 1993; Barnes *et al.* 2011). Its definition depends on the interactions existing between a planet and its host star, which are gravitational (tidal forces), magnetic (wind-planet interactions, hereafter referred to as SPMI) and radiative (*e.g.*, stellar EUV ionisation flux). Magnetized interactions between a star and its orbiting planets have recently been suggested to be at the origin of a possibly enhanced planet detectability (Jardine & Collier Cameron 2008; Fares *et al.* 2010; Miller *et al.* 2012). In the case of a close-in planet, these interactions may also be at the origin of anomalous stellar magnetic activity (Cuntz *et al.* 2000; Lanza 2008; Donati *et al.* 2008). It was also suggested that it could affect the star-planet rotational evolution (Laine *et al.* 2008; Pont 2009; Cohen *et al.* 2010; Vidotto *et al.* 2010; Lanza 2010). Theoretical work is needed to better understand SPMIs.

Based on a pioneering work done in the context of the satellites of Jupiter (Goldreich & Lynden-Bell 1969; Kivelson *et al.* 2004), Laine & Lin (2011) built an analytical model describing the various components of SPMIs in the case of unmagnetized planets. Pursuing the same goal, Lanza (2013) also developed semi-analytical models of SPMIs in the context of magnetized planets. However, a systematic numerical validation of those models still remains to be properly done (see Ip *et al.* 2004; Cohen *et al.* 2011, for first steps towards such a validation).

Focusing on close-in planets, the SPMIs include magnetic reconnection, magnetic field diffusion at the stellar surface and in the planet vicinity, radiation and ionisation processes in the planetary magnetosphere and magneto-sonic wave propagation. A numerical investigation of SPMI requires a careful description of those physical processes although it is generally not possible to treat all of them simultaneously with a unique model. Hence, specific strategies such as dedicated boundary conditions have to be developed to study SPMI from a global point of view. We detail in this work how to develop both stellar (section 2) and planetary (section 3) boundary conditions to globally model the different SPMI cases, within the MHD formalism.

## 2. Stellar boundary conditions

We model stellar winds following numerous previous analytical and numerical studies (Weber & Davis 1967; Washimi & Shibata 1993; Ustyugova *et al.* 1999; Keppens & Goedbloed 2000; Matt & Balick 2004; Matt *et al.* 2012). We use standard MHD theory to numerically model with the PLUTO code (Mignone *et al.* 2007) magnetized steady state flows anchored at the surface of a rotating star. We model winds driven by the thermal pressure of the stellar corona in a 2D axisymmetric cylindrical geometry (see Strugarek *et al.* 2012, for a more detailed description of the MHD model we use).

The steady-state wind solution can depend very sensitively on the type of boundary conditions that are imposed under the stellar surface. Because we want to use our model to study SPMIs, the stellar boundary conditions have to be able to both react and adapt to external stimuli originating from the orbiting planet. The design of a boundary

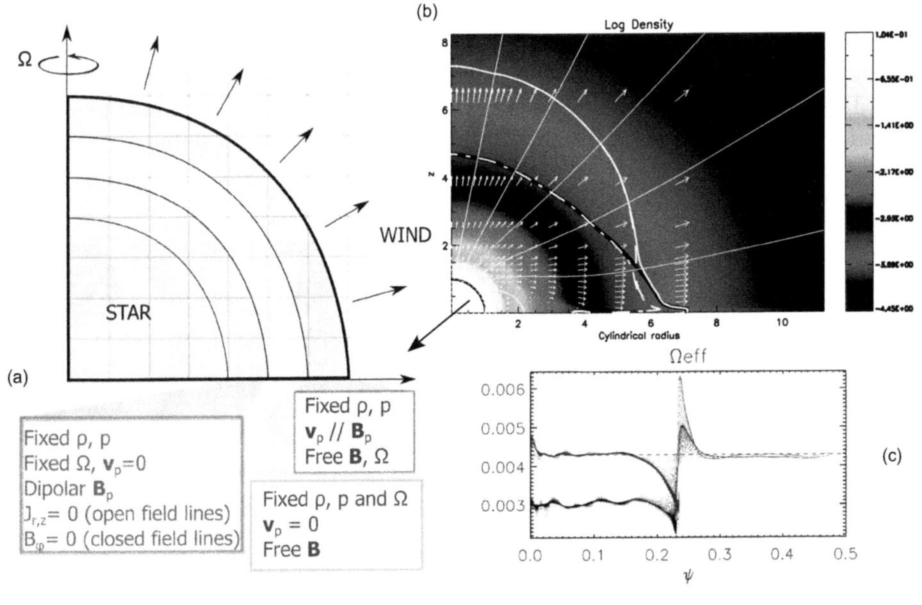


**Figure 1.** **(a)** Schematic of the multi-layer boundary condition ensuring good conservation properties of the MHD solution as well as reactivity to external stimuli. Fixed quantities are forced to the Parker wind solution. The subscript $p$ stands for the poloidal component $(\varpi, z)$ of vector in cylindrical coordinates. **(b)** Typical wind solution used for SPMI. The color map represents the logarithmic density, the white lines the poloidal magnetic field lines. The slow and fast Alfvèn surfaces are labeled by the dashed lines, and the arrows show velocity field. The stellar surface is labeled by a black quarter of a circle. The axes are in stellar radius units. **(c)** Effective rotation rate as a function of the streamfunction $\psi$ for good (blue dots) and bad (black dots) boundary conditions. The red dashed horizontal line labels the stellar rotation rate. Low values of $\psi$ correspond to open polar field lines and larger values of $\psi$ to closed equatorial field lines. Each dot corresponds to a grid point.

condition satisfying those two conditions, and its associated stellar wind solution, are displayed in panels (a) and (b) of fig. 1.

We developed a layered boundary condition over which the stellar wind characteristics are progressively enforced as we go deeper under the stellar surface. This boundary condition ensures very good conservation properties (Lovelace *et al.* 1986; Zanni & Ferreira 2009) along the magnetic field lines. This is exemplified in panel (c) of fig. 1. We display the effective rotation rate $\Omega_{\rm eff} \equiv \frac{1}{\varpi}(v_\phi - \frac{v_p}{B_p}B_\phi)$ as a function of the streamfunction $\psi$ generating the poloidal magnetic field. In a steady-state, ideal MHD wind, $\Omega_{\rm eff}$ should be constant along each field line and equal to $\Omega_\star$. The blue dots correspond to the boundary condition described in panel (a), and the black dots to a case where $B_\phi$ is set to 0 at all latitudes in the third boundary level. We observe that the target stellar rotation rate (dashed horizontal red line) is recovered only with the correct boundary conditions. Conservation errors exist at the open-closed field lines boundary ($\psi \sim 0.23$), but they remain confined to very few grid points in the simulation domain. Finally, this boundary condition is intrinsically able to react to a perturbation by a planet orbiting a star by, *e.g.*, modifying the stellar wind topology. We discuss now the importance of planetary boundary conditions when studying SPMIs.

## 3. Planetary boundary conditions

SPMIs are generally decomposed in two categories: the so-called unipolar and dipolar interactions (Zarka 2007), which refer to the cases of unmagnetized and magnetized planets. Both interactions can be modeled within the MHD formalism with an adequate boundary condition design. We detail in this section how to design such boundary conditions. The examples given here were all done for a planet with a radius of $r_p = 0.1\,r_\star$, a mass of $M_p = 0.01\,M_\star$, an orbital radius of $r_{\rm orb} = 3r_\star$ and a resolution of $0.03\,r_p$ at the planetary surface.

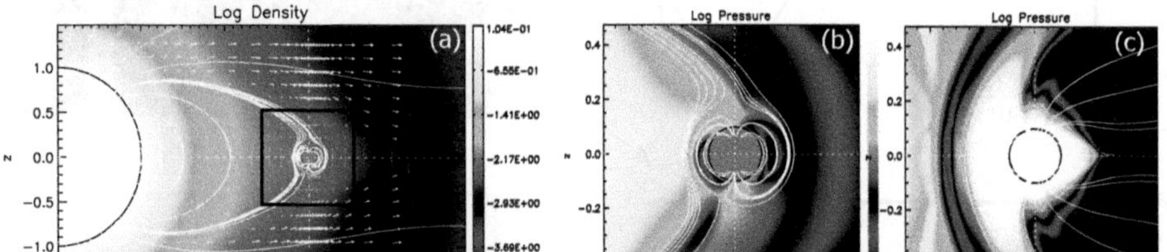

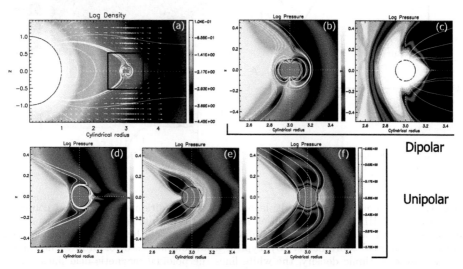


**Figure 2.** Zoom on planetary boundary conditions effects for dipolar (upper panels) and unipolar (lower panels) interactions. The color map represents the gas pressure in logarithmic scale, and the white lines the magnetic field lines. The planet surface is labeled by a black circle at 1 stellar radius. Panels (a) and (b) show the fiducial dipolar case, and panel (c) is the unrealistic case of a planet with a very high internal pressure. Panel (d) represents a Venus-like interaction and panels (e) and (f) two Io-Jupiter-like interactions.

We consider the planet itself as a boundary condition. The PLUTO code allows one to define internal domains as boundary conditions over which all variables can be altered during the model evolution. In all cases, we set the poloidal velocity to zero and the azimuthal velocity to the keplerian velocity inside the planet. We also set the density and pressure values inside the planet to fiducial values which are consistent with its gravity field. These value have to be carefully prescribed since they can trigger undesirable effects in the vicinity of the planet. We give an example of a dipolar case in panels (a) and (b) of fig. 2 (the planetary magnetic field is simply enforced in the planetary interior in this case). A stable configuration is obtained when the magnetic pressure and the gas pressure equilibrate at the interface between the planetary magnetosphere and the stellar wind. The ram pressure plays little role here because the planet we consider is in the so called *dead-zone* of the stellar wind, in which the poloidal velocity is negligible. We show in panel (c) the exact same simulation for an extreme case where we multiplied the internal pressure of the planet by a factor of 20. The former pressure balance then fails and a wind is driven from the planet itself. The planetary dipole opens up and a shock eventually creates at the interface between the two "winds". Such undesirable effects may also be obtained by varying the density of the interior of the planet. Hence, any SPMI model must be developed to minimize such undesirable effects in the final solution.

Modeling a planet in the unipolar case is a bit more complex than in the dipolar case. Two classes of unipolar interactions can indeed be distinguished: Venus-like interaction (case V) and Io-Jupiter like interaction (case IJ). Note however that in both cases, we consider a planet located inside the stellar wind dead-zone, at $r_p = 3\,R_\star$.

In case V, the ionisation of the planetary atmosphere by the stellar EUV radiation flux allows the creation of a ionosphere which acts as a barrier between the stellar wind magnetic field and the unmagnetized interior of the planet (Russell 1993). Depending on the stellar wind conditions around the planet, an induced magnetosphere may then be sustained on secular time scales. We show case V in panel (d) of fig. 2. The ionosphere is modeled as a very thin ($<0.2\,r_p$) highly conductive boundary layer under the planetary surface. The wrapping of the magnetic field lines around planet (Russell 1993) is naturally recovered.

In case IJ, no ionosphere is created and the stellar wind magnetic field pervades inside the planet. The SPMI then depends on the ratio of electrical conductivities between the planetary interior and the stellar surface where the magnetic field lines are *a priori* anchored. This ratio sets the effective drag the planet is able to induce on the stellar wind magnetic field lines. We use the ability of the PLUTO code to add extra ohmic diffusion in the planet interior to model it and show in figure 2 two extreme cases in which magnetic field lines are dragged (panel e) or diffused (panel f) by the planet. In all cases, we obtain a statistical steady state in which the SPMI can be analyzed in details.

## 4. Conclusions

We showed in this work that is it possible to model the global, magnetized and non-linear interactions between a star and a planet, within the MHD formalism. It requires a careful development of adequate boundary conditions to represent the various interaction cases. We showed that boundary conditions play a very important role both at the stellar surface and in the planetary interior. Steady state solutions could be found in the dipolar case as well as in both the Venus-like and Io-Jupiter-like unipolar cases.

The SPMI model we developed will be useful for exploring stable interaction configurations between a close-in planet and its host star. In addition, It will enable quantitative predictions of rotational evolution of star-planet systems due to the effective magnetic torques which develop in the context of dipolar and unipolar interactions (Strugarek *et al.*, in prep). Finally, such models could also be used to study potential SPMI induced emissions, which we will analyze in a future work.

**Acknowledgements**

We thank A. Mignone and his team for making the PLUTO code open-source. We thank A. Cumming, R. Pinto, C. Zanni and P. Zarka for inspiring discussions on star-planet magnetized interactions. This work was supported by the ANR TOUPIES and the ERC project STARS2. We acknowledge access to supercomputers through GENCI project 1623 and Prace infrastructures. A. Strugarek acknowledges support from the Canada's Natural Sciences and Engineering Research Council.

## References

Barnes, R., Meadows, V. S., Domagal-Goldman, S. D., *et al.* 2011, *16th Cambridge Workshop on Cool Stars*, 448, 391
Cohen, O., Drake, J. J., Kashyap, V. L., Sokolov, I. V., & Gombosi, T. I. 2010, *ApJ*, 723, L64
Cohen, O., Kashyap, V. L., Drake, J. J., *et al.* 2011, *ApJ*, 733, 67
Cuntz, M., Saar, S. H., & Musielak, Z. E. 2000, *ApJ*, 533, L151
Donati, J.-F., Moutou, C., Fares, R., *et al.* 2008, *MNRAS*, 385, 1179
Fares, R., Donati, J.-F., Moutou, C., *et al.* 2010, *MNRAS*, 406, 409
Goldreich, P. & Lynden-Bell, D. 1969, *ApJ*, 156, 59
Ip, W.-H., Kopp, A., & Hu, J.-H. 2004, *ApJ*, 602, L53
Jardine, M. & Collier Cameron, A. 2008, A&A, 490, 843
Kasting, J. F., Whitmire, D. P., & Reynolds, R. T. 1993, *Icarus*, 101, 108
Keppens, R. & Goedbloed, J. P. 2000, *ApJ*, 530, 1036
Kivelson, M. G., Bagenal, F., Kurth, W. S., *et al.* 2004, *In: Jupiter. The planet*, 513
Laine, R. O. & Lin, D. N. C. 2011, *ApJ*, 745, 2
Laine, R. O., Lin, D. N. C., & Dong, S. 2008, *ApJ*, 685, 521
Lanza, A. F. 2008, *A&A*, 487, 1163
—. 2010, *A&A*, 512, 77
—. 2013, *A&A*, 557, 31
Lovelace, R. V. E., Mehanian, C., Mobarry, C. M., & Sulkanen, M. E. 1986, *ApJS*, 62, 1
Matt, S. & Balick, B. 2004, *ApJ*, 615, 921
Matt, S. P., MacGregor, K. B., Pinsonneault, M. H., & Greene, T. P. 2012, *ApJL*, 754, L26
Mignone, A., Bodo, G., Massaglia, S., *et al.* 2007, *ApJS*, 170, 228
Miller, B. P., Gallo, E., Wright, J. T., & Dupree, A. K. 2012, *ApJ*, 754, 137
Pont, F. 2009, *MNRAS*, 396, 1789
Russell, C. T. 1993, Reports on Progress in Physics, 56, 687
Strugarek, A., Brun, A. S., Matt, S. P., & Reville, V., in preparation
Strugarek, A., Brun, A. S., & Matt, S. 2012, *in SF2A-2012: Proceedings of the Annual meeting of the French Society of Astronomy and Astrophysics*. Eds.: S. Boissier, 419–423
Ustyugova, G. V., Koldoba, A. V., Romanova, M. M., Chechetkin, V. M., & Lovelace, R. V. E. 1999, *ApJ*, 516, 221
Vidotto, A. A., Opher, M., Jatenco-Pereira, V., & Gombosi, T. I. 2010, *ApJ*, 720, 1262
Washimi, H. & Shibata, S. 1993, *MNRAS*, 262, 936
Weber, E. J. & Davis, L. J. 1967, *ApJS*, 148, 217
Zanni, C. & Ferreira, J. 2009, *A&A*, 508, 1117
Zarka, P. 2007, *Planetary and Space Science*, 55, 598

*Nature of Prominences and their role in Space Weather*
*Proceedings IAU Symposium No. 300, 2013*
*B. Schmieder, J.-M. Malherbe & S. T. Wu, eds.*

doi:10.1017/S1743921313011174

# Stellar CME activity and its possible influence on exoplanets' environments: Importance of magnetospheric protection

**Maxim L. Khodachenko[1,2], Yury Sasunov[1], Oleksiy V. Arkhypov[1], Igor I. Alexeev[2], Elena S. Belenkaya[2], Helmut Lammer[1], Kristina G. Kislyakova[1], Petra Odert[3], Martin Leitzinger[3], and Manuel Güdel[4]**

[1]Space Research Institute, Austrian Academy of Sciences, 8042 Graz, Austria
email: maxim.khodachenko@oeaw.ac.at

[2]Institute of Nuclear Physics, Moscow State University, 119992 Moscow, Russia

[3]Institute of Physics, Karl-Franzens-University, 8010 Graz, Austria

[4]Institute of Astronomy, University of Vienna, 1180 Vienna, Austria

**Abstract.** CMEs are large-scale magnetized plasma structures carrying billions of tons of material that erupt from a star and propagate in the stellar heliosphere, interacting in multiple ways with the stellar wind. Due to the high speed, intrinsic magnetic field and the increased plasma density compared to the stellar wind background, CMEs can produce strong effects on planetary environments when they collide with a planet. The main planetary impact factors of CMEs, are associated interplanetary shocks, energetic particles accelerated in the shock regions, and the magnetic field disturbances. All these factors should be taken into account during the study of evolutionary processes on exoplanets and their atmospheric and plasma environments. CME activity of a star may vary depending on stellar age, stellar spectral type and the orbital distance of a planet. Because of relatively short range of propagation of majority of CMEs, they impact most strongly the magnetospheres and atmospheres of close orbit ($< 0.1$ AU) exoplanets.

**Keywords.** magnetic fields, plasmas, stars:planetary systems, stars:winds, stars:activity

## 1. Introduction

The constantly growing number of discovered exoplanets and accumulation of data regarding their physical and orbital characteristics provide an empirical platform for a more detailed study of general principles and major trends of the formation and evolution of planets and planetary systems (including the planetary potential habitability aspect). More than a half of known exoplantes have orbits around their host stars shorter than 0.6 AU. By this, an evident maximum in the orbital distribution of exoplanets takes place in the vicinity of 0.05 AU, with two well pronounced major sub-populations there corresponding to the giant type planets ($0.2M_{\rm J} < m_{\rm p} < 8M_{\rm J}$), so called "Hot Jupiters", and less massive ($0.008M_{\rm J} < m_{\rm p} < 0.08M_{\rm J}$), Neptune- and Super-Earth type planets. Here $M_{\rm J}$ stays for the mass of Jupiter. Altogether the Hot Jupiters comprise about 30% of the total number of known exoplanets.

Close location of the majority of known exoplanes to their host stars results in intensive heating, ionization, and chemical modification of their upper atmospheres by the stellar X-ray/EUV (XUV) radiation with the subsequent expansion of the ionized atmospheric material and its loss due to interaction with the stellar wind (Lammer *et al.*, 2009; Khodachenko *et al.*, 2007a,b). A number of actual questions regarding the evolutionary paths of planetary systems and influencing them key factors is nowadays under continuous tackling. Among these questions a prominent position belongs to the problem of stellar - planetary interactions, including consideration of influences of stellar radiation and plasma flows, e.g., stellar wind, coronal mass ejections (CMEs), on planetary environments and evolution of planets. Magnetic fields, those connected with the planetary intrinsic magnetic dipole $\mathcal{M}$, as well as the magnetic fields associated with the electric current systems induced in the planetary close surroundings, play here an important role. They the planetary magnetosphere which appears as an obstacle (magnetospheric obstacle) interacting with a stellar wind and protecting the internal planetary environments (ionosphere, atmosphere, surface) against of direct impact of stellar plasmas and energetic particles (e.g., cosmic rays).

The plasma of stellar CMEs colliding with a planet, interacts with the planetary magnetosphere, and in the case of a weak magnetospheric protection (i.e., weak or no intrinsic planetary magnetic dipole), the magnetosphere is compressed down to the planetary surface, resulting in strong erosion of the planetary atmosphere. Sufficiently large magnetospheres are known to protect the underlying planetary environments, e.g. ionosphere, atmosphere, and surface against of stellar XUV/EUV and stellar wind factors. These usually require strong enough intrinsic planetary magnetic fields and/or extended magnetospheric current systems such as magnetodisks. Below we discuss the role of such factors like activity of a host star and intrinsic magnetic field of a planet and show how the account of these factors may influence the scaling of the planetary magnetosphere and its protecting capabilities.

## 2. Impact of stellar radiation and plasma flows on planets

Interaction of short-periodic exoplanets with the stellar wind plasma and high XUV flux at close orbital distances plays a crucial role regarding the ionization and ion loss processes of atmospheric species. The action of intensive stellar radiation and stellar winds on planetary environments consists of the following effects.

1) XUV radiation of the host star affects the the planetary thermosphere heat budget, resulting in the heating and expansion of the upper atmosphere, which under certain conditions could be so large that the majority of light atmospheric constituents overcome the gravitational binding and escape from the planet in the form of a hydrodynamic wind. This effect is called as a hydrodynamic or *thermal* escape (Tian *et al.*, 2008; Penz *et al.*, 2008; Erkaev *et al.*, 2013). Simultaneously with the direct radiational heating of the upper atmosphere, the processes of ionization with the consequent production of energetic neutral atoms (ENAs) by various photo-chemical and charge exchange reactions take place (Lammer *et al.*, 2008; Lichtenegger *et al.*, 2009). Such processes, together with the thermal escape, result in the formation around planets of extended (in some cases) coronas, filled with hot neutral atoms.

2) The expanding upper planetary atmospheres and/or hot neutral coronas may reach and even exceed, the boundaries of the planetary magnetospheres. In this case they will be directly exposed to the plasma flows of the stellar wind and CMEs with the consequent loss due to ion pick-up, as well as sputtering, and different kinds of photo-chemical energizing mechanisms which all contribute to the so-called *non-thermal* atmospheric mass-loss process (Lichtenegger *et al.*, 2009). As a crucial parameter here appears the

size of the planetary magnetosphere. Altogether, this makes the planetary magnetic field, as well as the parameters of the stellar wind (mainly density $n_{sw}$ and speed $v_{sw}$) to be very important for the processes of atmospheric erosion and mass-loss of a planet, affecting finally the whole evolution of its environments. By this, the size of magnetosphere and its planetary protecting role should be always considered in context with the fact that the stellar radiation and plasma flows may vary significantly throughout the lifetime of the host star, as its luminosity and activity evolve. This evolution is different for different star types, and depends also on their age.

In that respect, the interaction of close-orbit exoplanets with stellar CMEs appears to be an important process, which is central to a better understanding of the non-thermal mass loss mechanism. High speed, intrinsic magnetic field and the increased density as compared to the stellar wind background, make CMEs an active factor which strongly influences the planetary environments and magnetospheres. Often collisions of the close-orbit exoplanets with massive stellar CME plasmas should compress planetary magnetospheres much deeper towards the surface of the exoplanet. This would result in much higher ion loss rates than that expected during the usual stellar wind conditions.

### 2.1. *Stellar activity*

The relevant physical phenomena of stellar activity on late-type stars (i.e., spectral classes G, K, M) and their observational manifestations include modulations of the stellar photospheric light due to stellar spots, intermittent and energetic flares, coronal mass ejections (CMEs), stellar cosmic rays, enhanced XUV emissions (see Scalo *et al.* (2007) and references therein). Evaluation of flaring rates and intensities usually require long-duration monitoring. So, the proxies for the flaring activity are used, such as optical Ca H and K emission cores, H-alpha and Mg II emission, soft X-ray continuous emission, and a large number of UV-to soft X-ray emission lines (*Ayres* 1997; *Gershberg* 2005).

Observations of stars in clusters have revealed that single late-type stars spin down monotonically with their age because of angular momentum loss (*Skumanich* 1972]). For a given age of star $t$, the stellar rotation period can be estimated as (*Newkirk* 1980): $P_{\rm rot} \propto \left(1+\frac{t}{\tau}\right)^{0.7}$, where $\tau = 2.56 \times 10^7$ yrs is a time constant calculated by *Newkirk* (1980)]. At the same time, already early studies pointed out a strong correlation between the rotation rate of a star and its activity level (*Wilson* 1966). This correlation means that there must be dependence between stellar activity and age. For solar-type stars this has been studied within the "Sun in Time" project (Ribas *et al.*, 2005). Based on the analysis of a large amount of X-ray, EUV and UV observations of a homogeneous sample of single nearby G0-5 stars with known rotation periods, luminosity and ages, it has been concluded that during the fist 100 Myr after the Sun arrived at the Zero-Age Main-Sequence (ZAMS), the integrated XUV flux was up to 100 times higher than today. After this very active stage, XUV flux of a solar-type star decreases with the time: $\propto t[Gyr]^{-1.72}$ (Ribas *et al.*, 2005).

According to the currently accepted paradigm, the wide range of activity levels and related phenomena observed in different stars is directly connected with operation of the stellar magnetic dynamo. By this, two basic parameters: (*i*) stellar rotation rate and (*ii*) depth of the convective zone, are believed to control the stellar dynamo efficiency, which increases with increasing of any, or both of these quantities. Since the stellar convective envelope becomes thicker with decreasing stellar mass, it is straightforward to infer that, at a given rotation period (i.e. age), the low-mass M- and K- stars should be more active than a solar-type G- star. This fact has many observational confirmations. For example, a relatively old ($\sim$ 5.5 Gyr) dwarf M- star, Proxima Centauri, experiences measurable flares at a rate of about one flare per hour (*Walker* 1981).

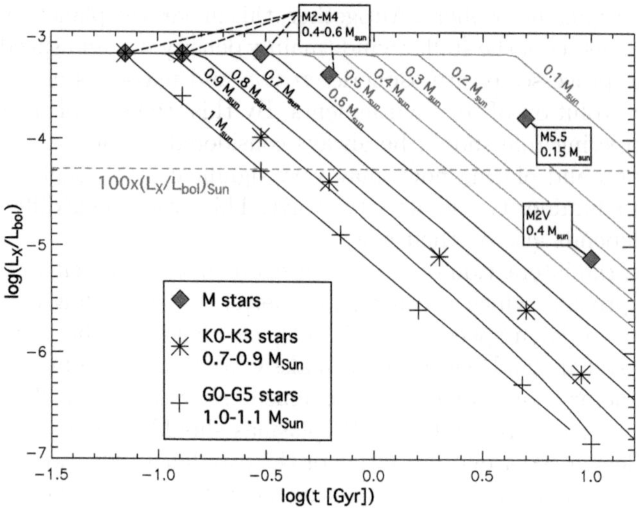


**Figure 1.** $L_X/L_{bol}$ as a function of age for stars with masses $< M_{Sun}$. Symbols represent stars from the "Sun in Time" program (adopted from Scalo *et al.* (2007).

Audard *et al.*, (2000) found that the energy of flares correlates with the stellar activity, characterized by $L_X/L_{bol}$, where $L_X$ and $L_{bol}$ are X-ray and bolometric luminosities of a star, respectively. The evolution of $log(L_X/L_{bol})$ with time for stars of various masses is shown in Figure 1, provided by Scalo *et al.* (2007). According to this activity-age diagram, the solar-type G- stars stay at saturated emission levels only until ages of $\sim$ 100 Myr, and then their XUV luminosities rapidly decrease with age: $\propto (t[Gyr])^{-1.72}$. On the other hand, M- stars have saturated emission periods up to 0.5–1 Gyr, and then their luminosity decreases in a way similar to the solar-type stars. According to Audard *et al.*, (2000), the rate of high-energy ($E > 10^{32}$ erg) flares per day as $logN|_{E>10^{32}}$ erg $= -26.7+0.95logL_X$, which in the case of M- stars with a saturated activity level $L_X = 7 \times 10^{28}$ erg/s implies $\sim$ 6 strong flares per day. Therefore, the powerful flares occur more often in X-ray bright stars. Altogether it has been found (Ribas *et al.*, 2005; Scalo *et al.*, 2007) that early K-stars and early M-stars may have XUV emissions level, and therefore flaring rates, of $\sim$(3–4) and $\sim$(10–100), respectively, times higher than solar-type G-stars of the same age.

### 2.2. *Stellar winds and CMEs*

In addition to being exposed to electromagnetic radiation from their host stars, exoplanets are also exposed to high-speed outflows of particles from the stellar atmosphere. For cool main sequence stars like the Sun, stellar winds arise in the hot coronas that represent the outermost atmospheres of the stars. Although the mechanisms of coronal heating and coronal wind acceleration remain hot topics of research, *Parker* (1958) demonstrated long ago that if once you have a hot corona, a wind much like that of the Sun arises naturally through thermal expansion. Thus, any star known to have a hot corona can be expected to possess a coronal wind. Observations from X-ray observatories such as *Einstein*, *ROSAT*, *Chandra*, and *XMM-Newton* have demonstrated that X-ray emitting coronas are ubiquitous among cool main sequence stars, so coronal winds can be expected to be a common feature as well. Unfortunately, detecting and studying these winds is much harder than detecting and studying the coronas in which they arise. Current observational capabilities do not yet allow us to directly detect solar-like coronal winds emanating from other stars.

Recently, there have been important developments towards indirect detections of stellar winds through their interactions with the surrounding interstellar medium. In particular, the stellar mass loss rates and related stellar wind parameters have been estimated by observing astrospheric absorption features of several nearby G- and K- stars. Comparison of the measured absorption to that calculated by hydrodynamic codes made it possible to perform empirical estimation of the evolution of the stellar mass loss rate as a function of stellar age (Wood *et al.*, 2002; 2005) and to conclude about the dependence of $n_{\rm sw}$ and $v_{\rm sw}$ on the age of the stellar system. In particular, the younger solar-type G- stars appeared to have much denser and faster stellar winds as compared to the present Sun. Combining the stellar mass loss measurements of (Wood *et al.*, 2005) with the results of (*Newkirk*, 1980) for the age-dependence of stellar wind velocity, *Grießmeier et al.* (2007) proposed a method for calculation of stellar wind density $n_{\rm sw}$ and velocity $v_{\rm sw}$ at a given orbital location of an exoplanet $d$ for a given mass $M_*$, radius $R_*$ and age $t_*$ of a star. As an example, the values of stellar wind plasma parameters for a solar-analog G-type star ($M_* = M_{\rm Sun}$, age: 4 Gyr) at orbital distances of 0.045 AU, 0.1 AU, and 0.3 AU are given in Table 1.

Furthermore, it is known from observations of our Sun that flaring activity of a star is accompanied by eruptions of coronal mass (e.g. CMEs), occurring sporadically and propagating in the stellar wind as large-scale plasma-magnetic structures. Traveling outward from the star at high speeds (up to thousands km/s), CMEs create major disturbances in the interplanetary medium and produce strong impacts on the planetary environments and magnetospheres. Since CMEs can be directly observed only on the Sun, the current knowledge on them comes from the study of the Sun and the heliosphere. On the Sun, CMEs are associated with flares and prominence eruptions and their sources are usually located in active regions and prominence sites. The likelihood of CME-events increases with the size and power of the related flare event. Generally, it is expected that the frequent and powerful flares on magnetically active flaring stars should be accompanied by an increased rate of CME production. By considering the Sun as a typical representative of G- stars, it seems reasonable to assume a similarity of the basic parameters of the stellar winds of G- stars and those known for the Sun. Such a solar-stellar analogy principle is widely considered for the investigation of basic processes of the stellar wind - planet interaction. Based on the estimations of solar CME plasma density $n_{\rm CME}$, using the in-situ spacecraft measurements (at distances $> 0.4$ AU) and the analysis of white-light coronagraph images (at distances $\leqslant 30 R_{\rm Sun} \approx 0.14$ AU), Khodachenko *et al.* (2007a) provided general power-law interpolations of $n_{CME}$ dependence on the distance to a star:

$$n_{\rm CME}^{min}(d) = 4.88(d[\mathrm{AU}])^{-2.3}, \quad n_{\rm CME}^{max}(d) = 7.10(d[\mathrm{AU}])^{-3.0}, \tag{2.1}$$

Equations (2.1) identify a typical maximum-minimum range of $n_{\rm CME}$. The dependence of stellar CME speed $v_{\rm CME}$ on the orbital distance $d$ can be approximated by the formula:

$$v_{\rm CME} = v_0 \left(1 - e^{\frac{2.8 R_{\rm Sun} - d}{8.1 R_{\rm Sun}}}\right)^{1/2}, \tag{2.2}$$

proposed in Sheeley *et al.* (1997) on the basis of tracking of several solar wind density enhancements at close distances ($d < 0.1$ AU). For the approximation of average- and high- speed CMEs one may take in (2.2) $v_0 = 500$ km/s and $v_0 = 800$ km/s, respectively. Besides of that, the average mass of CMEs is estimated as $10^{15}$ g, whereas their average duration at distances $\sim 0.05$ AU is close to 8 hours. Table 1 provides an example of stellar CME plasma parameters for a solar-analog G-type star at orbital distances of 0.045 AU, 0.1 AU, and 0.3 AU. Because of the relatively short range of propagation of majority

**Table 1.** Stellar wind and CME parameters for a solar-analog G-type star ($M_* = M_{\rm Sun}$, age: 4 Gyr) at different orbital distances. The values of $\tilde{v}_{\rm CME}^{fast}$ and $\tilde{v}_{\rm CME}^{av}$ are obtained using (2.2) with $v_0 = 800$ km/s and $v_0 = 500$ km/s, respectively. All the velocities include a contribution of the Keplerian planetary orbital velocity $V_K$.

| Orbital distance | $n_{\rm sw}$ [cm$^{-3}$] | $\tilde{v}_{\rm sw}$ [km/s] | $n_{\rm CME}^{min}$ / $n_{\rm CME}^{max}$ [cm$^{-3}$] | $\tilde{v}_{\rm CME}^{av}$ / $\tilde{v}_{\rm CME}^{fast}$ [km/s] |
|---|---|---|---|---|
| 0.045 AU | 9.1e3 | 210 | 6.1e3/7.8e4 | 520/810 |
| 0.1 AU | 1.2e3 | 260 | 1.0e3/7.1e3 | 510/810 |
| 0.3 AU | 92 | 340 | 78/2.6e2 | 500/800 |

of CMEs, they should strongly impact first of all the planets at close orbits ($\lesssim 0.3$ AU). Khodachenko *et al.* (2007a) have found that for a critical CME production rate $f_{\rm CME}^{cr} \approx 36$ CMEs per day (and higher) a close orbit exoplanet appears under continuous action of the stellar CMEs plasma, so that each next CME collides with the planet during the time when the previous CME is still passing over it. This means in general the harder conditions for the planetary environments than those in the case of a regular stellar wind. Therefore, the investigation of evolutionary paths of close-orbit exoplanets in potentially habitable zones around young active stars, besides of the higher XUV radiation, should take also into account the effects of "short range" (in astrophysical scales) planetary impacting factors of stellar activity such as relatively dense stellar winds and frequent magnetic clouds (MCs) and CMEs.

## 3. The problem of magnetospheric protection of exoplanets

For an efficient magnetospheric protection of a planet, the size of its magnetosphere characterized by the magnetopause stand-off distance $R_{\rm s}$ should be much larger than the height of the exobase. By this, the value of $R_{\rm s}$ is determined from the balance between the stellar wind ram pressure and the planetary magnetic field pressure at the substellar point (*Grießmeier et al.*, 2004; Khodachenko *et al.*, 2007a). In the most of studies so far, the investigation of an exoplanetary magnetospheric protection is performed within a highly simplifying assumption of a planetary *dipole-dominated* magnetosphere. This means that only the intrinsic magnetic dipole moment of an exoplanet $\mathcal{M}$ and the corresponding magnetopause electric currents (i.e., "screened magnetic dipole" case) are considered as the major magnetosphere forming factors. In this case, i.e. assuming $B(r) \propto \mathcal{M}/r^3$, the value of $R_{\rm s}$ has been defined by the following expression:

$$R_{\rm s} \equiv R_{\rm s}^{(dip)} = \left[\frac{\mu_0 f_0^2 \mathcal{M}^2}{8\pi^2 \rho_{\rm sw} \tilde{v}_{\rm sw}^2}\right]^{1/6}, \qquad (3.1)$$

where $\mu_0$ is the diamagnetic permeability of free space, $f_0 \approx 1.22$ is a form-factor of the magnetosphere caused by the account of the magnetopause electric currents, $\rho_{\rm sw} = n_{\rm sw} m$ is the mass density of the stellar wind, and $\tilde{v}_{\rm sw}$ is the relative velocity of the stellar wind plasma which includes also the planetary orbital rotation velocity. For the tidally locked close orbit exoplanets with weak magnetic moments exposed to a dense and/or fast stellar wind plasma flows, (3.1) yields rather small values for sizes of dipole-dominated magnetospheres, $R_{\rm s} = R_{\rm s}^{(dip)}$, compressed by the stellar wind plasma flow, which in the most extreme cases may even shrink down to the planetary radius $r_{\rm p}$. Therefore, the approach to estimation of the magnetosphere size based on (3.1) resulted in the commonly accepted conclusion, that in order to have an efficient magnetic shield, a planet needs a strong intrinsic magnetic dipole $\mathcal{M}$.

Khodachenko *et al.* (2007b) studied the mass loss of the Hot Jupiter HD 209458b due to the ion pick-up mechanism caused by stellar CMEs, colliding with the planet. In spite of the sporadic character of the CME-planetary collisions, in the case of a moderately active host star of HD 209458b, it has been shown that the integral action of the stellar CME impacts over the exoplanet's lifetime can produce significant effect on the planetary mass loss. The estimates of the non-thermal mass loss of the weakly magnetically protected Hot Jupiter, HD 209458b, due the stellar wind and CMEs ion pick-up, lead to significant and sometimes unrealistic values – up to several tens of planetary masses $M_{\rm p}$ lost during a planet life time (Khodachenko *et al.*, 2007b). In view of the fact that multiple close-in giant exoplanets, comparable in mass and size with the Solar System Jupiter exist, and that it is unlikely that all of them began their life as ten times, or even more massive objects, one may conclude that additional factors and processes have to be taken into consideration in order to explain the protection of close-in exoplanets against of destructive non-thermal mass loss. In the following sub-section we introduce a more complete model of magnetosphere of a giant gas exoplanet, which due to its consequent account of the specifics of close orbit Hot Jupiters provides under similar conditions larger sizes for the planetary magnetospheric obstacles, then those given by the simple screened magnetic dipole model, traditionally considered so far in the literature.

## 4. Magnetodisk-dominated magnetosphere of a Hot Jupiter

The investigation of exoplanetary magnetospheres and their role in evolution of planetary systems forms a new and fast developing branch. Magnetosphere of a close orbit exoplanet is a complex object, which formation depends on different external and internal factors. These factors may be subdivided on two basic groups: (a) *stellar factors*, e.g., stellar radiation, stellar wind plasma flow, stellar magnetic field and (b) *planetary factors*, e.g., type of planet, orbital characteristics, escaping material flow, and planetary magnetic field. The structure of an exoplanetary magnetosphere depends also on the speed regime of the stellar wind plasma relative the planet (Erkaev *et al.*, 2005; Ip *et al.*, 2004). In particular, for an exoplanet at sufficiently large orbital distance when the stellar wind is super-sonic and super-Alfvénic, i.e. when the ram pressure of the stellar wind dominates the magnetic pressure, a Jupiter-type magnetosphere with a bow shock, magnetopause, and magnetotail, is formed. At the same time, in the case of an extremely close orbital location of an exoplanet (e.g., $d < 0.03$ AU for the Sun analogue star), where the stellar wind is still under acceleration and remains to be sub-magnetosonic and sub-Alfvénic (Ip *et al.*, 2004; Preusse *et al.*, 2005), an Alfvénic wing-type magnetosphere without a shock in the upstream region is formed. The character of the stellar wind impact on the planetary nearby plasma environment and inner atmosphere is different for the super- and sub- Alfvénic types of the magnetosphere and in each particular planet case it has to be properly taken into account. In the present paper, however, we do not consider the Alfvénic wing-type magnetospheres, aiming at moderately short orbit giant planets near solar-type stars, under the conditions of a super-Alfvénic stellar wind flow, i.e., with the magnetospheres having in a general case a bow shock, a magnetopause, a magnetotail, similar to the case of the solar system Jupiter.

### 4.1. *Magnetodisk - a key element of Hot Jupiter magnetosphere*

To explain an obvious survival and sufficient magnetospheric protection of close orbit Hot Jupiters under the extreme conditions of their host stars Khodachenko *et al.* (2012)

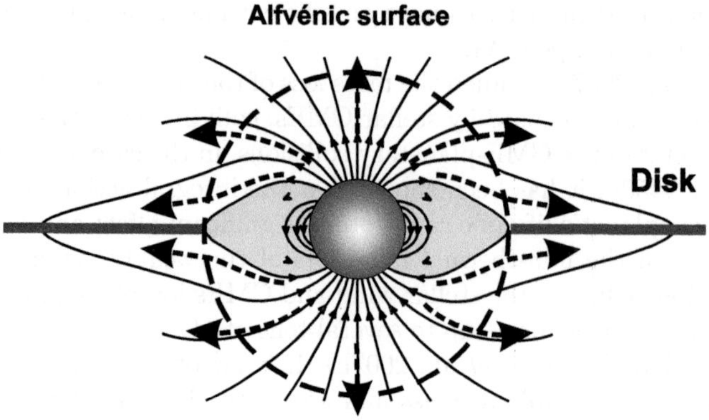


**Figure 2.** Schematic view of magnetodisk formation (adopted from Khodachenko *et al.* (2012).

proposed a more generic view of a Hot Jupiter magnetosphere. A key element in the proposed approach consists in taking into account of the upper atmosphere of a planet as an expanding dynamical gas layer heated and ionized by the stellar XUV radiation (Johansson *et al.*, 2009; Koskinen *et al.*, 2010, 2012). Interaction of the outflowing plasma with the rotating planetary magnetic dipole field leads to development of a current-carrying magnetodisk surrounding the exoplanet. The inner edge of magnetodisk is located at the so called Alfvénic surface ($r = R_A$) where the kinetic energy density of the moving plasma becomes equal to the energy density of the planetary magnetic field. This condition is equivalent also to the equality of the plasma ram pressure and magnetic pressure, or the Alfvén Mach number $M_A^2 = 1$. Beyond the Alfvénic surface the expanding plasma is not guided any more by the dipole magnetic field. It deforms the field lines leading to creation of a current-carrying magnetodisk which in turn entirely changes the topology of planetary middle and outer magnetosphere.

According to Khodachenko *et al.* (2012), a Hot Jupiter's magnetodisk can be formed by different mechanisms, acting simultaneously: 1) the thermal expansion of the escaping planetary plasma wind, heated by the stellar radiation, and 2) the centrifugal acceleration of plasma by rotating planetary magnetic field in the co-rotation region, with subsequent release of material in the vicinity of the Alfvénic surface (so called "sling" mechanism). A self-consistent description of both these mechanisms represents an important and complex physical problem. So far only a qualitative insight into origin and interconnection of the inner (dipole dominated) and outer (magnetodisk-dominated) parts of the magnetosphere of a Hot Jupiter was suggested in Khodachenko *et al.* (2012). Two major regions with the different topology of magnetic field can be distinguished in the magnetosphere of a Hot Jupiter driven by the escaping plasma flow (*Mestel*, 1968). The first region corresponds to the inner magnetosphere, or so-called "dead zone", filled with closed dipole-type magnetic field lines. The magnetic field in the "dead zone" is strong enough to keep plasma locked with the planet. In the second region, so-called "wind zone", the expanding plasma drags and opens the magnetic field lines. These two regions are separated by Alfvénic surface $r = R_A$ (see Figure 2). The plasma escaping along field lines beyond the Alfvénic surface not only deforms and stretches the original planetary dipole field, but also creates a thin disk-type current sheet in the equatorial region. Altogether, this leads to development of a new type of magnetodisk-dominated magnetosphere of a Hot Jupiter, which has no analogues among the solar system planets (Khodachenko *et al.*, 2012).

### 4.2. *Scaling of a magnetosphere with magnetodisk*

The proposed by Khodachenko *et al.* (2012)more complete view of the Hot Jupiter magnetosphere structure is based on the Paraboloid Magnetospheric Model (PMM). PMM is a semi-analytical approach to the modeling of planetary magnetosphere structure (Alexeev *et al.*, 2003; *Alexeev and Belenkaya*, 2005; Alexeev *et al.*, 2006; Khodachenko *et al.*, 2012). The name of the model is derived from its key simplifying assumption that the magnetopause of a planet may be represented by a paraboloid surface co-axial with the direction of the ambient stellar wind plasma. The PMM calculates the magnetic field generated by a variety of current systems located on the boundaries and within the boundaries of a planetary magnetosphere. Besides of the intrinsic planetary magnetic dipole and magnetopause currents, the PMM has, among the main sources of magnetic field, also the electric current system of the magnetotail, and the induced ring currents of the magnetodisk. The model works without any restrictions imposed on the values of interplanetary medium parameters, enabling therefore the description of the whole variety of possible magnetosphere configurations caused by different intrinsic magnetic fields of exoplanets and various stellar wind conditions. As applied to the Hot Jupiters, PMM reveals that the electric currents induced in the plasma disk produce an essential effect on the overall magnetic field structure around the planet, resulting in the formation of a *magnetodisk-dominated* magnetosphere of a Hot Jupiter. Due to certain extension of the plasma disks around close-in exoplanets, the sizes of their magnetodisk-dominated magnetospheres are usually larger than those, followed from the traditional estimates with the equation (3.1), based on the account of only the screened planetary magnetic dipoles (*Griemeier et al.*, 2004; Khodachenko *et al.*, 2007a). In general, the role of magnetodisk may be attributed to an expansion of a part of the dipole magnetic flux from the inner magnetosphere regions outwards and a resulting increase of the magnetosphere size. The magnetic field produced by magnetodisk ring currents, dominates above the contribution of intrinsic magnetic dipole of a Hot Jupiter and finally determines the size and shape of the whole magnetosphere. Khodachenko *et al.* (2012) provided an approximate formula for estimation of the magnetopause stand-off distance taking into account the contribution of the magnetodisk:

$$\frac{R_{\mathrm{s}}^{(dip+MD)}}{r_{\mathrm{p}}} \sim \frac{B_{d0J}^{1/2}(1+\kappa^2)^{1/4}}{(2\mu_0 p_{\mathrm{sw}})^{1/4}} \left(\frac{R_{\mathrm{AJ}}}{r_{\mathrm{p}}}\right)^{-1/2} \times \left(\frac{\omega_{\mathrm{p}}}{\omega_{\mathrm{J}}}\right)^{\frac{3k+1}{10}} \left(\frac{\mathrm{d}M_{\mathrm{p}}^{(th)}/\mathrm{d}t}{\mathrm{d}M_{\mathrm{J}}/\mathrm{d}t}\right)^{\frac{1}{10}} . \tag{4.1}$$

where $R_{\mathrm{AJ}}$, $\frac{\mathrm{d}M_{\mathrm{J}}}{\mathrm{d}t}$, and $B_{\mathrm{d0J}}$ are the known values corresponding to the Alfvénic radius, mass load to the disk, and surface magnetic field for the solar system Jupiter. The coefficient $\kappa \approx 2.44$ is an amplifying factor of the inner magnetospheric field at the magnetopause (Alexeev *et al.*, 2003), which is required to take into account the contribution of the Chapman-Ferraro field at the substellar point. It is connected with the form-factor $f_0$ from (3.1) as $\kappa = 2f_0$. Therefore, according to (4.1), for a given kinetic pressure of stellar wind, $p_{\mathrm{sw}}$, the size of magnetosphere increases for the increasing planetary angular velocity $\omega_{\mathrm{p}}$ and/or thermal mass loss rate $\mathrm{d}M_{\mathrm{p}}^{(th)}/\mathrm{d}t$.

A slower, than the dipole-type decrease of magnetic field with distance comprises the essential specifics of the magnetodisk-dominated magnetospheres of Hot Jupiters. This results in their $40-70\%$ larger scales, as compared to those traditionally estimated with taking into account of only the planetary dipole. Such larger magnetospheres, extending well beyond the planetary exosphere height, provide better protection of close-in planets against of the erosive action of extreme stellar winds (Khodachenko *et al.*, 2007a). Table 2 summarizes the values for a Hot Jupiter magnetopause stand-off distance at different

**Table 2.** Hot Jupiter Alfvénic radius, $R_A$, and magnetopause stand-off distance for only a dipole controlled magnetosphere, $R_s^{(dip)}$, and a magnetosphere with magnetodisk, $R_s^{(dip+MD)}$, given by PMM. Full analog of the solar system Jupiter orbiting the Sun analog star at different orbits is considered. [1]: Tidally locked. [2]: Not tidally locked. [3]: Jupiter

| $d$ [AU] | $R_s^{(dip+MD)}$ [$r_p$] | $R_s^{(dip)}$ [$r_p$] | $R_A$ [$r_p$] |
|---|---|---|---|
| 0.045[1] | 8.0 | 5.76 | 3.30 |
| 0.1[1] | 8.27 | 6.16 | 4.66 |
| 0.3[2] | 24.2 | 15.0 | 7.30 |
| 5.2[3] | 71.9 | 41.8 | 19.8 |

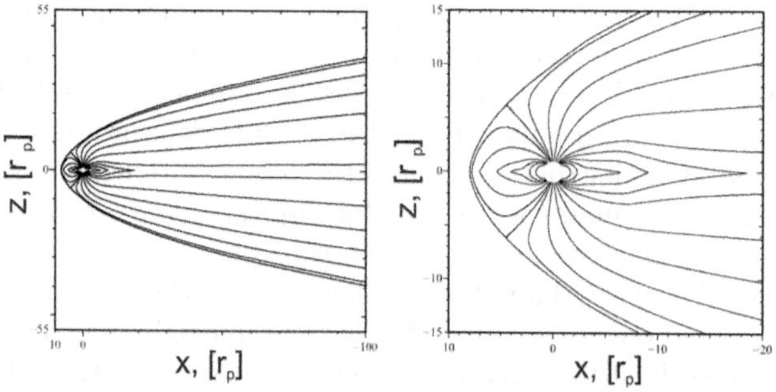


**Figure 3.** Typical view of a magnetodisk-dominated magnetosphere.

orbits around a Sun full analogue star and gives for the comparison the stand-off distance values, obtained with equation (3.1), i.e. in the case when the contribution of magnetodisk is ignored (e.g., a pure dipole case).

A typical example of the magnetic field structure in the in the magnetosphere of a Hot Jupiter, obtained with PMM, is shown in Figure 3.

## 5. Conclusions

To summarize this paper we would like to emphasize that stellar XUV radiation and stellar wind plasma flow strongly impact the environments of close-orbit exoplanets. Given the fact that the complete or partial tidal locking of such short periodic exoplanets may lead to relatively weak intrinsic planetary magnetic moments, the encountering stellar wind and CMEs will push the planetary magnetospheres down to the heights at which the ionization and pick-up of the planetary neutral atmosphere by the stellar plasma flow takes place. This makes the stellar activity and planetary magnetospheric protection to play a crucial role for the whole complex of planetary evolution processes, including atmosphere erosion and mass loss. Large enough extended magnetospheres are needed to protect the upper atmospheric environments against of stellar XUV and stellar wind/CMEs impacts.

The expanding and escaping upper atmospheric gas heated and ionized by the stellar radiation contributes to the build-up of the magnetodisk around the planet, which constitutes the major specifics of a Hot Jupiter magnetosphere considered in this work. The magnetic field produced by magnetodisk ring currents, dominates above the contribution of intrinsic magnetic dipole of a Hot Jupiter and finally determines the size and shape of the whole magnetosphere. A more realistic structure of the magnitodisk-dominated magnetosphere of a Hot Jupiter predicted by the Paraboloid Magnetospheric Model and its significantly larger size, as compared to a dipole-type magnetosphere, have important consequences for the study of magnetospheric protection of close orbit exoplanets.

## Acknowledgements

This work was supported by the Austrian Science Foundation (FWF) (projects S11606-N16 and P25587-N27). The authors acknowledge the EU FP7 project IMPEx for support of numerical modelling work and providing collaborative environment for research and communication.

## References

Alexeev, I. I., Belenkaya, E. S., Bobrovnikov, S. Yu., & Kalegaev, V. V. 2003, *Space Sci.Rev.*, 107, 7

Alexeev, I. I., Kalegaev, V. V., Belenkaya, E. S., Bobrovnikov, S. Yu., Bunce, E. J., Cowley, S. W. H., & Nichols, J. D. 2006, *Geophys.Res.Lett.*, 33, L08101

Alexeev, I. I., & Belenkaya, E. S. 2005, *Ann. Geophys.*, 23, 809

Audard, M., Güdel, M., Drake, J. J., & Kashyap, V. L. 2000, *ApJ*, 541, 396

Ayres, T. R. 1997, *JGR*, 102, 1641

Erkaev, N. V., Penz, T., Lammer, H., Lichtenegger, H. I. M., Biernat, H. K., Wurz, P., Grießmeier, J.-M., & Weiss, W. W. 2005, *ApJ. Suppl. Ser.*, 157, 396

Erkaev, N. V., Lammer, H., Odert, P., Kulikov, Yu. N., Kislyakova, K. G., Khodachenko, M. L., Güdel, M., Hanslmeier, A., & Biernat, H. 2013, *Astrobio.*, (in press)

Gershberg, R. E. 2005, *A Solar-Type Activity in Main-Sequence Stars*, (Berlin,Heidelberg,New York: Springer)

Grießmeier, J.-M., Stadelmann, A., Penz, T., Lammer, H., Selsis, F., Ribas, I., Guinan, E. F., Motschmann, U., Biernat, H. K., & Weiss, W. W. 2004, *A & A*, 425, 753

Grießmeier, J.-M., Preusse, S., Khodachenko, M. L., Motschmann, U., Mann, G., & Rucker, H. O. 2007, *Planet. & Space Sci.*, 55, 618

Ip, W.-H., Kopp, A., & Hu, J.-H. 2004, *Astrophys. J.*, 602, L53

Johansson, E. P. G., Bagdonat, T., & Motschmann, U. 2009, *A & A*, 496, 869

Khodachenko, M. L., Ribas, I., Lammer, H., Grießmeier, J.-M., Leitner, M., Selsis, F., Eiroa, C., Hanslmeier, A., Biernat, H., Farrugia, C. J., & Rucker, H. 2007a, *Astrobio.*, 7, 167

Khodachenko, M. L., Lammer, H., Lichtenegger, H. I. M., Langmayr, D., Erkaev, N. V., Grießmeier, J.-M., Leitner, M., Penz, T., Biernat, H. K., Motschmann, U., & Rucker, H. O. 2007b, *Planet.Space Sci.*, 55, 631

Khodachenko, M. L., Alexeev, I. I., Belenkaya, E., Leitzinger, M., Odert, P., Grießmeier, J.-M., Zaqarashvili, T. V., Lammer, H., & Rucker, H. O. 2012, *ApJ*, 744, 70

Koskinen, T., Yelle, R. V., Lavvas, P., & Lewis, N. K. 2010, *ApJ*, 723, 116

Koskinen, T. T., Harris, M. J., Yelle, R. V., & Lavvas, P. 2012, *Icarus*, (in press), http:// arXiv:1210.1535.

Lammer, H., Kasting, J. F., Chassefière, E., Johnson, R. E., Kulikov, Yu. N., & Tian, F. 2008, *Space Sci Rev.*, 139, 399

Lammer, H., Odert, P., Leitzinger, M., Khodachenko, M. L., Panchenko, M., Kulikov, Yu. N., Zhang, T. L., Lichtenegger, H. I. M., Erkaev, N. V., Wuchterl, G., Micela, G., Penz, A., Biernat, H. K., Weingrill, J., Steller, M., Ottacher, H., Hasiba, J., & Hanslmeier, A. 2009,)*A & A*, 506, 399

Lichtenegger, H. I. M., Gröller, H., Lammer, H., Kulikov, Yu. N., & Shematovich, V. 2009, *Geophys. Res. Lett.* ,36 , CiteID L10204

Mestel, L. 1968, *MNRAS*, 138, 359

Newkirk, G., Jr. 1980, *Geochim. Cosmochim. Acta Suppl.*, 13, 293

Parker, E. N. 1958, *ApJ*, 128, 664

Penz, T., Erkaev, N. V., Kulikov, Yu. N., Langmayr, D., Lammer, H., Micela, G., Cecchi-Pestellini, C., Biernat, H. K., Selsis, F., Barge, P., Deleuil M., & Léger, A. 2008, *Planet. Space Sci.*, 56, 1260

Preusse, S., Kopp, A., Büchner, J., & Motschmann, U. 2005, *A & A*, 434, 1191

Ribas, I., Guinan, E. F., Güdel, M., & Audard, M. 2005, *ApJ*, 622, 680

Scalo, J., Kaltenegger, L., Segura, A. G., Fridlund, M., Ribas, I., Kulikov, Yu. N., Grenfell, J. L., Rauer, H., Odert, P., Leitzinger, M., Selsis, F., Khodachenko, M. L., Eiroa, C., Kasting, J., & Lammer, H. 2007, *Astrobiol.*, 7, 85

Sheeley, N. R., Jr., Wang, Y.-M., Hawley, S. H., Brueckner, G. E., Dere, K. P., Howard, R. A., Koomen, M. J., Korendyke, C. M., Michels, D. J., Paswaters, S. E., Socker, D. G., St. Cyr, O. C., Wang, D., Lamy, P. L., Llebaria, A., Schwenn, R., Simnett, G. M., Plunkett, S., & Biesecker, D. A. 1997, *ApJ*, 484, 472

Skumanich, A. 1972, *ApJ*, 171, 565

Tian, F., Kasting, J. F., Liu, H., & Roble, R. G. 2008, *J. Geophys. Res.*, 113, E05008

Walker, J. C. G., Hays, P. B., & Kasting, J. F. 1981, *J. Geophys. Res.*, 86, 9776

Wood, B. E., Müller, H.-R., Zank, G. P., & Linsky, J. L. 2002, *ApJ*, 574, 412

Wood, B. E., Müller, H. -R., Zank, G. P., Linsky, J. L., & Redfield, S. 2005, *ApJ*, 628, L143

# Session IV

## INSTRUMENTATION, MISSIONS AND TECHNIQUES

*Nature of Prominence and their role in Space Weather*
*Proceedings IAU Symposium No. 300, 2013*
*B. Schmieder, J.-M. Malherbe & S. T. Wu, eds.*

doi:10.1017/S1743921313011186

# The Chinese Giant Solar Telescope

**Zhong Liu[1], Yuanyong Deng[2] and Haisheng Ji[3]**

[1]Yunnan Observatories, CAS,
650011, Kunming, P.R. China
email: lz@ynao.ac.cn

[2]National Astronomical Observatories, CAS,
100012, Beijing, P.R. China

[3]Purple Mountain Observatory, CAS,
210008, Nanjing, P.R. China

**Abstract.** Chinese Giant Solar Telescope is the next generation ground-based solar telescope. The main science task of this telescope is to observe the ultra fine structures of the solar magnetic field and dynamic field. Due to the advantages in polarization detection and thermal controlling with a symmetrical circular system, the current design of CGST is a 6∼8 meter circular symmetrical telescope. The results of simulations and analysis showed that the current design could meet the demands of most science cases not only in infrared bands but also in near infrared bands and even in visible bands. The prominences and the filaments are very important science cases of CGST. The special technologies for prominence observation will be developed, including the day time laser guide star and MCAO. CGST is proposed by all solar observatories and several institutes and universities in China. It is supported by CAS and NSFC (National Natural Science Foundation of China) as a long term astronomical project.

**Keywords.** Telescopes, High angular resolution, Magnetic fields, Prominence, Filaments

## 1. Introduction

Chinese Giant Solar Telescope (CGST) is the main facility of a future solar observatory, the Chinese Advanced Ground-based Solar Observatory. The Chinese Advanced Ground-based Solar Observatory consists of two huge facilities, the CGST and a large aperture coronagraph. The large coronagraph is a 1m refracting telescope similar to COSMO. According to the current plan, this 1m refracting coronagraph will be developed by a Sino-American cooperation group. This paper mainly introduces the CGST which is proposed by Yunnan Astronomical Observatory CAS, National Astronomical Observatories CAS, Purple Mountain Observatory CAS, Nanjing University, Nanjing Institute of Astronomical Optical Technology and Beijing Normal University.

The primary science goals of the next generation solar telescopes are similar, that is to push the human's understanding of the sun and the space weather to a new level. As is well known, the knowledge of the sun strongly depends on the very high precision observations of the solar magnetic and dynamic fields. That is the reason why solar physicists wish to develop the next generation solar telescopes, such as Solar-C, ATST (Keil *et al.* 2003), EST (Schmidt *et al.* 2012) and CGST. As large ground-based solar telescopes are more powerful in very high resolution and can do more complex observations, the ongoing space solar telescopes such as Solar-C and Chinese DSO cannot replace the ground-based solar telescopes in the near future.

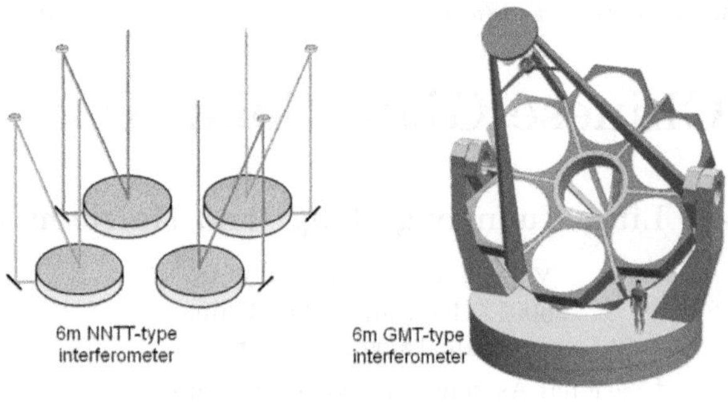


**Figure 1.** Sketches of NNTT-type CGST and GM- type CGST.

## 2. Current conceptual designs of CGST

Although the vacuum telescopes have many advantages in solar observation, it is very difficult to manufacture a vacuum solar telescope if the aperture diameter increases over 1 meter. Another disadvantage of vacuum solar telescope is the opaqueness of the vacuum window in the infrared range. So the next generation large solar telescopes are all designed to be open telescopes. The open solar telescopes are roughly classified into two kinds, off-axis and on-axis telescopes. An off-axis telescope is comparatively easy to do thermal control, but its unsymmetrical structure will result in polarization cross talk. It is a trouble for magnetic field measurements.

The CGST will focus on the ultra fine structures of magnetic and dynamic fields in different layers of the solar atmosphere. Many sciences cases were suggested to observe and study the physical phenomena from convection zone to corona. Compared with the current Chinese solar telescopes, the spectral range of CGST will be expanded to 15 micron or even farther. In order to meet the requirements of scientific goals, Chinese solar astronomers (Fang 2011) wish to combine the advantages of both off-axis and on-axis systems. Several symmetrical off-axis systems have been proposed. The first conceptual system consists of four telescopes. Each telescope is an independent off-axis solar telescope with 2 or 3 meters diameter. All the four telescopes need co-phasing with each other. As it looks like the former NNTT telescope, we named it NNTT-type. The second system consists of more primary sub-mirrors. The diameter of each mirror is 2 meters. All the mirrors are off-axis parabolic and are combined into a circular co-phasing interferometric telescope. It looks like the GMT telescope without the central mirror. We call it Aperture Ring Telescope or simply GMT-type.

The third design is an 8 meter ring telescope (Liu *et al.* 2012). In this design, the primary mirror is an 8 meter ring mirror with 1 meter width (Table 1, Figure 2). This telescope is named Ring Solar Telescope (RST). As a typical RIT (Ring Interferometric Telescope) (Liu *et al.* 2006), the resolution diameter of RST is 8 meters and the collecting area is 22 square meters, just equal to the collecting area of a traditional 5 meter telescope. No matter a whole ring or a segmented ring, the primary mirror will reach its diffraction limit at least at 1 micron band. We have done many simulations of the 8 meter RST, such as the PSF properties (Liu *et al.* 2011), the finite element analysis of the mechanical structures (Dai *et al.* 2012), the active optics and adaptive optics (Dai *et al.* 2011; Yuan *et al.* 2011). High resolution solar observations by using a real ring aperture (Figure 4) were also carried out on the 1 meter New Vacuum Solar telescope (Liu & Xu 2011). All the above simulations and experiments indicate the feasibility of the 8 meter RST. Now, the 8 meter RST is the relatively most complete conceptual design in all the proposed candidates of CGST.

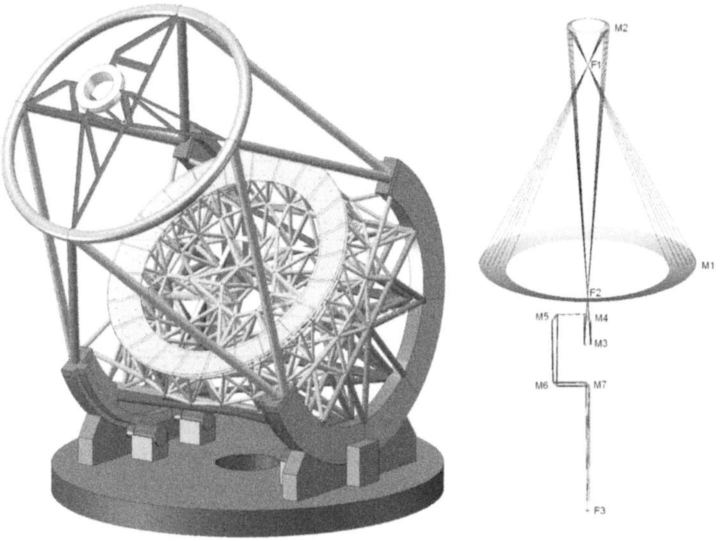


**Figure 2.** 3D sketch of 8m RST, the primary mirror is an 8 meter ring with 1 meter width.

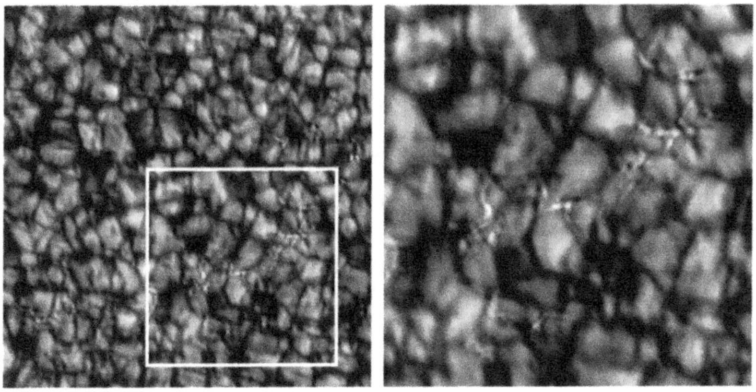

**Figure 3.** High resolution imaging by ring aperture, the diameter of the ring is 1 meter, the width is 0.15m.

## 3. Prominence observation with CGST

With the progress of modern solar observations and the developments of new technologies, fine structures smaller than 10 0km can be resolved by current ground-based solar telescopes. The recent researches also indicate that there should be ultra fine structures smaller than 0.1 arc-second in the solar photosphere. These ultra fine structures are not only statistically significant fundamental structures but also a part of solar activity itself. New observation results from NST showed the inherent relationship between the photosphere bright points and the obvious activities in solar corona (Ji *et al.* 2012). The recent simulations also predicted 20 km significant structures in the photosphere (Stein *et al.* 2006; Stein *et al.* 2011). The current solar telescopes cannot resolve these ultra fine features and their evolutions, such as the micro magnetic reconnection, the structure of bright points and the evolution of the tiny flux tubes. So, the next generation large solar telescopes all choose 4 meter or more as their aperture size. The scientific goals of CGST will be introduced in other papers. In this paper, prominence observations with CGST are introduced as an important case.

**Table 1.** Optical parameters of CGST.

| | |
|---|---|
| Diameter of primary mirror | 8m |
| Width of primary mirror | 1m |
| F/D of primary mirror | 1 |
| Collecting area of primary mirror | $22m^2$ |
| Diameter of secondary mirror | 1.6m |
| Spectral range | 0.3 ∼ 15 micron |
| Spatial resolution | 0.03 arc sec |
| Polarization accuracy | $\sim 10^{-4}$ |

Many cases of prominences and filaments have been discussed in CGST science group. Some of them are based on the high resolution observations from the 1 meter NVST (Liu & Xu 2011). Figure 4 shows the obvious changes of active filaments during a flare eruption. After flare, an active filament connected with another one in a short time and exchanged some mass and energy. Some observation data from SDO and the other solar telescopes also showed the filament reconnections (Jiang *et al.* 2013). Are these real reconnections between filaments or only visual effects? The deeper problems in such a case include the fine magnetic structures of filaments, the reconnection process between filaments, the magnetic energy transmission and storage. Although a one meter telescope could resolve many details inside a filament, it is still too small to resolve the individual flux structures and their evolutions. Direct evidences are needed to demonstrate these reconnections result from more fundamental magnetic reconnections or consist of more fine combinations of tiny structures. The relationship between filaments reconnection and the reconnection of flux ropes or flux tubes (Linton *et al.* 2001) is still open. It is one of the key problems in filament eruption and is also an important science case of prominence high resolution observation. According to the simulations, CGST will be able to observe filament reconnection with very high angular resolution and very high polarization accuracy.

Another important case is a perennial open question, that is, the quiescent prominences. We don't know well the structures and the origin of the quiescent prominence, especially the magnetic field in it or below it. Some works, such as the dynamic structures of quiescent prominence could be done by using a current 1 m-class telescope. Figure 5 shows a high resolution quiescent prominence taken by the 1 meter NVST. As the magnetic field of quiescent prominence is very weak, most problems are waiting for more powerful observational facilities. For example, some current observations showed that quiescent prominences do not have obvious footpoints in the photosphere. On the other hand, the recent observations and simulations demonstrate that tornados are related to the photospheric magnetic field and have obvious footpoints in the photosphere (Wedemeyer-Bohm *et al.* 2012). The relationship between quiescent prominences and tornados implies that many quiescent prominences may have footpoints in the photosphere and the photospheric magnetic field. Only a large aperture telescope with high polarization sensitivity could give the final answer. In all the remaining problems of quiescent prominence, the magnetic field is the most important part. It is directly relevant to the nature of prominence.

Unlike the objects on the solar disk, the observations of prominence need some special methods and new technologies. For example, the normal adaptive optics could not observe the prominence on the edge of the solar disk as it is very difficult to do the wave-front

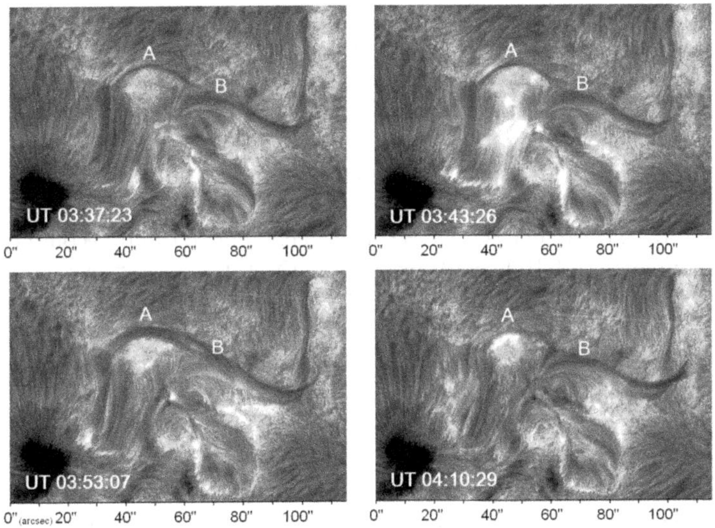


**Figure 4.** The change of active filaments during a flare on October 25, 2012 (AR11598).

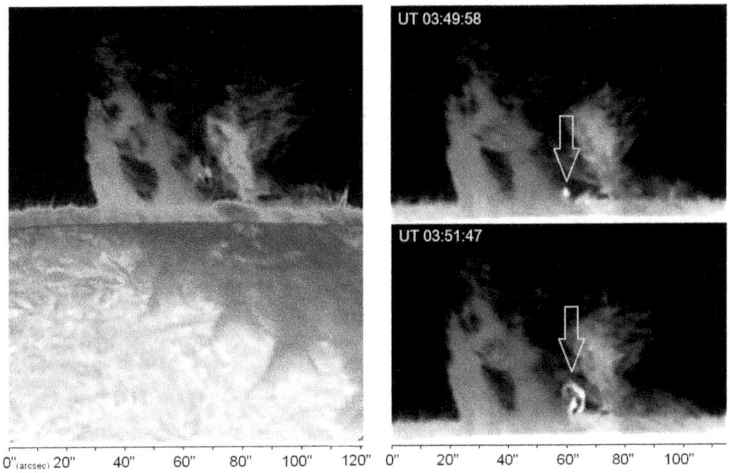


**Figure 5.** Quiescent prominence on January 15, 2013.

sensing. Normally, we could do wave-front sensing in photosphere bands and do AO observations in the weak chromosphere bands. But prominences on the edge of solar disk do not have such a photosphere background. The idea of using a chromospheric wave-front sensor encounters the problem of low photon numbers. An alternative choice is to use the Day-time Laser Guide Star (LGS) to do wave-front sensing, for example the sodium LGS. MCAO (Multi Conjugated Adaptive Optics) is also a very important technology for prominence observations, as the size of a prominence is much larger than the isoplanatic area of the Earth atmosphere.

## 4. Current situations of CGST

A site survey group is working on the plateau of southwest China. This work began in 2010 not only for CGST but also for the big aperture Sino-American cooperative coronagraph. GIS (Geography Information System) and the satellite meteorological information have been used in the general survey. The testing instruments include the portable image motion monitor, solar differential image motion monitor, sky brightness monitor, scintillometer, integrated water vapor detector, robot weather station and other useful instruments. Now, after three years' hard work, the site survey group has reduced the candidates from dozens to several lake sites and mountain sites.

Besides the progress in science and technology, CGST also got some progress for project approval. In 2010, a committee has been set up to push CGST to be a normal National Science Project. In the same year, CGST has been selected and recommended to the National Development and Reform Commission (NDRC) as a "National major basic scientific project for 2016–2030". In 2012, CGST was confirmed as one of the major national science projects by NDRC of China. The first budgets for preliminary research work and site survey are mainly funded by NSFC (National natural Science Foundation of China) and CAS (Chinese Academy of Sciences).

## 5. Acknowledgements

The authors are grateful to all colleagues in the Chinese solar community. Many thanks to the site survey group and all the people involved in CGST. These works are funded by the MOST (2011CB811400) of China, NSFC (11078004) and CAS (KJCX2-EW-T07).

### References

Dai, Y. C. & Lin, J. 2011, *Proc. SPIE*, 8336, 833607-1
Dai, Y. C., Yang, D. H., Zago, L., & Liu, Z. 2012, *Proc. SPIE*, 8449, 84491A-1
Fang,C. 2011, *RAA*, 11(12), 1377
Ji, H. S., Cao, W. D., & Goode, P. 2012, *ApJL*, 750, L25
Jiang, Y. C., Hong, J. C., Yang, J. Y., Bi, Y., Zheng, Y. S., Yang, B., Li, H. D., & Yang, D. 2013, *ApJ*, 764, 68
Keil, S., Rimmele, T., & Keller, C., ATST Team 2003, *Astronomische Nachrichten*, 324, 303
Linton, M. G., Dahlburg, R. B., & Antiochos, S. K. 2001, *ApJ*, 533, 905
Liu, Z., Jin, Z. Y., Li, Y., Lin, J., & Tan, H. S. 2006, *Proc. SPIE*, 6267, 62672L-1
Liu,Z., Xu,J. 2011, *1st APSP ASICS*, 2, 9
Liu, Z. & Jin, Z. Y. 2011, *Proc. SPIE*, 8336, 833609-1
Liu, Z., Deng, Y. Y., Jin, Z. Y., & Ji, H. S. 2012, *Proc. SPIE*, 8444, 844405-1
Schmidt,W., et.al. 2012, *ASPC*, 463, 365S
Stein, R. F., Nordlund,Å. 2006, *ApJ*, 642, 1246
Stein, R. F., Lagerfjard, A., Nordlund, Å., & Georgobiani, D. 2011, *Solar Phys.* , 268, 271
Wedemeyer-Bohm, S., Scullion, E., Steiner, O., van der Voort, L. R., de La Cruz Rodriguez, J., Fedun, V., & Erdelyi, R. 2012, *Nature*, 486, 505
Yuan, S. & Lin, J. 2011, *Proc. SPIE*, 8336, 83360I-1

*Nature of Prominences and their role in Space Weather*
*Proceedings IAU Symposium No. 300, 2013*
*B. Schmieder, J.-M. Malherbe & S. T. Wu, eds.*

doi:10.1017/S1743921313011198

# Scientific Programmes with India's National Large Solar Telescope and their contribution to Prominence Research

**S. S. Hasan**

Indian Institute of Astrophysics,
Bangalore 560034, India
email: hasan@iiap.res.in

**Abstract.** The primary objective of the 2-m National Large Solar Telescope (NLST) is to study the solar atmosphere with high spatial and spectral resolution. With an innovative optical design, NLST is an on-axis Gregorian telescope with a low number of optical elements and a high throughput. In addition, it is equipped with a high order adaptive optics system to produce close to diffraction limited performance.

NLST will address a large number of scientific questions with a focus on high resolution observations. With NLST, high spatial resolution observations of prominences will be possible in multiple spectral lines. Studies of magnetic fields, filament eruptions as a whole, and the dynamics of filaments on fine scales using high resolution observations will be some of the major areas of focus.

**Keywords.** Solar telescope, Solar magnetic fields, Solar observations, Solar prominences

## 1. Introduction

The Indian Institute of Astrophysics (IIA) has recently proposed a 2 m state-of-the-art National Large Solar Telescope (NLST) for carrying out high spatial and spectral observations of the Sun with a view to obtaining a better understanding of the fundamental nature of magnetic fields and other phenomena in the solar atmosphere. Observations have established that the solar magnetic field is structured in the form of flux tubes that can be as small as a few kilometers. Several numerical simulations indicate that crucial physical processes like vortex flows, dissipation of magnetic fields and the generation of magnetohydrodynamic (MHD) waves can occur efficiently on length scales even as small as 10 km (e.g. Vögler *et al.* 2005, Stein & Nordlund 2006, Shelyag *et al.* 2011). Such waves are likely candidates for transporting energy to the upper atmosphere of the Sun. Spatially resolved observations are, therefore, essential to shed light on the different physical processes involved. Unfortunately, even the largest current solar telescopes are limited by their apertures to resolve solar features to this level at visible wavelengths. On the global scale, the energy stored in magnetic fields is eventually dissipated in the higher layers of the solar atmosphere, for instance in the form of flares and coronal mass ejections (CMEs) that release energetic solar plasma into the interplanetary medium.

Currently the 1.5 m German GREGOR and the 1.6 m US New Solar Telescope (NST) in Tenerife and Big Bear respectively, are the largest international facilities for solar observations. The 4 m Advanced Technology Solar Telescope (ATST) will be commissioned around 2020. This has provided a window of opportunity for India to build a 2 m solar telescope in the next few years.

With an innovative optical design, NLST is a 3 mirror on-axis Gregorian telescope with a low number of optical elements (6 mirrors) to reduce the number of reflections and yield a high throughput with low polarization. It will be equipped with a high order adaptive optics package to produce close to diffraction limited performance (technical details on the telescope design and focal plane instruments can be found in Hasan *et al.* 2010 and Hasan 2012). In addition to the requirement of good angular resolution, a high photon throughput is also necessary for spectropolarimetric observations to accurately measure vector magnetic fields in the solar atmosphere with a good signal to noise ratio. With an aperture of 2 m, NLST will be able to resolve structures with sub-arc sec resolution in the solar atmosphere as well as carry out spectropolarimetry with a high time cadence. A novel feature of NLST is that it will also be possible to perform night time observations of stars.

## 2. Broad Scientific Objectives

NLST is envisaged as a versatile instrument that will enable a broad class of problems to be investigated, to shed light on the complex interaction of magnetic fields with plasma in the solar atmosphere. It is well known that the magnetic field plays a crucial role in modulating solar variability and activity. Understanding how magnetic fields are generated and maintained on the Sun is basic to understanding the origin and nature of solar cycle and variability, and to predict in advance its behaviour both on short and long time scales. Another key issue that NLST will address is to study magnetic coupling between the interior and solar atmosphere that involves an investigation of dynamical processes in magnetic elements that enable the transport of energy to the chromosphere and corona.

Other areas of focus include local helsioseimology, measurement of weak magnetic fields in the internetwork where a significant fraction of field is below 400 G (Khomenko *et al.* 2003), the thermal structure of the chromosphere, particularly cool pockets with temperatures as low as 3600 K (Solanki, Livingston & Ayres 1994); and energetic phenomena and activity. An important goal of NLST will be to observe the magnetic field and dynamic changes that occur during solar flares. NLST will also contribute towards a better understanding of CMEs by combining optical data with radio to determine the connection between the changes that occur at chromospheric layers through $H_\alpha$ observations and type III radio bursts during flares.

An important focus of NLST will be the study of quiescent and active prominences.

## 3. Scientific programmes related to Prominences

It is well known that prominences exhibit a range of morphology when viewed above the limb consisting of nearly horizontal threads in some cases and almost vertical threads in other instances (Mackay *et al.* 2010). The reason for this is still not fully understood. Presently there are very few observations of magnetic fields in prominences.

### 3.1. *Magnetic Field Measurements in Filaments*

Balasubramaniam *et al.* (2006) obtained simultaneous observations of a filament structure observed on April 7, 2003 with the Advanced Stokes Polarimeter (ASP, Elmore *et al.* 1992) in $H_\alpha$, Mg I 517.27 nm and the Diffraction Limited Spectro-Polarimeter (DLSP, Sigwarth *et al.* 2001) in Fe I 630.15/630.25 nm. They inferred the 3-D magnetic geometry, using linear force-free reconstruction based on the line of sight measurements of the magnetic field. They concluded that the filament material is located in regions where

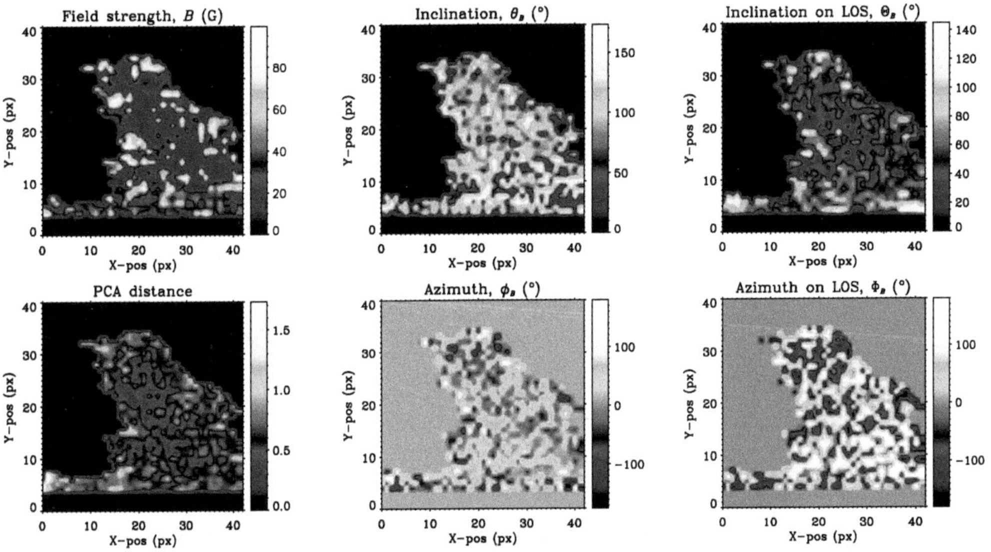


**Figure 1.** Magnetic map of a prominence observed on May 25, 2002 using the Dunn Solar Telescope (DST) at NSO derived from a PCA inversion of the four Stokes profiles of the He I D3 line. Both geometries of the field, respectively, in the reference frame of the local vertical, $(\theta_B, \phi_B)$, and in the reference frame of the LOS, $(\Theta_B, \Phi_B)$, are presented (from Casini *et al.* 2003).

the chromospheric field dips, similar to the Kippenhahn-Schlüter model (Kippenhahn & Schlüter 1957).

Figure (1) shows the magnetic map of a prominence observed on May 25, 2002 using the Dunn Solar Telescope (DST) derived from inversion of the four Stokes profiles of the He I D3 line (Casini *et al.* 2003). Vector magnetic fields were mapped down to the chromospheric limb. The analysis indicates that the average magnetic field in prominences

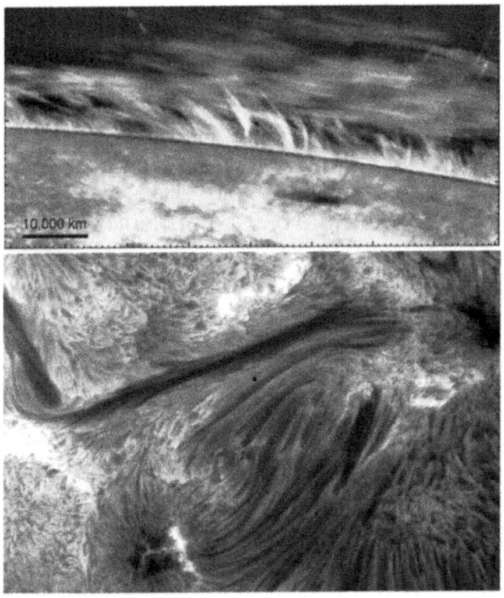


**Figure 2.** Upper panel: Thin threads in an active region prominence seen in Ca II H. The image was obtained with Hinode/SOT on 2006 November 9 (Okamoto *et al.* 2007). Lower panel: Thin threads of an active region filament seen in $H_\alpha$. The image was obtained from the Swedish 1 m Solar Telescope on 2003 August 22, with a field of view of 83000 × 56000 km (from Lin 2011).

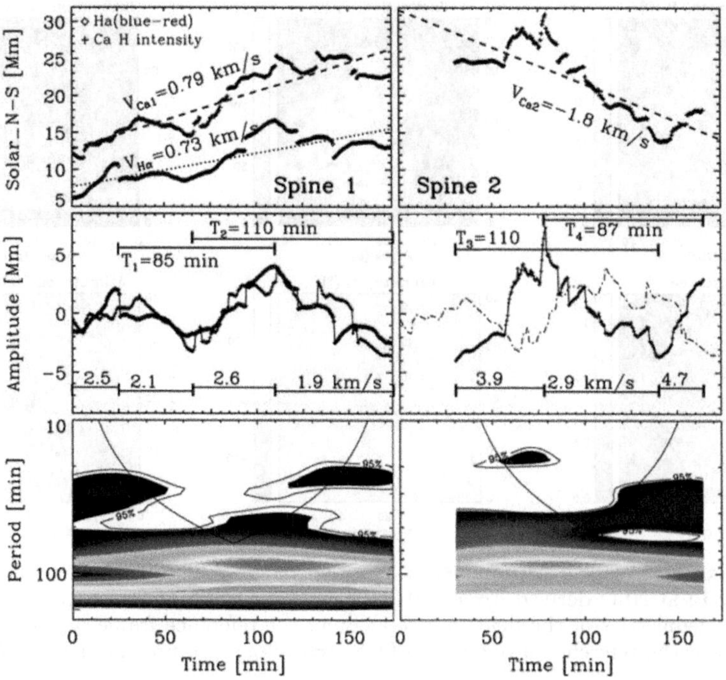


**Figure 3.** Quiescent filament observed in $H_\alpha$ by Hinode/SOT. Top panel: time evolution of two spine structures in Doppler and Ca II H intensity. Middle panel: oscillatory pattern of the two spine structures. Bottom panel: wavelet diagram of the oscillations of the two spines structures. (from Ning *et al.* 2009)

is mostly horizontal and varies between 10 and 20 G, thus confirming previous findings. However, these maps show that fields significantly stronger than average, even as large as 60 or 70 G, can often be found in clearly organized plasma structures of the prominence. NLST will routinely provide vector magnetic maps in prominences with higher accuracy.

### 3.2. *Fine Structure*

High resolution observations indicate the presence of fine structure down to the resolution limit of present day instruments (typically 100 km). Very thin dark fibrils visible along the spine of a quiescent prominence are seen as short structures inclined to the filament axis. Longer fibrils can be seen within the barbs or connecting various parts of the filament body. Figure (2) shows thin threads in an active region prominence in (a) Ca II H (Okamoto *et al.* 2007) as seen with Hinode (top panel), and (b) $H_\alpha$ (Lin 2011) using the 1 m Swedish Vacuum Solar Telescope. NLST, with its higher resolution, will be in a better position to study in greater detail the nature of the fine threads that make up spines, barbs and ends.

### 3.3. *Oscillations & Dynamics*

High resolutions observations (from the ground as well as space) reveal the presence of oscillatory motions in individual thin filaments. Small amplitude oscillations with velocity amplitudes 0.13 km $s^{-1}$ in filaments are detected as periodic variations in line-of-sight velocity, line intensity, line width, as well as spatial displacement (swaying) of the threads. A range of oscillatory periods have been reported ranging from very short periods $\leqslant$1 min, short periods 1- 20 min, intermediate periods 20 - 40 min, and long periods 40 -100 min (Lin 2011). Ning *et al.* (2009) examined oscillations in two spine structures of a quiescent

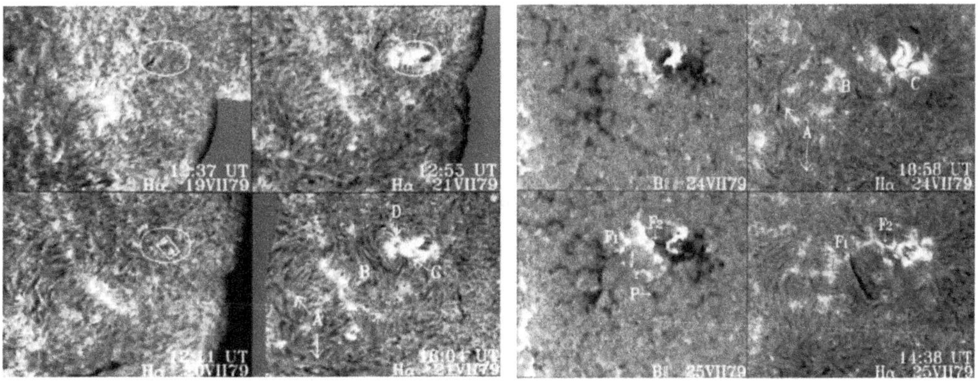


**Figure 4.** $H_\alpha$ image (right) and magnetogram (left) showing the final stages of the formation of an Intermediate Filament. The Intermediate Filament forms on the 25th July after flux convergence and cancellation occur at F1 in the bottom right panel (from Gaizauskas *et al.* 1997)

prominence. Figure (3) shows the time evolution of the $H_\alpha$ Doppler and Ca II H intensity using Hinode observations.

### 3.4. *Formation of filament channels*

Currently, there are two scenarios related to the filament channel (FC) formation. In the first one (Gaizauskas *et al.* 1997, Wang & Muglach 2007) channel formation takes place through surface motions acting on pre-existing fields. Figure (4) depicts the early stages (left panel) in the formation of an Intermediate Filament between an old remnant region (bright plages A & B in bottom right images) and an emerging activity complex (inside the oval).

The second scenario (Lites & Low 1997 and Okamoto *et al.* 2008) involves the emergence of a sheared flux rope that either replaces a pre-existing filament or reconnects with this filament. Figure (5) shows a time series of Hinode SP data to illustrate this phenomenon.

### 3.5. *Eruption & Disappearance of filaments*

STEREO/EUVI 304 Å observations of prominences show a helical twist in the spine during eruption (Joshi & Srivastava 2011). During eruption, the prominence exhibits non-radial motions. The non-radial motion and helical twist in spines can be used to determine the propagation dynamics of prominences. These results show that the acceleration of prominences is higher in the prominence leg where these two forces act in the same direction, and lower in the leg where they act in opposite directions. NLST will be able to carry out studies of the initiation of filament eruption or disappearance. It will carry out observations of filaments prior to their activation and identify some of the precursors.

## 4. Prominence Studies with NLST

Observations of prominences are ideally suited for a large aperture solar telescope such as NLST. In order to detect weak fields in quiescent prominences, a high polarimetric sensitivity with a good signal to noise is necessary. The aim is to map the vector magnetic field in the photosphere and the chromosphere accurately with high spatial resolution and high time cadence to study a range of dynamical processes. This will be carried out principally using a spectropolarimeter using multiple spectral lines simultaneously.

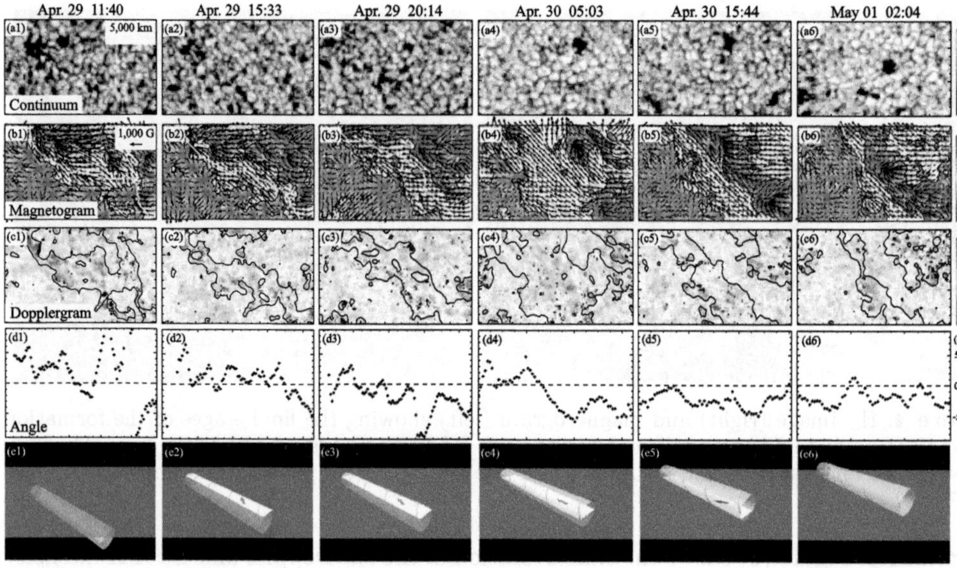


**Figure 5.** Time series of Hinode SP data. The field of view is 32" × 19". The red (blue) region indicates the plage (sunspot side) on the photosphere. The white tube is a helical flux rope. The yellow line shows a magnetic field line on the surface of the flux tube. Black arrows indicate the orientation of horizontal magnetic fields of this rope crossing with the photosphere (from Okamoto *et al.* 2008)

This instrument will seek to achieve a polarimetric accuracy better than $5 \times 10^{-4}$, a wavelength coverage from 380 nm to 2.5 $\mu$ and low instrumental polarization ($<$ 1% before modulation and $<$ 10% before demodulation).

In addition, NLST will carry out high resolution 2-D observations of Doppler shifts required for carrying out prominence seismology investigations that seek to accurately determine physical conditions in prominences as well as the role of barbs in the evolution of filaments.

Presently, there are very few observations of the formation of filament channels. NLST will seek to examine FC formation in action regions using high cadence observations along with combined high resolution $H_\alpha$ and line of sight magnetic field observations.

## 5. Current Status

After four years of site characterization, two superb sites in the Ladakh region of Jammu & Kashmir, India have been located. A detailed project report has been submitted to the Indian Government and formal approval is awaited. Fabrication of NLST is expected to begin in 2014, with first light in 2018. An international consortium has been identified for fabricating the telescope. The backend instruments will be made indigenously.

### Acknowledgements

I am grateful to Drs. Anand Joshi, Nandita Srivastava and K. S. Sankarasubramanian for kindly providing me useful reference material. Part of the work was carried out when the author was at DAMTP and Clare Hall College, Cambridge. His visit to Cambridge was supported through a Hamied Fellowship.

## References

Balasubramanium, K. S., Sankarasubramanian, K & Pevtsov, A. A. 2006, in: R. Casini & B. W. Lites (eds.), *Solar Polarization 4*, ASP Conf. Ser., Vol. 358, p. 68

Casini, R, López Ariste, Tomczyk, S. & Lites, B. W. 2003, *ApJ*, 598, L67

Elmore, D. F, Lites, B. W., Tomczyk, S., *et al.* 1992, in: D. H. Goldstein & R. A. Chipman (eds.), *Polarization Analysis and Measurement*, Proc. SPIE Vol. 1746, p. 22

Gaizauskas, V., Zirker, J. B., Sweetland, C., & Kovacs, A. 1997, *ApJ*, 479, 448

Hasan, S. S., Doltau, D., Kärcher, Süss, Berkfeld, T. 2010,*AN*. 331, 628

Hasan, S. S. 2012, in: T. Rimmele, M. Collados Vera, T. Berger *et al.* (eds.), *Magnetic Fields from the Photosphere to the Corona*, ASP Conf. Ser., Vol. 463, p. 395

Joshi, A. & Srivastava, N. 2011, *ApJ*, 739, 1

Khomenko, E., Collados, M., Solanki, S. K., Lagg, A., & Trujillo Bueno, J. 2003, *A&A*, 408, 1115

Kippenhahn, R. & Schlüter, A. 1957, *ZfA*, 43, 36

Lin, Y. 2011, *Space Sci. Rev.*, 158, 237

Lites, B. W. & Low, B. C. 1997, *Solar Phys.*, 174, 91

Mackay, D. H., Karpen, J. T., Ballester, J. L., Schmieder, B., & Aulanier, G. 2010, *Space Sci. Rev.*, 151, 333

Ning, Z, Cao, W. & Goode, P. R. 2009, *ApJ*, 707, 1124

Okamoto, T. J., Tsuneta, S., Berger, T. E. *et al.* 2007, *Science*, 318, 1577

Okamoto, T. J., Tsuneta, S., Lites, B. W., Kubo, M., Yokoyama, T., Berger T. E. *et al.*, 2008, *ApJ*, 673, L215

Shelyag, S., Keys, P., Mathioudakis, M., & Keenan, F. P. 2011, *A&A*, 526, A5

Sigwarth, M., Berst, C., Gregory, S., *et al.* 2001, in: M. Sigwarth (ed.), *Advanced Solar Polarimetry: Theory, Observation, and Instrumentation*, ASP Conf. Ser. Vol. 236, p. 57

Solanki, S. K., Livingston, W. L. & Ayres, T., *Science*, 263, 4

Stein, R. F. & Nordlund, A. 2006, *ApJ*, 642, 1246

Vögler, A., Shelyag, S., Schüssler, M., Cattaneo, F., Emonet, T., & Linde, T. 2005, *A&A*, 429, 335

Wang, Y.-M & Muglach, K. 2007, *ApJ*, 666, 1284

*Nature of Prominences and their role in Space Weather*
*Proceedings IAU Symposium No. 300, 2013*
*B. Schmieder, J.-M. Malherbe & S. T. Wu, eds.*

doi:10.1017/S1743921313011204

# Prominence Science with ATST Instrumentation

**Thomas Rimmele[1], Thomas Berger[1], Roberto Casini[2], David Elmore[1], Jeff Kuhn[3], Haosheng Lin[3], Wolfgang Schmidt[4] and Friedrich Wöger[1]**

[1]National Solar Observatory, Sunspot, NM-88349,
PO Box 62,USA
email: rimmele@nso.edu

[2]High Altitude Observatory, Boulder, CO,
[3]Institute for Astronomy, University of Hawaii,
[4]Kiepenheuer Institute für Sonnenphysik, Freiburg, Germany

**Abstract.** The 4m Advance Technology Solar Telescope (ATST) is under construction on Maui, HI. With its unprecedented resolution and photon collecting power ATST will be an ideal tool for studying prominences and filaments and their role in producing Coronal Mass Ejections that drive Space Weather. The ATST facility will provide a set of first light instruments that enable imaging and spectroscopy of the dynamic filament and prominence structure at 8 times the resolution of Hinode. Polarimeters allow high precision chromospheric and coronal magnetometry at visible and infrared (IR) wavelengths. This paper summarizes the capabilities of the ATST first-light instrumentation with focus on prominence and filament science.

**Keywords.** Sun, Prominence, Filaments, Corona, Magnetic Fields, Instrumentation

---

## 1. Introduction

Prominences are elevated structures observed at the limb of sun. On the disk these features are seen as dark filaments in chromospheric lines such as H$\alpha$. Prominences and filaments are large-scale structures of cool plasma embedded in an otherwise hot coronal environment. The cold prominence plasma is suspended by a complex magnetic fields. Prominences and filaments are drivers of space weather. The contribution by Gibson in this volume provides an overview of modeling efforts that describe the complex magnetic field configurations of prominences. Recent high-resolution time sequences of narrow band images exhibit intriguing, highly dynamic and variable properties (see e.g. contribution by Berger in this volume). These new data challenge conventional views of the nature of prominences. However, detailed diagnostics of the thermodynamic and magnetic properties of the prominences are still lacking due to the limitations of the imaging instruments used to obtain these observations.

The ATST with its 4m aperture and a set of highly capable first light instruments offers unique capabilities for ground-breaking observations and quantitative analysis of prominences and filaments. The significantly increased resolution and the photon collecting area provided by the large aperture open up new discovery space. One of the most challenging goals of the ATST is to perform measurements of the prominence (filament) magnetic field and the coronal background field, in which the prominence is embedded. ATST instrumentation will provide the spectroscopic analysis tools that cover a wide temperature range and allow measurements of the chromospheric and coronal magnetic field with sufficient sensitivity. The high photon flux of ATST is essential to accurately measure Stokes-V, which provides crucial diagnostics needed to distinguish between magnetic field models, for example, between a potential field extrapolation and a non-linear force

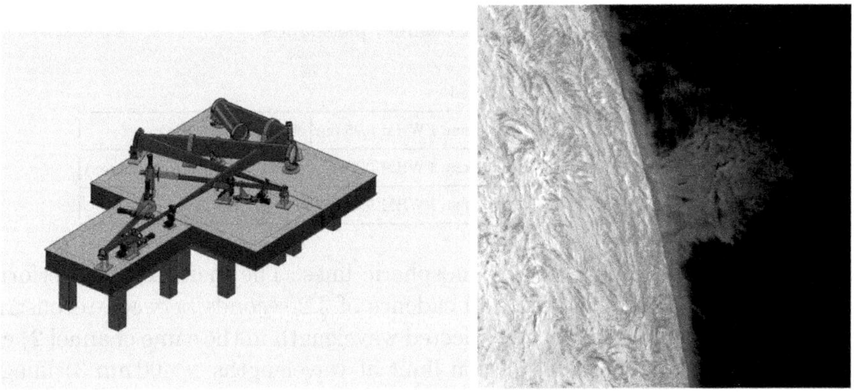

**Figure 1.** Right: VBI instrument layout. Left: Image of a prominence observed at the DST with the VBI H$\alpha$ filter.

free extrapolation (Wiegelmann 2004). Spectral diagnostic in the infrared is particularly attractive for coronal spectroscopy and magnetometry due to low sky and instrumental background and relatively bright mid-IR coronal emission lines.

This paper reviews the design and capabilities of the ATST first light instruments. These instruments can be operated individually or as a system and will provide a wealth of observational information about prominences that promises to significantly advance our knowledge of these important drivers of space weather.

## 2. ATST First Light Instruments

The suite of five first light instruments consists of the Visible Broad-band Imager (VBI), the Visible Tunable Filter (VTF), the Visible Spectro-Polarimeter (ViSP), the Diffraction-Limited Near-Infrared Spectro-Polarimeter (DL-NIRSP) and the Cryogenic Near-Infrared Spectro-Polarimeter (Cryo-NIRSP). The VBI and the VTF provide broad-band and narrowband imaging while the ViSP and the NIRSP instruments are spectrograph-based instruments. All instruments share a camera and data handling system provided by the facility. Polarimetry calibration and modulation are also functions of facility subsystems. Making adaptive optics function using the faint prominence structure that can be observed only with narrow-band (0.5Å) filters (e.g. H$\alpha$) is a challenge. A prototype system has recently been tested successfully at the Dunn Solar Telescope (DST) (Taylor *et al.* 2013). The ATST instruments build on significant heritage of instruments operated at current facilities, such as the DST. The goal is to operate these instruments in Service Mode (Uitenbroek& Tritschler 2012) for the majority of available observing time. The data collected by all instruments will be processed to remove instrument and residual atmospheric signatures (Level 1) at the NSO-ATST Data Center located at NSO Headquarters in Boulder. Distribution of data products for a broad user base is a function of the Data Center. The Science Working Group is defining standard data products for each of the instrument that will be available to the community.

### 2.1. *Visible Broadband Imager (VBI)*

The ATST Visible Broadband Imager (VBI) is a high-cadence imaging instrument that resolves structures within a 2×2 arcmin$^2$ field in the solar photosphere and chromosphere at the diffraction limit of ATST's aperture. The VBI consists of two imaging channels that can be synchronized to simultaneously observe targets at two different wavelengths in the visible light wavelength range between 390 and 800 nm selected by interference

**Table 1.** VBI channel passbands.

**VBI blue**

| | |
|---|---|
| 393.3 nm, FWHM 0.1 nm | chromosphere (Ca II K) |
| 430.5 nm, FWHM 0.5 nm | photosphere (G-band) |
| 450.4 nm, FWHM 0.4 nm | photosphere (blue continuum) |
| 486.1 nm, FWHM 0.05 nm | chromosphere (H $\beta$) |

**VBI red**

| | |
|---|---|
| 656.3 nm, FWHM 0.05 nm | chromosphere (H $\alpha$) |
| 668.4 nm, FWHM 0.4 nm | photosphere (blue continuum) |
| 705.8 nm, FWHM 0.6 nm | photosphere (TiO) |

filters as narrow as 0.05 nm for its chromospheric lines. The main science performance requirements of the VBI are: 1) temporal cadence of 3.2 seconds between reconstructed, high-resolution images observed at any selected wavelength in the same channel 2) critical sampling (or better) of ATST's diffraction limit at wavelengths > 400 nm 3) imaging of eight prominent lines and bands covering the photosphere and chromosphere, with high throughput.

For performance reasons, the VBI channel wavelengths are separated between 'blue' and 'red' wavelengths. Both channels contain interference filters that image either photospheric or chromospheric layers in the solar atmosphere. Table 1 lists the available interference filters and their passband.

The high-resolution, high-cadence capabilities of the two channels of the VBI will allow recording simultaneous images of, e.g., the chromospheric Ca II K (393.3 nm) line and the H$\alpha$ (656.3 nm) line (see Fig. 1) to investigate the dynamics and morphology of prominences within multiple layers of the solar atmosphere. These well-known and common diagnostics for prominence observations can be complemented by reconstructed images of the H$\beta$ (486.1 nm) line that might allow to infer further information about the temperature characteristics of the prominence. The ability to switch to any wavelength combination achievable by the interference filter configuration in each channel, and to acquire two simultaneous, reconstructed high-spatial resolution images at a fixed cadence enables accurate measurements of oscillations on-disk, or proper motions of prominence fine structure through the different geometric heights at the limb. In addition, the movies created will deliver important context for ATST instruments that have spectro-polarimetric capabilities but either operate at lower spatial resolution or within a smaller field of view.

### 2.2. *Visible Tunable Filter (VTF)*

The VTF is an imaging spectro-polarimeter based on tunable Fabry-Perot Etalons. It can observe virtually all spectral lines (one at a time) within the wavelength range of the instrument. It produces narrowband filtergrams with a field of view of about one arc minute. Several spectral lines can be observed in rapid sequence, where the cadence ranges between 1 s and 30 s per line, depending on the spectral sampling and on the desired signal quality. The VTF offers the several measurement modes:

- Imaging Stokes Polarimeter to derive magnetic field information from the polarization properties of spectral lines. This mode includes the possibility to measure full Stokes data set to retrieve the magnetic field, or to record Stokes-V magnetograms.
- Doppler imager to measure flow fields in the solar atmosphere
- Narrowband filter imager to record monochromatic filtergrams

The VTF covers a spectral range from 520 nm to 870 nm, with a spectral resolution of 3 pm (R=185,000 at 550 nm) and a circular field of view of 60 arcsec. The optical design uses three Fabry-Perot Etalons with spacing ratios of 1:1.581:3.412 in a telecentric configuration. The VTF has a pixel scale of 0.0125 arcsec and thus allows for diffraction-limited imaging with the ATST. The integration time for individual images is 25 ms,

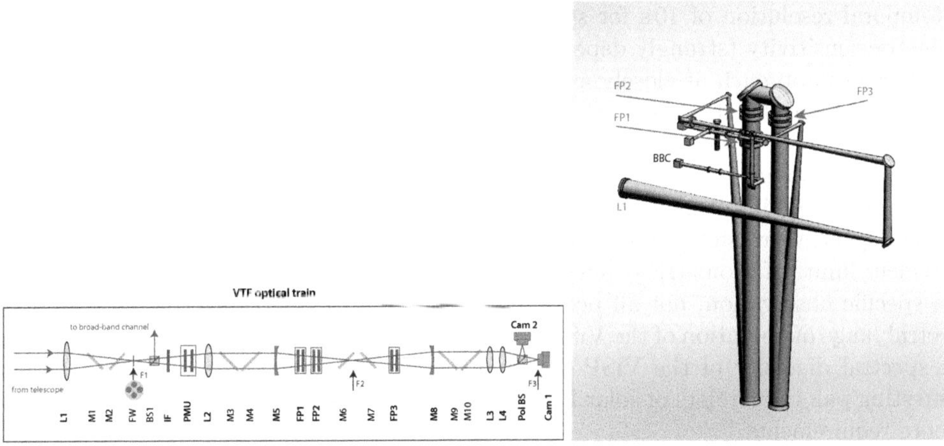


**Figure 2.** Left: Optical train of the VTF. Lenses and mirrors are labeled with Lx and Mx, respectively. FW is the device with the line selecting filters, PMU is the polarization modulation unit. The three Etalons are labelled FP1,2,3. A polarizing beam splitter (PolBS) serves as analyzer and sends two polarized beams to the detectors, Cam 1 and Cam2. Right: Optical configuration of the VTF. The three Etalons are marked in red (*FP1,2,3*). The broad-band channel is labeled *BBC*. Lens *L1* images the field of view to the VTF entrance focal plane.

corresponding to a signal-to-noise of 90 at disk center. The signal quality is improved through accumulating several images per wavelength position per Stokes component. Depending on the strength of the line and on the number of image accumulations, a signal-to noise-ratio of 300 is achieved. The spectral field of view can be chosen to cover line shifts up to 50 $\mathrm{kms}^{-1}$. Line selection is done with interference filters with a width of about 1 nm. The optical train of the VTF is shown in Fig.2, left. It shows the optical elements of the narrow-band channel. In the opto-mechanical configuration (Fig. 2,right) the Etalons will be placed horizontally, i.e. with the light beam passing vertically, to minimize gravitational deformation of the air gap between the Etalon plates. The Etalons for the VTF will have clear apertures of 260 mm.

The spectral window of the VTF includes the Mg b lines, the He D3 line, the Na D lines, H$\alpha$, and the infrared Calcium lines. All these lines have been used to study the NLTE behavior of the prominence atmosphere (Stellmacher& Wiehr 2005; Stellmacher *et al.* 2013). The VTF is therefore well suited for diagnostic spectroscopy of prominences, however with the restriction that only one line at a time can be recorded. In quiescent prominences, the cadence of multiline observations should be fast enough to combine the information obtained in different lines. The VTF records an area of about 30,000 km$\times$30,000 km instantaneously at each wavelength point. Magnetic fields estimates should be possible from He D3 Stokes spectra. Due to the low intensity of prominences compared to on-disk measurements, the quality of the Stokes measurements will be significantly reduced. It may be necessary to sacrifice some spatial resolution to improve the sensitivity.

### 2.3. *Visible Spectro-Polarimeter (ViSP)*

The Visible Spectro-Polarimeter (ViSP) of the ATST is a wavelength versatile instrument, whose design concept is specifically to provide a continuous coverage of the instrument's spectral range of operation (380 to 900 nm) over a field of view of $2 \times 2\,\mathrm{arcmin}^2$. The main science performance requirements of the ViSP are: 1) spectral resolution better than 180,000 ($\sim 3.5$ pm at 630 nm) over the entire spectral range; 2) spatial (sampling) resolution matching the diffraction limit of the ATST at the wavelength of observation;

3) temporal resolution of 10 s for spectro-polarimetric observations achieving $10^{-3}$ polarimetric sensitivity (strongly dependent on the target's brightness).

In order to approach as closely as possible the required spatial sampling condition at the observed wavelength, the ViSP is equipped with three photo-etched slit apertures, matching the Airy radius of the ATST respectively at 450, 650, and 850 nm. The ViSP is capable to observe up to three different spectral lines simultaneously. In order to do this, the ViSP implements low-order, high-dispersion diffraction gratings, and three movable spectral arms to pick up the different diffracted wavelengths (Fig. 3, right). Due to evident limits of geometric configurability, and depending on the desired performance of a specific observation, not all possible combinations of three wavelengths within the spectral range of operation of the ViSP may be attainable by the instrument. Nonetheless, the spectral diversity of the ViSP is significant, and allows the observations of many interesting pairs or triplets of solar lines, with good enough efficiency to meet the ATST science requirements.

The ViSP instrument covers the shorter end of the spectral region accessible to the ATST. This reaches down to 380 nm on the blue end, in order to allow spectro-polarimetric observations of the strong chromospheric resonance lines of Ca II at 393.4 nm (K line) and 396.9 nm (H line). On the red end of the "visible" solar spectrum, the ViSP spectral coverage extends to 900 nm, thus encompassing also the Ca II IR triplet ($\lambda\lambda$ 849.8, 854.2, 866.2 nm). The ViSP optics are optimized up to 1.1 $\mu$m, and so observations of the bright chromosperic resonance line of He I at 1083.0 nm are also possible, although the high quantum efficiency of a dedicated IR detector may be necessary to perform high-sensitivity spectro-polarimetric observations of this line with the ViSP. The Ca II H and K resonance lines are among the strongest spectral features of the quiet-Sun spectrum, and they represent important tracers of small-scale magnetic activity in the solar atmosphere. They have also been regularly observed in solar prominences for more than a century (Belopolsky 1908; Berger *et al.* 2010),and through an accurate wavelength calibration of spectral observations they allow high precision Doppler measurements of prominence plasma motions. Theoretical studies of the NLTE formation of Ca II and H I lines in prominences Gouttebroze, Vial & Heinzel (1997) have suggested that line-ratio techniques applied to various lines of these two different ion species can reliably be used for temperature diagnostics of prominence plasmas (Gouttebroze & Heinzel 2002).

For the magnetic diagnostics of prominences, the two chromospheric lines of He I at 587.6 (D3) and 1083.0 nm represent an almost ideal combination, because of the sensitivity of their polarization to a large range of magnetic strengths(approximately, 1 to $10^3$ gauss). This is possible because of the diverse diagnostic capabilities that are offered by the Hanle, Zeeman, and Paschen-Back effects of helium (Bommier 1980; Landi Degl'Innocenti 1983; López Ariste & Casini 2002; Asenio Ramos *et al.* 2008). Indeed, these two lines have almost exclusively dominated all attempts of the solar scientific community to reliably measure the magnetic field of quiescent and active-region prominences and filaments (Leroy 1981; Querfeld *et al.* 1985; Bommier *et al.* 1994; Paletou 2001; Trujillo-Bueno *et al.* 2002; Casini *et al.* 2003; Kuckein *et al.* 2009). Based on both line formation modeling and the reported observations described above, a potential list of interesting solar lines for prominence diagnostics accessible with the ViSP is proposed in Table 3.

### 2.4. *Diffraction-Limited Near-Infrared Spectro-Polarimeter (DL-NIRSP)*

Unlike other ATST spectropolarimeters based on the conventional slit spectrograph design, DL-NIRSP will be equipped with two fiber-optic integral field units (IFUs) that will allow it to record the polarized spectra of a 2D field simultaneously in up to three

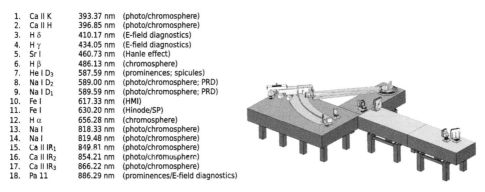

| | | | |
|---|---|---|---|
| 1. | Ca II K | 393.37 nm | (photo/chromosphere) |
| 2. | Ca II H | 396.85 nm | (photo/chromosphere) |
| 3. | H $\delta$ | 410.17 nm | (E-field diagnostics) |
| 4. | H $\gamma$ | 434.05 nm | (E-field diagnostics) |
| 5. | Sr I | 460.73 nm | (Hanle effect) |
| 6. | H $\beta$ | 486.13 nm | (chromosphere) |
| 7. | He I $D_3$ | 587.59 nm | (prominences; spicules) |
| 8. | Na I $D_2$ | 589.00 nm | (photo/chromosphere; PRD) |
| 9. | Na I $D_1$ | 589.59 nm | (photo/chromosphere; PRD) |
| 10. | Fe I | 617.33 nm | (HMI) |
| 11. | Fe I | 630.20 nm | (Hinode/SP) |
| 12. | H $\alpha$ | 656.28 nm | (chromosphere) |
| 13. | Na I | 818.33 nm | (photo/chromosphere) |
| 14. | Na I | 819.48 nm | (photo/chromosphere) |
| 15. | Ca II $IR_1$ | 849.81 nm | (photo/chromosphere) |
| 16. | Ca II $IR_2$ | 854.21 nm | (photo/chromosphere) |
| 17. | Ca II $IR_3$ | 866.22 nm | (photo/chromosphere) |
| 18. | Pa 11 | 886.29 nm | (prominences/E-field diagnostics) |

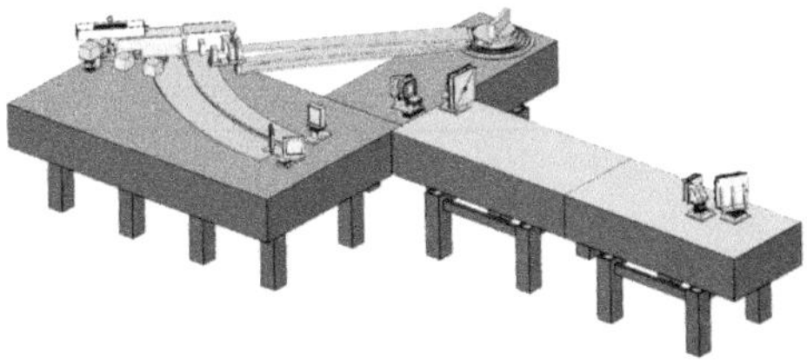

**Figure 3.** Left: Sample prominence diagnostics accessible with the ViSP. PRD stands for partial re-distribution. Right: Design of the Visible Spectro-Polarimeter (VISP).

spectral lines. This approach will enable observations of the dynamic evolution of the gas and magnetic fields of the prominences with high spatial and temporal resolution while also retaining the full spectral diagnostic capabilities of the conventional spectrographs. The DL-NIRSP will be equipped with a feed optical system that provides three different image scales, three spectral arms, and two diffraction gratings to provide a very versatile observing capability that can meet the observing requirements of programs with disparate science objectives. The DL-NIRSP can also observe three spectral lines simultaneously, with flexible line combination, which will provide the data needed for comprehensive spectral diagnostic to fully explore the prominence atmosphere in a broad temperature, density, and magnetic field ranges. This instrument allows to perform trades between spatial (and temporal) resolution and sensitivity. Figure 4, right, shows the field of view of DL-NIRSP in three observing modes: 1) a 7.2" $\times$ 5.4" (0".03/pix) field obtained with a 9 $\times$ 9 mosaic using the F/62 high-resolution mode, 2) a 19.2" $\times$ 14.4" (0".08/pix) field obtained with a 3 $\times$ 3 mosaic in the F/24 mid-resolution mode, and 3) a 80" $\times$ 60" field with 1" spatial sampling without mosaic. The IFU Imaging Array Format is 80 $\times$ 60. The spectral resolution $\frac{\lambda}{\Delta\lambda}$ is 250,000 (F/62, F24) and 80,000 (F/4). Figure 4, left, shows the optical layout of the 3-arm DL-NIRSP spectrograph.

DL-NIRSP will split the solar spectrum into three wavelength windows. The first window covers the spectral range of 500 to 750 nm, the second spectral windows covers the range of 750 nm to 1,000 nm, and the third windows covers 1,000 nm to the cutoff

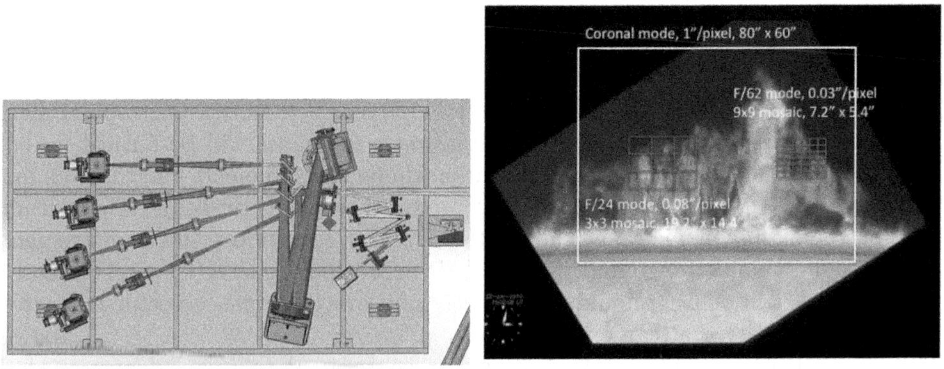


**Figure 4.** Left: optical design of DL-NIRSP. Right: FOV of DL-NIRSP for different feed optics configurations.

wavelength of the IR array detector. Table 2 shows some possible line combinations, limited by the availability of bandpass isolation filters for the spectral lines.

**Table 2.** DL-NIRSP possible Line Combinations

| | Spectral Arm 1 (500 - 750 nm) | Spectral Arm 2 (750 -1,000 nm) | Spectral Arm 3 (1,000 - 2,500 nm) |
|---|---|---|---|
| 1 | H$\alpha$ 656.3 nm | CaII 854.2 nm | HeI 1083.0 nm |
| 2 | HeI D3 587 nm | CaII 854 nm | HeI 1083 nm |
| 3 | HeI D3 587 nm | CaII 854 nm | FeXIII 1075 nm |

### 2.5. *Cryogenic Near-Infrared Spectro-Polarimeter (Cryo-NIRSP)*

The ATST and its IR instrumentation bring spectacular advantages for observing prominences because of its low-scattered light performance and broad wavelength coverage. Off-limb prominence observations depend on the suppression of scattered light from the solar disk and a reduced sky background. The ATST achieves both goals because of its dark-sky operating conditions from Haleakala and its all-reflecting coronagraphic optical design. While the ATST field-of-view is limited to about 4', this is adequate for observations of most off-limb magnetic structures. The extended wavelength coverage of the ATST allows instrumentation that can further improve the scattered light dynamic range of prominence measurements. The sky brightness decreases with wavelength. In particular, in the wavelength window between 3 and 4 microns that the terrestrial "blue" sky can be even fainter than the solar corona. Instrumental scattered light also declines in the IR, for example from mirror microroughness, where the background "noise" brightness decreases like $1/\lambda^2$.

| | |
|---|---|
| Grating (primary) | Echelle: 32 line/mm |
| Relay, focal ratio | 18 |
| Coll/Cam. Focal Ratio | 18/8 |
| Plate scale | 0.16 mm/arcsec |
| Slit Width | 52/175$\mu$ |
| Pixel scale (at 1 $\mu$) | 165 $m\AA$ |
| Pixel scale (arcsec) | 0.12" (18$\mu$ pix) |
| Design/Diffraction limited FOV | 240/90" |
| Cold blocking filter | BW 1% |
| System QE | 5% |
| Max spectral resolution at 4$\mu$ | 100000 |

**Table 3.** CryoNIRSP Optical Specifications

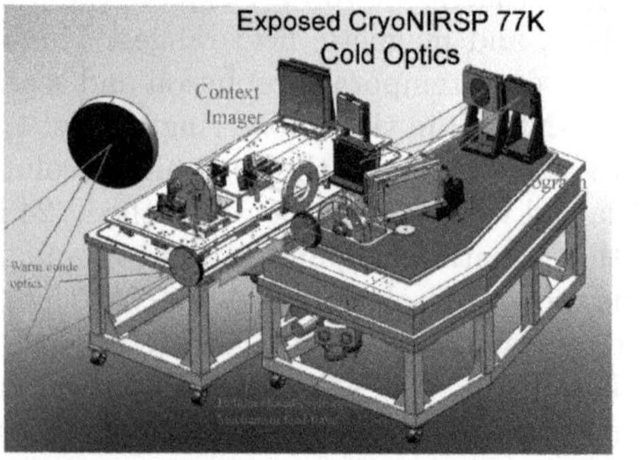


**Table 4.** The CryoNIRSP optics are cooled to approximately 80K in order to achieve photon-noise limited performance from visible to IR wavelengths of 5$\mu$ for bright and faint infrared sources.

Prominence magnetometry depends on very high signal-to-noise spectropolarimetry. For example the important HeI IR triplet-state resonance lines will be critical for both permitted-line, non-saturated linear polarization Hanle measurements and the even more difficult Stokes-V magnetometry of off-limb prominence magnetic fields. These studies depend on the high dynamic range capabilities of ATST and its post-focus instrumentation. The high-dynamic range and wide-field IR spectropolarimetry that ATST and its instrumentation will deliver should significantly improve prominence vector magnetometry with a sensitivity that can exceed 0.1G.

The CryoNIRSP imaging spectropolarimeter is designed to take advantage of the ATST's unmatched scattered light suppression, broad wavelength coverage, and field-of-view. It is thus a fully reflective, essentially photon-limited spectropolarimeter that observes over a 0.5 (goal) to 5$\mu$ wavelength band. It is designed to take full advantage of the ATST's coronagraphic properties while providing slit-scanning spectropolarimetry of a 3'x4' field with a spectral resolution of 100,000. To achieve high polarimetric accuracy it simultaneously observes orthogonal polarization states using a cryogenic polarizing beamsplitter. CryoNIRSP is the largest FOV ATST instrument and does not use system AO facilities. Under good seeing conditions the ATST/CryoNIRSP will be diffraction limited near 4.6$\mu$ wavelengths. To achieve photon-noise limited performance the essential instrument optics and detectors are cryogenically cooled. Instrument properties are summarized in Table 3.

The CryoNIRSP context imager and spectrograph can be operated independently for observations of the near-limb or far-limb solar atmosphere. Both modes are well suited to observing the origin and evolution of cool magnetic prominence loops into the hot coronal surroundings. Figure 4 shows how the optical components of the instrument will appear at the coude level of the ATST with dewar thermal shields removed. The instrument's wide FOV means it is uniquely suited for both Zeeman and Hanle effect prominence vector magnetometry using, for example, the IR Helium lines near 1$\mu$. The system's low scattered light performance allows direct measurement of the imbedding coronal magnetic field that is critical to understanding the stability and evolution of the cooler prominence material. This is possible because the CryoNIRSP can measure 1G-level coronal fields over a broad range of plasma temperatures. For this it relies on the sensitivity of forbidden coronal emission lines ranging in wavelength from the green line (0.53$\mu$) to SiX near 4$\mu$ where the magnetic Zeeman sensitivity is also a maximum.

## References


Wiegelmann, T. 2004, *Sol. Phys.*, 219, 87

Uitenbroek, H. & Tritschler, A. 2012, *IAU Special Session*, 6

Taylor, G. *et al.*, 2013, *Proceddings SPIE, Optics and Photonics 2013: Astronomical Optics and Instruments, in press.*

Stellmacher, G. & Wiehr, E. 2005, *Astron. Astrophys.*, 431, 1069

Stellmacher, G., Wiehr, E., & Dammasch, I. E. 2013, arXiv:1303.1126

Asensio Ramos, A., Trujillo Bueno, J., & Landi Degl'Innocenti, E. 2008, *ApJ*, 683, 542

Belopolsky, A. A. 1908, MiPul *(Bull. Pulkovo Obs.)*, 2, 239

Berger, T. E., Slater, G., Hurlburt, N., *et al.* 2010, *ApJ*, 716, 1288

Bommier, V. 1980, *Astron. Astrophys.*, 87, 109

Bommier, V., Landi Degl'Innocenti, E., Leroy, J.-L., & Sahal-Bréchot,S. 1994, *Solar Phys.*, 154, 231

Casini, R., López Ariste, A., Tomczyk, S., & Lites, B. W. 2003, *ApJ*, 582, 51L

Gouttebroze, P. & Heinzel, P. 2002, *Astron. Astrophys.*, 385, 273

Gouttebroze, P., Vial, J.-C., & Heinzel, P. 1997, *Solar Phys.*, 172, 125

Kuckein,C., Centeno, R., Martínez Pillet, V., *et al.* 2009, *Astron. Astrophys.*, 501, 1113

Landi Degl'Innocenti, E. 1982, *Solar Phys.*, 79, 291

Leroy, J.-L. 1981, *Solar Phys.*, 71, 285

López Ariste, A., & Casini, R. 2002, *ApJ*, 575, 529

Paletou, F., López Ariste, A., Bommier, V., & Semel, M. 2001, *Astron. Astrophys.*, 375, 39

Querfeld, C. W. , Smartt, R. N. , Bommier, V., *et al.* 1985, *Solar Phys.*, 96, 277

Trujillo Bueno, J., Landi Degl'Innocenti, E., Collados, M., *et al.* 2002, *Nature*, 415, 403

*Nature of Prominences and their role in Space Weather*
*Proceedings IAU Symposium No. 300, 2013*
*B. Schmieder, J.-M. Malherbe & S. T. Wu, eds.*

doi:10.1017/S1743921313011216

# Instrument concepts for the observation of prominences with future ground-based telescopes

**A. López Ariste**

THEMIS - CNRS UPS 853
C/ Vía Láctea s/n 38205 - La Laguna, Spain
email: arturo@themis.iac.es

**Abstract.** The observation of prominences with ground-based telescopes suffers from poor image quality due to atmospheric turbulence when compared with space-borne instruments which, for solar observations, are of similar apertures. To make ground-based instruments competitive, they should rely on spectropolarimetry and the measurement of prominence magnetic fields, a task which no foreseeable space instrument will perform. But spectropolarimetry alone does not suffice, and we argue that future instrumentation should combine it with imaging in a large field of view and good temporal resolution. We place numbers on those requirements and give examples of instrumental accomplishments already at work today that forecast a new generation of instruments for the observation of prominences from ground-based telescopes.

**Keywords.** Instrumentation; Polarimetry; Imaging

## 1. Introduction

The difficulties of observing the Sun force solar telescopes to small apertures, limited by the ability of materials and optical design to control heat. This constraint has a curious downside for ground-based solar telescopes, and it is that space solar telescopes have similar sizes too. Ground-based solar telescopes do not have the advantage of size over their space counterparts that night telescopes may enjoy.

What is left for ground-based solar telescopes to do when the advantage of size disappears?

This crucial question underlines the almost day-to-day prospective of future instrumentation for ground-based solar telescopes. It is clear that space telescopes, observing from beyond the disturbing effects of the atmospheric turbulence, enjoy an almost-perfect image of the Sun, constant in quality over the observing time. Until recently it could be asserted that the best spots on Earth could still take advantage of the larger mirror sizes and of performant adaptive optics systems to, from time to time, produce the highest-resolution images possible of the solar atmosphere. The best images and movies from the Swedish Solar Telescope in La Palma with its 1m aperture apparently outperformed the best images coming from the largest space solar telescope, Hinode, with its 0.5m aperture. Recent results by van Noort(2012) and Ruiz Cobo and Asensio Ramos (2013) deconvolving images from the SOT instrument aboard Hinode cast a doubt over this general idea and future space solar telescopes (such as the projected Solar-C with its 1.5 aperture) will put further pressure on the performance of adaptive optics for new ground-based solar telescopes like ATST (Advanced Technology Solar Telescope) and EST (European Solar Telescope), a performance already pushed to the limit by the large aperture (4m) of these future telescopes. For the observation of solar prominences, the advantage in image quality falls almost completely on the side of space telescopes. Adaptive optics systems require some light and a contrasted image to work. Those requirements are fulfilled over the solar disk where the observation of filaments with ground-based solar telescopes can still enjoy some good image quality under exceptional conditions (Lin *et al.*, 2003 and Lin *et al.*, 2005). Off-limb prominences on the other hand do not have the conditions for image correction, and in spite of some impressive examples when contrasted objects could be seen in the adjacent solar disk under conditions of large isoplanatic angles (Martínez González *et al.*, 2014), high resolution images of prominences are the the realm of space solar telescopes (Berger *et al.*, 2008, Chae *et al.*, 2008).

The question returns therefore to what is left for ground-based telescopes to do when observing prominences, in a time when the Hinode/SOT instrument has provided spectacular observations of prominences that have renewed the debate as to how prominences are formed and about the relation of the observed dynamics with its magnetic structure.

## 2. Spectra, polarization and imaging

A first answer is spectropolarimetry. At this date the measurement of magnetic fields in prominences and filaments through spectropolarimetry of the He lines is a exclusivity of just 3 telescopes around the world, all ground-based: THEMIS, the DST and the VTT through its TIP instrument. Spectropolarimetry is still a complex technique. It requires a combination of powerful spectrographs and reliable and highly sensitive polarimeters. However, its complexity arises mostly from the difficulty of its interpretation. The constraint to only two He lines (the He $D_3$ line and the He line at 1083nm) is due to the fact that only those allow a simple enough treatment of the polarized line formation in the presence of quantum coherences, a requirement for the Hanle effect on which the diagnostic is based (López Ariste & Casini, 2002). These difficulties make the technique not a subject of choice for space instruments which are characterized by their request for reliability and the absence of doubts on the interpretation of the raw measurement. Because of this no space instrument has ever carried a spectropolarimeter capable of these measurements, however important they are. Ground based solar telescopes observing prominences can profit from the ability to do spectropolarimetry in these He lines in order to be competitive when compared with space solar telescopes. A recent and dramatic example of this is the result published by Schmieder *et al.* (2013). In their work the authors use the Hinode SOT imaging instrument to provide them with a time series of high quality images of a prominence in which a wave-like perturbation appears to propagate upwards. This observation would not be surprising or new were it not for the fact that, simultaneously, THEMIS measured the magnetic field over the prominence revealing a perfectly horizontal field 7G strong. The image quality of the THEMIS data leaves much to be desired. But this is not a problem since the interest on THEMIS data is NOT on its image quality, which is provided by the space-borne Hinode and almost impossible to improve upon with existing ground-based instruments. The strength of THEMIS data rests upon its spectropolarimetric capabilities added to the ability to infer from them the magnetic field vector all over the prominence. This is where the ground-based solar telescope, THEMIS in that case, outperformed the space instrument and justified its use.

Spectropolarimetry is therefore an essential ingredient on future instruments for the observation of prominences with ground-based telescopes.

Many advances in prominence observations have resulted from the ability to measure and interpret the dynamics of the plasma confined in the prominence (as examples, Berger *et al.*, 2008, and Chae *et al.*, 2008). The ability to do imaging in adequate spectral ranges quick enough to allow for the follow up of the movements and perturbations

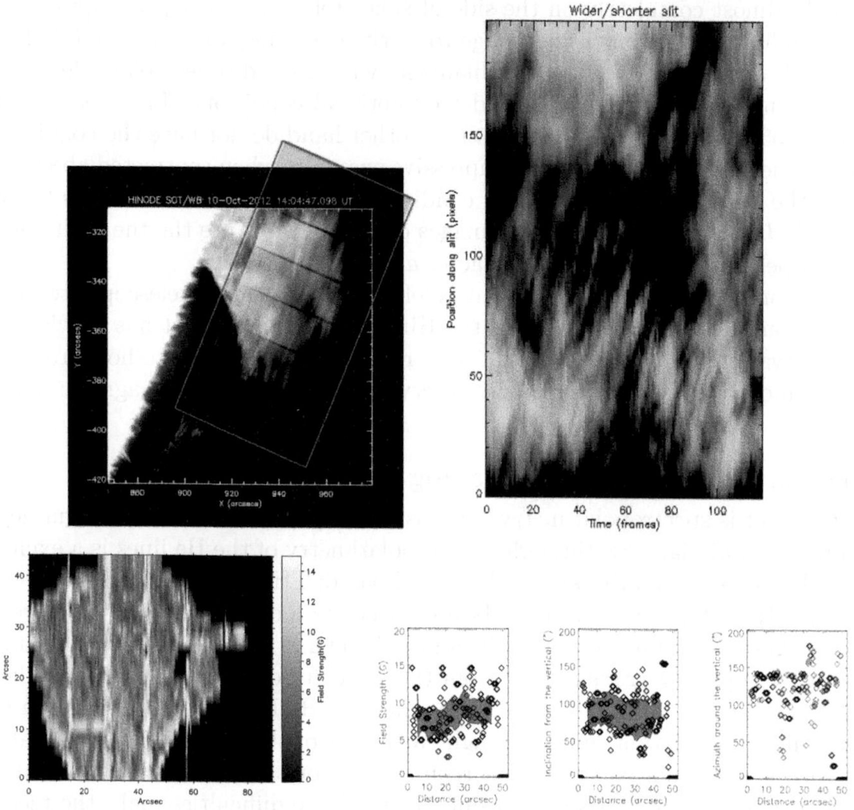


**Figure 1.** Collage of figures from Schmieder *et al.* (2013) illustrating space vs ground-based observations of prominences. The top images come from Hinode and show the prominence as seen in H$\alpha$ and a time-distance diagram of a cut through the foot of the prominence. The red rectangle on the left image shows the region scanned with THEMIS. The bottom images show the magnetic field measurements made with the He $D_3$ line at THEMIS: the field strength over the prominence at left and the values of strength, inclination and azimuth of the magnetic field vector along the same cut through the prominence foot on which the time-distance diagram was made.

of the prominence plasma has shown the presence of counterstreaming threads, prominence oscillations of different kinds, or the presence of bubbles and plumes deforming the magnetic structure of the prominence. A very recent instrument focusing in this kind of observations from the ground is ROSA (Jess *et al.*, 2010). Rather than giving the details of this instrument and its performances, it is interesting for the purpose of our present exercise to just look into the acronym that makes the name of this instrument: ROSA stands for Rapid Oscillations in the Solar Atmosphere. High quality imaging, we said, is the realm of space instruments. But as ROSA, and those recent observational results point out any future instrument must be capable of high cadence high quality imaging too. With this in mind, we have collected the three requirements for new instruments observing prominences: spectropolarimetry (in particular of the He lines) combined with imaging at sufficiently high cadence and quality as to follow the evolution, dynamics and oscillations of the prominence plasma. We should now give some numbers to make some sense of those three requirements.

Our first requirement for new instruments observing prominences was spectropolarimetry. The He lines $D_3$ (at 587nm) and at 1083nm are today the ones whose formation and

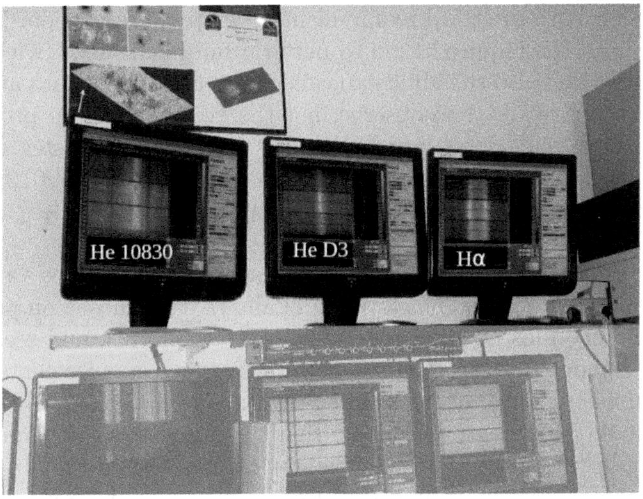


**Figure 2.** Observing prominences simultaneously in the 2 He lines at 587 and 1083nm plus Hα with THEMIS.

polarization in the conditions of prominences and in the presence of magnetic fields is better understood. Our prospective instrument should therefore be able to observe either one of those spectral ranges. A much better constraint on the measured magnetic fields and other physical conditions is obtained if both of these lines are simultaneously observed and, furthermore, if a line like Hα is put in the set as a means to obtain a good feeling on the prominence morphology thanks to its large brightness. Observations with the three lines cited have already been performed, e.g. at THEMIS (see Fig. 2). Concerning spectral resolution, all recent measurements of magnetic fields in prominences have been done at spectral resolutions $\frac{\lambda}{\Delta\lambda}$ of around 200,000 though probably one can relax this constraint and retrieve reliable magnetic fields from polarimetry of those lines with values as low as 100,000. The linear polarization expected in these lines can be as high as 10%, but it has been demonstrated time and again the importance of a reliable diagnostic for measuring the weak circular polarization in these lines, either due to the Zeeman effect or, as in the $D_3$ line, the subtle transfer of atomic alignment into orientation. Those weak polarization signals peak at most at 0.1% of the intensity. To observe them, a requirement of a signal-to-noise ratio approaching 10,000 must be imposed on the instrument.

Our second requirement was imaging. The spatial resolution will be limited by the performance of the adaptive optics systems and, in the case of off-limb observations, the ability to use them off-limb. It is clear that space instruments will almost always outperform any ground-based instrument combining imaging with spectropolarimetry. Furthermore, signal-to-noise requirements, as given above, may require spatial resolution to be dropped in exchange for the accumulation of photons in every single detector pixel. Because of those reasons it is advisable not to set stringent requirements on spatial resolution. Another story is the field of view. The typical size of prominences must be observed to correctly interpret the morphology and, eventually, the temporal evolution. A minimum requirement of $1 \times 1$ arcmin$^2$ with a goal to $2 \times 2$ arcmin$^2$ appears as a reasonable choice.

Finally, we are left with the third requirement of temporal resolution. Here our limitations come from both the required time to perform full polarimetry with the requested signal-to-noise ratios and also the ability to collect and handle the data at high rates. On the other hand we are required to observe, not the evolution of the prominence topology which more often than not happens in times of hours, but faster events as waves and oscillations which occur in periods of minutes (Ballester *et al.*, 2012, Arregui *et al.*, 2012). A prospective instrument should aim at temporal cadences of less than one minute therefore.

The two previous paragraphs almost look like a specification list for a new instrument. And in a basic sense they are so. The level of details given is chosen on purpose. Requiring spectropolarimetry plus imaging plus temporal resolution would seem at first sight as an impossible dream. Putting numbers into those broad requirements helps us identify priorities and where instrumental trade-offs can help solve the problem of a do-it-all instrument. It should be stressed once again that the purpose of this paper is not to consider what new observations of solar prominences can be done in the absolute, but what observations can be done from ground-based telescopes that are competitive respect to its space counterparts. The requirement for spectropolarimetry plus imaging plus temporal resolution is justified by these comparisons. Only by satisfying those requirements can we expect instruments in ground-based solar telescopes to offer new observations on prominences.

But can those requirements be satisfied at all by a single instrument? We think that, with the figures given, the answer is almost afirmative. The ROSA instrument is an obvious answer to almost all of those requirements assuming that another choice of observed spectral regions can be implemented to better suit prominence observations. Here we should end with another attempt in the same direction:TUNIS, being commissioned at this time on THEMIS. TUNIS stands for Tunable Universal Narrowband Imaging Spectrograph. It is an imaging instrument with a field of view of $2 \times 2$ arcmin$^2$ and a pixel sampling of 0.2 arcsec. It is based upon the concept of subtractive double pass, on which a spectrograph is used to select the spectral range and resolution. This concept has been used before very successfully in the Multi-Subtractive Double Pass (MSDP) suite of instruments (Mein, 2002). The usual long and narrow slit at the entrance of classic spectrographs disappears, allowing the said field of view into the spectrograph. In the first pass each point of that field of view is dispersed in wavelength. A spectral slit selects the appropriate spectral bandpass for each point. Because of the different angle of entrance into the spectrograph there will be a linear variation across the image of what wavelength is selected, but the bandwidth will be fixed by the spectral slit. In the present TUNIS instrument this bandwidth is set at 20 mÅ resulting in a maximum of 300,000 spectral resolution. The second pass through the spectrograph in subtractive mode ensures that the residual dispersion of 20 mÅ does not degrade image quality. Spectral coverage at each point has to be then performed by some wavelength scanning. Rather than just scan the spectrum, though, TUNIS relies on wavelength multiplexing with a Hadamard mask. With this method, each point of the image sees in one exposure a linear combination of the intensities at 63 different wavelengths 20 mÅ wide. By making different linear combinations of the kind in subsequent exposures one can solve the linear system and retrieve the intensity of each individual wavelength during data reduction. In the present TUNIS, this can be done in about 2 seconds without polarimetry. More details of this instrument have been given by López Ariste *et al.* (2010) and López Ariste *et al.* (2011). The work in progress with instruments like ROSA and TUNIS demonstrates in our view that the requirements given above can be satisfied, aside from small concessions concerning, for example, spectral scanning.

## 3. Conclusion

Summarizing, the observation of solar prominences with ground-based solar telescopes should rely mostly on spectropolarimetry and magnetic field measurements, the realm on which space instruments cannot compete. Added to that requirement, they should of course aim at also adding imaging and temporal resolution. Since in neither of these two aspects ground-based instruments can hardly outperform space instruments, they cannot be the priority of our ground-based instruments but should be seen as the perfect addition to spectropolarimetry to give ground-based instruments a place in the upcoming evolutions of our understanding of solar prominences. If the addition of spectropolarimetry plus imaging plus temporal resolution may seem an impossible task to achieve, when actually giving numbers on the actual requirements we see that the task can be actually accomplished with few concessions. Actually we think that two recent instruments as ROSA and TUNIS on the DST and THEMIS telescopes respectively are actually almost satisfying those requirements.

### References

Arregui, I., Oliver, R., & Ballester, J. L., 2012. Prominence Oscillations. Living Reviews in Solar Physics 9, 2.

Ballester, J. L., Arregui, I., Oliver, R., Terradas, J., Soler, R., Lin, Y., Engvold, O., Langagen, O., & Rouppe van der Voort, L. H. M., 2012. Prominence seismology using ground- and space-based observations, pp. 169–174.

Berger, T. E., Shine, R. A., Slater, G. L., Tarbell, T. D., Title, A. M., Okamoto, T. J., Ichimoto, K., Katsukawa, Y., Suematsu, Y., Tsuneta, S., Lites, B. W., & Shimizu, T., 2008. Hinode SOT Observations of Solar Quiescent Prominence Dynamics. The Astrophysical Journal Letters 676, L89–L92.

Chae, J., Ahn, K., Lim, E. K., Choe, G. S., & Sakurai, T., 2008. Persistent Horizontal Flows and Magnetic Support of Vertical Threads in a Quiescent Prominence. The Astrophysical Journal Letters 689, L73–L76.

Jess, D. B., Mathioudakis, M., Christian, D. J., Keenan, F. P., Ryans, R. S. I., & Crockett, P. J., 2010. ROSA: A High-cadence, Synchronized Multi-camera Solar Imaging System. Solar Physics 261, 363–373.

Lin, Y., Engvold, O., Rouppe van der Voort, L., Wiik, J. E., & Berger, T. E., 2005. Thin Threads of Solar Filaments. Solar Physics 226, 239–254.

Lin, Y., Engvold, O. R., & Wiik, J. E., 2003. Counterstreaming in a Large Polar Crown Filament. Solar Physics 216, 109–120.

López Ariste, A., & Casini, R., 2002. Magnetic Fields in Prominences: Inversion Techniques for Spectropolarimetric Data of the He I D3 Line. Astrophysical Journal 575, 529–541.

López Ariste, A., Le Men, C., & Gelly, B., 2011. Double-pass spectroimaging with spectral multiplexing: TUNIS. Contributions of the Astronomical Observatory Skalnate Pleso 41, 99–105.

López Ariste, A., Le Men, C., Gelly, B., & Asensio Ramos, A., 2010. Double-pass spectro-imaging: TUNIS. Astronomische Nachrichten 331, 658.

Martínez González, M., Manso Sainz, R., Asensio Ramos, A., Beck, C., & De La Cruz Rodríguez, J., 2014. Observation of the magnetic field in solar tornadoes.

Mein, P., 2002. The MSDP of THEMIS: Capabilities, first results and prospects. Astronomy and Astrophysics 381, 271–278.

van Noort, M., 2012. Spatially coupled inversion of spectro-polarimetric image data. I. Method and first results. Astronomy and Astrophysics 548, 5.

Ruiz Cobo, B., & Asensio Ramos, A., 2013. Returning magnetic flux in sunspot penumbrae. Astronomy and Astrophysics 549, L4.

Schmieder, B., Kucera, T., Knizhnik, K., Luna, M., López Ariste, A., & Toot, D., 2013. Propagating Waves Transverse to the Magnetic Field in a Solar Prominence. Astrophysical Journal 777, 108.

# V - Conclusion

## GENERAL CONCLUSION OF THE SYMPOSIUM

*Nature of Prominences and their role in Space Weather*
*Proceedings IAU Symposium No. 300, 2013*
*B. Schmieder, J.-M. Malherbe & S. T. Wu, eds.*

doi:10.1017/S1743921313011228

# Prominences: Conference Summary and Suggestions for the Future

**Eric R Priest**

Mathematics Institute, University of St Andrews
St Andrews KY16 9SS, UK
email: eric@mcs.st-and.ac.uk

**Abstract.** In this conclusion to the conference, I shall attempt to summarise what we knew before about solar prominences and what we have learnt during the conference (mainly from the review talks), as well as to make suggestions for their future study.

**Keywords.** Sun: prominences, Sun: filaments, Sun: activity, Sun: coronal mass ejections (CMEs), Sun: magnetic fields.

## 1. Introduction: Prominence Formation

We have been treated to an exciting and well delivered set of talks. Clearly, prominences are intriguing and fascinating, with many secrets to reveal to us (Fig. 1). I would in particular like to thank my Paris friends such as Pierre and Nicole Mein, Zadig Mouradian, Mme Martres and Serge Koutchmy, for introducing me to prominences when I first visited in the 1970's, and to Brigitte Schmieder and Jean-Marie Malherbe for continuing this interest when I met them at Hvar and La Palma in the 1980's.

Three mechanisms for prominence formation have been proposed, namely, condensation by radiative instability or non equilibrium, levitation and injection by reconnection. At this meeting, Tom Berger gave a convincing observation from SDO/AIA of radiative condensation and also proposed a new mechanism of magneto-thermal convection, inspired by amazing movies from SDO and numerical experiments. Also, Manuel Luna presented a talk prepared by Judith Karpen of impressive multi-thermal models for prominences.

The question "How do prominences form?" is still open and needs detailed modelling as well as new ideas. A related problem is to determine how they are maintained. What is the mass circulation, both the supply and the loss? What is the engine or driver for the flow? Jack Carlyle has made a start towards addressing this issue and is to be congratulated on winning the poster prize.

## 2. What is the Plasma/Magnetic Structure?

It was known before that the prominence density and temperature are typically a hundred times larger and smaller, respectively, than the surrounding corona. Also, most prominences consist of a vertical sheet suspended in a large horizontal flux rope of inverse polarity above a polarity inversion line (Fig. 2). One way of forming such a flux rope is for it to emerge already twisted near a polarity inversion line (PIL). Another is for the twist to be built up by flux cancellation below a prominence at the PIL (van Ballegooijen and Martens 1989).

An important realisation has been the importance of a coronal cavity around a prominence, as reviewed by Sarah Gibson (this volume) (Fig. 3). She described the range of

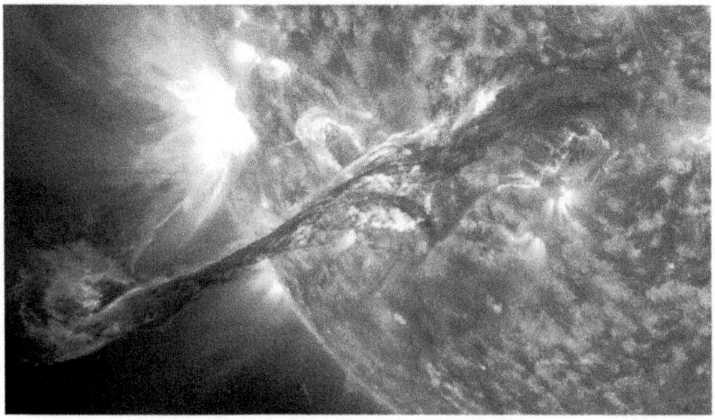

**Figure 1.** An erupting prominence seen by SDO/AIA (courtesy NASA/SDO science team).

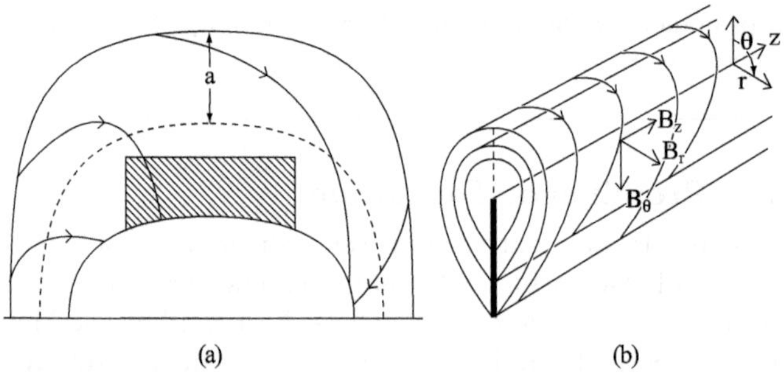


**Figure 2.** The overall structure of a prominence sheet within a large horizontal flux rope, from (a) the side and (b) the end, according to the Flux-Rope Model (Priest *et al.* 1989).

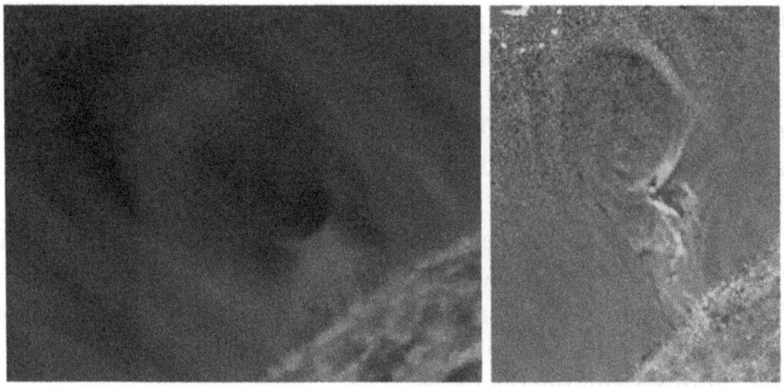

**Figure 3.** (a) A prominence cavity and (b) an inner flux rope observed by Alan Title (this volume), courtesy NASA/SDO science team.

sizes, the fact that they form a funnel, the density depletion by 25–30%, the substructure and the multi-thermal dynamic nature, with flows of 5–10 km $s^{-1}$.

Future questions are: how much mass is provided by the flows? How does this compare with the best estimate of mass loss from a prominence, taking account of the net flows on the limb and on the disc?

Alan Title (this volume) gave some fabulous evidence for inner flux ropes within the core of the cavity above erupting prominences (Fig. 4). They represent the continuation of the horns that are sometimes seen (e.g., Fig. 3) and represent the inner part of the classical three-part structure of a coronal mass ejection. Why do they appear? Is it that the eruption compresses the plasma in this part of the overlying flux rope and so makes it visible?

Jean-Claude Vial gave a coherent review (this volume, prepared by Susanna Parenti) of the prominence-corona transition region, which needs in future to be incorporated more into prominence modelling. Just as the transition region in a coronal loop is not a thin static region sitting between the chromosphere and corona, so its equivalent in a prominence represents dynamic plasma that is either heating up or cooling down and happens to be passing through $10^5$ K.

Bruce Lites (this volume) gave his customary authoritative review of magnetic field observations. He stressed that in prominences the Hanle effect is the best way of measuring magnetic fields and summarised results from many years by Leroy, Bommier, Lopez Ariste and Kuckein. He concluded by describing a comprehensive study by Orozco Suárez *et al.* (2013) that has given magnetic field strengths of 2–30 G and horizontal field inclinations to the prominence axis of $15–25^o$. Also, Zhi Xu (this volume) described measurements from the new Chinese telescope of photospheric and chromospheric measurements of magnetic fields near an active-region filament.

Two thoughts occur. The first is that the prominence magnetic field probably consists of two parts, a large-scale field (which is what we normally measure) together with a small-scale turbulent field (whose existence is implied by the small-scale plasma structure in prominences). The second thought is that in future we need new non-force-free techniques to extrapolate from observed magnetic fields in the photosphere (which are not force-free) up through the chromosphere to the corona.

Jose-Luis Ballester (this volume) gave a comprehensive review of prominence seismology, which is a promising way of determining physical properties in prominences, such as field strength, plasma density and filling factor. However, this field is very much in its infancy and so more realistic models are expected in future. For example, although this model is a useful beginning, it is a gross oversimplification to regard a coronal loop or a prominence fibril as an isolated one-dimensional flux tube, and so much more complex models of such structures need to be built in future.

## 3. Why Barbs and Feet?

A key model for barbs was proposed by Aulanier and Démoulin (1998). It consists of a force-free flux rope with a series of parasitic polarities on both sides of the polarity inversion line. Fig. 4a shows the photospheric polarities viewed from above, including the parasites, together with locations where the prominence plasma is expected to accumulate in magnetic dips: in particular, a series of barbs is produced above the parasites. Vertical cross sections across the prominence reveal an O-type topology with a bald patch at locations between the parasites and a flat field joining two X-points above the parasites (Fig. 4b, c).

Aad Van Ballegooijen gave a superb review of prominence magnetic structure, concluding that a flux rope model works well for explaining many aspects of prominences except for the vertical threads: for example, predicting barbs, describing the formation of the flux rope and producing the horns and inner flux rope above a prominence observed by SDO/AIA. The magnetic field of a prominence is distorted by gravity when the field is weak (Hillier and van Ballegooijen 2013) and the flux-rope insertion model is useful for modelling particular prominences (Su and van Ballegooijen 2013).

In future, it will be interesting to determine whether prominences are located in current sheets with the field either side being vertical or possessing a horizontal component. Also,

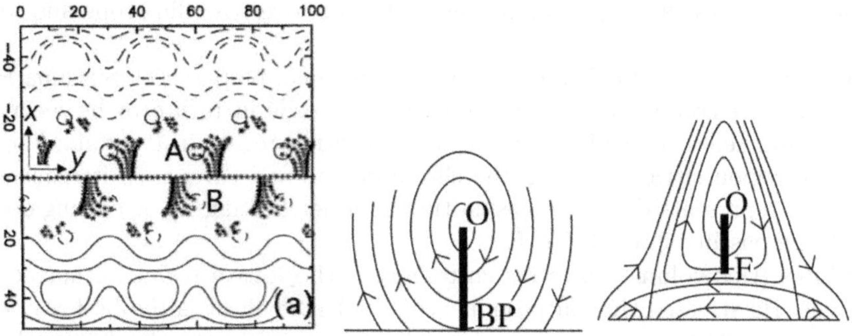


**Figure 4.** A force-free flux rope model with parasitic polarities, showing (a) the photospheric flux pattern viewed from above together with the locations (crosses) of dips and (b), (c) vertical sections across the prominence at two locations (Aulanier and Démoulin 1998).

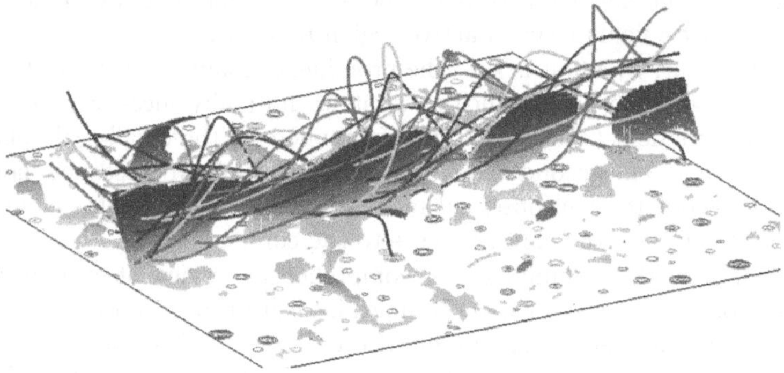

**Figure 5.** A twisted flux-rope model due to Su and van Ballegooijen (2013).

it is important to ensure that models are consistent with both disc and limb observations, which sometimes appear contradictory at first sight.

## 4. The Formation of Flux Ropes along a Polarity Inversion Line

Prominences possess a global chirality pattern, being mainly dextral in the northern hemisphere and sinistral in the southern hemisphere (Martin *et al.* 1994). Also, the shear is concentrated around the polarity inversion line (Schmieder *et al.* 1996). A way in which the chirality of a prominence is produced was suggested in the Dextral and Sinistral Model (Priest *et al.* 1996) and this was later developed by van Ballegooijen *et al.* (2000) in the mean-field model for filament channel formation. The model includes a flux-transport model for the evolution of the radial photospheric magnetic field in response to flux emergence, differential rotation, meridional flow and supergranular diffusion. It also shows how the coronal magnetic field evolves through a series of nonlinear force-free fields in response to the photospheric evolution.

Applying this to observed magnetic fields was highly effective (Mackay and van Ballegooijen 2005; Yeates *et al.* 2008). In particular, the model predicts the locations along the polarity inversion line where large flux ropes form and they agree in over 95% of cases with observed filament locations (Fig. 6).

What is clear from these impressive results is that that magnetic helicity transport over months and years is a fundamental part of coronal evolution. Thus, the coronal

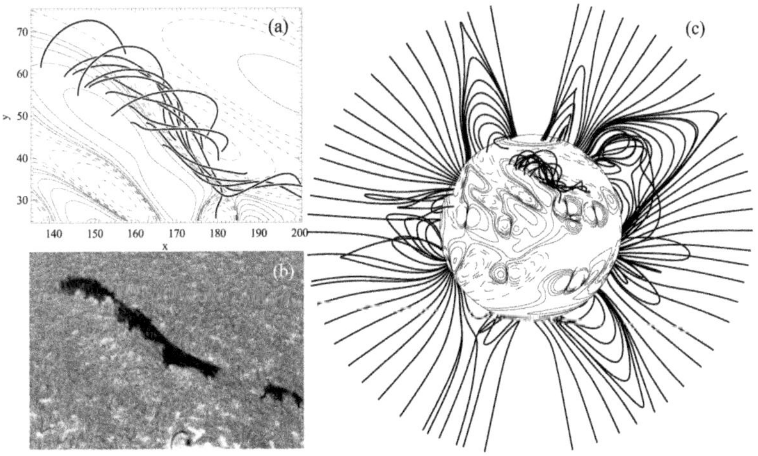


**Figure 6.** (a) A simulated nonlinear force-free field together with (b) an H$\alpha$ image of a filament at the same location and (c) the global context in which the filament lies (Yeates *et al.* 2008).

magnetic field is certainly not potential and is very much a global system with rapid communication between its different parts.

## 5. What is the Cause of Fine Structure (Bubbles, Plumes, Threads and Tornadoes)?

Schmieder *et al.* (1984) first observed quantitatively the dynamic nature of prominences, by mapping the upflows and downflows. Later, Schmieder *et al.* (2010) demonstrated that apparent vertical motions in hedgerow prominences have a substantial component out of the plane of the sky, so they are inclined to the vertical. Here, Tom Berger (this volume) has demonstrated the dynamics in a series of dramatic movies from SDO and Hinode. He suggested that prominence bubbles are probably hot, and one idea is that they are caused by emerging flux (Dudík *et al.* 2012). This is suggestive, but more evidence is needed: the creation of a loop-like lower boundary could be a response to the evolution of fields in the filament channel and not necessarily flux emergence.

For plumes and threads, we are moving towards an explanation in terms of Rayleigh-Taylor instability, and the properties of plumes have been used to infer a plasma beta (Hillier *et al.* 2012). Including partial ionisation (Khomenko *et al.* 2013) has been a crucial step that is greatly welcomed (Fig. 7). Questions include: what is a physical explanation for the upflows and downflows and for the widths of threads? Can the observed properties of threads be quantified and then explained? Is the cause of threads a simple magnetic Rayleigh-Taylor instability or does it include resistive and radiative effects too?

David Orosco Suarez gave a promising observation of the magnetic field of threads, but it was puzzling that the field was uniform in threads. Stano Gunnar reviewed models for fine structure: he presented the possibility of tangled magnetic fields (van Ballegooijen and Cranmer 2010) and also of multi-thread models with a local dip and radiative transfer (Gunár *et al.* 2008).

Tornadoes are puzzling. Maria Martinez Gonzalez (this volume) used observations of Stokes parameters to deduce that a tornado is a rotating double helix with magnetic field 20–60 G. But what is the cause of tornadoes? How do they compare with barbs? Are tornadoes just barbs that are rotating?

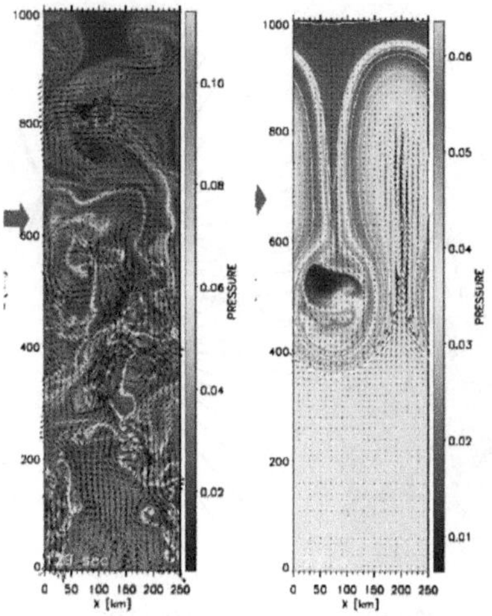


**Figure 7.** A numerical experiment to determine the effect of partial ionisation on magnetic Rayleigh-Taylor instability, showing contours of pressure (Khomenko *et al.* 2013).

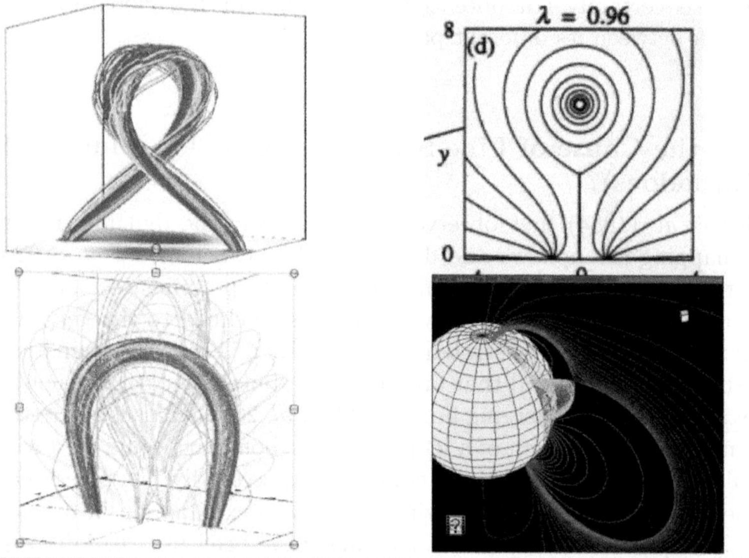


**Figure 8.** Four possible models for the eruption of coronal magnetic fields in a two-ribbon flare.

Masumi Shimojo (this volume) showed how prominence activation varies with the solar cycle. At present, even though the number of sunspots is half of the previous solar maximum, the number of prominence activations is almost as great. Also a butterfly diagram of prominence activity shows a rise in maximum latitude followed by a decline after solar maximum. At present, the activity is normal in the northern hemisphere but anomalous in the southern.

## 6. Why Do Prominences Erupt?

Previously, four mechanisms for prominence eruption had been suggested, namely, kink instability, nonequilibrium or catastrophe, torus instability and breakout (Fig. 8).

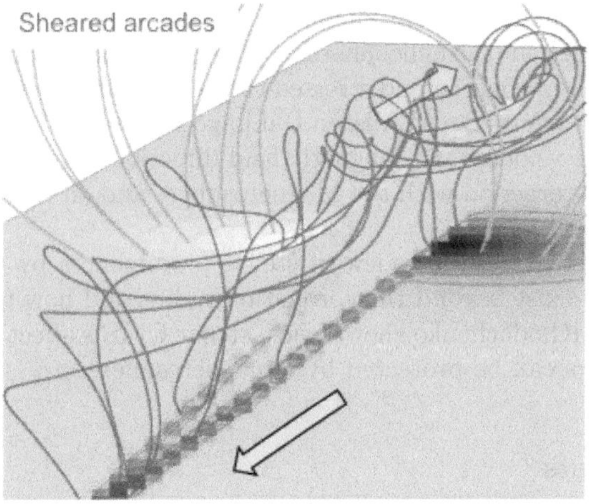


**Figure 9.** A sheared arcade model of a pre-eruptive magnetic configuration (Aulanier *et al.* 2006).

Kazunari Shibata (this volume) described his unified model (Shibata 1999) with plasmoid-induced reconnection and fractal current sheets (Nishizuka and Shibata 2013), triggered by emerging flux (Kusano *et al.* 2012).

Shibata also found 365 super flares ($10^{35}$ erg) on solar-type stars. On the Sun this would be preceded by superspots and superprominences, so we don't need to start worrying until we see a superspot.

Guillaume Aulanier (this volume) discussed the physical processes for eruption and compared curved wire with MHD models, in which there is a balance between an upwards magnetic pressure force and a downwards magnetic tension force. Eruption could occur due to an increase in magnetic pressure or a decrease in tension due to reconnection or evolution. He concluded that eruption is usually due to torus instability or non equilibrium, and only rarely by breakout. Also, evolution to the critical point could occur in different ways.

One point to note is that torus instability is exactly the same as lateral kink instability, and another is that some flares have the reconnection occurring at separators (Longcope *et al.* 2007), some at null points (Masson *et al.* 2009) and some at quasi-separatrix layers (Mandrini *et al.* 1997; Aulanier *et al.* 2006).

## 7. Coronal Mass Ejections

Pascal Démoulin (this volume) described the properties of magnetic clouds, which are probably part of all interplanetary coronal mass ejections (ICME's). They show up as a rotation in the magnetic field and a low proton temperature in a one-dimensional spacecraft track. It is difficult to recognise flux ropes, but he showed how to find the axis and boundaries of a rope, and also how to calculate the density expansion. A flux rope can lose up to 50% of its flux by reconnection.

Noe Lugaz then showed how the whole CME can be imaged by Stereo and how complex the propagation of from Sun to Earth can be. Many properties can be determined, such as CME rotation, expansion, mass increase (by up to 50% by a snowplow effect), and interaction with other CME's. In all this, his numerical simulations are helping our understanding. Then Bob Wimmer-Schweingruber discussed the effects of CME's on space weather, starting with a wide-ranging history of ideas. He stressed that space weather matters throughout the heliosphere, since it affects outer planets and there can be a global reaction throughout the heliosphere to a CME. He showed that the radiation for a manned trip to Mars is of concern and stressed the need for multi-spacecraft observations in future. Also, Benoit Lavraud and Alisson Dal Lago showed how the geo-effectiveness of CMEs depends on the velocity and magnetic field direction, on their flux erosion during propagation, on the interaction with the Magnetosphere and on the accompanying shock waves.

The meeting concluded with a review of stellar prominences by Gaitee Hussain that showed how they can exist beyond the coronation radius and how they are occasionally ejected. Also, Maxim Khodachenko showed how stellar CME's affect planetary formation and how close Jupiters can be protected by a magnetosphere.

## 8. Final Comments

As an aside, some amusing comments during the meeting concerned the Huntsville express (Shi Tsan Wu), coronal *magnetic* eruptions (Tom Berger),"It is too good to be true" and "I am very conservative" (Jean-Claude Vial), "Here is a crazy idea" (Aad Van Ballegooijen, when describing his own work), "I am pretending to be Piet Martens" (Duncan Mackay) "or Judy Karpen" (Manuel Luna), "Twisted fields breed bunnies" (Sarah Gibson),"I like star trek and science fiction" (Guillaume Aulanier and Noe Lugaz), "I like superprominences" (Kazunari Shibata), "I like Hagar the Horrible" (Bob Wimmer Schweingruber) and "We need to cooperate with the Sun" (Pascal Démoulin).

Finally, the person whose kindly presence we have been remembering this week is Einar Tandberg-Hanssen, and the person whom we thank most of all for a fantastic conference is La Reine Brigitte.

## References

Aulanier, G. & Démoulin, P. (1998). 3-D magnetic configurations supporting prominences. I. The natural presence of lateral feet. *Astron. Astrophys.* 329, 1125–1137

Aulanier, G., Pariat, E., Démoulin, P., & DeVore, C. R. (2006). Slip-running reconnection in quasi-separatrix layers. *Solar Phys.* 238, 347–376

Dudík, J., Aulanier, G., Schmieder, B., Zapiór, M., & Heinzel, P. (2012). Magnetic topology of bubbles in quiescent prominences. *Astrophys. J.* 761, 9

Gunár, S., Heinzel, P., Anzer, U., & Schmieder, B. (2008). On Lyman-line asymmetries in quiescent prominences. *Astron. Astrophys.* 490, 307–313

Hillier, A., Hillier, R., & Tripathi, D. (2012). Determination of prominence plasma $\beta$ from the dynamics of rising plumes. *Astrophys. J.* 761, 106

Hillier, A. & van Ballegooijen, A. (2013). On the support of solar prominence material by the dips of a coronal flux tube. *Astrophys. J.* 766, 126

Khomenko, E., Diaz, A., de Vicente, A., Luna, M., & Collados Vera, M. (2013). Rayleigh-Taylor instability in prominences from numerical simulations including partial ionization effects. *IAU Symp.* 300,

Kusano, K., Bamba, Y., Yamamoto, T. T., Iida, Y., Toriumi, S., & Asai, A. (2012). Magnetic field structures triggering solar flares and coronal mass ejections. *Astrophys. J.* 760, 31

Longcope, D. W., Beveridge, C., Qiu, J., Ravindra, B., Barnes, G., & Dasso, S. (2007). Modeling and measuring the flux reconnected and ejected by the two-ribbon flare/CME event on 7 November 2004. *Solar Phys.* 244, 45–73

Mackay, D. H. & van Ballegooijen, A. A. (2005). New results in modeling the hemispheric pattern of solar filaments. *Astrophys. J. Letts.* 621, L77–L80

Mandrini, C. H., Démoulin, P., Bagala, L. G., van Driel-Gesztelyi, L., Henoux, J. C., Schmieder, B., & Rovira, M. G. (1997). Evidence of magnetic reconnection from H$\alpha$, soft X-ray and photospheric magnetic field observations. *Solar Phys.* 174, 229–240

Martin, S. F., Bilimoria, R., & Tracadas, P. (1994). Magnetic field configurations basic to filament channels. In *Solar Surface Magnetism*, R. Rutten and C. Schrijver, eds. (Springer-Verlag, New York), 303

Masson, S., Pariat, E., Aulanier, G., & Schrijver, C. J. (2009). The nature of flare ribbons in coronal null-point topology. *Astrophys. J.* 700, 559–578

Nishizuka, N. & Shibata, K. (2013). Fermi acceleration in plasmoids interacting with fast shocks of reconnection via fractal reconnection. *Phys. Rev. Letts. 110,* 5 (Feb.), 051101

Orozco Suárez, D., Asensio Ramos, A., & Trujillo Bueno, J. (2013). Measuring vector magnetic fields in solar prominences. In *Highlights of Spanish Astrophysics VII.* ), 786–791

Priest, E. R., Hood, A. W., & Anzer, U. (1989). A twisted flux-tube model for solar prominences. I. General properties. *Astrophys. J.* 344, 1010–1025

Priest, E. R., van Ballegooijen, A. A., & Mackay, D. H. (1996). A model for dextral and sinistral prominences. *Astrophys. J.* 460, 530–543

Schmieder, B., Chandra, R., Berlicki, A., & Mein, P. (2010). Velocity vectors of a quiescent prominence observed by Hinode/SOT and the MSDP (Meudon). *Astron. Astrophys.* 514, A68

Schmieder, B., Demoulin, P., Aulanier, G., & Golub, L. (1996). Differential magnetic field shear in an active region. *Astrophys. J.* 467, 881

Schmieder, B., Malherbe, J. M., Mein, P., & Tandberg-Hanssen, E. (1984). Dynamics of solar filaments. III - Analysis of steady flows in H-alpha and C IV lines. *Astron. Astrophys.* 136, 81–88

Shibata, K. (1999). Evidence of magnetic reconnection in solar flares and a unified model of flares. *Astrophys. and Space Sci.* 264, 129–144

Su, Y. & van Ballegooijen, A. (2013). Rotating motions and modeling of the erupting solar polar-crown prominence on 2010 December 6. *Astrophys. J.* 764, 91

van Ballegooijen, A. A. & Cranmer, S. R. (2010). Tangled magnetic fields in solar prominences. *Astrophys. J.* 711, 164–178

van Ballegooijen, A. A. & Martens, P. C. H. (1989). Formation and eruption of solar prominences. *Astrophys. J.* 343, 971–984

van Ballegooijen, A. A., Priest, E. R., & Mackay, D. H. (2000). Mean field model for the formation of filament channels on the Sun. *Astrophys. J.* 539, 983–994

Yeates, A. R., Mackay, D. H., & van Ballegooijen, A. A. (2008). Modelling the global solar corona. II. Coronal evolution and filament chirality comparison. *Solar Phys.* 247, 103–121

# Poster session I

## PROMINENCES

*Nature of Prominences and their role in Space Weather*
*Proceedings IAU Symposium No. 300, 2013*
*B. Schmieder, J.-M. Malherbe & S. T. Wu, eds.*

doi:10.1017/S174392131301123X

# On Critical Heights and Longitudinal Magnetic Field Strength in Prominences

**I. V. Alexeeva and I. S. Kim**

Lomonosov Moscow State University, Sternberg Astronomical institute,
119992 Universitetsky pr-t, 13, Moscow, Russia
email: kim@sai.msu.ru

**Abstract.** The distributions of the measured longitudinal magnetic field strength, $B_{//}$, and maximum height observed, $h$, are presented. 50 Mm, 30–35 G and 50 G respectively are found to be the critical $h$ and $B_{//}$ for the pre-eruption of quiescent prominences.

**Keywords.** Sun: prominences, Sun: magnetic fields, techniques: polarimetric

## 1. Histograms on $B_{//}$ and $h$

The former magnetic measurements by "the Fabry-Perot magnetograph + the 50 cm coronagraph" assembly (Nikolsky *et al.* 1982, Stepanov 1990, Klepikov 1990) with a "magnetic resolution" (the diameter of a pinhole) of $4 - 8''$ (arc sec) and an accuracy of $3 - 5$ G are used. When the "magnetic resolution" is ten times worse than the angular one, it is appropriate to use statistical analysis to find $h$ and $B_{//}$ typical for different classes of prominences. The preliminary analysis was made with measurements in 145 prominences observed in 1975–1985 (Kim 1990). Here, an analysis is presented for 312 prominences observed in 1975–1990. Selection of quiescents (Q) was based on the location far from active regions (AR) and the fine vertical structure, activated quiescents (AQ) represent Q with quickly increasing $h$, active region filaments (ARF) are observed as fine dark filaments in AR at the disk and fine horizontal structures on the limb.

### 1.1. *Distributions on the longitudinal magnetic fields strengths*

Each of prominences underwent measuring in 3–15 "points", each "point" was measured at least 3 times to determine the averaged $B_{//}$. Fig. 1 (left) shows the histograms of $B_{//}$.

– The upper distribution represents the whole set of 312 prominences. The y-axis indicates the quantity both in %, n (left), and actual number, N′(right). An asymmetry in the distribution and the relatively long "tail" indicate a possible multimodality. $B_{//}$ varies from 0 to 65 G with a significant peak at 10–15 G.

– The Q distribution (N = 159) is the unimodal one with an asymmetric maximum centered at 10 G, and seems to correspond the stability of quiescent prominences.

– The AQ distribution (N = 31) has maxima at $10 - 15$ G (an initial stage of activation), $30 - 35$ G, and a broad one centered at 50 G. There is not enough data to estimate the probabilities of the maxima.

– The bimodal ARF distribution (N = 122) has maxima at 10–15 G and about 30 G.

### 1.2. *Distribution on prominence heights*

To our knowledge, Leroy *et al.* (1984) were the first to identify the maximum prominence height observed, $h$, as a "magnetic" parameter. Later, Makarov *et al.* (1992) used the idea that the height of prominences characterizes the strength of the background magnetic

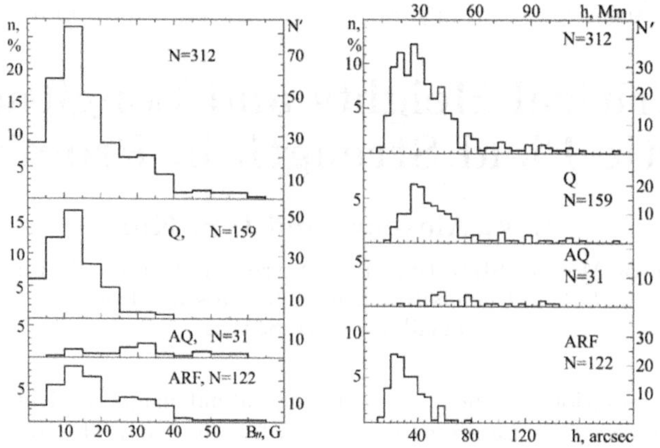


**Figure 1.** Prominence distributions on $B_{//}$ (left) and $h$ (right).

field. We used the filtergrams of the Fabry-Perot magnetograph to obtain the distribution on $h$ with a step of $5''$. The x-axis in Fig. 1 (right) indicate $h$ in arc sec and Mm.

– The upper distribution represents the whole set. It is a multimodal one with 2 maxima centered at $25''$ and $40''$.

– The Q distribution (N = 159) is a unimodal one with a maximum centered at $40''$ and corresponds to the stability of quiescent prominences.

– The distribution for AQ (N = 31) is a broad one with $h = 40 - 135''$, with the averaged $h$ of $70''$ ($\approx 50$ mM). We note the agreement with the critical height calculated by Filippov *et al.* (2006) and based on disk observations of filaments and magnetic fields.

– The ARF distribution (N = 122) is a unimodal one with a maximum at $20 - 30''$.

## 2. Summary

The above mentioned allows us to determine the averaged $B_{//}$ and $h$ typical for different classes of prominences: $B_{//} = 10$ G and $h = 40''$ for the stable Q; $B_{//} = 30 - 35$ or 50 G (depending on the latitude) and $h = 70''$ ($\approx 50$ mM) for activated or pre-erupting Q; $B_{//} = 10 - 15$ G or 30 G (depending on the latitude) and $h = 20 - 30''$ for ARF.

**Acknowledgements.** The reported study was partially supported by RFBR (research project No. 11-02-00631), IAU, SCOSTEP, SF2A and KLSA/CAS.

## References

Filippov, B. P., Zagnetko, A. M., Ajabshirizadeh, A., & Den, O. G. 2006, *Solar System Research*, 40, p. 319

Kim, I. S. 1990, *Lecture Notes in Physics*, 363, p. 49. Springer-Verlag Series France

Klepikov, V. Y. 1990, *Ph.D. thesis*, IZMIRAN, Moscow

Leroy, J.-L., Bommier, V., & Sahal-Breshot, S. 1984, *Astron. Astrophys.*, 131, p. 33

Makarov V. I., Tavastsherna K. S., Davydova E. I., & Sivaraman K. R., *Solar Data*, No 3, p. 90

Nikolsky, G. M., Kim, I. S., & Koutchmy, S. 1982, *Solar Physics*, 81, p. 81

Stepanov, A. I. 1989, *Ph.D. thesis*, IZMIRAN, Moscow

*Nature of Prominences and their role in Space Weather*
*Proceedings IAU Symposium No. 300, 2013*
*B. Schmieder, J.-M. Malherbe & S. T. Wu, eds.*

doi:10.1017/S1743921313011241

# The promise of Bayesian analysis for prominence seismology

**Iñigo Arregui, Andrés Asensio Ramos, and Antonio J. Díaz**

Instituto de Astrofísica de Canarias, E-38205 La Laguna, Tenerife, Spain
Departamento de Astrofísica, Universidad de La Laguna, E-38205 La Laguna, Tenerife, Spain
email: `iarregui@iac.es`

**Abstract.** We propose and use Bayesian techniques for the determination of physical parameters in solar prominence plasmas, combining observational and theoretical properties of waves and oscillations. The Bayesian approach also enables to perform model comparison to assess how plausible alternative physical models/mechanisms are in view of data.

**Keywords.** MHD, Sun: prominences, Sun: oscillations, methods: statistical

---

Prominence seismology aims to determine physical parameters in prominence plasmas by a combination of observed and theoretical properties of waves and oscillations (Joarder *et al.* 1997). The technique has been successful in the determination of a number of parameters using prominence fine structure oscillations (Arregui *et al.* 2012). Yet, solving the inversion problem is not an easy task, because observational information is always incomplete and uncertain. As a consequence, extracting information from model parameters by comparison of their predictions with observed data has to be carried out in a probabilistic framework. The Bayesian formalism is the only fully correct way we have to obtain information about physical parameters from observations (inference) and to compare the performance of alternative models to explain observed data (model comparison) (see e.g., Gregory 2005; von Toussaint 2011).

Two Bayesian data analysis tools are here used:

• the marginal posterior, $p(\theta_i|d) = \int p(\theta|d)d\theta_1 \ldots d\theta_{i-1}d\theta_{i+1} \ldots d\theta_N$, provides us with the most probable values of a given parameter, $\theta_i$, compatible with observed data $d$, in the form of a conditional probability distribution, $p(\theta_i|d)$.

• the marginal likelihood, $p(d|M) = \int p(d,\theta|M)d\theta = \int p(d|\theta,M)p(\theta|M)d\theta$, provides us with the probability of the observed data $d$, given that the model $M$ is true. It tells us how well the observed data are predicted by model $M$, with parameter set $\theta$.

**Determination of field strength and transverse inhomogeneity.** We used observations of period ($P$) and damping time ($\tau_d$) of transverse thread oscillations to obtain information on magnetic field strength ($B_0$) and transverse density inhomogeneity length scale ($l/R$). Figure 1 shows the marginal posteriors for these two parameters for given observed oscillation data. The posteriors provide a well constrained fully consistent solution to the inverse problem, with correct propagation of uncertainty.

**Discrimination of damping mechanisms.** We used our Bayesian model comparison tool to quantify the ability of different mechanisms to explain the damping of transverse thread oscillations in prominences. We considered three proposed mechanisms and computed their evidence in view of data (the damping ratio $\tau_d/P$). The mechanisms were: (1) Alfvénic resonance; (2) slow resonance; and (3) Cowling's diffusion (see Soler 2010). The computation of the marginal likelihood for each model – the probability of the data given that the model is true – enables us to assess how well each one reproduces the observed damping ratio (Figure 2). Alfvénic resonance (model M1) is able to properly

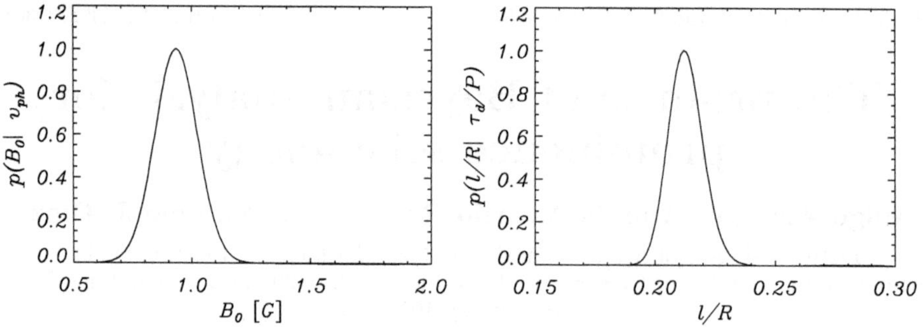


**Figure 1.** Posterior probability distributions for field strength ($B_0$) and transverse density length-scale ($l/R$, in units of the thread radius) for a thread oscillation with $P = 3$ min, $\tau_d = 9$ min, and phase speed $v_{ph} = 16$ km s$^{-1}$. Uncertainty of 10% in data has been considered.

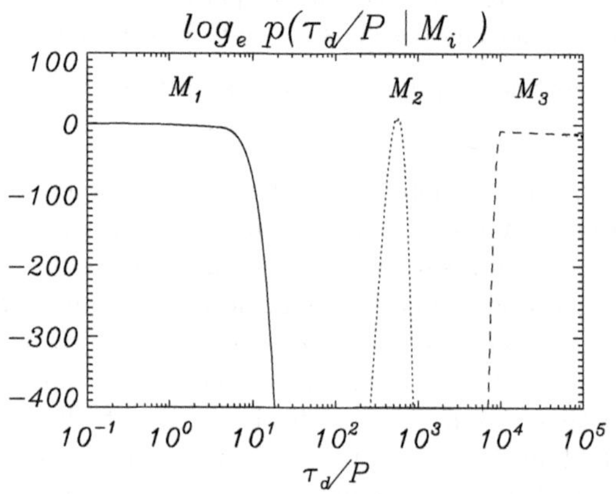


**Figure 2.** Graphical representation of the validity of each considered damping mechanism as a function of the observable damping ratio, in the form of marginal likelihoods in natural logarithm.

explain low damping ratios, such as the observed ones. Cowling's diffusion (M2) has the ability to reproduce damping ratios of the order of $10^3$. Model M3 (slow resonance) can only account for damping rations above $10^4$, even for plasma-$\beta$ values near unity.

These and additional examples (omitted for brevity) show that Bayesian analysis techniques exhibit great promise for prominence seismology.

## Acknowledgements

We acknowledge the Spanish MICINN/MINECO and FEDER funds for the funding provided under projects AYA2010-18029 and AYA2011-22846. IA and AAR acknowledge the Spanish MINECO for a Ramón y Cajal fellowship.

## References

Arregui, I., Oliver, R., & Ballester, J. L. 2012, *Living Reviews in Solar Phys.*, 9, 2

Gregory, P. C. 2005, *Bayesian Logical Data Analysis for the Physical Sciences: A Comparative Approach with 'Mathematica' Support* (Cambridge University Press)

Joarder, P. S., Nakariakov, V. M., & Roberts, B. 1997, *Solar Phys.*, 173, 81

Soler, R. 2010, *PhD thesis*, Universitat de les Illes Balears

von Toussaint, U. 2011, *Reviews of Modern Physics*, 83, 943

*Nature of Prominences and their role in Space Weather*
*Proceedings IAU Symposium No. 300, 2013*
*B. Schmieder, J.-M. Malherbe & Shi Wu, eds.*

doi:10.1017/S1743921313011253

# The spatial relation between EUV cavities and linear polarization signatures

**Urszula Bąk-Stęślicka[1], Sarah E. Gibson[2], Yuhong Fan[2], Christian Bethge[2], Blake Forland[3] and Laurel A. Rachmeler[4]**

[1]Astronomical Institute, University of Wroclaw, ul. Kopernika 11, 51-622 Wrocław, Poland
email: bak@astro.uni.wroc.pl

[2]High Altitude Observatory, NCAR, P.O. Box 3000, Boulder, CO 80307, USA
email: (sgibson,yfan,bethge)@ucar.edu

[3]Metropolitan State College of Denver, P.O. Box 173362, Denver, CO 80217-3362, USA
email: bforland@msudenver.edu

[4]Royal Observatory of Belgium, Avenue Circulaire 3, 1180 Brussels, Belgium
email: rachmeler@oma.be

**Abstract.** Solar coronal cavities are regions of rarefied density and elliptical cross-section. The Coronal Multi-channel Polarimeter (CoMP) obtains daily full-Sun coronal observations in linear polarization, allowing a systematic analysis of the coronal magnetic field in polar-crown prominence cavities. These cavities commonly possess a characteristic "lagomorphic" signature in linear polarization that may be explained by a magnetic flux-rope model. We analyze the spatial relation between the EUV cavity and the CoMP linear polarization signature.

**Keywords.** Sun: corona; Sun: filaments, prominences; Sun: infrared; Sun: magnetic field

## 1. Data Analysis and Results

In our previous paper (Bąk-Stęślicka *et al.* 2013) we have surveyed daily images from the *Solar Dynamics Observatory*/Atmospheric Imaging Assembly (*SDO*/AIA 193 Å, Lemen *et al.* 2012) for polar-crown cavities, and examined CoMP (Tomczyk *et al.* 2008) data to establish cavity signatures in linear polarization. Observations in linear polarization showed characteristic structures we termed lagomorphic, due to its resemblance to a rabbit-head seen in silhouette. We found 68 cases visible over the course of 78 days monitored. Using forward modeling we calculated synthetic CoMP-like data and we showed that lagomorphic structures are consistent with the flux rope model (e.g. Fan 2010).

In this paper we present an initial test of whether size of the cavity and size of the lagomorph are comparable, using a subset of cases including cavities with a good signal to noise in both AIA and CoMP and were large enough so that their center lay above the CoMP occulter. We compared the width of the EUV cavity and corresponding structures in the linear polarization. For this purpose we examined polar-angle cuts in *SDO*/AIA 193 Å and CoMP L/I (degree of linear polarization) (see Figure 1 B-C) at the same height as the height of the center of the cavity seen on AIA 193 Å images. We measured the average signal outside the cavity at the same height and defined a width using an area where the signal decreased more than $3\sigma$ (black solid line on Figure 1 B-C). The relation between the width of the EUV cavity and the width of the lagomorph structures (for 16 cases) is presented on Figure 2. Analysis of the intensity profiles shows that the size of the lagomorph structure scales with the size of the cavity.

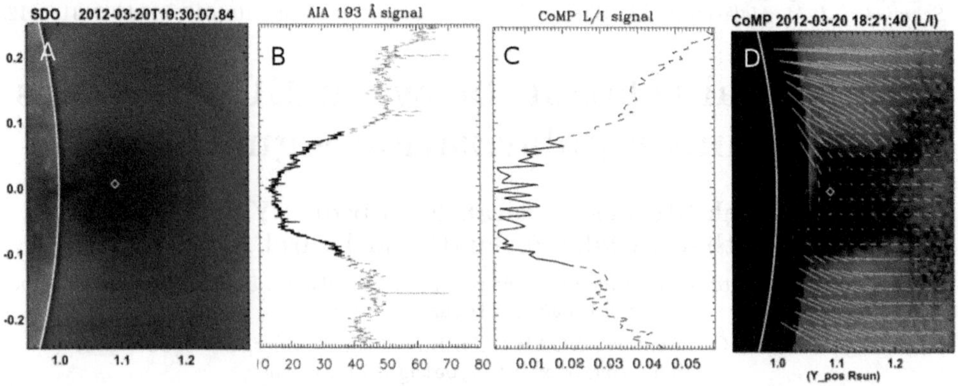


**Figure 1.** A) Example of cavity observed by SDO/AIA 193 Å. B) AIA 193 Å intensity profile across polar-angle cuts at the height 1.09 $R_\odot$. Solid line shows $3\sigma$ depletion. C) L/I profile across polar-angle cuts at the height 1.09 $R_\odot$. Solid line shows $3\sigma$ depletion. D) LOS-integrated L/I for CoMP observations. Direction of Stokes linear polarization vectors (integrated through the LOS) is shown as green lines. The edge of the solar disk is indicated by the curved yellow lines. White diamonds show the center of the cavity seen on AIA images.

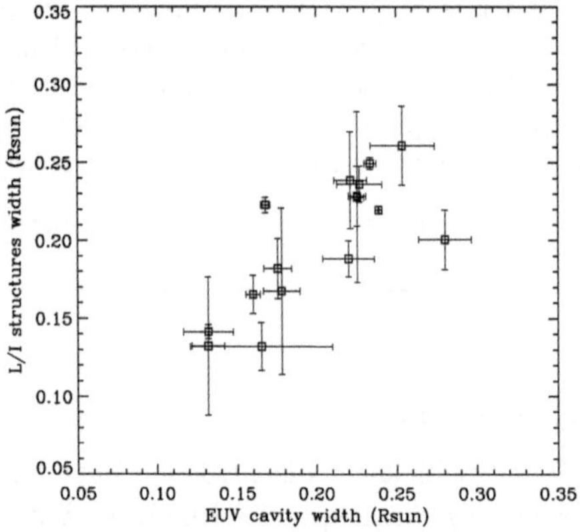


**Figure 2.** Relation between the width of the cavity seen on AIA 193 Å images and the width of the lagomorph structure seen on CoMP L/I images (see text for details).

## Acknowledgements

We thank Mark Miesch for an internal review of this manuscript. This work was supported in part by NASA LWS grant NNX09AJ89G to NCAR. UBS acknowledges financial support from the Polish National Science Centre grant 2011/03/B/ST9/00104. This work was enabled by participation of several of the authors in the International Space Science Institute (ISSI) working group on coronal magnetism. The CoMP data was provided courtesy of the MLSO, operated by the HAO, as part of the NCAR. NCAR is supported by the National Science Foundation.

## References

Bąk-Stęślicka, U., Gibson, S. E., Fan, Y., Bethge, C., Forland, B., & Rachmeler, L. A. 2013, *ApJ* (Letters), 770, 28

Fan Y. 2010, *ApJ*, 719, 728

Lemen, J. R., Title, A. M., Akin, D. J., *et al.* 2012, *Solar Phys.*, 275, 17

Tomczyk, S., Card, G. L., Darnell, T., Elmore, D. F., Lull, R. *et al.* 2008, *Solar Phys.*, 247, 411

*Nature of Prominences and their role in Space Weather*
*Proceedings IAU Symposium No. 300, 2013*
*B. Schmieder, J.-M. Malherbe & S. T. Wu, eds.*

doi:10.1017/S1743921313011265

# 24 synoptic maps 1974-1982 (ascending phase of cycle XXI) of 323 prominence average magnetic fields measured by the Hanle effect

**Véronique Bommier**

LESIA, Observatoire de Paris, CNRS-INSU-UMR8109, UPMC Univ. Paris 06, Université Paris Diderot-Paris 7; 5, Place Jules Janssen, 92190 Meudon, France

**Abstract.** The poster was made of 323 average prominence magnetic fields reported on 24 synoptic maps. The paper first resumes the methods for the field derivation, and the different results of the whole program of these second generation Hanle effect observations. From their conclusions, it was possible to derive a unique field vector for each of the 323 prominences. The maps put in evidence a large scale structure of the prominence magnetic field, probably distorted by the differential rotation, which leads to a systematically small angle (on the order of 30°) between the field vector and the prominence long axis.

**Keywords.** Prominences, Magnetic Field, Hanle effect, Polarization

## 1. Introduction

The observational discovery of the Hanle effect in prominences is due to Bernard Lyot. Lyot (1934, 1936) clearly and repeatedly observed the rotation of the polarization direction, but its first interpretation in terms of the Hanle effect is due to Hyder (1965). Fourty years after Lyot, the observation polarimetric accuracy had increased by one order of magnitude, which has enabled quantitative interpretation of the measurements in terms of magnetic field diagnostics. This required the density matric formalism to be developed in order to generalize the statistical equilibrium equations of the mean atom, to the off-diagonal coherences responsible for the Hanle effect. At the colloquium, it was several times claimed that "only about thirty prominences were completely analyzed" from these observations. On the one hand this is the case with one unique 3D average field vector determined for only 14 prominences (Bommier *et al.* 1994), and 18 prominences with 3D field vector determination but ambiguous by Athay *et al.* (1983), but on the other hand more partial studies, yet leading to important results, were previously led on much larger samples of prominences. In order to help the reader to disentangle the subtly different classes of results, and to follow how the method was developed and was step by step complemented, Sect. 2 is devoted to describe the thread of these Hanle effect studies of the second generation. An unpublished result is given at the end of the section, about the discovery of Inverse Polarity prominences from the analysis of two following days observations, in 1980. The content of the poster, which was made of 24 synoptic maps on which the final magnetic field vector was reported for 323 prominences observed at the Pic-du-Midi during the ascending phase of Cycle XXI, is described in Sect. 3. A unique field vector was obtained for each of these 323 prominences by applying the conclusions of the whole story.

## 2. Thread of the cycle XXI prominence Hanle effect studies

A campaign of prominence Hanle effect observations was runned at the Pic-du-Midi coronagraph from 1974 to 1982 (Leroy *et al.* 1977), along the ascending phase of solar cycle XXI. The polarimeter was equipped with a filter and two photomultipliers (Ratier 1975), without any spectral resolution, unnecessary for the Hanle effect. The aperture was a pinhole of 3 arcsec. A dozen of exposures in different locations were typically taken for each prominence, but curiously the polarization was found rather constant (see Fig. 1 of Leroy *et al.* 1977), so that the polarization was finally averaged over the whole prominence for interpretation to increase the accuracy. The observed line was He I $D_3$, but this was insufficient to fully determine the field vector, because two linear polarization parameters were measured (the linear polarization degree and direction), and three vector coordinates were searched for. Accordingly, two lines at least are necessary for a complete field vector determination. Moreover, these two lines must have different Hanle sensitivity. The Hanle sensitivity of a line is given by $\omega\tau = 1$, where $\omega = 2\pi\nu$ where $\nu$ is the Larmor frequency, and $\tau$ is the line upper level lifetime. Typically, a visible permitted line is sensitive to fields of about 10 G (see a table of typical sensitivities in Sahal-Brechot 1981, Table I). $D_3$ is sensitive to 6 G, also the typical prominence field.

Two-line observations were runned at the same moment with the experiment Stokes II at Sac. Peak. Stokes II was a spectropolarimeter and the two lines were the two components of the He I $D_3$ line. One is made of five unresolved fine-structure components $3d^3D_{3,2,1} \rightarrow 2p^3P_{2,1}$ whereas the other is made of the single $3d^3D_1 \rightarrow 2p^3P_0$. The condition of different sensitivity is fulfilled, and the field vector horizontality (i.e. the field vector lying in the solar horizontal plane) was obtained in 18 (Athay *et al.* 1983) and 2 (Querfeld *et al.* 1985) individual prominences. The field horizontality was also statistically visible in the Pic-du-Midi data by superimposing the observed polarization with the horizontal field diagram computed from the new theory by Bommier & Sahal-Brechot (1978). The superposition is visible in Fig. 5 of Sahal-Brechot *et al.* (1977), and the statistical derivation of the field horizontality from this figure is discussed in Bommier (2009).

Once the field horizontality was proven, it was possible to interpret each Pic-du-Midi observation in terms of field strength and azimuth. More than 400 prominences were observed during the whole campaign, which finally result in 323 interpreted prominences by considering only quiescent objects, and by retaining only those prominences that could be identified with a filament observed eight days before (W limb) or after (E limb). 110 prominences were Polar Crown prominences and their magnetic field was presented and discussed in Leroy *et al.* (1983). In particular, it was obtained that the field strength increases with altitude with a vertical gradient of $+0.5 \times 10^{-4}$ G/km. The 213 remaining prominences of the sample are lower latitude objects, which have various orientations of their filament with respect to the line-of-sight. Advantage was taken of this variety and of the mirror effect of the symmetry with respect to the line-of-sight, to derive a statistical solution of the fundamental ambiguity (Leroy *et al.* 1984) from the idea that the angle between the field vector and the filament long axis should be rather constant through the sample. Indeed, the Hanle effect is submitted to the fundamental ambiguity, which originates in the very small atomic size with respect to the radiation wavelength, so that the atom cannot discriminate between the two propagation directions possible along a given ray path. The result is that two field vectors symmetrical with respect to the line-of-sight cannot be discriminated. The symmetry is rigorous when the prominence is located in the plane of the sky, i.e. when the scattering angle is 90°. For different scattering angles, there are two ambiguous field vectors but they are not strictly symmetrical with

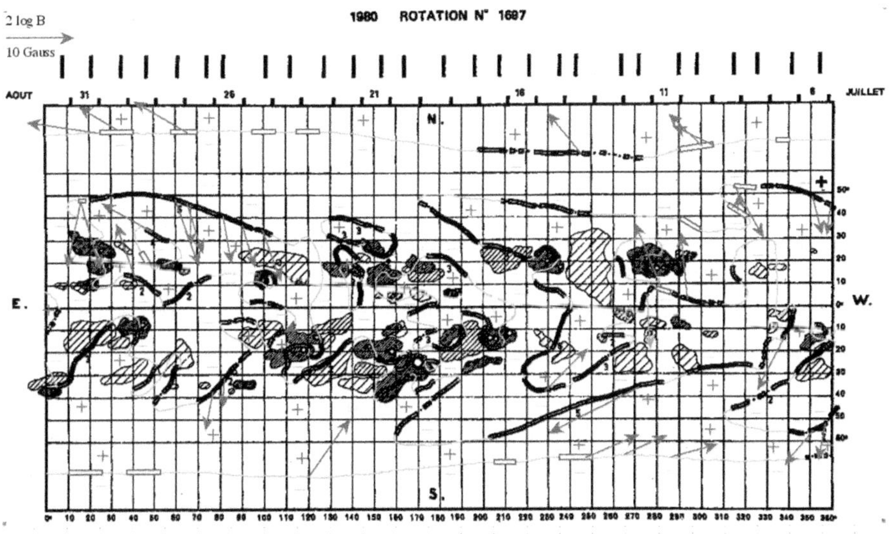


**Figure 1.** Example of the Meudon Observatory synoptic map of Carrington Rotation 1697 (July 1980), with prominence average magnetic field vectors superimposed. Photospheric neutral lines and polarities, and Polar Crown prominences have been added taken from the McIntosh map.

respect to the line-of-sight. The statistical resolution led to 75% of prominences of the Inverse Polarity type (Kuperus-Raadu type), whereas the 25% remaining ones were of the Normal Polarity type (Kippenhahn-Schlüter type).

Two-lines observations were also runned during the last years of the campaign, after the modification of the polarimeter to enable multiline observations. Hydrogen H$\beta$ or H$\alpha$ were observed together with D $_3$. Hydrogen H$\beta$ is optically thin in prominences, and its Hanle sensitivity is different from the $D_3$ one, so that the full vector was determined and the field horizontality was again obtained in 18 prominences (Bommier *et al.* 1986b). Four polarization parameters were measured with the two lines, and the field vector is three coordinates only, so that a parameter remained free, which was devoted to the electron density determination via the collisional depolarization. The electron density was found to range from $10^9$ to $4 \times 10^{10}$ cm$^{-3}$ in these 18 prominences (Bommier *et al.* 1986a). H$\alpha$ is optically thick and a model of its radiative transfer was developed by Landi Degl'Innocenti *et al.* (1987). The effect of the optical thickness is that there is a non-negligible incident radiation along the filament axis, and that the underlying photospheric radiation is differentially absorbed on both sides of the prominence. The scattering geometry is then much more complicate than in the optically thin case, and the result is that two ambiguous field vectors are derived, which do not have the same symmetry as in the optically thin case. Advantage may then be taken of this difference to solve the fundamental ambiguity. This was done in 14 prominences in Bommier *et al.* (1994), leading to 12 Inverse and only 2 Normal Polarity prominences. Besides, the ensemble of results on the prominence magnetic field is summarized in Table I of that paper, with in particular the angle of about 35° found between the field vector and the filament long axis.

A third method was explored for solving the ambiguity, which is the comparison of the results of prominences observed on two following days. This was investigated by Bommier *et al.* (1981) and found very efficient, but the result remained unpublished because too surprising at that moment. A large majority of Inverse Polarity prominences were found in the sample of about twenty analyzed prominences.

## 3. Content of the presented poster

Once taken the results on the field horizontality and the prominence polarity, it is possible to infer a unique field vector for each of the 323 prominences observed at the Pic-du-Midi, and to plot this field vector on the synoptic maps. This was the object of the poster, where 24 synoptic maps were presentend that contained the 323 prominences. One of those is given here in Fig. 1. In the few cases where the two ambiguous field vectors were pointing on the same side of the filament and then were of the same polarity so that the ambiguity could not be solved by the polarity law, the ambiguity was removed by selecting the one of the two vectors that had the same orientation of the field component along the filament, as another prominence observed along the same neutral line. A more detailed presentation of this systematic ambiguity solution is presented in Bommier (2009), with figures.

On the map given in example in Fig. 1, the continuity of the filament long axis field component is visible along the neutral line, and in addition its sign changes when crossing successive neutral lines from north to south, or from east to west. It appears also that the prominence field is in accordance with alignment along a north-south general line distorted by the differential rotation. Such results were visible in each of the 24 maps.

## 4. Conclusion

For page number limit reasons, it was not possible to attach the 24 synoptic maps to this publication. This will be the object of a forthcoming paper. Nevertheless the large scale structure of the prominence magnetic field, which seems to be strongly linked to the one of the neutral line, is already visible in the example map given in Fig. 1 of this paper. The small angle (on the order of 30°) between the field vector and the prominence long axis appears as systematic. It could be the result of a north-south general field line distorted by the differential rotation. The smallness of the angle leads to the necessity of 3D MHD modeling. The third generation of Hanle effect observations will be spatially resolved prominence observations. They have presently begun, but the preliminary investigation by Leroy *et al.* (1977) in their Fig. 1 displays a rather homogeneous field, which seems to contradict the highly structured aspect of the H$\alpha$ observations. This feature presently remains to be explained.

### References

Athay, R. G., Querfeld, C. W., Smartt, *et al.* 1983, *Solar Phys.*, 89, 3
Bommier, V. 2009, in Lecture Notes in Physics, Berlin Springer Verlag, Vol. 765, 231–259
Bommier, V., Landi Degl'Innocenti, E., Leroy, J.-L., & Sahal-Brechot, S. 1994, *Solar Phys.*, 154, 231
Bommier, V., Leroy, J. L., & Sahal-Brechot, S. 1986a, *Astron. Astrophys.*, 156, 90
Bommier, V. & Sahal-Brechot, S. 1978, *Astron. Astrophys.*, 69, 57
Bommier, V., Sahal-Brechot, S., & Leroy, J. L. 1981, *Astron. Astrophys.*, 100, 231
Bommier, V., Sahal-Brechot, S., & Leroy, J. L. 1986b, *Astron. Astrophys.*, 156, 79
Hyder, C. L. 1965, *Astrophys. J.*, 141, 1374
Landi Degl'Innocenti, E., Bommier, V., & Sahal-Brechot, S. 1987, *Astron. Astrophys.*, 186, 335
Leroy, J. L., Bommier, V., & Sahal-Brechot, S. 1983, *Solar Phys.*, 83, 135
Leroy, J. L., Bommier, V., & Sahal-Brechot, S. 1984, *Astron. Astrophys.*, 131, 33
Leroy, J. L., Ratier, G., & Bommier, V. 1977, *Astron. Astrophys.*, 54, 811
Lyot, B. 1934, *Compt. Rend. Acad. Sci.*, 198, 249
Lyot, B. 1934, *Compt. Rend. Acad. Sci.*, 202, 392
Querfeld, C. W., Smartt, R. N., Bommier, V., *et al.* 1985, *Solar Phys.*, 96, 277
Ratier, G. 1975, *Nouvelle Revue d'Optique*, 6, 149
Sahal-Brechot, S. 1981, *Space Sci. Rev.*, 29, 391
Sahal-Brechot, S., Bommier, V., & Leroy, J. L. 1977, *Astron. Astrophys.*, 59, 223

*Nature of Prominences and their role in Space Weather*
*Proceedings IAU Symposium No. 300, 2013*
*B. Schmieder, J.-M. Malherbe & S. T. Wu, eds.*


doi:10.1017/S1743921313011277

# Density evolution of in-falling prominence material from the 7th June 2011 CME

**Jack Carlyle[1,2], David Williams[1], Lidia van Driel-Gesztelyi[1,3,4] and Davina Innes[2]**

[1]UCL-Mullard Space Science Laboratory, UK
email: j.carlyle@ucl.ac.uk

[2]Max Planck Institute for Solar System Research, Germany
[3]Observatoire de Paris, LESIA, UMR 8109 (CNRS), France
[4]Konkoly Observatory, Budapest, Hungary

**Abstract.** This work investigates the density of in-falling prominence material following the $7^{th}$ June 2011 eruption. Both the evolution and the distribution of the density is analysed in five discreet "blobs" of material. The density appears to be remarkably uniform, both spatially within the blobs, and temporally over the course of the descent of each, although a slight concentration of material towards the leading edge is noted in some cases. Online material is available at bit.ly/jackblob

## 1. Introduction

The Coronal Mass Ejection (CME) on the $7^{th}$ June 2011 was one of the largest captured by the Solar Dynamics Observatory Atmospheric Imaging Assembly (SDO/AIA), with a large amount of the ejected matter falling back to the solar surface, suggesting high densities (in order for gravitational forces to overcome the magnetic forces, thought to drive CMEs).

As the ejected prominence material fell, it formed discreet blobs, appearing to undergo the magnetic Rayleigh-Taylor (mRT) instability (Innes *et al.* 2012). By achieving a quantitative assessment of the in-falling material, the physical conditions for such mRT instabilities may be investigated. This investigation assumes photoionisation is the predominant process by which photons are removed from the line of sight, and uses a temporal-interpolative approach to determine the column density of the cool filament material when in absorption against the disc of the Sun (Williams *et al.* 2013).

## 2. Method

Using 94, 131, 171, 193 and 211 Å AIA images of a blob and unocculted co-spatial frames 5 minutes prior, optical depth of the blob is determined from the observed difference in intensity. These measures are then used to constrain the parameters of the model developed by Williams *et al.* (2013) which uses a Levenberg-Marquardt least-squares minimisation algorithm to find the best fit for the two free parameters, column density $N_H$ and geometric filling factor $G$, in each pixel. For a more thorough description of the event and method, please refer to Carlyle *et al.* (2013).

## 3. Results and Analysis

Figure 1 shows maps of column density and $G$ for the first blob analysed. All analysed blobs displayed column densities of approximately $2 \times 10^{19}$ cm$^{-2}$, remaining constant over their descent. The value of $G$ is approximately 0.8 (with the edge of the blob defined

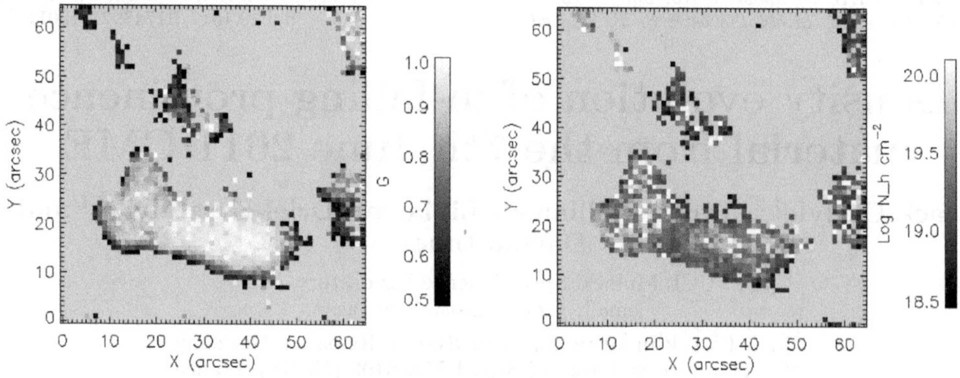


**Figure 1.** Values obtained for G (left) and $N_H$ (right) of one blob as it first appeared against the solar disc at 07:06 UT. The direction of travel is approximately 45° to the negative x-axis.

by $G > 0.5$, as beyond this, the majority of emission in the pixel is unattenuated). The constant value of $N_H$ suggests the blobs are being held together, with the material possibly carrying with it a bundle of magnetic field. Higher densities are observed towards the leading edge just before the blobs split, suggesting that we may be observing the mRT instability. The fact that $G$ decreases is likely due to a greater proportion of emission being in the foreground as the blobs fall through the solar atmosphere.

## 4. Discussion and Conclusion

This work has found column density values for the discreet blobs of material which fell back to the Sun after the June $7^{th}$ 2011 CME and shown that these are relatively high values. Gilbert *et al.* (2005) calculated the column density of a prominence to be $1.6 \pm 1 \times 10^{19}$ cm$^{-2}$ *before* erupting – a value comparable with those calculated for material in this study, which was seen to expand greatly in the initial eruption.

The distribution of the density is consistent with the behaviour of the mRT instability, and the large separation width between successively divided blobs (relative to their length) implies suppression of the higher-$k$ modes of the instability, consistent with the presence of magnetic fields.

### Acknowledgements

JC thanks UCL and MPI for Impact PhD Studentship. The research leading to these results has received funding from EC FP7 under grant agreement No. 284461 (eHEROES). LvDG's work was supported by Hungarian Research grant OTKA K-081421.

### References

Carlyle, J., Williams, D., van Driel-Gesztelyi, L., Innes, D., Hillier, A., Matthews, S., 2013, *ApJ*, submitted

Gilbert, H., Holzer, T. E., & MacQueen, R. M., 2005, *ApJ*, 618, 524

Innes, D. E., Cameron, R. H., Fletcher, L., Inhester, B., & Solanki, S. K., 2012, *A&A* 540, L10

Williams, D. R., Baker, D., & van Driel-Gesztelyi, L., 2013, *ApJ*, 764, 165

*Nature of Prominences and their role in Space Weather*
*Proceedings IAU Symposium No. 300, 2013*
*B. Schmieder, J.-M. Malherbe & S. T. Wu, eds.*

doi:10.1017/S1743921313011289

# Observations of Overlying Extreme-ultraviolet Arches confining the eruption of a Filament

**Huadong Chen[1,2], Suli Ma[1], and Jun Zhang[2]**

[1]College of Science, China University of Petroleum, Qingdao 266580, China
email: hdchen@upc.edu.cn

[2]Key Laboratory of Solar Activity, National Astronomical Observatories, Chinese Academy of Sciences, Beijing 100012, China

**Abstract.** Using the multi-wavelength data from AIA/SDO, we report a failed filament eruption, which was associated with an X1.9 flare, but without any distinct CME, coronal dimming or EUV wave. Some magnetic arches above the filament was observed distinctly in EUV channels, especially in 94 Å and 131 Å, before and during the filament eruption. Our results show that the overlying arcades expanded along with the ascent of the filament at first until they reached a projected height of about 49 Mm above the Suns surface, where they stopped. The following filament material was observed to be confined by the stopped EUV arches and not to escape from the Sun. These results support that the overlying arcades play an important role in preventing the filament to erupt outward successfully.

**Keywords.** Sun: activity — Sun: filaments, prominences — Sun: flares — Sun: UV radiation

---

On 2011 November 3, a filament erupted in NOAA AR 11339. According to GOES-15 observations, the associated flare is an X1.9 class flare. However, no coronal dimming, EUV wave or CME was related to the event according to the observations of SECCHI/STEREOB. These results indicate that it is a failed filament eruption.

The AIA 171 Å images (the top panels of Fig. 1) show the main eruption process of the filament "A" (see Fig. 1(a)). During the eruption, A broke into two major segments – "$Ab_1$" and "$Ab_2$" (see Fig. 1(c) and (d)). At about 20:23 UT, $Ab_1$ reached its maximum projected height of about 49 Mm (see Fig. 2) and then moved laterally. As for $Ab_2$, it only reached a maximum projected height of about 35 Mm (also see Fig. 2) and then returned. The bottom panels of Fig. 1 show the failed eruption at 131 Å. Obviously, there existed some bright arch structures ("Ar") overlying the filament before the flare

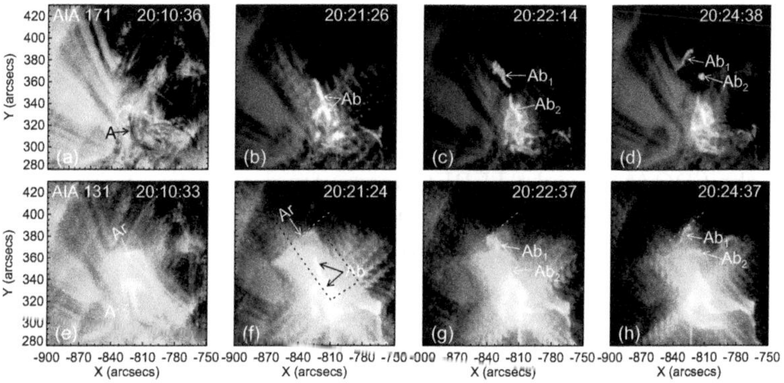


**Figure 1.** AIA 171Å and 131 Å images showing the filament eruption.

(see Fig. 1(e)). According to the observations in the other AIA passbands, it was found that these arches were only visible as emission in the 131 Å and 94 Å channels. As the flare commenced, Ar started to expand upwardly. After a rapid expanding for about 2 minutes, at 20:21 UT, the apex of Ar quickly stopped at a projected height of ~49 Mm (indicated by the dotted lines in Fig. 1(f)-(g)). When the following filament $Ab_1$ reached the same altitude about one minute later, it did not keep rising but moved laterally under and along the apex of Ar, as shown in Fig. 1(g) and (h).

The detailed kinetics of Ar, $Ab_1$, and $Ab_2$ are shown in Fig. 2. Panels (a) and (b) are the time-slit maps along the wide slit (indicated by the dotted box in Fig. 1(f)) from the AIA 131 Å and 171 Å images, respectively. From these time-distance diagrams, we can clearly see that Ar was invisible at 171 Å, but present in the 131 Å wavebands. The time variations of the mean projected heights and the derived velocities of Ar (plus, dashed), $Ab_1$ (asterisk, dash dot), and $Ab_2$ (diamond, dash dot dot) are plotted in Fig. 2(c) and (d), respectively. During the rapid expanding phase, Ar was accelerated with an increased velocity from tens of km $s^{-1}$ to more than 200 km $s^{-1}$. Then, the apex of Ar quickly stopped at a projected height of about 49 Mm. $Ab_1$ firstly erupted from the lower corona with a mean upward velocity of about 320 km $s^{-1}$. When it approached the apex of Ar, it was prevented to rise up by the overlying Ar, with a mean deceleration of about 2.7 km $s^{-2}$. This value is near to ten times that of the solar gravitational deceleration and is comparable to that reported by Ji *et al.* (2003). As for $Ab_2$, its velocity rapidly declined from more than 350 km $s^{-1}$ to zero during a period of about 2 minutes, which derives a mean deceleration of about 2.9 km $s^{-2}$ similar to that of $Ab_1$.

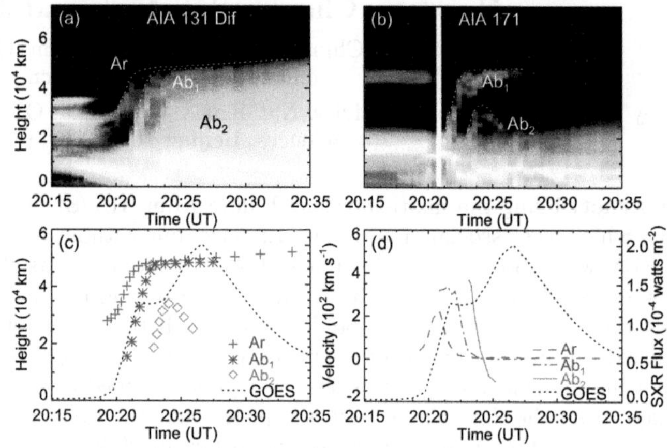


**Figure 2.** Time-slit maps from AIA 131 Å (a) and 171 Å (b) images; Time variations of the mean projected heights (c) and the derived velocities (d) of Ar, $Ab_1$, and $Ab_2$.

According to these results, it is very likely that the closed overlying EUV arches played a crucial role in preventing the successful eruption of the filament (e.g., Liu *et al.* 2009, Schmieder *et al.* 2013).

## Acknowledgements

We thank the SDO/AIA and STEREO teams for data support. The work was supported by the NSFC (11103090, 41204124, 40825014 and 41331068), SPNSF (ZR2011AQ009) and CAS/NAO (KLSA201209).

## References

Ji, H., Wang, H., Schmahl, E. J., Moon, Y.-J., & Jiang, Y. 2003, *ApJ*, 595, L135
Liu, Y., Su, J., Xu, Z., Lin, H., Shibata, K., & Kurokawa, H. 2009, *ApJ*, 696, L70
Schmieder, B., Démoulin, P., & Aulanier, G. 2013, *Adv. Sp. Res.*, 51, 1967

*Nature of Prominences and their role in Space Weather*
*Proceedings IAU Symposium No. 300, 2013*
*B. Schmieder, J.-M. Malherbe & S. T. Wu, eds.*

doi:10.1017/S1743921313011290

# Estimation of Plasma Properties and Magnetic Field in a Prominence-like Structure as Observed by SDO/AIA

**B. N. Dwivedi[1], A. K. Srivastava[2] and Anita Mohan[1]**

[1]Department of Physics, IIT (BHU), Varanasi, India
email: bnd.app@iitbhu.ac.in

[2]Aryabhatta Research Institute of Observational Sciences (ARIES), Nainital, India

**Abstract.** We analyze a prominence-like cool plasma structure as observed by Atmospheric Imaging Assembly (AIA) onboard the Solar Dynamics Observatory (SDO). We perform the Differential Emission Measure (DEM) analysis using various filters of AIA, and also deduce the temperature and density structure in and around the observed flux-tube. In addition to deducing plasma parameters, we also find an evidence of multiple harmonics of fast magnetoacoustic kink waves in the observed prominence-like magnetic structure. Making use of estimated plasma parameters and observed wave parameters, under the baseline of MHD seismology, we deduce magnetic field in the flux-tube. The wave period ratio P1/P2 = 2.18 is also observed in the flux-tube, which carries the signature of magnetic field divergence where we estimate the tube expansion factor as 1.27. We discuss constraints in the estimation of plasma and magnetic field properties in such a structure in the current observational perspective, which may shed new light on the localized plasma dynamics and heating scenario in the solar atmosphere.

**Keywords.** Solar Prominence, MHD Waves, Magnetic Fields

## 1. Introduction

In this brief paper, we report the differential emission measure (DEM) analysis using various AIA filters and deduce temperature structure of the emitting region using the method of Aschwanden *et al.* (2013). We also estimate local magnetic field in the tube from MHD seismology. We have carried out emission measure and temperature distribution analysis as well as tracing and estimating the density along the prominence by extensively using the automated method. SDO/AIA data provide information about the emission measure and estimation of the average density and temperature. We use the full-disk SDO/AIA in all EUV channels around 13:21 UT on 7 March 2011. We calibrate and clean the data using the aia_prep subroutine of SSWIDL and co-align the AIA images as observed in its various AIA filters using the co-alignment test. We obtain emission measure and temperature maps for six AIA filter full-disk and co-aligned images (304 Å, 171Å, 193 Å, 94 Å, 335 Å, and 211 Å) in the temperature between 0.4 MK and 9.0 MK. The partial FOV analysis in the form of the emission measure and temperature measurements has been performed. The prominence structure is not fully resolved in multi-temperature emissions as it only contributes to the information of 304 Å He II emission. However, the cool plasma streaks are evident in the prominence sub-FOV maintained at temperature below 1.0 MK.

## 2. Results

The main results obtained are briefly described below:
[1] The automated emission analyses, using Aschwanden *et al.* (2013) method, clearly

indicate the presence of cool plasma streaks in the prominence structure. The electron density decreases in the highly tangled prominence structure up to its apex which lies at a distance of $\approx$ 85 Mm from the left-footpoint of the flux-tube. It shows some increment towards right foot-point side along the tube. The density near the apex of the prominence is Ne = $1.0\times10^9$ cm$^{-3}$. The density near the left foot-point lies in the range of Ne = $3.16\times10^9-1.0\times10^{10}$ cm$^{-3}$.

[2] The temperature estimations with best least chi square show a uniform electron temperature (Te$\sim5.0\times10^5$ K) below 1.0 MK along the prominence-like cool loop system. However, the calculated average temperature is 1.0 MK, because a few temperature points are estimated at > 1.0 MK. The average temperature of about 1.0 MK is two orders of magnitude large than the cool prominence plasma temperatures (below $10^4$ K) and most likely represent a very hot part of the prominence-corona transition region.

[3] Srivastava *et al.* (2013) observed the multiple harmonics of the fast magnetoacoustic kink waves in this prominence-like cool loop system.

[4] Srivastava *et al.* (2013) have found $\approx$ 667 s and $\approx$ 305 s periods respectively near the apex and foot-point of the prominence-like cool loop system. The phase speed of fundamental mode period (within tube length of $\sim$ 170 Mm) lies in the fast regime (510 km s$^{-1}$) of the tube MHD waves, which can either be magnetoacoustic sausage or kink waves. However, under the given morphology and plasma conditions (typical coronal density inside the prominence), sausage modes are unlikely because of longer wavelength cut-off issue. Therefore, the most likely fast tubular mode is the non-linear magneto-acoustic kink waves that can weakly modulate the density near the boundary of the flux-tube (and thus intensity) during the modulation of plasma column depth in the obliquely oriented flux-tube in accordance with Cooper *et al.* (2003) theory.

[5] The period ratio shift of the fundamental mode to the first harmonics is > 2.0, which indicates the flux-tube expansion and the magnetic field stratification in the loop. This is dominant over the longitudinal density structuring. For the first time, the expansion factor of the loop was estimated to be 1.27. It was also suggested as the first clues of the shift of fundamental mode period by a factor of 0.85 in the duration of 600 s. This provides an observational clues to further develop the appropriate theory (Morton and Erdlyi, 2009 , Ruderman 2011).

[6] The average density near the apex of the observed filament is Ne = 1.0 $\times10^9$ cm$^{-3}$. Therefore, we estimate the magnetic field of the prominence near its apex as 9.0 Gauss based on the theory of MHD seismology and observational signature of the fundamental kink waves (Nakariakov and Ofman, 2001). Therefore, we estimate the density (Ne = 1.0 $\times10^9$ cm$^{-3}$), temperature ($\sim$ 1.0 MK), and magnetic field (9.0 Gauss) values in the prominence-like flux-tube observed by SDO/AIA.

## 3. Acknowledgements.

We thank the referee for his/her suggestions. BNDs participation in the IAUS300 is enabled by the financial support from the ESA. We acknowledge SDO/AIA-NASA for data, and E. OShea for the randomlet software, and Torrence and Compo for the wavelet software. We also acknowledge the use of the procedures for automated DEM and Te - analyses developed by M.J. Aschwanden. AKS acknowledges the support from DST-RFBR and Indo-Austrian Projects (INT/RFBR/P-117;INT/AUA/BMWF/P-18/2013), and also thanks Shobhna Srivastava for her patient encouragement during research.

### References

Aschwanden, M. J., Boerner, P., Schrijver, C. J., & Malanushenko, A. 2013, *ApJ*, 283, 5

Cooper, F. C., Nakariakov, V. M., & Tsiklauri, D. 2003, *A&A*, 397, 765
Morton, R. J. & Erdélyi, R. 2009, *ApJ*, 707, 750
Nakariakov, V. M. & Ofman, L. 2001, *A&A*, 372, 53
Ruderman, M. S. 2011, *Sol. Phys.*, 271, 41
Srivastava, A. K., Dwivedi, B. N., & Kumar, Mukul 2013, *Ap&SS*, 345, 25

*Nature of Prominences and their role in Space Weather*
*Proceedings IAU Symposium No. 300, 2013*
*B. Schmieder, J.-M. Malherbe & S. T. Wu, eds.*

doi:10.1017/S1743921313011307

# Kappa-distributions and Temperature Structure of the Prominence-Corona Transition Region

**Elena Dzifčáková[1], Šimon Mackovjak[2,1], & Petr Heinzel[1]**

[1]Astronomical Institute of the Academy of Sciences of the Czech Republic,
Fričova 298, 251 65 Ondřejov, Czech Republic,
email: `elena.dzifcakova@asu.cas.cz, petr.heinzel@asu.cas.cz`

[2]Faculty of Mathematics Physics and Informatics, Comenius University,
Mlynska Dolina F2, 842 48 Bratislava, Slovakia, email: `mackovjak@fmph.uniba.sk`

**Abstract.** The influence of the electron $\kappa$ - distributions on the differential emission measure (DEM) of the prominence-corona transition region (PCTR) derived from observed line intensities has been investigated. An important consequence of the $\kappa$ - distribution is formation of the emission lines in much wider temperature ranges. The implications for the formation temperature of the observed SDO/AIA band emissions are shown.

**Keywords.** DEM, PCTR, SUMER spectral lines, SDO/AIA

---

A strong gradient of temperature and density in the PCTR can form non-Maxwellian distributions with an enhanced number of particles with high energies - the $\kappa$-distribution. Therefore we have studied the influence of the $\kappa$-distributions on calculated DEM. The observed line intensities from SOHO/SUMER spectrometer listed in Parenti & Vial (2007) as A_1 were used in our analysis. The ionization equilibria for $\kappa$-distributions were taken from Dzifčáková & Dudík (2013) and the excitation equilibrium was calculated for atomic data corresponding to the CHIANTI 6 (Dere *et al.*, 2009). The Withbroe-Sylwester method (Withbroe, 1975; Sylwester *et al.*, 1980) was employed for the calculation of DEM. The differences between our reconstructed DEM for the Maxwellian distribution with original DEM by Parenti & Vial (2007) (Fig. 1, top left) should be a result of different calculation methods only. The DEM's calculated for the $\kappa$-distributions are wider and flatter in comparison with DEM for the Maxwellian distribution (Fig. 1, top right; bottom left). This is mainly the result of changes in the ionization equilibrium for the $\kappa$-distributions, where the ionization peaks are wider and shifted in comparison with the Maxwellian distribution (Dzifčáková & Dudík, 2013). For the $\kappa$-distributions, the spectral lines are formed in wider temperature ranges and the maxima of contribution to the line intensities can be substantially shifted to lower $T$, especially in transition region (Fig. 1, bottom right). These changes could affect the temperature region that is visible in SDO/AIA filters. For the AIA 171 and 193 filters, and Fe X - XII lines (Fig. 2), the maxima of contributions to the line intensity have similar positions for both Maxwellian and $\kappa$-distributions. However, the Fe IX and Fe VIII lines show enhanced low-temperature contributions and the contributions from O V and O VI lines, formed at even lower temperatures, are significantly widened. We can conclude that the $\kappa$-distributions allow to see wider temerature range in AIA filters than the Maxwellian one.

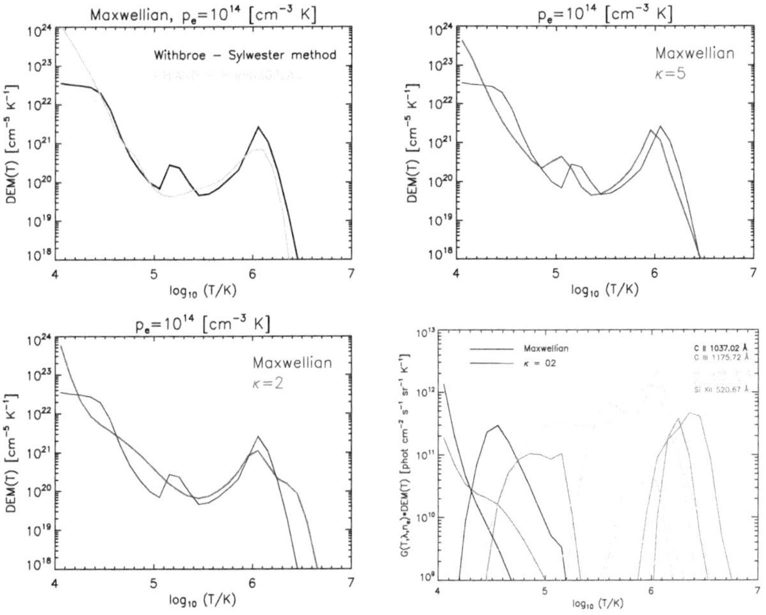


**Figure 1.** *Top left:* The PCTR DEM for the Maxwellian distribution by Parenti & Vial (2007) (green line) and by Withbroe-Sylwester method (black line). *Top right:* DEM for the $\kappa$-distribution with $\kappa = 5$ (blue line). *Bottom left:* DEM for the $\kappa$-distribution with $\kappa = 2$ (red line). *Bottom right:* $G(T,\lambda,n_e) * \mathrm{DEM}$ for the Maxwellian (full lines) and for the $\kappa$-distributions with $\kappa = 2$ (dot-dot-dot dashed lines) for six lines formed in different temperature ranges.

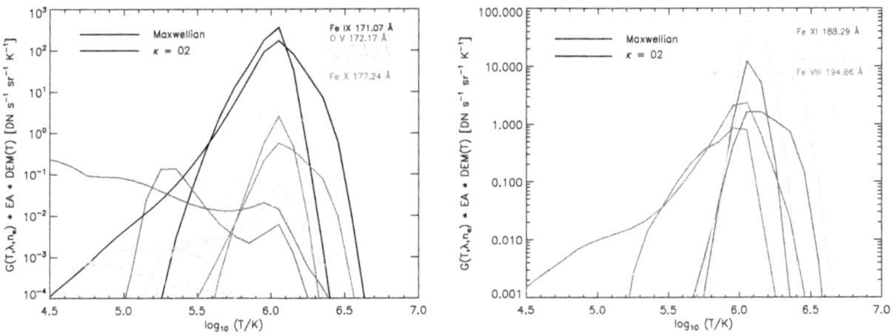


**Figure 2.** $G(T,\lambda,n_e) * \mathrm{DEM}$ multiplied by the effective area (EA) of 171 (*left*) and 193 band (*right*) for the Maxwellian (full lines) and $\kappa$ - distribution with $\kappa = 2$ (dot-dot-dot dashed lines).

## Acknowledgements

This work has been supported by Grant No. 209/12/1652 of the Grant Agency of the Czech Republic, grant No. 1/0240/11 of Scientific Grant Agency VEGA Slovakia, and the bilateral project APVV CZ-SK-0153-11 (7AMB12SK154) involving the Slovak Research and Development Agency and the Ministry of Education of the Czech Republic.

## References

Dzifčáková, E. & Dudík J., 2013, *ApJS*, 206, 6

Dere, K. P., Landi, E., Young, P. R., Del Zanna, G., Landini, M., & Mason, H. E., 2009, *A&A*, 498, 915

Parenti, S. & Vial J. C., 2007, *A&A*, 469, 1109

Sylwester, J., Schrijver, J., & Mewe, R., 1980, *SoPh*, 67, 285

Withbroe, G. L., 1975, *SoPh*, 45, 301

*Nature of Prominences and their role in Space Weather*
*Proceedings IAU Symposium No. 300, 2013*
*B. Schmieder, J.-M. Malherbe & S. T. Wu, eds.*

doi:10.1017/S1743921313011319

# Modeling Prominence Formation in 2.5D

**X. Fang, C. Xia and R. Keppens**

Centre for mathematical Plasma Astrophysics, Department of Mathematics, KU Leuven

**Abstract.** We use a 2.5-dimensional, fully thermodynamically and magnetohydrodynamically compatible model to imitate the formation process of normal polarity prominences on top of initially linear force-free arcades above photospheric neutral lines. In magnetic arcades hosting chromospheric, transition region, and coronal plasma, we perform a series of numerical simulations to do a parameter survey for multi-dimensional evaporation-condensation prominence models. The investigated parameters include the fixed angle of the magnetic arcade, the strength and spatial range of the localized chromospheric heating.

**Keywords.** magnetohydrodynamics(MHD) — Sun: corona — Sun: filaments, prominences

---

Prominences, a common feature in active and quiet solar regions, represent huge structures of cold ($\approx 10^4$ K) and dense ($10^{10} - 10^{11}\mathrm{cm}^{-3}$) plasma in the solar atmosphere (Tandberg-Hanssen (1995)). They are hosted by strong and complex dip-shaped magnetic field configurations, usually above the magnetic polarity inversion lines. However, the magnetic field topology of prominences is still poorly understood, although observations indicate it is mainly horizontal, with an acute angle with respect to the main axis of prominences (Bommier & Leroy (1998)). Prominences attracted plenty of theoretical studies to address different aspects of them, such as formation and eruption. Especially considering the formation of prominences, recently Xia *et al.* (2012) realized a 2.5D simulation of in situ formation of a filament in a sheared magnetic arcade, with chromospheric evaporation plus coronal condensation, using the MPI-parallelized Adaptive Mesh Refinement (AMR) Versatile Advection Code (Keppens *et al.* (2012)).

Here we present a parameter study based on Xia *et al.* (2012). We follow the setup in Xia *et al.* (2012) as a 2.5D thermodynamic MHD model on a 2D domain of size 40 by 50 Mm (in $x - y$), but now adopt a linear force-free magnetic field characterized by a constant angle $\theta_0$ (as in Fang *et al.* (2013)) as the initial magnetic field topology. The background heating rate decays exponentially with height, which helps to obtain a self-consistent thermally structured corona at first, and a relatively strong additional heating near the chromosphere injects energy and evaporates the plasma. We choose different angles ($\theta_0 = 30°, 45°$) for the initial linear force-free magnetic field topology. In the simulations, we regulate the energy input from additional chromospheric heating to reach the same value among the models with different angles. We also study models with different magnetic field strength, different energy input heating scale and different spatial ranges of the additional chromospheric heating.

## Results and discussion

In the left panel of Fig.1, simulations of two representative models with different angles for the arcade magnetic field are compared, showing the evolution of the prominence mass. They indicate that after the appearance of cool plasma, an approximate linear relationship with time is found and the growth rates of the condensations in these models are similar. We infer this is because of the regulated same energy input from the additional chromospheric heating. By analyzing the growth rates of accumulated prominence mass

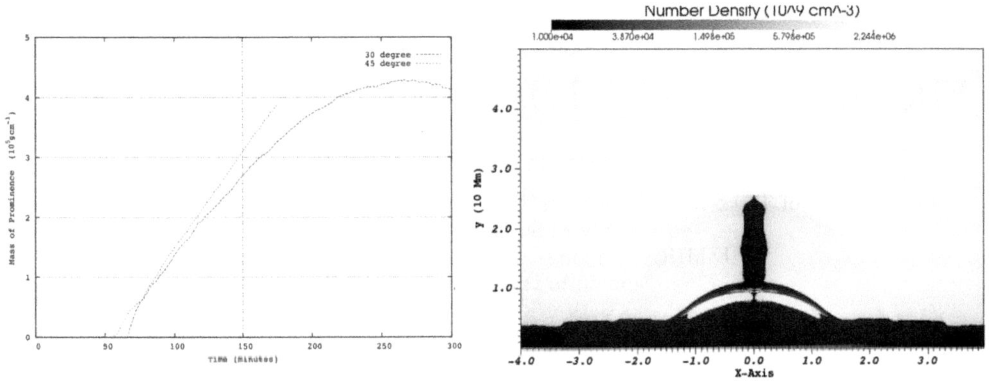


**Figure 1.** Left Panel: Temporal evolution of the prominence mass of models with two different angles of magnetic field. The vertical dashed line at t=150 min shows the moment when prominences begin to drag extra mass.; Right Panel: The density snapshot at t = 240 min.

in models with different parameters (a total of 10 models were intercompared, not shown in Fig. 1), we infer that these growth rates are basically determined by the energy input from additional chromospheric heating, although the time of formation and the heights of the first condensations can differ. Furthermore, by adopting different heating lengthscales and spatial ranges of the additional chromosphere heating with the same angle ($\theta_0 = 30°$) of magnetic field topology, simulations still demonstrate that the growth of condensations display nearly linear relationship with time and positively correlate with total energy inputs from the additional heating. In models hosting different magnetic field strength, we did not find any obvious relationship between the growth rates of the condensations and the magnetic field strength.

Some of our modeled prominences develop additional internal structure, with the side boundaries of the prominence resembling sawteeth, when the magnetic field of the arcade is strong. Indeed, when the lateral growing prominence can not bend the arched loops fast enough, segments of the prominence body residing in self-created magnetic dips fall down to the chromosphere along the arched loops. This drags extra mass from inside the magnetic dips to stream down until all prominence mass in the affected loops drains to the chromosphere. Consecutively, the evacuated loops reform condensations, and this phenomenon propagates from lower to higher loops. This realizes a down-streaming channel adjacent to an up-streaming channel, reforming the prominence as it rises, and we suggest these long-lived streams connecting the prominence and the chromosphere resemble the barbs of prominences (Fig. 1, right panel). They also shed light on the mass recycling puzzle of prominences in general.

## References

Bommier, V. & Leroy, J. L. 1998, IAU Colloq. 167: *NPSP*, 150, 434
Fang, X., Xia, C., & Keppens, R. 2013, *Ap. Lett.*, 771, L29
Keppens, R., *et al.* 2012, *JCP*, 231, 718
Xia, C., Chen, P. F., & Keppens, R. 2012, *Ap. Lett.*, 748, L26
Tandberg-Hanssen, E. 1995, *Science*, 269, 111

*Nature of Prominences and their role in Space Weather*
*Proceedings IAU Symposium No. 300, 2013*
*B. Schmieder, J.-M. Malherbe & S. T. Wu, eds.*

doi:10.1017/S1743921313011320

# Filament Connectivity and "Reconnection"

**Boris Filippov**

Pushkov Institute of Terrestrial Magnetism, Ionosphere and Radio Wave Propagation, Russian Academy of Sciences (IZMIRAN)
IZMIRAN, Troitsk, Moscow, 142190, Russia
email: bfilip@izmiran.ru

**Abstract.** Stable long lived solar filaments during their lives can approach each other, merge, and form circular structures. Since filaments follow large scale polarity inversion lines of the photospheric magnetic field, their evolution reflects changes of the photospheric field distribution. On the other hand, filament interaction depends on their internal magnetic structure reviled in particular by filament chirality. Possibility of magnetic field line reconnection of neighbor filaments is discussed. Many examples of connectivity changes in a course of photospheric field evolution were found in our analysis of daily H$\alpha$ filtergrams for the period of maximum activity of the solar cycle 23.

**Keywords.** Sun: chromosphere, Sun: filaments, Sun: prominences, Sun: magnetic fields

---

The axial component of the filament magnetic field defines two classes of filaments, depending on the direction of the axial component: a filament is called dextral if this component is directed toward the right the filament is viewed from the side of the positive background polarity, and sinistral if the direction of the axial component is opposite to this (Martin *et al.* 1994). Analysis of the fine structure of filaments shows that the thin threads are rotated through a small angle clockwise to the axis in dextral and counterclockwise in sinistral filaments. This makes it possible to determine the class of a filament (the filament chirality) from its visual appearance, without information on the magnetic fields (Pevtsov *et al.* 2003).

When the ends of filaments with the same chirality approach each other, the "positive end" of one rope (a source of field lines) approaches the "negative end" of the other rope (a sink of field lines). The filament ends located on opposite sides of the neutral line have the same directions for both the axial and transverse fields. However, the vertical components can have antiparallel segments, where reconnection can occur. This results in the formation of loops overlapping the rope), which connect the regions of earlier rooting of each rope, so that the ropes themselves become "sewn" into a single object. The electric currents close at the photosphere at the filament ends before reconnection, and so form a single circuit from the right end of the right filament to the left end of the left filament after reconnection.

At a contact of filaments with opposite chiralities, their ends contact each other on one side of the polarity inversion line. Because the directions of the azimuthal components are the same and the directions of the axial components are opposite, a typical cusp is formed.

The approach of two antiparallel dipoles results in a quadrupolar magnetic configuration. At the initial time, the two neutral lines are initially separated by some polarity. These lines can sometimes contact (intersect) each other, but, strictly speaking, such a state is unstable, and the lines will diverge again in the case of arbitrary small changes in the fields. The connectivity of the neutral lines can change, and new lines will be separated by another polarity. If filaments of the same chirality exist at both polarity

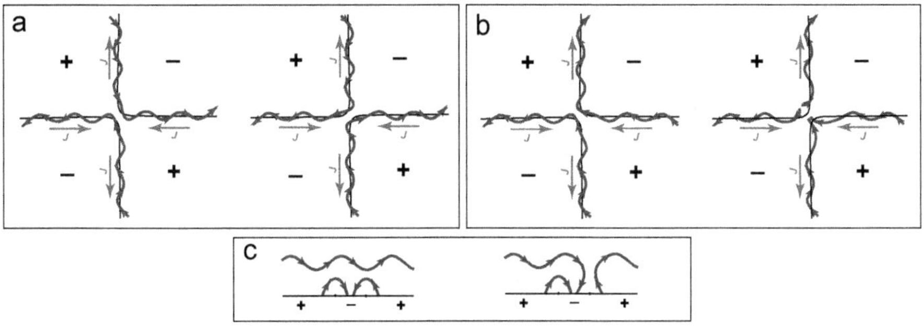


**Figure 1.** Schematics for (a) the reconnection of filaments with the same chirality; (b) the rupture of filaments with opposite chiralities during reconnection of neutral lines; (c) the rupture of a rope during reconnection with the photospheric field. Red arrows show the directions of electric currents $J$.

inversion lines, these filaments can "reconnect," exchanging their halves at the contact point (see Fig. 1(a)). The reconnection of filaments with the same chirality occurs at a null point of the field, where the dominant antiparallel axial components reconnect. The azimuthal components of the new halves are easily joined, since they have the same direction, and the currents reconnect. Of course, this idealized scenario is applicable only if the magnitudes of the currents and longitudinal fields in the approaching filaments are equal.

In the case of opposite chiralities, the axial components of the filament halves located at the reconnected neutral line are antiparallel and cannot be joined (see Fig. 1(b)). Therefore, they must reconnect with photospheric fields approaching the contact point; i.e., they must close at the photosphere. The ends of ruptured filaments should form cusps, as is typical for contacts of filaments with opposite chiralities. Figure 1(c) presents a schematic illustrating the reconnection between one field line of a magnetic rope and a loop of photospheric field, as viewed in the vertical plane passing through the diagonal of Fig. 1(b) (right), i.e., from the lower left to the upper right corner.

Examples of alterations of the filament connectivity occuring during the evolution of photospheric fields are presented in Kumar *et al.* (2010), Su *et al.* (2007), Filippov (2011). Theoretical and observational aspects of interaction of filaments, which show the same or opposite signs of magnetic helicities, are discussed in Schmieder *et al.* (2004), Aulanier & Schmieder (2008), Török *et al.* (2011).

This work was supported in part by the Russian Foundation for Basic Research (grants 12-02-00008 and 12-02-92692) and the Program 22 of the Russian Academy of Sciences.

## References

Aulanier, G. & Schmieder, B. 2008, in: C. Charbonnel, F. Combes, & R. Samadi (eds.), *SF2A-2008: Proceedings of the Annual meeting of the French Society of Astronomy and Astrophysics*, http://proc.sf2a.asso.fr, p. 543

Filippov, B. P. 2011, *Astron. Rep.*, 55, 541

Martin, S. F., Billimoria, R., & Tracadas, P. W. 1994, in: R. J. Rutten & C. J. Schrijver (eds.), *Solar Surface Magnetism* (Dordrecht: Kluwer Acad.), p. 303

Kumar, P., Manoharan, P. K., & Uddin, W. 2010, *ApJ*, 710, 1195

Pevtsov, A. A., Balasubramaniam, K. S., & Rogers, J. W. 2003, *ApJ*, 595, 500

Schmieder B., Mein N., Deng Y., *et al.* 2004, *Solar Phys.* , 223, 119

Su, J., Liu, Y., & Kurokawa, H., *et al.* 2007, *Solar Phys.* , 242, 53

Török, T., Chandra, R., Pariat, E., Démoulin, P., & Schmieder, B., *et al.* 2011, *ApJ*, 728, 65

*Nature of Prominences and their role in Space Weather*
*Proceedings IAU Symposium No. 300, 2013*
*B. Schmieder, J.-M. Malherbe & S. T. Wu, eds.*

doi:10.1017/S1743921313011332

# The solar physics FORWARD codes: Now with widgets!

**Blake Forland[1,2]†, Sarah Gibson[1], James Dove[2] and Therese Kucera[3]**

[1]High Altitude Observatory/National Center for Atmospheric Research
3080 Center Green Dr. Boulder, CO, 80027, USA
email: lyon.bcf@gmail.com,sgibson@ucar.edu

[2]Dept of Physics & Metro State College Denver
NC 3123 Campus Box 69 1201 5th Street, Denver, CO, 80204-2005, USA
email: dove@msudenver.edu

[3]NASA Goddard Space Flight Center & Greenbelt, MD
20771, USA
email: therese.a.kucera@nasa.gov

**Abstract.** We have developed a suite of forward-modeling IDL codes (FORWARD) to convert analytic models or simulation data cubes into coronal observables, allowing a direct comparison with observations. Observables such as extreme ultraviolet, soft X-ray, white light, and polarization images from the Coronal Multichannel Polarimeter (CoMP) can be reproduced. The observer's viewpoint is also incorporated in the FORWARD analysis and the codes can output the results in a variety of forms in order to easily create movies, Carrington maps, or simply observable information at a particular point in the plane of the sky. We present a newly developed front end to the FORWARD codes which utilizes IDL widgets to facilitate ease of use by the solar physics community. Our ultimate goal is to provide as useful a tool as possible for a broad range of scientific applications.

**Keywords.** Sun: prominences, Sun: corona, Sun: magnetic fields.

---

Solar coronal observables depend upon physical plasma properties such as density, temperature, velocity, and magnetic field, all of which are important constraints on theoretical models, but which in general are not themselves directly observable. Comparing models to data is not always straightforward, however, since each observable depends on coronal plasma properties in different ways. Moreover, the optically-thin corona results in observables that are integrated along the line of sight. A useful and general approach is forward modeling: taking a model-defined three-dimensional distribution of coronal plasma properties and performing line-of-sight integrals specific to a given coronal observable's dependence on these properties.

The "FORWARD" suite of forward-modeling IDL codes is designed to convert analytic models or simulation data cubes into coronal observables. It includes a set of analytic magnetohydrodynamic models and the capability to incorporate the output from any numerical MHD simulation or potential-field extrapolation, enabling modeling of magnetic structures ranging in size from that of active regions to the global corona. The output of these models can be translated into a broad range of observables, from white light to EUV to coronal Stokes polarimetry parameters in the infrared. These codes have broad applicability to coronal observations in general, and present a powerful tool for distinguishing between theoretical models through side-by-side comparison of their predictions to coronal data.

† NCAR is supported by the National Science Foundation

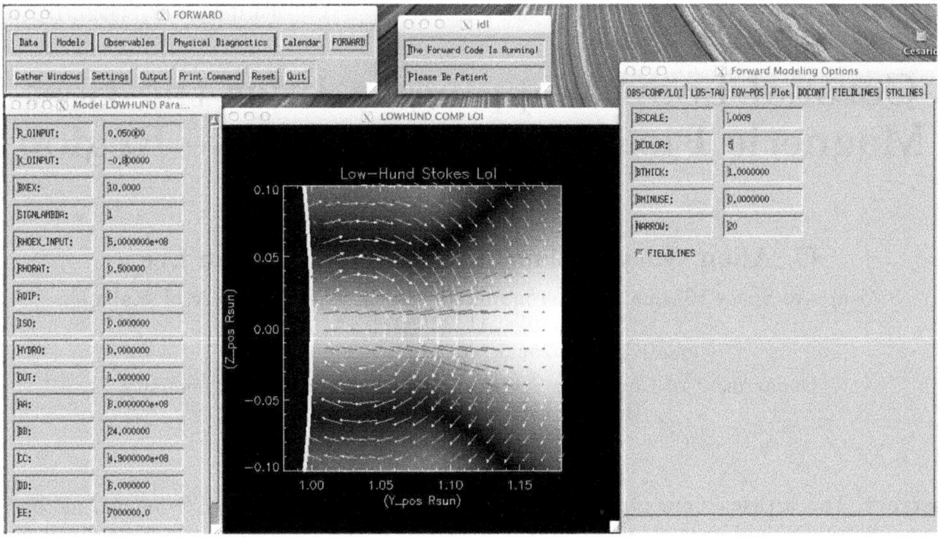

**Figure 1.** Example of widget interface of the FORWARD codes. Further information about how to access and install these codes available at *http : //people.hao.ucar.edu/sgibson/ FORWARD/*

The FORWARD codes were recently used to extract the 3D morphology, density, and temperature of a coronal cavity from multiwavelength observations (Gibson *et al.* (2010), Schmit & Gibson (2011), and Kucera *et al.* (2012)). They have also been used to analyze coronal magnometric properties in coronal cavities (Dove *et al.* (2011), Rachmeler *et al.* (2013), and Bąk-Stęślicka *et al.* (2013)).

Recently, we have upgraded FORWARD to possess a widget interface (Figure 1). Our goal was to make this interface dynamic enough to adapt to parameter choice changes made by the user, but also to ongoing additions made to the base FORWARD codes. Structured default files for the models, observables, and associated parameters were set up within the FORWARD code so that the widget interface could automatically incorporate such changes. Overall, the goal was to allow users to apply the FORWARD codes to a broad range of scientific applications, without requiring a comprehensive understanding of its capabilities *a priori.*

The FORWARD codes are designed for ongoing expansion, with the ultimate goal of inclusion as a SolarSoft package. FORWARD enables model-data comparison at multiple wavelengths, enabling the user to effectively choose between models and ultimately reconstruct the three-dimensional coronal field. To this end, efforts are underway to expand FORWARD to incorporate radio and visible wavelength magnetometric diagnostics.

## References


Bąk-Stęślicka, U., Gibson, S. E., Fan, Y., Bethge, C., Forland, B., & Rachmeler, L. A. 2013, *Astrophys. J.* in press; Arxiv 1304.7388

Dove, J., Gibson, S., Rachmeler, L. A., Tomczyk, S., & Judge, P. 2011, *Astrophys. J.*, 731, 1

Gibson, S. E., Kucera, T. A., *et al.* 2010, *Astrophys. J.*, 723, 1133

Kucera, T. A., Gibson, S. E., Schmit, D. J., Landi, E., & Tripathi, D. 2012, *Astrophys. J.*, 757, 73

Rachmeler, L. A., Gibson, S. E., Dove, J. B., DeVore, C. R., & Fan, Y. 2013, *Solar Phys.* in press, Arxiv 1304.7594

Schmit, D. J. & Gibson, S. E. 2011, *Astrophys. J.*, 733, 1

*Nature of Prominences and their role in Space Weather*
*Proceedings IAU Symposium No. 300, 2013*
*B. Schmieder, J.-M. Malherbe & S. T. Wu, eds.*

doi:10.1017/S1743921313011344

# Coronal Loop Mapping to Infer the Best Magnetic Field Models for Active Region Prominences

**G. Allen Gary[1], Qiang Hu[1] and Jong Kwan Lee[2]**

[1]Center of Space Plasma and Aeronomic Research, The University of Alabama, Huntsville, AL 35899, USA
email: `gag0002@uah.edu, qh0002@uah.edu`

[2]Department of Computer Science, Bowling Green State University, Bowling Green, OH 43403, USA
email: `leej@bgsu.edu`

**Abstract.** This article comments on the results of a new, rapid, and flexible manual method to map on-disk individual coronal loops of a two-dimensional EUV image into the three-dimensional coronal loops. The method by Gary, Hu, and Lee (2013) employs cubic Bézier splines to map coronal loops using only four free parameters per loop. A set of 2D splines for coronal loops is transformed to the best 3D pseudo-magnetic field lines for a particular coronal model. The results restrict the magnetic field models derived from extrapolations of magnetograms to those admissible and inadmissible via a fitness parameter. This method uses the minimization of the misalignment angles between the magnetic field model and the best set of 3D field lines that match a set of closed coronal loops. We comment on the implication of the fitness parameter in connection with the magnetic free energy and comment on extensions of our earlier work by considering the issues of employing open coronal loops or employing partial coronal loop.

**Keywords.** Magnetic fields, Models; Magnetic fields, Corona; Active Regions

---

## 1. 2D Cubic Bezier Splines, 3D Extensions, & Misalignment Angles

Gary, Hu, and Lee (2013) developed a rapid manual method to map individual coronal loops of a 2D EUV image as Bézier curves using only four points per loop. Their article outlined explicitly how the coronal loops can be employed in constraining competing magnetic field models by transforming 2D coronal loop images into 3D field lines. Their method uses the minimization of the misalignment angles between the magnetic field model and the best set of 3D field lines that match a set of closed coronal loops. The article considered only closed loops with given photospheric foot points; here we comment on the consequences of considering open loops and loop segments in the analysis.

The center of coronal loops is a locus of a point moving with one degree of freedom along the curve. One class of parametric curves is the Bézier curves. Cubic Bézier splines provide a curve that can fit most coronal loops with four number control points and hence provides a rapid matching process. Four 2D control points $[\mathbf{P}_1, \mathbf{P}_2, \mathbf{P}_3, \mathbf{P}_4]$ can be manipulated to provide a rapid fit to an EUV coronal loop image. In the previous study, the first and last control points were associated with the coronal-loop foot points. The four control points of the Bézier splines that are used to fit a coronal loop image are 2D, but these can be extended to 3D by the addition of z-components (Fig. 1). The non-zero z-components of the control points provide a 3D curve (psuedo field line) that will remain true to the projected coronal image. The first panel of Fig. 1 is for close loops and other two panels extend this concept to open field lines and a segment of a field line.

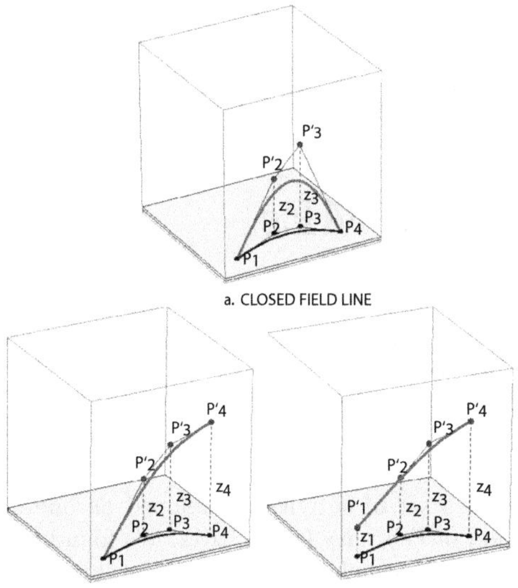


**Figure 1.** The extentsion of 2D cubic Bézier splines to 3D for closed, open, and partial loops. The thick lower (photospheric) line: a 2D cubic Bézier curve in the $z = 0$ plane, defined by the four control points $\mathbf{P}_i(x_i, y_i)$. The 3D cubic Bézier curve (thick upper line) is obtained through the introduction of parameters: $z_i$. The two curves are related by having points lying at the same location in the image plane.

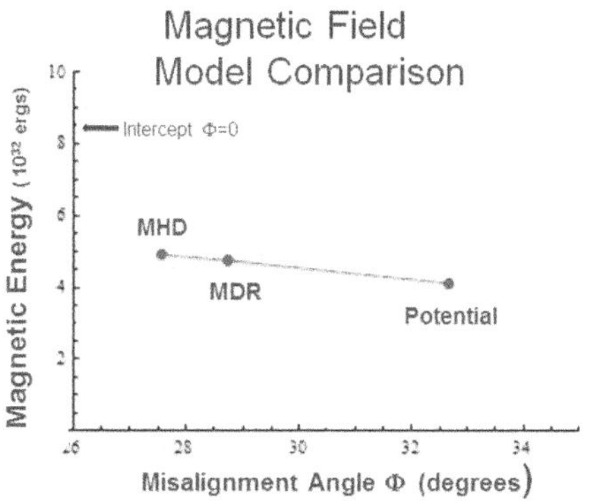


**Figure 2.** For the three magnetic-field models studied (potential, MDR, and MHD), the respective misalignment angles are 32.7°, 28.8°, and 27.6° for active region AR 11117. The resulting magnetic-energy values in the volume are 4.12, 4.47 and 4.90 $\times 10^{32}$ ergs. There is an almost inverse linear correspondence between the energy and the misalignment angle, with an extrapolated zero misalignment angle having an energy of about twice the potential energy.

## 2. Concluson

We can adjust the z-components of the 2, 3, or 4 control points to generate a best fit of the extended 3D spline to a magnetic field line. The requirement is the sum of the misalignment angles between the normalized tangents of the spline and the normalized magnetic field be a minimum. We have compared the average misalignment angle [Φ] to determine which magnetic field model best fits the mapped coronal images (see Gary, Hu, & Lee (2013) for details). Fig. 2 shows volume magnetic energy versus the misalignment angle. The results are for a potential field extrapolation compared with Wu's 3D time-dependent data-driven MHD solution, and, for comparison, the NLFFF minimum dissipative rate (MDR) model is included. This analysis assumed the identified loops were closed, i.e. both ends of the loop were coronal foot points at $z = 0$. The MHD field is the best fit to the coronal loops, *i.e.* the lowest misalignment angles. It also has the highest magnetic energy. When assuming open field lines or segment of field lines, the minimization process becomes problematic, as a result of the misalignment angle having stronger inter-relationships near the photosphere.

## References

Gary, G. A., Hu, Q., & Lee, J. K. 2013, *Solar Physics*, August 2013 (on line), DOI: 10.1007/s11207-013-0359-8.

*Nature of Prominences and their role in Space Weather*
*Proceedings IAU Symposium No. 300, 2013*
*B. Schmieder, J.-M. Malherbe & S. T. Wu, eds.*

doi:10.1017/S1743921313011356

# Evolution of a Group of Coronal Holes Associated with Eruption of Nearby Prominences and CMEs

**Heidy Gutiérrez[1], Lela Taliashvili[1] and Zadig Mouradian[2]**

[1]Space Research Center, University of Costa Rica, 2060 San José, Costa Rica
email: heidy.gutierrez@cinespa.ucr.ac.cr
[2]Observatoire de Paris-Meudon, CNRS, UPMC, Universit Paris Diderot, 5 Place Jules Janssen, 92190 Meudon, France

**Abstract.** We present the results of detailed study of a set of activities developed on one of three enclosed sectors of solar region during the period of February 07–13, 2012. We found the sequence of certain topological perturbations of whole coronal holes (CHs) and their surroundings associated to the eruption of nearby prominence and subsequent Coronal Mass Ejections (CMEs). Especially, we observe the emergence of small bright points (BPs) and the formation of dimming regions (DRs) close to the filament's channel associated with a pre–evolution of filament/prominence eruption, whereas BPs disappearance and the shrinkage of CH we found associated with the post-eruption evolution of prominence and of CME.

**Keywords.** coronal hole, prominence, magnetic field, CME.

## 1. Description of Event

This study is based on multispectral analyze of images provided by Observatory of Paris–Meudon, GONG, WSO, STEREO and SDO linked to a set of activities observed during Feb. 07–13, 2012. This set of activities involves: a) the evolution of group of CHs, b) two active regions, c) the disappearance of filaments, located $< 20°$ distance from the boundary of CHs, d) pre-/post-evolution of associated CMEs (Figure 2,3). We analyzed this set considering its development along three sectors (Figure 2) as well as a system of inter–correlated activities, but in present study we concentrate only on the sector 1 (S1).

## 2. Discussion and Preliminary Results

The starting processes of F1 and F2 eruptions are associated with the formation of BPs, and DR1 and DR2 (respectively) close to the filament's channels (Figure 2). After ~2 h from the start-time of F2 eruption first CME1 and after additional ~2 h the second CME2 formed (both CMEs related with only F2 eruption). Additionally, the disappearance of BPs, DR1 and DR2 correspond to the post–eruption evolution of F1 and F2. BPs formation/disappearance is related with the interchange reconnection (Fisk (2005)) near to CH1 boundaries, topological changes of CH1 and reorganization of magnetic field, contributing in this way to it destabilization and eruption (Taliashvili *et al.* (2009), Gutiérrez *et al.* (2013)). Post–eruption evolution of F1 and post–CME2 are also related with the topological changes and shrink of CH1. However, in case of F1 eruption there is no subsequent formation of CME, probably due to the few material that constitutes F1 body and its maximum height attained (Gopalswamy *et al.* (2006)). Therefore, the CME1 formation is a direct consequence of F2 eruption, while CME2 is related with

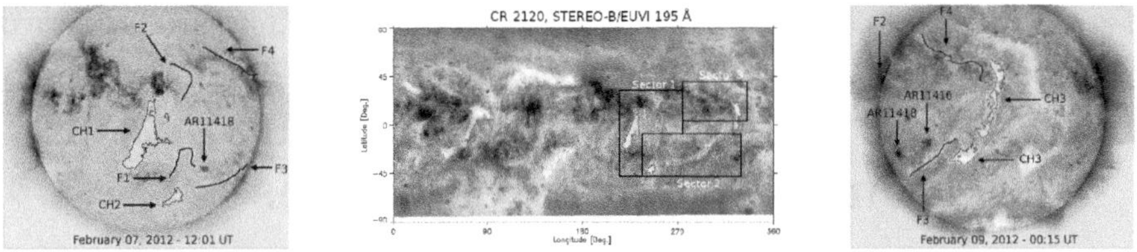


**Figure 1.** STEREO-B/EUVI-195Å (left) and SDO/AIA-193Å (right) images. STEREO /EUVI-195Å map with three sectors (central). The activities are indicated by arrows. CH boundaries are enhanced, filaments are hand-drawn and the colors are inverted for clarity.

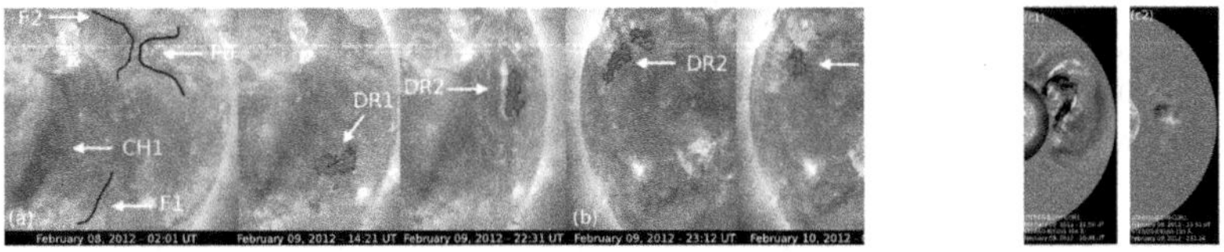


**Figure 2.** (a)STEREO-B/EUVI-195Å and (b)SDO/AIA-193Å images sequence (Feb. 08-13); filaments (hand–drawn). (c) Compositions of STEREO images: (c1)CME1 (COR1 diff/EUVI-304Å), (c2)CME2 (COR1diff/EUVI-195Å). We enhanced the DRs boundaries.

the evacuated material of F2 (that appears after F2 eruption) and with DR2 formation, which decreases and finally disappears after CME2.

Three activity sectors are inter-related in time, space and magnetically. The whole region that encloses these sectors is negative and delimited mainly by filaments. The possibility of magnetic reconnection between the field lines associated with BPs and the magnetic field lines close to CH1 boundaries, triggers the reconfiguration of magnetic field around CH1 (and CH1 topological changes too), including the foot points of filaments ($< 20°$), contributing in this way to their perturbation, eruption and subsequent formation of CME. DRs that formed after the onset of eruption of filaments disappear and CH adjacent to the filament shrinks after eruption of filaments and formation of CME. The association between DR and CME could be related with this magnetic reconfiguration (Jiang *et al.* (2011)) with BPs as indicators of this reconfiguration. In this study we gave the results regarding of primary evolution of the large region, starting mainly from S1. Based on WSO magnetograms we conclude that a small-scale magnetic reconfiguration associated with the set of activities starts also mainly from S1. This conduces to the large–scale magnetic reconfiguration through magnetic diffusion after filaments eruption, which converts the whole region in magnetically disintegrated sectors, persisted over the next CR 2121. Continuous magnetic observations are important to conclude regarding the magnetic evolution of CHs and filaments that conduces the formation of CMEs.

**Acknowledgements.** We acknowledge to Observatoire Paris-Meudon, GONG, WSO, SDO, STEREO and COR1 Preliminary Events List for open access to their data sets. H.G and L.T. are gratefully acknowledged the financial supports by IAU/LOC.

## References

Fisk, L. A. *ApJ* 626, 563, 2005.

Gopalswamy, N., Mikić, Z., Maia, D., Alexander, D., Cremades, H., Kaufmann, P. Tripathi, D., Wang, Y.-M. *Space Science Rev.* 123, 303, 2006.

Gutiérrez, H., Taliashvili, L., Mouradian, Z. *Ad. Space Res.* 51 , 1824, 2013.

Jiang, Y., Yang, J., Hong, J., Bi, Y. and Zheng, R. *ApJ* 738, 179, 2011.

Taliashvili, L., Mouradian, Z., Páez, J. *Sol. Phys.* 258, 277, 2009.

*Nature of Prominences and their role in Space Weather*
*Proceedings IAU Symposium No. 300, 2013*
*B. Schmieder, J.-M. Malherbe & S. T. Wu, eds.*

doi:10.1017/S1743921313011368

# Mapping prominence plasma parameters from eclipse observations

**Sonja Jejčič[1], Petr Heinzel[2], Maciej Zapiór[2,3], Miloslav Druckmüller[4], Stanislav Gunár[2], and Pavel Kotrč[2]**

[1]University of Ljubljana, Department of Physics, Ljubljana, Slovenia
email: sonja.jejcic@guest.arnes.si

[2]Astronomical institute of the Academy of Sciences, Ondřejov, Czech Republic
email: pheinzel@asu.cas.cz, email: gunar@asu.cas.cz email: pkotrc@asu.cas.cz

[3]Departament de Física, Universitat de les Illes Balears, Palma de Mallorca, Spain
email: maciej.zapior@uib.es

[4]Institute of Mathematics, Brno University of Technology, Brno, Czech Republic
email: druckmuller@fme.vutbr.cz

**Abstract.** Using the eclipse observations, we construct the maps of quiescent prominence temperatures, electron densities, pressures and geometrical thicknesses. For this we use the RGB signal of prominence visible-light emission detected during the total solar eclipse on August 1, 2008 in Mongolia, and quasi-simultaneous H$\alpha$ spectra taken at Ondřejov observatory. The method of disentangling the electron density and effective geometrical thickness was described by Jejčič & Heinzel (2009) and is used here for the first time to analyse the spatial variations of various prominence parameters.

**Keywords.** Eclipse Observations; Prominences, Quiescent; Spectral Line, Intensity and Diagnostics

## 1. Determination of temperature and electron densities from eclipse observations

In this study we used the H$\alpha$ line spectral observations of a quiescent prominence together with RGB eclipse observations from August 1, 2008 (Figure 1). After coalignment between H$\alpha$ line spectra and visible-light prominence from the eclipse observations, we used calibrated H$\alpha$ profiles to compute the temperature. The temperature is derived from simple relation $\Delta\lambda_{\rm D} = F \cdot E_{\rm H\alpha}/(I_{\rm max}\ \sqrt{\pi})$, where $\Delta\lambda_{\rm D}$ is the Doppler width (in Å), $E_{\rm H\alpha}$ the H$\alpha$ integrated intensity and $I_{\rm max}$ the maximal (peak) intensity of the H$\alpha$ emission profile. $F$ is the opacity correction factor (Heinzel *et al.* (1994)). The kinetic temperature is derived assuming a uniform value of the microturbulent velocity 5 km s$^{-1}$. The electron density was determined using a generalized approach of Jejčič & Heinzel (2009). We processed the eclipse observations in RGB channels of the digital Canon camera and used the ratio between two channels (RGB signals are not calibrated), together with the H$\alpha$ integrated intensity, to derive the electron density and the effective geometrical thickness along the line of sight.

## 2. Summary of results

We obtained 2D maps of the kinetic temperature and electron density of the studied prominence (Figure 2). The temperature varies between 6000 and 15000 K and is consistent with recent measurements of Park *et al.* (2013). The central part of the prominence has lower temperature than the edges, between 7000 and 8000 K. The electron density

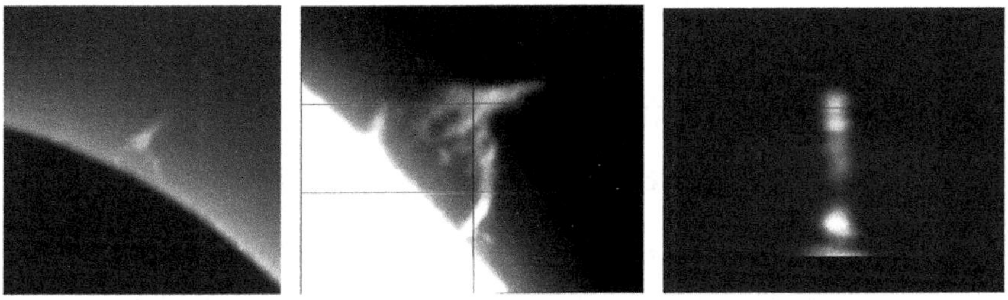

(a) (b) (c)

**Figure 1.** a)Total solar eclipse observed on August 1, 2008 at 11:05:29.6 UT in Mongolia with the studied prominence. b) The slit-jaw image of the studied prominence obtained at Ondřejov observatory on August 1, 2008 at 08.10.42.0 UT. c)The H$\alpha$ line spectra along the vertical slit shown in (b).

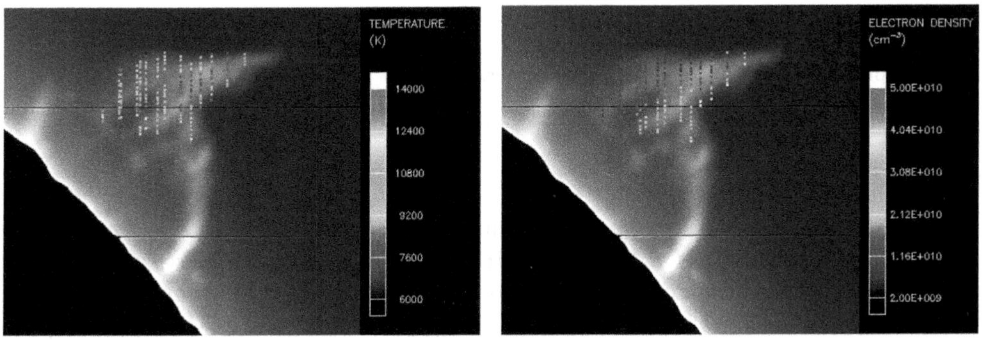


(a) (b)

**Figure 2.** a) Map of the kinetic temperature of the studied prominence. b) Map of the electron density.

varies between $5 \times 10^9 - 10^{11}$ cm$^{-3}$ and is quite comparable with the values from Bommier *et al.* (1994). The electron density is systematically increasing towards the bottom of the prominence which could be explained by an enhanced photoionization by the incident solar radiation. We also obtained 2D map of the effective thickness which varies between about 200 and 15000 km. Typical value is around 10000 km and is decreasing towards the bottom. Using the obtained maps of the electron density and kinetic temperature we also computed the prominence gas pressures, assuming a uniform ionization degree of hydrogen $i = 0.5$ in cool parts. The prominence seems to be a low-pressure structure with a typical gas pressure around 0.05 dyn cm$^{-2}$.

New prominence observations obtained during solar eclipses or using the space coronagraphs are required together with simultaneous detection of the Balmer lines and other lines falling into the wavelength range of detected visible-line continuum.

## References

Bommier, V., Landi Degl'Innocenti, E., Leroy, J.-L., & Sahal-Brechot, S. 1994 *Solar Phys.*, 154, 231

Heinzel, P., Gouttebroze, P., & Vial, J. C. 1994, *Astron. Astrophys.* , 292, 656

Jejčič, S. & Heinzel, P. 2009 *Solar Phys.*, 254, 89

Park, H., Chae, J., Song, D., Maurya, R. A., Yang, H., Park, Y.-D., Jang, B.-H., Nah, J., Cho, K.-S., Kim, Y.-H., Ahn, K., Cao, W., & Goode, P. R. 2013 *Solar Phys.*, doi:10.1007/s11207-013-0271-2

*Nature of Prominences and their role in Space Weather*
*Proceedings IAU Symposium No. 300, 2013*
*B. Schmieder, J.-M. Malherbe & S. T. Wu, eds.*

doi:10.1017/S174392131301137X

# A Statistical Study on Characteristics of Disappearing Prominences

**Anand D. Joshi**[1], **Su-Chan Bong**[1] **and Nandita Srivastava**[2]

[1]Korea Astronomy and Space Science Institute, Daejeon, South Korea
email: `janandd@gmail.com; scbong@kasi.re.kr`

[2]Udaipur Solar Observatory, Physical Research Laboratory, Udaipur, India.
email: `nandita@prl.res.in`

**Abstract.** Real-time monitoring of filaments is essential for the prediction of their eruption and the ensuing coronal mass ejection (CME). We apply an automated algorithm for the detection and tracking of filaments in full-disc H$\alpha$ images to obtain their physical attributes. This provides an accurate onset time of the eruption, and also allows us to study the physical characteristics of the erupting filaments in an objective manner.

## Introduction

A disappearing prominence, most of the times, is associated with a coronal mass ejection (CME) (Gopalswamy *et al.* 2003). A CME, if directed towards the Earth and carrying a southward magnetic field, can lead to a geomagnetic storm that can cause considerable damage (Srivastava & Venkatakrishnan 2004). If one has to predict the onset of such CMEs, it is essential to monitor filaments seen on the solar disc. For this purpose, an automated algorithm has been developed for detection and tracking of filaments observed in full-disc H$\alpha$ images (Joshi *et al.* 2010). The algorithm tracks the filaments through the full period of observation to generate their physical attributes such as size and length. In this paper we apply this algorithm to seven filament eruptions and observe their characteristics based on the attributes derived from the algorithm.

## Data Selection and Analysis

For this study, we have used full-disc H$\alpha$ images of the Sun for seven filament eruptions observed from different ground-based observatories. The algorithm is flexible enough to be used on any dataset with only a few minor modifications. On applying a median-based local threshold to the greyscale H$\alpha$ images we get binary images wherein the filaments are identified as black features on a white disc. The grouping criterion is used to determine the number of filaments in each image, and also the number of fragments that a single filament is split into. Each filament is identified with a unique label in this step. Thereafter, we use labelling criterion which tracks the filaments over successive images, and maintains the labels consistently.

## Results

The algorithm provides length and size of the filament, along with the number of fragments that each filament is broken into. Based on the area and length, we derive three more parameters, which are the average area and length before the eruption began, its percentage left behind after the eruption, and the duration of the eruption. This has

| Date | Observatory | Area | | | Length | | |
|---|---|---|---|---|---|---|---|
| | | average ($10^{-3}$ disc) | % after eruption | duration (min) | average ($R_\odot$) | % after eruption | duration (min) |
| 20 Jul 2004 | MLSO | 1.87 | 44 | 93 | 0.38 | 48 | 102 |
| 28 Jul 2005 | MLSO | 0.54 | 14 | 83 | 0.03 | 15 | 60 |
| 19 May 2007 | KSO | 0.46 | 22 | 159 | 0.14 | 27 | 137 |
| 01 Aug 2010 | KSO | 1.03 | 52 | 270 | 0.19 | 78 | 201 |
| 08 Aug 2012 | SMART | 1.60 | 18 | 180 | 0.34 | 23 | 105 |
| 31 Aug 2012 | GONG | 1.47 | 10 | 38 | 0.33 | 9 | 34 |
| 19 Feb 2013 | KSO | 1.10 | 58 | 82 | 0.28 | 65 | 69 |

**Table 1.** Summary of the seven events analysed using the automated filament detection algorithm. The H$\alpha$ images used are taken from Mauna Loa Solar Observatory (MLSO), Kanzelhöhe Solar Observatory (KSO), Solar Magnetic Activity Research Telescope (SMART), and Global Oscillation Network Group (GONG).

been summarised in Table 1. To maintain uniformity across different datasets, area is shown as a fraction of the area of solar disc, while the length is shown as a fraction of the solar radius, $R_\odot$. The start of eruption can be approximately determined from the time-lapse movies of H$\alpha$ images. However, in this table, the start of eruption is chosen as the time when the filament area and length start to decrease from their average values, thereby providing an accurate value based on the attributes from the algorithm.

## Summary

The automated filament detection and tracking algorithm was applied on seven filaments observed from different observatories. Qualitatively, we find that the changes in area and length of a filament are proportional. However, we observe that the quantitative measures, duration of eruption and percentage left behind after eruption, are unequal for area and length. It implies that while one parameter may undergo a certain degree of variation, the other may vary slightly differently. We also clearly see from the data that when an eruption/disappearance begins, the number of fragments increase, i.e. the filament breaks up into smaller parts. The breaking up indicates that the whole of the filament becomes unstable at the same time, and there is no preferential point at which the eruption starts. An increase in the number of fragments would prove to be very significant during real-time monitoring of filament.

We would like to extend this work to several more cases. The analysis would also include applying the algorithm to a day before the eruption, thereby allowing a comparison of attributes between a quiet and an eruptive filament. From this study, we would try to establish threshold criterion for the filament eruptions.

## References

Gopalswamy, N., Shimojo, M., Lu, W., Yashiro, S., Shibasaki, K., & Howard, R. A. 2003, *Astrophys. J.*, 586, 562

Joshi, A. D., Srivastava, N., & Mathew, S. K. 2010, *Sol. Phys.*, 262, 425

Srivastava, N. & Venkatakrishnan, P. 2004, *J. Geophys. Res.*, 109, 10103, A18

*Nature of Prominences and their role in Space Weather*
*Proceedings IAU Symposium No. 300, 2013*
*B. Schmieder, J.-M. Malherbe & S. T. Wu, eds.*

doi:10.1017/S1743921313011381

# Signatures of magnetic reconnection during the evolutionary phases of a prominence eruption and associated X1.8 flare

**Bhuwan Joshi[1], Upendra Kushwaha[1], KyungSuk Cho[2] and Astrid Veronig[3]**

[1]Udaipur Solar Observatory, Physical Research Laboratory, Udaipur 313004, India
[2]Korea Astronomy and Space Science Institute, Daejeon 305-348, Korea
[3]IGAM/Institute of Physics, University of Graz, Universitätsplatz 5, A-8010 Graz, Austria

**Abstract.** In this paper, we present RHESSI and TRACE observations of multiple flare activity that occurred in the active region NOAA 10656 on 2004 August 18. Out of four successive flares, there were three events of class-C while the final event was a major X1.8 solar eruptive flare. During localized C-class flares, the filament undergoes slow yet crucial morphological evolution. The filament eruption is accompanied with an X1.8 flare during which multiple HXR bursts are observed up to 100–300 keV energies. From the location, timing, strength, and spectrum of HXR emission, we conclude that the prominence eruption is driven by the distinct events of magnetic reconnection occurring in the current sheet below the erupting prominence. These multi-wavelength observations also provide evidence for tether-cutting reconnection as the triggering mechanism for filament eruption and associated X-class flare.

**Keywords.** flares, filaments, X-rays, magnetic reconnection.

---

## 1. Introduction

The association of eruptions of solar prominences with coronal mass ejections and solar flares is well known. It is widely accepted that the fundamental processes responsible for these phenomena are closely related and of magnetic origin. Multi-wavelength measurements are crucial to infer the characteristics of physical processes responsible for impulsive energy release and eruptions (Joshi *et al.* 2012). In this paper, we report RHESSI and TRACE observations of the eruption of an active region filament (Cho *et al.* 2009, Joshi *et al.* 2013) occurred in NOAA 10656 on 2004 August 18. We present observations of four flares (there were three flares of class C while the last events was an X1.8 eruptive flare) that occur during the evolutionary phases of the filament toward its eruption (Fig. 1).

## 2. Observations and results

**1.** The pre-eruption phase is characterized by three localized flares (i.e., events I, II and III) which are associated with heating, activation, and rise of the filament (Fig. 2(a)-(d)). During the first event, we observed plasmoid eruption (Fig. 1(a)) which was followed by the rise of filament (Fig. 2(b)). Events II and III occurred close to the footpoints of the rising filament. The pre-eruption flares and subsequent changes in the morphology of the prominence provide evidence for the tether-cutting mechanism for solar eruptions (Moore & Roumeliotis 1992).

**2.** The filament eruption was accompanied by an X1.8 flare (Fig. 2(e)-(f)) during which multiple HXR bursts are observed up to 100–300 keV energies. We observed a bright and

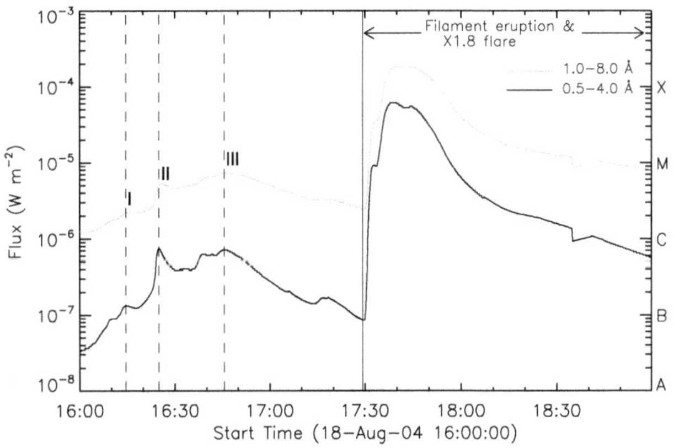


**Figure 1.** GOES soft X-ray time profiles in the 0.5–4 and 1–8 Å wavelength bands. The vertical solid line differentiates the pre-eruption phase from the eruptive X1.8 flare. Three C-class flares observed before the prominence eruption are denoted as I, II, and III.

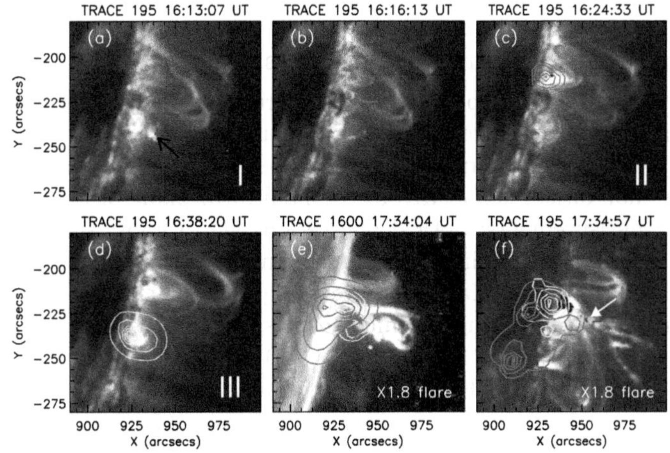


**Figure 2.** TRACE observations of pre-eruption events (I, II, and III; panels (a)-(d)) and X1.8 class flare associated with prominence eruption (panels (e)-(f)). Arrow in panel (b) indicates the rise of prominence after event I. Contours represent RHESSI PIXON images in 12–25 keV (yellow), 25–50 keV (red), and 50–100 keV (blue) energy bands.

extended coronal source simultaneously in EUV and 50–100 keV HXR images underneath the expanding filament which provides strong evidence for the ongoing magnetic reconnection. This phase is accompanied with very high plasma temperatures of ~31 MK which is followed by the detachment of the prominence from the solar source region.

## References

Cho, K.-S., Lee, J., Bong, S.-C., Kim, Y.-H., Joshi, B., & Park, Y.-D. 2009, *ApJ*, 703, 1

Joshi, B., Veronig, A., Manoharan, P. K., & Somov, B. V. 2012, *in Astrophysics and Space Science Proceedings, Vol. 33, Multi-scale Dynamical Processes in Space and Astrophysical Plasmas*, ed. M. P. Leubner & Z. Vörös (Springer-Verlag Berlin Heidelberg), 29

Joshi, B., Kushwaha, U., Cho, K.-S., & Veronig, A. M., *ApJ*, 771, 1

Moore, R. L. & Roumeliotis, G. 1992, *in Lecture Notes in Physics, Berlin Springer Verlag*, Vol. 399, IAU Colloq. 133: Eruptive Solar Flares, ed. Z. Svestka, B. V. Jackson, & M. E. Machado, 69

*Nature of Prominences and their role in Space Weather*
*Proceedings IAU Symposium No. 300, 2013*
*B. Schmieder, J.-M. Malherbe & S. T. Wu, eds.*

doi:10.1017/S1743921313011393

# On magnetic measurements in prominences

**I. S. Kim and O. I. Bugaenko**

Lomonosov Moscow State University, Sternberg Astronomical institute,
119992 Universitetsky pr-t, 13, Moscow, Russia
email: kim@sai.msu.ru

**Abstract.** The successes of magnetic measurements in faint objects located near very bright ones are strongly depending on the stray light in the telescope. We propose a mask with a variable transmission placed on the primary optics of a telescope. Our computations of the stray light in such a telescope indicate that the calculated coronagraphic factor of improvement, $K$, would increase at least by 2 orders of magnitude compared to the Lyot-type coronagraph.

**Keywords.** Sun:prominences, Sun: magnetic fields, techniques: polarimetric

## 1. Reducing the stray light in telescopes

In prominences the Zeeman splitting is 3 – 4 orders of magnitude less than the line width. Therefore, in spite of available advanced polarimeters and recording systems, "weak" magnetic fields measurements are still complicated because of the stray light, $I_{stray}$, caused by non-object signatures in the final focal plane of a telescope. Different effects affect the measurements: optical aberrations (1), a ghost solar image produced by multiple reflections in the primary lens (2), random inhomogeneities in the glass of the primary lens (3), diffraction of the solar disk light at the entrance aperture, $I_{dif}$ (4), scattering at micro-roughness of the primary optics, $I_{sc}$ (5), the sky brightness, $I_{sky}$ (6), the continuum corona, $I_{cont}$ (7). The items (2 - 5) are negligible during the totality of solar eclipses and dominate for the non-eclipse coronagraphic observations.

An ideal super-smooth primary optics is assumed below, e.g. $I_{stray} = I_{dif}$. The diffraction of a bright round source at the round-shape aperture was treated by Nagaoka (1920) and Sazanov (1968). The latter suggested the simplified expression for the range $R < 1.3R_{\odot}$.

$$lgI_{dif}(R) = lg\frac{2\lambda}{\pi^3 D\gamma_0} + lg\left(\frac{\sqrt{1+(\nu-1)^2}}{\nu-1} - \frac{\sqrt{1+(\nu+1)^2}}{\nu+1}\right) - 0.27(\nu-1) - 0.017, \tag{1.1}$$

where $R$ (the distance from the solar disk center) $= \nu a$, $a = \frac{\pi D}{\lambda}\gamma_0$ is the radius of the assumed round source in arbitrary units, $D$ is the diameter of the primary lens, $\gamma_0$ is the angular radius of the source. Intensities are given in units of 1 Å nearby the solar disk continuum. The $I_{dif}(R)$ for H$\alpha$ and near IR lines for 0.5 m and 4 m apertures, and the acceptable level of the stray light for "weak" magnetic measurements were presented recently (Kim *et al.* 2012). For a 0.5 m aperture of a non-coronagraphic telescope, $I_{dif}$ in H$\alpha$ reaches $10^{-3}$ at the prominence heights ($\approx 40''$) and is $> 5\cdot 10^{-3}$ at the chromosphere level ($\approx 4''$).

### 1.1. *Internal occulting technique (the Lyot method)*

The coronagraphic technique suggested by Lyot (1930) is based on putting a mask both in the primary focal plane and in the plane of the exit pupil to minimize the input of items (2 - 4). The Lyot-type coronagraph has a primary single lens, a mask in the primary focal plane (an artificial Moon), a field lens, a mask in the plane of the exit pupil (the Lyot-stop), the relay optics, and a final focal plane. In practice, the correct use of the Lyot method results in reducing $I_{dif}$ by $1-2$ orders of magnitude which is the coronagraphic factor, $K$, depending on the size of the mask in the primary focal plane and in the plane of the Lyot stop. For a 0.5 m aperture, $K \approx 20 - 100$ for prominences and $< 20$ for the chromosphere depending on the height observed.

1.2. *Apodizing with a special mask in the plane of an entrance aperture*

The diffraction pattern in the focal plane is a result of the discontinuity of the transmission function, $G$, (or its derivatives) of the entrance aperture. $I_{dif}$ at the optical axis is defined by $I_{dif}(0) = 1 - J_0^2(a) - J_1^2(a)$, where $J_1(x)$ is the Bessel function of the first kind, $a = (\pi D sin\gamma_0)/\lambda$, $\lambda$ is the wavelength, $\gamma_0$ is the angular radius of solar disk, and $D$ is the diameter of the aperture. If $a \gg 1$, then $I_{dif}(0) \approx 1-2/(\pi a)$ (Kim *et al.* 1995). For $\gamma_0 = 960''$, $\lambda = 600$ nm, $D = 200$ mm, we obtain $a = 4874$ and $I_{dif}(0) \approx 1.3) \cdot 10^{-4}$. A mask with variable transmission $G(\rho) = 1 - \rho^2$ has discontinuities in the first derivative. For $\gamma = (1 + \epsilon)\gamma_0$, the apodized ($I^a_{dif}$) and non-apodized ($I_{dif}$) are related as following:

$$I^a_{dif}(\epsilon) \simeq 2/3\pi^4 [I_{dif}(\epsilon)]^3. \tag{1.2}$$

Let's estimate the $K$ factor for the chromosphere and prominence heights and $a \approx 5000$.
– Quiescent prominence: $h = 40''$, $\epsilon = 0.04$ ($\epsilon\gamma_0 = 40''$). Then $I_{dif}(0.04) = 10^{-3}$, $I^{(a)}_{dif}(0.04) = 0.7 \cdot 10^{-7}$, and $K = 10^4$.
– The upper chromosphere: $h = 4''$, $\epsilon = 0.004$ ($\epsilon\gamma_0 = 4''$). Then $I_{dif}(0.004) = 10^{-2}$, $I^a_{dif}(0.004) = 0.7 \cdot 10^{-4}$, and $K = 10^2$.
We note that the mask $(1 - \rho^2)$ reduces the output by 3 times.

**Summary.** Current state of technology suggests to put on the primary optics a mask with a variable transmission $G(\rho) = 1 - \rho^2$ for having a low-scattered light telescope. At prominence heights, the *calculated* coronagraphic factor, $K$, is at least 2 orders of magnitude higher as compared with the Lyot-type coronagraph.

**Acknowledgements.** The reported study was partially supported by RFBR (research project No. 11-02-00631), IAU, SCOSTEP, SF2A and KLSA/CAS.

## References

Kim, I. S., Bugaenko, O. I., & Bruevich, V. V. and Evseev, O. A., 1995, *Bull. of Rus. Acad. of Sci.*, 59, 153.
Kim, I. S, Alexeeva, I. V, & Suyunova, E. Z., 2012, *ASP Conference Series*, T. R. Rimmele, M. Collados Vera *et al.*, eds, Vol. 463, 337.
Lyot, B., *Comptes Rendus Acad. Sci.*, 1931, 1169.
Nagaoka, H., 1920 *Astrophysical Journal*, 51, 73.
Sazanov, A. A., 1968, *Ph.D. thesis*, IZMIRAN, Moscow.

*Nature of Prominences and their role in Space Weather*
*Proceedings IAU Symposium No. 300, 2013*
*B. Schmieder, J.-M. Malherbe & S. T. Wu, eds.*

doi:10.1017/S174392131301140X

# Observational Study of Large Amplitude Longitudinal Oscillations in a Solar Filament

**Kalman Knizhnik[1,2], Manuel Luna[3], Karin Muglach[2,4]**
**Holly Gilbert[2], Therese Kucera[2], and Judith Karpen[2]**

[1]Department of Physics and Astronomy
The Johns Hopkins University, Baltimore, MD 21218
email: kalman.knizhnik@nasa.gov

[2]NASA/GSFC, Greenbelt, MD 20771, USA
[3]Instituto de Astrofísica de Canarias, E-38200 La Laguna, Tenerife, Spain
[4]ARTEP, Inc., Maryland, USA

**Abstract.** On 20 August 2010 an energetic disturbance triggered damped large-amplitude longitudinal (LAL) oscillations in almost an entire filament. In the present work we analyze this periodic motion in the filament to characterize the damping and restoring mechanism of the oscillation. Our method involves placing slits along the axis of the filament at different angles with respect to the spine of the filament, finding the angle at which the oscillation is clearest, and fitting the resulting oscillation pattern to decaying sinusoidal and Bessel functions. These functions represent the equations of motion of a pendulum damped by mass accretion. With this method we determine the period and the decaying time of the oscillation. Our preliminary results support the theory presented by Luna and Karpen (2012) that the restoring force of LAL oscillations is solar gravity in the tubes where the threads oscillate, and the damping mechanism is the ongoing accumulation of mass onto the oscillating threads. Following an earlier paper, we have determined the magnitude and radius of curvature of the dipped magnetic flux tubes hosting a thread along the filament, as well as the mass accretion rate of the filament threads, via the fitted parameters.
**Keywords.** solar prominences, oscillations, magnetic structures

## 1. Procedure

LAL oscillations consist of periodic motions of the prominence threads along the magnetic field that are disturbed by a small energetic event close to the filament (see Luna *et al.* paper in this volume). Luna and Karpen (2012) argue that prominence oscillations can be modeled as a damped oscillating pendulum, whose equation of motion satisfies a zeroth-order Bessel function. In their model, a nearby trigger event causes quasi-stationary preexisting prominence threads sitting in the dips of the magnetic structure to oscillate back and forth, with the restoring force being the projected gravity in the tubes where the threads oscillate (e.g. Luna *et al.* (2012)). In this paper, we report preliminary results of comparisons of observations of prominence oscillations with the model presented by Luna and Karpen (2012). More details will be available in the forthcoming paper by Luna *et al.* (2013).

In this analysis, we place slits along the filament spine and measure the intensity along each slit as a function of time. Fig. 1 (left) shows the filament in the AIA 171Å filter with the slits overlaid. Each slit is then rotated in increments of 0.5° from 0° to 60° with respect to the filament spine. We select the best slit according to the following criteria: (a) continuity of oscillations, (b) amplitude of the oscillation is maximized, (c) clear transition from dark to bright regions, (d) maximum number of cycles.

The oscillation for a representative slit is shown in Figure 1 (right), which corresponds to the grey slit in Figure 1 (left). We identify the position of the center of mass of the

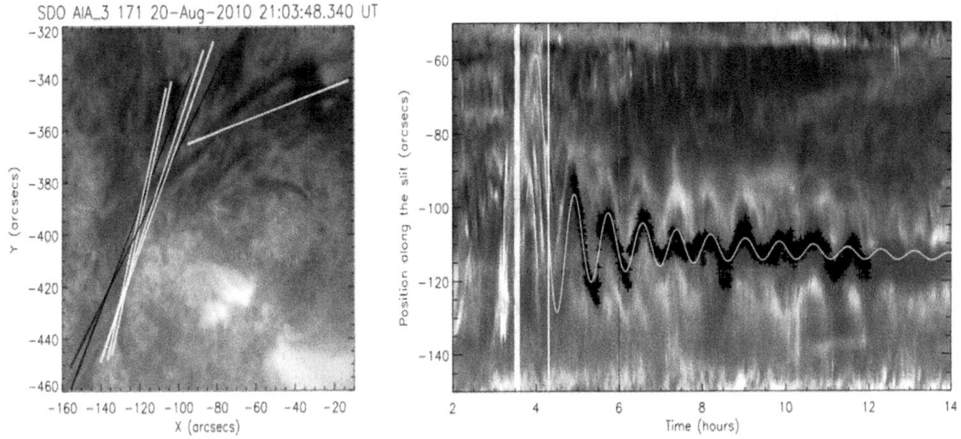


**Figure 1.** *Left*: Filament seen in AIA 171 with best slits overlaid. *Right*: An intensity distance–time slit, showing an oscillation with the Bessel fit (white curve) to equation (2.1) in Luna *et al.* (this volume). The sinusoidal fit was not as good as the Bessel fit and is not shown.

thread by finding the intensity minimum along the slit, indicated by black crosses in Figure 1 (right). These points are then fit to equation (2.1) of Luna *et al.* (this volume), and the resulting fit is shown in white.

## 2. Results

Fitting our data to equation (2.1) of Luna *et al.* (this volume) yields values of $\chi^2$ ranging between 1-13. Using equation (2.2) of Luna *et al.* (this volume), we find the average radius of curvature of the magnetic field dips that support the oscillating threads. We find it to be approximately 60 Mm. We also calculate a threshold value for the field itself that would allow it to support the observed threads. Using equation (3.1) of Luna *et al.* (this volume), we find an average magnetic field of $\sim$ 20 $G$, assuming a typical filament number density of $10^{11}$ cm$^{-3}$, in good agreement with measurements (e.g. Mackay *et al.* 2010). On average, the oscillations form an angle of $\sim 25^o$ with respect to the filament spine, and have a period of $\sim$ 0.8 hours. To explain the very strong damping mass must accrete onto the threads at a rate of about $60 \times 10^6$ kg/hr.

## 3. Conclusions

We conclude that the observed oscillations are along the magnetic field, which forms an angle of $\sim 25^\circ$ with respect to the filament spine (Tandberg-Hanssen & Anzer, 1970). We find that both the curvature and the magnitude of the magnetic field are approximately uniform on different threads. Both the Bessel and sinusoidal functions are well fitted, indicating that mass accretion is a likely damping mechanism of LAL oscillations, and that the restoring force is the projected gravity in the dips where the threads oscillate. The mass accretion rate agrees with the theoretical value (Karpen *et al.*, 2006, Luna, Karpen, & DeVore, 2012).

## References

Karpen, J. T., Antiochos, S. K., & Klimchuk, J. A. 2006, *ApJ*, 637, 531

Luna, M., Karpen, J. T., & Devore, C. R. 2012a, *ApJ*, 746, 30

Luna, M. & Karpen, J. 2012, *ApJ*, 750, L1

Luna, M., Knizhnik, K., Muglach, K., Gilbert, H, Kucera, T., & Karpen, J., *this volume*, 2014

Luna, M., Knizhnik, K., Muglach, K., Gilbert, H, Kucera, T., & Karpen, J., *ApJ*, 2013, *in prep.*

Mackay, D., Karpen, J., Ballester, J., Schmieder, B., & Aulanier, G. 2010, *Sp. Sci. Rev.*, 151, 333

Tandberg-Hanssen, E. and Anzer, U. 1970, *Solar Physics* 15, 158T

*Nature of Prominences and their role in Space Weather*
*Proceedings IAU Symposium No. 300, 2013*
*B. Schmieder, J.-M. Malherbe & S. T. Wu, eds.*

doi:10.1017/S1743921313011411

# 3D dynamical structuring of a high latitude erupting prominence: I- Analysis of the cool plasma flows before the eruption

**Serge Koutchmy[1], Boris Filippov[2], Ehsan Tavabi[3], Cyril Bazin[1] and Sylvain Weiller[1]**

[1]IAP UMR 7095, INSU-CNRS and UPMC, 98 bis Bd Arago 75014 Paris, France
email: koutchmy@iap.fr, bazin@iap.fr, sweiller@free.fr

[2]Pushkov Institute of Terrestrial Magnetism, Ionosphere and Radio Wave Propagation, Russian Academy of Sciences, Troitsk, Moscow Region 142190, Russia
email: boris_filippov@mail.ru

[3]Payame Noor University of Tehran, 14155-6466,I.R. of Iran
email: etavabi@gmail.com

**Abstract.** Both the origin of the quiescent prominences and their eruption related to CMEs are still a matter of extended studies. The small scale dynamic aspects like vortex structures and counter- flows are now seriously taken into account having in mind that the flows are a good proxy of the line of force of the omnipresent but rather unknown in detail force free or not magnetic field. Large scale vortex has been detected in a high latitude prominence observed on November 13- 14, 2011 before its eruption.

**Keywords.** prominence eruption, quiescent, flows, turbulence, counter- flows, vortices, tornado

## 1. Observations and data processing

A high latitude prominence was observed on November 13- 14, 2011 see Fig. 1 (left panels), using high resolution H$\alpha$ fast imaging (60 fps; .5/px) refractor (diameter=100mm) equipped with a Coronado, to look at the fine scale structure and its dynamics. This hedgerow quiescent type prominence with many vertical flow lines, indeed erupted the following day (Koutchmy II in this issue). The filter bandwidth (.07 nm) is marginally sensitive to both transverse motions (Doppler-Fizeau effect) and optical thickness effects. Image processing was used to reconstruct images at a 15 s cadence with overlaps to remove seeing effects. Each reconstructed image is the consequent result of processing 1800 individual images resulting in a resolution significantly better than the SDO/AIA resolution achieved with the latest 304 filtergrams of 4Kpx size. The very high signal/noise ratio permits an analysis of proper motions inside the entire prominence keeping an excellent and homogeneous resolution (Fig. 1). The origins of vertical structures with elongated features moving up and down and vortices of different scales are considered.

## 2. Results

Using a Fourier local correlation tracking algorithm combined with the 3D analysis, our results demonstrate that prior to the prominence eruption, radial counter-flows with

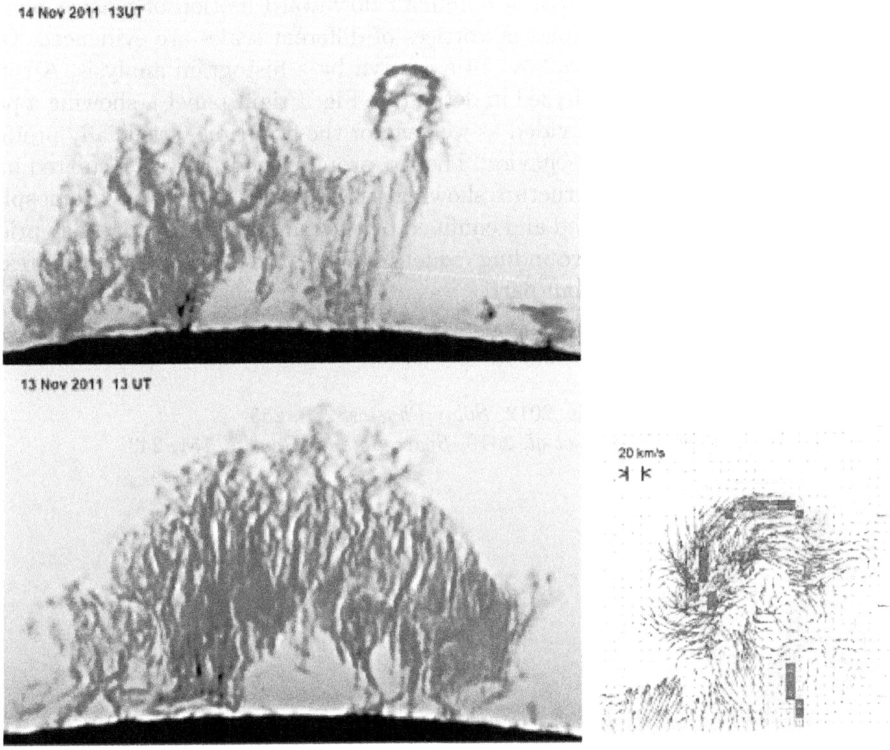


**Figure 1.** (*left panels*) Sample of a reconstructed negative image taken in H$\alpha$ at 24 hours interval. (*right panel*) Instantaneous flow map computed using the LCT code from images taken 1 min apart at a time near the beginning of the sequence. The scale is given at top in a zoomed-in view of the part with a well developed vortex structure.

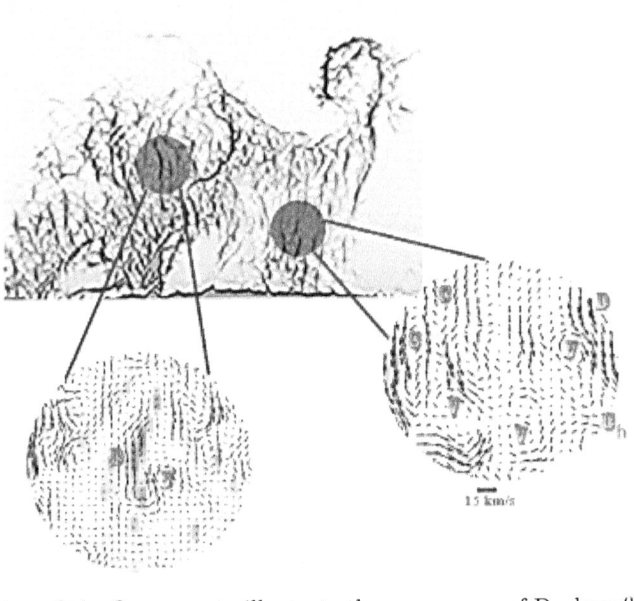

**Figure 2.** Zooming of the flow map to illustrate the occurrence of D- downflows; V- vortices; C- counterflows in vertical direction (Ch- in horizontal direction)

vortices are observed in threads with a dominant downward motion of plasma blobs or plasmoids see Fig. 2. Many examples of vortices of different scales are evidenced. Down flows are clearly still dominant on Nov. 14 as shown by a histogram analysis. A typical large-scale vortex structure is analyzed in detail (see Fig. 1 right panel ), showing a possible 3D behavior. A movie was provided as well as, for the first time, artificially produced stereograms to illustrate the 3D behavior. The H$\alpha$ prominence is indeed insulated inside a complex coronal temperature structure showing strong vortex motion of chromospheric temperature plasma still suspended and confined by the magnetic field half a day prior to the eruption. The 3D coronal surrounding evidenced using the AIA images is impressive, including hot plasma in the bottom part.

## References

Bazin, C. Koutchmy, S. & Tavabi, E. 2012, *Solar Physics*, 286, 255
Labrosse, N. Heinzel, P. Vial, J.-C. *et al.* 2010, *Space Science Review*, 151, 243

*Nature of Prominences and their role in Space Weather*
*Proceedings IAU Symposium No. 300, 2013*
*B. Schmieder, J.-M. Malherbe & S. T. Wu, eds.*

doi:10.1017/S1743921313011423

# 3D dynamical structuring of a high latitude erupting prominence: II- Analysis of the coronal context and eruption

**Serge Koutchmy**[1], **Boris Filippov**[2], **Ehsan Tavabi**[3], **Cyrille Bazin**[1] **and Sylvain Weiller**[1]

[1]IAP UMR 7095, INSU-CNRS and UPMC, 98 bis Bd Arago 75014 Paris, France
email: koutchmy@iap.fr, bazin@iap.fr, sweiller@free.fr

[2]Pushkov Institute of Terrestrial Magnetism, Ionosphere and Radio Wave Propagation, Russian Academy of Sciences, Troitsk, Moscow Region 142190, Russia
email: boris_filippov@mail.ru

[3]Payame Noor University of Tehran, 14155-6466,I.R. of Iran
email: etavabi@gmail.com

**Abstract.** Both the origin of the quiescent prominences and their eruption related to CMEs event are still a matter of extended studies. The case of high latitudes quiescent prominences producing slow CMEs can be considered as a potential component of the slow wind. A high latitude prominence was observed on November 13 - 14, 2011. A schematic representation of flux rope is proposed to describe the magnetic structure of the prominence prior to its eruption.

**Keywords.** prominence eruption, flows, counter- flows, vortices, tornado, CME

## 1. Observations

A high latitude prominence (see Fig. 1 of Koutchmy 2013a this issue) was observed on November 13 - 14, 2011, using high resolution Hα fast imaging.

Prior to the prominence eruption, radial counter-flows with vortices are observed in threads with a dominant downward motion of plasma blobs or plasmoids. Some blobs might move along field lines and at the same time could be seen as moving vertically. However, we cannot exclude that some blobs might fall through the dominantly horizontal field under the action of gravity and due to plasma instabilities. We primarily used SDO/AIA coronal and transition region (TR) filtergrams processed to improve the signal/noise ratio, see Fig. 1 left panel, and SoHO/LASCO images, to look at the details of the erupting southern high-latitude prominence on 14 - 15 November 2011 and the corresponding very low speed CME event (v = 163 km/s from LASCO data processed at NRL). The intermingled cool chromospheric and TR details (in white in Fig. 1) with the hot more stretched coronal details (in dark in Fig. 1) are particularly striking, in apparent contrast with the usual picture of a cavity surrounding the prominences. The prominence is here a part of a southern polar crown filament which is too faint and possibly too high to derive its chirality from the observed internal fine structure as it is usually possible for active region filaments, although internal motions observed within the prominence in Hα suggest negative helicity of the flux rope, which is not unusual but not typical for the southern hemisphere. However, the prominence is not stretched strictly in longitudinal direction but turns to the North in its left-hand side, see Fig. 1 right panel. Therefore, we believe the prominence is seen from the Earth at the angle about 45° to its long axis or direction of the polarity inversion line.

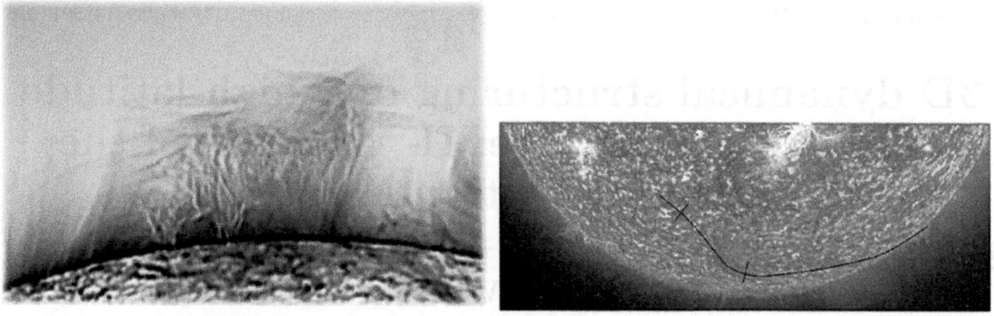

**Figure 1.** (*left panel*) Negative processed coronal image from a superposition of 200 elementary images taken with the FeXII 195 filter of AIA/SDO during our observations. (*right panel*) Filament channel (black line) deduced from the STEREO/B 304 SECCHI images on 01-15 November 2011 drawn on the image on 11 November at 06 UT.

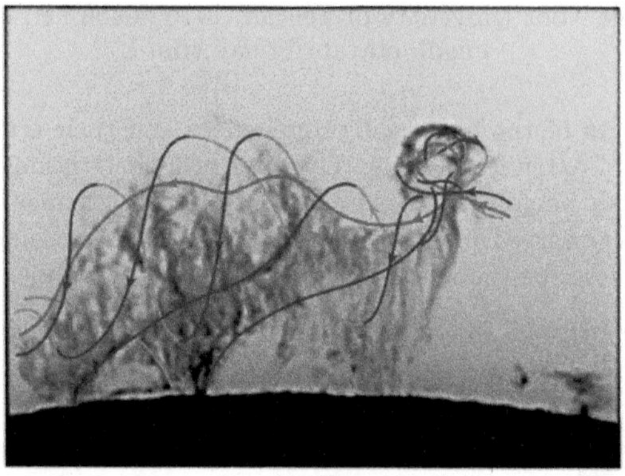

**Figure 2.** Negative H$\alpha$ image of the prominence on 14 November with schematic representation of flux-rope helical field lines. Parts of the field lines located before (behind of) the prominence symmetry plane have blue (green) color. Red line shows the flux-rope axis or the filament spine.

## 2. Results

A 3D structure is proposed to represent the prominence (Fig. 2). The details of the eruption are analyzed using additional SECCHI (STEREO mission of NASA) simultaneous filtergrams, see Fig. 1 right panel and the LASCO coronagraph data. The magnetic context is discussed, including the use of a computed PFSS map, although photospheric magnetic fields patches inferred from HMI magnetograms of SDO are rather weak near the filament location and do not show a signature. A likely scenario is suggested taking into account the 3D peculiarities for explaining this very low speed CME.

### References

Engvold, *et al.* 1989, *Hvar Obs. Bull.*, 13, 205

Filippov, B. & Koutchmy, S. 2008, *Ann. Geophys.*, 26, 3025

Koutchmy , S. Filippov, B. Tavabi, E. Weiller, S., & Bazin, C. 2013, *IAUS-300 Proc.* Poster I

Regnier, S. Walsh, R. N., & Alexander, C. E. 2011, *Astron. Astrophys.*, 533, L1

*Nature of Prominences and their role in Space Weather*
*Proceedings IAU Symposium No. 300, 2013*
*B. Schmieder, J.-M. Malherbe & S. T. Wu, eds.*

doi:10.1017/S1743921313011435

# Propagating waves transverse to the magnetic field in a solar prominence

**Therese Kucera[1], Brigitte Schmieder[2], Kalman Knizhnik[3,1], Arturo Lopez-Ariste [4], Manuel Luna[5] and David Toot[6]**

[1]NASA/GSFC, Greenbelt, MD, USA, email: therese.a.kucera@nasa.gov
[2]Observatoire de Paris, LESIA, Meudon, 92195, France,
email: brigitte.schmieder@obspm.fr
[3]Johns Hopkins University, Baltimore, MD USA [4]THEMIS, CNRS , E38205 LaLaguna, Spain
[5]Instituto de Astrofsica de Canarias, E38205 LaLaguna, Spain
[6]Alfred University, Alfred, NY, USA

**Abstract.** We have observed a quiescent prominence with the Hinode Solar Optical Telescope (SOT) (Ca II and Hα lines), Sacramento Peak Dunn Solar Telescope using the Universal Birefringent Filter (DST/UBF, in Hα, Hβ and Sodium-D lines), THEMIS (Télescope Héliographique pour l Etude du Magnétisme et des Instabilités Solaires/MTR (Multi Raies) spectromagnetograph (He $D_3$), and the Solar Dynamics Observatory Atmospheric Imaging Assembly (SDO/AIA) in EUV over a 4 hour period on 2012 October 10. The small fields of view of the SOT, DST, and MTR are centered on a large prominence footpoint extending towards the surface. This feature appears in the larger field of view of the AIA/304 Å filtergram as a large, quasi-vertical pillar with loops on each side. The THEMIS/MTR data indicate that the magnetic field in the pillar is essentially horizontal and the observations in the optical domain show a large number of horizontally aligned features in the pillar. The data are consistent with a model of cool prominence plasma trapped in the dips of horizontal field lines. The SOT and DST data show what appear to be moving wave pulses. These pulses, which include a Doppler signature, move vertically, perpendicular to the field direction, along quasi-vertical columns of horizontal threads in the pillar. The pulses have a velocity of propagation of about 10 km/s, a wavelength about 2000 km in the plane of the sky, and a period about 280 sec. We interpret these waves in terms of fast magnetosonic waves.

**Keywords.** Sun: prominences, magnetic field, oscillations

## 1. Prominence observed with Hinode/SOT, DST and THEMIS

The prominence is observed by SDO/AIA (Fig. 1 left panel) in 304 Å. The fields of view of Hinode/SOT in Ca II H (Fig. 1 middle panel) and of the THEMIS vector magnetograph working in He $D_3$ are pointed at the central footpoint of the prominence. The raw data of the THEMIS/MTR mode was reduced with the DeepStokes procedure (Lopez *et al.* 2009). The Stokes profiles are fed to an inversion code based on Principal Component Analysis (Casini *et al.* 2003). The database used contains 90,000 profiles. The inferred magnetic field strength is in the range 5-10 Gauss, with an uncertainty of 2 Gauss, the inclination is estimated to be $90° \pm 10°$, and the azimuth is about 110° from the plane defined by the line of sight and local vertical. Thus we conclude that the magnetic field in the locations of the cool prominence material is horizontal and about 30° from the plane of the sky (Schmieder *et al.* 2013).

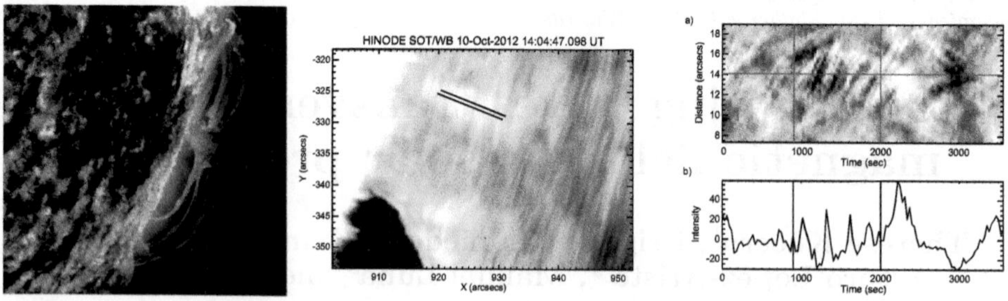


**Figure 1.** Prominence of 2012 October 10, observed by SDO/AIA in 304 Å (left), by Hinode/SOT in Ca II H (middle), and an intensity map as a function of time along a long narrow box through the brightest column (top right), and a cut along a horizontal line in the time vs. distance plot (bottom right).

## 2. Oscillations: Observations and Theory

We integrated the intensity across a 0.55″(5-pixel) wide area positioned across the oscillating region (Fig. 1 middle panel). The distance between each intensity peak (Fig. 1 right panels) is approximately 2000 km. The slopes of the intensity peaks in these plots of time vs. distance correspond to the upwards velocity of the moving features, approximately 10 km/s. Fourier and wavelet analyses give a wave period of $277 \pm 50$ sec, which appears to be approximately constant. We have modeled the oscillations as a fast magnetosonic wave in an uniform atmosphere. The driver oscillates three times with a period of 300 sec, similar to the observed one. The prominence-slab has a total width of 10 Mm. The magnetic field is taken to be uniform and horizontal ($B$ = 7.5 Gauss). It is a similar equilibrium configuration to that of Joarder and Roberts (1992). The model prominence electron density ($10^{11}\mathrm{cm}^{-3}$) is 200 times larger than that of the surrounding corona and also on the high end of measured prominence densities (Labrosse *et al.* 2010). The temperature of the plasma is 8000 K. We impose pressure perturbations.

## 3. Conclusion

The field strength in the bright columns is 5-10 Gauss and the magnetic field vector is mainly horizontal, confirming previous results. Propagating waves with periods around 300 sec are detected. The Doppler shifts measured in H$\beta$ show maxima during the passage of the wave. The driver of these waves is unknown. A simulation of a train of waves traveling upward producing disturbances of the gas pressure, the velocity and the magnetic field was performed. The wave front is almost planar in the prominence, oscillating vertically and horizontally in the corona with larger perturbations in the cool plasma than in the corona. The phase speed is the fast magnetosonic velocity of a uniform medium. This simulation demonstrates that the propagating waves are consistent with a fast magnetosonic wave confined to the prominence. However, the phase speed is much larger than the one observed. This could be due to projection effects.

## References

Casini R., Lopez Ariste A., Tomczyk S. & Lites B. W. 2003, *ApJ*, 598, L67

Labrosse N., Heinzel P., Vial J.-C., Kucera T., Parenti S., Gunár S., Schmieder, B. & Kilper G. 2010, *Space Sci. Rev.*, 151, 243

Lopez-Ariste A., Asensio Ramos A., Manso Sainz R., *et al.* 2009, *ApJ*, 501, 729

Joarder P. S. & Roberts B. 1992, *A&A* 261, 625

Schmieder B., Kucera T., Knizhnik K., Lopez-Ariste A., Luna M. & Toot D. 2013, *ApJ*, 777, 108

*Nature of Prominences and their role in Space Weather*
*Proceedings IAU Symposium No. 300, 2013*
*B. Schmieder, J.-M. Malherbe & S. T. Wu, eds.*

doi:10.1017/S1743921313011447

# High-resolution spectroscopy of a giant solar filament

**Christoph Kuckein, Carsten Denker and Meetu Verma**
Leibniz-Institut für Astrophysik (AIP), An der Sternwarte 16, 14482, Potsdam, Germany
email: ckuckein@aip.de

**Abstract.** High-resolution spectra of a giant solar quiescent filament were taken with the Echelle spectrograph at the Vacuum Tower Telescope (VTT; Tenerife, Spain). A mosaic of various spectroheliograms (H$\alpha$, H$\alpha \pm 0.5$ Å and Na $D_2$) were chosen to examine the filament at different heights in the solar atmosphere. In addition, full-disk images (He I 10830 Å and Ca II K) of the Chromspheric Telescope and full-disk magnetograms of the Helioseismic and Magnetic Imager were used to complement the spectra. Preliminary results are shown of this filament, which had extremely large linear dimensions ($\sim 740''$) and was observed in November 2011 while it traversed the northern solar hemisphere.

**Keywords.** Sun: filaments, Sun: chromosphere, Sun: photosphere, techniques: spectroscopic, techniques: high angular resolution

## 1. Introduction

Filaments are large structures observed in the solar corona or chromosphere. On the disk, they are seen as elongated dark features whereas above the limb they appear bright against the dark background and are called prominences. They have been observed for many centuries and are largely classified into two groups: quiescent (QS) filaments and active region (AR) filaments. The so-called polar crown filaments are long QS filaments that lie at high latitudes ($> 45°$) and usually form a crown around the pole (Cartledge *et al.* 1996). The observed filament in this work does not fit into the scheme of a polar crown filament. It is rather a QS filament which is particularly interesting owing to its large linear dimensions and its location across the northern solar disk. In the literature, only a few examples of giant QS filaments are found (e.g., Yazev & Khmyrov 1988).

## 2. Observations

The ground-based observations of the filament were acquired in 2011 November 15 with the Echelle spectrograph at the VTT. The good seeing conditions made it possible to scan with the slit the whole filament along the northern hemisphere of the Sun. The observing strategy was to divide the filament into ten pieces. For each piece the scanned area was $100'' \times 182''$ and consecutive scans slightly overlapped to assure the continuity of the filament and to facilitate the subsequent reconstruction of the mosaic. The whole filament was scanned between 11:38 UT and 13:11 UT, starting at heliographic coordinates (59° E, 48° N) and ending at (14° W, 18° N). The filament extends roughly $740''$ (536 Mm) from solar East to West.

Two CCD cameras were mounted at the Echelle spectrograph to acquire two different spectral regions. The first one was centered at the chromospheric H$\alpha$ line at 6562.8 Å and spanned a spectral range of 8 Å. The second region was centered at the Na $D_2$ line at 5889.9 Å and covered a spectral range of 7 Å.

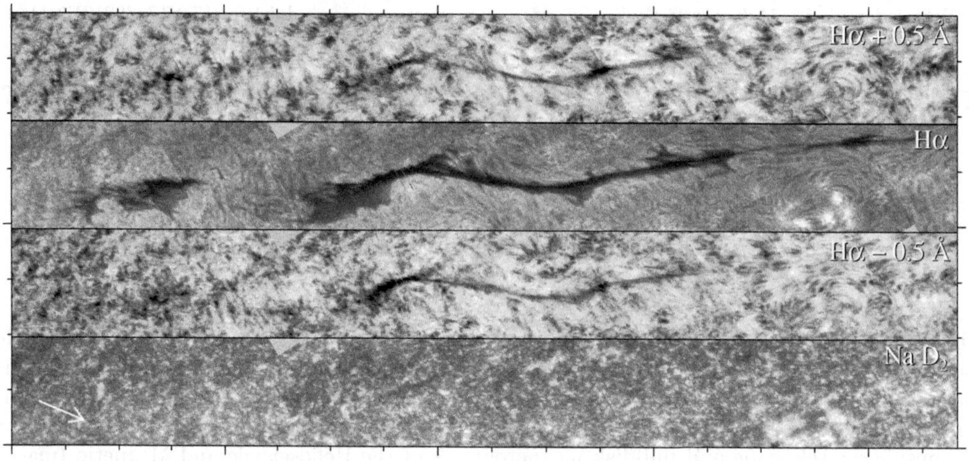


**Figure 1.** The spectroheliograms were assembled from ten partially overlapping slit-reconstructed images: Hα + 0.5 Å, Hα line core, Hα − 0.5 Å and Na $D_2$ (*top to bottom*). On the right hand side an emerging flux region appears close to the filament. Major tick marks are separated by 200″. The white arrow points towards the disk center.

In addition, full-disk images of the Chromospheric Telescope (ChroTel; Bethge *et al.* 2011) attached to the VTT building were used for this study. ChroTel acquires images at several wavelengths (Hα, He I 10830 Å and Ca II K) with a cadence of 3 min. Furthermore, full-disk magnetograms of the Helioseismic and Magnetic Imager instrument (HMI; Schou *et al.* 2012) were used to complement the spectra. The magnetograms confirm that the filament lies on top of the polarity inversion line (PIL). Positive (negative) polarity is above (below) the filament as shown in Fig. 1.

## 3. Results

The filament with extraordinary linear dimensions shows a gap on the left hand side in the Hα panel of Fig. 1. A few hours before, plasma was removed from that side by a coronal mass ejection (CME). However, thin threads appear to link both ends of the separated filament. Barbs, i.e., groups of thin threads that protrude from the side of the main body of the filament (e.g., Martin 1998), are detected. The spine of the filament is also seen at Hα ± 0.5 Å. In these areas the Hα line is extremely broad. Very little absorption at the filament can be seen in the Na $D_2$ line core panel.

### Acknowledgements

CK greatly acknowledges the travel support received from the IAU. CD was supported by grant DE 787/3-1 of the German Science Foundation (DFG). MV thanks the German Academic Exchange Service (DAAD) for its support in the form of a PhD scholarship.

### References

Bethge, C., Peter, H., Kentischer, T. J., *et al.* 2011, *Astron. Astrophys* 534, A105
Cartledge, N. P., Titov, V. S., & Priest, E. R. 1996, *Solar Phys.* 166, 287
Martin, S. F. 1998, *Solar Phys.* 182, 107
Schou, J., Scherrer, P. H., Bush, R. I., *et al.* 2012, *Solar Phys.* 275, 229
Yazev, S. A., & Khmyrov, G. M. 1988, *Adv. Space Res.* 8, 199

*Nature of Prominences and their role in Space Weather*
*Proceedings IAU Symposium No. 300, 2013*
*B. Schmieder, J.-M. Malherbe, & S. T. Wu, eds.*

doi:10.1017/S1743921313011459

# Prominences in SDO/EVE spectra: contributions from large solar structures

**Nicolas Labrosse**[1], **Hugh Hudson**[1,2] **and Maria Kazachenko**[2]

[1]SUPA, School of Physics and Astronomy, University of Glasgow, Glasgow G12 8QQ, UK
email: Nicolas.Labrosse@glasgow.ac.uk

[2]Space Sciences Laboratory, University of California, USA

**Abstract.** The EVE instrument on SDO is making accurate measurements of the solar spectral irradiance in the EUV between 30 and 1069 Å, with 1 Å spectral resolution and 10 s sampling rate. These data define solar variability in the "Sun-as-a-star" mode and reveal many interesting kinds of variation. Its high sensitivity also makes it suitable for spectroscopic diagnostics of solar features such as flares. Here we present EVE's potential contribution to the diagnostics of large-scale, slowly evolving features such as prominences and active regions, and what we can learn from this.

**Keywords.** Sun: activity, Sun: prominences

## 1. Prominences in EVE spectra

We use AIA and EVE to study the Sun-as-a-star variability in the time range from 2010-06-19 20:00UT until 2010-06-20 06:00UT. Two active regions (NOAA 11082 and 11083) were visible on the disk, though they did not produce any flare in this time period. Two prominence eruptions occurred on the East limb during these observations.

On Fig. 1, a quiescent prominence can be seen at location (-921″, 371″). It erupts between 23:48UT and 3:40UT. This eruption is associated with a GOES A6.4 disturbance. The left panel of Fig. 2 shows that a clear signature in AIA 304 of the NE prominence eruption can be seen, well correlated with an increase in irradiance at 304 Å observed by EVE. Another prominence eruption takes place between 19:47UT and 23:10UT on 19th June on the South East limb (-939″,-407″) but its signature in EVE or AIA signals is less clear. The increasing activity near 06:00UT is due to a filament activation in an active region. Note the overall correlation between AIA intensity across its entire FOV and EVE irradiance. The right panel of Fig. 2 shows that the NE prominence event is not only detected by EVE at 304 Å but also in four Fe lines from Fe XIII to Fe XVI, revealing plasma heating during the eruption.

## 2. Active regions in EVE spectra

A single large active region can temporarily dominate the EVE spectra (Fig. 3). In July 2012, NOAA AR11520 became the dominant sunspot group, and we believe that the time-series data can be corrected to permit the use of many EVE lines to characterize the coronal luminosity of this region. During more confused times, with many regions, we can resort to correlation analysis based upon magnetic proxy indices.

The left panel of Fig. 3 shows irradiance evolution in one of the many EVE emission lines, Fe XV 284 Å. The right panel shows the total unsigned magnetic flux of the whole solar disk derived from the HMI/SDO. Dashed vertical lines show moments when the

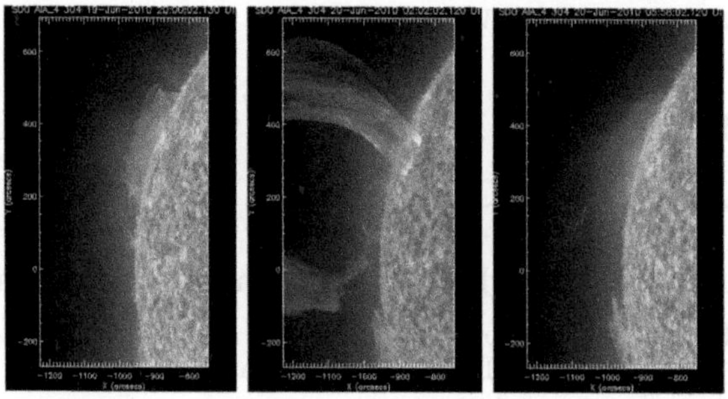

**Figure 1.** AIA 304 Å observations of the NE eruption at (from left to right) 20:00UT (2010-06-19), 02:02UT, and 05:58UT (2010-06-20).

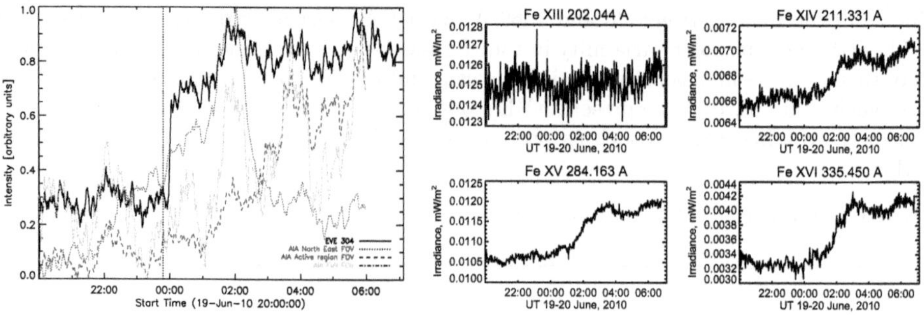


**Figure 2. Left:** Light curves from EVE and AIA at 304 Å for the 20 June 2010 event. The sudden increase in EVE 304 Å at 00:00UT is an artefact, but the irradiance increase before 02:00UT correlates with the AIA signal from the NE prominence. The vertical dotted line shows the start of the eruption. **Right:** Four EVE lines, at different Fe ionization states, that are well-observed and show hints of the prominence activity.

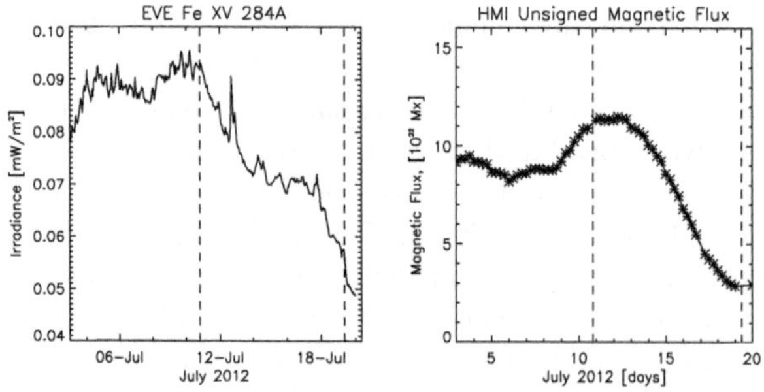


**Figure 3.** Irradiance in Fe XV 284 Å, and total unsigned magnetic flux of the whole solar disk.

AR NOAA 11520 was at the disk center and away from the visible disk. The latter can be used as a quiet sun level for studying the AR properties.

## 3. Conclusions

EVE data can be used to analyse individual large-scale events, such as prominences and active regions. We believe that other and much better examples will be found in future surveys. The signatures are difficult to interpret without AIA, but the combination of these instruments should provide good characterizations of different solar features, for which the full EVE spectroscopy can be deployed for physical characterizations.

*Nature of Prominences and their role in Space Weather*
*Proceedings IAU Symposium No. 300, 2013*
*B. Schmieder, J.-M. Malherbe, & S.-T. Wu, eds.*

doi:10.1017/S1743921313011460

# Coronal Condensation in Funnel Prominences as Return Flows of the Chromosphere-Corona Mass Cycle

**Wei Liu**[1], **Thomas E. Berger**[2] **and B. C. Low**[3]

[1]Stanford-Lockheed Institute for Space Research, HEPL Solar Physics, 452 Lomita Mall, Stanford, CA 94305-4085, USA; email: weiliu@sun.stanford.edu
[2]National Solar Observatory, 950 N. Cherry Avenue, Tucson, AZ 85719, USA
[3]High Altitude Observatory, P.O. Box 3000, Boulder, CO 80307, USA

**Abstract.** We present *SDO*/AIA observations of a potentially novel type of prominence, called "funnel prominence", that forms out of coronal condensation at magnetic dips.
They can drain a large amount of mass (up to $\sim 10^{15}$ g day$^{-1}$) and may play an important role as return flows of the chromosphere-corona mass cycle.

**Keywords.** Sun: chromosphere, Sun: corona, Sun: filaments, prominences, Sun: magnetic fields

---

It has recently been recognized that a significant portion of the mass of the solar atmosphere is cycled between the hot, tenuous corona and the underlying cool, dense chromosphere (Berger *et al.* 2011; McIntosh *et al.* 2012; see a review by T. Berger in this proceedings). Like the water cycle on Earth, hot plasma is transported upward into the corona in such forms as spicules and flux emergence manifested as prominence bubbles and plumes, while cool plasma condenses out of the corona and drains back to the chromosphere via vertical prominence threads and transient coronal rain. One likely cooling mechanism is the radiative cooling instability (e.g., Karpen & Antiochos 2008) in magnetized plasma that effectively inhibits across-field thermal conduction.

Recent *SDO*/AIA observations have revealed a potentially new type of prominence, called "funnel prominence", that is an evident manifestation of this cooling process. As shown in Fig. 1, its distinct feature is a funnel shape formed by the combination of dipped coronal loops best seen at 171 Å and cool 304 Å material appearing and draining from the dips. Usually an emission cloud lasting for hours progressively appears at lower heights and in cooler EUV channels from 211 Å ($\sim$2.0 MK) to 193 Å ($\sim$1.6 MK), and then 171 Å ($\sim$0.8 MK), indicating a continual cooling process that eventually leads to the in-situ condensation of the 304 Å ($\sim$0.08 MK) prominence material (Liu *et al.* 2012; Berger *et al.* 2012). Field extrapolation from *SDO*/HMI magnetograms confirmed the existence of such magnetic dips at the initial condensation site (Liu *et al.* in prep.), which provide access to a large coronal volume for cooling mass to be collected and channeled to its lower portion for condensation. The condensed mass subsequently drains along meandering paths down to the chromosphere, suggestive of cross-field slippage of cool, poorly ionized mass that can involve magnetic reconnection (Low *et al.* 2012a,b). We found an average drainage velocity of $30 kmps$ at a slower than free-fall acceleration of g/ 6 (Liu *et al.* 2012). A moderate-sized funnel prominence can drain a significant mass $\sim 10^{15}$ g day$^{-1}$, comparable to a fraction of the entire corona or a typical CME mass, suggestive of its important role as return flows of the chromosphere-corona mass cycle.

In a preliminary survey of the first three years (2010–2013, near the solar maximum) of *SDO*/AIA data, we have found 13 funnel prominences as summarized in Table 1. They appear off-limb at various locations predominantly at high latitudes $>50°$, consisting of a single or multiple columns on the order of 70 Mm tall and 20 Mm wide

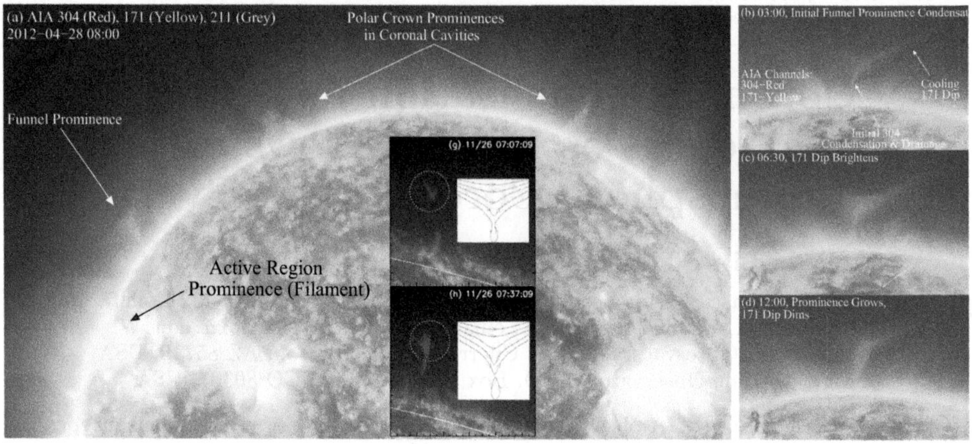


**Figure 1.** (a) Composite *SDO*/AIA 304 Å (red), 171 Å (yellow), and 211 Å (grey) images showing a funnel prominence and two classical types of prominences. The inset, taken from Fig. 2 of Liu *et al.* (2012), shows initially a V-shaped 304 Å condensation in another funnel prominence and a possible magnetic field configuration. (b)–(d) Enlarged view of the funnel prominence in (a) rotated to the local vertical up showing a cooling/condensation sequence indicated by the changing dip brightness and prominence size.

**Table 1.** Preliminary sample of 13 funnel prominences detected by *SDO*/AIA.

| Year | 2010 | 2011 | 2012 | | | | | | | | 2013 | | |
|---|---|---|---|---|---|---|---|---|---|---|---|---|---|
| Date | 11/26 | 10/14 | 03/15 | 03/17 | 04/28 | 06/24 | 06/27 | 06/29 | 07/13 | 09/30 | 05/11 | 06/21 | 07/13 |
| Location | W86S13 | E100N61 | W79N57 | W40N80 | E83N39 | W88S42 | E91N58 | E92N68 | E90N75 | W95N63 | W82N52 | E87N77 | E96N68 |
| Eruptive? | No | No | No | Yes | No | No | No | No | No | Yes | No | Yes | No |

(e.g., Fig. 1) with lifetime of hours to a day or two. Some funnel prominences disappear probably because of depletion by drainage, while others erupt together with their hosting magnetic dips. These characteristics distinguish funnel prominences from active region prominences and quiescent polar-crown prominences within coronal cavities situated along polarity inversion lines (PILs) (see Fig. 1); they may or may not require a PIL or flux-rope cavity. Meanwhile, funnel prominences share some morphological similarities with quiescent prominences, such as the conical shapes at the tops of funnels and the "horns" in quiescent prominences (Berger 2012), both seen as 171 Å emission. Funnel prominences are also similar to "cloud prominences" (Lin *et al.* 2006) in that they both involve downflows streaming out of high-altitude condensations. However, some spider-shaped cloud prominences show drainage along well-defined curved paths at close to free-fall speeds (Allen *et al.* 1998), resembling coronal rain sliding down coronal loops, while funnel prominences drain along meandering paths at much lower speeds (yet 2–3 times faster than those of quiescent prominences) possibly due to cancellation of gravity by a Lorentz force at the magnetic dips. Further analysis is underway to validate funnel prominences as a new prominence type and to investigate their physical nature.

### References

Allen, U. A., Bagenal, F., & Hundhausen, A. J. 1998, in *ASP-CS*, Vol. 150, ed. D. F. Webb, *et al.* 290
Berger, T. 2012, in *ASP-CS*, Vol. 463, ed. T. R. Rimmele, A. Tritschler, F. Wöger, *et al.* 147
Berger, T. E., Testa, P., Hillier, A., *et al.* 2011, *Nature*, 472, 197
Berger, T. E., Liu, W., & Low, B. C. 2012, *ApJ*, 758, L37
Karpen, J. T. & Antiochos, S. K. 2008, *ApJ*, 676, 658
Lin, Y., Martin, S. F., & Engvold, O. 2006, *BAAS*, 38, 219; AAS/SPD Meeting #37, #1.21
Liu, W., Berger, T. E., & Low, B. C. 2012, *ApJ*, 745, L21
Low, B. C., Berger, T. E., Casini, R., & Liu, W. 2012a, *ApJ*, 755, 34
Low, B. C., Liu, W., Berger, T. E., & Casini, R. 2012b, *ApJ*, 757, 21
McIntosh, S. W., Tian, H., Sechler, M., & De Pontieu, B. 2012, *ApJ*, 749, 60

*Nature of Prominences and their role in Space Weather*
*Proceedings IAU Symposium No. 300, 2013*
*B. Schmieder, J.-M. Malherbe & S. T. Wu, eds.*

doi:10.1017/S1743921313011472

# Two distinct peculiar "dimming channels" observed by SDO/AIA

**Suli Ma and Huadong Chen**

College of Science, China University of Petroleum,
Qingdao 266580, China
email: sma@upc.edu.cn

**Abstract.** In this work, we report two distinct peculiar "dimming channels" observed in all the seven EUV wavelengths around AR 11520 by SDO/AIA on July 12, 2012. Our results show that: (1) the two dimming channels are very narrow and the intensity in them dropped fierce; (2) specially, some flare ribbons appeared at the edge and prior to the appearance of dimming channels, which is a rare phenomenon; (3) the dimming channels seem to be located at the boundaries of some magnetic networks (or supergranules).

**Keywords.** Sun: Dimming, Sun: Flare, Sun: Filament, Sun: Solar Eruption, Sun: CME

## 1. Introduction

Coronal dimming sometimes referred to as transient coronal holes are common phenomena associated with coronal mass ejections (CMEs). They are usually observed as decrease in intensity in soft X-rays (e.g., Sterling and Hudson, 1997) and extreme ultra-violet (EUV) data (e.g., Thompson *et al.*, 1998). However, dimmings are infrequent in the chromosphere (e.g., Neidig *et al.*, 1997 and Jiang *et al.*, 2003). In this paper, we report two well-confined dimmings observed from chromosphere to corona.

## 2. Observations and Results

On July 12, 2012, a solar eruption occurred in the active region AR 11520, which was associated with an X1.4 class flare, a partial halo CME and an EUV wave. Unusually, two distinct narrow dimming channels, marked by "Dim1" and "Dim2" in the panels c through e in Figure 1, are observed during this eruption. The "dimming channels" appeared in seven EUV wavelengths simultaneously for several hours. They came into forth gradually from the center of the solar eruption to the surrounding area. Specially, the dimming channels are embraced by bright flare ribbons at the beginning of the dimming channels appeared, which is distinctly different from the traditional dimmings.

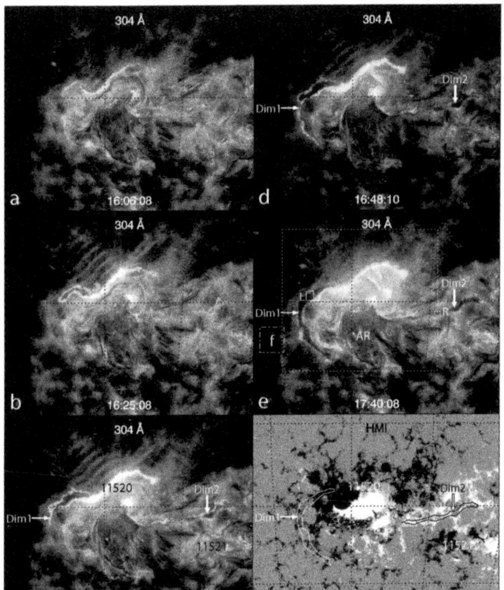


**Figure 1.** SDO/AIA and SDO/HMI observation showing the formation process of the two dimming channels and the surrounding magnetic field. The four boxes marked with "L", "R", "f" and "AR" indicate the areas which we chose to measure the variation in intensity.

The panel f of Figure 1 is SDO/HMI image, the white and black contours indicate the position of Dim1 and Dim2, respectively. It shows that Dim1 is located in the negative magnetic field, while Dim2 is in the positive one. Furthermore, the panel f also indicates that the dimming channels are located at the boundaries of some magnetic networks (or supergranules).

In Figure 2 are plotted intensity profiles in different EUV wavelengths in the given areas indicated by the boxes in the panel e of Figure 1. In the rising phase of the X1.4 flare, the intensities of the active region at all the seven wavelengths are increasing, especially at 94Å and 131Å , which increase suddenly like GOES flux. In the surrounding reference area, the intensity at 94Å, 131Å and 171Å showed increase while at other wavelengths decrease. The intensity in the Dim1 increased first and then decrease except in 335Å, which decrease monotonously. The intensity at 193Å, 211Å and 335Å dropped more deep, which can be dropped upto 80% of its value, than at the other wavelengths. The intensity in Dim2 displayed a distinct drop at 171Å, 193Å and 211Å as compared to other wavelengths.

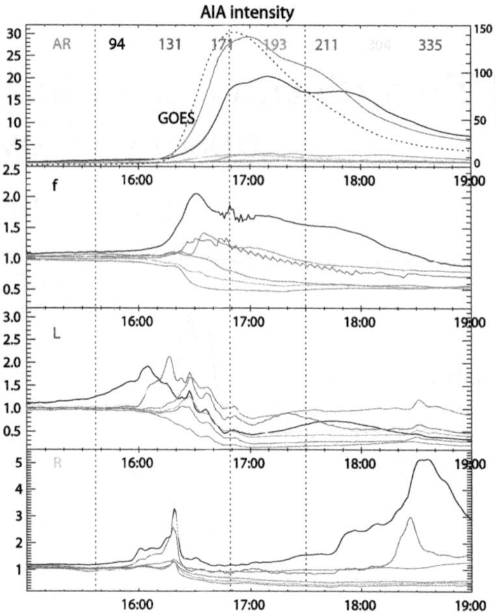


**Figure 2.** The time intensity profile of the dimming areas as well as with the active region and reference area at different EUV wavelengths. The dotted line in the top panel indicates GOES 1-8 Å flux and the three vertical lines represent start, peak and end time of the corresponding flare.

## 3. Summary

In this paper, we report two dimming channels associated with a coronal sigmoid structure eruption. Our results show that: (1) the dimming channels appeared in all the seven EUV wavelengths observed by AIA, which supports that the idea that the dimming channel are density depletion (e.g., Hudson *et al.*,1996); (2) the appearance of the dimming channels at 304Å and the appearance of flare ribbons might imply the solar eruption is strong enough to disturb or change the magnetic field in the low corona and chromosphere; (3) the borders of the dimming channels coincide with the boundaries of some magnetic networks, where the magnetic field is stronger than the surrounding areas; (4) the gradual formation of the dimming channels and flare ribbons might be due to a slipping magnetic reconnection. A detailed study is needed to establish/confirm this.

**Acknowledgements.** We thank the SDO/AIA teams for data support. The work was supported by the NSFC (41204124, 11103090, 40825014 and 41331068), SPNSF (ZR2011AQ009) and CAS/NAO (KLSA201209).

## References

Jiang, Y., Ji, H., Wang, H., & Chen, H. 2003, *ApJ*, 597, L161
Hudson, H. S., Acton, L. W., & Freeland, S. L. 1996, *ApJ*, 470, 629
Neidig, D. F., *et al.* 1997, *Sol. Phys.*, 170, 321
Sterling, A. C. & Hudson, H. S. 1997, *ApJ*, 491, L55
Thompson, B. J., Plunkett, S. P., Gurman, J. B., *et al.* 1998, *Geophys. Res. Lett.*, 25, 2465

*Nature of Prominences and their role in Space Weather*
*Proceedings IAU Symposium No. 300, 2013*
*B. Schmieder, J.-M. Malherbe & S. T. Wu, eds.*

doi:10.1017/S1743921313011484

# Where Do Solar Filaments Form?

**Duncan H Mackay[1], Victor Gaizauskas[2] and Anthony R. Yeates[3]**

[1]School of Mathematics and Statistics, University of St Andrews, St Andrews, KY16 9SS, UK
email: `duncan@mcs.st-and.ac.uk`

[2]HIA, CNRC, 100 Sussex Drive, Ottawa, Ontario, Canada

[3]Department of Mathematical Sciences, Durham University, Durham, UK

**Abstract.** In the present study, we consider where large, stable solar filaments form relative to underlying magnetic polarities. We find that 92% of all large stable filaments form in magnetic configurations involving the interaction of two or more bipoles. Only 7% form above the Polarity Inversion Line (PIL) of a single bipole. This indicates that a key element in the formation of large-scale stable filaments is the convergence of magnetic flux, resulting in either flux cancellation or coronal reconnection.

**Keywords.** Sun: filaments, Sun: magnetic fields, Sun: prominences

## 1. Introduction

Solar filaments (prominences) form over a wide range of latitudes on the Sun. One early classification scheme splits them into two categories. Tandberg-Hanssen (1995) describes these two categories as Type A (bipolar region filament) and Type B (between bipolar region filament) respectively. Observations by Tang (1987) have shown that over 60% of all filaments form between bipolar regions (Type B; see also Gaizauskas & Zwaan(1997)). We re-examine where large, stable solar filaments form relative to underlying magnetic polarities, extending the study of Tang (1987) to cover the full solar cycle.

## 2. Classification Scheme and Data Sets

To full categorise the magnetic interactions that lead to the formation of filaments we introduce 4 categories. These are,

(*a*) **Interior BR Filament (IBR)**: forms above the PIL of a single magnetic bipole.
(*b*) **Exterior BR Filament (EBR)**: forms above the PIL that lies between two or more magnetic bipoles.
(*c*) **Interior/Exterior BR Filament (I/EBR)**: forms above both the internal and external PIL surrounding a bipole.
(*d*) **Diffuse BR Filament (DBR)**: a filament that lies above a PIL that sits in a bipolar distribution at high latitudes, where the bipolar distribution at high latitudes was produced from earlier emergences and cancellations of flux at lower latitudes.

Four distinct periods are analysed over Solar Cycle 21 where in total 603 filaments are considered. The periods studied are: Set 1: 1977, CR1653-1658, 101 Filaments; Set 2: 1979, CR1680-1685, 234 Filaments; Set 3: 1982, CR1720-1725, 149 Filaments; Set 4: 1984, CR1747-1752, 119 Filaments. For each time period the classification of each filament is carried out through a 2 stage process: (1) Use synoptic data to determine filament locations and identify underlying magnetic flux and (2) Use high resolution data and simulations to classify each filament.

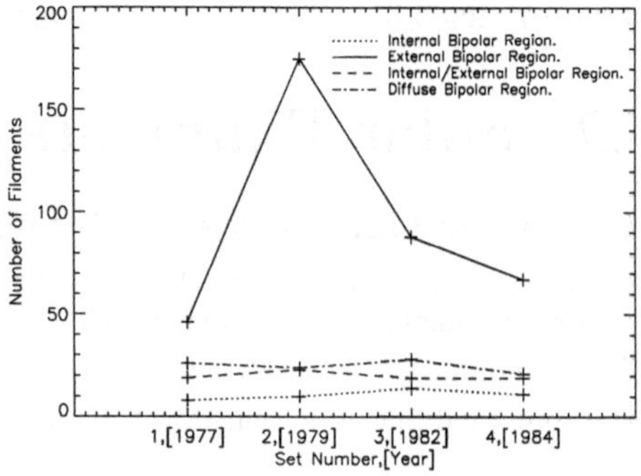


**Figure 1.** Number of filaments in each category as a function of time in Cycle 21.

**Table 1.** Percentage of filaments in each category

| Type | IBR | EBR | I/EBR | DBR | U |
|---|---|---|---|---|---|
| % | 7 | 63 | 13 | 16 | 1 |

## 3. Results

Figure 1 shows the number of filaments in each category as a function of data set number. The plus signs denote the number of filaments in each category, while the varying line-style denotes the overall cycle variation of each category. Cycle maximum lies midway between Sets 2 and 3. For each of the four data sets a similar behaviour is found. EBR filaments (solid line) dominate. This graph clearly shows that filaments prefer to form in flux systems involving multiple bipole interactions rather than just a single magnetic bipole. Throughout the cycle only EBR filaments show a strong variation. This indicates that the formation mechanism for these must be closely related to the number of bipoles on the Sun throughout the solar cycle. Table 1 gives the percentage of filaments in each category calculated over the combined data sets (where "U" is unclassified). Over 92% of all filaments (EBR, I/EBR, DBR) occur in flux distributions that are non-bipolar in nature and require the interaction of two or more bipoles.

## 4. Conclusions

Over 92% of all filaments occur in flux distributions that are non-bipolar in nature and require the interaction of two or more bipoles. Since the vast majority of large-scale stable filaments occur along PILs that are external to any one bipole, we conclude that a key element in the formation of the majority of these solar filaments must be the convergence of magnetic flux between bipoles, resulting in either flux cancellation (van Ballegooijen & Martens(1989)) or coronal reconnection (Galsgaard & Longbottom(1999)). The consequences of this for theoretical models is discussed in Mackay *et al.* (2008).

### References

Gaizauskas, V. & Zwaan, C. 1997, *Bulletin of the American Astronomical Society*, 29, 902
Galsgaard, K. & Longbottom, A. W. 1999, *ApJ*, 510, 444
Mackay, D. H., Gaizauskas, V., & Yeates, A. R. 2008, *Solar Phys.*, 248, 51
Tang, F. 1987, *Solar Phys.*, 107, 233
Tandberg-Hanssen, E. 1995, *Astrophysics and Space Science Library*, 199,
van Ballegooijen, A. A. & Martens, P. C. H. 1989, *ApJ*, 343, 971

*Nature of Prominences and their role in Space Weather*
*Proceedings IAU Symposium No. 300, 2013*
*B. Schmieder, J.-M. Malherbe & S. T. Wu, eds.*

doi:10.1017/S1743921313011496

# Spectral Observations of Filament Activation

**G. Mashnich**

Institute of Solar-Terrestrial Physics, SB RAS, P. O. Box 4026, Irkutsk, 664033 Russia
email: mashnich@iszf.irk.ru

**Abstract.** Studies of solar filament (prominence) activation and eruption are often based on measurements of intensity fluctuations in various solar emission bands and rarely on Doppler velocity measurements. The goal of this paper is to analyze the process of quiescent filament activation, using spectral data, and its associated events in solar UV band. Motions have been examined in a small southern fragment of a quiescent, extended filament in the northern hemisphere prior to and during its activation on June 14 2012. A part of the fragment disappeared after the filament activation.

## 1. Observations and method of data processing

The quiescent filament was observed by Sayan Solar Observatory's horizontal solar telescope prior to and during the activation on June 14 2012 (E11N27). A series of spectra in the $H\beta$ line region (including $\lambda$ 486.1 nm and Fe I $\lambda$ 489.7 nm in the chromosphere and photosphere respectively) was registered from 01:43 UT to 04:18 UT with a cadence of 10 s between the consecutive spectra. The mirror spectrograph's slit and an auxiliary device with an $H\alpha$-filter provide $H\alpha$-filtergrams together with an actual slit position against the filament body. The spectral slit position covered the region of filament bifurcation. The analysis of dynamics of motions in the filament also involves simultaneous events in SDO/AIA, SDO/HMI magnetograms and BBSO $H\alpha$-images. The SDO magnetograms show a complex configuration of weak polarities in the photospheric magnetic field under the region of the filament bifurcation. Velocities perpendicular to the line of sight are measured by the method of lowest intensity of $H\beta$ line (Mashnich *et al.* (2012)) and the Doppler velocity by the bisector method. Wavelet analysis for the prior to activation time interval (between 01:43UT to 03:05 UT) carried out to investigate time behaviors of the filament and the photosphere under filament Doppler velocity oscillations. The sequence of images SDO/AIA images in He II (30.4 nm) is processed by subtracting the first image prior to the filament activation from each subsequent image.

## 2. Results and discussions

Short-period (about 5 minutes) oscillations occurred across the whole width of the filament 25 minutes before the filament activation. Such oscillations are typical only for fine structures which are well-resolved at filament edges. For a given time period the profile of the velocity distribution along the slit in the photosphere was shown to reoccur in the filament with a delay of 260-280 sec. Assuming the filament height to be at 50 Mm above the photosphere, we estimated the travel velocity of the disturbance from the photosphere to the filament to be between 110 km/s and 180 km/s.

The filament activation is illustrated by the spectra (Fig. 1). In the first phase, the core filament structure rose; after that time, the edges of the filament channel were being traced by bright points in He II (30.4 nm). With the increasing rising velocity, the filament structure broke off, a bright region flared up in He II (30.4 nm), and, after the

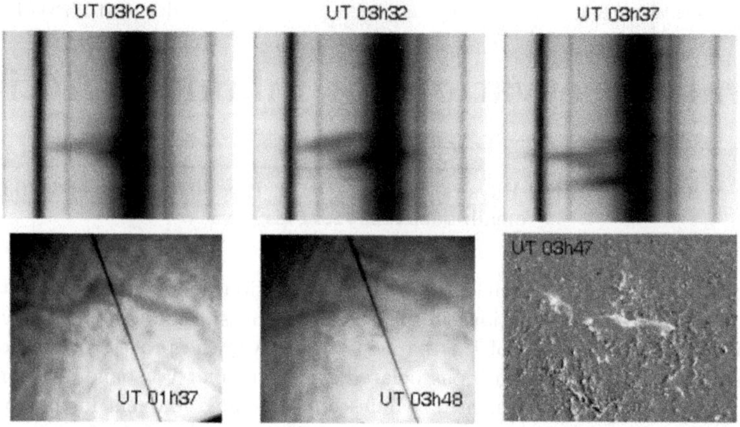


**Figure 1.** Top panel shows the sequence of spectra in the $H\beta$ line during the filament activation. The bottom panel shows the filament position on the mirror spectrograph's slit (black line) at the beginning of observations (left) and during the filament activation (middle) in the blue wing of the $H\alpha$ line. The difference image in He II (30.4 nm) during the filament activation, is on the right. Each image is labelled with the universal time (UT) at which it was taken.

maximum of filament rise velocity (75-95 km/s), bright arches began to spread from this point to opposite sides of the filament channel (Fig. 1). The bright arches were visible in other bands of SDO/AIA. This process repeated after about 50 minutes. The $H\alpha$-filament split into fragments for a short time during the filament activation and some of them disappeared (Fig. 1). The reason for the filament destabilization and acceleration is still unclear. It should be noted that there is a time correlation between the activation of this filament located in the North hemisphere with flare events in AR 11504 located in the South hemisphere visible in He II (30.4 nm) and other bands of SDO/AIA. The activation of this filament can be compared with the dynamics observed in a large quiescent polar crown filament observed with STEREO (Gosain *et al.* (2009)). Gosain *et al.* (2009) measured apparent velocities of the same order. We are measuring plasma Dopplershifts.

I am thankful to Dr. B. Schmieder for useful comments and discussion of results. I would also like to thank Mashnich V.I. for his help in data processing. This study was supported by the Programme of the Basic Fundamental Research of SB RAS No. 16.1.

## References

Gosain, S., Schmieder, B., Venkatakrishnan, R. Chandra, & Artzner, G. 2009, *Solar Phys*, 259, 13

Mashnich, G. P., Bashkirtsev, V. S., & Khlystova, A. I. 2012, *Astronomy Reports*, v. 56, 3, 241

*Nature of Prominences and their role in Space Weather*
*Proceedings IAU Symposium No. 300, 2013*
*B. Schmieder, J.-M. Malherbe & S. T. Wu, eds.*

doi:10.1017/S1743921313011502

# Column Density Measurements of a Prominence Observed by AIA

**Patrick I. McCauley, Yingna Su, Edward DeLuca, and Adriaan van Ballegooijen**

Harvard-Smithsonian Center for Astrophysics, 60 Garden street, Cambridge, MA 02138, USA

**Abstract.** We present column density measurements of a polar crown prominence observed on March 9th, 2012 by the Atmospheric Imaging Assembly (AIA) aboard the Solar Dynamics Observatory. The structure was viewed on the east limb by AIA and erupted about 30 hours after the observations shown here. We estimate column density by approximating the obscured background emission to obtain an optical depth. This can then be combined with the absorption cross sections of neutral hydrogen and helium, along with the He:H abundance ratio, to calculate column density. We perform this calculation for the 171, 193, 211, and 335 Å AIA passbands.

**Keywords.** Sun: prominences – Sun: corona – Sun: coronal mass ejections

## 1. Observations and Analysis

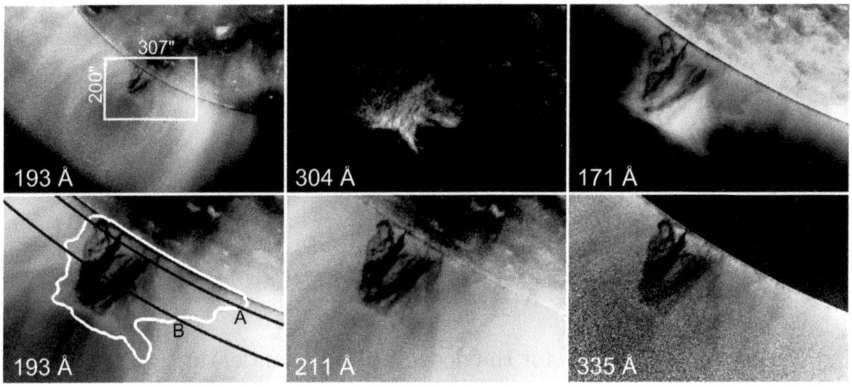


**Figure 1.** AIA prominence observations taken on 2013/03/09 at 19:35 UT. The white contour on the 193 Å image denotes the 304 Å extent, and the black arcs denote the cuts used in Fig 2.

To estimate neutral hydrogen column density ($N_{\rm HI}$), we employ a simple technique used in several past studies (e.g. Daw *et al.* 1995, Kucera *et al.* 1998, Anzer & Heinzel 2005). Given the observed intensity ($I_{obs}$) across the EUV absorption features in Fig 1, we approximate the obscured background ($I_b$) and overlying foreground ($I_f$) using a linear fit to the emission at the intersections of the white contour and concentric slices shown in the lower-left panel. We take $I_f$ to be 29% of this fit based on the deepest absorption. If a larger portion is attributed to $I_f$, then $I_{obs}$ is over-subtracted and Eqn 1.1 is NaN at certain positions. This leads to the optical depth ($\tau$), which is combined with the ratio of He:H ($r$=15%; Del Zanna *et al.* 2004) and their absorption cross sections ($\sigma_{\rm HI}$ & $\sigma_{\rm He\ I}$; Anzer & Heinzel 2005, West & Marr 1976) to yield $N_{\rm HI}$ along the line-of-sight.

$$I_{obs} = I_b e^{-\tau} + I_f \Rightarrow \tau = -\ln \frac{I_{obs} - I_f}{I_b} \tag{1.1}$$

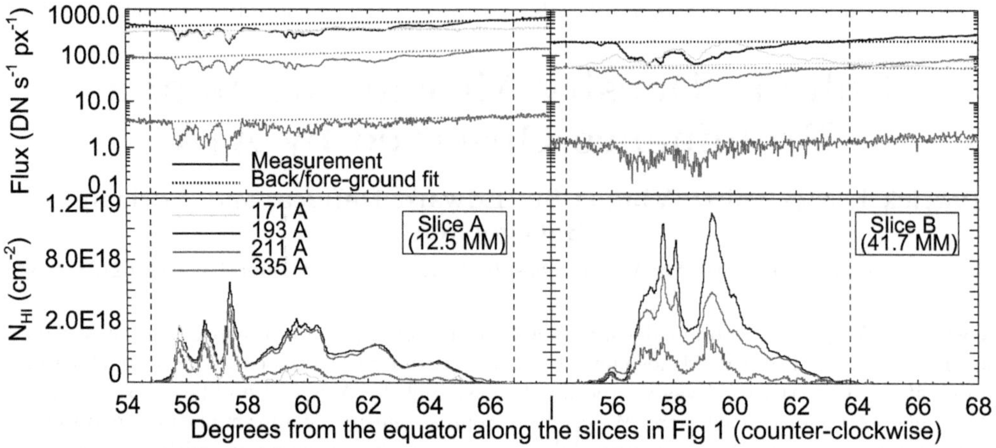


**Figure 2.** *Upper:* observed flux. *Lower:* estimated column densities. Dashed lines denote intersections of white & black contours in Fig 1, used for linear back/fore-ground fits (dotted).

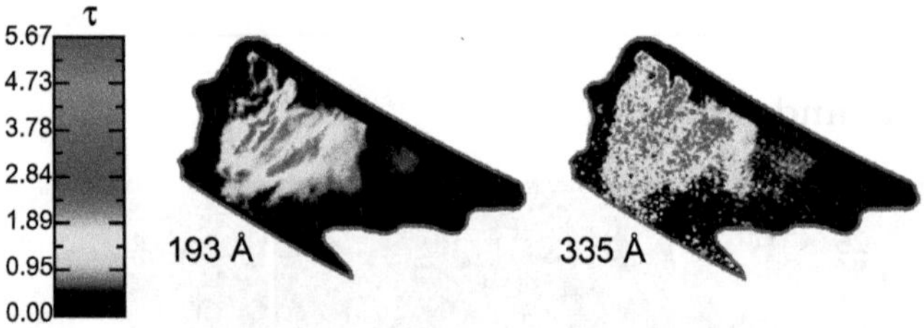


**Figure 3.** Maps of the optical depth, $\tau$.

$$\tau = \sigma_{\rm HI} N_{\rm HI} + \sigma_{\rm He\ I} N_{\rm He\ I} + \sigma_{\rm He\ II} N_{\rm He\ II} \approx (\sigma_{\rm HI} + r\sigma_{\rm He\ I}) N_{\rm HI} \quad (1.2)$$

$$\Rightarrow N_{\rm HI} \approx \frac{\tau}{\sigma_{\rm HI} + r\sigma_{\rm He\ I}} \quad (1.3)$$

We find $N_{\rm HI}$ to be $\sim 10^{19}$ cm$^{-2}$ for the deepest absorption features, which is consistent with similar work (Labrosse *et al.* 2011). Note that we are sensitive only to the neutral hydrogen, and the ionized fraction likely varies, particularly with height. This, along with a varying filling factor, may explain the discrepancy between the 335 Å channel and the other bands. See Su *et al.* (2013) for additional details on this event.

*Acknowledgements:* This project is supported by NASA grant NNX12AI30G and NASA contract SP02H1701R from LMSAL to SAO.

## References

Anzer, U. & Heinzel, P. 2005, *ApJ*, 622, 714
Daw, A., DeLuca, E. E., & Golub, L. 1995, *ApJ*, 453, 929
Del Zanna, G., Chiuderi Drago, F., & Parenti, S. 2004, *A&A*, 420, 307
Kucera, T. A., retta V, & Poland, A. I. 1998, *SoPh*, 183, 107
Labrosse, N., Schmieder, B., Heinzel, P., & Watanabe, T. 2011, *A&A*, 531, A69
Su, Y., Reeves, K. K., McCauley, P., van Ballegooijen, A., & DeLuca, E. 2013, *Proceedings of IAUS 300: Nature of Prominences and their role in Space Weather*
West, J. B. & Marr, G. V. 1976, *Proc. R. Soc. London A*, 349, 397

*Nature of Prominences and their role in Space Weather*
*Proceedings IAU Symposium No. 300, 2013*
*B. Schmieder, J.-M. Malherbe & S. T. Wu, eds.*

doi:10.1017/S1743921313011514

# Dynamics in the filament of september 17 2010 and in its channel

**Nicole Mein[1], Pierre Mein[2], Brigitte Schmieder[2], Jean-Marie Malherbe[2] and Thierry Roudier[3]**

[1]Observatoire de Paris, Meudon, 92195, France
[2]Observatoire de Paris, LESIA, Meudon, 92195, France
[3]Observatoire Midi-Pyrénées, Toulouse, France

**Abstract.** Dynamics of a filament is investigated in Hα. Counterstreaming flows are observed along the filament. Photospheric horizontal motions have been computed by using a Coherent Structure Tracking algorithm in the filament environment.

**Keywords.** Sun: prominences, granulation, velocity field

## 1. Multi-wavelength observations

We investigate the dynamics of a large filament in the central part of the active region NOAA 11106 observed near the central meridian on Sept 17 2010. This filament was observed in Hα with the THEMIS telescope in Tenerife in both modes (MTR and MSDP) and in 304 Å with the EUV imager (AIA: Lemen *et al.* 2012) on board of the Solar Dynamics Observatory (SDO).

Figures (1) and (2) show two sets of Hα intensity and dopplershift images obtained with the Multichannel Subtractive Double Pass (MSDP) spectrograph operating on the THEMIS telescope (Tenerife). The velocities are calculated by the bisector method. If we assume that the flows are horizontal, and taking into account the position on the disk, velocities may be multiplied by a factor 2. An other possibility is to suppose that the filament is like a cloud over the chromosphere. In that case, the velocity may still be multiplied by an additional factor 2. If we apply both corrections, the velocities could be estimated to be between -10 km/s and 20 km/s. The most striking feature is the upward velocity in the filament itself and the presence of downward velocities in the filament channel showing perhaps the signature of the counterstreaming flow detected also in 304 Å movies. But the order of magnitude, between 50 to 150 km/s, is very different. Perhaps this is due to the filament direction or to the mixing of many threads in Hα observations along the line of sight. Figure (3a) shows the environment of the filament seen by AIA and Figure (3b) the magnetic field given by SDO/HMI. We can see that the filament lies between positive and negative polarities, as usual.

## 2. Supergranule velocities computed with the CST algorithm

The photospheric velocity flow has been calculated by the Coherent Structure Tracking algorithm (CST) (Roudier *et al.* 2009).

The cork distribution map in the filament environment has been derived by using the photospheric flow averaged over three hours. The corks are gathered at the edge of supergranules in the quiet Sun. In the facular regions, they are located on the brightness maxima in 1600 Å (Figure 3c). The mean proper motion vectors over 3 hours calculated by CST are shown in Figure (3d) together with the contour of the filament and its channel (full lines). The field of view covers the two different fields of view of the MSDP. Two important features can be noticed: converging zones associated with barbs (one of them

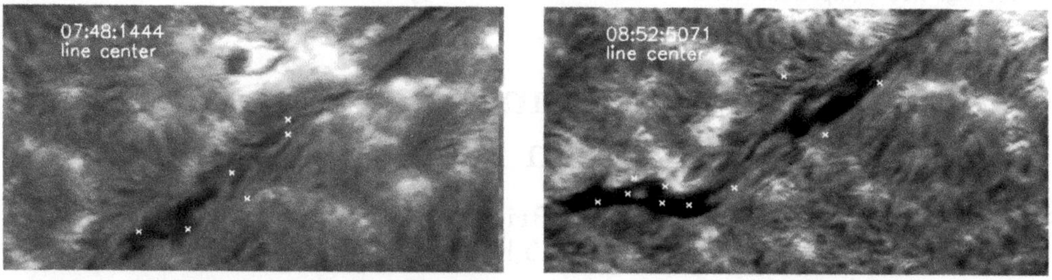


**Figure 1.** Intensities measured with MSDP at THEMIS

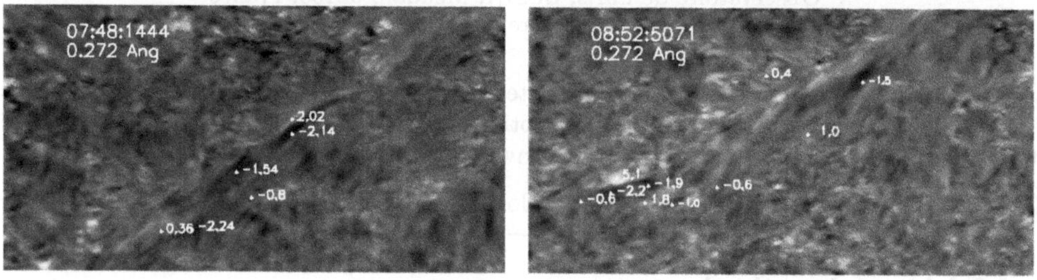


**Figure 2.** Dopplershifts (km/s) measured with MSDP at THEMIS

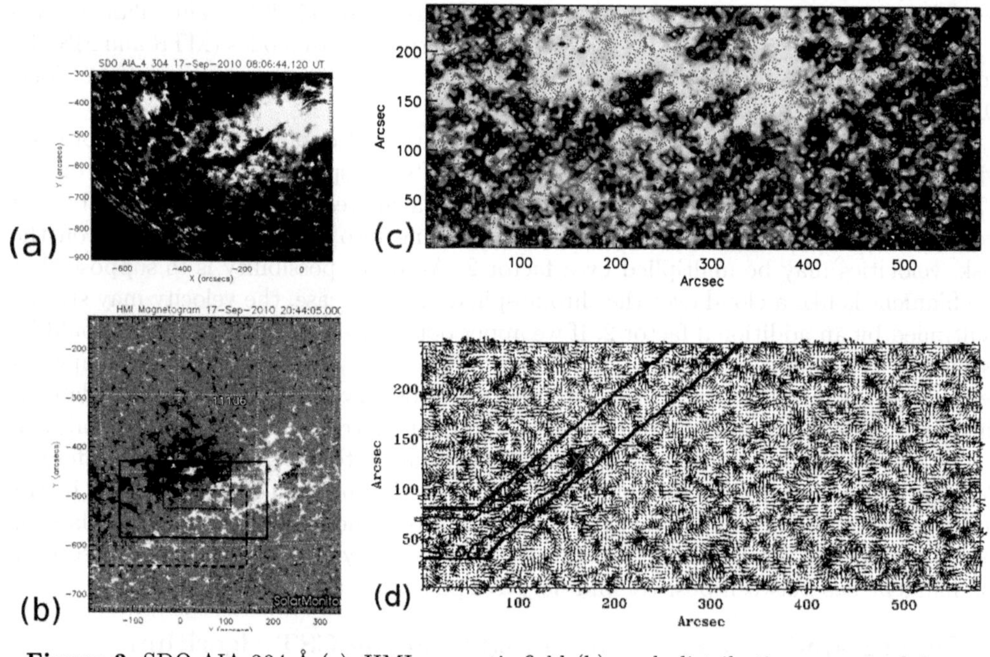


**Figure 3.** SDO AIA 304 Å (a), HMI magnetic field (b), cork distribution map overlying an AIA image at 1600 Å (c) and proper motion velocity vectors (d)

is indicated by a red/dark star), and a flow crossing the filament in its middle part (x = 200 arc sec, y = 150 arc sec).

## References

Roudier, Th., Rieutord, M., Brito, D., Rincon, F., Malherbe, J. M., Meunier, N., Berger, T., & Frank, Z., 2009, *A & A]*, *496*, *945*

Lemen, J. R., Title, A. M., Akin, D. J., *et al.* 2012, *Solar Phys.* 275, 17

*Nature of Prominences and their role in Space Weather*
*Proceedings IAU Symposium No. 300, 2013*
*B. Schmieder, J.-M. Malherbe & S. T. Wu, eds.*

doi:10.1017/S1743921313011526

# Multidimensional and inhomogeneity effects on scattering polarization in solar prominences

**Ivan Milić[1,2] and Marianne Faurobert[1]**

[1]UMR 7293 J.L. Lagrange Laboratory, Université de Nice Sophia Antipolis, CNRS, Observatoire de la Côte d'Azur, Campus Valrose, 06108 Nice, France
email: `milic@aob.rs`

[2]Astronomical observatory Belgrade, Volgina 7, 11060 Belgrade, Serbia
email: `marianne.faurobert@oca.eu`

**Abstract.** Measurements of magnetic fields in solar prominences via Hanle effect usually assume either single scattering approximation or simple, one-dimensional, slab model in order to perform an inversion and find the unknown magnitude and the orientation of the magnetic field from spectropolarimetric observations. Here we perform self-consistent NLTE modeling of scattering polarization in inhomogeneous 2D slab, illuminated from its sides by the solar continuum radiation. We show that even in the absence of a magnetic field, in the non-optically thin regime, significant non-zero Stokes U is to be expected. Neglecting these effects, in principle, could cause systematic errors in spectropolarimetric inversions, in the case when the prominence is optically thick.

**Keywords.** line formation, scattering polarization, radiative transfer

## 1. Introduction

In order to measure magnetic fields pervading solar prominences one usually performs spectropolarimetric inversion. One of the most common choices for such procedure is He1083 line as demonstrated by, for example, Asensio Ramos *et al.* (2008). To find the unknown magnitude and the orientation of the magnetic field, one usually assumes that the emergent spectral line is formed by single scattering (optically thin regime) of the incident continuum radiation, or, in a bit more sophisticated approach, that the prominence behaves as a vertical, one-dimensional slab in which polarized radiative transfer takes place. In both of these approaches, in the absence of a magnetic field, the emergent Stokes $U$ is zero. Thus, non-zero Stokes U implies the presence of a magnetic field and the ratio between Stokes $U$ and $Q$ gives information on the orientation of the field. In this paper the positive reference direction for Stokes $Q$ is parallel to solar limb.

However, prominences are sometimes optically thick in diagnostically important lines and they are far from one-dimensional. It is our assumption that a more complex and realistic prominence model would create additional polarization signatures. Here we investigate scattering polarization in a hypothetical line, formed by scattering in a two-dimensional, inhomogeneous slab, in the presence of velocity fields.

## 2. Prominence model and method of solution

We consider a 2D slab, with arbitrary opacity distribution in the $x, y$ plane and homogeneous along $z$ ($z$ is the atmospheric normal). The slab is placed 20 "above the solar surface and illuminated by the limb-darkened solar radiation where the limb darkening

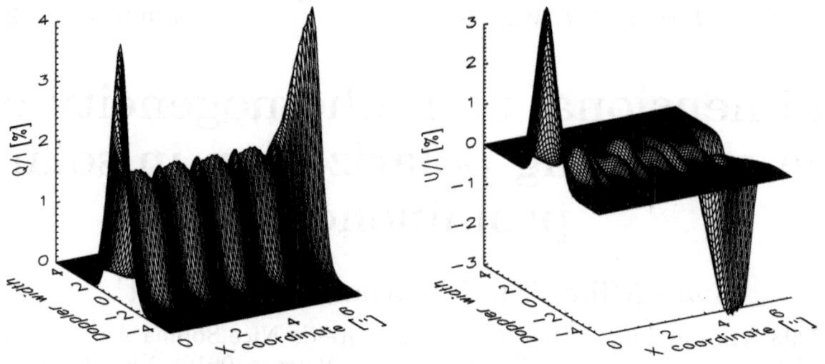


**Figure 1.** Spatial distribution of line profiles in the simple prominence model. Line of sight is parallel to $y$ axis, that is, corresponds to the prominence seen exactly on the limb.

coefficients correspond to the wavelength of $\approx 1100\,\text{nm}$. The slab is 1000 km "thick" and 5000 km "wide". In order to emulate vertical threads, we adopt the following distribution of the line-integrated opacity:

$$\chi(x,y) = \chi_0 \sum_{i}^{9} \exp(-(x - x_{0,i})^2/\sigma^2)\exp(-(y - y_{0,i})^2/\sigma^2). \tag{2.1}$$

Here $\sigma = 150$ km and $\chi_0$ is chosen so that the mean, line-integrated, prominence optical thickness along $x$ is equal to 5. Spacing in $x$ between the peaks of the opacity distribution is 500 km. The plasma is also pervaded by a velocity field which has sinusoidal variations along $x$ and has two components:

$$v_{y,z} = v_{max} \sin\left((x - 500\,\text{km})\frac{2\pi}{1000\,\text{km}}\right), \tag{2.2}$$

where $v_{max}$ is equal to one Doppler velocity, which, for He1083 line, corresponds to about 10 km/s.

To self-consistently solve the NLTE polarized radiative transfer problem for a two-level atom and compute the emergent Stokes vector ($I$,$Q$,$U$), we use the reduced intensity basis as shown by Anusha & Nagendra(2011). The formal solution is based on short characteristics approach by Olson & Kunasz(1988) while the self-consistent NLTE solution follows the approach given by Milić(2013). We neglect collisional depolarization and use $W_2 = 0.3$ for intrinsic line polarizability. This is similar to the red wing of He1083, which consists of two lines, with $W_2 = 0.25$ and 0.35.

## 3. Results and the discusion

Fig. 1 shows that, even in the absence of a magnetic field, the $Q/I$ and $U/I$ polarization profiles have complex shapes which vary strongly with the line of sight. Note that the spatial distribution has been smeared with a 0.5"-wide mean filter in order to account for the finite resolving power of today's ground based instruments. It is important to notice that Stokes $Q$ is enhanced near the edges of the prominence due to the multidimensional radiative transfer effects. Stokes U is also non-zero all over the prominence, again with strong peaks near the edges due to the additional anisotropy.

These simple, "toy-model" results show that, in complex prominence models there are much more effects which contribute to emergent scattering polarization then just the Hanle effect acting on the scattering of the anisotropic incident radiation. In subsequent work we will explore the interplay between these effects and the Hanle effect resulting from large scale magnetic field pervading the prominence.

## References

Anusha, L. S. & Nagendra, K. N. 2011, *ApJ*, 726, 6
Asensio Ramos, A., Trujillo Bueno, J., & Landi Degl'Innocenti, E. 2006, *ApJ*, 683, 542
Milić, I. 2013, *A&A*, 555A, 130M
Kunasz, P. B. & Olson, G. L. 1988, *JQSRT*, 39, 1

*Nature of Prominences and their role in Space Weather*
*Proceedings IAU Symposium No. 300, 2013*
*B. Schmieder, J.-M. Malherbe & S. T. Wu, eds.*

doi:10.1017/S1743921313011538

# The polar belts of prominence occurence as an indicator of the solar magnetic field reversal

**Teodor Pintér, Milan Rybanský and Ivan Dorotovič**

Slovak Central Observatory, Komárňanská 134, SK-94701 Hurbanovo, Slovak Republic
email: teodor.pinter@suh.sk, rybansky.milan@gmail.com, ivan.dorotovic@suh.sk

**Abstract.** The global magnetic field of the Sun is the determining parameter of spreading the solar wind in the interplanetary space. The global field changes the polarity synchronically with the cycle of solar activity. The interesting indicator of the polarity change are the occurence so-called polar belts of the prominences. The article shows the performance of these belts on observational work from 1975 to 2009. A coordinated effort is suggested for the compilation of data from different observers following the method described by Rušin *et al.*, 1988.

**Keywords.** Prominence occurence, polar belts, solar magnetic field reversal

## 1. Introduction

Although the magnetic field of the Sun is measured for a long time, the existence of polar field was often questioned until the 1950s. An argument for the existence of such field was mainly based on observations of polar coronal rays during total solar eclipses. However, it failed to be confirmed by measurements. Svalgaard *et al.* (1978) published, for the first time, a methodology for calculating the polar fields using measurements of background magnetic field of the Sun from the heliographic latitudes ±55° to the poles. These measurements have been carried out at Stanford since 1975. The resulting data are published at http://wso.stanford.edu.

The maximum magnetic field strength occurs at the poles during the minimum of the solar activity cycle and reaches values about 0.1−0.2 mT. It is peculiar, that in the last three cycles the absolute value of magnetic field strength continues to decrease. Another interesting fact is that the polarity reversal of the poloidal magnetic field, according to the data does not occur at the same time. Therefore, during a certain period of time the Sun has the same polarity on the both poles.

## 2. Initial data, method of processing, and results

In this contribution we would like to point out a fact that a concomitant event of the polarity reversal of the magnetic field of the Sun is an 'arrival' of polar branches of prominences to the solar poles. It seems that prominences occur randomly across the surface of the Sun. This applies only to heliographic latitudes of ±40°, and even this is not valid everywhere. In higher latitudes is clearly observable the so called polar branch of prominences (Fig. 1 and 2).

The Figures 1 and 2 are drawn on the basis of the catalogue of prominences compiled according the observations in the coronal station of the Astronomical Institute at Lomnický Štít (Rušin *et al.*, 1988 and subsequent supplements up to 2009). Each marker in the figure depicts observation of one prominence at a given time and at a given latitude.

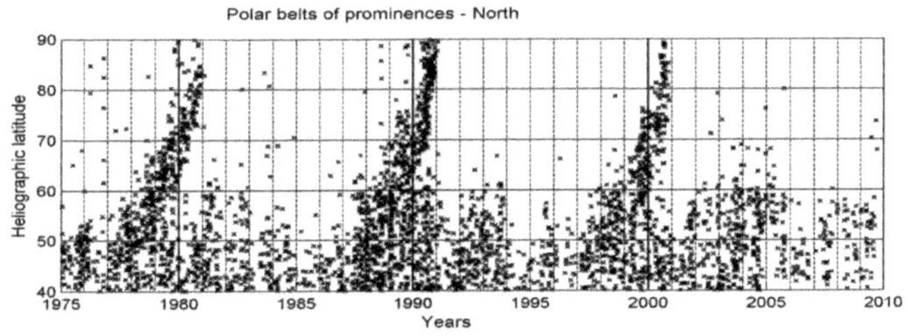


**Figure 1.** Polar branches of prominences near the North Pole of the Sun.

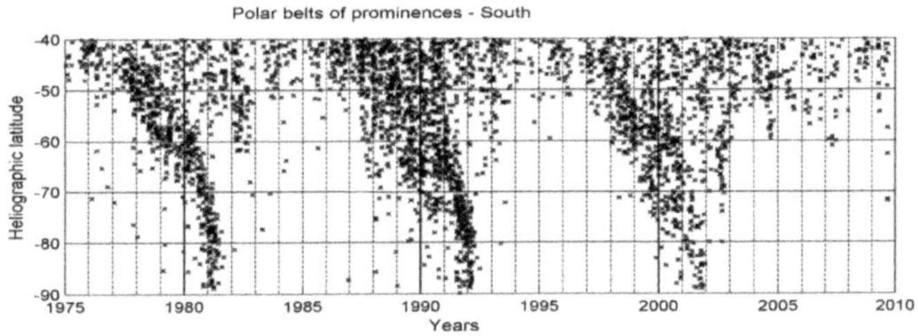


**Figure 2.** Polar branches of prominences near the South Pole of the Sun.

Only prominences the area of which is larger than 50 units are considered. Area is in the catalogue indicated in units of 1° x 1″, the range in positional angles is given in degrees, and the height of prominences in arc seconds. Catalogue in the electronic form can be requested via emailaddressed to suh@suh.sk.

The method of observation and hence the catalogue itself was created upon the initiative of Dr. Gnevyshev (Russia) and therein are published observations from above mentioned observatory from 1967 to 2009. From the images we can see that apart from the period around the polarity reversal prominences (of the given kind) did not occur at latitudes above 65°. This fact can be used to predict the time of the polarity reversal.

| Prominences reached the pole: | Polarity reversal occured: |
|---|---|
| 1981.2 N and 1981.2 S in the 21st cycle; | 1979.9 N and 1980.5 S |
| 1990.8 N and 1992.0 S in the 22nd cycle; | 1990.3 N and 1991.3 S |
| 2000.7 N and 2001.8 S in the 23rd cycle; | 2000.5 N and 2000.0 S. |

## 3. Conclusions

From these data one would assume that the polarity reversal occurs at a time when the polar branches of prominences reach a latitude around ±70°. Observation of prominences is performed at many professional and amateur institutes. The authors believe that it would be appropriate to consolidate the method of processing of observations and to publish them electronically in a standardized form by one institute. This contribution uses observations only from one observatory, thus there are big gaps caused by the weather. Inspite of that also from such observations can be possible to contribute to reveal a mechanism of solar cycle which is not well understood, yet.

## References

Rušin, V., Rybanský, M., Dermendjiev V., & Stavrev K. Y. 1988, *Contrib. Astron. Obs. Skalnaté Pleso*, 17, 63

Svalgaard, L., Duvall, T. L. jr., & Scherrer, P. H. 1978, *Solar Phys.*, 58, 225

*Nature of Prominences and their role in Space Weather*
*Proceedings IAU Symposium No. 300, 2013*
*B. Schmieder, J.-M. Malherbe & S. T. Wu, eds.*

doi:10.1017/S174392131301154X

# Total mass loading of prominences estimated from their multi-spectral observations

**Pavol Schwartz[1,2], Peter Heinzel[2], Pavel Kotrč[2], František Fárník[2], Yurij A. Kupryakov[2,3], Edward E. DeLuca[4] and Leon Golub[4]**

[1]Astronomical Institute of Solvak Academy of Sciences, 05960 Tatranská Lomnica
email: `pschwartz@ta3.sk`

[2]Astronomical Institute, Academy of Sciences of the Czech Republic, 25165 Ondřejov
email: `pheinzel pkotrc ffarnik kupry @asu.cas.cz`

[3]Sternberg Astronomical Institute, 119899 Moscow, Russia

[4]Harvard-Smithsonia Center for Astrphysics, Cambridge, MA 02138, USA
email: `edeluca lgolub @cfa.harvard.edu`

**Abstract.** The total mass of several quiescent prominences observed in EUV by the AIA instrument on board SDO, in soft X-rays by XRT on Hinode and in H$\alpha$ and CaII H by the SLS and HSFA spectrographs of the Ondřejov observatory, was estimated. Values of asymmetry of coronal emissivity obtained during the mass computations are compared with those estimated from 193 Å intensities measured at the disk edge and just above the limb.

**Keywords.** solar prominences, spectroscopy, EUV coronal radiation, X-rays

---

## 1. Introduction

Total mass of a prominence can be estimated from amount of EUV coronal radiation that it absorbs (see works of Kucera *et al.* (1998); Gilbert *et al.* (2005, 2006); Williams *et al.* (2013). In these works only absorption in resonance continua of hydrogen and helium was taken, but decrease of intensities of hot corpnal lines due to lack of coronal emission from cool prominence or low-dense cavity, was not taken into account. Moreover, interpolations in time or space were used to estimate background coronal radiation. In this work we take into account the blocking and fraction of emissivity behind a prominence is estimated from simultaneous coronal EUV observations at three different wavelengths.

## 2. Observations

More than 30 quiescent prominences were observed during a campaign held from April through June 2011. Prominences were observed from space in EUV by the AIA instrument on board the SDO , in soft X-rays by *X-ray Telescope* (XRT) (Golub *et al.* 2007) on Hinode and by ground-based spectrographs *Solar Laboratory Spectrograph* (SLS) (`http://radegast.asu.cas.cz/MFS/prominence_archiv/sls.html`) and *Horizontal--Sonnen-Forschung-Anlage* (HSFA) (Kotrč 2009) in H$\alpha$ and CaII H, respectively. Example of prominence observations in EUV at 304 Å and 193 Å X rays and H$\alpha$ from 18 May 2011 is shown in Fig. 1. In this work we chose from the campaign six prominence observations made between 19 April and 18 May 2011. More about these observations can be found in Schwartz *et al.* (2013).

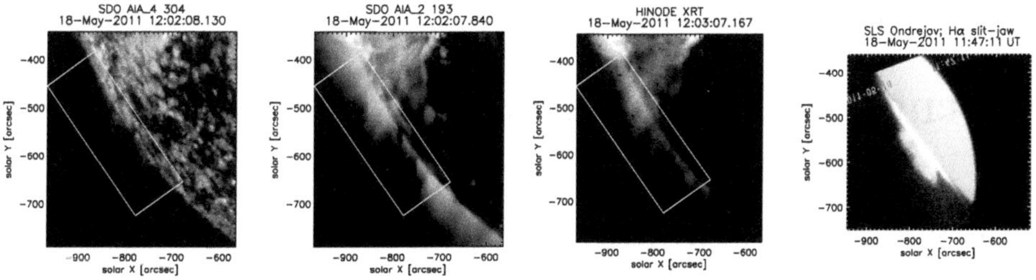


**Figure 1.** Multi-spectral observations of prominence on 18 May 2011. Panels from left to right: images in the 304 Å and 193 Å AIA channels, X-ray image by XRT and Hα slit-jaw image of the SLS spectrograph.

## 3. Method

For estimation of the optical thickness at wavelength below 912 Å we used improved method of Heinzel *et al.* (2008). The method is based on the fact that both X-ray and EUV intensities are decreased only by the emissivity blocking in the cavity and in a prominence the EUV intensity is lowered by both absorption and blocking while X-ray emission is not absorbed by the prominence plasma. More details about the method can be found in Schwartz *et al.* (2013).

## 4. Results and discussion

The total mass ranges from $2.9\times10^{11}$ up to $1.7\times10^{12}$ kg that is close to values estimated by other authors (Gilbert *et al.* 2005, 2006; Williams *et al.* 2013). The $\alpha$ values are varying for each prominence from zero to almost unity. Such wide intervals of $\alpha$ variations shows that emissivity of the corona is rather inhomogeneous. This fact is supported also by the indirect $\alpha$ measurements: When a radial cut is made close to a prominence partly crossing the disk and partly going off limb, interpolation to disk edge of quiet-Sun (outside brightenings) 193 Å intensities along disk part of the cut, gives at least rough estimation of foreground intensity. An intensity from the whole line of sight at the limb close to a prominence is estimated by fitting off-limb intensities along the cut by exponential function. Then $\alpha$ is calculated simply by subtraction from the unity of ratio of the foreground to the whole line-of-sight intensity. Such value of $\alpha$ is strongly sensitive to position in which the radial cut is made. Thus, background coronal emission obtained by interpolation from prominence vicinity, cannot be reliable. More detailed discussions about the method and results can be found in Schwartz *et al.* (2013).

### References


Gilbert, H. R., Falco, L. E., Holzer, T. E., & MacQueen, R. M. 2006 *ApJ*, 641, 606
Gilbert, H. R., Holzer, T. E., & MacQueen, R. M. 2005 *ApJ*, 618, 524
Golub, L., DeLuca, E. E., Austin, G., *et al.* 2007 *Sol. Phys.*, 243, 63
Harrison, R. A., Sawyer, E. C., Carter, M. K., Cruise, A. M., *et al.* 1995 *Sol. Phys.*, 162, 233
Heinzel, P., Schmieder, B., Fárník, F., Schwartz, P. *et al.* 2008 *ApJ*, 686, 1383
Kotrč, P. 2009, *Central European Astrophysical Bulletin*, 33, 327
Kucera, T. A., Andretta, V., & Poland, A. I. 1998 *Sol. Phys.*, 183, 107
Lemen, J. R., Title, A. M., Akin, D. J., *et al.* 2012 *Sol. Phys.*, 275, 17
Schwartz, P., Heinzel, P., Kotrč, P., Fárník, Kupryakov, Yu. A., DeLuca, E. E., & Golub, L. 2013, *A&A*, in preparation
Williams, D. R., Baker, D., & van Driel-Gesztelyi, L. 2013 *ApJ*, 764, 165

*Nature of Prominences and their role in Space Weather*
*Proceedings IAU Symposium No. 300, 2013*
*B. Schmieder, J.-M. Malherbe & S. T. Wu, eds.*

doi:10.1017/S1743921313011551

# Structure and Dynamics of an Eruptive Prominence on the Quiet Sun

**Yingna Su, Katharine K. Reeves, Patrick McCauley, Adriaan A. van Ballegooijen and Edward E. DeLuca**

Harvard-Smithsonian Center for Astrophysics, 60 Garden street, Cambridge, MA 02138, USA
email: ysu@cfa.harvard.edu

**Abstract.** We present preliminary results on the investigation of one polar crown prominence that erupted on 2012 March 11. This prominence is viewed at the east limb by SDO/AIA and displays a simple vertical-thread structure. A bright U-shape (double horn-like) structure is observed surrounding the upper portion of the prominence before the eruption and becomes more prominent during the eruption. When viewed on the disk, STEREO_B shows that this prominence is composed of series of vertical threads and displays a loop-like structure during the eruption. We focus on the magnetic support of the prominence by studying the structure and dynamics before and during the eruption using observations from SDO and STEREO. We will also present preliminary DEM analysis of the cavity surrounding the prominence.

**Keywords.** Sun: coronal mass ejections (CMEs), Sun: filament eruptions, Sun: magnetic fields.

## 1. Observations

What is the magnetic structure supporting hedgerow prominences? Are they supported by the dips in a twisted flux rope or tangled fields in a current sheet (e.g., van Ballegooijen & Cranmer 2010, Berger 2012, Fan 2012, and references therein)? To address this question, we study a polar crown prominence that erupted on 2012 March 11.

Figure 1 shows AIA emission maps of the filament and surroundings at different temperature bands before the eruption. The maps are derived from the DEMs calculated using six AIA EUV channels in each pixel. The filament contains vertical threads surrounded with a bright inverted triangle structure and a bright U-shape structure on top (Figures 1a-1b). The DEM study suggests that the bright emission surrounding the vertical threads has relatively low temperature ($<$ 1 MK), and the cavity has lower density than the background corona. There is also a hot core (concentrated substructure at 2–3 MK in Figure 1c) located on top of the threads within the cavity.

Figure 2 shows AIA and STEREO_B observations of the erupting prominence on 2012 March 12. The filament displays a slow-rise phase (V$\sim$ 0.2 km s$^{-1}$) and a fast-rise phase (V $\sim$ 27 km s$^{-1}$) starting around 17:00 UT and 22:40 UT on March 11, 2012, respectively (Figure 2b). Upon exiting the AIA field of view, the filament has velocity of 64.8 km s$^{-1}$, and acceleration is 0.02 km s$^{-2}$. The eruption is associated with a partial halo CME with a linear speed of 638 km s$^{-1}$ as provided by the SOHO LASCO CME catalog. The timing of the filament appearance in LASCO suggests additional acceleration. AIA 171 Å observations show that during the eruption, more and more bright U-shape structures appear to go across the vertical filament threads starting from the upper to the lower portion. The filament displays an asymmetric arch-like structure during the eruption as shown in STEREO_B (Figure 2f). We also observe brightenings of prominence material at the inner edge of the filament arch. A clear U-shape structure is also seen at 171 Å in STEREO_B during eruption (Figure 2g). More comprehensive analysis

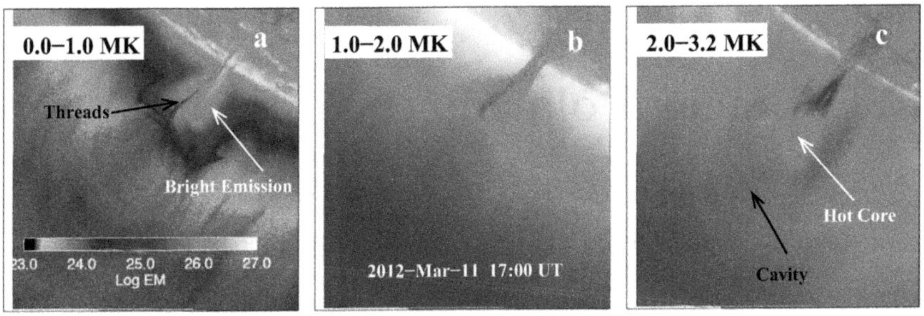


**Figure 1.** AIA Emission Measure maps at 17:00 UT on 2012 March 11.

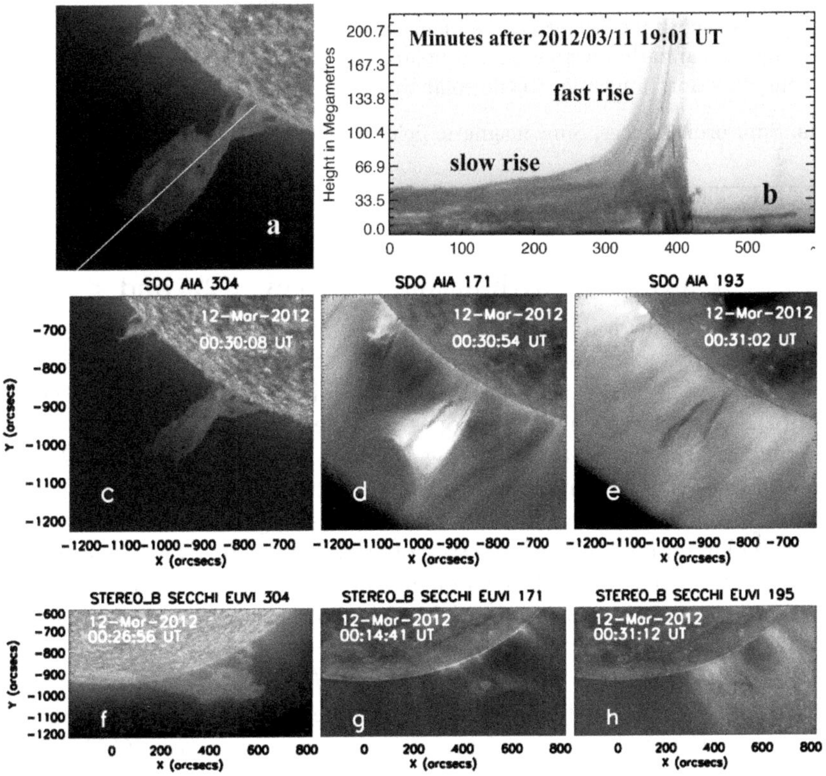


**Figure 2.** SDO and STEREO_B observations of the erupting prominence on March 12, 2012.

of the magnetic structure and thermodynamics of the erupting prominence is under investigation. For detailed column density study of this prominence, please see the paper by McCauley *et al.* in this proceedings.

Acknowledgements: This project is supported by NASA grant NNX12AI30G as well as NASA contract SP02H1701R from LMSAL to SAO.

## References

Berger, T. 2012, *Second ATST-EAST Meeting: Magnetic Fields from the Photosphere to the Corona.*, 463, 147

Fan, Y. 2012, *ApJ*, 758, 60

van Ballegooijen, A. A. & Cranmer, S. R. 2010, *ApJ*, 711, 164

*Nature of Prominences and their role in Space Weather*
*Proceedings IAU Symposium No. 300, 2013*
*B. Schmieder, J.-M. Malherbe & S. T. Wu, eds.*

doi:10.1017/S1743921313011563

# On 2D Linear Polarimetry in Prominences

**V. V. Popov, I. S. Kim, and E. Z. Suyunova**

Lomonosov Moscow State University, Sternberg Astronomical institute,
119992 Universitetsky pr-t, 13, Moscow, Russia
email: kim@sai.msu.ru

**Abstract.** An approach for high-precision 2D linear polarimetry is briefly described. The key components are reducing random errors, reducing systematic errors, and obtaining 2D distributions of the linear polarization degree, $p$, and polarization angle, $\chi$ (deviation of the polarization plane from the direction tangential to the solar limb).

**Keywords.** Sun: prominences, Sun: magnetic fields, techniques: polarimetric

## 1. 2D linear polarimetry with actual accuracy $< 2\%$ and $< 2°$

Significant noise is inevitable for the near-limb filter linear polarimetry. We suggest an approach aimed at obtaining the polarization "images": 2D distributions of $p$, $\chi$, and the sign of $\chi$. The main components are the following (Kim *et al.* 2011): low sky brightness (a), low-scattered light telescopes (b), identity of the polarizer performance for any "point" of the image (c), reducing random errors by the use of 24 successive orientations of the polarizer corresponding to its full turnover instead of traditional three (d), reducing systematic errors by the use of a special algorithm of data reduction (e), obtaining the polarization "images" (f). We note that items $(d)-(f)$ are the key ones. The polarization $p$- and $\chi$-images of $H_\alpha$-prominence of March 29, 2006 (centered at position angle of 307°) were presented in Figure 2 by Suyunova *et al.* (2013), they illustrate the relative accuracies $< 2\%$ and $< 2°$.

***Main "steps" of the final corrected version of the algorithm (e)*** are given here. The use of $I$, $p$, and $\chi-$parameters with different dimensions before statistical procedures results in systematic errors. Parameters of the Stokes vector $(I, U, Q)$ have the same dimension, the intensity. $I = I_*$(non-polarized) + $I_\updownarrow$(linearly polarized). The alternating-sign $Q$ and $U$ are the projections of $I_\updownarrow$ on the axes of the Cartesian coordinates chosen randomly in the sky plane. $I_\updownarrow = \sqrt{Q^2 + U^2}$ and is oriented at an angle $\varphi$: $\tan\varphi = U/Q$. So, $p = I_\updownarrow/I$, and $\chi = \varphi/2$. The intensity of partially linearly polarized light passed through an ideal polarizer is described by the expression

$$S \sim I_* + 2I_\updownarrow \cos^2(\chi - \alpha) = (I_* + I_\updownarrow) + I_\updownarrow \cos(2\chi - 2\alpha) = I + Q\cos 2\alpha + U \sin 2\alpha, \quad (1.1)$$

where $\alpha$ is the orientation of the polarization plane in the $\chi$ coordinate system. The intensities $I$, $Q$ and $U$ depend on the coordinates $(i, j)$, i.e., the frames $\mathbf{I} \equiv \mathrm{I(i,j)}$, $\mathbf{Q} \equiv \mathrm{Q(i,j)}$, and $\mathbf{U} \equiv \mathrm{U(i,j)}$. $\alpha$ differs in successive frames by $2\pi/24 = 15°$, and $\alpha_k = \alpha_0 + \pi k/12$ in the $k-$frame, where $\alpha_0$ is the orientation in the initial frame. Thus, there are 24 equations for each pixel with the $(i, j)$ coordinates based on 24 successive frames $\mathbf{S_k}$:

$$\mathbf{S_k} \equiv \mathrm{S_k(i,j)} = \mathbf{I} + \mathbf{Q}' \cos\frac{\pi \mathrm{k}}{6} + \mathbf{U}' \sin\frac{\pi \mathrm{k}}{6}, \qquad \mathrm{k} = 0, 1, \ldots, 23, \quad (1.2)$$

where the $(\mathbf{Q}', \mathbf{U}')$ vector is related to the $(\mathbf{Q}, \mathbf{U})$ vector by rotation by $2\alpha_0$:

$$\begin{pmatrix} \mathbf{Q}' \\ \mathbf{U}' \end{pmatrix} = \begin{pmatrix} \cos 2\alpha_0 & \sin 2\alpha_0 \\ -\sin 2\alpha_0 & \cos 2\alpha_0 \end{pmatrix} \begin{pmatrix} \mathbf{Q} \\ \mathbf{U} \end{pmatrix}; \quad \begin{pmatrix} \mathbf{Q} \\ \mathbf{U} \end{pmatrix} = \begin{pmatrix} \cos 2\alpha_0 & -\sin 2\alpha_0 \\ \sin 2\alpha_0 & \cos 2\alpha_0 \end{pmatrix} \begin{pmatrix} \mathbf{Q}' \\ \mathbf{U}' \end{pmatrix}. \tag{1.3}$$

The solution of the overdetermined system of 24 equations $\mathbf{S_k}$, with respect to $\mathbf{I}, \mathbf{Q}'$ and $\mathbf{U}'$ by the least square method minimizes the systematic errors and is given by

$$\mathbf{I} = \frac{1}{24}\sum_{\mathrm{k}=0}^{23} \mathbf{S_k}, \qquad \mathbf{Q}' = \frac{1}{12}\sum_{\mathrm{k}=0}^{23} \mathbf{S_k} \cos\frac{\pi \mathrm{k}}{6}, \qquad \mathbf{U}' = \frac{1}{12}\sum_{\mathrm{k}=0}^{23} \mathbf{S_k} \sin\frac{\pi \mathrm{k}}{6}. \tag{1.4}$$

The possible input of the steep brightness gradient is reduced by the use of the $\mathbf{q}'$ and $\mathbf{u}'$ frames which are alternating-sign components of $p$, so that the module of $\mathbf{p} = \sqrt{\mathbf{q}'^2 + \mathbf{u}'^2}$.

$$\mathbf{q}' \equiv \mathrm{q}'(\mathrm{i},\mathrm{j}) = \mathbf{Q}'/\mathbf{I}, \qquad \mathbf{u}' \equiv \mathrm{u}'(\mathrm{i},\mathrm{j}) = \mathbf{U}'/\mathbf{I}. \tag{1.5}$$

The polarization vector $(\mathbf{q}', \mathbf{u}')$ transfer to local coordinates is carried out by turning by $2\psi \equiv 2\psi(i,j)$: $\cos 2\psi = \frac{i^2-j^2}{\sqrt{i^2+j^2}}$ and $\sin 2\psi = \frac{2ij}{\sqrt{i^2+j^2}}$. Tangential $\mathbf{t}'$ and radial $\mathbf{r}'$ frames are derived from $\mathbf{q}'$ and $\mathbf{u}'$ cartesian projections by the following transformations

$$\begin{pmatrix} \mathbf{t}' \\ \mathbf{r}' \end{pmatrix} = \begin{pmatrix} \cos 2\psi & \sin 2\psi \\ -\sin 2\psi & \cos 2\psi \end{pmatrix} \begin{pmatrix} \mathbf{q}' \\ \mathbf{u}' \end{pmatrix}. \tag{1.6}$$

The angle $\alpha_0$ is determined with sufficient accuracy under the assumption that the radial component of polarization, $\langle \mathbf{r} \rangle$, averaged around the limb equals zero: $\tan 2\alpha_0 = \langle \mathbf{r}' \rangle / \langle \mathbf{t}' \rangle$. Turning the $(\mathbf{t}')$ and $(\mathbf{r}')$ frames by $2\alpha_0$ results in the $(\mathbf{t})$ and $(\mathbf{r})$ frames with the averaged $\mathbf{r}$ of zero. Finally the $(\mathbf{t})$ and $(\mathbf{r})$ frames are used to obtain $p$ and $\chi$.

The described method was applied to H$\alpha$-prominence to obtain 2D distributions of polarization parameters (Suyunova *et al.* 2013). On the average, our $p$-values are 1-6% more than the values of non-eclipse coronagraphic filter linear polarimetry in H$\alpha$ quiescent prominences for the same height range (Bommier *et al.* 1994). Several factors can be responsible for the difference: the underestimation of the observed $p$ in the presence of large-angle stray light (Chae *et al.* 1998) inevitable for non-eclipse polarimetry, the input of the continuum corona for total solar eclipse polarimetry, our accuracies, and a possible change of [I, Q, U] during 5 seconds (the polarizer full turnover period). The latter factor was indicated by V. Bommier (private communication during IAU S300). This will be discussed in a separate article.

**Summary.** The application of the approach to non-eclipse coronagraphic filter linear polarimetry in two lines looks promising for study of magnetic structure of prominences.

***Acknowledgements.*** This work was partially supported by RFBR (research project No. 11-02-00631), IAU, SCOSTEP, SF2A and KLSA/CAS. The authors are very indebted to the referee for the review with important comments.

## References

Bommier, V., Leroy, J. L., & Sahal-Brechot, S. 1994, *Solar Phys.*, 154, p. 231
Chae, J., Yun, H. S., Sakurai, T., & Ichimoto, K. 1998, *Solar Phys.*, 183, p. 229
Kim, I. S., Lisin, D. V., Popov, V. V., & Popova, E. V., 2011, *ASP-CS*, 437, p. 181
Suyunova, E. Z., Kim, I. S., Popov, V. V., & Bugaenko, O. I. 2013, *CEAB*, 37.

*Nature of Prominences and their role in Space Weather*
*Proceedings IAU Symposium No. 300, 2013*
*B. Schmieder, J.-M. Malherbe & S. T. Wu, eds.*

doi:10.1017/S1743921313011575

# Spectroscopic measurements of EUV ejecta in a CME: a high-blueshift trailing thread

**David Williams[1], Deborah Baker[1], Lidia van Driel-Gesztelyi[1,2,3] and Lucie Green[2]**

[1]UCL Mullard Space Science Laboratory, UK
email: d.r.williams@ucl.ac.uk
[2]Observatoire de Paris, LESIA, UMR 8109 (CNRS), France
[3]Konkoly Observatory, Budapest, Hungary

**Abstract.** The mass of erupting prominence material can be inferred from the obscuration of emission behind this mass of cool plasma thanks to the rapid cadence of *SDO*/AIA images in the short EUV wavelength range (Carlyle *et al.* 2013, these proceedings). In comparing this approach with spectral observations from *Hinode*/EIS, to monitor contributions from emission seen around the erupting prominence material, we have found an intriguing component of blue-shifted emission, trailing the erupting prominence, with Doppler shifts on the order of 350 km $s^{-1}$ in bright lines of both He II and Fe XII.

**Keywords.** Sun: coronal mass ejections (CMEs), Sun: activity, Sun: prominences, Sun: UV radiation

## 1. Introduction

While observations of prominence eruptions are increasingly common with imaging instruments such as *SDO*/AIA, observations of these events by slit spectrometers are far more rare because of the small instantaneous field of view (IFOV), *i.e.*, the slit size. Co-temporal, co-spatial observations from both imager and spectrometer, observing common ions are even rarer. Fortunately, *Hinode*/EIS and *SDO*/AIA each observe strong lines emitted by Fe XII and He II, and this allows us to describe not only the emission/absorption of plasma and its plane-of-sky motion, but also its line-of-sight motion where the slit IFOV overlaps with that of the imager.

## 2. Observations & Method

A filament eruption from the south-east portion of the Sun was observed on 31-Aug-2012 and observed by *Hinode*/EIS as it scanned its slit in 60-s exposures across a nearby portion of the off-limb corona. The compact line list contained observations of the spectrum around the Fe XII 195.12 Å and He II 256.32 Å lines. These same ions are also chiefly responsible for emission in the 193 Å and 304 Å bands of *SDO*/AIA, and so it was possible to carefully align the data of EIS to those of AIA and examine the Doppler shift of the parts of the filament that passed under the EIS slit. Conspicuously bright components of each of these lines were seen strongly blue-shifted from the mean, quasi-static off-limb components, in areas of the spectrum unoccupied by other bright lines (Brown *et al.* 2008; Labrosse *et al.* 2011), and so single-Gaussian fits to these components were made, allowing us to estimate the Doppler shift of this fast-moving material.

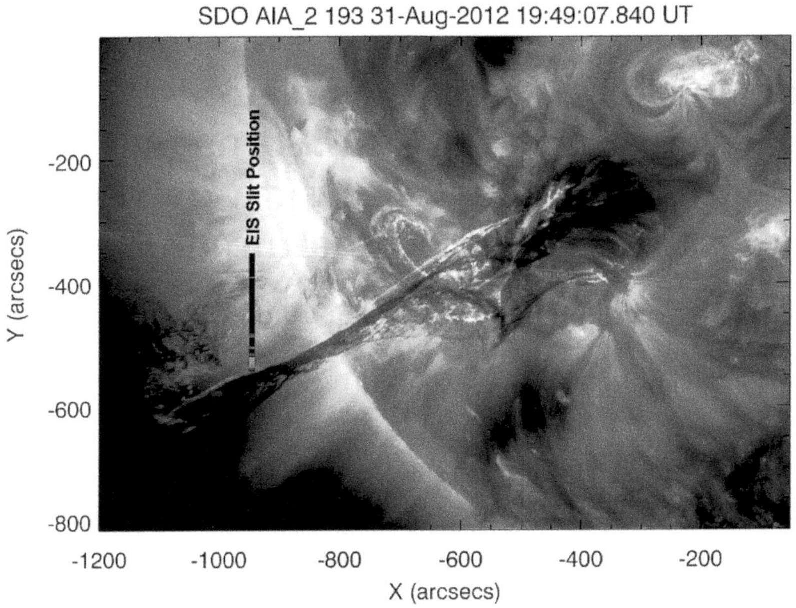


**Figure 1.** *Hinode*/EIS measurements of fitted intensity for only the blue-shifted component of Fe XII 195.12 Å (overlaid on an image from *SDO*/AIA 193 Å). The blue-shifted component appears as a bright section of the *Hinode*/EIS slit (indicated), north of the dark "leg" of the prominence. This high-velocity component persists in both lines for several minutes.

## 3. Results & Summary

The blue-shifted component of the Fe XII and He II lines (Fig. 1) marks the intersection of the slit with a thin bright feature visible in AIA, running behind the entire prominence and anchored in the flare ribbons. In both EIS lines mentioned, it is centred on positions slightly offset in $y$, indicating that temperature and velocity are a function of position. Nevertheless, the bulk motion towards the observer is supersonic ($v_{Doppler} > 300$ km s$^{-1}$). Since the bulk velocity of this component is non-zero, it may indicate an asymmetric outflow of material from this region behind the prominence. Co-spatial and co-temporal observations with EIS and AIA reveal a high-blueshift component of weaker emission trailing the dark prominence material of a CME. Although bright features have been noted in other EIS observations of CMEs, such high Doppler velocities have not. This rare combination of same-ion eruption data with a spectrometer and imager reveals that heating and acceleration of plasma is proceeding along and behind the erupting structure.

### Acknowledgement

The research leading to these results has received funding from the EC's FP7 under grant agreement No. 284461 (eHEROES project). LvDG's is supported by Hungarian Research grant OTKA K-081421, and LG by a Royal Society University Fellowship.

### References

Brown, C. M., Feldman, U., Seely, J. F., Korendyke, C. M., & Hara, H., 2008, *ApJS* 176, 511
Carlyle, J., Williams, D. R., van Driel-Gesztelyi, L., & Innes, D. E., 2013, *Proceedings IAU Symposium No. 300*
Labrosse, N., Schmieder, B., Heinzel, P., & Watanabe, T., 2011, *A&A* 531, 69

*Nature of Prominences and their role in Space Weather*
*Proceedings IAU Symposium No. 300, 2013*
*B. Schmieder J.-M. Malherbe & S. T. Wu, eds.*

doi:10.1017/S1743921313011587

# Simulation of sigmoid structure and filament eruption of AR11283 using a three-dimensional data-driven magnetohydrodynamic model

**S. T. Wu[1], Chaowei Jiang[2], Xueshang Feng[2], Qiang Hu[1], and Yang Liu[3]**

[1]Center for Space Plasma and Aeronomic Research, University of Alabama in Huntsville, Huntsville, AL 35805, USA
email: wus@uah.edu

[2]SIGMA Weather Group, State Key Laboratory for Space Weather, Center for Space Science and Applied Research, Chinese Academy of Sciences, Beijing, 100190, China

[3]W.W. Hansen Experimental Physics Laboratory, Stanford University, Stanford, CA 99305, USA

**Abstract.** This paper describes an MHD simulation of an observed Sigmoid in AR 11283 from its formation to eruption. The Non-linear Force Free MHD model (Jiang and Feng, 2012) and the data-driven active region evolution model (Wu *et al.*, 2006; Jiang *et al.* 2013) together with the SDO/HMI magnetograms are used. We show the successful simulation results of the eruption of a flux-rope structure.

**Keywords.** Sigmoid, MHD Simulation, Active Region, Magnetogram

Sigmoids are generally observed in the solar corona in soft X-ray. It has been shown observationally (Canfield *et al.* 1999, 2007) that active regions with sigmoidal morphology are more likely to produce eruptive events. In this scenario, we will analyze a recent observed active region (AR11283) using a three-dimensional data-driven magnetohydrodynamic (MHD) active region evolution model (Wu *et al.* 2006; Jiang *et al.* 2012) together with a nonlinear force-free model (Jiang & Feng, 2012) for the construction of the initial state with the presence of a sigmoidal feature. It has been recognized that the AR11283 is a very productive source region, producing a number of flares and CMEs when it was located near disk center. In the present study, we will focus on the initiation process of a CME on 6 September 2011 accompanied by a sigmoid eruption around 22:20UT. To carry out this simulation, we first have input the measured vector magnetic field on the photosphere

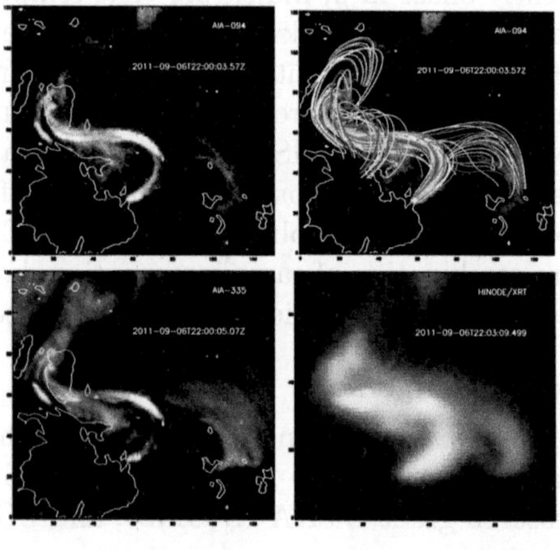


**Figure 1.** Pre-eruption MHD equilibrium state in comparison with observations for AR11283. *Counterclock-wise from upper left panel:* AIA-094, AIA-335, and Hinode/XRT images, and the simulated magnetic field lines with the AIA-094 image as the background.

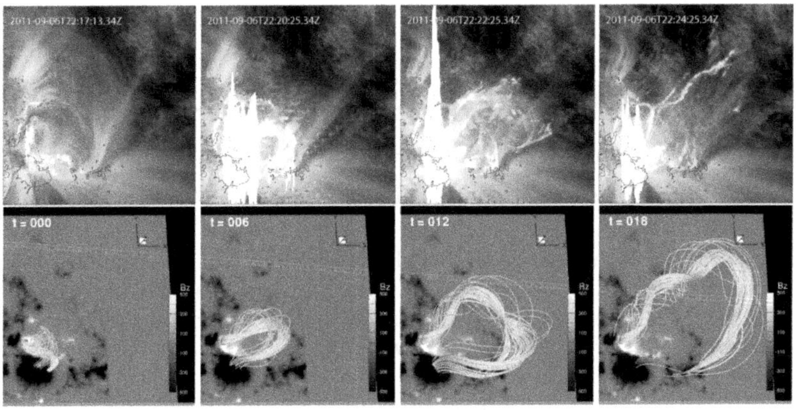


**Figure 2.** The MHD simulation result of the erupting flux-rope (twisted magnetic field-line) structure (bottom row) and the corresponding AIA-171 images at four different times.

into a nonlinear force-free model (Jiang & Feng, 2012) to obtain the sigmoid structure before the eruption. Fig. 1 shows the simulated sigmoid at 2011-09-06 22:00UT immediately before the eruption. It is obvious that the simulated sigmoid matches the observations well. Then we embed this solution into our data-driven CESE-MHD model (Jiang *et al.* 2012), which results in an eruption as shown sequentially in Fig. 2. There the simulated eruption of the sigmoid compares remarkably well with the corresponding AIA images†. The cause of the eruption is due to twist instability (Kliem & Török, 2006; Török & Kliem, 2007). We have ruled out the kink instability because we found that immediately above the sigmoid the magnetic field structure corresponding to part of a flux rope is strongly sheared and slightly twisted. Detailed discussion was given by Jiang *et al.* (2013a,b).

In summary, this is the first time an observed sigmoid from its birth to eruption has been successfully simulated based on vector magnetograms from SDO/HMI. A detailed description of the simulation results can be found in Jiang *et al.* 2013a,b.

## Acknowledgements

This work is jointly supported by NSF (AGS-1153323; AGS-1062050) and NNSFC (41204126, 41231068, 41274192, 41031066, and 41074122). Data from observations are courtesy of NASA SDO/AIA and HMI science teams.

## References

Canfield, R. C., Hudson, H. S., & McKenzie, D. E. 1999, *GRL*, 26, 627
Canfield, R. C., Kazachenko, M. D., Acton, L. W., *et al.* 2007, *ApJL*, 671, L81
Jiang, C., Feng, X., Wu, S. T., & Hu, Q. 2013a, *ApJL*, 771, L30
Jiang, C., Wu, S. T., Feng, X., & Hu, Q. 2013b, *ApJ*, in press
Jiang, C., Feng, X., Wu, S. T., & Hu, Q. 2012, *ApJ*, 759, 85
Jiang, C. & Feng, X. 2012, *ApJ*, 749, 135
Kliem, B. & Török, T. 2006, *Physical Review Letters*, 96, 255002
Török, T. & Kliem, B. 2007, *Astronomische Nachrichten*, 328, 743
Wu, S. T., Wang, A. H., Liu, Y., & Hoeksema, J. T. 2006, *ApJ*, 652, 800

† See `https://dl.dropboxusercontent.com/u/96898685/Eruption_AR11283.gif` for an animation.

*Nature of Prominences and their role in Space Weather*
*Proceedings IAU Symposium No. 300, 2013*
*B. Schmieder, J.-M. Malherbe & S. T. Wu, eds.*

doi:10.1017/S1743921313011599

# Prominence Formation and Destruction

**Chun Xia, Patrick Antolin and Rony Keppens**

Centre for mathematical Plasma Astrophysics, KU Leuven, email: `chun.xia@wis.kuleuven.be`

**Abstract.** In earlier work, we demonstrated the in-situ formation of a quiescent prominence in a sheared magnetic arcade by chromospheric evaporation and thermal instability in a multi-dimensional MHD model. Here, we improve our setup and reproduce the formation of a curtain-like prominence from first principles, while showing the coexistence of the growing, large-scale prominence with short-lived dynamic coronal rain in overlying loops. When the localized heating is gradually switched off, the central prominence expands laterally beyond the range of its self-created magnetic dips and falls down along the arched loops. The dipped loops recover their initially arched shape and the prominence plasma drains to the chromosphere completely.

**Keywords.** Sun: prominences, filaments — MHD

---

The present work brings great improvements to our earlier work (Xia *et al.* 2012), such as a larger domain, higher resolution, more accurate numerical scheme, and more realistic localized heating restricted in finite chromospheric regions. Our computational box is in a vertical plane perpendicular to the polarity inversion line (PIL, taken in the $z$-direction) and reduced to only the right half of the simulated area using the symmetry of the model. Our model starts from a force-free sheared magnetic arcade permeated by stratified solar atmosphere in overall force-balance. We numerically solve the MHD equations with optically thin radiative cooling, thermal conduction, coronal heating, and gravity terms using the Adaptive Mesh Refinement Versatile Advection Code (MPI-AMRVAC, Keppens *et al.* 2012). An equivalent resolution of $1024 \times 1280$ is achieved by a 5-level AMR grid with the smallest spacing of 39 km. To get an equilibrium state, we start the simulation with a background heating $\propto |B|$ until the system relaxes to a quasi-equilibrium (see Fig. 1(a)). Starting from this state, a relatively strong localized heating $H_1$ is added in selected regions (see the contours in Fig. 1(b)).

As $H_1$ is introduced, chromospheric plasma of the heated region gets evaporated into the coronal loops increasing the density and the temperature. At $t = 6$ mins, the temperature near the top of the heated coronal loops reaches a maximum value of 2.2 MK and starts to decrease slowly. At $t = 79$ mins, the temperature decreases significantly in an extensive region in the loops with height around 20 Mm where the increase of density is less prominent. At $t = 86$ mins, first condensations appear off center as a pair of horizontally extended thin sheets at 20 Mm height. Soon the condensation sheets extend and connect at the center below where a verical condensation appears along central axis sheets after the collision of strong counter flows driven by the loss of pressure from thermal instability (see Fig. 1(b)). At the same time, a smaller condensation forms at the center of lower loops. Then successive condensations happen on the top of neighbouring loops, and cool dense plasma assembles along the $y$-axis forming a curtain-like prominence.

Along with the growing prominence, an assembly of condensations resembling a zigzag belt appears at the shoulders of large loops. At this moment, the inner part of the belt connects with the top of the prominence and the zigzag condensation has three sharp cusps. The one closer to the top of the arcade is drawn uphill and merges with its symmetric twin at the top of their loop. The other two fall down along their loops. The belt of condensed plasma is stretched to a very long extension before fragmentation. The

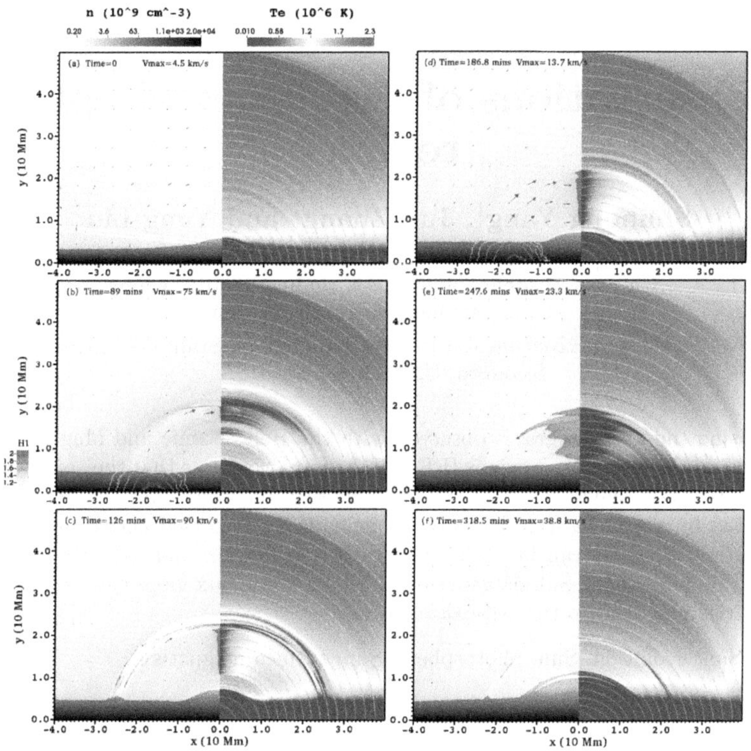


**Figure 1.** Snapshots of the prominence lifecycle. In each panel, the left halves show the density distribution with the projected velocity in blue arrows and the localized heating $H_1$ in contours (not in panel (c)). The right halves show the projected magnetic field lines on top of the temperature distribution.

falling blobs of condensations resemble coronal rain (see Fig. 1(c)). In those loops with coronal rain, the localized heating lengthscale over the loop length is small enough so that condensations occur off-center (Xia *et al.* 2011). Due to the arched shape of the loops, the off-center condensations have to fall down due to gravity unless an upward pressure gradient force dominates.

The coronal rain stops and the prominence sits stably in self-created bended dips at the top of the loops (see Fig. 1(d)). All processes mentioned above are under steady localized heating $H_1$. We then decrease $H_1$ linearly to zero over 29 minutes. In the prominence-hosting loops, the temperature in the hot corona decreases but remains the same inside the prominence, which leads to a gas pressure gradient causing the prominence to swell. As the prominence expands laterally, the plasma density in the dips decreases and the Lorentz force overcomes the gravity restoring the field line from dipped to flat and finally to arched shape. Without magnetic dips, the prominence plasma falls down along the loops into the chromosphere (see Fig. 1(e)-(f)). 150 minutes after $H_1$ stopped, the prominence drains completely and the system recovers to its initial state.

## References


Keppens, R., *et al.* 2012, *J. Comput. Phys.*, 231, 718

Xia, C., Chen, P. F., Keppens, R., & van Marle, A. J. 2011 *ApJ*, 737, 27

Xia, C., Chen, P. F., & Keppens, R. 2012, *ApJ*, 748, L26

*Nature of Prominences and their role in Space Weather*
*Proceedings IAU Symposium No. 300, 2013*
*B. Schmieder, J.-M. Malherbe & S. T. Wu, eds.*

doi:10.1017/S1743921313011605

# HMI observations of two types of ephemeral regions

**Shuhong Yang**[1], **Jun Zhang**[1] **and Yang Liu**[2]

[1]Key Laboratory of Solar Activity, National Astronomical Observatories,
Chinese Academy of Sciences, Beijing 100012, China
email: shuhongyang@nao.cas.cn

[2]W.W. Hansen Experimental Physics Laboratory, Stanford University,
Stanford, CA 94305-4085, USA

**Abstract.** Using the magnetograms observed with the Helioseismic and Magnetic Imager, we statistically study the ephemeral regions (ERs) of the Sun. we notice that the areas with locations around S15° and N25° have larger ER number density, implying that the generation of ERs may be affected by the large-scale background fields from dispersed active regions. According to their evolution, the ERs can be classified into two types, i.e., normal ERs (2798 ones) and self-canceled ERs (190 ones). Submergence of initial magnetic flux loops connecting the opposite dipolar polarities may lead to the self-cancellation.

**Keywords.** Sun: evolution, Sun: photosphere, Sun: surface magnetism

## 1. Introduction

The dipolar magnetic field regions in the solar photosphere ranges from smaller than $10^{18}$ Mx to larger than $10^{23}$ Mx, and the small short lived ones are named ephemeral regions (ERs; Harvey & Martin 1973). With the magnetograms from the Solar and Heliospheric Observatory (SOHO), Schrijver *et al.* (1998) noticed that the mean total unsigned flux per ER is $1.3 \times 10^{19}$ Mx. In the quiet Sun, ERs continuously emerge and replenish the magnetic flux loss due to the dispersion and cancellation (Schrijver *et al.* 1998; Hagenaar *et al.* 2003). The Helioseismic and Magnetic Imager (HMI; Scherrer *et al.* 2012) onboard the Solar Dynamics Observatory (SDO; Pesnell *et al.* 2012) uninterruptedly measures the full-disk magnetic fields with a 45 s cadence and a pixel size of 0.5 arcsec. These advantages are very helpful for us to statistically investigate the ERs in the quiet Sun.

## 2. Observations and Results

In this study, we adopt the HMI line-of-sight magnetograms observed in a four-day period, i.e., from 2010 June 11 12:00 UT to June 15 12:00 UT. We only consider the pixels with heliocentric angle ($\alpha$) smaller than 60° (delineated by the red circle in Figure 1 (a)). All the magnetograms are differentially rotated to a reference time (2010 June 13 12:00 UT). The blue curve outlined the target with $\alpha < 60°$ during the four days. The area $S$ and the magnetic flux density $B$ of each pixel in the derotated magnetograms are calculated as $S/\cos(\alpha_0)$ and $B/\cos(\alpha_1)$, respectively. Note that $\alpha_0$ is the heliocentric angle of the pixel, while $\alpha_1$ is the heliocentric angle at the observation time.

We identify 2988 ERs and their spatial distribution are presented in Figure 1 (a) (marked with blue dots). The target area after projection correction is $8.0 \times 10^5$ Mm$^2$, and thus the mean number density is $9.32 \times 10^{-4}$ day$^{-1}$ Mm$^{-2}$. We display the latitudinal

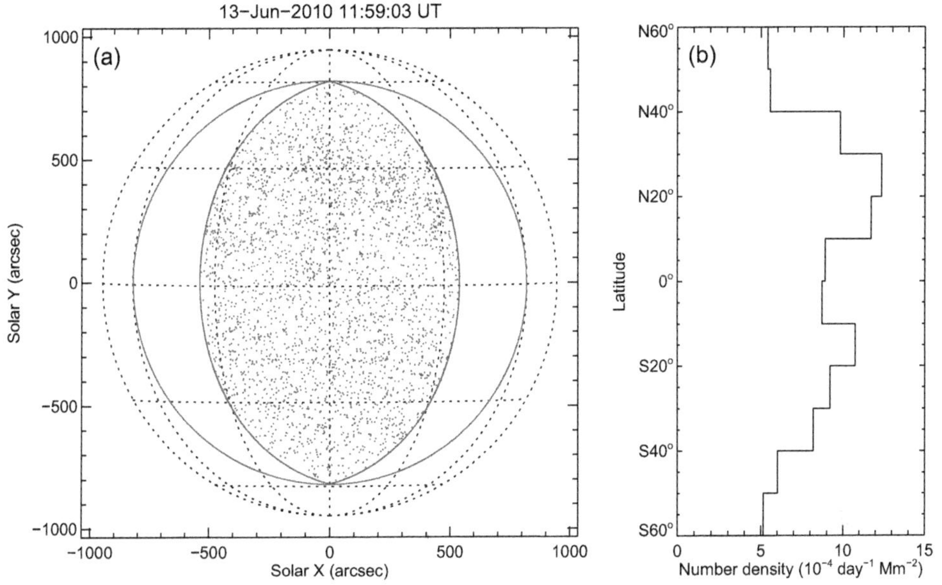


**Figure 1.** Spatial distribution (left panel) and latitudinal distribution (right panel) of the ERs. The red circle marks the place with heliocentric angle $\alpha$ of 60°, and the blue curve outlines the area with $\alpha < 60°$ during the whole four days.

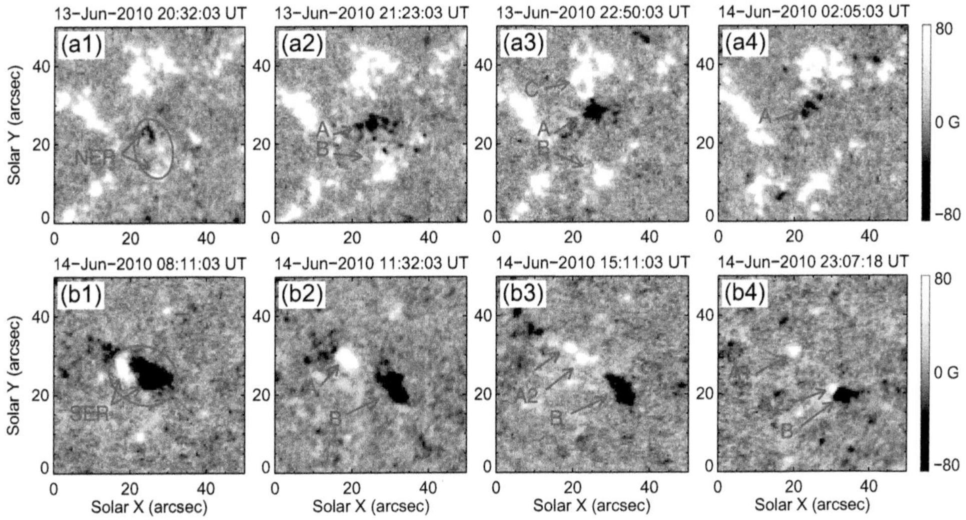


**Figure 2.** Sequence of magnetograms showing the evolution of an NER (upper panels) and an SER (lower panels). The ellipses outline the areas where the ERs emerged.

distribution of these ERs in Figure 1 (b), and find that the ERs are not distributed uniformly. Two regions have larger number density: one region is around S15° and the other one is around N25°, where the mumber density exceeds $10 \times 10^{-4}$ day$^{-1}$ Mm$^{-2}$. We measure the unsigned magnetic flux of the ERs and find that their mean magnetic flux is $9.27 \times 10^{18}$ Mx.

According to their performance, these ERs can be classified into two types, i.e., normal ERs (NERs) and self-canceled ERs (SERs), and their numbers are 2798 and 190, respectively. We find that 9.8% of the total flux of ERs disappeared due to the self-cancellation.

The upper panels in Figure 2 show the evolution of an NER. The NER emerged as a dipolar region (marked by arrows in panel (a1)). Then the opposite polarities (denoted by "A" and "B" in panel (a2)) separated, and the negative polarity encountered and canceled with the pre-existing positive field (denoted by arrow "C" in panel (a3)). At 02:05 UT on June 14, most magnetic flux of patch "A" disappeared (panel (a4)). The lower panels display an example of SERs. The ellipse outlines the location of the SER (see panel (b1)). The two patches of the SER (denoted by arrows "A" and "B" in panel (b2)) separated, and the positive patch "A" split gradually into elements "A1" and "A2" (panel (b3)). Patch "A2" moved toward to patch "B" and canceled with it (see panel (b4)).

## 3. Discussion

As shown in this study, the average unsigned magnetic flux of ERs is $9.3 \times 10^{18}$ Mx, smaller than that ($1.3 \times 10^{19}$ Mx) determined with the SOHO magnetograms (Schrijver *et al.* 1998). Figure 1 (b) shows that, instead of the high latitudinal and the equatorial regions, the areas located at around S15° and N25° (the general latitudes of active regions) have larger ER number density, implying that the generation of ERs may be affected by the large-scale background magnetic fields from decayed and dispersed active regions.

Zwaan (1978, 1987) illustrated that the retraction of initial magnetic flux loops connecting the two poles into the sub-photosphere can lead to magnetic flux cancellation, an observational phenomenon. Besides the theory, the submergence of a sunspot group was also observed by Zirin (1985). We suggest that the self-cancellation of SERs results from the submergence after emergence of magnetic flux loops connecting the opposite dipolar polarities. When dipolar patches with opposite polarities cancel with the surrounding magnetic fields, magnetic reconnection takes place accompanied with energy release, the flux connection is changed and the magnetic configuration is restructured. While when the initial flux loops submerge after emergence, no magnetic flux reconnection occurs and thus no magnetic energy is released during this process.

**Acknowledgements**

This work is supported by the Outstanding Young Scientist Project 11025315, the National Basic Research Program of China under grant 2011CB811403, the National Natural Science Foundations of China (11203037, 11221063, 11373004, and 11303049), and the CAS Project KJCX2-EW-T07.

## References

Hagenaar, H. J., Schrijver, C. J., & Title, A. M. 2003, *ApJ*, 584, 1107
Harvey, K. L. & Martin, S. F. 1973, *Solar Phys.*, 32, 389
Pesnell, W. D., Thompson, B. J., & Chamberlin, P. C. 2012, *Solar Phys.*, 275, 3
Scherrer, P. H., Schou, J., Bush, R. I., *et al.* 2012, *Solar Phys.*, 275, 207
Schrijver, C. J., Title, A. M., Harvey, K. L., *et al.* 1998, *Nature*, 394, 152
Zirin, H. 1985, *ApJ*, 291, 858
Zwaan, C. 1978, *Solar Phys.*, 60, 213
Zwaan, C. 1987, *ARAA*, 25, 83

*Nature of Prominences and their role in Space Weather*
*Proceedings IAU Symposium No. 300, 2013*
*B. Schmieder, J.-M. Malherbe & S. T. Wu, eds.*

doi:10.1017/S1743921313011617

# Solar wind fluctuations and solar activity long-term swing: 1963-2012

**Zerbo, J. L.[1,2], Amory-Mazaudier, C.[2] and Ouattara, F.[3]**

[1]UFR/ST, Université Polytechnique de Bobo Dioulasso, Burkina Faso

[2]LPP-Laboratoire de Physique des Plasmas/UPMC/Polytechnique/CNRS, UMR 7648, 4 Avenue de Neptune 94 107 Saint-Maur-des-Fossés, France

[3]Université de Koudougou, BP 376, Koudougou, Burkina Faso

email: jeanlouis.zerbo@gmail.com

**Abstract.** In this study we investigate the time variation of several solar activity, geomagnetic indices, and solar wind parameters (B, V). It is well known that solar wind is one of the main contributing factors to geomagnetic activity and his topology is strongly affect by solar events such as CMEs and coronals. For these two solar events, we study the correlation between PCI and BV during solar cycle phases and point out the close link between PCI and the occurring of CMEs and high wind speed flowing from coronal holes.

**Keywords.** solar wind, solar activity indices-Joule heating, IMF reversal, PCI

## 1. Introduction

The Sun interacts with the Earth environment throughout radiation and solar wind. This permanent and variable interaction can cause sometimes severe damages to our society. Therefore the Sun-Earth study is so important for general science and the life on Earth as well. Since it is well known that the CMEs occur in majority on the maximum phase of the solar cycle and the high stream solar wind following from coronal holes occur during the declining of solar cycle, we choose in this paper to investigate the natural links between solar wind parameters (B, V) and the Polar Cap Index (PCI) introduced by Troshichev *et al.*, 1988, and used in a statistical sense as a measure of electric field. Our interest is to learn more about the link between the solar dynamo and the energy dissipated in interplanetary medium in the context of geomagnetic activity (Legrand and Simon, 1989; Zerbo *et al.*, 2012).

## 2. Data presentation

Based on the temporal distribution of solar events (CMES, coronal holes) and the fact that he Polar Cap Index can be used as a proxy of Joule heating effect (Francis K. Chun *et al.*, 1999), we investigate the level of the PCI during solar activity. The Figure presented in this work shows the time variations of the Interplanetary Magnetic Field (B), the sunspot number (Rz), the solar wind speed (V) and the Polar Cap Index during the period 1975-2012. We can remark that the peaks in PCI profile occur near the sunspot maximum and more frequently on the declining of solar cycle. We can assume that the peaks observed in PC index are closely linked with rapid solar wind streams from the coronal holes and CMEs. Using these observations, we investigate the correlation between the PCI and the BV and find the most interesting results, about $\approx 70\%$., on the maximum and declining phase of solar cycle. These results show that PCI is closely

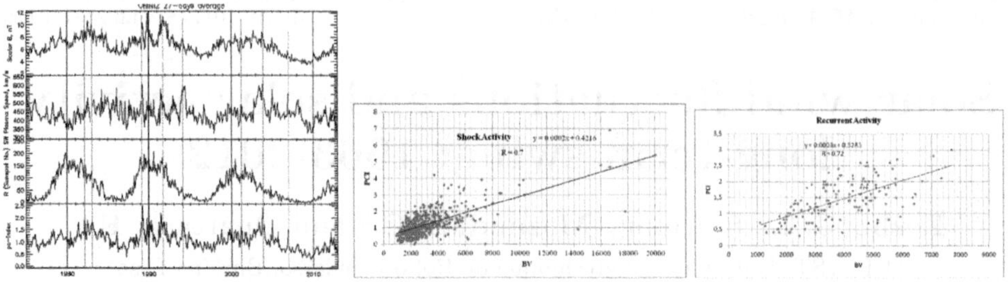

**Figure 1.** *left panel*: Time profiles of the Interplanetary Magnetic Field (B) , the sunspot number (Rz), the solar wind speed (V) and the Polar Cap Index during the period 1975-2012, *middle panel*: Correlation between PCI and BV for Shock activity, *right panel*: Correlation between PCI and BV for Recurrent activity.

linked to BV and inform at the same time on the effects of solar wind and CMEs in the interplanetary medium and terrestrial atmosphere during the two solar cycle phases when the most important energy (Shock and high stream solar wind, Legrand and Simon, 1989) is bring on the interplanetary medium by solar radiation.

## 3. Conclusion

The important remark is that the highest energy is produced near the sunspot maximum and especially during the declining phase of solar cycle. This study shows that to investigate the interconnection between solar activity indices is very important for forecasting the solar activity and to learn how the terrestrial atmosphere reacts during solar events. Our results point out the necessity to do studies for each class of solar events in order to know the different characteristics of the coupling between solar wind and magnetosphere.

## References

Francis, K., *et al.* 1999, Polar cap index as a proxy for hemispheric Joule heating, *J. Geophys. Letter*, vol. 26, NO. 8, p. 1101–1104.

Legrand, J. P. & P. A. Simon 1989, Solar cycle and geomagnetic activity: A review for geophysicists. Part I. The contributions to geomagnetic activity of shock waves and of the solar wind. *Annales geophysicae*, 7, (6).

Ouattara, F. & C. Amory-Mazaudier 2009, Solar geomagnetic activity and Aa indices toward a Standard, *J. Atmos. Solar-Terr. Phys*, 71, pp. 1736–1748.

Troshichev, O. A., V. G. Andrezen, S. Vennerstrm, & E. Friis-Christensen 1988, Magnetic activity in the polar cap A new index, *Planet. Space Sci.*, 36, 1095.

Zerbo, J.L., Amory-Mazaudier, C., Ouattara, F., & Richardson, J. 2012, Solar wind and geomagnetism, toward a standard classification 18682009. *Ann Geophys.* 30, 4216.

*Nature of Prominences and their role in Space Weather*
*Proceedings IAU Symposium No. 300, 2013*
*B. Schmieder, J.-M. Malherbe & S. T. Wu, eds.*

doi:10.1017/S1743921313011629

# Shearing motions and torus instability in the 2010 April 3 filament eruption

**F. P. Zuccarello[1,2], P. Romano[2], F. Zuccarello[3] and S. Poedts[1]**

[1]Centre for mathematical Plasma Astrophysics, KU Leuven, Belgium
[2]INAF - Osservatorio Astrofisico di Catania, via S. Sofia 78, 95123 Catania, Italy
[3]Dipartimento di Fisica e Astronomia, Sezione Astrofisica, Via S. Sofia 78, 95123 Catania, Italy

**Abstract.** The magnetic field evolution of active region NOAA 11059 is studied in order to determine the possible causes and mechanisms that led to the initiation of the 2010 April 3 coronal mass ejection (CME).

We find (1) that the magnetic configuration of the active region is unstable to the torus instability and (2) that persistent shearing motions characterized the negative polarity, resulting in a southward, almost parallel to the meridians, drift motion of the negative magnetic field concentrations.

We conclude that these shearing motions increased the axial field of the filament eventually bringing the flux rope axis to a height where the onset condition for the torus instability was satisfied.

**Keywords.** Sun: filaments, Sun: magnetic fields, instabilities.

## 1. Introduction

On 2010 April 3 at 09:05 UT in NOAA 11059 (S25 W03) a filament eruption occurred, resulting in a geoeffective CME. Figure 1(a) shows an EUV image of the active region taken by SWAP on board PROBA2. The white contour outlines the filament observed in the $H_\alpha$ images taken at the Kanzelhöhe Observatory (KSO). A detailed description of the chromospheric and photospheric evolution of the active region can be found in Zuccarello *et al.* (2012). This event has also been studied by Seaton *et al.* (2011). These authors used PROBA2/SWAP and STEREO/EUVI data to reconstruct the three-dimensional trajectory of the eruption. At the moment of the eruption the top of the expanding loop system had a height of about 0.2 $R_\odot$. If the flux rope fills the entire volume between the photosphere and the top of the loops, at the moment of the eruption the flux rope axis is located at about 70 Mm from the photosphere. The three-dimensional threshold render of the SWAP intensity image, where the z direction represents the height above the photosphere of the flux rope axis, is shown in Fig. 1(b).

This work is aimed at investigating the role that the overlying magnetic field and the shearing motions played in the initiation of the 2010 April 3 CME.

## 2. Results

For a given magnetic field configuration, the decay index is defined as $n = -R\, d(\ln B_{ex})/dR$, where $R$ is the flux rope major radius and $B_{ex}$ is the external magnetic field (Kliem & Török 2006). When $n$ is larger than a critical value ($n_c$) the system becomes unstable. Demoulin & Aulanier (2010) found $n_c \simeq 1.1 - 2$. Figure 1(b) shows the spatial distribution of the computed decay index (rainbow color scale) together with the three-dimensional rendering of the estimated height of the flux rope axis (see Introduction). It

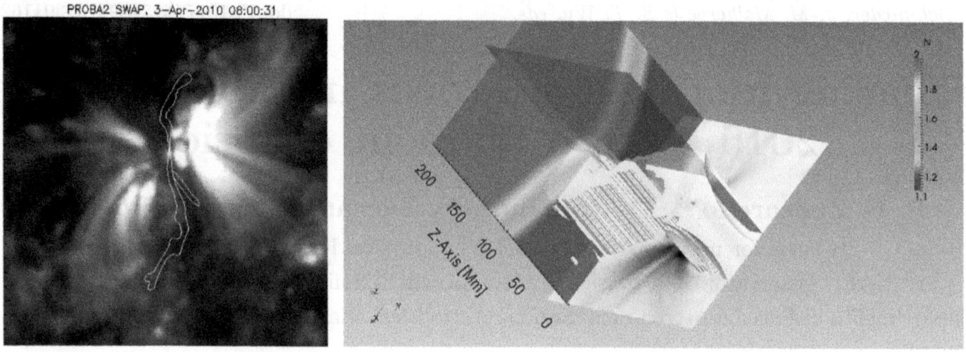


(a) (b)

**Figure 1.** (a) SWAP EUV image taken at 174 Å: the white contour represents the boundaries of the $H_\alpha$ filament. (b) Spatial distribution of the computed decay index (rainbow color scale) together with a three-dimensional rendering of the estimated height of the flux rope axis (see text for more details). A color version of the figure is available in the online version.

is evident that at the moment of the eruption the flux rope was nominally torus unstable. However, it is interesting that the southern part of the dark EUV sigmoid (left in the view of Fig. 1(b)), i.e. the one that underwent the eruption, is actually the least unstable one.

Using MDI magnetograms, we determined the photospheric velocity maps (see Zuccarello *et al.* (2012) for details). We found that persistent southward-directed shearing motions, with average velocity of 0.2-0.3 km $s^{-1}$, characterized the fragmented negative polarity during the 24h that preceded the eruption. As a result, the negative flux concentrations moved southward by about 16-20 arcsec.

## 3. Conclusions

The shearing motions discussed above may have played a significant role in the eruption. In fact, even though the system was nominally torus unstable, the filament was observed for at least two days before the eruption. Therefore, other effects that are not taken into account in the estimation of the decay index, such as the line tying (Olmedo & Zhang 2010) and a non-zero toroidal component of the ambient field, may have played a role in stabilizing the filament. However, the observed shearing motions may have contributed to reduce these stabilising effects and they also resulted in the increase of the axial flux of the filament and as a consequence in the increase of its magnetic pressure. This increase in the magnetic pressure lifted up the flux rope slowly, eventually bringing its axis to a height where the condition for the torus instability was satisfied resulting in the filament eruption.

## References

Démoulin, P. & Aulanier, G. 2010, *ApJ*, 718, 1388
Kliem, B. & Török, T. 2006, *Physical Review Letters*, 96, 255002
Olmedo, O. & Zhang, J. 2010, *ApJL*, 718, 433
Seaton, D. B., Mierla, M., Berghmans, D., Zhukov, A. N., & Dolla, L. 2011, *ApJL*, 727, L10
Zuccarello, F. P., Romano, P., Zuccarello, F., & Poedts, S. 2012, *A&A*, 537, A28

# Poster session II

CORONAL MASS EJECTIONS

INTERPLANETARY CORONAL MASS EJECTIONS

SPACE WEATHER

*Nature of Prominences and their role in Space Weather*
*Proceedings IAU Symposium No. 300, 2013*
*B. Schmieder, J.-M. Malherbe & S. T. Wu, eds.*

doi:10.1017/S1743921313011630

# Topological study of active region 11158

**Jie Zhao[1,2], Hui Li[1,2], Etienne Pariat[3], Brigitte Schmieder[3], Yang Guo[4] and Thomas Wiegelmann[5]**

[1]Purple Mountain Observatory, CAS, 2 West Beijing Road, Nanjing 210008, China
[2]Key Laboratory of Dark Matter and Space Astronomy, CAS, Nanjing 210008, China
[3]LESIA, Observatoire de Paris, Section de Meudon, F-92195,Meudon Principal Cedex, France
[4]School of Astronomy & Space Science, Nanjing University, Nanjing 210093, China
[5]Max-Planck-Institut für Sonnensystemforschung, Max-Planck-Strasse 2, 37191 Katlenburg-Lindau, Germany
email: nj.lihui@pmo.ac.cn

**Abstract.** With the cylindrical equal area (CEA) projection data from the Helioseismic and Magnetic Imager (HMI) onboard the Solar Dynamics Observatory (SDO), we reconstructed the three-dimensional (3D) magnetic fields in the corona ,using a non-linear force-free field (NLFFF) extrapolation method every 12 minutes during five days, to calculate the squashing degree factor $Q$ in the volume. The results show that this AR has an hyperbolic flux tube (HFT) configuration, a typical topology of quadrupole, which is stable even during the two large flares (M6.6 and X2.2 class flares).

**Keywords.** corona, flares, magnetic topology

## 1. Introduction

The different behaviors of solar atmosphere in response to the plasma motions and instabilities are thought to be mainly due to the different topologies in the coronal magnetic field (Berger 1991). Like null point exists at cross section of separatrices, there is hyperbolic flux tube (HFT) at cross section of quasi-separatrix layers (QSLs). Strong electric currents, which dynamically form in non-potential magnetic field, are physically responsible for flaring (Aulanier *et al.* 2005). QSLs, especially the HFTs, are crucial for a region to flare as they are preferential places for narrow current sheet formation (Titov *et al.* 2002). We focus on the topological analysis of active region (AR) 11158 in this work. The method used to calculate Q values is introduced in Section 2 and we discuss our results in Section 3.

## 2. Computational method of QSLs

We use the equations proposed in Pariat *et al.* (2012)

$$Q_{m3} = \begin{array}{l} ((d_{X2x_c}d_{y1y_c} - d_{X2y_c}d_{y1x_c})^2 \\ +(d_{X2y_c}d_{x1x_c} - d_{X2x_c}d_{x1y_c})^2 \\ +(d_{Y2x_c}d_{y1y_c} - d_{Y2y_c}d_{y1x_c})^2 \\ +(d_{Y2y_c}d_{x1x_c} - d_{Y2x_c}d_{x1y_c})^2)f \end{array} \tag{2.1}$$

with

$$f = \frac{|B_{z,1}(x_1, y_1)B_{Z,2}(X_2, Y_2)|}{|B_{n,c}(x_c, y_c)|^2\delta^4} \tag{2.2}$$

to calculate the Q value at ($x_c$,$y_c$) on a fix plane.

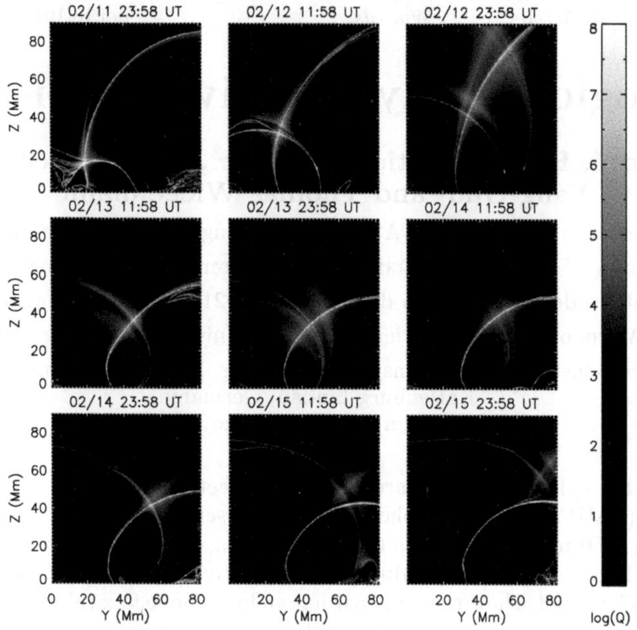


**Figure 1.** Evolution of Q maps in a vertical cut. The Q value is displayed in logarithm scale and time cadence between two successive panels is 12 hours.

The reference boundary here is chosen to be the photosphere. The parameters $d_{x1x_c}$, $d_{y1x_c}$, $d_{x1y_c}$, $d_{y1y_c}$, $d_{X2x_c}$, $d_{Y2x_c}$, $d_{X2y_c}$, $d_{Y2y_c}$ are the components of the field lines footpoint distance on the photosphere. These field lines pass through the four surrounding points of ($x_c$,$y_c$) on the fix plane. $B_{z,1}(x_1, y_1)$ and $B_{Z,2}(X_2, Y_2)$ are the vertical components of magnetic field at the two footpoints ($x_1$,$y_1$) and ($X_2$,$Y_2$), $B_{n,c}(x_c, y_c)$ is the value of the magnetic field component normal to the fix plane at ($x_c$,$y_c$).

## 3. Results and discussion

From the temporal series of Q maps in 3D, we noticed that the quadrupole is relatively stable, which was there even during the large flares. Large-scale HFT always exists and is stable for 36 hours since 11:58 UT of 13 February. Titov *et al.* (2003) has studied this kind of large-scale HFT structure and suggested that the magnetic pinching inside caused by the large-scale shearing motion on the photosphere could produce a large flare. However, in our case, we do not find large flare directly related to this HFT. We suggest that it will be hard to be distinguished from the visible flare ribbons although the possible reconnection at this place may induce the ultra-violet (UV) enhancement at the intersection of HFT configuration and the photosphere.

### References

Aulanier, G., Pariat, E., & Démoulin, P. 2005, *A&A*, 444, 961
Berger, M. A. 1991, *Advances in Solar System Magnetohydrodynamics*, 241
Pariat, E. & Démoulin, P. 2012, *A&A*, 541, A78
Titov, V. S., Hornig, G., & Démoulin, P. 2002, *JGR*, 107, 1164
Titov, V. S., Galsgaard, K., & Neukirch, T. 2003, *ApJ*, 582, 1172

*Nature of Prominences and their role in Space Weather*
*Proceedings IAU Symposium No. 300, 2013*
*B. Schmieder, J.-M. Malherbe & S. T. Wu, eds.*

doi:10.1017/S1743921313011642

# Constraints on the Release History of Solar Energetic Particles by Flux-tube Variations

**N. Agueda[1] and K.-L. Klein[2]**
[1]DAM/ICC, Universitat de Barcelona, Spain
[2]Observatoire de Paris, LESIA, CNRS-UMR, France

**Abstract.** We present the analysis of a large solar near-relativistic (>50 keV) electron event observed by the *Wind* spacecraft on 1998 April 20. In-situ data show variations of the local magnetic field direction accompanied by changes in the topology of the observed electron pitch-angle distributions. These suggest changes in the magnetic flux tubes scanned by *Wind*. Using simulations of the interplanetary particle transport, we model the early rising phase of the electron event, from 10:30 to 11:00 UT, and infer the propagation conditions and the injection history of the first arriving 50–82 keV solar electrons. The results reveal a prompt ($\leqslant$1 min) release in coincidence with the soft X-ray and type III radio bursts, suggesting that the first arriving electrons were flare-accelerated.

**Keywords.** Sun: particle emission, interplanetary medium, Sun: flares

## 1. Introduction

The solar energetic particle (SEP) event on 1998 April 20 was observed during quiescent interplanetary (IP) conditions: no IP coronal mass ejections (CMEs) passed Earth during the previous ~10 days (Cane & Richardson 2003) and the in-situ particle intensities were near solar-quiet levels for roughly a week before the event (Tylka *et al.* 1999).

The SEP event was presaged by a prominence eruption at the southwest limb of the Sun, where an M1.4 X-ray flare was reported at 10:21 UT. Distinct radio emission episodes included a moving type IV burst (9:40–10:00 UT), followed by a series of decametric-hectometric (DH) type III bursts (10:03–10:33 UT), and a type II radio burst at 5-10 MHz after 10:25 UT (Maia *et al.* 2000; Klassen *et al.* 2002). The DH type III bursts were weak at frequencies above 2 MHz, suggesting that the parent active region was occulted behind the solar limb. A partial halo fast (1863 km/s) CME was seen above 3 solar radii at 10:04 UT (Vourlidas *et al.* 1999). The expanding CME loops were imaged directly at radio wavelengths, and the emission ascribed to non-thermal synchrotron emission from ~0.5–5 MeV electrons in the CME magnetic field (Bastian *et al.* 2001).

## 2. In-situ Observations

Fig. 1 (left) shows the 50–82 keV electron omni-directional intensities observed by *Wind*/3DP on 1998 April 20, and the pitch-angle distributions (PADs). The following panels show the measured solar wind speed and the IP magnetic field strength and direction in the GSE coordinate system. It can be seen that during the rising phase of the event the magnetic field direction changed significantly. We identified four periods of time in Fig. 1, with the following $\vec{B}$ mean latitude and longitude: (−35°,242°); (−66°,80°), (−30°,245°), and (32°,230°). These four periods showed different PADs topologies, that is, antisunward PADs with a positive polarity, bidirectional PADs, antisunward PADs with a negative polarity and antisunward PADs with a positive polarity, respectively.

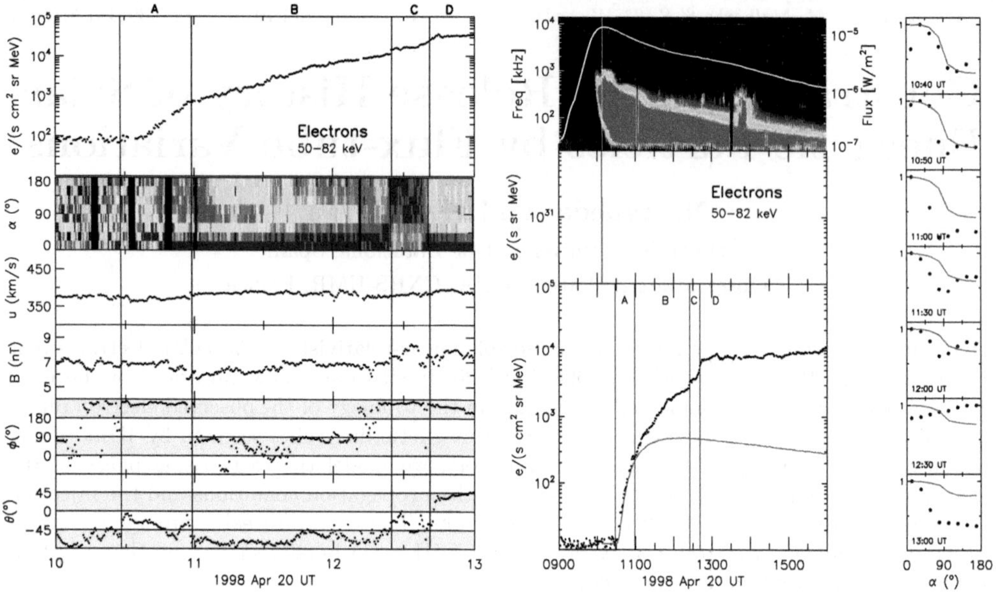


**Figure 1.** *Left:* Electron 50–82 keV omni-directional intensities and PADs; solar wind speed; magnetic field strength and direction (longitude and latitude) in the GSE coordinate system. *Center:* 20 kHz–14 MHz Radio spectrum and soft X-ray flux profile (white curve) expressed in terms of solar release time ($t_{\rm SRT} = t_{\rm AU} - 500$s); Best fit injection function; Omni-directional intensities (black: observations, red: fit). *Right:* Observed and modeled normalized PADs.

## 3. Results

We used an inversion method (Agueda *et al.* 2008) to fit the PADs observed by *Wind* from 10:30 to 11:00 UT. The method utilizes Green's functions of IP transport (Agueda *et al.* 2012) to infer the best fit injection function and the value of the electron mean free path in the IP medium. The method reveals a prompt ($\leqslant$1 min) release of electrons (indicated by the red vertical line in Fig. 1, central panel) at the time of the soft X-ray and type III radio bursts. The inferred value of the radial mean free path is 0.05 AU. Beyond 11:00 UT, the modeled intensities are unable to explain the observations, since the event is dominated by changes in the magnetic flux tube at least until 12:41 UT. We conclude that the duration and timing of the injection suggest the first arriving electrons were flare-accelerated, and got access to the field line magnetically connecting *Wind* to the Sun along diverging open field lines in the parent active region.

## References

Agueda, N., Vainio, R., Lario, D., & Sanahuja, B. 2008, *ApJ*, 675, 1601

Agueda, N., Vainio, R., & Sanahuja, B. 2012, *ApJ* (Supplement), 202, id: 18

Bastian, T. S., Pick, M., Kerdraon, A., & Maia, D., Vourlidas A. 2001, *ApJ* (Letters), 558, L65

Cane, H. V. & Richardson, I. G. 2003, *J. Geophys. Res.*, 108, 1156

Klassen, A., Bothmer, V., Mann, G., Reiner, M. J., Krucker, S., Vourlidas, A., & Kunow, H. 2002, *A&A*, 385, 1078

Maia, D., Pick, M., Vourlidas, A., & Howard, R. 2000, *ApJ*, 528, L49

Tylka, A. J., Reames, D. V., & Ng, C. K. 1999, *Geophys. Res. Lett.*, 26, 2141

Vourlidas, A., Maia, D., Pick, M., & Howard, R. A. 1999, in *Magnetic Fields and Solar Processes* (ESA SP-448), ed. A. Wilson (Noordwijk: ESA), 1003

*Nature of Prominences and their role in Space Weather*
*Proceedings IAU Symposium No. 300, 2013*
*B. Schmieder, J.-M. Malherbe & S. T. Wu, eds.*

doi:10.1017/S1743921313011654

# Galactic cosmic ray decreases associated with non-interacting magnetic clouds in the $23^{rd}$ solar cycle

**J. J. Masías-Meza[1] and S. Dasso[2,3]**

[1]Departamento de Física (FCEN-UBA-IFIBA), BsAs, Argentina
[2]Departamento de Física (FCEN-UBA), BsAs, Argentina
[3]Instituto de Astronomía y Física del Espacio (UBA-CONICET), BsAs, Argentina
email: `(masiasmj, dasso)@df.uba.ar`

**Abstract.** Sudden Galactic Cosmic Ray (GCR) intensity decreases are related to the passage of Interplanetary Coronal Mass Ejections (ICMEs). These phenomena are also known as Forbush Decreases (FDs). The deepest FDs are associated with the passage of Magnetic Clouds (MCs). In this preliminary study we select "non-interacting" MCs associated with FDs observed from ground Neutron Monitors in the period 1996-2009, with the aim of reducing the complexity and the number of parameters involved in the GCR-MC interactions. We introduce a method to determine properties of the "*ejecta* component" of the FD. We analyze properties of the *ejecta* component in combination with properties of MCs. From the resulting selection of events, we find that those FDs containing *ejecta* components show stronger correlations with MC parameters than our total sample of events.

**Keywords.** Sun: magnetic fields, Sun: coronal mass ejections (CMEs), cosmic rays

---

*Introduction:* Forbush decreases (FDs) are depletions of fluxes of Galactic Cosmic Rays (GCRs), typically observed at Earth using ground Neutron Monitors (NMs). Magnetic Clouds (MCs) are a particular subset of Interplanetary Coronal Mass Ejections (ICMEs), and the deepest FDs are associated with the passage of a MC through the terrestrial environment [Richardson & Cane (2011)].

In many cases, FDs present two different phases: (a) a gradual one during the passage of the post-shock turbulent region, and (b) a steeper one during the passage of the MC or *ejecta*. They are said to have a two-step FD profile. The decomposition of the effects of shock and *ejecta* components is very important, in particular for comparisons of FD observations with model predictions.

In the present study we select "clean" ICMEs associated with FDs in the period 1996-2009. We then introduce a method to determine the *ejecta* components for those FDs showing a two-step profile. Finally, we present the correlations between the *ejecta* components and MC properties.

*Selection and decomposition:* From the ICME list in Richardson & Cane (2011), we select those classified as MCs, because their structure are better understood and/or modeled [e.g., Lepping *et al.* (2003)]. If the leading edge of the analyzed MC shows larger velocity than the MC trailing edge, and the velocity profile is linear in the MC temporal window, we consider it an unperturbed expansion and select it, otherwise we discard it.

Richardson & Cane (2011) report a total of 322 ICMEs, from which 99 are MCs. Four of them are reported as multiple MCs. From the 95 remaining, we determined that 35 have an unperturbed expansion, and 13 of these have an associated Forbush Decrease in the NM of Rome ($Rc = 6.3GV$, http://cr0.izmiran.rssi.ru/rome). So these events correspond to significant shielding affecting GCRs with $R \gtrsim 6\ GV$. In order to analyze effects of MCs on GCRs with lower values of $R$, in this work we analyze the FDs observed from the McMurdo NM ($Rc = 0.01\ GV$, http://neutronm.bartol.udel.edu/realtime/mcmurdo.html) and *in situ* observations from the OMNI data compilation (http://omniweb.gsfc.nasa.gov/html/HROdocum.html) for MC properties. The MC parameters were determined according to the start and end times reported in Richardson & Cane (2010).

From our selection of 13 events, only 6 show a two-step FD profile and only 5 show no other interplanetary transients in $\sim 2 days$ before the event. We select these 5 events to make component decompositions in the cosmic ray data.

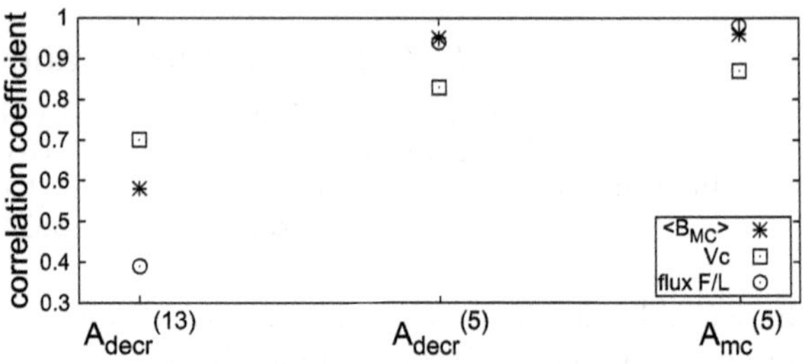


**Figure 1.** Summary of correlations coefficients between MC and FD parameters. The upper indexes in FD parameters (horizontal axis) indicate the number of events taken into account. The highest correlation (very close to 1) is obtained between $A_{mc}^{(5)}$ and each MC parameter, with better correlation for $F/L$.

To determine the shock component, we make a polynomial fit of degree $n = 6$, $P(t) = \sum_{i=0}^{n} a_i t^i$ to the observed time structure of the FD, excluding the time range of the cloud passage. We chose a polynomial degree $n = 6$ because it is the minimum value that simultaneously captures the trends before and after the cloud in the 5 analyzed events. The *ejecta* component is determined by substracting the shock profile from the original data.

*Results, Discussion, and Conclusions:* After we make the shock-cloud decompositions, we determine the peak amplitudes of the observed total FD ($A_{decr}$) and of the *ejecta* component ($A_{mc}$) profiles. Following Dasso *et al.* (2006), we use the definition of the cumulative magnetic flux per unit length $F/L = \int_{X_{in}}^{x} B_{y,cloud}(x')dx'$, as expressed in their eq. (19), but using the module of the magnetic field instead of a component. For the present study, $x'$ is the coordinate parallel to the spacecraft path through the MC, and the limits of integration are the MC borders. We determine this quantity as a proxy of the flux traversed by CRs when they pass through the MC. The properties determined from spacecraft obsevations we analyze here are: the mean magnetic field strength inside the MC, $< B_{MC} >$, the bulk plasma velocity at the MC center, $V_c$, and the magnetic flux per unit length $F/L$. The correlation coefficients between the FD parameters ($A_{decr}$ and $A_{mc}$) and MC properties ($< B_{MC} >$, $V_c$, and $F/L$) are summarized in Figure 1.

We find that the correlation coefficients are higher for FDs with two-step profiles. From the comparison of the correlation coefficients of MCs parameters with $A_{decr}$ and $A_{mc}$, we find a larger correlation with $A_{mc}$, the decomposed cloud peak amplitude.

These preliminary results suggest that the parameters we define here better represent the MC effects on GCR transport.

*Acknowledgments:* Financial support is from the Argentinian grant UBACyT 20020120100220.

## References

Dasso, S., Mandrini, C. H., Démoulin, P., & Luoni, M. L. 2006, *A&A*, 455, 349
Gulisano, A. M., Démoulin, P., Dasso S., Ruiz, M. E., & Marsch E. 2010, *A&A*, A39, 509
Lepping, R. P., Berdichevsky, D. B., & Ferguson, T. J. 2003, *J. Geophys Res*, 108, 1356
Richardson, I. G. & Cane, H. V. 2011, *Solar Phys*, 270, 609
Richardson, I. G. & Cane, H. V. 2010, *Solar Phys*, 264, 189

*Nature of Prominences and their role in Space Weather*
*Proceedings IAU Symposium No. 300, 2013*
*B. Schmieder, J.-M. Malherbe & S. T. Wu, eds.*

doi:10.1017/S1743921313011666

# Characterization of intermittent structures in the solar wind

**M. S. Nakwacki**[1,2]**, M. E. Ruiz**[1,2] **and S. Dasso**[2,1]

[1]Instituto de Astronomía y Física del Espacio,
CC 67, Suc. 28, CP 1428, CABA, Argentina
email: sole@iafe.uba.ar

[2]Departamento de Física, FCEyN, UBA, Argentina

**Abstract.** The solar wind (SW) is a suitable natural scenario to study the intermittent nature of magnetohydrodynamic (MHD) turbulence for systems with low dissipation rate. In particular, nonlinear wave-wave interactions can be characterized by the degree of phase correlation and by departures from Gaussianity of the magnetic field. In this work, we study *in situ* observations of magnetic field intensity from the spacecraft ACE, which is located near one astronomical unit from the Sun, in the SW near Earth. We compute the phase coherence index analyzing two sets of observations, each one consisting of approximately three months during 2008 and 2012, respectively. From these sets of data we characterize intermittent features of the magnetic field intensity corresponding to a solar maximum and a solar minimum.

**Keywords.** Solar Wind, Intermittency, Magnetohydrodynamic turbulence.

## 1. Introduction

Intermittent turbulence is characterized by fluctuations in different spatial scales. Such multiscale interactions are localized regions of plasma with phase synchronization giving place to the presence of phase-coherent intermittent structures which dominate fluctuations at small scales and show departures from Gaussianity. The solar wind (SW) is considered a natural laboratory for observing intermittent turbulence (Chian *et al.* (2009)), and the presence of these phase-coherent structures have been shown previously by Chian *et al.* (2009), Koga *et al.* (2007), Hada *et al.* (2003).

In this work, we study the intermittent fluctuations of the SW taking into account the proton to magnetic presure ratio ($\beta_p = P_{proton}/P_{mag}$) as a proxy for distinguishing between the usual SW and its transient component during two conditions of the solar cycle, namely, solar maximum and solar minimum. We aim at determining which condition leads to a more frequent ocurrence of phase syncronization by computing a phase coherence index obtained from first order structure functions.

## 2. Observations of solar wind at 1 AU and results

We study three months of magnetic field data with a temporal cadence of 1 second from ACE spacecraft for two different solar conditions: close to solar minimum (Jan-Mar 2008) and close to solar maximum (Jan-Mar 2012). We perform a daily analysis for both period of observations.

We characterize the spatial fluctuation of the magnetic field intensity ($B = |\vec{B}|$) by computing the structure function of first order $S(\tau) = \sum_{i=1}^{n} |B_{i+\tau} - B_i|$ for each temporal scale, $\tau$, and over the $n$ elements of the data series. In order to quantify the degree of phase syncronization we derive from $S(\tau)$ the phase coherence index as $C_\phi(\tau) = \frac{S_{prs}(\tau)-S_{org}(\tau)}{S_{prs}(\tau)-S_{pcs}(\tau)}$.

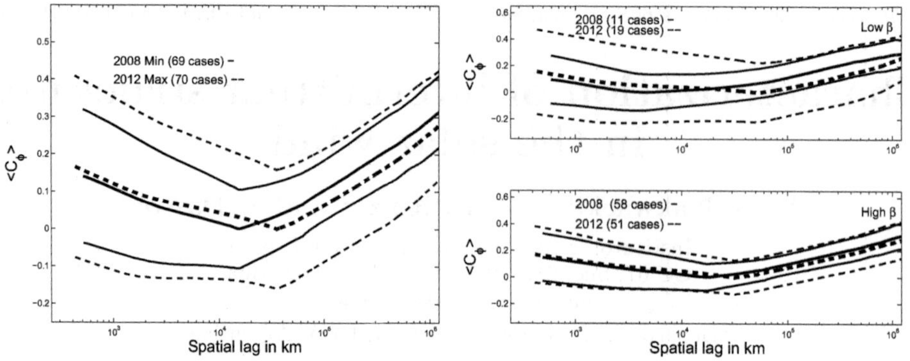


**Figure 1.** Left panel shows the mean variation of the coherence index $< C_\phi >$ and its standard deviation ($\sigma$ in thinner lines) as a function of the spatial scale during solar minimum (full line) and solar maximum (dashed line). Right panels separate the cases shown in left panel for $\beta_p \leqslant 1.50$ (upper panel), and for $\beta_p \geqslant 1.50$ (lower panel).

Here $org$, $prs$, and $pcs$ stand for the original data series, random phase series and constant phase series, respectively, each one obtained from the original series (Chian *et al.* (2009)). We present our analysis for solar maximum (2012) and solar minimum (2008), and we also consider $\beta_p = 1.50$ as a threshold value for separating low and high $\beta_p$ regimes.

The left panel of Fig. 1 shows the mean coherence index $< C_\phi >$ and its standard deviation ($\sigma$) during both solar activity periods. Although the tendency is similar in both periods showing more coherence towards smaller spatial scales ($< C_\phi > \sim 0.2$ for $r < 10^3$ km), it is noticeable that during solar maximum the values of $< C_\phi >$ are more spread than during solar minimum, in particular, at smaller scales.

The right panels of Fig. 1 show $< C_\phi >$ during both solar activity periods but comparing high and low $\beta_p$ regimes (lower and upper panels, respectively). In this case, the phase syncronization is slightly more enhanced during high $\beta_p$ periods, indicating that phase syncronyzation would be not only more common in higher solar activity periods, but also in higher $\beta_p$ regimes.

## 3. Discussion, preliminary results and future work

We study intermittency in the solar wind through the phase coherence comparing both regimes of solar activity for high $\beta_p$ and low $\beta_p$ cases using *in situ* observations at 1 AU in the ecliptic plane. We find that phase syncronization can be found more frequently during solar maximum. However, for $\beta_p < 1.5$, this feature is less noticeable. For $\beta_p > 1.5$, the values of $< C_\phi >$ are more spread being more coherent at smaller spatial scales indicating more cases with phase syncronization as a signature of intermittency.

Phase syncronization seems to be related to high solar activity, in agreement with the higher level of fluctuations, and with high $\beta_p$ regimes, making $\beta_p$ a parameter for ordering the fluctuations properties in the SW. However, further studies must be done.

**Acknowledgments:** This research has made used of NASA's Space Physics Data Facility (SPDF). We thank CONICET for finnancial support. We acknowledge support by UBACyT 20020120100220.

## References


Chian, A. C. L. & Miranda, R. A. 2009, *Ann. Geophys.*, 27, 1789–1801.

Hada, T., Koga, D., & Yamamoto, E. 2003, *Space, Science Reviews*, 107, 463–466.

Koga, D., Chian, A. C. L., & Miranda, R. A. 2007, *Phys. Rev. E*, 75, 046401.

*Nature of Prominences and their role in Space Weather*
*Proceedings IAU Symposium No. 300, 2013*
*B. Schmieder, J.-M. Malherbe & S. T. Wu, eds.*

doi:10.1017/S1743921313011678

# Statistical relationship between CME speed and soft X-ray peak flux of the associated flare during solar cycle 23

**C. Salas-Matamoros[1], K.-L. Klein[2] and L. Taliashvili[1]**

[1]Space Research Center, University of Costa Rica (CINESPA)
email: carolina.salas@planetario.ucr.ac.cr

[2]LESIA - UMR 8109, Observatoire de Paris, CNRS, Univ. Paris 6 & 7
email: ludwig.klein@obspm.fr

**Abstract.** The relationship between the speed of coronal mass ejections and the peak soft X-ray flux of the associated flares is studied for events occurring near the solar limbs between 1996 and 2008. An improved, though still moderate, correlation between the two parameters is found.

**Keywords.** Coronal mass ejections, flares, X-ray emission

## 1. Introduction

Coronal Mass Ejections (CME), i.e. expulsions of huge masses of plasma and magnetic field into the heliosphere, are often associated with soft X-ray (SXR) bursts (Tandberg-Hanssen & Emslie 1988), which are routinely observed by the GOES spacecraft. Statistical studies of correlations between the parameters of CMEs (such as the linear velocity) and the associated SXR bursts have led to conflicting results, ranging from no significant correlation (Aggarwal *et al.* 2008) to moderate correlations between SXR flux and CME speed (Moon *et al.* 2003, Vrsnak *et al.* 2005, Bein *et al.* 2012), or SXR flux and CME kinetic energy (Hundhausen, 1997, Burkepile *et al.* 2004). Since one can only measure the projected CME speed, an existing correlation may be smeared out if one considers CMEs irrespective of their location on the Sun. This can be one of the reasons for the low correlations shown in these studies. Moon *et al.* (2003), indeed, found a higher correlation when restricting their sample to limb events, and so did Burkepile *et al.* (2004), considering the kinetic energy of CMEs. In this work we investigate if the entire SOHO/LASCO data set since 1996 reveals a more significant correlation between CME speed and SXR peak flux near the limbs, where projection effects are minimized.

## 2. Analysis and preliminary results

This study is based on (1) a listing of the measured position angles (PA), widths, heights and speeds of CMEs provided in the LASCO/SOHO CME catalogue during the period 1996 – 2008; (2) the X-ray key parameters from GOES satellites obtained from the National Oceanic and Atmospheric Administration (NOAA) and National Geophysical Data Center (NGDC) during the same time period. We selected CMEs on the western (PA = 60°-120°) and eastern (PA = 240°-300°) solar limb with velocities ⩾ 300 km/s. The associated SXR bursts were identified from the timing association and the linear speed of the CME. From this new list of CME/flare events, we selected those with heliolongitudes near the limb (70°-85° east and west) to evaluate the correlation between CME speed and SXR peak flux. A total of 77 CME events fulfilled the selection criteria.

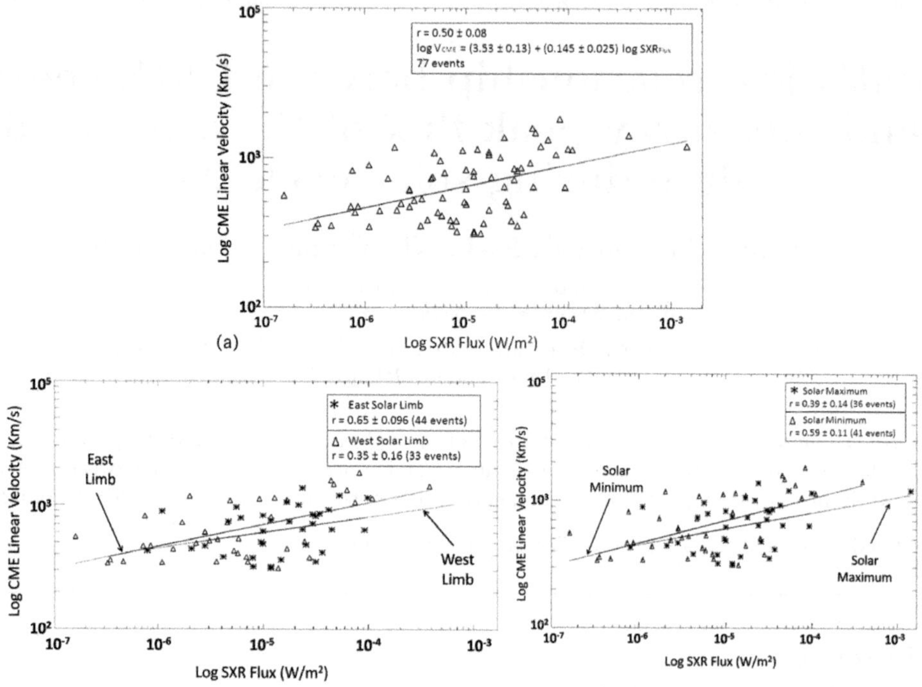


**Figure 1.** Logarithmic plot of CME velocity versus SXR peak flux of the associated flare for events originating near the eastern and western solar limb. (a) All events during the interval 1996-2008. (b) Separate plots for eastern and western events. (c) Separate plots for events near solar minimum and maximum.

We found a correlation of $r = 0.50$ and a weak dependence (very similar to other authors) of the CME speed on SXR peak flux (Fig. 1a). The correlation is lower than found by Moon *et al.* (2003) for a small data set (24 events), but higher than that of Bein *et al.* (2012; $r = 0.32$ for 70 events) who considered CMEs anywhere on the disk.

Higher correlations were found for (1) the east limb events (Fig. 1b) and (2) events around solar minimum as compared to solar maximum (Fig. 1c). It is presently not clear if result (1) is due to a selection effect or if it is physically relevant. Result (2) might be related to the clearer temporal separation of events in times of low solar activity.

## Acknowledgements

C. S.-M. and L.T. acknowledge financial support for attending the symposium. We are grateful to the teams providing the LASCO CME and GOES X-ray catalogues.

## References

Aggarwal, M., *et al.* 2008, *J. Astrophys. Astr.*, 29, 195

Bein, B. M., *et al.* 2012, *Astrophys. J.*, 755, 44

Burkepile, J. T., *et al.* 2004, *J. Geophys. Res.*, 109, A03103

Hundhausen, A. J. 1997, in: J. R. Jokipii *et al.* (eds.), *Cosmic Winds and the Heliosphere* (Tucson: Univ. Arizona Press), p. 259

Moon, Y. J., *et al.* 2003, *J. Korean Astron. Soc.*, 36, 61

Tandberg-Hanssen, E. & Emslie, A. G. 1988, *The Physics of Solar Flares* (Cambridge: Cambridge Univ. Press)

Vrsnak, B., *et al.* 2005, *Astron. Astrophys.*, 435, 1149

*Nature of Prominences and their role in Space Weather*
*Proceedings IAU Symposium No. 300, 2013*
*B. Schmieder, J.-M. Malherbe & S. T. Wu, eds.*

doi:10.1017/S174392131301168X

# Recurrent filament eruptions and associated CMEs

**Brigitte Schmieder**[1], **Hebe Cremades**[2], **Cristina Mandrini** [3], **Pascal Démoulin**[1] **and Yang Guo**[4]

[1]Observatoire de Paris, LESIA,Meudon, 92195, France
email: brigitte.schmieder@obspm.fr

[2]FRM-UTN, Mendoza, Argentina and CONICET

[3]IAFE, UBA-CONICET, Buenos Aires, Argentina

[4]School of Astronomy and Space Science, Nanjing University, Nanjing, China

**Abstract.** We investigate the violent events in the cluster of two active regions (ARs), NOAA numbers 11121 and 11123, observed on 11 November 2010 by the Solar Dynamics Observatory (SDO). Within one day the magnetic field intensity increased by 70% with the emergence of new groups of bipoles in AR 11123, where three filaments are seen along the complex inversion line. The destabilization of the filaments led to flares and CMEs. The CMEs around 08:24 UT and 17:00 UT are directly related to the partial eruption of one filament in the new AR, as shown by a topology computation and analysis. The other CMEs on this day are due to either other ARs or to the destabilization of the global magnetic configuration of the two ARs. This conclusion can be only reached by using the three eyes of SOHO, STEREO and SDO.

**Keywords.** Sun: active region, Sun:magnetic field, Sun: CME, Sun: filament

## 1. Active regions observed by SDO and THEMIS

AR 11123 emerged rapidly within the following negative polarity of AR 11121 from 9 to 11 November 2010. It consisted of several bipoles that formed a nest by 11 November (Fig. 1, left). Several flares accompanied by filament eruptions occurred in the two active regions. In the center of AR 11121 a long filament was oriented NW-SE, while in AR 11123 two well-identified small filaments erupted at the time of the flares (Fig. 1, center), i.e. 07:15 UT and 15:54 UT. A magnetic topology analysis (Fig. 1, right) showed that the flare ribbons overlaid the quasi-separatrix layers (QSLs; Mandrini *et al.* 2013).

## 2. CMEs on 11 November 2010

According to the SOHO/LASCO CME Catalog, six CMEs occurred on 11 November 2010. The AR complex was nearly at S21W00 (Fig. 2, left). Associations between the addressed filament eruptions and the cataloged CMEs are not straightforward from H$\alpha$, SDO/AIA, and SOHO/LASCO data. On this date, STEREO-A and -B were $\sim$90° away from the Sun-Earth line, and thus able to observe the AR complex on the solar limb. A systematic joint analysis of multiple viewpoint data provided by SDO, SOHO, and STEREO yielded the associations summarized in Table 1. The first column is the first CME observation by SOHO/LASCO, position angle (PA) and speed (V) are projected values listed by the Catalog, and flare time is the GOES start time. A CME at 00:12 UT is not listed because it is unlikely related to the AR complex. The CME at 08:24 UT by SOHO/LASCO (Fig. 2, center), seen from the STEREOs, is well associated with a small filament eruption from AR 11123 and with the 07:15 UT flare. The SOHO/LASCO CMEs

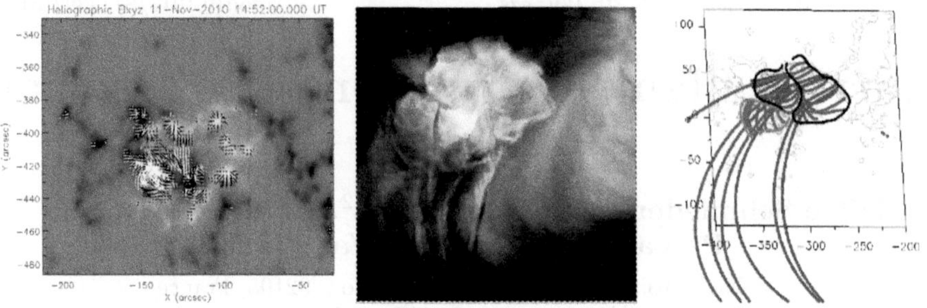


**Figure 1.** AR 11123: (*left*) Vector magnetogram of the emergence of nested bipoles (THEMIS/MTR), (*center*) Filament eruption and bright ribbons at 07:28 UT (SDO/AIA 304-171-211Å), (*right*) Topology showing the QSLs and field lines computed from both sides.

at 11:00 and 17:00 UT only left some traces of material moving high in the corona as viewed in the STEREO EUVIs images. The CME at 14:00 UT, seen from the STEREOs, agrees with the eruption of the filament crossing AR 11121. The CME at 20:12 UT is directly related to AR 11124.

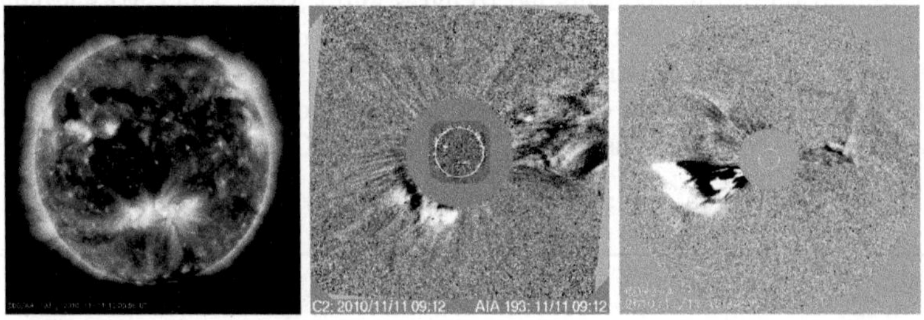


**Figure 2.** (*left*) Full disk SDO/AIA 193 Å showing the AR complex towards the south. (*center*) The SOHO/LASCO CME of 08:24 UT seen to the SE at 09:12 UT. (*right*) The SOHO/LASCO CME of 17:00 UT, here from the quadrature perspective of STEREO COR2-A at 18:24 UT.

Table 1 : CMEs and their sources.

| CME (UT) | PA (deg) | V(km $s^{-1}$) | Flare (UT) | Source |
|---|---|---|---|---|
| 08:24 | 147 | 250 | 07:15 | small filaments in AR 11123 |
| 11:00 | 142 | 315 | 10:12 | high above AR 11121 + AR 11123 |
| 14:00 | 150 | 266 | 12:58 | large central fil. in AR 11121 |
| 17:00 | PH | 419 | 15:54 | high above AR 11121 + AR 11123 |
| 20:12 | 112 | 155 | 19:23 | AR 11124 |

## 3. Conclusion

The analysis of the CME sources shows the importance of observations from the quadrature perspective of STEREO. Most of the CMEs on this day arose from global instability in the AR complex. They appear completely differently from the STEREOs and LASCO, very impressive from a sideway view and weak events with LASCO (Fig. 2).

### Reference

Mandrini C., Schmieder B., Démoulin P., & Guo Y. 2013 *Solar Physics*, in press.

*Nature of Prominences and their role in Space Weather*
*Proceedings IAU Symposium No. 300, 2013*
*B. Schmieder, J.-M. Malherbe & S. T. Wu, eds.*

doi:10.1017/S1743921313011691

# Evolution of the 5 January 2005 CMEs associated with eruptive filaments in inner heliosphere

**Rahul Sharma**[1], **Nandita Srivastava**[2], **Bernard V. Jackson**[3], **D. Chakrabarty**[4], **Nolan Luckett**[3], **Hsiu-Shan Yu**[3], **Qiang Hu**[5,6] **and Christian Möstl**[7,8,9]

[1]3, Indra Nagar, North Sunderwas, Udaipur, India.
email: sharmarahul20@googlemail.com

[2]Udaipur Solar Observatory, Physical Research Laboratory, Udaipur, India. [3]Center for Astrophysics and Space Sciences, University of California at San Diego, CA, USA. [4]Space and Atmospheric Sciences Division, Physical Research Laboratory, Ahmedabad, India. [5]Department of Physics, The University of Alabama in Huntsville, Huntsville, AL, USA. [6]Center for Space Plasma and Aeronomic Research, The University of Alabama in Huntsville, Huntsville, AL, USA. [7]Space Sciences Laboratory, University of California, Berkeley, CA, USA. [8]Kanzelhöhe Observatory-IGAM, Institute of Physics, University of Graz, Graz, Austria. [9]Space Research Institute, Austrian Academy of Sciences, A-8042 Graz, Austria.

**Abstract.** On 5 January 2005, SoHO/LASCO observed two CMEs associated with eruptive filaments with different initial velocities and acceleration. The second CME accelerates much faster than the previous and the resulting interaction has been revealed in in-situ spacecraft measurements by the presence of magnetic holes at the border of the two distinct magnetic clouds. At their interface region, these magnetic clouds have embedded filament plasma that shows complex magnetic structures with a distinct magnetic flux rope configuration; these have been modeled by the Grad - Shafranov reconstruction technique. The geomagnetic consequences of these structures have been associated with substorms in recovery phase of a storm and detailed analysis is presented in Sharma *et al.* (2013). In the present paper, we highlight the comparison of shape and extent of two filament plasma remnants in magnetic clouds as revealed by three - dimensional (3D) reconstruction and analysis from the Solar Mass Ejection Imager (SMEI) data. The results provide an overview of the two eruptive filaments on 5 January 2005 and their interplanetary propagation.

**Keywords.** Solar filaments, ICMEs, geomagnetic storms, substorms

## 1. Introduction

Despite the fact that around 70% of the Coronal Mass Ejections (CMEs) in interplanetary medium are associated with eruptive filaments (Webb & Hundhausen, 1987), there remain a number of questions concerning identification of their remnants in in-situ data and propagation patterns in the heliosphere. As the CME propagates, the associated magnetic fields may reconnect with solar wind fields or those of other CMEs creating complicated signatures in spacecraft data. Also, CMEs kinematically elongate in angular extent, expand from high internal pressure, and may distort due to inhomogeneous ambient solar wind environment. We focus on the evolution of two CMEs associated with eruptive filaments on 5 January 2005, in the inner heliosphere with identification of filament remnants. The angular extents and general shapes of the associated structures were reconstructed using SMEI data while the magnetic topology of embedded flux ropes is modeled using the Grad-Shafranov reconstruction technique.

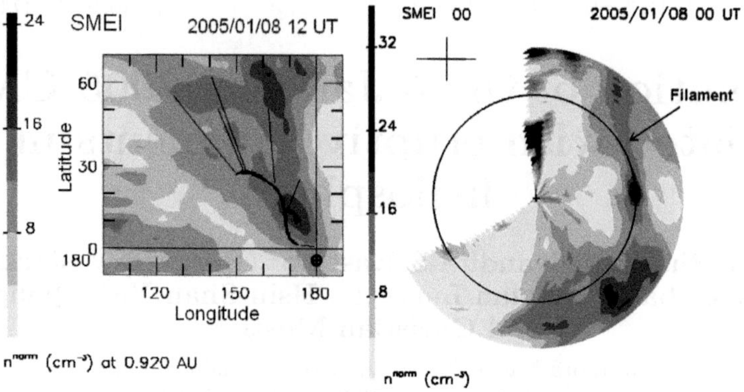


**Figure 1.** Left: Portions of the filament shown expanded to the larger size in the SMEI reconstructions by black thin lines traced from the surface filament to the expanded structure at 1 AU. Right: 3D reconstructed ecliptic cuts viewed from the north with the Sun at the center and Earth (⊕) shown to the right on its elliptical orbit.

## 2. Data, Analysis and Results

Solar disk observations were taken from ground-based (KSO and YNAO) and space-borne (SoHO-EIT/MDI/LASCO) instruments. The in-situ properties were investigated through measurements from ACE and Wind spacecraft. 3D reconstruction of the plasma structures and their extents were done using SMEI data. Filament plasma remnants were identified in magnetic clouds by techniques described by Sharma & Srivastava (2012). The magnetic topology of embedded magnetic flux ropes is modeled using Grad-Shafranov reconstruction technique (Hu & Sonnerup, 2002). 3D rendering of associated plasma structures is done by the use of time dependent reconstruction algorithm (Jackson *et al.*, 2002, and references therein).

Sharma *et al.* (2013) reported the event by the study of remote and in-situ observations of two CMEs associated with an active region and quiescent filaments on 5 January 2005. The first CME was associated with an east - west oriented filament and was slow in velocity and acceleration as compared to the second north - south oriented filament. Due to this difference, the two CMEs interacted in the interplanetary medium through magnetic reconnection which resulted in complex magnetic structures at the interface region, which was further confirmed by modeling of magnetic configuration of flux rope structures using G-S technique. The in-situ signatures of solar wind for the events studied also provide evidence for filament plasma remnants.

3D reconstructions from SMEI data (Fig. 1) reveal that the filament structure(s) near 1 AU have expanded in their latitude and longitude extent on the solar surface to nearly twice the size and volume at 1 AU. We identify the CME associated with first filament eruption to the west and the in-situ remnant as a knotted structure enveloped in the cloud nearer to the Earth. The second filament is observed as a dense structure extending to the northwest to a distance over half an AU from the Earth. A qualitative comparison with in-situ data suggests that spacecraft (ACE/Wind) encountered a dense, flank edge of the filament structures, associated with 5 January 2005 solar eruptions.

### References

Hu, Q. & Sonnerup, B. U.Ö. 2002, *J. Geophys. Res.*, 107, 1142
Jackson, B. V., Hick, P. P., & Buffington, A. 2002, *Proc. SPIE*, 4853, 23
Sharma, R. & Srivastava, N. 2012, *J. Space Weather Space Clim.*, 2, A10
Sharma, R., Srivastava, N., Chakrabarty, D., Möstl, C., & Hu, Q. 2013, *J. Geophys. Res.*, 118
Webb, D. F. & Hundhausen, A. J. 1987, *Sol. Phys.*, 108, 383

*Nature of Prominences and their role in Space Weather*
*Proceedings IAU Symposium No. 300, 2013*
*B. Schmieder, J.-M. Malherbe & S. T. Wu, eds.*

doi:10.1017/S1743921313011708

# Role of filament plasma remnants in ICMEs leading to geomagnetic storms

**Rahul Sharma[1], Nandita Srivastava[2] and D. Chakrabarty[3]**

[1]3, Indra Nagar, North Sunderwas, Udaipur, India.
email: sharmarahul20@googlemail.com

[2]Udaipur Solar Observatory, Physical Research Laboratory, Udaipur, India.
email: nandita@prl.res.in

[3]Space and Atmospheric Sciences Division, Physical Research Laboratory, Ahmedabad, India.
email: dipu@prl.res.in

**Abstract.** We studied three interplanetary coronal mass ejections associated with solar eruptive filaments. Filament plasma remnants embedded in these structures were identified using plasma, magnetic and compositional signatures. These features when impacted the Earth's terrestrial magnetosphere - ionosphere system, resulted in geomagnetic storms. During the main phase of associated storms, along with high density plasma structures, polarity reversals in the Y-component (dawn-to-dusk) of the interplanetary electric field seem to trigger major auroral substorms with concomitant changes in the polar ionospheric electric field. Here, we examine the cases where plasma dynamics and magnetic structuring in the presence of the prompt penetration of the electric field into the equatorial ionosphere affected the space weather while highlighting the complex geomagnetic storm-substorm relationship.

**Keywords.** Solar filaments, ICMEs, geomagnetic storms, substorms

---

## 1. Introduction

Nearly 70% of Coronal Mass Ejections (CMEs) are associated with erupting filaments (Webb & Hundhausen, 1987) and identification of source region remnants are rare in spacecraft measurements. Low-$\beta$ magneto-plasmas injected into the solar wind during these energetic events plays an important role in causing geomagnetic storms. While the geomagnetic storms are externally driven by solar wind drivers, the magnetospheric substorms involve a sequence of processes in the Earth's magnetosphere during which energy is extracted from the solar wind and deposited in the magnetosphere and the ionosphere (McPherron, 1979). To understand the magnetospheric substorm it is necessary to determine whether substorm onset is always externally triggered by the interplanetary sources or it occurs spontaneously as a result of internal processes. Indicators for a near-Earth location of the substorm onset are poorly understood and the relationship between the magnetic storm and substorm is controversial (Kamide *et al.*, 1998). The research presented here investigates the role of solar filament material in triggering the substorms in the main/recovery phase of 3 geomagnetic storms related to CMEs associated with eruptive filaments.

## 2. Data and Analysis

The three reported interplanetary CME events associated with eruptive filaments arrived at 1 AU on 21 Feb 2000, 21 Oct 2001 and 7 Jan 2005. The $in-situ$ properties were investigated through magnetic (MFI/Wind, MAG/ACE) and plasma (3DP/Wind,

SWE/Wind, SWEPAM/ACE and SWICS/ACE) instruments onboard the ACE and Wind spacecraft. The geomagnetic response has been studied using indices obtained from the NASA/GSFC CDAWeb (*www.cdaweb.gsfc.nasa.gov/istp_public/*). Filament plasma were identified using signatures described in Sharma & Srivastava (2012), while the resulting geomagnetic effects of events are studied by a combination of interplanetary and geomagnetic indices as reported by Sharma *et al.* (2013).

## 3. Results and Conclusion

*21 Feb 2000 event*: The interplanetary CME arrived at 1 AU with a shock at 9:50 UT on 20 Feb, followed by a magnetic cloud at 5:37 UT on 21 Feb. Filament plasma located over magnetic flux rope structure arrived during the interval 16:13 UT - 19:17 UT on 21 Feb, preceded by a low temperature ($1.8 \times 10^4$ K) and high density (31 n/cc) pressure pulse at 15:00 UT. The cloud had a Dst index of -26 nT along with two substorms (A1 and A2) associated with the pressure pulse and later with filament material. The former (A1) is observed with sharp polarity changes in Z- component of interplanetary magnetic field (IMF Bz) and the Y- component interplanetary electric field (IEFy), while the changes in these parameters are accompanied with increase in ram pressure for A2 which is correlated with the filament remnant.

*21 Oct 2001 event*: The magnetic cloud is observed during the interval from 19:23 UT on 21 Oct to 00:47 UT on 22 Oct with a high density (65 n/cc) structure with temperature of the order of $10^5$ K at the rear boundary of the cloud with depressed ionic thermal velocities, RMS fluctuations, charge states and elevated $\alpha$/p ratio. Upon comparing the arrival time of this structure (23:15 UT on 21 Oct) to the sudden impulse in AL and PC indices (23:21 UT on 21 Oct), we found insignificant propagation delay ($\approx 6min$) of interplanetary feature to the magnetosphere, highlighting the potential role of high density plasma in triggering mechanisms of substorms.

*7 Jan 2005 event*: Two interacting magnetic clouds (MC1 and MC2) associated with two distinct erupted filaments arrived at 1 AU and resulted into three (C1, C2 and C3) significantly sharp polarity changes (C1 $\approx$ 12:00 UT, C2 $\approx$ 22:00 UT on 7 Jan and C3 $\approx$ 07:00 UT on 8 Jan) in IMF Bz (southward transitions) and IEFy (eastward transitions) during 7-9 January 2005. C1 and C2 brought in proportionate changes in the PC index and AL while the effect of C3 on PC index was disproportionately small and its effect on AL was absent. Upon comparison of the arrival times of MC1 and MC2 at 1 AU with the development of the main phase of the storm and triggering of substorms, it is clear that MC2 has almost negligible impact on the terrestrial magnetosphere-ionosphere system.

This study concludes that there is a possibility that filament plasma remnants in interplanetary CMEs can contribute to the triggering of substorms. Out of the three reported cases, two (Feb, 2000 and Oct, 2001) suggest triggering through compression by the filament plasma and further a "$cause - result$" relationship. The third (Jan, 2005) event highlights the role of magnetic cloud interactions and that the triggering might be caused or enhanced by discontinuities at the interface region between two clouds.

### References

Kamide, Y., *et al.* 1998, *J. Geophys. Res.*, 103, 17705
McPherron, R. L. 1979, *Rev. Geophys. Space Phys.*, 17(4), 651
Sharma, R. & Srivastava, N. 2012, *J. Space Weather Space Clim.*, 2, A10
Sharma, R., Srivastava, N., Chakrabarty, D., Möstl, C., & Hu, Q. 2013, *J. Geophys. Res.*, 118
Webb, D. F. & Hundhausen, A. J. 1987, *Sol. Phys.*, 108, 383

*Nature of Prominences and their role in Space Weather*
*Proceedings IAU Symposium No. 300, 2013*
*B. Schmieder, J.-M. Malherbe & S. T. Wu, eds.*

doi:10.1017/S174392131301171X

# On the onset of recurrent eruptions of a filament observed during August 2012

**Nandita Srivastava[1], Anand D. Joshi[2] and Shibu K. Mathew[1]**

[1]Udaipur Solar Observatory, Physical Research Laboratory, Udaipur, India.
email: nandita@prl.res.in; shibu@prl.res.in

[2]Korea Astronomy and Space Science Institute (KASI), Daejeon, Korea
email: anand@kasi.re.kr

**Abstract.** We report observations of a long filament that underwent recurrent partial eruptions on August 4, 6, and 8, 2012. The filament reappeared in the subsequent rotation of the Sun, and disappeared completely on August 31, 2012. We implemented an automated filament detection algorithm developed by us for estimating different attributes of these filaments few hours prior to its disappearance in H$\alpha$ and studied their evolution. Based on these attributes, we determine the onset time of the disappearance of H$\alpha$ filaments. We then compared these onset times with that of the associated CMEs observed by LASCO/SOHO coronagraphs. This is also useful to understand temporal relationship of EUV and X-ray flux variation associated with filament disappearances in H$\alpha$. Our results show the importance of such studies in understanding the mechanism of CME initiation, particularly the role of eruptive filaments, in this process.

**Keywords.** Filaments, CMEs, Onset time, etc.

---

## 1. Introduction

Coronal Mass Ejections (CMEs) are the key drivers of space weather and are known to be the main cause of major geomagnetic storms at the Earth (Gosling, 1993). They are often associated with flares and eruptive filaments or EFs (Webb *et al.* 1976; Webb and Hundhausen, 1987). Munro *et al.* (1979) found that more than 70% of the CMEs were associated with EFs. Those originating from around centre of solar disc are potential candidates for geo-effective CMEs as shown by Srivastava & Venkatakrishnan (2004). CMEs associated with flares and eruptive prominences differ in their properties. A recent study by Joshi and Srivastava (2011) has shown that flare associated CMEs show bimodal acceleration while filament associated CMEs do not. As per our present understanding, same mechanism drives all CMEs, and the two types of CMEs, lie at the two extremes of the energy range that is driving them. One of the major constraints in investigating the driving force of CMEs is the difficulty in estimating their correct onset time. The onset time is currently estimated by back extrapolating the projected height-time plot of the leading edge of a CME using LASCO/SoHO observations (http://cdaw.gsfc.nasa.gov). In the events where the source region involves a filament, monitoring its activation is crucial for forewarning of its disappearance in H$\alpha$, in EUV and also the associated soft X-ray (SXR) flare and CME lift off. Determining precise temporal relationship between the eruption of H$\alpha$ and EUV filament, SXR emission and appearance of CME leading edge (in white light) is therefore of utmost importance in understanding the CME initiation mechanism. To achieve this objective, we examine, recurrent eruptions of a single large filament and the associated CMEs that occurred on August 4, 6, 8 and in the next rotation of the sun on August 31, 2012.

## 2. Events, Analysis and Summary

A long filament appeared on the Sun on August 1 at the SE limb as observed in H$\alpha$ images. On August 4, it showed activations leading to its disappearance in H$\alpha$ at around 12:20 UT. This was associated with a C3.5 class flare observed by GOES and enhancement in EUV flux observed by LYRA aboard PROBA2 (Dominique *et al.* 2013). The filament eruption was associated with a white light halo CME with projected speed of 850 km/s as observed by LASCO-C2 coronagraph at 13:22 UT. Following this, the filament reformed and underwent subsequent eruptions on August 6 and 8 accompanied by C-class flares and slow and accelerating CMEs. It may be noted that all the three eruptions were not powerful enough to drive a geomagnetic storm although the location of the filaments was favourable (Srivastava and Venkatakrishnan, 2004). In the next rotation of the Sun, the filament reappeared on August 30, on the SE limb and erupted completely on August 31 at 19:40 UT in H$\alpha$ with an associated C8.4 class flare and EUV flux peaking around 20:43 UT and 21:13 UT. Concurrently, LASCO-C2 observed a partial halo CME with a speed of 1440 km/s. Although this CME was in SE direction, it gave rise to a moderate geomagnetic storm on September 3 (Dst $\sim -78$ nT). As described above, the chain of recurrent eruptions of a long quiescent filament at intervals of few days, provides an excellent data-set to understand the initiation, and propagation of slowly accelerating CMEs during the minimum phase of the solar cycle, which under suitable circumstances may prove to be geoeffective. We implemented an algorithm developed by Joshi, Srivastava and Mathew (2010) on the full disk H$\alpha$ images obtained for all the 4 events. This algorithm detects, tracks and estimates the location, length, area and number of fragments of disappearing filaments. From the temporal evolution of these attributes, we estimated the time of start and end of H$\alpha$ filament disappearance, compared these with the time of SXR flux peak, EUV flux peak and CME lift off. Our analysis shows that the start time of the filament disappearance in H$\alpha$ precedes the time of peak of SXR flux by more than 1.5 hr. The EUV flux peaks 15 to 30 min after the SXR flux. Further, the H$\alpha$ filament disappearance starts at least an hour earlier than the onset time of CME. These clearly demonstrate the usefulness of implementation of the automated filament detection technique to forewarn potential geo-effective eruptions, based on full disk H$\alpha$ observations, prior to the launch of a CME.

## 3. Acknowledgements

This work contributes to the research for European Union Seventh Framework Programme (FP7/2007-2013) for the COMESEP project under Grant Agreement No. 263252. We thank the GONG & PROBA2 team for providing H$\alpha$ and LYRA data, respectively.

### References

Dominique, M., Hochedez, J.-F., Schmutz, W., Dammasch, I. E., Shapiro, A. I., Kretzschmar, M., Zhukov, A. N., Gillotay, D. *et al.* 2013, *Solar Phys.*, 286,1, 21–42

Gosling, J. T. *J. Geophys. Res*, 98, A11, 18937–18950

Joshi, A. D., Srivastava, N., & Mathew, S. K. 2010, *Solar Phys.*, 262, 425–436

Joshi, A. D. & Srivastava, N. 2011, *Astrophys. J.*,739, 1, 8

Munro, R. H., Gosling, J. T., Hildner, E., MacQueen, R. M., Poland, A. I., Ross, C. L.1979, *Solar Phys.* 61, 201–215

Srivastava, N. & Venkatakrishnan, P.2004, *J. Geophys. Res.* 109, A10, CiteID A10103

Webb, D. F., Krieger, A. S., & Rust, D. M. 1976, *Solar Phys.* 48, 159–186

Webb, D. F. & Hundhausen, A. 1987, *Solar Phys.*108, 2, 383–401.

*Nature of Prominences and their role in Space Weather*
*Proceedings IAU Symposium No. 300, 2013*
*B. Schmieder, J.-M. Malherbe & S. T. Wu, eds.*

doi:10.1017/S1743921313011721

# Different Stages of Evolution of Prominence and the Associated CMEs

**Lela Taliashvili[1], Zadig Mouradian[2] and Heidy Gutiérrez[1]**

[1]Space Research Center, University of Costa Rica, 2060 San Jos, Costa Rica
email: lela.taliashvili@cinespa.ucr.ac.cr
[2]Observatoire de Paris–Meudon, LESIA, 92190 Meudon, France

**Abstract.** We study the different evolutional stages of a large quiescent prominence, mainly considering its dynamic/thermal instabilities occurred close from the boundary of coronal hole (CH). We identify the critical conditions, such as the minimum distance between the CH's boundary and prominence channel and the emergence of a new magnetic flux linked to the prominence instability and its general evolution in connection to CH. Our observations indicate peculiar filament activations prior to its thermal/dynamic instabilities, suggesting the connection of nearby CH with the general evolution of prominence and vice versa. Additionally, we analyze each evolutional stage of prominence and the associated Coronal Mass Ejections (CMEs).

**Keywords.** Prominence, coronal hole, coronal mass ejection, magnetic flux, magnetic field

## 1. Introduction

An important issue for understanding the evolution of prominences/filaments is their thermal and dynamic instabilities (Mouradian *et al.* (1981), Mouradian *et al.* (1986), Tandberg–Hanssen (1995)). Observational results indicate the implication of CH at these different evolutional stages of prominence. Based on 42 quiescent prominences, the study by Taliashvili *et al.* (2009) have found that all thermal and 91% of dynamic eruptions were occurred within ~15° distance from CH's boundary; and 47% of thermal and 64% of dynamic eruptions were associated with CMEs. It is well known the relation between the prominence dynamic eruptions and CMEs due to magnetic reconnection, frequently anticipated by an emergence of new magnetic flux at one of filament's foot points (Mouradian *et al.* (1987)) while the energy is transported by waves along the flux tubes in the prominence feet from the pivot point (Mouradian *et al.* (1989)). Various authors have discussed different tools associated to the magnetic reconfiguration before/after prominence eruption and the formation of CMEs; Schmieder *et al.* (2013) (and their references) summarizing the causes of prominence eruption, the emergence of new flux and/or the external magnetic field dispersion and/or reconnection of field lines below/above the flux rope that reduces the downward magnetic tension, brings the flux rope to an unstable state; however they conclude that the most efficient mechanism for CME formation is torus instability. Yet these processes are still unclear step by step.

## 2. Event description and Discussion

We study the continuous evolution of a long–lived quiescent filament /prominence spread out from equator to −19° during two periods, Aug. 3–13 (CR 2126), Aug. 30–Sept. 4 of 2012 (CR 2127), located at ~15° distance from the equatorial CH; based on Hα images reported by Paris–Meudon Observatory and GONG. In order to study the evolutional stages of filament and the global interaction with CH evolution, we analyze:

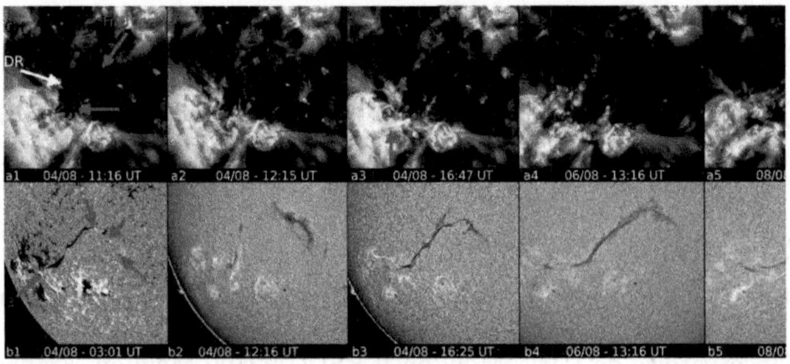


**Figure 1.** (a1-a6) EUV evolution of filament/CHs by SDO/AIA–193 Å. (b2-b6) H$\alpha$ evolution of filament by GONG. (b1) Superposition of GONG H$\alpha$ image and SDO/HMI magnetogram of 04/08–03:00UT, with four newly emerged fluxes.

STEREO and SDO/AIA EUV images, WSO, SDO/HMI magnetograms, and maps. In regard to the prominence eruption–CME association we consider their similar position angles, widths and their starting times. CMEs are identified using SOHO/LASCO and STEREO CME catalogs, images and movies.

The large ($\sim$45°) filament has an additional northern unstable $\sim$15° section, extended toward the flux 4 (Figure 1, b1), which is observed only at EUV images, showing that the prominence plasma reached coronal temperatures, while keeping its shape i.e. the magnetic field structure. Four small active regions (ARs) surround the southern foot point of filament. At the East side, parallel to the filament channel extends the dimming region (DR) and a small equatorial CH within $\sim$15° distance (Figure 1, a1). Every evolutional stage of filament successively involves three sections that compose the filament's body: southern (SS), central (CS) and northern (NS); each of them characterized by a different evolution Mouradian *et al.* (1989) and an independent and almost constant movement of its material that sometimes upraises, but the whole filament maintains stable until Aug. 8. During the evolution of this filament, additionally three peculiar motions are observed at Aug. 4, 6 and 7 in form of a directional plasma launch. These motions anticipate a consecutive emergence of certain new magnetic fluxes near and along the filament channel, especially close to the filament foot points (Figure 1, b1; additional fluxes emerge among flux 2/3, at the east side of filament channel). Moreover, C1.1–C3.5 flares, related with NOAA ARs 11539–40–41 observed 0.5 h–3 h before these launches. Then a directional plasma launch starts from southern toward the northern foot point that follows the same path of emerged fluxes ending with filament appearance at its original position. These processes develop according to the following three scenarios (Figure 1): (1) Directional plasma launch (Aug. 4. H$\alpha$: 11:15–13:30 UT; EUV: 11:15–18:00 UT) from a small EUV loop that reconnects with the southern foot point at 11:16 UT; At 13:30 UT filament returns to its original form in H$\alpha$ due to plasma cooling by radiation; then it raises and its SC expands due to an increasing energy input leading to its thermal disappearance in H$\alpha$ (13:45–14:45 UT); At 15:40 flux 3 cancelled and a small coronal loop arise from its location, at 16:40 UT, moving radially; the northern EUV hot section cools down; this NS that was visible only at EUV, appears in H$\alpha$ too. After $\sim$4 h, two consecutive narrow CMEs start at nearly position angle of the cancelled flux; after CMEs, filament and CH increase, reaching their maximum size at Aug. 6–7. (2) Directional plasma launch (Aug. 6, 09:15–22:45 UT) combined with filament plasma motions mainly inside its sections, accompanied by the whole filament expansion and the transient CH birth within DR, associated with evacuated filament plasma (20:45 UT); transient CH grows and starts the CS and SS thermal disappearance. (3) Directional plasma launch (Aug. 7, 23:01 UT–Aug. 8. 02:00 UT) is similar to the first launch; nonetheless is followed by a dynamic eruption of CS and SS (03:45 UT) and perturbation of filament's surroundings, covering the CH; after ~30 min. two consecutive CMEs start and both CHs disappear (14:02 UT). At Aug. 13, the SS and the CS's remnant erupt followed by consecutive CMEs (two west–limb and one Halo). Based on STEREO EUV evolution of filament we note the filament rebuild (Mouradian *et al.* 1987) from Aug. 14, almost with its original form. Lately, from Aug. 30 (CR 20127) it is observed again from Earth with two DRs, at east side (within ~15° distance) and around its NS. At 17:00 UT, Aug. 31 the whole filament eruption begins and at 20:00 UT a Halo CME occurs. A new small equatorial CH is formed within DR at the east side of filament channel few hours before this eruption, which grows during the eruption and disappears slowly after 3.5 days. We observed a part of the evacuated filament's material right after this eruption. From Sept. 4, H$\alpha$ filament is observed entirely with considerably less material.

As a summary of our observations, we interpret that new CH appears before and grows during the thermal/dynamic instabilities, whereas CH fading is a consequence of dynamic instabilities or/and after classical CMEs (Taliashvili *et al.* (2009)); we observe the CH growth only after CMEs associated with thermal instabilities. A stable magnetic filed supports the filament. Thermal instabilities and the associated CMEs prior to the filament eruption are related to a small–scale magnetic reconfiguration, but they might be indicators of upcoming destabilization of filament and its surroundings (including adjacent CH, within 15° distance from the filament channel) followed by a huge CMEs and a large-scale magnetic reconfiguration observed at the late evolutional stage of filament as longitudinal magnetic diffusion (Gutiérrez *et al.* (2013)), with CH fading from the closest region to the southern foot point toward the CH. These observational results indicate the involvement of a newly emerged magnetic flux near/along of filament channel and close to the foot points at the starting process of instabilities and also at the general evolution of filament/adjacent CH; ongoing series of magnetic reconnections due to flux cancelation leading to peculiar motions and different type of filament instabilities, birth and growth of adjacent CH and formation of CMEs. The flux rope that is progressively formed by photospheric reconnection and successively flux cancellation, approaches a critical point of the equilibrium curve driven by a constant increase of the twist and/or by changing the magnetic flux below/above the flux rope, removes the overlying arcades by coronal reconnection and erupts by developed torus instability (Aulanier *et al.* (2010) and their references).

**Acknowledgments.** We thank the Observatoire Paris-Meudon, GONG, WSO, SDO, SOHO/LASCO, STEREO for open access to their data sets. L.T and H.G. acknowledge for the financial supports by IAU/LOC.

## References

Aulanier, G., Török, T., Démoulin, P., & DeLuca, E. E. *Astrophys. J.* 708, 314, 2010.

Gutiérrez, H., Taliashvili, L., & Mouradian, Z. *Ad. Space Res.* 51 , 1824, 2013.

Mouradian, Z., Martres, M. J., & Soru–Escaut, I. *Proc. Japan–France Seminar on Solar Physics, Tokyo, ed., Moriyama, F., and Henoux, J.C.* 195, 1981.

Mouradian, Z., Martres, M. J., & Soru–Escaut, I. *Coronal and Prominence Plasmas, NASA conf. Public. ed., Poland A. I.* 221, 1986.

Mouradian, Z., Martres, M. J., Soru–Escaut, I., & Gesztelyi, L. *Astron. Astrophys.* 183, 129, 1987.

Mouradian, Z., & Soru–Escaut, I. *Astron. Astrophys.* 210, 410, 1989.

Schmieder, B., Dmoulin, P., & Aulanier, G. *Adv. Space Res* 51, 1967, 2013.

Tandberg–Hanssen, E. *The Nature of Solar Prominences, Kluwer Acad. Pub.* §6.2.1, 1995.

Taliashvili, L., Mouradian, Z., & Páez, J. *Sol. Phys.* 258, 277, 2009.

*Nature of Prominences and their role in Space Weather*
*Proceedings IAU Symposium No. 300, 2013*
*B. Schmieder, J.-M. Malherbe & S. T. Wu, eds.*

doi:10.1017/S1743921313011733

# From solar eruption to transformer saturation: the space weather chain

**Larisa Trichtchenko**

Canadian Space Weather Forecast Centre, Natural Resources Canada, 2617 Anderson Rd., Ottawa ON, K1A 0E7, Canada
email: `larisa.trichtchenko@nrcan-rncan.gc.ca`

**Abstract.** Coronal mass ejections (CME) and associated interplanetary-propagated solar wind disturbances are the established causes of the geomagnetic storms which, in turn, create the most hazardous impacts on power grids. These impacts are due to the large geomagnetically induced currents (GIC) associated with variations of geomagnetic field during storms, which, flowing through the transformer windings, cause extra magnetisation. That can lead to transformer saturation and, in extreme cases, can result in power blackouts. Thus, it is of practical importance to study the solar causes of the large space weather events. This paper presents the example of the space weather chain for the event of 5-6 November 2001 and a table providing complete overview of the largest solar events during solar cycle 23 with their subsequent effects on interplanetary medium and on the ground. This compact overview can be used as guidance for investigations of the solar causes and their predictions, which has a practical importance in everyday life.

**Keywords.** Space Weather, Coronal Mass Ejections, Magnetic Storms, Geomagnetically Induced Currents

---

The event, presented as a detailed example, started on November 4, 2001 with a halo CME and associated X-ray flare at about 16:20 UT (Fig. 1). Two days later the disturbance arrived at 1 AU as shown in variations of the solar wind magnetic field recorded by the ACE satellite (Fig. 1a). The southward (negative) component reached minimum value of 80 nT at $\sim$ 02:00 UT. On the same day, magnetic field variations in Western Canada (Fig. 1b) maximized at $\sim$ 04:00 UT and in Eastern Canada at $\sim$ 02:20 UT (Fig. 1d). Power grids in the West and East (Power site 1 and Power site 2) have responded with GIC maxima at $\sim$ 02:30 UT (site 1) and at $\sim$ 2:40 UT (site 2), as shown in Figs. 1c and 1e.

The complete overview of the most significant Space Weather Events of the Solar Cycle 23 is presented in Table 1 with their subsequent effects on geomagnetic field and power grids. The events were chosen based on the global index of geomagnetic activity $K_p > 8-$. The halo CME and location (approximate) of the source region, based on location of associated flare, are the key solar parameters. The southward interplanetary magnetic field has a controlling role in the interaction between the ICME and the Earth's magnetosphere. The effects on the ground magnetic field at specific region (eastern Canada) are defined by a peak local hourly range index of magnetic variations. Finally, the peak values of GIC recorded on the Nova Scotia Power system and by the Sunburst monitoring network (Electric Power Research Institute) at one of the power grids in the USA are shown. There were two short power outages during solar cycle 23 attributed to space weather impacts, in New Zealand (November 2001) and in Sweden(October 2003).

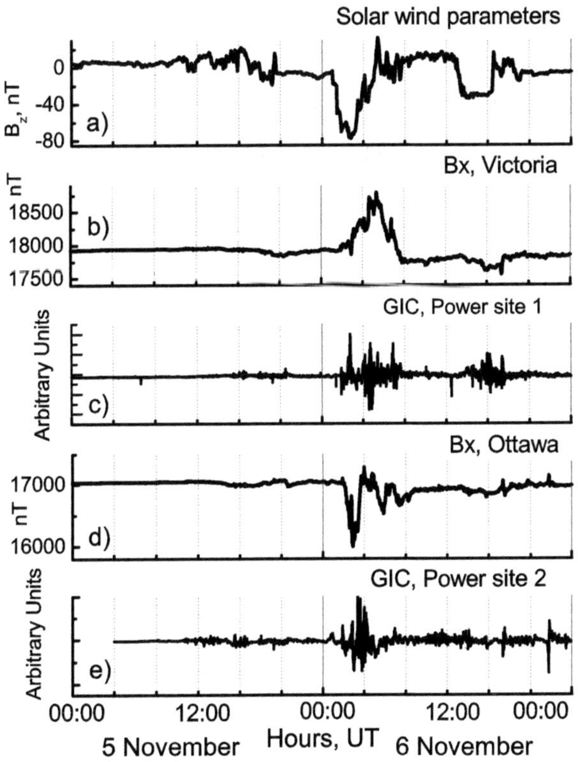


**Figure 1.** Magnetic field in the solar wind and on the ground and GIC on 5-6 November.

**Table 1.** Critical parameters of Space Weather Events from the Sun to the Earth.

| Dates of events | CME | Flare (assoc.) | Location (possible) | IMF $B_z$ nT | $K_p$ | HR local nT | Site 1 Max GIC,A | Site 2 Max GIC,A |
|---|---|---|---|---|---|---|---|---|
| 2-4 May 98 | Halo | X1, M6 | S17W15 | -32 | 9- | 898 | 70 | 74 |
| 23-25 Sep 98 | N/A | M6.9 | N18E09 | -27 | 8+ | 1270 | N/A | N/A |
| 21-22 Oct 99 | NE | M1 | N/A | -30 | 8 | 908 | 47 | N/A |
| 4-7 Apr 00 | Halo/DSF | C9 | N16W66 | -27 | 9- | 807 | 96 | 80 |
| 14-16 Jul 00 | Halo | X5 | N22W07 | -54 | 9 | 1729 | 92 | 76 |
| 17 Sep 00 | Halo | M2 | N13E09 | -34 | 8+ | 875 | 62 | N/A |
| 31 Mar 01 | Halo | X1.7 | N20W19 | -46 | 9 | 1236 | 100 | 76 |
| 11 Apr 01 | Halo | X2 | S23W09 | -27 | 8 | 806 | 67 | 27 |
| 06 Nov 01 | Halo | X1 | N06W18 | -68 | 9- | 1020 | 50 | 64 |
| 24 Nov 01 | Halo | M9.9 | S15W34 | -40 | 8 | 839 | 67 | 90 |
| 23 May 02 | Halo/DSF | C5 LDF | S22W53 | -43 | 8+ | 327 | 24 | 10 |
| 29-30 May 03 | ?? | X1, X3 | S07W17 | -33 | 8+ | 915 | 40 | 60 |
| 29-31 Oct 03 | 2Halo | X17/X10 | S16E08/ S15W02 | -48/ -35 | 9 | 1500 | 100 | 75 |
| 20-21 Nov 03 | Halo | M3 | S01E18(?) | -53 | 9- | 450 | 60 | 20 |
| 24-27 Jul 04 | Halo | M1.2 LDE | N03W27 | -21 | 9- | 1422 | N/A | 50 |
| 7-8 Nov 04 | Halo | M5/M9 | N11E11/ N10E08 | -48.5 | 9- | 1561 | N/A | 78 |
| 9-10 Nov 04 | Halo | X2/M9 | N09W17/ N07W51 | -25 | 9- | 1317 | N/A | 80 |
| 15 May 05 | Halo | M8 | N12E12 | -43 | 8- | 1924 | N/A | 83 |
| 24 Aug 05 | Halo | M2/M5 LDE | S08W50/ S12W60 | -55 | 9- | 662 | N/A | 55 |

DSF=Disappearing Solar Filament, LDE=Long Duration Event, N/A=data not available

*Nature of Prominences and their role in Space Weather*
*Proceedings IAU Symposium No. 300, 2013*
*B. Schmieder, J.-M. Malherbe & S. T. Wu, eds.*

doi:10.1017/S1743921313011745

# Magnetic reconnection driven by filament eruption in the 7 June 2011 event

**L. van Driel-Gesztelyi[1,2,3], D. Baker[1], T. Török[4], E. Pariat[2], L. M. Green[1], D. R. Williams[1], J. Carlyle[1], G. Valori[2], P. Démoulin[2], S. A. Matthews[1], B. Kliem[5] and J.-M. Malherbe[2]**

[1]University College London, Mullard Space Science Laboratory, Holmbury St. Mary, Dorking, Surrey RH5 6NT, UK
[2]LESIA, Observatoire de Paris, CNRS, UPMC, Université Paris Diderot, 5 place Jules Janssen, 92190 Meudon, France;email: Lidia.vanDriel@obspm.fr
[3]Konkoly Observatory, Hungarian Academy of Sciences, Budapest, Hungary
[4]Predictive Science Inc., 9990 Mesa Rim Rd., Suite 170, San Diego, CA 92121, USA
[5]Institut für Physik und Astronomie, Universität Potsdam, Karl-Liebknecht-Str. 24-25, 14476 Potsdam, Germany

**Abstract.** During an unusually massive filament eruption on 7 June 2011, SDO/AIA imaged for the first time significant EUV emission around a magnetic reconnection region in the solar corona. The reconnection occurred between magnetic fields of the laterally expanding CME and a neighbouring active region. A pre-existing quasi-separatrix layer was activated in the process. This scenario is supported by data-constrained numerical simulations of the eruption. Observations show that dense cool filament plasma was re-directed and heated in situ, producing coronal-temperature emission around the reconnection region. These results provide the first direct observational evidence, supported by MHD simulations and magnetic modelling, that a large-scale re-configuration of the coronal magnetic field takes place during solar eruptions via the process of magnetic reconnection.

**Keywords.** MHD, instabilities, Sun: activity, magnetic fields, coronal mass ejections (CMEs), filaments, methods: numerical, data analysis

---

A spectacular solar eruption occurred on 7 June 2011 observed by the Solar Dynamic Observatory's Atmospheric Imaging Assembly (SDO/AIA: Lemen *et al.* 2012). The CME originated in a complex of three adjacent active regions (ARs, see Figure 1) in the south-western quadrant and carried an unusually massive erupting filament in its core. We carried out a multiwavelength analysis of the event. Using SDO/HMI data we computed the magnetic topology, determining the locations of quasi-separatrix layers in the three-AR complex. We also carried out data-constrained MHD simulations of the eruption.

We found that the strong lateral expansion of the erupting magnetic structure led to flux pile-up, current sheet formation/intensification, and magnetic reconnection along a pre-existing quasi-separatrix layer in the three-AR complex. The onset of reconnection first became apparent in the SDO/AIA images when downward flowing dense, cool filament plasma, originally contained within the erupting flux rope, was re-directed towards a neighbouring active region, tracing the change of large-scale magnetic connectivity. Williams *et al.* (2013) estimated a lower limit of the electron density of the redirected plasma to be $10^{10}$ cm$^{-3}$, at least one order of magnitude larger than the typical coronal density. As a result of this unusually high density around the reconnection region, direct plasma heating took place there. The most prominent brightening was seen in the AIA 171 Å waveband ($6.3 \times 10^5$ K).

These SDO observations provide one of the first direct imaging observations of magnetic reconnection in the solar atmosphere. Furthermore, a combination of observations,

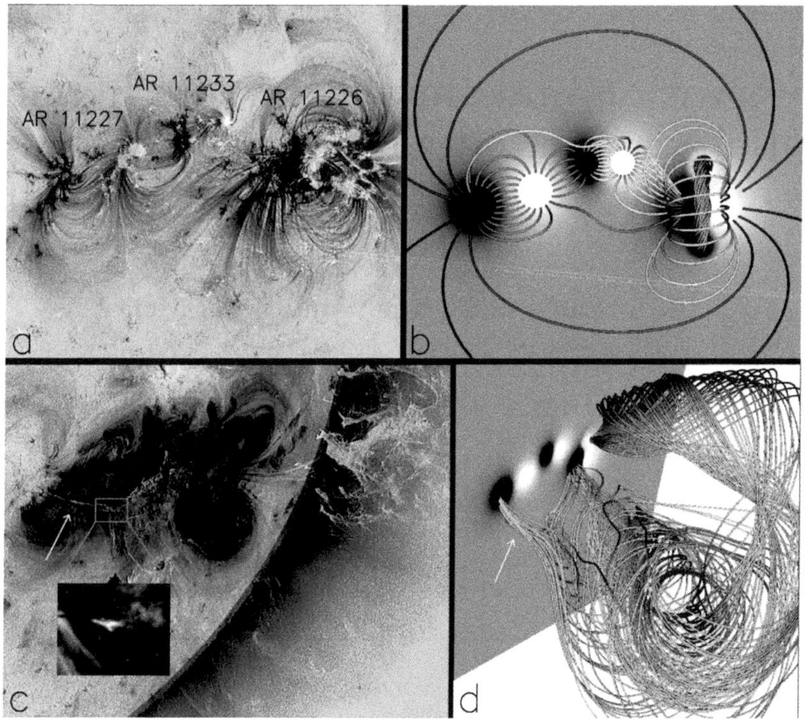


**Figure 1.** Observations and MHD numerical simulation of the magnetic configuration and the eruption. (**a**) SDO/AIA 171 Å reverse colour scale image over a co-aligned SDO/HMI magnetic field map of the three neighbouring active regions on 3 June 2011 and (**b**) corresponding magnetic configuration in the simulation. Black/white indicate negative/positive magnetic polarity, the pre-eruption magnetic flux rope in the rightmost active region is shown in gold. (**c**) SDO/AIA 171 Å reversed-colour image and (**d**) a simulation snapshot during the CME eruption on 7 June, by when the active regions have rotated close to the solar limb. The new connections, formed by magnetic reconnection between the magnetic flux rope erupting from AR 11226 and magnetic field lines of AR 11227, are indicated by white arrows in (**c**) and (**d**). The inset in (**c**) is a magnification of the in-situ heated bright reconnection region.

magnetic modelling, and MHD simulations (Figure 1) provides evidence that, during the expansion of a CME's magnetic structure, instantaneous magnetic reconnection can occur with ambient magnetic field leading to large-scale restructuring. For more details see van Driel-Gesztelyi *et al.* (2013).

## Acknowledgements

The authors acknowledge funding from the European Commission's Seventh Framework Programme under the grant agreement No. 284461 (eHEROES project) and the Hungarian Research grant OTKA K-081421. TT was supported by NASA's HTP, LWS, SR&T programs and NSF (AGS-1249270). LMG is grateful for a Royal Society Fellowship, and JC for a joint UCL-MPI Impact PhD Studentship. BK acknowledges support by the DFG.

## References

Lemen, J. R., Title, A. M., Akin, D. J., *et al.*, 2012, *Solar Phys.* 275, 17

van Driel-Gesztelyi, L., Baker, D., Török, T., Pariat, E., Green, L. M., Williams, D. R., Carlyle, J., Valori, G., Démoulin, P., Matthews, S. A., Kliem, B., & Malherbe, J.-M. 2013, *ApJ*, submitted.

Williams, D. R., Baker, D., & van Driel-Gesztelyi, L. 2013, *ApJ*, 764, 165

*Nature of Prominences and their role in Space Weather*
*Proceedings IAU Symposium No. 300, 2013*
*B. Schmieder, J.-M. Malherbe & S. T. Wu, eds.*

doi:10.1017/S1743921313011757

# Observation and simulation of a filament eruption associated with the contraction of the overlying coronal loops and the filament rotation

**X. L. Yan[1], Z. K. Xue[1,2] and Z. X. Mei[1]**

[1]Yunnan Astronomical Observatory, Chinese Academy of Sciences, Kunming 650011, China
email: `yanxl@ynao.ac.cn`

[2]University of Chinese Academy of Sciences, Zhongguancun Beijing 100049, China

**Abstract.** By using the data of Solar Dynamics Observatory (SDO), we present a case study of the contraction of the overlying coronal loop and the rotation motion of a sigmoid filament on 2012 May 22. At the beginning of the filament eruption, the overlying coronal loop experienced a significant contraction. In the following, the filament started to rotate counterclockwise. We also carried the simulation to investigate the process of the filament eruption.

## 1. Introduction

The contraction of the magnetic loop associated with flares and flarelike events has been shown by a number of authors (Forbes & Acton 1996; Švestka *et al.* 1997; Liu *et al.* 2010). The shrinkage of large-scale loop without flares was also reported by Wang *et al.* (1997) using data of the Yohkoh Soft X-ray telescope. Wang *et al.* (1997) suggested that the shrinkage is not an apparent motion, but a real contraction of the coronal loops as a result of the heating at the footpoints followed by gradual cooling. Hudson (2000) suggested that a magnetic implosion at preflare energy storage must occur simultaneously with coronal events such as flares or coronal mass ejections.

## 2. Observation

The Atmospheric Imaging Assembly (AIA; Lemen *et al.* 2012) on board Solar Dynamics Observatory (SDO) provides multiple simultaneous high resolution full-disk images of the corona and the transition region. The observational range of the AIA can extend to 1.3 $R_\odot$. We analyzed the contraction process of the overlying coronal loop and the rotation of the filament by using the 171 Å and 304 Å images of the AIA, which have high temporal (∼12 s) and spatial resolution (0.6″).

## 3. Result

The successive contraction of the coronal loops overlying the filament can be seen from a series of 171 Å images before the filament eruption (see the white arrows in Fig. 1). The maximum contraction speed was 45 km/s. After the contraction of the overlying coronal loops, the left part of the filament began to rotate counter-clockwise seen from the 304 Å observation (Fig. 2). Meanwhile, the counterclockwise rotation of the right foot of the filament was also observed during its eruption (Yan *et al.* 2013). We also carried out the simulation of filament eruption (Fig. 3). We started with a parameter set that is used

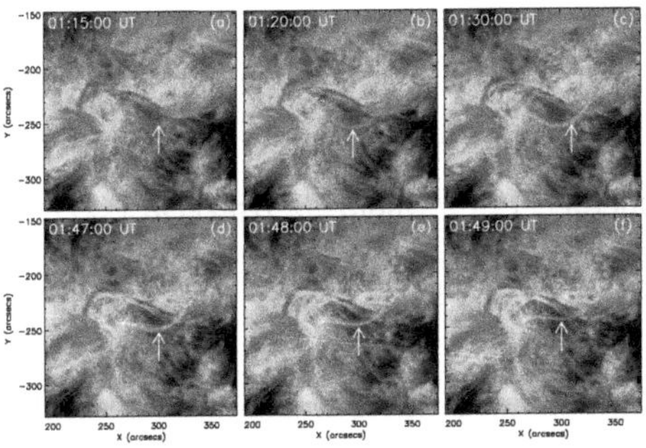


**Figure 1.** A sequence of images acquired at 171 Å to show the contraction process of the coronal loop overlying the sigmoid filament from 01:15:00UT to 01:49:00UT on 2012 May 22. The white arrows point to the coronal loop.

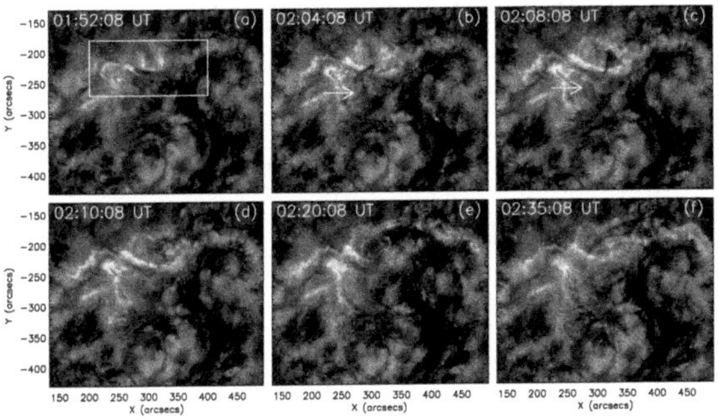


**Figure 2.** A sequence of images acquired at 304 Å to show the rotation process of the sigmoid filament from 01:52:08UT to 02:35:08UT on 2012 May 22. The white box and the arrows denote the filament.

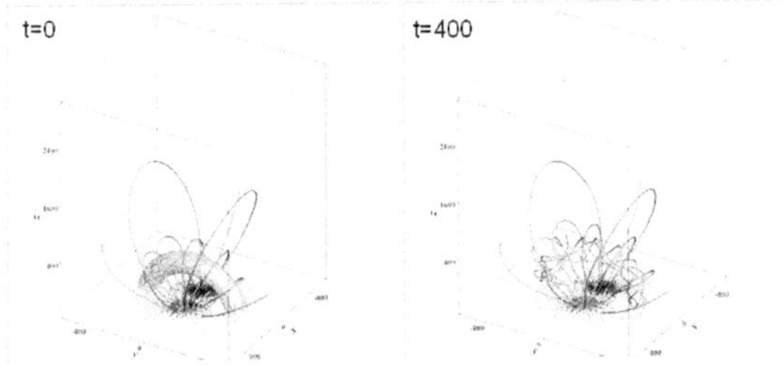


**Figure 3.** The 3D-MHD simulation of a filament eruption.

by Titov & Demoulin (1999). The IRVANA code was used to carry out the simulation (Ziegler 2008). At the initial phase of filament eruption, we find that the coronal loop overlying the filament began to contract due to the expansion of the filament in our simulation. During the filament eruption, the overlying coronal loops at the end of the filament present the contraction as the coronal loops start to contract toward the below of the filament.

## 4. Conclusion

During the filament eruption, the contraction of the overlying coronal loops and the rotation motion of the filament was observed. Through the MHD simulation of filament eruption, the contraction of overlying coronal loop is found at the initial stage of the filament eruption. We explain that the contraction of the overlying coronal loops was due to the decrease of magnetic pressure and magnetic energy release. The obvious rotation of the right foot of the filament is caused by the unwinding motion of the filament during its eruption.

### Acknowledgements

SDO is a mission of NASA's Living With a Star Program. This work is supported by the National Science Foundation of China (NSFC) under grant numbers 11373066, Key Laboratory of Solar Activity of CAS under number KLSA201303.

## References

Forbes T. G. 2000, *J. Geophys. Res.*, 105, 23153

Hudson, H. S. 2000, *ApJ* (Letters), 531, L75

Liu, R. & Wang, H. M. 2010, *ApJ*, 714, L41

Lemen, J. R., Title, A. M., Akin, D. J., Boerner, P. F., Chou, C., Drake, J. F., Duncan, D. W., *et al.* 2012, *Solar Phys.*, 275, 17

Švestka, Z. F., Fontenla, J. M., Machado, M. E., Martin, S. F., & Neidig, D. F. 1987, *Solar Phys.*, 108, 273

Titov, V. S. & Demoulin, P. 1999, *A&A*, 351, 707

Wang, J., Shibata, K., Nitta, N., Slater, G. L., Savy, S. K., & Ogawara, Y. 1997, *ApJ* (Letters), 478, L41

Yan, X. L., Pan, G. M., Liu, J. H., Qu, Z. Q., Xue, Z. K., *et al.* 2013, *AJ*, 145, 153

Ziegler, U. 2008, *Computer Physics Communications*, 179, 227

# Poster sessions III & IV

POSTER SESSION III

STAR EJECTA

&

POSTER SESSION IV

INSTRUMENTATION

*Nature of Prominences and their role in Space Weather*
*Proceedings IAU Symposium No. 300, 2013*
*B. Schmieder, J.-M. Malherbe & S. T. Wu, eds.*

doi:10.1017/S1743921313011769

# Stellar ejecta from falling comet-like bodies: young stars

**Firuz S. Ibodov**[1] **and Subhon Ibadov**[1,2]

[1]Lomonosov Moscow State University, Sternberg Astronomical Institute, Moscow, 119234 Russia
email: `mshtf@sai.msu.ru`

[2]Institute of Astrophysics, Tajik Academy of Sciences, Dushanbe, Tajikistan
email: `ibadovsu@yandex.ru`

**Abstract.** High-resolution spectral observations of young stars with dense protoplanetary discs like Beta Pictoris led to the discovery of variable emission lines of metal atoms, Na, Fe etc., that indicate the presence of fluxes of comet-like evaporating bodies falling onto the stars, FEBs. Assuming the presence of stellar atmospheres similar to the solar one, we show that passages of the FEBs through the stellar chromosphere and photosphere with velocities around 600 km/s will be accompanied by aerodynamic crushing of the nuclei, transverse expansion of the crushed matter, "explosion" of the flattened nuclei in a relatively very thin sub-photosphere layer due to sharp deceleration, and impulse production of a hot plasma. The impulsive rise of the layer's temperature and density lead to the generation of a strong "blast" shock wave and shock wave-induced ejection/eruption of hot plasma into space above the chromosphere. Observations of such impact-induced high-temperature phenomena are of interest for the physics/prognosis of stellar/solar flares as well as physics of comets.

**Keywords.** stars: individual ($\beta$ Pic); stars: flare; comets: general; shock waves; Sun: coronal mass ejections (CMEs)

## 1. Introduction

Coronagraph observations by SOLWIND, SMM, SOHO and STEREO missions, together with ground-based observations and celestial-mechanics calculations of the orbital evolution of comets, indicate the presence of a continuous comet flux passing close to the solar surface or colliding with the Sun (Weissman 1983; MacQueen & St.Cyr 1991; Bailey *et al.* 1992; Info 1998; `http://sungrazer.nrl.navy.mil/`).

At the same time, high-resolution spectral observations of young stars with dense protoplanetary discs, like $\beta$ Pictoris, led to the discovery of variable emission lines of metal atoms, Na, Fe, etc., that indicate the presence of fluxes of comet-like evaporating bodies falling onto the stars, FEBs (Lagrange *et al.* 1987; Beust *et al.* 1996; Grinin *et al.* 1996).

We are developing an analytical approach for investigating processes accompanying the passage of star/Sun impacting comet like bodies through their atmospheres and the relation of these processes to active ones on the stars, assuming the stellar atmospheres are similar to the solar atmosphere.

## 2. Impulse aerodynamic deceleration of crushed comet nuclei in the solar/stellar atmosphere: explosive photospheric mass ejection

The height range of basic deceleration of the aerodynamically fully fragmented and transversally expanding nucleus where the decrease of velocity from $V_1 = 0.9V_0$ to $V_2 = 0.1V_0 = 60$ km/s occurs is $\Delta h_\mathrm{d} \approx 0.7H = 140$ km (Grigorian *et al.* 1997, 1998, 2000; Ibadov *et al.* 2009).

The characteristic time for thermalization of the kinetic energy of the fragmented mass in the decelerating layer, $\tau_\mathrm{th} = \Delta h_\mathrm{d}/V_\mathrm{e} = 0.7eH/V_0 \approx 0.5$ s, indicates the strongly impulse/explosive character of the energy release process that leads to generation of a hot plasma and strong shock wave.

The initial velocity of "blast" shock wave in the explosive layer is estimated as

$$V_\mathrm{sh} = \left[\frac{kT_0}{2\pi A m_\mathrm{p}} + \frac{3k(1+z)T_0}{A m_\mathrm{p}}\right]. \tag{2.1}$$

Here $T_0$ is the initial plasma temperature, $T_0 = A m_\mathrm{p} V_\mathrm{e}^2 / \left[12k\left(1 + z + 2x_1/3\right)\right]$, $A$ is the mean atomic number for the falling comet nucleus material and matter of the solar photosphere, $m_\mathrm{p}$ is the proton mass, $k$ is the Boltzmann constant, $z$ is the mean multiplicity of charge of plasma ions, $x_1$ is the mean relative ionization potential (Ibadov 1986, 1990, 1996; Ibodov & Ibadov 2011 ).

The maximum height of the photospheric mass ejections due to cometary impacts may be estimated as

$$h_\mathrm{m} = \frac{V_\mathrm{sh}^2}{2g_0} = \frac{R_0^2 V_\mathrm{sh}^2}{2GM_0}, \tag{2.2}$$

where $g_0$ is the gravity acceleration on the solar/stellar surface, $G$ is the gravitational constant, and $M_0$ is the mass of the Sun/star.

Accepting realistic values of $A = 20$, $z = 5$, $x_1 = 3$ and using (2.1), (2.2) with $M_0 = 2 \times 10^{33}$ g we get $T_0 = 7 \times 10^6$ K, $V_\mathrm{sh} = 1.7 \times 10^7$ cm/s, $h_\mathrm{m} = 5 \times 10^9$ cm.

It is known that there is a variety of solar prominences having maximum heights in the range 30–50 thousand kilometers (Mackay *et al.* 2010 and references therein). It means that comet impact-generated photospheric mass ejections can form a certain type of solar/stellar prominences, too.

Application of results of our consideration is possible for studying, analytically, the collision of comet SL 9 with Jupiter in July 1994: "plumes" have been detected up to maximum heights around 3500 km by the HST (e.g., Hammel *et al.* 1995).

The physics of a similar explosive event in the Earth's atmosphere, known as the 1908 Tunguska phenomenon, has analytically been developed during several last decades (Grigorian 1979; Ibadov *et al.* 2008, 2010; Grigorian *et al.* 2009, 2013).

## 3. Conclusions

The passage of comet nuclei through the solar/stellar chromosphere is accompanied by intense aerodynamic crushing, transverse expansion of crushed matter, sharp aerodynamic stopping in a relatively very thin sub-photosphere layer, impulse production of a hot plasma, strong "blast" shock wave, ejection/eruption of a hot "plume" consisting of photospheric and cometary material to the heights reaching the lower solar corona.

## Acknowledgments

The authors are grateful to the IAUS 300 SOC for a grant to participate in the Symposium (Paris, 10-14 June 2013) and present the paper in the Session III "Stellar Ejecta and Impact on Exoplanets". Sincere thanks to the LOC of the meeting, DIC MSU, SAI MSU for hospitality, Drs. V.P. Arkhipova, M.A. Burlak and G.M. Rudnitskij for discussions.

## References

Bailey, M. E., Chambers, J. E., & Hahn, G. 1992, *A&A*, 257, 315

Beust, H., Lagrange, A.-M., Plazy, F., & Mouillet, D. 1996, *A&A*, 310, 181

Grigorian, S. S. 1979, *Cosmic Res.*, 17, 724

Grigorian, S. S., Ibodov, F. S., & Ibadov, S. 1997, *Dokl. Akad. Nauk*, 354, 187 [Engl. Transl.: *Doklady Physics*, 42, 262]

Grigorian, S. S., Ibadov, S., & Ibodov, F. S. 1998, in: *Cometary Nuclei in Space and Time, IAU Colloq. 168*, Abstracts, Nanjing, China

Grigorian, S. S., Ibadov, S., & Ibodov, F. S. 2000, *Dokl. Akad. Nauk*, 374, 40 [Engl. Transl.: *Doklady Physics*, 45, 463]

Grigorian, S. S., Ibodov, F. S., & Ibadov, S. I. 2009, *Vestnik RFFI*, 1–2 (61–62), 56 (In Russian), `www.rfbr.ru`

Grigorian, S. S., Ibodov, F. S., & Ibadov, S. I. 2013, *Sol. System Res.*, 47, 268

Grinin, V. P., Kozlova, O. V., The, P. S., & Rostopchina, A. N. 1996, *A&A*, 309, 474

Hammel, H. B., Beebe, R. F., Ingersoll, A. P., *et al.* 1995, *Science*, 267, 1288

Ibadov, S. 1986, *ESA SP-250*, 1, 377

Ibadov, S. 1990, *Icarus*, 86, 283

Ibadov, S. 1996, *Physical Processes in Comets and Related Objects*, Cosmosinform Publishing Company, Moscow (In Russian)

Ibadov, S., Ibodov, F. S., & Grigorian, S. S. 2008, in: *Internat. Conf. "100 Years Since Tunguska Phenomenon: Past, Present and Future" Presentations*, RAS, Moscow, `http://tunguska.sai.msu.ru/index.php?q=present`

Ibadov, S., Ibodov, F. S., & Grigorian, S. S. 2009, in: Gopalswamy, N. & Webb, D. F. (Eds.), *Universal Heliophysical Processes, Proc. IAU Symp. 257*, CUP, p. 341

Ibadov, S., Ibodov, F. S., & Grigorian, S. S. 2010, in: Fernandez, J. A., Lazzaro, D., Prialnik, D., & Schulz, R. (Eds.), *Icy Bodies of the Solar System, Proc. IAUS 263*, CUP, p. 269, `www.iau.org`

Ibodov, F. S. & Ibadov, S. 2011, in: Bonanno, A. & Kosovichev, A. (Eds.). *Advances in Plasma Astrophysics, Proc. IAUS 274*, CUP, p. 92

Info 1998, *COSPAR Inform. Bull.*, 142, 22 `www.cospar-assembly.org`

Lagrange, A.-M., Ferlet, R., & Vidal-Madjar, A. 1987, *A&A*, 173, 289

Mackay, D. H., Karpen, J. T., Ballester, J. L., Schmieder, B., & Aulanier, G. 2010, *Space Sci. Rev.*, 151, 333

MacQueen, R. M. & St. Cyr, O. C. 1991, *Icarus*, 90, 96 `http://sungrazer.nrl.navy.mil/`

Weissman, P. R. 1983, *Icarus*, 55, 448.

*Nature of Prominences and their role in Space Weather*
*Proceedings IAU Symposium No. IAUS300, 2013*
*B. Schmieder, J.-M. Malherbe & S. T. Wu, eds.*

doi:10.1017/S1743921313011770

# The Heliophysics Feature Catalogue, a tool for the study of solar features

**Xavier Bonnin[1], Nicolas Fuller[1], Christian Renié[1], Jean Aboudarham[1], Baptiste Cecconi[1], Robert D. Bentley[2] and André Csillaghy[3]**

[1]LESIA, Observatoire de Paris, CNRS, UPMC, Universit Paris-Diderot, 5 place Jules Janssen, 92195 Meudon, France
email: `xavier.bonnin@obspm.fr` email: `nicolas.fuller@obspm.fr`
email: `christian.renie@obspm.fr` email: `jean.aboudarham@obspm.fr`
email: `baptiste.ceconni@obspm.fr`

[2]MSSL, University College London, Hombury St. Mary, Dorking, Surrey RH5 6NT, U.K.
email: `rdb@mssl.ucl.ac.uk`

[3]Institute of 4D Technologies, FHNW, Steinackerstrass 5, CH-5210 Windisch, Switzerland
email: `andre.csillaghy@fhnw.ch`

**Abstract.** The behavior of filaments and prominences during the Solar Cycle is a signature of Sun's activity. It is therefore important to follow their evolution during the cycle, in order to be able to associate it with the various phases of the Solar Cycle as well as with other Solar features or events. The virtual observatory HELIO provides information that can be used for such studies, especially its Heliophysics Feature Catalogue gives a unique access to the description of various features during around one cycle. Features available are: filaments, prominences, photospheric and coronal active regions, coronal radio emission, type III radio bursts, coronal holes and sunspots. Web interfaces allow the user to query data for these features. Useful information can also be shared with other HELIO services, such as Heliophysics Event Catalogue, which provides access to dozens of tables of events such as flares, CMEs, ...

**Keywords.** Solar features, Automatic recognition, Image processing, Database, Data mining, Virtual observatory

## 1. Introduction

In the frame of the European project HELIO (Bentley *et al.* 2011), funded under European Commission's seventh Framework Program, (Project No. 238969) the Heliophysics Feature Catalogue (HFC) has been growing for the three last years. It is populated using automatic solar and heliospheric features detection codes which have been mainly developed at LESIA (Paris Observatory, France) and TCD (Dublin, Ireland). Currently, the HFC holds descriptions of filaments, prominences, active regions, coronal holes, sunspots, type III radio bursts and radio sources, which are searchable through dedicated user interfaces. From that, it is possible to study the behavior of several kinds of signatures of the solar activity, and possibly look for correlations between them. With the development of tracking capabilities it is also possible to derive events from the observations, like filaments disparitions brusques which are linked to CMEs. The HFC graphical user interface is accessible at: *http://voparis-helio.obspm.fr/hfc-gui/*, and more information about HELIO can be found at: *http://www.helio-vo.eu.*

Next Section provides an overview of the detection codes used to populated the HFC.

## 2. Detection Tools

**SoSoFT** (Fuller *et al.* 2005) is dedicated to the segmentation of solar filaments on full Sun H$\alpha$ images. After some cleaning steps (irregular background intensity, dust lines, etc.), seed pixels of dark areas are obtained using a windowed threshold. The seeds are then grown to catch the whole filament (region growing technique). The segmented filament are described by using morphological operators to retrieve shape and position parameters. **TrackFil** (Bonnin *et al.* 2013) is a filament tracking code that uses a curve-matching algorithm to compare the shape of filament skeletons on successive images.

**SoSoPro** has been developed together with SoSoFT to detect filaments, but on the limb, where they are called prominences. We use CaII K3 images from Meudon Observatory to detect them with a region growing technique. A raster scan of each prominence is then encoded (Run Length Encoding) and stored in the HFC, together with parameters like coordinates, height, etc.

**CHARM** (Krista *et al.* 2009) identifies coronal holes using SOHO/EIT (195Å) and SOHO/MDI observations. It uses a histogram-based intensity thresholding technique to detect low intensity regions on EIT and then classifies them as coronal holes or filaments using magnetograms.

**NRH2D**, developped by Christian Renie, detects and tracks radio sources on 2D images at 150 or 164 Mhz observed by Nancay radioheliograph. A gaussian fit is performed on the source to describe its shape.

**SMART** (Higgins *et al.* 2011) algorithm relies on consecutive image differencing to remove both quiet-Sun and transient magnetic features on MDI magnetograms to detect solar active regions. A region-growing technique allows to group flux concentrations into classifiable features. SMART extracts both position and magnetic properties of the features.

**RABAT3** (Bonnin *et al.* 2011) applies a specific Hough transform on a binary image of a dynamical spectrum to determine the type III main direction.

**SDOSS** is an upgrade of the code developed for EGSO to detect sunspots on SOHO/MDI (MDISS) (Zharkova *et al.* 2005), which have been superseded by SDO/HMI. The detection is based on the watershed operator and on the Canny edge technique. SDOSS efficiently discriminates umbra and penumbra in a given sunspot.

**SPOCA** (Barra *et al.* 2009) uses a multi-channel unsupervised spatially-constrained fuzzy clustering method, which automatically segments solar extreme ultraviolet (EUV) images into regions of interest. It is successfully applied on SOHO/EIT (195Å, 171Å) and SDO/AIA (193Å, 171Å) to detect active regions and coronal holes, and proposes an additional algorithm to track features with the solar rotation.

### References

Barra, V., Delouille, V., Kretzchmar, M, & Hochedez, J.-F. 2009, *A&A*, 505, 361

Bentley, R. D., Csillaghy, A., Aboudarham, J., Jacquey, C., Hapgood, M. A., Bocchialini, K., Messerotti, M., Brooke, J., Gallagher, P., Fox, P., Hurlburt, N., Roberts, D. A., & Duarte, L. 2011, *Adv. Sp. Res.*, 47, 2235

Bonnin, X., Aboudarham, J., Fuller, N., Csillaghy, A., & Bentley, R. D. 2013, *Solar Phys.*, 283, 49

Bonnin, X., Aboudarham, J., Fuller, N., Renie, C., Perez-Suarez, D., Gallagher, P., Higgins, P., Krista, L., Csillaghy, A., & Bentley, R. D. 2011, *SF2A-2011 Conf. Proceed.*, 373

Fuller, N., Aboudarham, J., & Bentley, R. D. 2005, *Solar Phys.*, 227, 61
Higgins, P., Gallagher, P., McAterr, R., & Bloomfield, D. 2011, *Adv. Sp. Res.*, 47, 2105
Krista, L. & Gallagher, P. 2009, *Solar Phys.*, 256, 87
Zharkova, V., Aboudarham, J., Zharkov, S., Ipson, S., Benkhalil, K., & Fuller, N. 2005, *Solar Phys.*, 228, 361

*Nature of Prominences and their role in Space Weather*
*Proceedings IAU Symposium No. 300, 2013*
*B. Schmieder, J.-M. Malherbe & S. T. Wu, eds.*

doi:10.1017/S1743921313011782

# Infrared Stokes Polarimeter at NAOJ/Mitaka

**Yoichiro Hanaoka, Takashi Sakurai, & IRMag Group**

National Astronomical Observatory of Japan,
2-21-1 Osawa, Mitaka, Tokyo 181-8588, Japan
email: hanaoka@solar.mtk.nao.ac.jp

**Abstract.** We have been operating an infrared Stokes spectro-polarimeter, whose observing wavelength bands include the He I 10830 Å and Fe I 15648 Å lines. A couple of full-Sun, full-Stokes maps in both wavelength bands are taken on a daily basis, with the polarization sensitivity better than $10^{-3}$. With this sensitivity, the helium polarization maps clearly show the atomic and Hanlé polarizations besides the Zeeman polarization, particularly in prominences/filaments. On these polarization maps, we can track the magnetic field signals of the prominences/filaments during their passages on the solar disk. Therefore, this instrument works as a 'synoptic prominence magnetograph'. Our preliminary study of the linear polarizations seen in quiescent filaments suggests that the magnetic field producing the polarization signals is located at the bottom part of a flux-rope, which supports a filament.

**Keywords.** Polarimetry, Prominences, Magnetic Field

## 1. Introduction and the Instrument

A full-Sun spectro-polarimeter, which aims at studying the evolution of the solar dynamo with a long-term operation, was installed onto the Solar Flare Telescope (Sakurai *et al.* 1995) at NAOJ/Mitaka (Hanaoka *et al.*, 2011). The observing wavelengths are He I 10830 Å/Si I 10827 Å (for the chromospheric and photospheric magnetic field) and Fe I 15648 Å (g=3)/15653 Å (for the photospheric magnetic field).

The solar light from the 15 cm objective lens enters the polarimeter, which consists of two ferroelectric liquid crystals and a polarizer (we adopt the single-beam polarimetry), before folding mirrors. Polarized light enters into an Echelle spectrograph, and a spectrum is imaged on an infrared camera (XENICS XEVA-CL-640) with 640 × 512 pixels. The exposure time is set to 20 msec, and 96 polarization modulated images are taken at one slit position for two seconds. The integration of 96 images reduces the polarization noise to less than $10^{-3}$. The slit covers half of the solar diameter, and two east-west scans, are performed to cover the full-Sun. A full-Sun scan takes about two hours.

## 2. Preliminary Results

Here we show examples of the observations of polarization in filaments. A He I 10830 Stokes I map taken on 2011 Dec 10 is shown in Figure 1. Enlargements of major filaments (labeled 1–4 in the full-disk image) and linear polarization signals of the red component of the He I 10830 line, displayed with short lines showing the orientation and the degree of the polarization, are also shown in Figure 1. Linear polarizations in filaments are produced by atomic level imbalance under the presence of magnetic field (see e.g. Trujillo Bueno *et al.*, 2002).

As seen in Figure 1, the polarization signals, namely the magnetic fields, are sheared. On the basis of the statistical study of the observed polarizations in filaments, the majority of the filaments are found to show the polarizations that the equator side shifts to

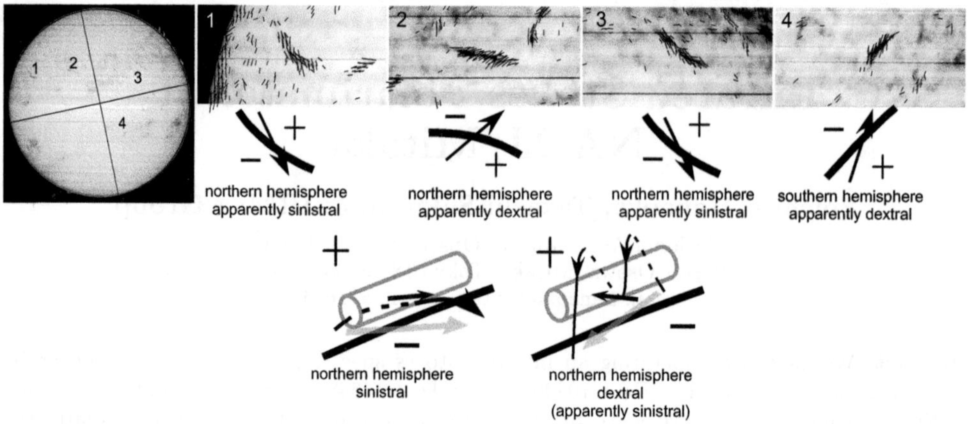


**Figure 1.** Top left: He I 10830 image on 2011 Dec 10. Major dark filaments are labeled 1–4, and their linear polarizations are shown in the top row with short black lines. Apparent magnetic connections presumed from the polarizations are shown below them. The cartoons in the bottom row show the possible interpretations of the observations. In the cartoons, black arrows show the magnetic field corresponding to the observed polarizations and gray arrows show the overall magnetic connection between the positive and negative magnetic regions.

the west (filaments 1, 3, and 4 show the polarization of the majority, and filament 2 is exceptional). This result seems to suggest that the majority of the filaments show sinistral (dextral) chirality in the northern (southern) hemisphere, as shown in the bottom-left cartoon in Figure 1. This is opposite to the interpretation by the former studies (Martin, 1998). However, if there is a flux rope with twisted magnetic field and filaments are located at the bottom of the flux rope as shown in the bottom-right cartoon in Figure 1, the apparent chirality of the magnetic field in the filament and the actual chirality of the overall magnetic connection are opposite to each other. In such a case, the overall magnetic connection corresponds to the dextral (sinistral) chirality in the northern (southern) hemisphere. Therefore, our result supports the view that filaments are supported by coronal cavity flux ropes (Berger, 2012).

This work was supported by a Grant-in-Aid for Scientific Research (No.17204014, 2005-2008, P.I.: T. Sakurai; and also partly by No.23244035, 2011-2014, P.I.: Y. Hanaoka) from the Ministry of Education, Culture, Sports, Science and Technology of Japan.

## References

Berger, T. 2012, in T. Rimmele, A. Tritschler, F. Woger, *et al.* (eds.) *The Second ATST-EAST Meeting: Magnetic Fields from the Photosphere to the Corona*, ASP. Conf. Ser. 463 (San Francisco: ASP), p. 147

Hanaoka, Y., Sakurai, T., Shinoda, K., Noguchi, M., Miyashita, M., Fukuda, T., Suzuki, I., Hagino, M., Arai, T., Yamasaki, T., & Takeyama, N. 2011, in J. R. Kuhn, D. M. Harrington, H. Lin, *et al.* (eds.) *Solar Polarization 6*, ASP. Conf. Ser. 437 (San Francisco: ASP), p. 371

Martin, S. F. 1998, in D. F. Webb, B. Schmieder, & D. M. Rust (eds.) *New Perspectives on Solar Prominences*, ASP. Conf. Ser. 150 (San Francisco: ASP), p. 419

Sakurai, T., Ichimoto, K., Nishino, Y., Shinoda, K., Noguchi, M., Hiei, E., Li, T., He, F., Mao, W., Lu, H., Ai, G., Zhao, Z., Kawakami, S., & Chae, J.-C. 1995, *PASJ*, 47, 81

Trujillo Bueno, J., Landi Degl'Innocenti, E., Collados, M., Merenda, L., & Manso Sainz, R. 2002, *Nature*, 415, 403

*Nature of Prominences and their role in Space Weather*
*Proceedings IAU Symposium No. 300, 2013*
*B. Schmieder, J.-M. Malherbe & S. T. Wu, eds.*

doi:10.1017/S1743921313011794

# Featuring dark coronal structures: physical signatures of filaments and coronal holes for automated recognition

**Judith Palacios[1], Consuelo Cid[1], Elena Saiz[1], Yolanda Cerrato[1], and Antonio Guerrero[1]**

[1]Space Reseach Group–Space Weather, Physics Dpt., University of Alcalá
University Campus, Sciences Building, P.O. 28871, Alcalá de Henares, Spain
email: `judith.palacios@uah.es`

**Abstract.** Filaments may be mistaken for coronal holes when observed in extreme ultraviolet (EUV) images; however, a closer and more careful look reveals that their photometric properties are different. The combination of EUV images with photospheric magnetograms shows some characteristic differences between filaments and coronal holes. We have performed analyses with 7 different SDO/AIA wavelengths (94, 131, 171, 211, 193, 304, 335 Å) and SDO/HMI magnetograms obtained in September 2011 and March 2012 to study coronal holes and filaments from the photometric, magnetic, and also geometric point of view, since projection effects play an important role on the aforementioned traits.

**Keywords.** Sun: solar filaments, Sun: coronal holes

---

## 1. Introduction: coronal holes and filaments

Dark features observed in EUV can be coronal holes (CHs) or filaments (Fs). CHs present a unipolar magnetic field. This unipolar magnetic field helps particles to escape, originating fast solar wind. Filaments are dark structures, with a very different magnetic field topology. Best observed in H$\alpha$, they are located mainly over neutral lines. The goal of this study is to check the photometric properties of CHs and filaments aiming at an automated recognition purpose.

## 2. Data

For this work we have used SDO data (Pesnell *et al.*, 2012); more precisely, EUV data from AIA instrument of the following wavelengths: 193, 211, 304, 94, 171, 94, 335 Å, and also longitudinal magnetic field data from SDO/HMI. These datasets are incomplete due to the eclipse season at the equinoxes. The cadence for the study is 30 min. The filaments studied are quiescent, away from active regions, and the low-latitude CHs are long-lived (>10 days) in the same field of view. Datasets from 2012, March 08-19 and 2011 Sept 06-15 are studied, but only the first one will be shown.

## 3. Photometric and magnetic properties of coronal holes and filaments

Some doubts may arise when automatically detecting dark features in AIA calibrated images. The coronal hole/filament photometric characteristics are found by studying the intensity histograms. First of all, a threshold is set for each wavelength. Then, these photometric histograms are studied in every wavelength for filaments and coronal holes.

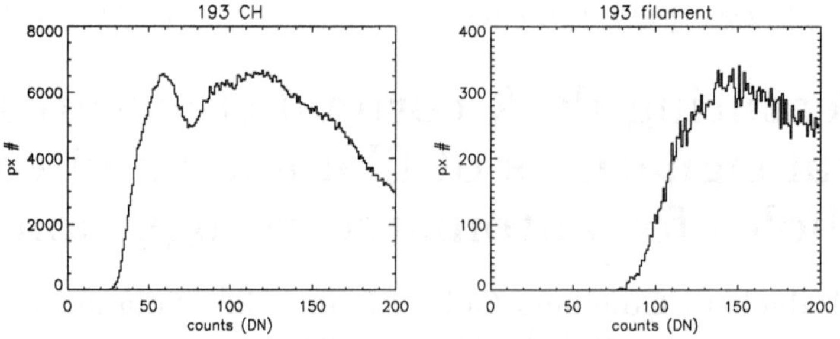


**Figure 1.** *Left* : Intensity histogram of a large CH in AIA 193 Å. X-axis shows the intensity, while the corresponding number of pixels is counted. *Right* : Intensity histogram of a filament in AIA 193 Å.

Running windows are set to follow the regions of interest across the solar disk. When necessary, the lambertian equal-area projection was used (Krista *et al.*, 2011).

In full-disk images, the intensity distribution can mark different features: bimodal intensity distributions show the existence of AR and CHs for 195 Å images (Krista & Gallagher, 2009). In this work we show different intensity distributions inside CH and filament areas. CH intensity follows a bimodal distribution, clearly marked in 193 and 211 Å (Fig. 1 *left*) while filaments usually display a unimodal distribution (Fig. 1 *right*). AIA wavelengths 94 Å and 131 Å are too noisy for dark regions, since their temperature response is more adequate for flaring regions (O'Dwyer *et al.*, 2010).

Regarding the magnetic field of these features, when filaments (and CHs) are located and contoured, their areas are superimposed to HMI magnetograms to relate the EUV image to the photospheric magnetic field. The filament mean magnetic field is around −0.5 G for all wavelengths.

Besides the HMI analyses, aiming at locating neutral lines, HMI images also can be used to pinpoint areas where quiescent filaments may be located, using a segmentation method.

## 4. Final remarks and conclusions

We have studied filaments and CHs. Generally, the mean magnetic field is higher for CHs than filaments, while intensity thresholds are usually higher for filaments than for coronal holes across the solar disk. The intensity histogram profiles are also different in filaments and CHs.

## 5. Acknowledgements

The authors want to acknowledge SDO/AIA and SDO/HMI Data Science Centers and Teams; and we would like to thank funding from the Spanish project PPII10-0183-7802 from "Junta de Comunidades de Castilla – La Mancha", and also from ESA, through the travel grant to participate in the IAUS300 Symposium.

### References

Krista L. & Gallagher P., 2009, *SolPhys*, 256, 87
Krista L. D., Gallagher P. T., & Bloomfield D. S., 2011, *ApJ*, 731, 26
O'Dwyer, B., Del Zanna, G., Mason H. E., Weber M. A., & Tripathi D., 2010, *A&A*, 521, A21
Pesnell, W. D., Thompson, B. J., & Chamberlin, P. C., 2012, *SolPhys*, 275, 3

*Nature of Prominences and their role in Space Weather*
*Proceedings IAU Symposium No. 300, 2013*
*B. Schmieder, J.-M. Malherbe & S. T. Wu, eds.*

doi:10.1017/S1743921313011800

# A system for near real-time detection of filament eruptions at Kanzelhöhe Observatory

**Werner Pötzi[1], Gernot Riegler[2], Astrid Veronig[1], Thomas Pock[2] and Ute Möstl[1]**

[1]University of Graz, Institue of Physics, IGAM-Kanzelhöhe Observatory, Austria
[2]Technical university of Graz, Institue for Computer Graphics and Vision, Austria

**Abstract.** Kanzelhöhe Observatory (kso.ac.at) performs regular high-cadence full-disk observations of the solar chromosphere in the H$\alpha$ and CaIIK spectral lines as well as the solar photosphere in white-light. In the frame of ESA's Space Situational Awareness (SSA) activities, a new system for near real-time H$\alpha$ image provision through the SSA Space Weather (SWE) portal (swe.ssa.esa.int) and for automatic alerting of flares and erupting filaments is under development. Image segmentation algorithms, based on optical flow image registration , for the automatic detection of solar filaments in real time H$\alpha$ images have been developed and implemented at the Kanzelhöhe observing system. We present first results of this system with respect to the automatic recognition and segmentation of filaments and filament eruptions on the Sun.

**Keywords.** H$\alpha$, filaments, solar activity, image processing

## 1. Introduction and Data

Full disc H$\alpha$ images provided by the H$\alpha$ telescope at Kanzelhöhe Observatory with a focal length of 2000 mm and an aperture of 100 mm are used for our purpose. The telescope is equipped with a Lyot filter centered at 656.28 nm and a FWHM of 0.07 nm. The time cadence of the 4 Megapixel images (resolution 1 arcsec/pixel) is 6 seconds and the system makes use of the frame selection technique, i.e. the camera software grabs the best image based on the image contrast during a preselected time interval. A fast computer is connected to the real-time observations, which analyses the incoming H-alpha images with respect to filaments and flares, and sends the output in near real-time to the SSA SWE portal (http://swe.ssa.esa.int).

## 2. The KSO SSA SWE subportal & Filament Eruption recognition

On the ESA SWE portal the observatory hosts a subportal (Fig. 1) displaying near real-time H$\alpha$ images and movies and lists of flares and filament eruptions. Every minute the page is refreshed and a new H$\alpha$ image is displayed. Detected flares and filament eruptions are shown in a table. Additionally a 360 degree view of a web-cam informs about the actual weather conditions at the observatory.

Filaments are detected by a highly parallelized feature recognition algorithm (Riegler *et al.* (2013)). The system is able to process up to 8 images per minute. Each feature is labeled and tracked during the whole observing day. For each filment, its ID, position and length are saved into log-files for further processing.

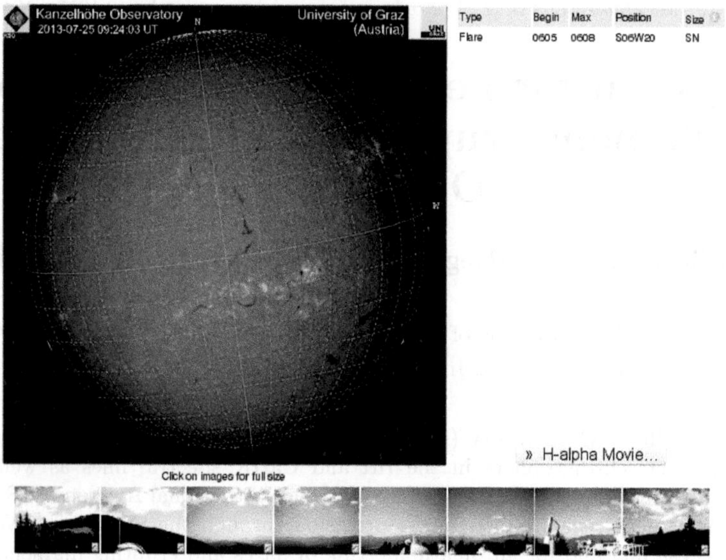


**Figure 1.** Sub portal design of the Kanzelhöhe H$\alpha$ services on ESA's SSA SWE portal. With a click on the image a new window pops up showing the H$\alpha$ image at 1024x1024 pixels and a daily history. The button "H-alpha movie" shows an interface, where a movie of the latest hour of observations is visible including the posiblity to control the movie speed and direction. Flares and filaments automatically detected in near real-time are displayed in a table. The interface is updated every minute.

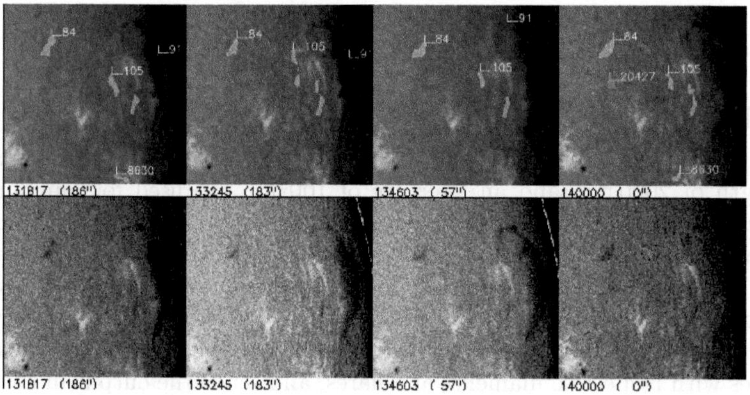


**Figure 2.** Bottom: sequence of KSO Halpha images, top: same sequence but with the output of the automated filament recognition overplotted. The filament close to the limb disappears from the $3^{rd}$ to the $4^{th}$ image. After 14:00 the filament was not detected anymore by the algorithm and it was therefore identified as filament eruption.

If a filament ID disappears for a certain time, a filament eruption alert is provided on the SWE H-alpha subportal.

*Acknowledgements*

This study was developed within the framework of ESA's SSA programme (SWE SN IV-2 activity).

## Reference

Riegler, G., Pock, T., Pötzi, W., & Veronig A. 2013, *arXiv:1304.7132*

*Nature of Prominences and their role in Space Weather*
*Proceedings IAU Symposium No. 300, 2013*
*B. Schmieder, J.-M. Malherbe & S. T. Wu, eds.*

doi:10.1017/S1743921313011812

# Coronal Multi-channel Polarimeter at the Lomnicky Peak Observatory

**P. Schwartz[1,2], J. Ambroz[1], P. Gömöry[1], M. Kozák[1], A. Kučera[1], J. Rybák[1], S. Tomczyk[3], S. Sewell[3], P. Aumiller[3], R. Summers[3], L. Sutherland[3] and A. Watt[3]**

[1]Astronomical Institute of Slovak Academy of Sciences, 05960 Tatranská Lomnica, Slovakia
email: pschwartz@astro.sk

[2]Astronomical Institute, Academy of Sciences of the Czech Republic, 25165 Ondřejov, Czech Republic

[3]High Altitude Observatory, National Center for Atmospheric Research, Boulder, CO 80307, USA

**Abstract.** Coronal Multi-channel Polarimeter (CoMP-S), developed by HAO/NCAR, has been introduced to regular operation at the Lomnicky Peak Observatory (High Tatras in northern Slovakia, 2633 m a.s.l.) of the Astronomical Institute of Slovak Academy of Sciences. We present here the technical parameters of the current version of the instrument and its potential for observations of prominences in the visual and near-IR spectral regions. The first results derived from observations of prominences in the H$\alpha$ emission line taken during a coordinated observing campaign of several instruments in October 2012 are shown here.

**Keywords.** solar instrumentation, spectroscopy, prominences

---

The Coronal Multi-channel Polarimeter (CoMP-S) was installed on the coronagraph (Lexa, 1966) of the Lomnicky Peak Observatory of the Astronomical Institute of SAS in March 2011 and nowadays it performs regular observations.

The CoMP-S instrument (Kučera *et al.*, 2011) is based on the concept of the CoMP instrument installed nowadays at the Maona Loa Observatory (Tomczyk *et al.*, 2008). Its core is a tunable 4-stage Lyot filter equipped with a Stokes polarimeter. It can operate in visible and near IR wavelength ranges from 500 to 1100 nm. This broad interval is allowed due to the latest achievements in polarizing materials: birefringent material VIS700BC4 by CODIXX and super-achromatic APSAW half-plates by ASTROPRIBOR. The following emission spectral lines of corona – Fe XIV 530.3 nm, Ca XV 569.5 nm, Fe X 637.5 nm, Fe XI 789.2 nm, Fe XIII 1074.7, 1079.8 nm – and prominences – He I 587.6 nm, H I 656.3 nm, Ca II 854.2 nm, He I 1083.0 nm can be observed. The Lyot filter bandpass width (FWHM) varies from 0.028 to 0.13 nm and its free spectral range is between 0.50 and 2.5 nm. Two orthogonal polarization states slightly shifted in wavelength are acquired simultaneously by two separate detectors. This allows subtraction of scattered light. The pco.edge sCMOS cameras by PCO are giving final image sampling of 0.33 arcsec/pixel and the FoV of 860×680 acrsecs (for wavelength of 656 nm).

Examples of the CoMP-S observations are shown in Fig. 1. The data were taken during the HOP186 coordinated campaign "Mass loading of quiescent prominences from multi-wavelength observations". A quiescent prominence was observed in the H$\alpha$ spectral line tuning the filter to 9 positions across the spectral profile with four individual polarizations per wavelength step. The detector exposure time was 50 ms. Reduction has been done only in the Stokes I profile with binning of 4×4 pixels leading to the final spatial sampling of 1.3 arcsecs. Each spectral profile detected with maximum intensity greater than 4000

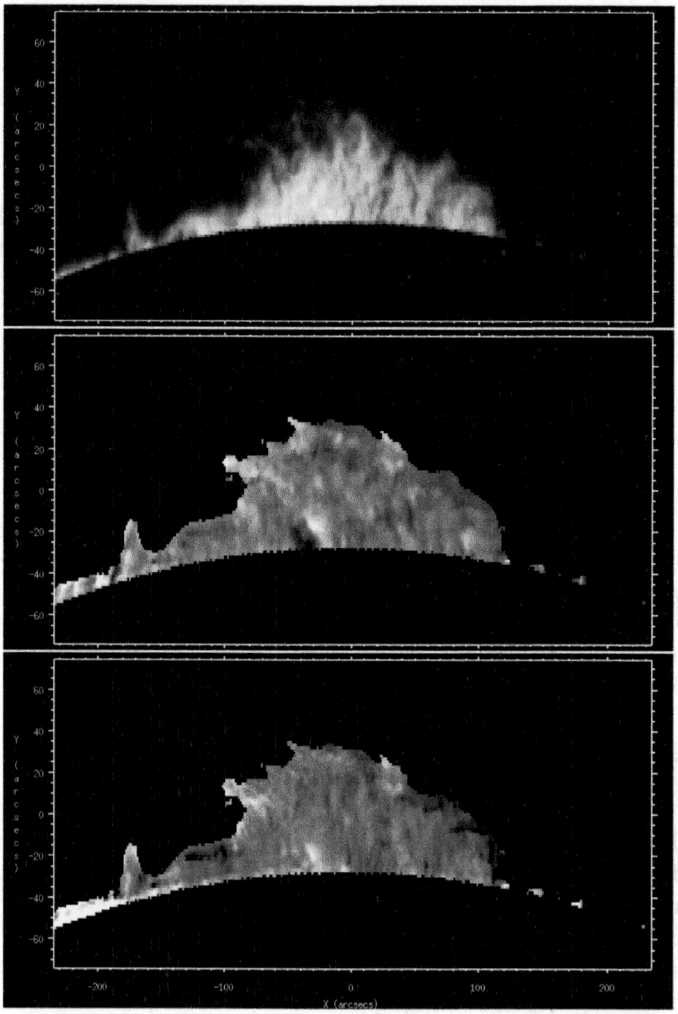


**Figure 1.** The line profile amplitude (top panel), the Doppler velocity (middle panel), and the Gaussian line width (bottom panel) of the H$\alpha$ spectral line derived for the quiescent prominence observed on 20/10/2012 at 07:06 UT at the position angle 170°. The displayed ranges are $\pm 12$ km/s for the Doppler shift range and 0.020-0.045 nm in case of the Gaussian line width.

counts above background was fitted by single Gaussian. The line profile amplitude, the Doppler shift (km/s), and the Gaussian line width (nm) were then derived.

Work on an extension of the original CoMP-S for data acquisition at wavelengths longer than 900 nm is now in progress preparing incorporation of IR detectors to the instrument (camera model Goodrich GA1280J by the Sensors Unlimited). Expected time of installation of this extension to the original instrument is January 2014 and full operation of the CoMP-S instrument is planned during summer 2014.

## References

Lexa, J., 1966, *BAC*, 14, 107

Kučera, A., Ambroz, J., Gömöry, P., Kozák, M., & Rybák, J., 2011, *Contributions of Astronomical Observatory Skalnate Pleso*, 40, 135

Tomczyk, S., Card, G., Darnell, T., Elmore, D. *et al.*, 2008, *Sol. Phys.*, 247, 411

*Nature of Prominences and their role in Space Weather*
*Proceedings IAU Symposium No. 300, 2013*
*B. Schmieder, J.-M. Malherbe & S. T. Wu, eds.*

doi:10.1017/S1743921313011824

# Solar Activity Monitoring of Flares and CMEs Precursors through Lyman-Alpha Imaging and Tracking of Filaments and Prominences

**Luc Damé[1] and Safinaz A. Khaled[2,1]**
[1]LATMOS/IPSL/CNRS/UVSQ, 11 Boulevard d'Alembert, 78280 Guyancourt, France
email: luc.dame@latmos.ipsl.fr
[2]Space Weather Monitoring Center, Helwan University, Helwan, Egypt
email: safinaz_1986@yahoo.com

**Abstract.** We investigate the advantages of imaging solar filaments and prominences in Lyman-Alpha, coupled to H-Alpha on ground, to develop more reliable precursors indicators for large flares, several hours before their occurrence.

**Keywords.** Solar flares, filaments, ultraviolet.

## 1. Rationale

Events preceding the onset of a flare are called 'precursors', and one of the prominent precursors is a newly emerging bipolar region at the surface, which may interact with pre-existing magnetic field in the corona and trigger a flare. Another well-known precursor is the activation, or eruption, of a filament that is composed of relatively cool plasma (around 10000 K), floating in the hot coronal plasma. Both emerging regions and filaments are very well observed in Lyman-Alpha (in Space) and H-Alpha (on ground and formed lower in the solar atmosphere), both on the disk and at the limb, and we expect that their combination can lead to better identification of changes at the origin of major eruptions and most important coronal mass ejections (CMEs).

Lyman-Alpha can provide early (precursor) detection on the disk, hours before the event, of filament/prominence eruptions (better than the He II line, well suited only for limb observations). As studied by Mierla *et al.* (2013), the imaging of an erupting prominence shows that the acceleration at the top of the prominence increases smoothly and continuously with height (slow rise), and this over hours (4–5) before the eruption. This indeed indicates that prominences are not accelerated immediately by local reconnection but are rather swept away (modifying shape, helicity) as part of a large-scale relaxation of the coronal magnetic field.

## 2. Observations and results

As shown in Fig. 1, the Lyman-Alpha signal (observed with LYRA/PROBA-2) peaks during the rising phase of the flare and has a very significant level. This M2.0 flare of February 8, 2010, was analyzed in detail by Kretzschmar *et al.* (2013). It indeed shows that Lyman-Alpha is very sensitive to flares, 1000 times more than H-Alpha formed lower in the atmosphere; the signature on the light curve of this M2 flare is reaching almost 1% in integrated light of the solar disk! The Lyman-Alpha line, the most intense solar line, is obviously very sensitive to flares and temperature variations in the chromosphere (see Milligan *et al.* 2011) but also to velocities and magnetic fields (Zeeman effect). It

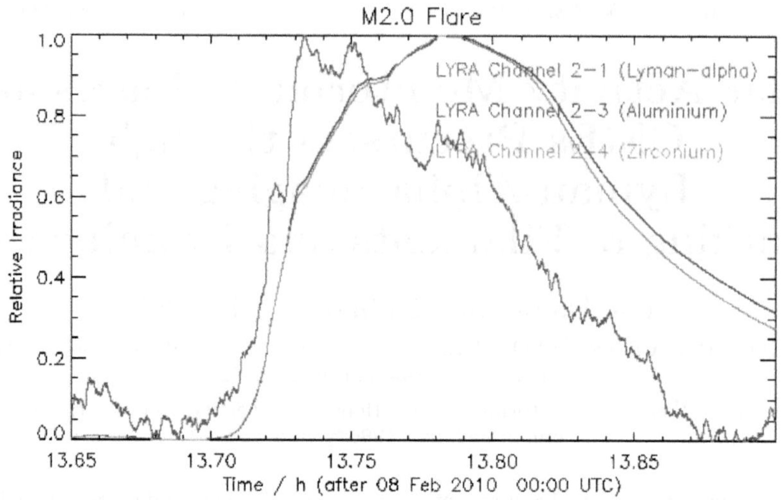


**Figure 1.** During a flare, the Lyman-Alpha signal, cooler, peaks during the rising phase, slightly earlier than X-rays or XUV (early LYRA/PROBA-2 data with proper Lyman-Alpha channel).

has the advantage to be much thicker than the H-Alpha line (cf. Labrosse *et al.* 2010) and to be formed higher in the solar atmosphere. By combining imaging (and velocities if possible) in both Lyman and H-Alpha, one can expect to better assess the filament 3D shape and its evolution up to the disruption/eruption "disparition brusque".

Lyman-Alpha has both the capacity of the early detection of a large flare/CME by imaging the filament change/acceleration, and the detection sensitivity of the flare/CME occurrence, even better than soft X-rays (GOES measurements) or XUV and EUV bands (as illustrated on Fig. 1).

Lyman-Alpha imaging, in that respect, would be a very high value Space Weather product complementing EUV imaging available on several satellites.

## 3. Conclusion

These considerations on Lyman-Alpha and H-Alpha observations are pointing the interest for future Space Lyman-Alpha irradiance measurements coupled, for precursors identification, to solar disk imaging, as proposed by the SWUSV *(Space Weather & Ultraviolet Solar Variability)* Microsatellite Mission (Damé and The SWUSV Team 2013).

### Acknowledgements

We are grateful to I. Dammasch and M. Dominique for access to and use of LYRA/PROBA-2 data, to S. UeNo for the H-Alpha data of Hida Observatory, and to R. Kariyappa and S.T. Kumara for help with the data analysis (image segmentation).

### References

Damé, L., The SWUSV Team 2013, *Journal of Adv. Research*, 4, 235

Kretzschmar, M., Dominique, M., & Dammasch, I. E. 2013, *Solar Phys.*, 286, 221

Labrosse, N., Heinzel, P., Vial, J.-C., Kucera, T., Parenti, S., Gunár, S., Schmieder, B., & Kilper, G. 2010, *Space Sci. Rev.*, 151, 243

Mierla, M., Heinzel, P., Seaton, D. B., Berghmans, D., Chifu, I., De Groof, A., Inhester, B., Rodriguez, L., Stenborg, G., & Zhukov, A. N. 2013, *Solar Phys.*, 286, 241

Milligan, R. O., Chamberlin, P., Hudson, H., Woods, T., Mathioudakis, M., Fletcher, L., Kowalski, A., & Keenan, F. 2011, *ApJ.*, 748, L14

*Nature of Prominences and their role in Space Weather*
*Proceedings IAU Symposium No. 300, 2013*
*B. Schmieder, J.-M. Malherbe & S. T. Wu, eds.*

doi:10.1017/S1743921313011836

# The Space Weather & Ultraviolet Solar Variability Microsatellite Mission (SWUSV)

**Luc Damé[1] and The SWUSV Team (Mustapha Meftah, Alain Hauchecorne, Philippe Keckhut, Alain Sarkissian, Marion Marchand, Abdenour Irbah, Éric Quémerais, Slimane Bekki, Thomas Foujols, Matthieu Kretzschmar, Gaël Cessateur, Alexander Shapiro, Werner Schmutz, Sergey Kuzin, Vladimir Slemzin, Sergey Bogachev, José Merayo, Peter Brauer, Kanaris Tsinganos, Antonis Paschalis, Ayman Mahrous, Safinaz A. Khaled, Ahmed Ghitas, Besheir Marzouk, Amal Zaki, Ahmed A. Hady, Rangaiah Kariyappa)**

[1]LATMOS/IPSL/CNRS/UVSQ, 11 Boulevard d'Alembert, 78280 Guyancourt, France
email: luc.dame@latmos.ipsl.fr

**Abstract.** We present a summary of the scientific objectives, payload and mission profile of the Space Weather & Ultraviolet Solar Variability Microsatellite Mission (SWUSV) proposed to CNES and ESA (small mission).

**Keywords.** Space Weather, solar flares, Coronal Mass Ejections, solar variability, ultraviolet.

---

## 1. Objectives

The SWUSV mission encompasses three major scientific objectives: (1) Space Weather including the prediction and detection of major eruptions and coronal mass ejections (Lyman-Alpha and Herzberg continuum imaging); (2) solar forcing on the climate through radiation and their interactions with the local stratosphere (UV spectral irradiance from 180 to 400 nm by bands of 20 nm, plus Lyman-Alpha and the CN bandhead); (3) simultaneous radiative budget of the Earth, UV to IR, with an accuracy better than 1% in differential.

SWUSV is targeting the observation of the Space environment, and in particular the premisses, the very early start ("precursor indicators"), of the coronal mass ejections, the interplanetary ones (ICMEs), that are the most important since the only ones with a destructive impact potential on Earth. With Lyman-Alpha (and H-Alpha on ground in complement, 2 major solar lines formed at different heights in the solar chromosphere), we should be able to better predict, hours in advance, the major flares and ICMES and their geoeffectiveness.

Concerning the solar influence on climate, the ultraviolet part of the spectra (<350 nm) is the only one having its energy completely absorbed in the high atmosphere (stratosphere), in the ozone layer and in the oxygen bands (Herzberg continuum at 200–220 nm). The large variability of UV wavelengths over the solar cycle (5–10% at 200–220 nm, factor 2 at Lyman-Alpha) is probably at the origin of a solar influence on climate, the UV affecting the temperatures and the stratospheric dynamic (the response to the solar heating extends to the poles and down, to the tropopause, modifying planetary waves and meteorological conditions). With these UV observations, and when coupled to a complete local radiative budget from UV to IR, we shall be able to characterize this solar influence to its full dynamical extent.

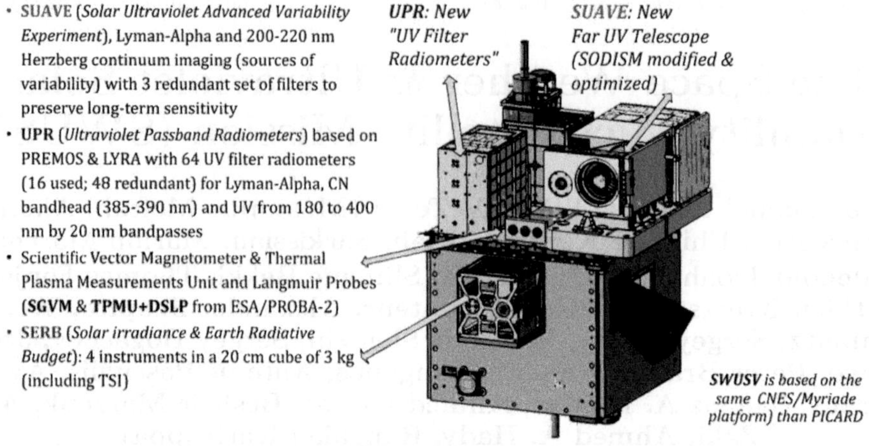
• SUAVE (Solar Ultraviolet Advanced Variability Experiment), Lyman-Alpha and 200-220 nm Herzberg continuum imaging (sources of variability) with 3 redundant set of filters to preserve long-term sensitivity
• UPR (Ultraviolet Passband Radiometers) based on PREMOS & LYRA with 64 UV filter radiometers (16 used; 48 redundant) for Lyman-Alpha, CN bandhead (385-390 nm) and UV from 180 to 400 nm by 20 nm bandpasses
• Scientific Vector Magnetometer & Thermal Plasma Measurements Unit and Langmuir Probes (SGVM & TPMU+DSLP from ESA/PROBA-2)
• SERB (Solar irradiance & Earth Radiative Budget): 4 instruments in a 20 cm cube of 3 kg (including TSI)
UPR: New "UV Filter Radiometers"
SUAVE: New Far UV Telescope (SODISM modified & optimized)
SWUSV is based on the same CNES/Myriade platform) than PICARD

**Figure 1.** Illustration of the possible accommodation of the SWUSV instruments on the CNES/Myriade microsatellite platform (dimensions 600 x 600 x 800 mm$^3$).

## 2. Model Payload & Mission Profile

SWUSV includes 5 instruments: SUAVE (*Solar Ultraviolet Advanced Variability Experiment*), an optimized telescope for FUV (Lyman-Alpha) and MUV (200–220 nm Herzberg continuum) imaging (sources of variability); UPR (*Ultraviolet Passband Radiometers*), with 64 UV filter radiometers; a vector magnetometer; thermal plasma measurements and Langmuir probes; and a total and spectral solar irradiance and Earth radiative budget ensemble (SERB, *Solar irradiance & Earth Radiative Budget*). The model payload and microsatellite (CNES/Myriade platform on a Sun synchronous orbit comparable to PICARD: altitude, 725 km; inclination, 98.29°; local time of ascending node, 6h00 +/- 10 minutes) are presented in Fig. 1.

## 3. Conclusion

The microsatellite investigation SWUSV is responding to the need to understand the influence of stratospheric dynamics on the climate by providing the tools to measure and quantify the influence of UV variability and determine its origin. It also carries, through Lyman-Alpha imaging, probably the best indicator for precursor signs of major Space Weather events. The program has the advantage of relying on technological developments made in very recent microsatellite missions of CNES (PICARD) and ESA (PROBA-2), and on a long expertise of the proposing team in far UV imaging (Lyman-Alpha in particular but also Herzberg continuum at 220 nm). The SWUSV mission is presented in detail in Damé and The SWUSV Team (2013).

### Acknowledgements

We are grateful to the CNES that supports the SWUSV initiative with a R&D program on far ultraviolet solar telescopes design and performances.

### Reference

Damé, L., The SWUSV Team (Meftah, M., Hauchecorne, A., Keckhut, P., Sarkissian, A., Marchand, M., Irbah, A., Quémerais, E., Bekki, S., Foujols, T., Kretzschmar, M., Cessateur, G., Shapiro, A., Schmutz, W., Kuzin, S., Slemzin, V., Urnov, A., Bogachev, S., Merayo, J., Brauer, P., Tsinganos, K., Paschalis, A., Mahrous, A., Khaled, S. A., Ghitas, A., Marzouk, B., Zaki, A., Hady, A. A., & Kariyappa, R.) 2013, *Journal of Adv. Research*, 4, 235

*Nature of Prominences and their role in Space Weather*
*Proceedings IAU Symposium No. 300, 2013*
*B. Schmieder, J.-M. Malherbe & S. T. Wu, eds.*

doi:10.1017/S1743921313011848

# Unpublished contributions

## 1. UNPUBLISHED ORAL CONTRIBUTIONS

### *Session I - 1.1 Prominence: fine structure, dynamics and seismology*

• Theoretical Models of the Origin of Prominence Mass (Judy Karpen, NASA Goddard Space Flight Center)

### *Session I - 1.3 Magnetic Field: Measurements and Models*

• On the magnetic topology of quiescent prominence bubbles (Jaroslav Dudik, Comenius University)
• Observation of the magnetic field in solar tornadoes (Maria Jesus Martinez Gonzalez, Instituto de Astrofisica de Canarias)

### *Session I - 1.4 Filament Environment*

• The role of prominences in defining the quiescent and dynamic large scale coronal structures (Shadia Habbal, Institute for Astronomy, University of Hawaii)
• Observations and simulations of longitudinal oscillations of an active region prominence (Qingmin Zhang, Purple Mountain Observatory)

### *Session II - 2.1 Prominence destabilization, CMEs, 3D reconstructions*

• Key Physics of Prominence Eruption: Models and Observations (Kazunari Shibata, Kwasan and Hida Observatories)
• Torus instability of a line-tied flux rope (Oscar Olmedo, Naval Research Laboratory)
• Evidence for Flux Ropes (Alan Title, Lockheed Martin Advanced Technology Center)
• Fractal Reconnection and Stochastic Particle Acceleration induced by a Prominence Eruption (Naoto Nishizuka, Japan Aerospace Exploration Agency)
• Low polarised emission from the core of coronal mass ejections (Marilena Mierla, Royal Observatory of Belgium)
• The 3-D NLFFF reconstruction of Active Region NOAA 11158 (Yihua Yan, Key Lab of Solar Activity, National Astronomical Observatories, Chinese Academy of Sciences)
• Dynamo driven coronal ejections (Joern Warnecke, Nordic Institute for Theoretical Physics, Department of Astronomy, Stockholm University)

### *Session II - 2.2 CMEs and Magnetic clouds in the Heliosphere and their impacts on Eart's Environment*

• The geoeffectiveness of ICMEs (Alisson Dal Lago, National Institute for Space Research)
• Statistical analysis of magnetic cloud erosion by magnetic reconnection (Alexis Ruffenach, Institut de recherche en astrophysique et planétologie)
• Coronal Mass Ejections and associated shocks: Build-up and propagation in a complex environment (Monique Pick, Observatoire de Paris)
• Filament Eruptions, Jets, and Space Weather (Ron Moore, NASA Marshall Space Flight Center)

***Session III - Stellar Ejecta and Impact on Exoplanets***

• Solar wind properties and coronal rotation during the activity cycle (Rui Pinto, Paris Observatory & CEA Saclay)
• Space observations of evaporating exoplanets (David Ehrenreich, University of Geneva)

***Session IV - Instrumentation, Missions and Techniques***

• Prominences observations from space: advances and future challenges (Frédéric Auchère, Institut d'astrophysique spatiale)
• The IRIS mission - prospects for prominence and filament science (Mats Carlsson, University of Oslo)
• Observation of the prominence eruptions and CME during the Interhelioprobe solar mission (Sergey Bogachev, Lebedev Physical Institute of Russian Academy of Sciences)
• ALMA Observations of Solar Prominences (Petr Heinzel, Astronomical Institute, Academy of Sciences of the Czech Republic)

## 2. UNPUBLISHED POSTER PAPERS

***Poster Session I - Prominences***

• Strategies for the inversion of He I 10830 A data (Asensio Ramos Andres)
• The Influence of Coronal Radiation on Prominence Plasma (Brown Gerrard)
• Automated detection, characterization, and tracking of filaments from SDO data (Buchlin Eric)
• Cut-off wavenumber of Alfvén waves in partially ionized plasmas of the Solar Atmosphere (Carbonell Marc)
• Coronal magnetic field modeling using stereoscopy constraints (Chifu Iulia)
• Detection of partial ionization effects in prominences with observed Doppler velocities (Diaz Antonio)
• Rayleigh-Taylor Instability in Prominence Partially Ionized Plasma (Diaz Antonio)
• Two-ribbon flare without a filament eruption: Slipping magnetic reconnection observed with SDO/AIA (Dudik Jaroslav)
• Total mass loading of prominences estimated from their multi-spectral observations (Farnik Frantisek)
• Effect of shear flow on damping of linear non-adiabatic MHD waves in a prominence medium (Kumar Nagendra)
• Measuring magnetic fields in prominences: the effect of multiple scattering on polarisation (Lopez Ariste Arturo)
• Partially ionized plasma downflows and vertical threads in solar prominences (Oliver Ramon)
• Magnetic properties of coronal pseudo-streamers (Rachmeler Laurel)
• A multi-spcecraft view of a giant filament eruption during 26/27 September 2009 (Sanjay Gusain)
• Rayleigh-Taylor unstable modes in filament threads (Terradas Jaume)
• Prominence MHD models and their eigenmodes (Terradas Jaume)
• Magnetic fields of an active region filament from full Stokes analysis of Si 10827 and He I 10830 A (Xu Zhi)
• Emerging dimmings of active regions observed by the Solar Dynamics Observatory (Zhang Jun)

### *Poster Session II - CME - ICME - Space Weather*

- Investigating the Initiation and dynamics of Flux-rope CMEs (Cheng Xin)
- Magnetic reconnection at the leading edge of a solar erupting loop and an ICME (Chian Abraham)
- Comparison of Helicity Signs in Interplanetary CMEs and Their Solar Source Regions (Cho Kyungsuk)
- Role of the terrestrial bow shock on magnetic clouds structure: 1. CLUSTER observations downstream of the bow shock (Fontaine Dominique)
- The dynamical behavior of Magnetic Clouds: From 0.3 to 5.4 astronomical units (Gulisano Adriana)
- SEP's during Halloween storms and space weather (Hady Ahmed)
- Multiwavelength Observations of Helical Kink Instability as a Trigger of Solar Flare and CME in AR NOAA 11163 (Kumar Pankaj)
- ICMEs Associated with Major Geomagnetic Storms Over the Solar Cycle 24 (Ontiveros Veronica)
- A Multi-wavelength Observational Study of Eruption Processes of Active Prominences in the Solar Active Region NOAA 11261 (Park Sung-Hong)
- Solar high energy observations within SEPServer project: spatially resolved X-ray observations of flares associated with SEP events (Rodríguez-Gasen Rosa)
- Statistics of magnetic autocorrelation lengths in the Solar Wind (Ruiz Maria Emilia)
- Analysis of SC23 major geomagnetic storms produced by CMEs (Stere Oana)
- Role of the terrestrial bow shock on magnetic clouds' structure: 2. 3D analytical MHD model (Turc Lucile)
- Statistical Investigation of Physical Parameters of Coronal Mass Ejection in 2002-2012 (Ajabshirizadeh Ali)
- EUV Solar Corona Sources of Geomagnetic Disturbances in the Solar Cycle 24 (Talebpour Sheshvan Nasrin)
- Correlation between Interplanetary Parameters and Geomagnetic Indices during Geomagnetic Storms in 2010-2011 (Nabizadeh Armin)

### *Poster session IV - Instrumentation*

- New Coronal Imaging Instrumentation (Golub Leon)
- Solar Radio Imaging-Spectroscopy Observations in cm-dm Wavelengths (Yan Yihua)

## Author Index